A Brief Table of Integrals

$$\int u^p\,du = \frac{1}{p+1}u^{p+1} + C,\ p \neq -1$$

$$\int e^u\,du = e^u + C$$

$$\int \frac{1}{u}\,du = \ln|u| + C$$

$$\int \cos u\,du = \sin u + C$$

$$\int \sin u\,du = -\cos u + C$$

$$\int \sec^2 u\,du = \tan u + C$$

$$\int \csc^2 u\,du = -\cot u + C$$

$$\int \sec u\ \tan u\,du = \sec u + C$$

$$\int \csc u\ \cot u\,du = -\csc u + C$$

$$\int \tan u\,du = -\ln|\cos u| + C$$

$$\int \cot u\,du = \ln|\sin u| + C$$

$$\int \sec u\,du = \ln|\sec u + \tan u| + C$$

$$\int \csc u\,du = \ln|\csc u - \cot u| + C$$

$$\int \sin^2 u\,du = \frac{1}{2}u - \frac{1}{4}\sin 2u + C$$

$$\int \cos^2 u\,du = \frac{1}{2}u + \frac{1}{4}\sin 2u + C$$

$$\int \sec^3 u\,du = \frac{1}{2}(\sec u \tan u + \ln|\sec u + \tan u|) + C$$

$$\int \csc^3 u\,du = -\frac{1}{2}(\csc u \cot u - \ln|\csc u - \cot u|) + C$$

$$\int \sin au\ \sin bu\,du = \frac{\sin((a-b)u)}{2(a-b)} - \frac{\sin((a+b)u)}{2(a+b)} + C$$

$$\int \cos au\ \cos bu\,du = \frac{\sin((a-b)u)}{2(a-b)} + \frac{\sin((a+b)u)}{2(a+b)} + C$$

$$\int \sin au\ \cos bu\,du = -\frac{\cos((a-b)u)}{2(a-b)} - \frac{\cos((a+b)u)}{2(a+b)} + C$$

$\int e^{au}\sin bu\,du$ [illegible] a^2+b^2 [illegible]

$$\int e^{au}\cos bu\,du = \frac{e^{au}}{a^2+b^2}(a\cos bu + b\sin bu) + C$$

$$\int \frac{du}{a^2+u^2} = \frac{1}{a}\tan^{-1}\frac{u}{a} + C$$

$$\int \frac{du}{a^2-u^2} = \frac{1}{2a}\ln\left|\frac{u+a}{u-a}\right| + C$$

$$\int \frac{du}{\sqrt{a^2-u^2}} = \sin^{-1}\frac{u}{a} + C$$

$$\int \sqrt{a^2-u^2}\,du = \frac{u}{2}\sqrt{a^2-u^2} + \frac{a^2}{2}\sin^{-1}\frac{u}{a} + C$$

$$\int \frac{du}{\sqrt{u^2 \pm a^2}} = \ln\left|u + \sqrt{u^2 \pm a^2}\right| + C$$

$$\int \sqrt{u^2 \pm a^2}\,du = \frac{u}{2}\sqrt{u^2 \pm a^2} \pm \frac{a^2}{2}\ln\left|u + \sqrt{u^2 \pm a^2}\right| + C$$

$$\int u\,dv = u\,v - \int v\,du$$

$$\int u^n e^u\,du = u^n e^u - n\int u^{n-1}e^u\,du$$

$$\int u^n \ln u\,du = \frac{1}{n+1}u^{n+1}\left(\ln u - \frac{1}{n+1}\right) + C$$

$$\int u^n \sin u\,du = -u^n \cos u + n\int u^{n-1}\cos u\,du$$

$$\int u^n \cos u\,du = u^n \sin u - n\int u^{n-1}\sin u\,du$$

Differential Equations
with Boundary Value Problems

DIFFERENTIAL EQUATIONS
WITH BOUNDARY VALUE PROBLEMS

SELWYN HOLLIS

Armstrong Atlantic State University

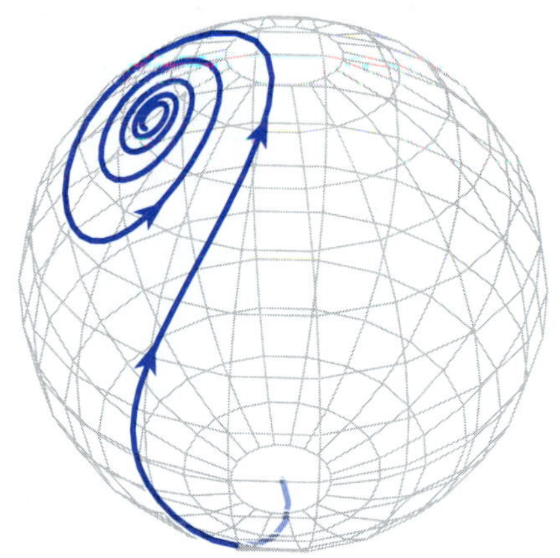

PRENTICE HALL, Upper Saddle River, NJ 07458

Library of Congress Cataloging-in-Publication Data

Hollis, Selwyn L.
Differential equations with boundary value problems/Selwyn Hollis.
p. cm.
Includes bibliographical references and index.
ISBN: 0-13-015927-1
1. Differential Equations. 2. Boundary value problems. II. Title
QA371. H67 2002 2001050072
515'.35–dc21

Acquisition Editor: *George Lobell*
Editor-in-Chief: *Sally Yagan*
Vice-President/Director of Production and Manufacturing: *David W. Riccardi*
Executive Managing Editor: *Kathleen Schiaparelli*
Senior Managing Editor: *Linda Mihatov Behrens*
Assistant Managing Editor: *Bayani Mendoza de Leon*
Production Editor: *Jeanne Audino*
Assistant Managing Editor, Math Media Production: *John Matthews*
Manufacturing Buyer: *Alan Fischer*
Manufacturing Manager: *Trudy Pisciotti*
Marketing Manager: *Angela Battle*
Marketing Assistant: *Rachel Beckman*
Editorial Assistant: *Melanie Van Benthuysen*
Art Director: *Jonathan Boylan*
Assistant to the Art Director: *John Christiana*
Interior Designer: *Marjory Dressler*
Cover Designer: *Jonathan Boylan*
Art Editor: *Thomas Benfatti*
Managing Editor, Audio/Video Assets: *Grace Hazeldine*
Creative Director: *Carole Anson*
Director of Creative Services: *Paul Belfanti*
Assistant Editor of Media: *Vince Jansen*
Cover Photo: *Obtained from www.eyewire.com*
Illustrations: *Laserwords Private Limited*

Prentice Hall

Upper Saddle River, NJ 07458

Printed in the United States of America

10 9 8 7 6 5 4 3 2 1

ISBN 0-13-015927-1

Pearson Education Ltd., *London*
Pearson Education Australia Pty., Limited, *Sydney*
Pearson Education Singapore, Pte. Ltd.
Pearson Education North Asia Ltd., *Hong Kong*
Pearson Education Canada, Ltd., *Toronto*
Pearson Education de Mexico, S. A. de C.V.
Pearson Education—Japan, *Tokyo*
Pearson Education Malaysia, Pte. Ltd.

CONTENTS

5 SECOND-ORDER LINEAR EQUATIONS I 133

6 SECOND-ORDER LINEAR EQUATIONS II 185

7 THE LAPLACE TRANSFORM 220

8 FIRST-ORDER LINEAR SYSTEMS 259

9 GEOMETRY OF AUTONOMOUS SYSTEMS IN THE PLANE 309

PREFACE

The introductory differential equations course plays an interesting role in the undergraduate mathematics curriculum. It is a required course for most science and engineering students, many of whom will take major courses that require certain knowledge of and skills related to differential equations. The course also plays the role of an introductory "applied course" for mathematics students, where modeling is often a primary focus.

Our point of view, which in no way diminishes the utility of the course's content, is that the differential equations course has two overarching functions: (1) It is where students are introduced to the central subject in all of applied mathematics, a subject whose questions have spawned most of the classical theory of analysis since the time of Newton and continue to create a rich field of mathematical activity today. (2) It is a place to reinforce and extend the student's (perhaps tenuous) understanding of calculus. It is the next step after calculus on the path that leads to real analysis. Whether the student will follow that path any farther makes little difference.

Throughout this book, we attempt to emphasize that the primary goal when studying a differential equation is always to understand the behavior of its solutions. Writing down formulas for solutions is just one means to that end. Qualitative and numerical methods are equally, if not more, important. Moreover, all of these "tools" are manifestations of theory, and we endeavor to emphasize that fact at a level which is appropriate for the course.

Times have changed. Amazingly fast computers are commonplace, and powerful software systems such as *Mathematica*, *Maple*, and MATLAB make it possible to solve highly complex problems and create wonderfully illuminating graphics. This computing power has enormous potential for enhancing the differential equations course—perhaps more so than in any other mathematics course. Indeed, the computer can facilitate the analysis of solutions algebraically, numerically, and graphically. Yet the computer cannot teach the conceptual and theoretical foundations that this book strives to convey—but neither can tedious algebraic manipulations done with paper and pencil. This book does not emphasize technology, and only occasionally do we mention it directly. However, we recommend the use of a computer to facilitate those computations that require more than a few routine algebraic steps. We are platform-neutral with respect to technology, and our belief is that the best way to address technology specifically is with separate companion manuals. Thus we have created *A Mathematica Companion for Differential Equations* and *A Maple Companion for Differential Equations* for those who use *Mathematica* or *Maple* in their courses, and we highly recommend their use.

Organization and Content

For the most part, this book covers the traditional topics in the introductory differential equations course, while enhancing the usual algebraic approach with more geometric ideas and interpretations. Naturally reflecting the biases of the author (an inkling of which should be provided by the preceding commentary), certain standard topics are omitted or given less emphasis, while others are given increased emphasis. A few of the topics covered are rather novel for an introductory text. The following are some of our more significant deviations from the "standard course":

- Systems and numerical methods are introduced early.
- Series methods are deemphasized, particularly the method of Frobenius.
- The method of undetermined coefficients is banished to near oblivion by the use of "exponential shift" and complex solutions.
- Linear equations of order three or higher are relegated to an appendix.
- An entire chapter is devoted to nonlinear systems in applications. We personally believe that this is where the subject of differential equations really becomes exciting and that students can greatly benefit from an introductory-level taste of some important models from biology, chemistry, and physics.

The book presupposes that the student has completed at least two, preferably three, semesters of calculus. Yet we realize that students at this stage have widely varying levels of mathematical maturity, and so we have taken some care to bring students along fairly slowly in the first few chapters. Overall, we have strive for a presentation that is neither overwhelming to weaker students nor patronizing to stronger ones.

We have aimed for a high degree of flexibility in terms of topic coverage. For instance, it is doubtful that all of the applications in Chapters 2 and 3 can be covered in any but the most applications-oriented course. Indeed, in the first four chapters, only Sections 1.3, 2.1, 3.1, 3.2, 4.2, 4.3 are truly essential. While the book as a whole is designed for a two-semester sequence, a single introductory course (or the first course in a sequence) can be taught in various ways, three of which are offered as follows:

- Thorough coverage of Chapters 1–6.
- Less thorough coverage of Chapters 1–7. (For instance, omit sections 2.2.3, 2.3, 3.5.3, 4.1, 4.5, and 4.6.)
- Minimal coverage of Chapters 1–4 followed by Chapters 8 and 9 plus topics from Chapter 10.

Linear algebra is often not a prerequisite for the differential equations course, though it should be. At some universities it is combined with differential equations to form a single course. Appendix I provides a reasonably complete development of elementary linear algebra, up to and including eigenvalues and eigenvectors. This material can be touched upon or referred to as needed, or it can be wholly integrated into the course just prior to Chapter 8 (Linear First-Order Systems).

There are nearly 1500 exercises in the book. The Hints and Answers section at the end of the book contains hints for many of the less routine problems and answers to most of the odd-numbered problems. The *Instructor's Solutions Manual* contains solutions to all of the problems, and the *Student's Solutions Manual* contains solutions to all of the odd-numbered problems.

The Website `www.prenhall.com/hollis` is the place to find *Mathematica* notebooks, *Maple* worksheets, animations, and general news and information, including the inevitable errata. Most importantly you will find applets for plotting direction fields and phase portraits, providing a convenient way for the reader to solve equations geometrically without access to a computer algebra system.

Acknowledgments

Primary thanks go to John Davis of Baylor University for cheerfully proof reading the manuscript, working all of the problems, and helping to smooth countless rough edges. I also wish to thank the reviewers, who provided much helpful criticism of the manuscript:

Johnny L. Henderson
Auburn University

V. Anne Noonburg
University of Hartford

Alejandro B. Engel
Rensselaer Polytechnic Institute

Philip S. Crooke
Vanderbilt University

M.A.M. Alwash,
University of California, Los Angeles

Krystyna Kuperberg
Auburn University

William L. Siegmann
Rensselaer Polytechnic Institute

Rennie Mirolo
Boston College

Hendrik J. Kuiper
Arizona State University

I am indebted to numerous professors and collaborators who in one way or another have influenced me and helped me become a better mathematician. To each of you, thanks—you know who you are. I thank Ed Wheeler and the rest of my friends and colleagues in, and connected to, the AASU mathematics department for being caring, patient, and dedicated. I also thank my editor George Lobell for his guidance and for wanting to publish the book. Much appreciation also goes to my production editor Jeanne Audino, and my copy editor Patricia M. Daly for their fine work.

Selwyn Hollis
`shollis@armstrong.edu`

To the Student

This book is meant to be adaptable to a variety of teaching styles. While one professor may teach a traditional course that emphasizes pencil-and-paper solution techniques, another may leave the calculations almost entirely to a computer algebra system (*CAS*). We believe that students in a technology-based or technology-enhanced course should not be completely sheltered from computational details, especially those which cast light upon important concepts or reinforce the central ideas of calculus. So we have attempted to present calculations in a useful way without belaboring tedious or insignificant details. Occasionally, we skip to the end of a calculation that is long, laborious, and error prone when done by hand, and we point out that the result was obtained by a *CAS*-aided computation. When reading this book, you should keep in mind your professor's particular emphasis with regard to this issue. It may be perfectly all right to gloss over certain details, accepting results more or less on faith and moving on, but even if not, never allow the trees to obscure your view of the forest.

Essentially the same advice applies with regard to proofs. We prove theorems whenever it is practical and instructive to do so. Often, however, a proof will require ideas from a more advanced course, and so we either omit the proof completely or attempt to explain the essence of it on an elementary level. In case you wish to investigate any of these "gaps" further—and we encourage you to do so—a number of references are suggested in the bibliography (Appendix V). The bibliography also suggests sources where you can pursue various topics in greater depth.

The problem set at the end of each section usually begins with a number of computational problems of a routine nature, particularly in the early chapters. The purpose of these is to facilitate understanding of basic results, and they are designed to be done "by hand." However, in a technology-based course, it may be acceptable for you to do these problems with a computer algebra system. Again, this is up to your professor. Other problems have you fill in certain omitted details in the section, direct you through the development of related concepts, or preview concepts that will be encountered later. Many of the problems will call upon your knowledge of calculus, which you will likely need to refresh from time to time with the help of the well-worn calculus book on your shelf.

The differential equations classroom is usually populated by students with diverse interests and goals. We hope that this book will be interesting, accessible, and relevant to each of you, regardless of your major. Differential equations are where calculus meets the real world—they are "calculus in action." So enjoy the course. This should be the payoff for the time and hard work you've invested in calculus.

DIFFERENTIAL EQUATIONS
WITH BOUNDARY VALUE PROBLEMS

CHAPTER

1

INTRODUCTION

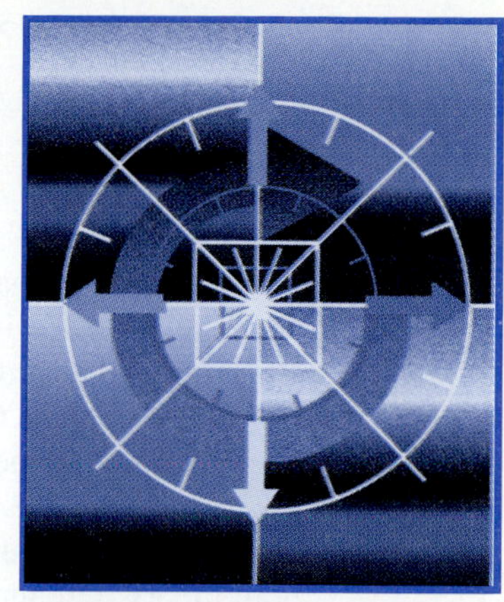

The purpose of this introductory chapter is to acquaint the student with many of the basic ideas related to the study of differential equations in an elementary and informal setting. Section 1.1 describes what differential equations are and attempts to give an indication of their wide range of application. Section 1.2 previews a few major themes in the context of four elementary applications. Section 1.3 gets down to serious business by introducing several fundamental concepts, including order, linearity, homogeneity, and general solution.

1.1 Prologue: What Are Differential Equations?

A differential equation is an equation that involves some unknown function and one or more of its derivatives. Derivatives are taken with respect to some independent variable, which may also appear explicitly in the equation. For example,

$$\frac{dy}{dt} + ty = 2 \quad \text{and} \quad y'' - 3y' + 2y = \sin t$$

are differential equations for the function y of the independent variable t, and

$$\frac{du}{dx} = u^2 \quad \text{and} \quad u'' - xu' + u = x^2$$

are differential equations for the function u of the independent variable x.

A function is said to be a **solution** of a given differential equation, if, upon substitution of the function and its derivatives into the equation, the equation is satisfied for each value of the independent variable in some interval. For example, one can verify that the function $y = te^t$ is a solution of the equation

$$y'' = 2y' - y$$

on the interval $(-\infty, \infty)$ as follows:

$$\begin{aligned}\text{If } y = te^t, \text{ then } y'' - 2y' + y &= e^t(2+t) - 2e^t(t+1) + te^t \\ &= e^t(2 + t - 2t - 2 + t) \\ &= 0.\end{aligned}$$

So we observe that the equation is satisfied for all t in $(-\infty, \infty)$. (Notice the advantage of viewing the equation rearranged as $y'' - 2y' + y = 0$.)

The study of differential equations dates back to Newton and Leibniz and the beginnings of the Calculus in the seventeenth century. Differential equations have occupied many other great minds in the history of mathematics: the Bernoullis, Euler, Lagrange, Laplace, Fourier, Cauchy, and Poincaré among many others. Today the broad field of mathematics known as differential equations is as lively as ever, in both theory and applications. Its numerous subfields together comprise, arguably, the nucleus of all of applied mathematics.

Since its beginning, the study of differential equations has been primarily motivated by applications. To illustrate the applied nature of differential equations, let's look at a basic physical principle with which you are probably familiar. Newton's second law, a fundamental principle in elementary physics, *is* a differential equation. It states that the rate of change in an object's momentum is equal to the net force acting on it; that is,

$$\frac{d}{dt}(mv) = F,$$

where m = mass, v = velocity, and $F = \sum F_i$, with F_i representing individual forces acting on the object. When m is constant, this becomes the perhaps more familiar statement that *mass times acceleration equals force*:

$$m\frac{dv}{dt} = F.$$

If the motion is along a straight path, then this is a single differential equation for the scalar-valued function $v(t)$; otherwise, it is a vector equation consisting of two or three scalar equations, one for each directional component of the velocity. Let us assume for this discussion that the mass is constant and the motion is along a straight path coordinatized by the variable x. Now, in principle, if F were a known function of the single variable t (or constant in the simplest case), then we could determine the velocity $v(t)$ in terms of its value at time $t = 0$ by straightforward integration of F with respect to t. Once the velocity is known, the position $x(t)$ of the object, in terms of its value at time $t = 0$, could also be recovered by integration. However, it may easily happen that F depends on the velocity v or the position x. For example, if air resistance is not ignored, then a model of the vertical motion of a projectile has $F = -mg - kv$, and so

$$m\frac{dv}{dt} = -mg - kv,$$

where g is the acceleration due to gravity and k is a positive "drag coefficient." This complicates matters considerably in that v cannot be found by simply integrating $\frac{dv}{dt}$

with respect to t. (We'll soon discover a fairly simple technique that allows us to find v almost as easily.)

Newton's second law is certainly not the only scientific principle that leads to a differential equation. Among many other elementary applications from physics and elsewhere, we will soon see that

- The decay of a radioactive substance is governed by the differential equation
$$\frac{dA}{dt} = -kA,$$
where k is a positive constant and A denotes the amount of the sample remaining at time t.
- The depth $y(t)$ of water draining through a spigot at the bottom of a cylindrical tank obeys
$$\frac{dy}{dt} = -k\sqrt{y},$$
where k is a positive constant.
- A differential equation that models the motion of a frictionless, simple pendulum is
$$L\theta'' = -g \sin\theta,$$
where $\theta(t)$ is the angle that the pendulum forms at time t with its downward vertical equilibrium position, and g and L are the gravitational acceleration constant and length of the pendulum, respectively.
- A differential equation that models the damped, unforced motion of a mass m attached to a spring is
$$my'' + ry' + ky = 0,$$
where $y(t)$ is the position of the mass, and r and k are damping and spring-stiffness constants, respectively.

Each of the above mentioned applications is based upon assumptions that are well-grounded in Newtonian physics. Other important differential equations arise in empirical models of complex natural phenomena. For instance, a common population model from mathematical biology involves the equation
$$\frac{dP}{dt} = k(1 - P/M)P,$$
where $P(t)$ is the population size at time t, and k and M are positive constants. This is known as the *logistic equation*, upon which there are numerous variations. Such models are typically not expected to accurately mirror the real world; yet they often provide useful qualitative information as well as important insight into the complex mechanisms involved.

The basic models mentioned so far only begin to suggest the crucial role played by differential equations today in wide-ranging areas of science and engineering, including such examples as weather and climate modeling, the design of aircraft and automobiles, and modeling the spread of disease and ground-water contamination.

Another important idea that should be pointed out at this early stage is that when we study a differential equation, actually *finding* a solution—in the sense of writing down an explicit expression involving elementary functions—may be extremely difficult, tedious, or even literally impossible. Consequently, the study of differential equations is about much more than learning "recipes" for deriving solutions. It is also about techniques for describing and understanding the *behavior* of solutions directly from the equation itself, especially when it is impractical, or impossible, to derive an explicit solution of the equation. Among these techniques are numerical approximation of solutions and graphical description of qualitative features of solutions. In some ways our notion of what it means to "solve" an equation will become much more general than that to which you're probably accustomed. If we can completely describe the behavior or the important qualities of the solution of an equation, then we can in some sense say that we have "solved" the equation.

> *The function of mathematics in providing answers is often less important than its function in providing understanding.*
>
> –Ben Noble

PROBLEMS

Remark. The following problems ask you to *verify* or *check* a given solution to some differential equation. The best strategy in general for verifying or checking a solution is to begin by substituting the supposed solution into the left side of the equation and then manipulating until the right side is obtained—*not* by manipulating both sides of the equation until arriving at some silly conclusion such as $0 = 0$! It is often helpful to rearrange the equation first so that the "target" right side of the equation is 0.

Verify that the given function is a solution of the differential equation beside it

1. $y = 5e^{-3t}, \quad \dfrac{dy}{dt} + 3y = 0$

2. $y = \cos 2x, \quad \dfrac{d^2y}{dx^2} + 4y = 0$

3. $u = t^2 + t^3, \quad u''' = 6$

4. $x = te^{-t}, \quad x'' + x = -2x'$

Verify that the given function is a solution of the differential equation beside it for any value of the constant c.

5. $y = ce^{-5t} + e^{-2t}, \quad \dfrac{dy}{dt} + 5y = 3e^{-2t}$

6. $y = \dfrac{c}{ct - 1}, \quad y' + y^2 = 0$

Verify that the given function is a solution of the differential equation beside it for any values of the constants c_1 and c_2.

7. $x = c_1 \cos 3t + c_2 \sin 3t, \quad x'' + 9x = 0$

8. $x = c_1 e^t + c_2 e^{-t}, \quad x'' = x$

9. $y = (c_1 + c_2 t)e^t, \quad y'' - 2y' + y = 0$

10. $x = c_1 t + c_2 t^3, \quad t^2 x'' = 3tx' - 3x$

1.2 Four Introductory Models

In this section we will discuss four elementary differential-equation models of real-world phenomena. Only basic calculus will be used in these discussions; no knowledge of the theory of differential equations is required. Our purpose is twofold:

(1) to illustrate the wide applicability of differential equations, and (2) to introduce informally a few ideas that are fundamental in the study of differential equations.

Radioactive Decay and Carbon Dating Atoms of a radioactive substance break down spontaneously, or decay, into atoms of other elements. The probability that a given atom will decay during a time interval of length Δt is assumed to depend only upon Δt. Let $p(\Delta t)$ denote this probability. Suppose that a sample of material contains a large number of these atoms. Then $p(\Delta t)$ can be interpreted as the proportion of the atoms that decay during any time interval of length Δt. Thus, with $A(t)$ denoting the amount of the radioactive substance in the sample at time t, we have

$$\frac{A(t) - A(t + \Delta t)}{A(t)} = p(\Delta t). \tag{1}$$

We assume further that

$$p(0) = 0 \text{ and } \lim_{\Delta t \to 0^+} \frac{p(\Delta t)}{\Delta t} = k > 0,$$

which essentially means that $p(\Delta t) \approx k\Delta t$ for small Δt. We then rearrange (1) to obtain

$$\frac{A(t + \Delta t) - A(t)}{\Delta t} = -\frac{p(\Delta t) - p(0)}{\Delta t} A(t),$$

which, in the limit as $\Delta t \to 0$, becomes

$$\frac{dA}{dt} = -kA. \tag{2}$$

Thus in a sample of material containing an amount $A(t)$ of a radioactive substance at time t, the rate at which A decreases is proportional to A itself. The positive constant k is called the *decay constant*.

An appropriate method for solving equation (2) is one that should be familiar to the reader from calculus, namely **separation of variables**. First we observe that (2) can be rewritten in the form

$$\frac{A'(t)}{A(t)} = -k,$$

so long as $A(t) \neq 0$. Then we antidifferentiate each side with respect to t:

$$\int \frac{A'(t)}{A(t)}\, dt = \int -k\, dt.$$

This produces

$$\ln |A(t)| = -kt + C,$$

where C is an arbitrary constant that comes from combining the two constants of integration from each side of the equation. Now, using the fact that $A(t)$ must be nonnegative, we solve for $A(t)$ and find

$$A(t) = e^{-kt+C} = e^C e^{-kt}.$$

Since C is an arbitrary constant, e^C is also an arbitrary constant that we can also refer to simply as C. So we rewrite the preceding equation as

$$A(t) = Ce^{-kt}$$

and conclude that every solution of interest has this form for some positive constant C. Substitution of $t = 0$ into the solution reveals that

$$A(0) = C;$$

thus, with $A_0 = A(0)$, we have

$$A(t) = A_0\, e^{-kt}.$$

Each radioactive isotope has a characteristic **half-life** H defined by

$$A(H) = \frac{1}{2}A_0,$$

which, for example, is about 1620 years for radium and roughly 4.5 *billion* years for uranium. The decay constant k can be determined from the half-life as follows. Since

$$A(H) = A_0e^{-kH} = \frac{1}{2}A_0,$$

it follows that

$$-kH = \ln\frac{1}{2}, \quad \text{and so } k = \frac{\ln 2}{H}.$$

Consequently,

$$A(t) = A_0e^{-\frac{\ln 2}{H}t}.$$

Carbon dating is an important method for estimating the age of organic material. The radioactive isotope C^{14} comprises some proportion of the carbon contained in all organic material. It is known to have a half-life of approximately 5570 years. Due to reactions in the upper atmosphere, there is a constant supply of C^{14} in the atmosphere. Plants absorb carbon directly from the atmosphere. Carbon then travels rapidly throughout the food chain into all living organisms. Consequently, the carbon within all living matter contains a near constant fraction of C^{14}, essentially the same as that of the carbon in the surrounding atmosphere. When an organism dies, the C^{14} in it stops being replaced, and the proportion of C^{14} decreases due to radioactive decay. Thus if $A(t)$ represents the amount of C^{14} per gram of carbon at time t, with $t_0 = 0$ representing time of death, then

$$A(t) = A_0e^{-kt}, \quad \text{where } k = \frac{\ln 2}{5570}.$$

Under the sensible assumption that when the organism was alive the atmosphere had essentially the same composition as it does today, we can use the amount of C^{14} per gram of carbon in present-day living matter for A_0.

Example 1

Suppose that a human bone found at an archeological site contains an amount of C^{14} per gram of carbon that is 32% that of presently living matter. We conclude that the bone contains 32% as much C^{14} as it did when it was "fresh." Using the value $\frac{\ln 2}{5570} = .000124$ for the decay constant, we compute the age of the bone from

$$0.32\, A_0 = A_0\, e^{-.000124\, t},$$

which results in

$$t = -\frac{\ln 0.32}{.000124} \approx 9190 \text{ years.}$$

In practice, carbon dating is based upon measurements of the *disintegration rate* of C^{14}, which we label as $r(t)$, measured in disintegrations (or "counts") per minute per gram of carbon. For living organisms this rate is about $r_0 = 15.3$ disintegrations per minute per gram of carbon. The amount of C^{14} per minute per gram of carbon is proportional to this disintegration rate. Thus

$$A(t)/A_0 = r(t)/r_0,$$

and so the disintegration rate satisfies

$$r(t)/r_0 = e^{-\frac{\ln 2}{5570} t}.$$

Solving for t and using $r_0 = 15.3$ gives the following formula for the age t:

$$t = 8040\Big(2.73 - \ln\big(r(t)\big)\Big) \text{ years.}$$

Integration between limits Let's look at a slightly different approach to solving (2) that involves *definite* integration, or *integration between limits.* As before, we begin by rewriting the equation as

$$\frac{A'(t)}{A(t)}\, dt = -k\, dt.$$

We then introduce a dummy variable s in place of t (so that we may use t as a limit of integration) and integrate both sides from $s = 0$ to $s = t$:

$$\int_0^t \frac{A'(s)}{A(s)}\, ds = \int_0^t -k\, ds.$$

The substitution $u = A(s)$ results in

$$\int_{A_0}^{A(t)} \frac{du}{u} = \int_0^t -k\, ds.$$

Now integrating each side gives

$$\ln|A(t)| - \ln|A_0| = \ln\left|\frac{A(t)}{A_0}\right| = -kt,$$

from which we get

$$A(t) = A_0 e^{-kt}.$$

Thus we arrive at the solution without ever seeing the arbitrary C.

Heat Transfer Suppose that some object is surrounded by air or water at some fixed temperature T^*—for instance, a mug of coffee on your desk or a potato baking in the oven. (Such a scenario is a reasonable approximation to reality if the object is small relative to its surroundings.) **Newton's law of cooling** states that the rate at which the temperature of the object changes is proportional to the difference between its temperature $T(t)$ and the temperature T^* of its surroundings. That is,

$$\frac{dT}{dt} = -k(T - T^*).$$

Let the initial temperature of the object be T_0. Separating variables and integrating over the time interval $[0, t]$ leads to

$$\int_0^t \frac{T'(s)}{T(s) - T^*}\, ds = \int_0^t -k\, ds.$$

By means of the substitution $u = T(s) - T^*$, we arrive at

$$\ln\left|T(t) - T^*\right| - \ln\left|T_0 - T^*\right| = -kt,$$

from which follows

$$\ln\left|\frac{T(t) - T^*}{T_0 - T^*}\right| = -kt$$

and then

$$T(t) = T^* + (T_0 - T^*)e^{-kt}.$$

Example 2

Suppose that we put a bottle of water, initially at a temperature of 75°F, into a freezer in which the temperature is 10°F. Naturally, we would like to know how long it will take the water to cool to the freezing temperature of 32°. Using $T_0 = 75$ and $T^* = 10$, our model for the temperature t minutes later is

$$T(t) = 10 + 65e^{-kt},$$

and so we solve the equation

$$32 = 10 + 65e^{-kt}$$

for t, obtaining

$$t = \frac{1}{k} \ln \frac{65}{22}.$$

This is not very useful, since we don't know the value of k. So we conduct an experiment: Ten minutes after we put the can in the freezer, we measure the temperature of the water and find that so far it has dropped to 55°. We put this information into our model and see that

$$55 = 10 + 65e^{-10k},$$

which allows us to solve for k:

$$k = \frac{1}{10}\ln\frac{13}{9} \approx 0.037.$$

Finally, we estimate that the water will reach 32° at

$$t = \frac{10\ln\frac{65}{22}}{\ln\frac{13}{9}} \approx 29.5 \text{ minutes.}$$

■

The Spread of Disease Consider a contagious disease—such as smallpox or measles—that results in either quick death or recovery with future immunity. To study the spread of the disease within a population, let us consider the proportion of individuals of age t (in years) who have not had the disease and are therefore still susceptible to it.* We assume that the age-t population, for $t \geq 0$, consists of

$$x(t) \text{ living individuals,}$$

among whom there are

$$y(t) \text{ \textit{susceptible} individuals}$$

(i.e., those who have not yet caught the disease). Then the *proportion* of the age-t population who are susceptible to the disease is $y(t)/x(t)$. To model this situation, we make the following assumptions regarding the changes in x and y during any short time interval $[t, t + \Delta t]$.

i. Due to all non-disease-related causes, the number of age-t individuals, or any subgroup thereof, decreases by some age-dependent factor $R(t)\Delta t$.
ii. The number of age-t susceptibles who contract the disease is jointly proportional to $y(t)$ and Δt—and thus given by $\alpha y(t)\Delta t$ for some positive constant $\alpha < 1$. (This means, for instance, that the *proportion* of age-t susceptibles who contract the disease each month is approximately $\alpha/12$.)
iii. The disease is fatal to some proportion δ of all those who contract it. So the number of age-t individuals who die from the disease during $[t, t + \Delta t]$ is proportional to the number of age-t susceptibles who contract the disease—and thus given by $\delta\alpha y(t)\Delta t$ for some positive constant $\delta < 1$.
iv. At age $t = 0$, the population consists entirely of susceptibles; i.e., $x(0) = y(0)$.

---

* The model presented here parallels Daniel Bernoulli's analysis of smallpox in 1760.

In light of assumptions (i) and (ii), we have the following statement about the change in the number of age-t susceptibles:

$$y(t+\Delta t) - y(t) \approx -\alpha y(t)\Delta t - R(t)\Delta t y(t).$$

In light of assumptions (i) and (iii), we have the following statement about the change in the overall number of living age-t individuals:

$$x(t+\Delta t) - x(t) \approx -\delta\alpha y(t)\Delta t - R(t)\Delta t x(t).$$

After dividing each of these by Δt and letting $\Delta t \to 0$, we arrive at a pair of differential equations for x and y:

$$\frac{dy}{dt} = -\alpha y - Ry$$
$$\frac{dx}{dt} = -\delta\alpha y - Rx.$$

The primary difficulty here is that we have no model for the the function R. However, note that we can eliminate R as follows:

$$y\frac{dx}{dt} - x\frac{dy}{dt} = -\delta\alpha y^2 + \alpha x y.$$

Dividing each side of this equation by y^2 produces

$$\frac{d}{dt}\left(\frac{x}{y}\right) = -\delta\alpha + \alpha\frac{x}{y}.$$

Thus the quantity $q = x/y$—the reciprocal of the proportion of age-t individuals still susceptible to the disease—satisfies

$$\frac{dq}{dt} = \alpha q - \delta\alpha, \quad q(0) = 1,$$

which can be solved easily by separation of variables. (The initial value of 1 results from assumption (iv).) So we divide by $q - \delta$ and integrate over $[0, t]$:

$$\int_0^t \frac{q'(s)}{q(s) - \delta}\, ds = \int_0^t \alpha\, ds.$$

Then we substitute $u(s) = q(s) - \delta$:

$$\int_{1-\delta}^{q(t)-\delta} \frac{du}{u} = \int_0^t \alpha\, ds.$$

Thus we find

$$\ln\frac{|q(t)-\delta|}{|1-\delta|} = \alpha t.$$

Since $q(t) \geq 1$ and $0 < \delta < 1$, this results in

$$q(t) = \delta + (1-\delta)e^{\alpha t}.$$

So the proportion of the age-t population still susceptible to the disease is

$$\frac{y(t)}{x(t)} = \frac{1}{q(t)} = \frac{1}{\delta + (1-\delta)e^{\alpha t}}.$$

Example 3

Suppose that 1% of all age-t susceptibles contract the disease each month. Then $\alpha \approx (.01)(12) = 0.12$. Suppose further that the disease is fatal to $\delta = \frac{1}{10}$ of all individuals who contract it. The proportion of the age-t population still susceptible to the disease is then

$$\frac{y(t)}{x(t)} = \frac{10}{1 + 9e^{0.12t}}.$$

According to this model, approximately 32% of age-10 individuals and approximately 10% of age-20 individuals will still be susceptible to the disease. ■

The Simple Pendulum Consider a simple pendulum consisting of a frictionless pivot and a bob with mass m attached to an arm of length L with negligible mass. If $\theta(t)$ is the angle that the arm of the pendulum makes with the vertical at time t, then the position of the mass is described by $s(t) = L\theta(t)$, and the component of the force of gravity in the direction tangent to the path is given by

$$F_{\tan} = -mg\sin\theta.$$

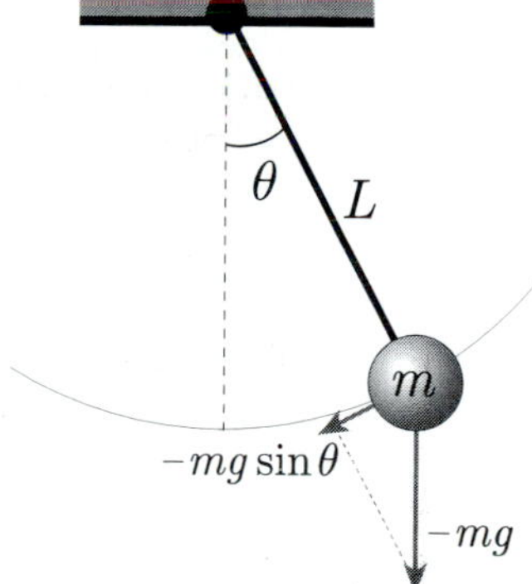

Thus, by Newton's second law, we have

$$m(L\theta)'' = -mg\sin\theta;$$

that is,

$$\theta'' = -\frac{g}{L}\sin\theta.$$

This differential equation differs from the preceding ones in this section in two fundamental ways. It is a *second-order* equation because of the presence of the second derivative. It is also a *nonlinear* equation. (These terms will be defined in the next section.) Moreover, it turns out that *we will never be able to express (nonconstant) solutions of this equation in terms of elementary functions.* However, if we assume that the angle θ is always small, we may use the approximation $\sin\theta \approx \theta$ to *linearize* the equation, obtaining

$$\theta'' = -\frac{g}{L}\theta.$$

For this equation, we simply observe that $\cos\omega t$ and $\sin\omega t$ are solutions, where $\omega = \sqrt{g/L}$, as is any *linear combination*

$$\theta(t) = A\cos\omega t + B\sin\omega t,$$

where A and B are constants. So we have found that the period of low-amplitude motion is (approximately) $2\pi/\omega = 2\pi\sqrt{L/g}$ and is therefore (approximately) proportional to $\sqrt{L}$ and (approximately) independent of the amplitude.

PROBLEMS

1. Consider the general exponential growth/decay problem

$$\frac{dy}{dt} = ky, \quad y(t_0) = y_0.$$

Use separation of variables and integration between limits:

$$\int_{t_0}^{t} \frac{y'(s)}{y(s)}\, ds = \int_{t_0}^{t} k\, dt$$

to arrive at the solution

$$y = y_0 e^{k(t-t_0)}.$$

Use the result of Problem 1 to write down by inspection the solution of each of the initial value problems in Problems 2 through 7.

2. $\dfrac{dy}{dt} = -6y, \; y(0) = 2$

3. $y' = 0.03y, \; y(0) = 1$

4. $\dfrac{dx}{dt} = x, \; x(1) = 2$

5. $\dfrac{du}{dx} = -\dfrac{u}{10}, \; u(3) = 3$

6. $\dfrac{du}{dt} = 1.01u, \; u(0) = 10$

7. $\dfrac{d\varphi}{dz} = -\dfrac{3}{8}\varphi, \; \varphi(-1) = 5$

8. Suppose that 5% of a radioactive substance has disintegrated after 2 years. Find the decay constant k and the half-life H of this material.

9. Suppose that the half-life of a radioactive isotope is 1000 years. What percentage of an initial amount would remain after 10, 100, and 500 years?

10. A sample from a wood artifact is analyzed and determined to contain C^{14} decaying at a rate of 9.8 counts per minute per gram of carbon. Date the sample.

11. A hat discovered in a Florentine flea market is claimed to have belonged to Leonardo da Vinci, who lived in the second half of the 15th century. Analysis reveals a C^{14} disintegration rate of 15 counts per second per gram of carbon. Could the hat have belonged to Leonardo?

12. U^{238} decays into Pb and has a half-life of 4.51×10^9 years. Suppose that a stone was formed long ago and originally contained no lead. The stone is analyzed and found to contain U^{238} atoms and Pb atoms in a ratio of 0.85. How old is the stone?

13. Suppose that a stone cools from 80°C to 70°C in 5 minutes in a room at a temperature of 20°C. How long until it cools to 30°C?

14. Suppose that a covered† pot of boiling water (212°F) began to cool some time ago in a room at 70°F. You measure temperatures of 110°F and 100°F fifteen minutes apart. When did the pot begin to cool?

---

† An *uncovered* pot of very hot water would cool more quickly, since much of its heat would be lost by the release of steam.

15. A murder victim's body is found in a room where the temperature is 70°F, and the temperature of the body is found to be 78°F. It is assumed that the temperature of the room has been roughly constant since the time of death. From the victim's body weight and type of clothing, you have an estimate of $k \approx .25$ hours^{-1} for the temperature decay rate. Estimate the time of death, given that the body temperature at that time was about 99°F.

16. Suppose that the number of bacteria $N(t)$ in a culture obeys the differential equation

$$\frac{dN}{dt} = kN.$$

Given that the population is observed to number 10,000 at time $t = 0$ and 12,000 at $t = 2$ hours, determine the constant k, and then find the amount of time T_2 that it takes for the population to double in size.

17. Suppose that the number of bacteria $N(t)$ in a culture obeys the differential equation $N' = kN$. Derive a formula for the tripling time T_3 of the population in terms of the doubling time T_2.

18. A motorboat is moving on still water at a speed of 30 mph (44 ft/s). At time $t = 0$, the motor is shut off, and thereafter its velocity obeys

$$\frac{dv}{dt} = -kv.$$

Given that after ten seconds the speed has dropped to 15 mph, solve for the velocity $v(t)$ and then the position $x(t)$. Calculate how far the boat will have drifted after 10, 20, and ∞ seconds.

19. Rework Example 3, assuming that 5% of the susceptible population contract the disease each month and that the disease is fatal to 1/2 of those who contract it. In particular, find the approximate proportion of the population still susceptible at age 10 and at age 20.

20. Consider the spread of disease in an age-t population. Modify the model to include inoculation as follows. Assume that the number of age-t susceptibles who become immune by inoculation during $[t, t + \Delta t]$ is jointly proportional to the number of age-t susceptibles and Δt—and thus given by $r.y(t)\Delta t$ for some positive constant $r < 1$. Then do the following.

(a) Show that the differential equations for $x(t)$ and $y(t)$ become

$$\frac{dx}{dt} = -\delta\alpha y - Rx, \qquad \frac{dy}{dt} = -(\alpha + r)y - Ry.$$

(b) Find $y(t)/x(t)$.

(c) Rework Example 3, assuming that 2% of the susceptible population become immune by inoculation each month. In particular, find the approximate proportion of the population still susceptible at age 10 and at age 20.

21. Consider the (approximate) small-amplitude motion of a simple pendulum:

$$\theta(t) = A\cos\omega t + B\sin\omega t, \qquad \omega = \sqrt{g/L}.$$

(a) Find L so that the period of the pendulum will be 2 seconds. (Use $g \approx 9.8$ m/s^2.)

(b) Using L from part (a), find $\theta(t)$ if the pendulum is gently released at time $t = 0$ (so $\theta'(0) = 0$) with an initial angle of $\theta(0) = \pi/18$.

(c) Again using L from part (a), find $\theta(t)$ if the pendulum is set in motion at time $t = 0$ with an initial angle $\theta(0) = 0$ and initial angular velocity $\theta'(0) = 1/2$ s^{-1}. What is the amplitude of the resulting motion in degrees?

22. A model of the small-amplitude motion of a *damped* pendulum leads to the differential equation

$$\theta'' = -\frac{r}{m}\theta' - \frac{g}{L}\theta,$$

where r is a positive coefficient of friction, and m is the mass of the bob. Show that, if r is sufficiently small and $k = \frac{r}{2m}$, then there is some $\omega > 0$ so that $e^{-kt}\cos\omega t$ and $e^{-kt}\sin\omega t$ are solutions.

23. Suppose that $y(t)$ satisfies $y' = ky$. Multiply both sides of the equation by dt and integrate over $[a, b]$ to find that

$$y(b) - y(a) = k\int_a^b y(t)\,dt, \quad \text{and thus} \quad \frac{y(b) - y(a)}{b - a} = k\tilde{y}$$

where $\tilde{y}$ is the average value of $y(t)$ over $[a, b]$. Interpret this result in the case where $k = 1$.

24. Consider a continuous function $y(t)$ on an interval $[a, b]$. Two of the more famous theorems from calculus tell us that

1) there is some $\tilde{t} \in (a, b)$ where $y(\tilde{t}) = \dfrac{1}{b-a}\displaystyle\int_a^b y(t)\,dt$, and

2) if y' exists on (a, b), then $y'(t^*) = \dfrac{y(b) - y(a)}{b - a}$ for some $t^* \in (a, b)$.

Show that if $y(t) = Ce^{kt}$, then there will be only one such t^* and one such $\tilde{t}$, *and furthermore* $t^* = \tilde{t}$. (*Suggestion*: Use Problem 23.) Interpret the result and illustrate with a picture.

1.3 Fundamental Concepts and Terminology

Differential equations are traditionally grouped into various categories. At the most general level, a differential equation is either an *ordinary* differential equation (ODE) or a *partial* differential equation (PDE). Ordinary differential equations involve an unknown function of one variable and one or more of its derivatives. Partial differential equations involve an unknown function of two or more variables and its partial derivatives.* The following are examples of typical ordinary and partial differential equations, respectively:

$$x' + 3t = \cos t, \qquad \frac{\partial^2 u}{\partial x^2} + \frac{\partial^2 u}{\partial y^2} = \sin x \cos y.$$

In any ordinary differential equation, all derivatives are with respect to a single independent variable; so prime notation for derivatives (as in x' or y'') is often used when the identity of the independent variable is not ambiguous. For instance, in the first of the preceding equations, x' is assumed to mean $\frac{dx}{dt}$.

Order The **order** of a differential equation is the order of the highest-order derivative appearing in the equation. Here are examples of first-, second-, and fourth-order

---

* This book deals mainly with ordinary differential equations; so by "differential equation" we usually mean *ordinary* differential equation.

equations, respectively:

$$x' + tx = 1, \quad y'' + t^3y' - y = 0, \quad y^{(4)} - y = x.$$

(Notice the notation $y^{(n)}$ for the nth derivative of y with respect to the independent variable.) All of the nth-order equations with which we will be concerned can be solved for $y^{(n)}$ in terms of a (single-valued) function of all other variables. Thus we write a generic nth-order differential equation (with dependent variable y and independent variable t) as

$$y^{(n)} = F(t, y, y', \dots, y^{(n-1)}),$$

where F is some function of $n + 1$ variables.† So first- and second-order equations may be written, respectively, in the forms

$$y' = F(t, y), \quad y'' = F(t, y, y'),$$

and so on.

Linearity All differential equations can be categorized as either *linear* or *nonlinear*. A differential equation is linear if it is linear in the unknown function and all its derivatives; otherwise it is nonlinear. Two examples of linear differential equations are

$$x' + 2x = t^3, \quad y'' + t^3y' - y = \sin t.$$

Notice that each of these linear equations is actually nonlinear in the independent variable t. These two differential equations are nonlinear:

$$x' + 2x^2 = t^3, \quad y'' + yy' = \sin t.$$

The first equation is nonlinear because of the x^2 term, and the second equation is nonlinear because of the yy' term.

To be more precise, a **linear differential equation** (for an unknown function y) is one that can be put into the form

$$p_n y^{(n)} + p_{n-1} y^{(n-1)} + \cdots + p_1 y' + p_0 y = f, \tag{L}$$

where $y^{(j)}$ denotes the jth-order derivative of y and $f, p_0, \dots, p_n$ are given functions of the independent variable. The equation is said to have *constant coefficients* if $p_0, \dots, p_n$ are constants. Note that (L) is an nth-order equation if $p_n \neq 0$.

Generally speaking, linear equations are much easier to deal with than nonlinear equations. In fact, it may be said that the theory of linear equations is thoroughly understood in general, while the theory of nonlinear equations is much more complicated and not nearly as well understood.

---

† This excludes, for example, equations such as $(y')^2 - y = t$ and $\tan(y'') - y^2 = 0$.

Homogeneity All differential equations may also be categorized as either *homogeneous* or *nonhomogeneous*. A differential equation is said to be **homogeneous** if, given any solution u, the function cu is also a solution for any constant c. Otherwise, the equation is **nonhomogeneous**.

Consider the two (nonlinear) equations

$$yy'' - (y')^2 = 0 \text{ and } y' - y^2 = 0.$$

The first of these is homogeneous because, given any solution u, the function cu satisfies the same equation for any constant c:

$$(cu)(cu)'' - ((cu)')^2 = c^2\left(uu'' - (u')^2\right) = c^2 \cdot 0 = 0.$$

The second equation is nonhomogeneous because, given a nonzero solution u, the function $2u$, for example, does not satisfy the same equation:

$$(2u)' - (2u)^2 = 2u' + 4u^2 = 2(u' + u^2 + u^2) = 2(0 + u^2) \neq 0.$$

Of the two (linear) equations

$$y' + t^3 y + \cos t = 0 \text{ and } y' + t^3 y = 0,$$

the first is nonhomogeneous, and the second is homogeneous, as you should verify yourself. Note that the only difference between the two equations is that the nonhomogeneous equation has a nonzero term that does not involve y or any of its derivatives.

It should be easy to see that *the general linear differential equation (L) is homogeneous precisely when $f = 0$ and nonhomogeneous otherwise.* Thus homogeneous linear differential equations are easy to recognize. (Can you describe a way to recognize homogeneous *nonlinear* differential equations?)

Superposition The notion of homogeneity plays an important role in understanding the set of solutions of any linear differential equation, even a nonhomogeneous one. In fact, from now on, homogeneity will enter our discussions only in the context of linear equations. The following theorem states a very important property of solutions to a homogeneous linear differential equation. Its proof is easy and instructive.

Theorem 1

If u and v are any two solutions of

$$p_n y^{(n)} + p_{n-1} y^{(n-1)} + \cdots + p_1 y' + p_0 y = 0, \tag{H}$$

then $c_1 u + c_2 v$ is also a solution for any constants c_1 and c_2. ▲

Proof Suppose that u and v satisfy (H) and that c_1 and c_2 are constants. Then

$$\begin{aligned} &p_n(c_1u + c_2v)^{(n)} + \cdots + p_1(c_1u + c_2v)' + p_0(c_1u + c_2v) \\ &\qquad = p_n(c_1u^{(n)} + c_2v^{(n)}) + \cdots + p_1(c_1u' + c_2v') + p_0(c_1u + c_2v) \end{aligned}$$

$$= c_1(p_n u^{(n)} + \cdots + p_1 u' + p_0 u) + c_2(p_n v^{(n)} + \cdots + p_1 v' + p_0 v)$$
$$= c_1 \cdot 0 + c_2 \cdot 0 = 0. \quad \bullet$$

Example 1

Consider the second-order equation $y'' + y = 0$. This is a homogeneous, linear equation of which it is easy to observe that $u = \cos t$ and $v = \sin t$ are solutions. Therefore, $y = c_1 \cos t + c_2 \sin t$ is a solution for any choice of constants c_1 and c_2. ■

Another important consequence of linearity and homogeneity is the following **superposition principle**.

Theorem 2

If u is a solution of (L) and v is a solution of (H), then $u + v$ is a solution of (L). Moreover, given any solution u of (L), every solution of (L) is of the form $u+v$ where v is some solution of (H). ▲

Proof The first assertion is left for the reader to verify. To see the second assertion, suppose that u is some particular solution of (L) and that w is any other solution of (L). Then the difference $w - u$ satisfies (H). Therefore, if we set $v = w - u$, then v is a solution of (H), and $w = u + v$. ●

Example 2

Consider the first-order, linear equation $y' + 3y = 6$. By inspection, the constant function $u = 2$ is a particular solution. Separation of variables reveals that every solution of $y' + 3y = 0$ is given by Ce^{-3t} for some constant C. Therefore, $y = Ce^{-3t} + 2$ is a solution of $y' + 3y = 6$ for any constant C, and *every* solution is of this form. ■

"Particular" Solutions We will often use the term **particular solution** to emphasize that we are talking about one solution rather than a family of solutions. For a number of reasons—including the superposition principle in Theorem 2—we are often interested in finding *some* particular solution of a differential equation. This can sometimes be done by looking for a solution in some special form. The most basic of these forms is a constant or *equilibrium* solution. Other simple forms include exponential solutions, trigonometric solutions, or powers of the independent variable.

Example 3

Substitution of $y = a$ into the linear equation

$$y'' + 2y' + 3y = 6$$

gives us $3a = 6$, or $a = 2$. Thus the equation has the equilibrium solution $y = 2$. (Note that this solution may just as easily have been obtained by inspection!) ■

Example 4

Substitution of $y = a$ into the nonlinear equation

$$y' = y^2 - 3y + 2$$

gives us $0 = a^2 - 3\,a + 2$, and so factoring produces $a = 1, 2$. Thus we have two equilibrium solutions $y = 1$ and $y = 2$. ■

Example 5

Consider the linear equation

$$y'' - 4y' + 3y = 0.$$

It is easy to see that this equation has no equilibrium solutions other than the trivial one, $y = 0$. Let's see if it has any exponential solutions. Substitution of $y = e^{kt}$ into the equation gives us

$$k^2e^{kt} - 4ke^{kt} + 3e^{kt} = 0.$$

Now dividing through by e^{kt} produces the quadratic equation $k^2 - 4k + 3 = 0$ for the constant k. Factoring shows two solutions $k = 1, 3$. Thus $y = e^t$ and $y = e^{3t}$ are two particular solutions. ■

Example 6

It is sensible to think that the linear equation

$$y'' + y' + y = 13\cos 2t$$

might have a trigonometric solution of the form $y = a\cos 2t + b\sin 2t$. (Why?) Substitution of this form into the equation gives

$$-4a\cos 2t - 4b\sin 2t - 2a\sin 2t + 2b\cos 2t + a\cos 2t + b\sin 2t = 13\cos 2t,$$

which when simplified becomes

$$(-3a + 2b)\cos 2t - (3b + 2a)\sin 2t = 13\cos 2t.$$

Equating coefficients gives two equations for a and b:

$$-3a + 2b = 13,\ 3b + 2a = 0;\ \text{thus } a = -3, b = 2.$$

So a particular solution is $y = -3\cos 2t + 2\sin 2t$. ■

Example 7

Here we'll show that the nonlinear equation

$$y''y - 2(y')^2 = 0$$

has solutions of the form $y = t^a$. Substitution into the equation gives us

$$a(a-1)t^{a-2}t^a - 2a^2t^{2(a-1)} = 0,$$

which is an identity if $a(a-1) - 2a^2 = 0$. A little algebra produces two solutions: $a = 0, -1$. Thus $y = 1$ and $y = t^{-1}$ are particular solutions of the differential equation. (Note that any constant is an equilibrium solution of this equation, and, since the equation is homogeneous, ct^{-1} is a solution for any constant c.) ■

General Solutions A **general solution** of an nth-order differential equation,

$$y^{(n)} = F(t, y, y', \ldots, y^{(n-1)}),$$

is an *n-parameter family* of solutions—that is, a family of solutions that can be expressed in terms of no fewer than n independent, unrestricted parameters. For instance, antidifferentiating twice shows that the simple second-order equation

$$y'' = 6t$$

has the general solution

$$y = t^3 + c_1t + c_2.$$

The parameters c_1 and c_2 in this case are just constants of integration.

Example 8

Consider the first-order equation

$$\frac{dy}{dt} = y^2.$$

Since the equation is of first order, any one-parameter family of solutions will be a general solution. After separating variables and antidifferentiating:

$$\int \frac{y'(t)}{y(t)^2}\,dt = \int dt,$$

we find that *nonzero* solutions of the differential equation satisfy

$$-\frac{1}{y(t)} = t + C$$

for some constant C. So a general solution given by

$$y = \frac{1}{-C - t}.$$

If we replace C with $-\frac{1}{c}$, we get a slightly different general solution:

$$y = \frac{c}{1 - ct},$$

which has the advantage of containing the zero solution (with $c = 0$) as well as all of the nonzero solutions. ■

Example 9

Consider the second-order equation

$$y'' - y = 0.$$

It is easy to verify that both e^t and e^{-t} are solutions. Since the equation is linear and homogeneous, it is also easy to check that

$$y = c_1 e^t + c_2 e^{-t}$$

is a solution for any constants c_1 and c_2, and, since we can describe this family with no fewer than two parameters, it is a general solution. One can also observe that $\cosh t$ and $\sinh t$ are solutions; thus the family described by

$$y = c_1 \cosh t + c_2 \sinh t$$

is also a general solution. These two general solutions are actually equivalent, since they contain precisely the same functions. ■

Initial-Value Problems An nth-order **initial-value problem** consists of an nth-order differential equation

$$y^{(n)} = F(t, y, y', \ldots, y^{(n-1)}),$$

along with n *initial conditions*

$$y(t_0) = y_0, y'(t_0) = y_1, \ldots, y^{(n-1)}(t_0) = y_{n-1},$$

where $t_0, y_0, y_1, \ldots, y_{n-1}$ are given real numbers. So a first-order initial-value problem has one initial condition:

$$y' = F(t, y), y(t_0) = y_0,$$

and an initial-value problem for a second-order differential equation has two initial conditions:

$$y'' = F(t, y, y'), y(t_0) = y_0, y'(t_0) = y_1.$$

A **solution** of an nth-order initial-value problem is an n-times differentiable function y that (1) satisfies the differential equation for all t in some interval I containing t_0, and (2) satisfies the initial conditions at t_0. The *domain* of the solution is understood to be the interval I; indeed, the domain *must* be an interval.[‡] If the solution cannot be extended so as to be valid on a larger interval that contains I, then the domain I is said to be *maximal*. A solution whose domain is maximal is called a **maximal solution**.

[‡] The interval I may be closed, open, or "half open." At any endpoint that I contains, limits and derivatives are understood in the one-sided sense from within the interval. For a more thorough discussion of this issue, see Appendix II.

A solution with domain I is **unique** on I if there are no other solutions defined on I. When we talk about *the* solution of an initial-value problem, we are referring to a unique, maximal solution.[§]

Notice that the number of initial conditions in an nth-order initial-value problem coincides with the number of independent parameters in a general solution of the differential equation. Thus if we have a general solution, the initial conditions become a system of n equations for the n parameters. The next two examples illustrate this.

Example 10

Consider the first-order initial-value problem

$$\frac{dy}{dt} = y^2,\ y(0) = y_0.$$

In Example 1.8, we found the general solution

$$y = \frac{c}{1 - ct}.$$

The initial condition $y(0) = y_0$ is met if we choose $c = y_0$. So a solution of the initial-value problem is given by

$$y = \frac{y_0}{1 - y_0 t} \quad \text{for } t \text{ in } I,$$

where I is any interval that contains $t = 0$ but *not* $t = 1/y_0$. Therefore, the maximal domain is

$$I = \begin{cases} (-\infty, 1/y_0), & \text{if } y_0 > 0; \\ (1/y_0, \infty), & \text{if } y_0 < 0; \\ (-\infty, \infty), & \text{if } y_0 = 0. \end{cases}$$

Note that, if $y_0 \neq 0$, the solution "blows up" (i.e., approaches $\pm\infty$) as t approaches $1/y_0$ from within I. (The graph of the solution with $y_0 = \frac{1}{2}$ is shown in Figure 1.) ■

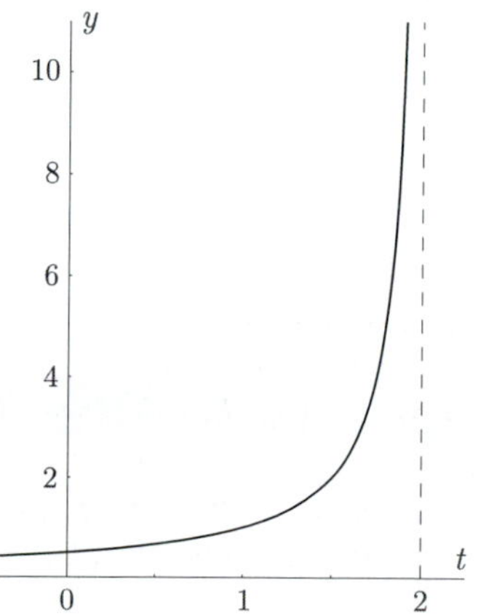

Figure 1

[§] The issues of uniqueness and maximality will be discussed in more detail in Section 3.2.

Example 11

Consider the second-order initial-value problem

$$y'' - y = 0,\ y(0) = y_0,\ y'(0) = y_1.$$

In Example 1.2 we found the general solution

$$y = c_1 e^t + c_2 e^{-t}.$$

The initial condition $y(0) = y_0$ is met if

$$c_1 + c_2 = y_0,$$

and the initial condition $y'(0) = y_1$ is met if

$$c_1 - c_2 = y_1.$$

Solving these two equations yields

$$c_1 = \frac{y_0 + y_1}{2} \text{ and } c_1 = \frac{y_0 - y_1}{2}.$$

So the function

$$y = \frac{1}{2}\left((y_0 + y_1)e^t + (y_0 - y_1)e^{-t}\right) \text{ for } t \text{ in } I$$

is a solution of the initial-value problem for any interval I containing $t = 0$—in particular, for the maximal domain $I = (-\infty, \infty)$. If we instead use

$$y = c_1 \cosh t + c_2 \sinh t$$

as the general solution, we easily find

$$c_1 = y_0 \text{ and } c_2 = y_1,$$

with which we write the solution equivalently as

$$y = y_0 \cosh t + y_1 \sinh t.$$

■

PROBLEMS

Classify each of the differential equations in 1 through 9

(a) according to order;

(b) as linear or nonlinear;

(c) as homogeneous or nonhomogeneous.

1. $\frac{dx}{dt} - x + t^3 = 0$

2. $y^{(4)}(x) = x^3 y$

3. $\frac{dx}{dt} = \sin x$

4. $x'' - xx' = 1$

5. $y' = y^2 - y - 1$

6. $y'' = y' + 2y$

7. $u''' + xu' + u = \cos x$

8. $\theta''' = t\theta\theta'$

9. $y'' = ((y')^3 + y^3)^{1/3}$

In Problems 10 through 14, verify that the given family of functions satisfies the linear equation beside it.

10. $y = ce^{-t} + \frac{1}{2}(\sin t - \cos t), \quad y' + y = \sin t$

11. $y = \left(c_1 - \frac{t}{2}\right)\cos t + c_2 \sin t, \quad y'' + y = \sin t$

12. $y = \left(c_1 e^{-t} + c_2 e^{t}\right)\cos t + \left(c_3 e^{-t} + c_4 e^{t} + \frac{1}{5}\right)\sin t,$
$y^{(4)} + 4y = \sin t$

13. $y = \frac{1}{t}(c_1 + c_2 \ln t), \quad t^2 y'' + 3ty' + y = 0$

14. $y = t\left(c_1 \cos(\ln t) + c_2 \sin(\ln t) + 1\right),$
$t^2 y'' - ty' + 2y = t$

Find all equilibrium solutions of each equation in 15–18.

15. $y'' + y' + y(6 - y) = 5$

16. $y' = (y - 1)(y - 2)(y - 3)$

17. $y' = \sin y$

18. $y'' = yy'$

Show that each equation in 19 through 21 has no nontrivial equilibrium solution.

19. $\dfrac{dy}{dt} + ty = 0$

20. $y'' + y' + y = 0$

21. $y''' - y' \sin t = y$

Show that each equation in 22 through 24 has no equilibrium solution.

22. $\dfrac{dy}{dt} + ty = 1$

23. $y'' + y' + y = t$

24. $y''' - y' \sin t = 1$

Find particular solutions of the form e^{kt} for the equations in 25–27.

25. $y'' - 4y = 0$

26. $y''' + 8y = 0$

27. $3y'''y - y''y' + 16y^2 = 0$

Find particular solutions of the form t^a for the equations in 28 through 31.

28. $t^2 y'' - 4ty' + 6y = 0$

29. $t^2 yy'' - (ty')^2 = 5y^2$

30. $t^2 yy'' + (ty')^2 + tyy' = 8y^2$

31. $y'' + (3y - ty')^3 \cos t = 6t$

32. Each of the equations in 28 through 31 has one or two solutions of the form t^a.

(a) The equation in 28 is homogeneous and linear. What can we conclude from the homogeneity and linearity of this equation?

(b) The nonlinear equations in 29 and 30 are homogeneous. What can we conclude from the homogeneity of those equations?

(c) Show that the equation in 31 is not homogeneous.

33. Find a general solution of each of the equations $y' = kt$ and $y' = ky$. Think about the difference between the two equations. Convince yourself that solving $y' = f(t)$ amounts to antidifferentiating $f(t)$, while solving $y' = f(y)$ amounts mainly to antidifferentiating $1/f(y)$.

34. Find the general solution of $\frac{dy}{dt} - 2y = 3$ by separation of variables.

35. Find the general solution of $\frac{dy}{dt} - 2y = 3$ by superimposing the general solution of $\frac{dy}{dt} - 2y = 0$ with a constant particular solution of $\frac{dy}{dt} - 2y = 3$.

36. Find the general solution of $y' - 2y = 3e^t$ by superimposing the general solution of $y' - 2y = 0$ with a particular solution of $y' - 2y = 3e^t$ in the form of ae^t.

37. Find the general solution of $y' - 2y = 3e^{2t}$ by superimposing the general solution of $y' - 2y = 0$ with a particular solution of $y' - 2y = 3e^{2t}$ in the form of ate^{2t}.

38. Solve the initial-value problem $\frac{dy}{dt} - 2y = 3$, $y(0) = 1$,

(a) by determining C in the general solution (see Problems 34, 35);

(b) by separating variables and integrating between limits.

39. Here we will construct the general solution of $y' - y = \sin t$ by superposition.

(a) Find a particular trigonometric solution by substituting $u(t) = a \sin t + b \cos t$ into the equation and solving for a and b.

(b) Add $u(t)$ to the general solution of $y' - y = 0$ to obtain the general solution of our problem.

(c) Check the result of (b).

Write down *by inspection* a *simple* particular solution of each of the equations in 40 through 42.

40. $x^2\dfrac{dx}{dt}+x^3=8$ **41.** $y''+y=1-x$

42. $y'''(t)=-8\cos 2t$

Solve the equation in each of 43 and 44 by separation of variables.

43. $\dfrac{dy}{dt}=-y^2,\ y(0)=1$

44. $x'=x^2\cos t,\ x(0)=1$

45. Verify that $y(t)=y_0e^{-kt}+e^{-kt}\int_0^t e^{ks}f(s)ds$ solves the initial-value problem

$$\frac{dy}{dt}+ky=f(t),\, y(0)=y_0,$$

where f is a given continuous function. You'll need the part of the Fundamental Theorem of Calculus that you may have purged from your memory, namely that if g is continuous then $\frac{d}{dt}\int_{t_0}^t g(s)\,ds=g(t)$.

46. Use the result of Problem 45 to write down the solution of the initial-value problem

$$\frac{dy}{dt}-2y=\cos(t^2),\, y(0)=1.$$

Is there any hope of expressing the integral in terms of elementary functions?

47. Verify that if u and v satisfy, respectively,

$$\frac{du}{dt}+ku=0,\, u(0)=y_0,$$
$$\frac{dv}{dt}+kv=f(t),\, v(0)=0,$$

then $y=u+v$ satisfies

$$\frac{dy}{dt}+ky=f(t),\, y(0)=y_0.$$

How does this relate to the result of Problem 45?

48. A basic theorem from calculus states that if $f'(x)=0$ for all x in an interval I, then $f(x)$ is constant on I. An important corollary is that any two antiderivatives of a function f on an interval I must differ by a constant—-or, equivalently, any two solutions of $y'=f(t)$ on an interval I must differ by a solution of $y'=0$. Prove the more general statement that any two solutions of $y'+p(t)y=f(t)$ on an interval I must differ by a solution of $y'+p(t)y=0$.

Solve the second-order initial-value problems in 49 through 52.

49. $y''(t)=12\cos 2t,\, y(0)=0,\, y'(0)=1$

50. $y''(t)=e^{-t},\, y(0)=1,\, y'(0)=0$

51. $y''+y=\sin t,\, y(0)=y'(0)=0$ (cf. Problem 11)

52. $t^2y''+3ty'+y=0,\, y(1)=y'(1)=0$ (cf. Problem 13)

Solve 53 and 54 by separating variables, and state the maximal domain of the solution.

53. $\dfrac{dy}{dt}=y^3,\ y(0)=1$

54. $\dfrac{dy}{dt}=1+y^2,\ y(0)=0$

55. Show that $y=\ln\cos t$ is a solution of the initial-value problem

$$y''-y'\tan t+2\,e^y=2\cos t-1,\, y(0)=y'(0)=0,$$

and find its maximal domain.

CHAPTER

2

LINEAR FIRST-ORDER EQUATIONS

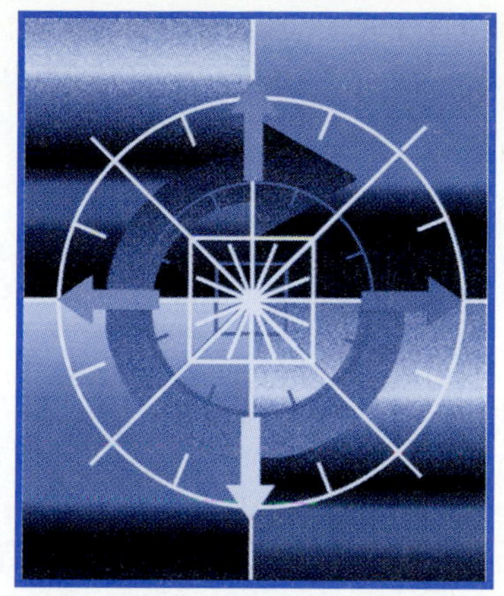

We consider in this chapter the class of differential equations of the form

$$y' + p\,y = f, \tag{1}$$

where p and f are given continuous functions on an interval I. Such equations arise in numerous elementary applications—a few of which we will explore in Section 2—and provide an elementary prototype for more complicated equations that we will encounter later.

2.1 Methods of Solution

This first section introduces two separate techniques for solving (1)—in the sense of finding a general solution that contains *all* solutions of (1) on the interval I. The second of these techniques, *variation of constants*, will prove to be a useful tool for studying other types of equations as well. Each of the techniques requires us to be able to solve the homogeneous version of (1) (i.e., (1) with $f = 0$). So it is there that we begin.

The Homogeneous Case The homogeneous version of (1),

$$y' + p\,y = 0, \tag{2}$$

can be solved easily by separation of variables. Observing first that the constant function $y = 0$ provides a trivial solution of the equation, we'll set about finding nontrivial solutions. So let's note that (2) can be rewritten in the form

$$\frac{1}{y}\frac{dy}{dt} = -p,$$

so long as $y(t) \neq 0$. We now antidifferentiate each side with respect to t and obtain

$$\ln|y| = -\int p + C,$$

where the symbol $\int p$ represents any antiderivative of p, and C is an arbitrary constant. Now solving for $|y|$, we get

$$|y| = e^{-\int p + C} = e^{C}\, e^{-\int p}.$$

Since C is an arbitrary constant, e^{C} is also an arbitrary (positive) constant that we may also refer to simply as C. So we may write the preceding equation as

$$|y| = Ce^{-\int p}.$$

From this and the necessary continuity of solutions, we can easily observe that any nonzero solution is either positive for all t or negative for all t. So the absolute value may be removed from y simply by allowing C to be negative. Note that the trivial solution also can be described this way by taking $C = 0$. Thus every solution of (2) is a member of the general solution described by

$$y = C\, e^{-\int p}, \tag{3}$$

where C may be any real number. Moreover, since p is continuous on I, any such solution is valid for all t in I.

The form of the general solution (3) should come as no surprise and, in fact, could even have been obtained by inspection. (How?) Having seen the derivation of (3), we indeed *should* solve specific instances of (2) by inspection! It is important, however, to realize that the above derivation brings with it a proof that the general solution (3) contains *all* solutions of (2) on I.

Example 1

The left column below contains six simple equations of the form in (2). Alongside each is a general solution obtainable by inspection.

$$\begin{aligned}
y' - y &= 0 & y &= C\, e^{t} \\
y' + 2\, y &= 0 & y &= C\, e^{-2\, t} \\
y' + 2t\, y &= 0 & y &= C\, e^{-t^2} \\
y' - y\, \sin t &= 0 & y &= C\, e^{-\cos t} \\
y' - \frac{y}{1+t^2} &= 0 & y &= C\, e^{\tan^{-1} t}
\end{aligned}$$

■

An Integrating Factor for the Nonhomogeneous Case The first method we'll present for solving (1) in the nonhomogeneous case consists of finding an appropriate *integrating factor* for the left side of the equation. This means that we seek a function φ for which the product $\varphi\,(y' + p\, y)$ is easily antidifferentiated.

So suppose that we multiply the expression $y' + p\, y$ by some function φ. The resulting expression, $\varphi\, y' + p\, \varphi\, y$, is recognizable as $(\varphi\, y)'$, provided that

$$\varphi' = p\, \varphi.$$

It is easy to see that a such a function φ is

$$\varphi = e^{\int p},$$

where $\int p$ means any antiderivative of p (the simpler the better). We now multiply each side of the equation $y' + p\,y = f$ by φ, producing

$$e^{\int p} y' + p\, e^{\int p} y = e^{\int p} f\,.$$

The left side is the derivative of the product $e^{\int p} y$; so

$$\left(e^{\int p} y\right)' = e^{\int p} f\,.$$

Antidifferentiating each side gives

$$e^{\int p} y = \int e^{\int p} f + C,$$

from which we get

$$y = e^{-\int p} \left(\int e^{\int p} f + C \right). \tag{4}$$

Let us emphasize that what should be remembered here is the procedure used, rather than the final "formula." The essence of the procedure lies in the fact that the equations

$$y' + p\,y = f \quad \text{and} \quad \left(e^{\int p} y\right)' = e^{\int p} f\,.$$

are equivalent, and the second one can be easily solved by antidifferentiation.

Example 2

Consider the equation

$$y' + 2\,t\,y = t\,e^{-2\,t^2}.$$

The appropriate integrating factor is

$$e^{\int 2\,t\,dt} = e^{\,t^2}.$$

Multiplying by this integrating factor turns the equation into

$$e^{\,t^2} y' + 2\,t\,e^{\,t^2} y = t\,e^{-t^2}; \quad \text{so} \quad (e^{t^2} y)' = t\,e^{-t^2}.$$

After antidifferentiating and dividing by $e^{\,t^2}$ we arrive at the general solution

$$y = -\frac{1}{2} e^{-2\,t^2} + C\,e^{-t^2}.$$ ■

Example 3

Consider the equation

$$y' + y \tan t = 2\,t\,\cos t.$$

The appropriate integrating factor is

$$e^{\int \tan t\, dt} = e^{-\ln|\cos t|} = |\sec t|.$$

Thus on any interval where $\sec t$ has constant sign, our integrating factor is either $\sec t$ or $-\sec t$. Multiplying both sides of our equation by either of these will produce the same result. So we just use $\sec t$ and obtain

$$y' \sec t + y \sec t \tan t = 2t; \quad \text{thus} \quad (y \sec t)' = 2t.$$

Antidifferentiation now gives $y \sec t = t^2 + C$; thus a general solution is

$$y = (t^2 + C)\cos t.$$ ■

It is clear, since p and f are continuous on I, that each solution given by (4) is valid for all t in I. However, we should wonder at this point whether the general solution described in (4) contains *every* solution of (1) on I. The key is the superposition principle. Let $\tilde{y}$ be the particular solution given by (4) with $C = 0$:

$$\tilde{y} = e^{-\int p} \int e^{\int p} f,$$

and suppose that y is any other solution. Since the difference $u = y - \tilde{y}$ satisfies the homogeneous equation $u' + p\,u = 0$, it follows that

$$y - \tilde{y} = C\, e^{-\int p}$$

for some constant C. Thus *every* solution y of (1) can be written in the form

$$y = \tilde{y} + C\, e^{-\int p} = e^{-\int p}\left(\int e^{\int p} f + C\right).$$

Variation of Constants Here we present a second approach to solving

$$y' + p\,y = f.$$

This approach is known as *variation of constants*.* Motivated by the fact that solutions of the homogeneous equation $y' + p\,y = 0$ are of the form $y = C\,e^{-\int p}$, where C is a constant, we look for solutions of the nonhomogeneous equation in the form

$$y = u(t)\, e^{-\int p},$$

where $u(t)$ is to be determined. (By replacing C with $u(t)$, we are allowing the *constant* C to *vary*.) So we compute

$$y' = u'\, e^{-\int p} - p\,u\, e^{-\int p} = \big(u' - p\,u\big)\, e^{-\int p}$$

and observe that y satisfies $y' + p\,y = f$, provided that u satisfies

$$(u' - p\,u)e^{-\int p} + p\,u(t)\, e^{-\int p} = f.$$

Simplification produces

$$u' = e^{\int p} f,$$

* Some prefer the phrase "variation of *parameters*."

which can be solved by antidifferentiation:

$$u = \int e^{\int p} f + C.$$

Thus we have

$$y = e^{-\int p} \left(\int e^{\int p} f + C \right).$$

Initial-Value Problems Suppose that p and f are continuous functions on an interval I and that t_0 is a number in I. Now consider the initial-value problem

$$y' + p\,y = f,\ y(t_0) = y_0. \tag{5}$$

If we define the function P on I by

$$P(t) = \int_{t_0}^{t} p(\tau)\,d\tau,$$

then $P' = p$ by the Fundamental Theorem of Calculus, and so $C\,e^{-P(t)}$ describes all solutions of the homogeneous equation $y' + p\,y = 0$. Furthermore, $P(t_0) = 0$; so $e^{-P(t_0)} = 1$. Using variation of constants as outlined above, we find that the solution of (5) has the form

$$y = u(t)\,e^{-P(t)},$$

provided that u satisfies

$$u' = e^{P(t)} f, \quad u(t_0) = y_0.$$

Integrating each side of the differential equation for u from t_0 to t produces

$$u(t) - y_0 = \int_{t_0}^{t} e^{P(s)} f(s)\,ds.$$

Therefore, the solution of (5) is

$$y = e^{-P(t)} \left(y_0 + \int_{t_0}^{t} e^{P(s)} f(s)\,ds \right). \tag{6}$$

This formula for the solution of (5) is called the *variation of constants formula*.

Example 4

Consider the initial-value problem

$$y' + 2t\,y = f(t), \quad y(0) = y_0.$$

Here we have $P(t) = \int_0^t 2s\,ds = t^2$. So, for any continuous function f, we can represent the solution via the variation of constants formula as

$$y(t) = e^{-t^2} \left(y_0 + \int_0^t e^{s^2} f(s)\,ds \right).$$

Most functions f will not allow the integral to be expressed in terms of elementary functions, but if, for example, $f(t) = t$, then the solution can be simplified as follows:

$$\begin{aligned} y(t) &= e^{-t^2}\left(y_0 + \int_0^t e^{s^2} s\,ds\right) \\ &= e^{-t^2}\left(y_0 + \frac{1}{2}(e^{t^2} - 1)\right) \\ &= y_0 e^{-t^2} + \frac{1}{2}(1 - e^{-t^2}). \end{aligned}$$

When f is not so accommodating—for example, if $f(t) = 1$—then the variation of constants formula still provides a means by which numerical values of the solution can be computed with a numerical integration procedure such as Simpson's rule. (See Problem 24.) ■

Uniqueness The solution given by the variation of constants formula (6) is indeed *the* (unique) solution of (5) on I. There are two ways to see this. One is to observe that the initial condition $y(t_0) = y_0$ uniquely determines C in

$$y = e^{-P(t)}\left(\int_{t_0}^t e^{P(s)} f(s)\,ds + C\right),$$

which is of the form in (4) and therefore contains every solution of the differential equation on I. Another proof of the uniqueness of (6) deals more directly with the initial-value problem (5) itself. Suppose that (5) has two solutions y_1 and y_2 on I. Then the difference $u = y_1 - y_2$ satisfies

$$u' + p\,u = 0, \quad u(t_0) = 0. \tag{7}$$

Since the differential equation $u' + p\,u = 0$ can be rewritten as

$$(e^{\int p} u)' = 0,$$

a basic theorem from calculus tells us that $e^{\int p} u$ is constant on I. Since $u(t_0) = 0$, that constant must be zero. Therefore, (7) has the unique solution $u = 0$ on I, which implies that $y_1 = y_2$; that is, the solution of (5) is unique.

PROBLEMS

Use an integrating factor to find a general solution for each equation in Problems 1 through 12.

1. $y' + y = e^{-t}$

2. $y' - 2\,y = e^{-t}$

3. $y' - \dfrac{2y}{t} = t$

4. $y' - \dfrac{2t\,y}{t^2 + 1} = 2t$

5. $y' + \dfrac{5y}{t - 3} = 4$

6. $y' + \dfrac{3y}{5 - t} = 1$

7. $y' - y\,\tan t = \sin t$

8. $y' + y\,\tan t = 1$

9. $y' + (\tan t + 1)\, y = \cos t$

10. $y' + \dfrac{y}{t \ln t} = \dfrac{1}{t}$

11. $2t\, y' - y = 2\, t^{3/2}$

12. $2t\, y' - y = 2t \cos \sqrt{t}$

Find the general solution for each equation in 13 through 15. No integrating factor is needed.

13. $t\, y' + y = 1$

14. $y' \sin t + y \cos t = 1$

15. $y'\, t \ln t + (1 + \ln t)\, y = 2t$

16. Solve, as in Example 4, the following initial-value problem on the interval $(-\frac{\pi}{2}, \frac{\pi}{2})$:

$$y' - y \tan t = \sin t,\ y(0) = y_0$$

For the initial-value problems in 17 and 18, write the solution as in Example 4.

17. $y' + y = f(t),\ y(0) = y_0$

18. $y' + y \cos t = \sin t,\ y(0) = y_0$

19. It is frequently the case that superposition is the simplest way to arrive at a general solution to a linear first-order equation $y' + p\, y = f$, particularly when p is constant. Show that if y and $\tilde{y}$ are any two solutions, then the difference $u = y - \tilde{y}$ satisfies the homogeneous equation $u' + p\, u = 0$. Conclude that

$$y - \tilde{y} = C\, e^{-\int p}$$

for some constant C, and therefore if $\tilde{y}$ is a known particular solution, then a general solution is given by

$$y = C\, e^{-\int p} + \tilde{y}.$$

20. Consider the equation $y' + k\, y = b$, where k and b are constants. Find a constant particular solution $y = a$ (i.e., an *equilibrium solution*), and use it to write a general solution as in Problem 19.

In Problems 21 through 23, find a particular solution $\tilde{y}$ of the suggested form, and use it to write a general solution of the differential equation.

21. $y' - y = t;\ \tilde{y} = a\, t + b$

22. $y' + y = e^t;\ \tilde{y} = a\, e^t$

23. $y' + y = 2 \cos t;\ \tilde{y} = a \cos t + a\ \sin t$

24. By the variation of constants formula, the function on the right is the solution of the initial-value problem on the left:

$$y' + 2t\, y = \cos t,\ y(0) = 0 \quad \Rightarrow \quad y = e^{-t^2} \int_0^t e^{s^2} \cos s\, ds.$$

(a) Use Simpson's rule with two subintervals,

$$\int_a^b f(x)\, dx \approx \frac{b-a}{6}\left(f(a) + 4f\left(\frac{a+b}{2}\right) + f(b)\right),$$

to compute approximations to each of the integrals

$$\int_0^{.25} e^{s^2} \cos s\, ds, \quad \int_{.25}^{.5} e^{s^2} \cos s\, ds, \quad \int_{.5}^{.75} e^{s^2} \cos s\, ds, \quad \int_{.75}^{1} e^{s^2} \cos s\, ds.$$

(b) Use the variation of constants formula and the computations from part (a) to compute approximations to $y(.25)$, $y(.5)$, $y(.75)$, and $y(1)$.

25. Let k be a constant, and consider the initial-value problem

$$y' + k\,y = f,\ y(t_0) = y_0.$$

Show that, for this problem, the variation of constants formula (6) becomes

$$y = y_0\,e^{-k\,t} + \int_{t_0}^{t} e^{k\,(s-t)}\,f(s)\,ds.$$

Let p and f be continuous on $[t_0, \infty)$. Then the solutions of (1) on $[t_0, \infty)$ are said to be **asymptotically stable** if, for any two solutions y and $\tilde{y}$,

$$\lim_{t\to\infty} (y(t) - \tilde{y}(t)) = 0.$$

Find a general solution of each equation in Problems 26 through 30, and determine whether the solutions are asymptotically stable. (These equations already appeared among Problems 1 through 12.)

26. $y' + y = e^{-t}$

27. $y' - 2y = e^{-t}$

28. $y' + \dfrac{3y}{5-t} = 1\ (t_0 > 5)$

29. $y' + \dfrac{5y}{t-3} = 4\ (t_0 > 3)$

30. $y' + \dfrac{y}{t\ln t} = \dfrac{1}{t}\ (t_0 > 0)$

31. Show that, if p and f are continuous on $[t_0, \infty)$, and if p has the property that

$$\lim_{t\to\infty} \int_{t_0}^{t} p(s)\,ds = \infty,$$

then the solutions of (1) are asymptotically stable. (*Hint*: Use the variation of constants formula.)

32. Consider the nonlinear first-order equation $y' = k\,y\,(1-y)$, where k is a constant. Show that the substitution $y = u^{-1}$ produces a linear first-order equation for u. (Note that $y' = -u^{-2}u'$.) Find a general solution of that equation, and use it to write a general solution of the nonlinear equation for y.

33. Consider the nonlinear first-order equation $y' = ay^2 + by + c$, where a, b, and c are continuous on some interval I. Show that if $\tilde{y}$ is any given particular solution, then the substitution $y = u^{-1} + \tilde{y}$ (i.e., $u = (y - \tilde{y})^{-1}$) leads to a linear first-order equation for u:

$$u' + (b + 2\,a\,\tilde{y})u = -a.$$

In Problems 34 through 37, use the given particular solution $\tilde{y}$ and the result of Problem 33 to find a general solution of the differential equation.

34. $y' = y^2 - y - 2;\ \tilde{y} = 2$

35. $y' = -2\,y^2 + y + 1;\ \tilde{y} = 1$

36. $y' = -\frac{1}{8}\,y^2 + \frac{1}{4}\,y + 1;\ \tilde{y} = 4$

37. $2ty' = -2y^2 + y + 2t;\ \tilde{y} = \sqrt{t}$

2.2 Some Elementary Applications

In this section we'll look at a few applications of linear first-order differential equations to problems from physics and engineering. We will first look at vertical projectile motion with air resistance, including the problem of finding the height attained by a fuel-burning rocket. A few of the problems following the first section deal with the path of a baseball in two dimensions, with and without air resistance. The remaining sections deal with applications such as pollution of a body of water and simple electrical circuits.

2.2.1 Projectile Motion with Resistance

Suppose that a projectile with mass m is moving vertically under constant gravitational force through air or some other more resistive medium such as water. The drag force experienced by the projectile may be modeled by $-\rho(v)$ where $\rho(v)$ is a continuous function of the velocity v with the following properties:

i) $\rho(0) = 0$, and ii) $\rho(v)$ has the same sign as v for all $v \neq 0$.

Property (i) means that there should be no drag force if the object is not moving, and property (ii) ensures that the drag force will always be in the direction opposite the velocity. If we agree that the positive direction is upward, then the force due to gravity will be $-mg$, where g is the acceleration due to gravity. With these considerations, Newton's second law, $\frac{d}{dt}(mv) = F$, takes the form

$$m\frac{dv}{dt} = -mg - \rho(v) \tag{1}$$

when m is constant. We will also assume that g is constant.

Linear Drag The simplest form that $\rho(v)$ can take is

$$\rho(v) = kv,$$

where $k > 0$ is a constant "drag coefficient." In this case, the equation for v is linear and first order. Since the coefficients in the equation are all constant, it is easy to write down a general solution by inspection:

$$\frac{dv}{dt} + \frac{k}{m}v = -g; \quad \text{therefore, } v = Ce^{-kt/m} - \frac{mg}{k}.$$

The constant C can be determined from an initial condition $v(0) = v_0$. This leads to the solution

$$v = \left(v_0 + \frac{mg}{k}\right)e^{-kt/m} - \frac{mg}{k}.$$

The height y of the projectile can be obtained by integrating v over $[0, t]$. The result is

$$y = y_0 + \frac{m}{k}\left(v_0 + \frac{mg}{k}\right)\left(1 - e^{-kt/m}\right) - \frac{mg}{k}t.$$

Note that the first term in the expression for the velocity approaches zero as $t \to \infty$, and consequently the second term, $-mg/k$, represents the limiting value of the velocity. Suppose that the initial velocity were $v_0 = 0$. Then the object's downward speed $|v|$ would never exceed mg/k. Consequently, $-mg/k$ is called the object's *terminal velocity*. The phenomenon of terminal velocity is not present if air resistance is ignored.

Linear Drag, Variable Mass, and Thrust If the mass of the projectile varies with time and an additional upward force f is present (such as with a fuel-burning rocket), then the differential equation is still linear and first order:

$$\frac{d}{dt}(mv) = -mg - kv + f, \quad \text{or} \quad \frac{dv}{dt} + \frac{1}{m}\left(k + \frac{dm}{dt}\right)v = -g + \frac{1}{m}f,$$

but it is somewhat more complicated to solve. Note that the appropriate integrating factor is now $e^{\int \frac{1}{m}(k+m')}$.

Example 1

A small rocket with mass 100 kg, including 30 kg of fuel, is fired vertically from rest at time $t = 0$. Its engine provides a thrust of 1000 N and burns fuel at a steady rate of 1 kg/s. Find the velocity and height attained by the rocket by the time its fuel is spent. Assume constant gravitational acceleration $g = 9.8$ m/s^2 and a known drag coefficient $k = 3$ N s/m.

The mass of the rocket satisfies $m'(t) = -1$, $m(0) = 100$, and therefore is given by $m(t) = 100 - t$, for $0 \le t \le 30$ seconds. Thus the equation of interest is

$$\frac{dv}{dt} + \frac{2}{100 - t}v = -9.8 + \frac{1000}{100 - t}, \qquad 0 \le t \le 30.$$

The integrating factor is $(100 - t)^{-2}$; therefore

$$\frac{d}{dt}\left(\frac{v}{(100 - t)^2}\right) = -\frac{9.8}{(100 - t)^2} + \frac{1000}{(100 - t)^3}.$$

Integration now yields

$$\begin{aligned} v &= (100 - t)^2\left(-\frac{9.8}{100 - t} + \frac{500}{(100 - t)^2} + C\right) \\ &= -9.8(100 - t) + 500 + C(100 - t)^2. \end{aligned}$$

The initial condition $v(0) = 0$ results in $C = 0.048$. Therefore the velocity of the rocket is

$$v = 9.8t - 480 + 0.048(100 - t)^2 = 0.048t^2 + 0.2\,t.$$

Now we integrate and use $y(0) = 0$ to obtain the height

$$y = 0.016\,t^3 + 0.1\,t^2.$$

Thus, after 30 seconds the rocket has reached a height of 522 m and has a velocity of 49.2 m/s. Once the fuel has been spent, the mass will remain constant, and there will be no more thrust; therefore, the velocity of the rocket for the remainder of its flight will obey the simpler equation

$$\frac{dv}{dt} + \frac{3}{70}v = -9.8.$$

■

PROBLEMS

1. Suppose that a ball with a mass of 1 kilogram is dropped from a height of 1000 meters above the ground. Assuming a linear drag model with $k = 0.1$, how long will it take the ball to hit the ground? (Use $g = 9.8$ m/s^2, and solve the necessary transcendental equation numerically.) What will be the ball's impact velocity?
2. How much higher will the rocket in Example 1 rise after its fuel is spent?
3. Consider a model rocket with mass 1 kg, including 0.1 kg of fuel, and a linear drag coefficient $k = 0.1$. The fuel burns completely in 10 seconds, and during that time the engine provides a constant thrust of 18 N.

   (a) What velocity and height will the rocket attain by the time its fuel is spent?

   (b) How much farther will the rocket rise after the fuel is spent?
4. Suppose that an object is dropped from a great height with zero initial velocity. Using a linear drag model, find (in terms of k/m) the time it takes for the object's velocity to become 95% of its terminal velocity.
5. An object with mass 1 kg is dropped from a height of 100 meters and hits the ground after 10 seconds. Find its linear drag coefficient k. (Use $g = 9.8$ m/s^2, and solve the necessary transcendental equation numerically.)
6. Consider the vertical motion of a projectile with initial height $y_0 = 0$ and positive initial velocity v_0.

   (a) Find a formula for the maximum height of the object if there is no air resistance.

   (b) Assuming a linear drag model, find a formula for the maximum height.

   (c) Show that the formula from part (b) reduces to the one from part (a) as $k \to 0^+$.
7. Suppose that a parachutist's terminal velocity is -40 ft/s if his parachute is not open and -5 ft/s if his parachute is open. The parachutist weighs 160 lb, which corresponds to a mass of about 5 slugs.

   (a) Let k_1 and k_2 be the linear drag coefficients of the parachutist while his parachute is closed and while it is open, respectively. Find k_1 and k_2.

   (b) If his parachute opens *instantaneously* while he is falling at -40 ft/s, find the force experienced by the parachutist at the instant when the parachute opens.
8. Consider again the parachutist in Problem 7, but suppose now that it takes 1 second for the parachute to open completely.

   (a) Construct a model of the form

   $$k(t) = \frac{a}{b + ct}, \qquad 0 \le t \le 1,$$

   for the drag coefficient while the parachute is opening. Use that model in the remainder of this problem.

   (b) If his parachute begins opening at $t = 0$, as he is falling at -40 ft/s, find the parachutist's velocity $v(t)$ for $0 \le t \le 1$.

   (c) Find the force experienced by the parachutist at the instant when the parachute becomes fully open (i.e., $t = 1$).
9. In a vacuum, the model for the vertical flight of a projectile, $\frac{dv}{dt} = -g$, leads to the conclusion that when an object is dropped from a height H the speed with which it hits the ground is the same as the initial velocity required to project the object from ground level to a maximum

height H. Argue that this is false if air resistance is present. (Suggestion: Consider the terminal velocity v_∞ of the object, and let H be the maximum height in Problem 6 with $v_0 = |v_\infty|$. What if the object is dropped from a height H or higher?)

10. This problem concerns the path of a baseball in two dimensions under constant gravitational force in a vacuum. Assume that the baseball is located at the point $(x(t), y(t))$ at time t and that $(x(0), y(0)) = (x_0, y_0)$ is the initial position. The velocity of the baseball is then the vector $V(t) = (x'(t), y'(t))$. Assume further that the initial velocity vector is $V_0 = (v_1, v_2)$, so that the initial speed is its length:

$$\|V_0\| = \sqrt{v_1^2 + v_2^2}.$$

Also, the initial angle that the path makes with the horizontal satisfies $\tan\theta = v_2/v_1$. By considering forces in the horizontal and vertical directions and applying Newton's second law, we arrive at second-order differential equations for x and y:

$$mx''(t) = 0, \quad my''(t) = -mg.$$

(a) Solve the equations separately to obtain the parametric solution

$$(x(t), y(t)) = \left(v_1 t + x_0, -\frac{1}{2}gt^2 + v_2 t + y_0\right).$$

(b) Taking $(x_0, y_0) = (0, 0)$, eliminate t to obtain the parabolic path of the baseball:

$$y = x\tan\theta - \frac{g\sec^2\theta}{2\,\|V_0\|^2}x^2.$$

(c) Assuming that $y = 0$ represents ground level, solve in terms of θ for the horizontal distance to the point where the baseball lands.

11. Assume the scenario of Problem 10. If $\theta = \pi/6$, what should be the initial velocity in order for the ball to fly 400 feet? Use $g \approx 32.2$ ft/s^2 and round all calculations to three significant figures.

12. Assume the scenario of Problem 10. If $\theta = \pi/6$, what should be the initial velocity in order for the ball to clear a 30-ft wall that is 350 feet from home plate?

13. This problem concerns the path of a baseball in two dimensions under constant gravitational force and linear drag. The set-up is the same as in Problem 10, except that the inclusion of a linear drag force, $-kV(t)$, leads to the equations

$$mx''(t) = -kx'(t), \quad my''(t) = -mg - ky'(t).$$

Reduce the order of the two equations by setting $u = x'$ and $v = y'$. Solve for u and v; then integrate to find $x(t)$ and $y(t)$.

14. Assume the scenario of Problem 13. Using $g = 32$ ft/s^2, $v_1 = v_2 = 104$ ft/s (100 mph at a 45° angle), find the horizontal distance travelled (before hitting the ground) for each of the cases: $k/m = 0.150$ and $k/m = 0.175$. (Each will require solving a transcendental equation numerically.)

15. Suppose that a cannon is positioned at the top of a cliff, and another identical cannon is positioned on the ground below, some distance away from the bottom of the cliff. Suppose further that the two cannons are aimed directly at each other, loaded in exactly the same way, and fired at exactly the same instant. Ignoring air resistance, investigate the question of whether the two cannonballs will collide somewhere along their respective paths. (*Hint*: Use the parametric solution in Problem 10.) Would your answer be the same if air resistance were not ignored?

2.2.2 Mixing Problems

Suppose that a container of some type contains a volume V gallons of water with some dissolved impurity at an initial concentration of k_0 grams per gallon. At time $t = 0$, water with dissolved impurity at a different concentration k_{in} begins to flow into the container at a rate of R gallons per minute. We assume that the water in the container stays thoroughly mixed at all times, and that water runs out of the tank at the same rate R gallons per minute, resulting in a constant volume V.

The problem is to determine the concentration of impurity in the container as a function of time. To do this, let's let $A(t)$ denote the *amount* of impurity (e.g., in grams) at time t. Then the concentration is simply $A(t)/V$. The form of the differential equation will be

$$\frac{dA}{dt} = \text{rate in} - \text{rate out}.$$

The key here is to take care to have proper units in each term. In our current set-up, each rate should be in units of grams per minute. The rate of impurity inflow is

$$k_{\text{in}} \frac{\text{grams}}{\text{gallon}} \cdot R \frac{\text{gallons}}{\text{minute}} = k_{\text{in}} R \frac{\text{grams}}{\text{minute}},$$

and the rate of impurity outflow is

$$A(t) \frac{\text{grams}}{V \text{ gallons}} \cdot R \frac{\text{gallons}}{\text{minute}} = \frac{R}{V} A(t) \frac{\text{grams}}{\text{minute}}.$$

Also, the initial value of A is $A_0 = k_0 \frac{\text{grams}}{\text{gallon}} \cdot V \text{gallons} = k_0 V$ grams. Thus we have the initial-value problem

$$\frac{dA}{dt} = k_{\text{in}} R - \frac{R}{V} A, \quad A(0) = k_0 V$$

for $A(t)$. The differential equation is linear and first order. Rearranging and using the integrating factor $e^{Rt/V}$ gives

$$(e^{Rt/V} A)' = k_{\text{in}} R e^{Rt/V}.$$

We now integrate and use the the initial condition to obtain

$$\begin{aligned} A(t) &= e^{-Rt/V} \left(k_0 V + k_{\text{in}} V (e^{Rt/V} - 1) \right) \\ &= \left(k_{\text{in}} + (k_0 - k_{\text{in}}) e^{-Rt/V} \right) V. \end{aligned}$$

Thus the concentration is

$$A(t)/V = k_{\text{in}} + (k_0 - k_{\text{in}}) e^{-Rt/V}.$$

Example 1

Suppose that a 1000-gallon water tank initially contains 5 kilograms of dissolved salt. Pure water begins to flow in at a rate of 5 gallons per minute. The tank is kept well-mixed, and the salt water flows out also at the rate of 5 gallons per minute. Find the concentration of salt in the tank as a function of time, and determine how long it will take for the salt concentration to reach one-tenth of its initial value.

SOLUTION: The inflow rate of salt is zero. The outflow rate is

$$\frac{A(t)\text{kg}}{1000 \text{ gallons}} \cdot \frac{5 \text{ gallons}}{\text{minute}} = .005\, A(t) \text{ kg/minute}.$$

Thus the initial-value problem of interest is

$$A' = -.005\, A, \quad A(0) = 5,$$

whose solution is $A(t) = 5e^{-.005\,t}$. The concentration will be one-tenth its initial value when $A(t) = 0.5$. Solving for t in

$$0.5 = 5\, e^{-.005\,t}$$

gives us $t = \frac{\ln 10}{.005} \approx 460.5$ minutes.

■

Example 2

Water polluted with tritium at a level of 6 parts per million begins to flow at a rate of 2000 cubic feet per day from a nuclear plant into a lake that initially contains no tritium. The lake maintains a constant volume of 5×10^6 cubic feet. Assuming thorough mixing in the lake at all times, how long will it take for the level of tritium in the lake to reach 1 part per million?

SOLUTION: The inflow rate of tritium is $(6 \times 10^{-6})(2000) = 0.012 \text{ ft}^3/\text{day}$. The outflow rate is

$$\frac{A(t)\text{ft}^3}{5 \times 10^6\text{ft}^3} \cdot \frac{2000 \text{ ft}^3}{\text{day}} = 4 \times 10^{-4}\, A(t) \text{ ft}^3/\text{day}.$$

The initial amount of tritium in the lake is zero. Thus our initial-value problem for the amount $A(t)$ of tritium in the lake is

$$A' = 0.012 - 4 \times 10^{-4}\, A, \quad A(0) = 0.$$

The solution is

$$A(t) = 30\,(1 - e^{-4\times 10^{-4}\,t}),$$

from which we obtain the tritium concentration $\frac{A(t)}{V} = 6 \times 10^{-6}(1 - e^{-4\times 10^{-4}\,t})$. Now, solving for t in

$$1 \times 10^{-6} = 6 \times 10^{-6}\left(1 - e^{-4\times 10^{-4} t}\right)$$

gives $t = -\frac{\ln(5/6)}{4\times 10^{-4}} \approx 456$ days.

■

Variable Volume We now consider the mixing problem with variable volume. The difference is that we assume different flow rates in and out of the container. Let R_{in} be the flow rate of water into the container and let R_{out} be the flow rate of water out of the container. The general form of the initial-value problem for the amount of impurity in the tank at time t is

$$\frac{dA}{dt} = k_{\text{in}} R_{\text{in}} - \frac{A}{V} R_{\text{out}}, \qquad A(0) = A_0. \tag{1}$$

The flow rates R_{in} and R_{out} may be constants or vary with t. Moreover, the concentration k_{in} of the impurity in the inflowing liquid may also vary with t. In any case, (1) is *coupled* to another initial-value problem for the volume V, which has the form

$$\frac{dV}{dt} = R_{\text{in}} - R_{\text{out}}, \qquad V(0) = V_0. \tag{2}$$

Notice that (1) is always a first-order linear equation for A; however, V must be known explicitly before we can attempt to solve for A by the integrating factor method.

When R_{in} and R_{out} are constants, (2) is easy to solve by inspection:

$$V(t) = V_0 + (R_{\text{in}} - R_{\text{out}})\,t.$$

The initial-value problem for $A(t)$ is now

$$\frac{dA}{dt} = k_1 R_{\text{in}} - \frac{R_{\text{out}}\,A}{V_0 + (R_{\text{in}} - R_{\text{out}})\,t}, \qquad A(0) = k_0\,V_0.$$

The differential equation here is still first-order linear. Since solving this in its general form will become somewhat messy, we will just illustrate with a specific example.

Example 3

A tank initially contains 100 gallons of pure water. Water with a dissolved salt concentration of 1 gram per gallon begins to flow in at a rate of 2 gallons per minute. At the same time, the well-mixed salt solution in the tank is being pumped out at a rate of 3 gallons per minute. Find the salt concentration in the tank up to the time when the tank becomes empty.

SOLUTION: Since $R_{\text{in}} - R_{\text{out}} = -1$, the volume of water in the tank is

$$V(t) = 100 - t \quad \text{for } 0 \le t \le 100.$$

So the initial-value problem for the amount of salt $A(t)$ during $0 \le t \le 100$ is

$$\frac{dA}{dt} = 2 - \frac{3}{100 - t} A, \qquad A(0) = 0.$$

The appropriate integrating factor is

$$e^{\int \frac{3}{100-t}\,dt} = e^{-3\ln(100-t)} = (100 - t)^{-3}.$$

The differential equation becomes

$$((100 - t)^{-3} A)' = 2(100 - t)^{-3},$$

and thus

$$A(t) = (100 - t)^3((100 - t)^{-2} + C) = (100 - t)(1 + C(100 - t)^2).$$

Using the initial condition $A(0) = 0$, we find $C = -\frac{1}{100^2}$, and so the solution of the initial-value problem for $A(t)$ is

$$A(t) = (100 - t)\left(1 - \left(\frac{100 - t}{100}\right)^2\right) \quad \text{for } 0 \le t \le 100.$$

Thus the concentration is

$$\frac{A(t)}{V(t)} = 1 - \left(\frac{100 - t}{100}\right)^2 \quad \text{for } 0 \le t \le 100.$$

■

PROBLEMS

1. A 20-gallon aquarium initially contains ammonia at a toxic level, say k_0 units per gallon. Pure water is pumped in and polluted water is pumped out, simultaneously, at a rate of 0.1 gallon per minute, and the water in the tank is kept well-mixed at all times. Set up and solve the initial-value problem for the amount of ammonia (in terms of k_0) in the tank after t minutes. How long until the concentration of ammonia is one-tenth of its initial value?
2. A 100-gallon tank is full of water containing a pollutant at a concentration of 0.7 grams/gallon. Cleaner water, in which the pollutant is present at a concentration of 0.1 grams/gallon, is pumped into the tank at a rate of 2 gallons/minute. The tank is kept well-mixed, and water leaves through an overflow valve at the same rate as the cleaner water is pumped in. Set up and solve the initial-value problem for the amount of pollutant in the tank at time t minutes. When will the concentration of pollutant in the tank be 0.2 grams/gallon?
3. A tank initially contains 40 gallons of brine with a salt concentration of 1 lb/gallon. Pure water is pumped into the tank at a rate of 2 gallons/minute, and the well-stirred mixture is pumped out at the same rate. Find the salt concentration as a function of $t \ge 0$, and find the time t at which the salt concentration is 0.05 lb/gallon.
4. Suppose that brine at a concentration of c pounds of salt per gallon is pumped into a tank, initially containing 50 gallons of pure water, at a rate of 2 gallons per minute and that the well-mixed contents flows out at the same rate. Find c so that the concentration of salt is 0.1 lb/gallon after 30 minutes.
5. A 60-gallon tank is initially half full of pure water. Brine with a salt concentration of 10 grams/gallon is pumped into the tank at a rate of 5 gallons/hour. The well-mixed contents of the tank is pumped out at a rate of 2 gallons/hour.

   (a) Set up and solve the initial-value problem for the amount of salt in the tank up to the time when the tank becomes full.

   (b) Assume that, once the tank becomes full, overflow causes the volume to remain constant thereafter. Set up and solve an initial-value problem for the amount of salt in the tank after the tank becomes full.
6. A 100-gallon tank is initially full of brine with a salt concentration of 10 grams/gallon. Pure water is pumped into the tank at a rate of 5 gallons/hour, while the well-mixed contents of the tank is pumped out at a rate of 10 gallons/hour. Set up and solve the initial-value problem

for the amount of salt in the tank up to the time when the tank becomes empty. What is the salt concentration at that instant?

7. A 100-gallon tank is initially full of pure water. Brine from another container, with a salt concentration of 10 grams/gallon, flows into the tank (by force of gravity) at a rate of $10 - t$ gallons/minute for $0 \leq t \leq 10$ minutes, while the well-mixed contents of the tank leaves through an overflow valve at the same rate. Set up and solve the initial-value problem for the amount of salt in the tank for $0 \leq t \leq 10$ minutes. What is the salt concentration at time $t = 10$?

8. Two 100-gallon tanks are initially full of pure water. Brine with a salt concentration of 20 grams/gallon is pumped into the first tank at a rate of 10 gallons/hour. The well-mixed contents of the first tank is pumped into the second tank, and the well-mixed contents of the second tank is pumped out, both at the same rate of 10 gallons/hour. Find the salt concentration in the second tank as a function of $t \geq 0$.

9. Suppose that brine with a salt concentration of 10 grams per cubic foot flows into a marsh at a rate of 100 cubic feet per day. Pure water is removed by evaporation at the same rate, causing the marsh to maintain a steady volume of 5×10^5 cubic feet of water. Assume that the marsh initially contains pure water and that no rainfall occurs to return pure water to the marsh. Find the concentration of salt in the marsh as a function of time.

2.2.3 Circuits

In this section we will look at a few simple electrical circuits. We assume that the student is at least vaguely familiar with the basic concepts of electricity such as current, voltage, and charge. Any introductory physics textbook may be consulted to supplement the brief discussion here. The circuits that we will consider are comprised of a single closed loop containing basic passive devices: resistors, capacitors, and inductors.

Resistors A resistor is a device that resists the flow of current and has a *resistance* R in units of *ohms* defined by *Ohm's law*:

$$V_R = I\,R,$$

where I is the current through the resistor in units of *amperes* (or *amps*) and V_R is the voltage across the resistor in units of *volts*. In other words, the ratio of voltage to current is a device characteristic called resistance. When R is independent of I and Ohm's law holds, the resistor is said to be linear. For a linear resistor at a near constant temperature, R can safely be assumed constant.

Capacitors A capacitor consists essentially of two metal plates separated by a thin layer of some insulator. It has the ability to hold an electrical charge, measured by its *capacitance* C, which is defined by the relation $Q = CV_C$ where Q is the charge (in *coulombs*) and V_C is the voltage across the capacitor. In other words, the ratio of the charge to the voltage is a device characteristic called capacitance. The basic unit of capacitance

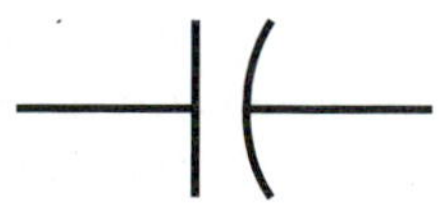

is the *farad*. The rate of change in charge is the current I; that is, $I = \frac{dQ}{dt}$. If C is constant (which is usually assumed), then $Q = CV_C$ leads to

$$C\frac{dV_C}{dt} = I.$$

This is the relationship between voltage across a capacitor and current.

Inductors An inductor is essentially a coil of wire. When a steady current passes through it, it offers essentially no resistance, and so there is essentially no voltage across it. However, an inductor tends to oppose *changes* in current; that is, change in current produces voltage across an inductor, given by

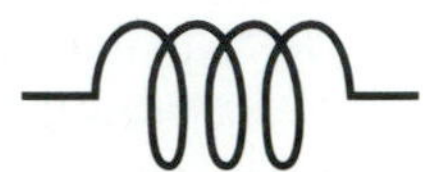

$$V_L = L\frac{dI}{dt}.$$

Here L is a device characteristic called *inductance*, measured in *henrys*.

Analysis of simple circuits centers on **Kirchhoff's*** **voltage law**:

When measured with a consistent clockwise or counterclockwise orientation, the voltages across circuit components in a closed loop always sum to zero.

In what follows, we will look at three circuits consisting of single loops containing two of the three passive devices just described. These three circuits are represented schematically in the following figure.

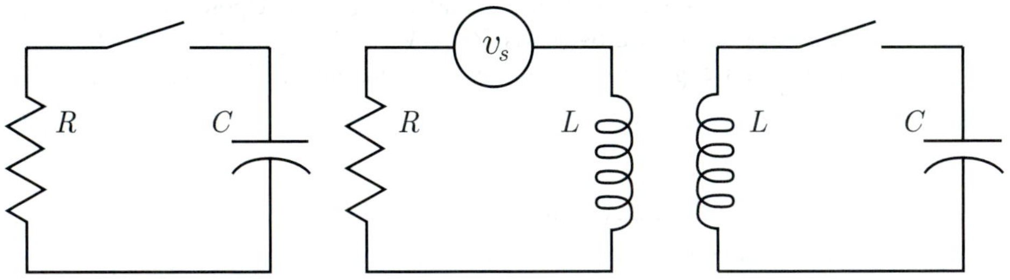

An *RC* Circuit Here we have the simple loop, consisting of a capacitor and a resistor, represented schematically by the left-most circuit in the preceding figure, and we assume that the capacitor holds some initial charge Q_0. Upon closure of the switch in the circuit, application of Kirchhoff's law results in $V_R + V_C = 0$, where V_R and V_C denote the voltages across the resistor and capacitor, respectively. Now, using $V_R = IR = \frac{dQ}{dt}R$ and $V_C = Q/C$, we obtain the differential equation

$$R\frac{dQ}{dt} + \frac{1}{C}Q = 0$$

* Pronounced (approximately) *Keerk-hoff* or *Keersh-hoff*, with *ee* as in "beer."

for the capacitor's charge. Thus,

$$Q(t) = Q(0)e^{-t/(RC)}.$$

So it is the quantity $\frac{1}{RC}$ that governs the rate at which the capacitor discharges. The voltage across the capacitor is now given by $V_C = Q(t)/C$ and the current in the circuit is given by $I(t) = Q'(t)$.

An *RL* Circuit with Source Here we have a simple loop consisting of an inductor, a resistor, and some voltage source $v_s(t)$. The schematic diagram of this arrangement is the middle circuit in the preceding figure. We assume that, when measured with consistent orientation, the voltages across the resistor, inductor, and voltage source are V_L, V_R, and $-v_s$, respectively. Application of Kirchhoff's law then gives us $V_L + V_R - v_s = 0$, which becomes

$$L\frac{dI}{dt} + RI = v_s(t),$$

a first-order linear equation for the current I. When v_s is constant (think of a battery), this can be solved easily with the integrating factor $e^{Rt/L}$ to show that

$$I(t) = I(0)e^{-Rt/L} + \frac{v_s}{R}(1 - e^{-Rt/L}).$$

If v_s is not constant, the same integrating factor (or the variation of constants formula) lets us write

$$I(t) = e^{-Rt/L}\left(I(0) + \frac{1}{L}\int_0^t e^{R\tau/L}v_s(\tau)\,d\tau\right).$$

An *LC* Circuit Here we have the simple loop consisting of an inductor and a capacitor indicated in the right-most circuit in the preceding figure, and we assume that the capacitor holds some initial charge Q_0. Upon closure of the switch in the circuit, Kirchhoff's Law gives $V_L + V_C = 0$, which becomes $LI' + V_C = 0$. Since $I = CV_C'$, we get the following *second-order* differential equation for the voltage $V = V_C$:

$$LCV''(t) + V(t) = 0.$$

It is easy to check that solutions of this equation include both $\sin\frac{t}{\sqrt{LC}}$ and $\cos\frac{t}{\sqrt{LC}}$. It is also easy to check that a two-parameter family of solutions is given by

$$V(t) = c_1 \sin\frac{t}{\sqrt{LC}} + c_2 \cos\frac{t}{\sqrt{LC}},$$

which therefore provides a general solution. Thus the voltage oscillates with a natural frequency of $\frac{1}{2\pi\sqrt{LC}}$ cycles per second, or *hertz* (Hz).

PROBLEMS

1. Consider an RC circuit with $R = 10$ and $C = 2$. Suppose that the initial charge on the capacitor is $Q(0) = Q_0$. Find the charge $Q(t)$ for $t \geq 0$, and determine how long it takes for the charge to reach 10% of its initial value.

Consider an RL circuit with a constant source voltage v_S and initial current $I(0) = 0$. For the parameter values in each of Problems 2 through 4,

- **(a)** find the current $I(t)$ for $t \geq 0$;
- **(b)** find $\lim_{t\to\infty} I(t)$;
- **(c)** find the time t at which $I(t) = \frac{9}{10} \lim_{t\to\infty} I(t)$.

2. $R = 100$, $L = 0.1$

3. $R = 10$, $L = 0.1$

4. $R = 10$, $L = 1$

5. Suppose that the switch in the RC circuit is closed at time $t = 0$ when the capacitor has a charge of Q_0. The current in the loop for time $t > 0$ is given by

$$I(t) = I_0\, e^{-t/(RC)},$$

where $I_0 = I(0^+) = \lim_{t\to 0^+} I(t)$ is the "jump" in the current at $t = 0$. What is value of I_0? (*Hint*: $I(t) = Q'(t)$ for all $t > 0$, where $Q(t) = Q_0 e^{-t/(RC)}$.)

6. Consider an RC circuit with $R = 100$ and $C = 0.01$. Suppose that the initial charge on the capacitor is $Q(0) = 50$ and that the switch is closed at time $t = 0$.

- **(a)** Find the charge $Q(t)$ for $t \geq 0$.
- **(b)** Compute the current $I(t) = Q'(t)$ for $t > 0$.
- **(c)** What is the "jump" in the current at $t = 0$?

7. Consider an RC circuit with a constant source voltage v_S and initial charge $Q_0 = 0$. Set up and solve the differential equation for the charge on the capacitor.

8. Consider an RL circuit with $R = 10$, $L = 5$, $I(0) = 0$, and source voltage $v_S(t) = e^{-t}$. Find the current $I(t)$.

9. Consider an RL circuit with source voltage $v_S(t) = \cos \omega t$ and $I(0) = 0$. Find the current $I(t)$ in terms of R, L, and ω, and phase shift ϕ.

10. Suppose that a LC circuit has an inductance of 1 mH (*millihenry*). What capacitance C will cause the circuit to have a natural frequency of 1000 Hz? Express your answer in *micro-farads* (μF).

11. **(a)** Derive the following differential equation for the current in an RLC circuit:

$$I''(t) + \frac{R}{L} I'(t) + \frac{1}{LC} I(t) = 0.$$

(b) Under what condition on R, L, and C will there be solutions of the form e^{at} with a real? (It is assumed that R, L, and C are positive.)

(c) Under what condition on R, L, and C will there be a solution of the form $e^{-kt} \cos(\omega t - \phi)$ with k, ω, ϕ real and $\omega > 0$? What then are k and ω?

2.3 Generalized Solutions

It is fairly common in applications to encounter initial-value problems involving first-order linear equations

$$y' + p\,y = f, \quad y(t_0) = y_0, \tag{1}$$

where at least one of p and f is a discontinuous function of t. Our purpose here is to introduce a generalized notion of solution that permits (1) to have a unique solution for a large class of discontinuous functions p and f. We begin with a standard definition.

Definition A function φ is **piecewise continuous** on a bounded interval I with endpoints a and b, if there are numbers $x_0 = a < x_1 < x_2 < \cdots < x_{n-1} < x_n = b$ such that

i) φ is continuous on each open subinterval (x_{i-1}, x_i), $i = 1, 2, \ldots, n$,
ii) the left-sided limit $\lim_{x \to x_i^-} \varphi(x)$ exists for all $i = 1, 2, \ldots, n$, and
iii) the right-sided limit $\lim_{x \to x_i^+} \varphi(x)$ exists for all $i = 0, 1, \ldots, n-1$.

A function is piecewise continuous on $(-\infty, \infty)$ if it is piecewise continuous on every bounded interval I. ◆

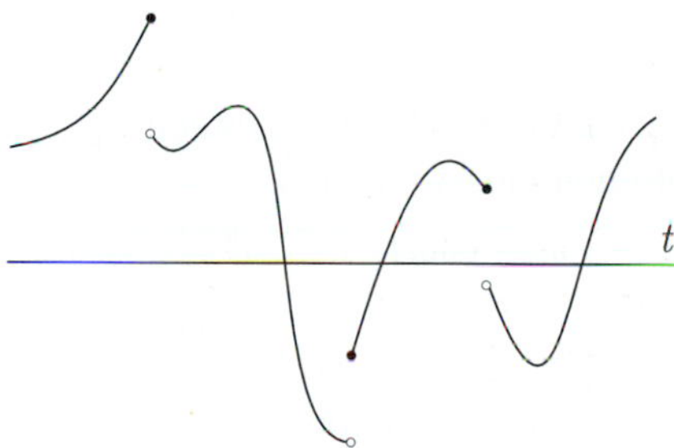

Figure 1

The graph in Figure 1 is essentially indicative of the nature of piecewise continuous functions. Calculus tells us that if a function φ is piecewise continuous on $[a, b]$, then φ is integrable on $[a, b]$, and for any t_0 in $[a, b]$ the function

$$\Phi(t) = \int_{t_0}^{t} \varphi(s)\, ds$$

is continuous on $[a, b]$. Moreover, Φ is differentiable, with $\Phi'(t) = \varphi(t)$, at each t where φ is continuous.

Let us assume now that the functions p and f in (1) are piecewise continuous on an interval I containing t_0. Then the variation of constants formula from Section 2.1, applied to (1), results in the function

$$y(t) = e^{-P(t)} \left(y_0 + \int_{t_0}^{t} e^{P(s)} f(s) ds \right), \text{ where } P(t) = \int_{t_0}^{t} p(\tau) d\tau. \qquad (2)$$

This function y is continuous on I, because our assumptions on p and f guarantee that $\int_{t_0}^{t} e^{P(s)} f(s) ds$ is continuous on I. Also, y is differentiable at each t in I where p and q are both continuous, and at those points y satisfies the differential equation in (1). However, y may fail to be differentiable at points where either p or f is discontinuous. So y may not be a solution of (1) in the usual sense.

A solution of (1), in the "usual" sense, is a continuously differentiable function $y(t)$ on an interval I containing 0 for which $y(t_0) = y_0$ and the differential equation in (1) holds at each point t in I. A solution in this sense is called a **classical solution**. The difficulty here is that if p and/or f are not continuous on I, then (1) may not have a classical solution. Indeed, it is easy to see that if p is continuous and f has a discontinuity in I, then there cannot be a classical solution of (1) on I. This is because a discontinuity of f would force a discontinuity in either y or y', by virtue of the differential equation.

So we need a weaker definition of solution if (1) is to have a solution when p and/or f are bounded and piecewise continuous. Such a definition should reduce to the usual definition of classical solution when p and q are continuous, as well as allow us to prove uniqueness under reasonable assumptions on p and f. The following is the definition we want.

Definition A function y is a **generalized solution** of (1) on an interval I containing t_0, if

i) y is continuous on I,
ii) at each point t in I where p and f are both continuous, y is differentiable and satisfies the differential equation in (1), and
iii) $y(t_0) = y_0$.

Note that if p and q are continuous on I, then a function is a generalized solution of (1) on I if and only if it is a classical solution of (1) on I. Another important observation to make here is that a generalized solution is independent of the values of p and f at their respective discontinuities. ◆

Theorem 1

Suppose that p and f are piecewise continuous on a bounded interval I containing t_0. Then (1) has a unique generalized solution on I, and this solution is given by (2). ▲

The existence of a generalized solution under the hypotheses of the theorem is easy to prove by checking that the function y given by (2) satisfies the three requirements.

To prove uniqueness, we suppose that y_1 and y_2 are generalized solutions of (1) on the interval I. Then the difference $u = y_1 - y_2$ is a generalized solution of $u' + pu = 0$, $u(t_0) = 0$ on I. Suppose that p is discontinuous at a number $t_1 > t_0$ in I, and assume that t_1 is the least such number. Since p is continuous on (t_0, t_1) and $\lim_{t \to t_0^+} p(t)$ exists, we can redefine $p(t_0)$ if necessary so that p is continuous on $[t_0, t_1)$. Then the solution of $u' + pu = 0$, $u(t_0) = 0$, is unique on $[t_0, t_1)$; therefore, $u(t) = 0$ for all t in $[t_0, t_1)$. Continuity then demands that $u(t_1) = 0$. If p is discontinuous at some $t_2 > t_1$ in I, then much the same argument shows that $u(t) = 0$ for all t in $[t_1, t_2]$. Continuing in this way, we eventually exhaust all the discontinuities of p in $I \cap (t_0, \infty)$, showing that $u(t) = 0$ for all t in $[t_0, t_j]$, where

t_j is the greatest of the discontinuities of p in I. A final, similar step then shows that $u(t) = 0$ for all $t > t_j$ in I. A similar argument shows that $u(t) = 0$ for all $t \le t_0$ in I. We therefore conclude that $y_1 = y_2$; i.e., the generalized solution of (1) on I is unique.

Example 1

Without the continuity requirement in the definition of generalized solution, there could be many, in fact infinitely many, functions satisfying the other two requirements. For example, consider the simple problem

$$y' = f(t),\ y(0) = 0, \quad \text{where } f(t) = \begin{cases} 2t, & \text{if } t \le 1 \\ 0, & \text{if } t > 1. \end{cases}$$

The only discontinuity of f is at $t = 1$. For any constant C, the function

$$y(t) = \begin{cases} t^2, & \text{if } t < 1 \\ C, & \text{if } t \ge 1 \end{cases}$$

satisfies $y' = f(t)$ except at $t = 1$, but only $C = 1$ makes this function continuous. Thus the choice of $C = 1$ gives the generalized solution, whose graph is seen in Figure 2. ■

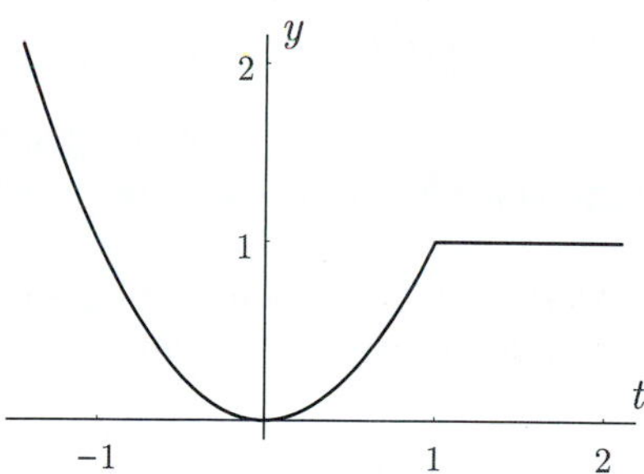

Figure 2

Example 2

Consider the initial-value problem

$$y' + p(t)y = 0,\ y(0) = 1, \quad \text{where } p(t) = \begin{cases} 1, & \text{if } 0 < t < 1 \\ 0, & \text{elsewhere.} \end{cases}$$

Rather than appeal to (2), let's just construct the solution from scratch. On the interval (0, 1), y satisfies $y' + y = 0$, and so $y = Ce^{-t}$ for some constant C. To satisfy $y(0) = 1$, we take $C = 1$ and get $y = e^{-t}$ for $0 \le t < 1$. Clearly $y(t)$ is constant for $t < 0$ and for $t > 1$, since $y' = 0$ on each of those intervals. So in order

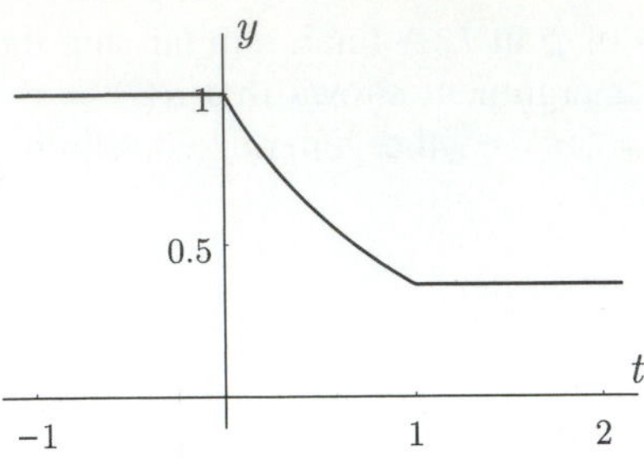

Figure 3

that y be continuous, we take $y(t) = 1$ if $t < 0$ and $y(t) = e^{-1}$ if $t \geq 1$. Thus

$$y(t) = \begin{cases} 1, & \text{if } t < 0 \\ e^{-t}, & \text{if } 0 \leq t \leq 1 \\ e^{-1}, & \text{if } t > 1 \end{cases}$$

is the generalized solution, whose graph is seen in Figure 3. ■

Example 3

Consider the initial-value problem

$$y' = f(t), \quad y(0) = 0$$

for $t \geq 0$, where f is the step function defined by

$$f(t) = (-1)^k \quad \text{when } k \leq t < k+1 \quad \text{with } k \text{ an integer.}$$

The graph of f for $t \geq 0$ appears in Figure 4a. The graph of the generalized solution can be drawn from purely geometric considerations, and this "sawtooth" graph appears in Figure 4b.

Let's make sure that the solution given by (2) corresponds to the sawtooth graph. Since $p(t) = 0$ for all t, we have the simple generalized solution

$$y(t) = \int_0^t f(s)ds.$$

This can be explicitly evaluated on each interval on which $f(t)$ is constant. Suppose now that $n \leq t \leq n+1$ where n is a nonnegative integer. Then $\int_0^t = \int_0^1 + \int_1^2 + \cdots +$

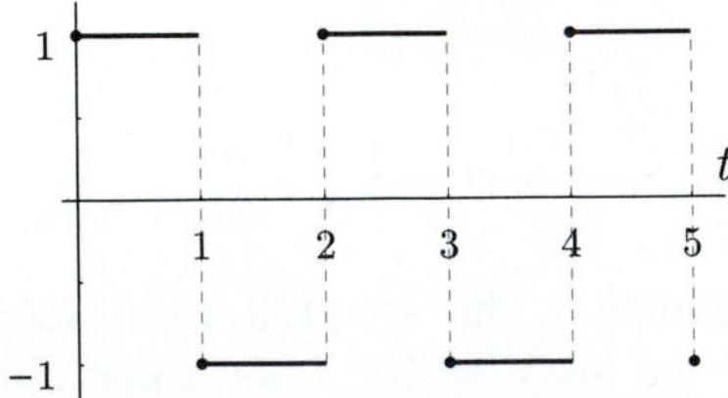

Figure 4a

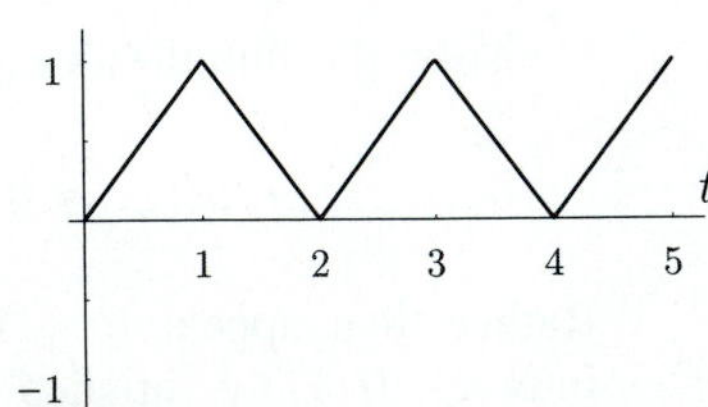

Figure 4b

$\int_{n-1}^{n} + \int_{n}^{t}$, and so

$$y(t) = \sum_{k=0}^{n-1} \int_{k}^{k+1} f(s)\,ds + \int_{n}^{t} f(s)\,ds = \sum_{k=0}^{n-1} \int_{k}^{k+1} (-1)^k\,ds + \int_{n}^{t} (-1)^n ds.$$

Evaluating the integrals and simplifying gives us

$$y(t) = \sum_{k=0}^{n-1} (-1)^k + (-1)^n (t-n) = \begin{cases} n+1-t, & \text{if } n \text{ is odd} \\ t-n, & \text{if } n \text{ is even.} \end{cases}$$

This does indeed correspond to the sawtooth graph. ■

Example 4

Consider the initial-value problem

$$y' + y = f(t), \quad y(0) = 0,$$

for $t \geq 0$, where f is the same step function as in Example 3. Since $p(t) = 1$,

$$y(t) = e^{-t} \int_{0}^{t} e^{s} f(s)\,ds.$$

Suppose now that $n \leq t \leq n+1$, where n is a nonnegative integer. Then

$$\begin{aligned} y(t) &= e^{-t} \left(\sum_{k=0}^{n-1} \int_{k}^{k+1} e^{s} f(s)\,ds + \int_{n}^{t} e^{s} f(s)\,ds \right) \\ &= e^{-t} \left(\sum_{k=0}^{n-1} \int_{k}^{k+1} (-1)^k e^{s}\,ds + \int_{n}^{t} (-1)^n e^{s}\,ds \right) \\ &= e^{-t} \left(\sum_{k=0}^{n-1} (-1)^k (e^{k+1} - e^{k}) + (-1)^n (e^{t} - e^{n}) \right), \end{aligned}$$

where the sum is zero if $n = 0$. Further simplification of this eventually gives

$$y(t) = (-1)^n - e^{-t}(1 + 2(-e + e^2 - e^3 + \cdots + (-1)^n e^n)), \text{ if } n \leq t \leq n+1,$$

whose graph appears in Figure 5. ■

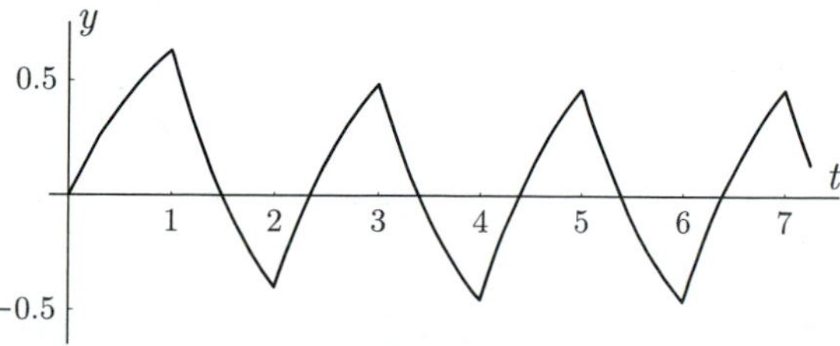

Figure 5

PROBLEMS

Each of Problems 1 through 6 deals with one of these three functions:

$$f(t) = \begin{cases} -1, & \text{if } t < 0 \\ 1, & \text{if } t \geq 0 \end{cases} \quad g(t) = \begin{cases} 1, & \text{if } t < -1 \\ 0, & \text{if } -1 \leq t \leq 1 \\ 1, & \text{if } t > 1 \end{cases}$$

$$h(t) = \begin{cases} -1, & \text{if } t < 1 \\ 0, & \text{if } t \geq 1 \end{cases}$$

For each equation in Problems 1 through 3,

(a) sketch the graph of the generalized solution satisfying $y(0) = 0$ *without solving the equation*;

(b) Write a formula for the solution you graphed in part (a).

(c) Check that your formula from part (b) agrees with that given by (2).

1. $y' = f(t)$ **2.** $y' = g(t)$ **3.** $y' = h(t)$

For each equation in Problems 4 through 6, use (2) to find the generalized solution satisfying $y(0) = 0$. Then sketch its graph.

4. $y' + y = f(t)$ **5.** $y' + y = g(t)$

6. $y' + y = h(t)$

7. Find and sketch the graph of the generalized solution of

$$y' + py = 1,\ y(0) = 0,\ \text{where}$$

$$p(t) = \begin{cases} -1, & \text{if } t < 1, \\ 1, & \text{if } t \geq 1. \end{cases}$$

In Problems 8 and 9, find the generalized solution of the initial-value problem, where the function p is defined by

$$p(t) = \begin{cases} -1, & \text{if } t < 0, \\ 1, & \text{if } t \geq 0. \end{cases}$$

8. $y' + py = 1,\ y(0) = 0$

9. $y' + py = -p,\ y(0) = 0$

10. Consider the vertical motion of a projectile with velocity governed by

$$\frac{dv}{dt} + \frac{k}{m}v = -g.$$

Find and graph the generalized solution $v(t)$ for $t \geq 0$, given that $v(0) = 0$ m/sec, $m = 100$ kg, and

$$k(t) = \begin{cases} 1, & \text{if } 0 \leq t < 10, \\ 10, & \text{if } t \geq 10. \end{cases}$$

11. A 100-gallon tank is initially full of pure water. Brine with a salt concentration of 10 grams/gallon is pumped into the tank at a rate of 20 gallons/minute for $0 \leq t \leq 5$ minutes, after which pure water is pumped in at the same rate. All the while, the contents of the tank is kept well-mixed and is pumped out at a rate of 20 gallons/minute. Set up and solve the initial-value problem for the amount of salt in the tank for $t \geq 0$. Sketch a graph of the solution.

12. Consider an *RL* circuit with $R = L = 1$, $I(0) = 0$, and source voltage

$$v_S(t) = \begin{cases} 0, & \text{if } t < 0, \\ 1, & \text{if } 0 \leq t < 1, \\ -1, & \text{if } 1 \leq t < 2, \\ 0, & \text{if } 2 \leq t. \end{cases}$$

Find and graph the current $I(t)$ that solves the initial-value problem in the generalized sense.

CHAPTER

3

Nonlinear First-Order Equations I

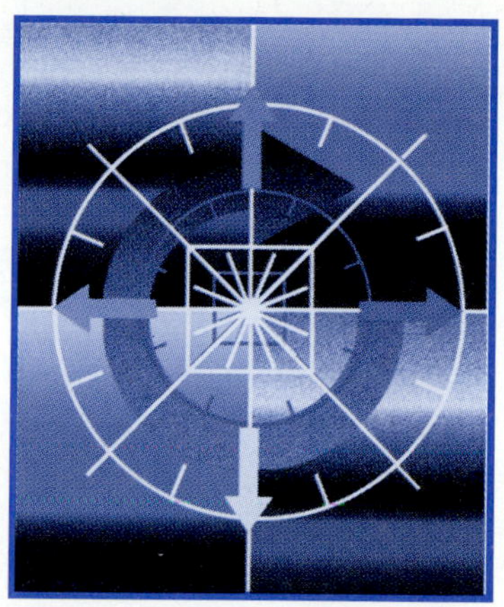

In this chapter we will survey some elementary solution techniques for nonlinear first-order differential equations and then present a few basic applications. We start with a brief, introductory discussion of graphical and numerical techniques before going on to look at some "analytical" methods for deriving formulas for solutions. The chapter concludes with a look at a few applications that involve nonlinear first-order equations: motion with nonlinear resistance, the draining-tank problem, and electrical circuits with nonlinear resistors.

3.1 Direction Fields and Numerical Approximation

We have seen that solutions of linear first-order differential equations can always be represented explicitly by means of the variation of parameters formula, if not more simply in terms of elementary functions. We will also study several important techniques for finding formulas for solutions of linear second-order equations as well as certain types of nonlinear first-order equations. However, such techniques apply to only a small set of nonlinear equations, and "real-world" problems rarely fall into these solvable categories. Therefore, it is important to be able to get both qualitative and numerical information about solutions of a differential equation from the equation itself. Remember that our primary goal always is to understand the *behavior* of solutions, even when formulas for solutions are unobtainable.

Consider an equation of the general form

$$y' = f(t, y), \tag{1}$$

where f is a given continuous function of two variables. From a geometric point of view, such an equation specifies the slope of the graph of any solution at each point through which it passes in the (t, y)-plane. It is therefore useful to associate with f a *direction field* that "guides" *solution curves* in the plane as t increases. We visualize this direction field as an array of arrows with uniform length, each located at a point (t, y) and having slope $f(t, y)$.

The following examples illustrate direction fields. Each deals with a simple, linear differential equation for the sake of simplicity and solvability; however, the construction of the direction field and the information gained from it depend in no way upon linearity.

Example 1

The direction field for the simple linear equation $y' = y$, along with one solution curve passing through (0, 1), is shown in Figure 1a. Note that this solution is simply $y = e^t$. Figure 1b shows several solution curves, corresponding to various initial values, for the same equation. In both figures, the slope of each arrow is equal to the y-coordinate at its base, since the function specifying the slope is $f(t, y) = y$. ■

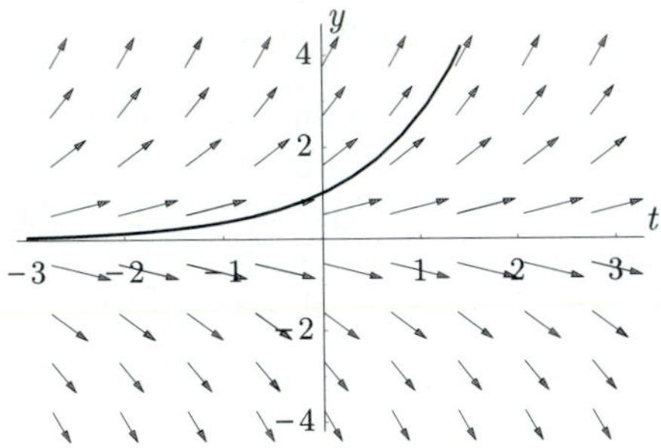

Figure 1a

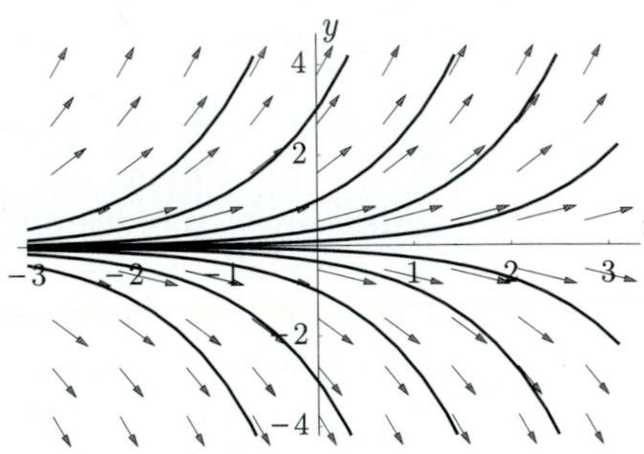

Figure 1b

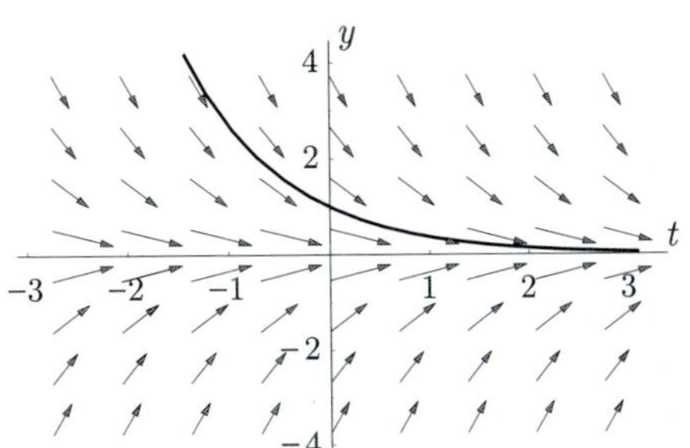

Figure 2a

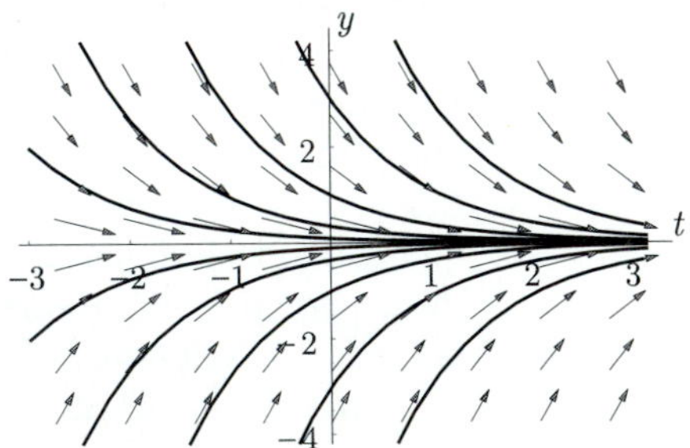

Figure 2b

Example 2

Figures 2a and 2b show the direction field for the equation $y' = -y$, first with one solution curve through (0, 1) (which is the graph of $y = e^{-t}$) and then with several solution curves. In both figures, the slope of each arrow is equal to the negative of the y-coordinate at its base, since the function specifying the slope is $f(t, y) = -y$. ■

An important thing to notice about the direction fields in all four of the preceding figures is the effect of having the right side of the differential equation depend only on y. This causes all direction arrows on any given horizontal line to be parallel, since the slope depends only on y and not on t. These lines are simple examples of what are called *isoclines*. An **isocline** is a curve with the property that every solution curve that crosses it does so with the same slope. In Figures 1 and 2, horizontal lines have this property.

In general, isoclines for solutions of $y' = f(t, y)$ are described by

$$f(t, y) = m,$$

where m is an arbitrary constant that represents the slope with which solution curves cross a particular isocline. Notice that these may be thought of as the level curves of the surface $z = f(t, y)$ in tyz-space.

Example 3

Figure 3 shows the direction field and several solution curves for

$$y' = t - y.$$

The direction field here depends in a simple way on both y and t. The isoclines (shown as dashed lines in the figure) are the lines

$$t - y = m,$$

each of which has slope 1, crosses the t-axis at $t = m$, and crosses the y-axis at $y = -m$. Notice, for example, that the line $t - y = 0$ is an isocline along which tangents to solution curves are all horizontal. ■

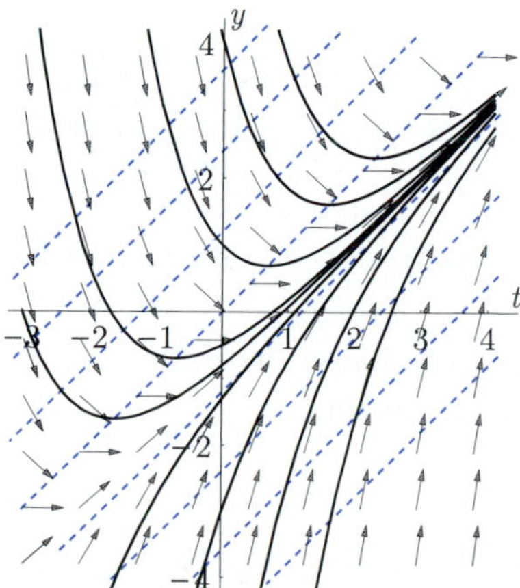

Figure 3

Example 4

The direction field and several solution curves for the equation

$$y' = -ty$$

are shown in Figure 4. Here the isoclines (the dashed curves in the figure) are the hyperbolas

$$-ty = m, \quad m \neq 0,$$

together with the coordinate axes, corresponding to $m = 0$. ■

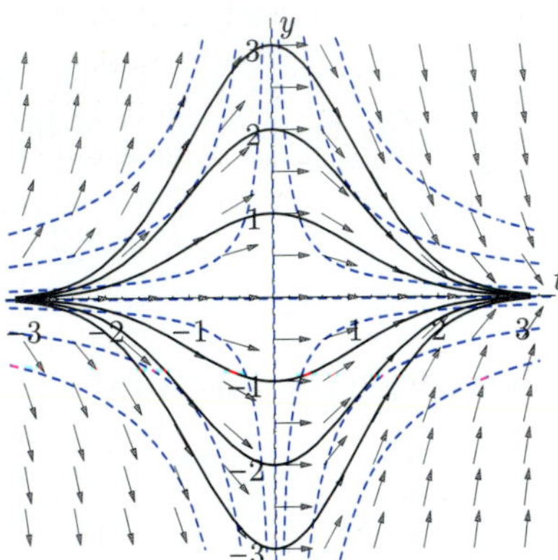

Figure 4

The equations in Examples 3 and 4 are fairly simple to solve explicitly, and you'll be asked to do so in Problem 1. The point, however, is that an equation's direction field can help us understand the behavior of solutions directly from the equation itself.

Numerical Approximation Solution curves such as those in Figures 1 through 4 can be generated—without explicitly solving the equations—with the aid of a numerical approximation procedure. Here we will only scratch the surface of the subject by looking at two fairly simple methods, beginning with the most basic one. A somewhat more detailed study of numerical methods comprises Section 4.5. In particular, more attention is given there to estimation of the errors in these numerical approximations.

Euler's Method Consider the equation $y' = f(t, y)$. By integrating each side of this equation over an interval $[t, t + h]$, we obtain

$$y(t + h) = y(t) + \int_t^{t+h} f(s, y(s))\, ds.$$

A left-endpoint approximation to the integral on the right side gives

$$y(t + h) \approx y(t) + h\, f(t, y(t)).$$

This formula describes a simple way of approximating the value of the solution at $t+h$ from knowledge of the solution at t. The resulting iterative procedure is called **Euler's method**. We start with an initial value $y_0 = y(t_0)$ and a small "stepsize" h. We then iteratively compute approximations

$$y_1 \approx y(t_0+h), \quad y_2 \approx y(t_0+2h), \quad y_3 \approx y(t_0+3h), \ldots,$$

using the formula

$$y_{n+1} = y_n + hf(t_n, y_n), \qquad \text{where } t_n = t_0 + nh.$$

As one would expect, the size of h greatly affects the quality of the approximations. Generally speaking, the smaller h is, the better the approximations will be. However, smaller values of the stepsize require more steps to reach a given t-value from the starting point t_0, and with each step a new error can compound errors made at previous steps. Section 4.5 contains a more careful discussion of these ideas. For now we will simply proceed with the awareness that numerical approximations can be inaccurate and therefore should be viewed with a healthy skepticism.

Example 5

Consider the initial value problem

$$y' = -ty, \quad y(0) = 1.$$

The general "Euler step" for this problem is

$$y_{n+1} = y_n + h(-t_n y_n).$$

Five steps with $h = 0.1$ are computed (to four decimal places) as follows:

$$\begin{aligned}
y(0.1) &\approx y_1 = 1 + (0.1)(-0)(1) = 1 \\
y(0.2) &\approx y_2 = 1 + (0.1)(-0.1)(1) = 0.99 \\
y(0.3) &\approx y_3 = 0.99 + (0.1)(-0.2)(0.99) = 0.9702 \\
y(0.4) &\approx y_4 = 0.9702 + (0.1)(-0.3)(0.9702) = 0.9411 \\
y(0.5) &\approx y_5 = 0.9411 + (0.1)(-0.4)(0.9411) = 0.9035.
\end{aligned}$$

■

Example 6

Figure 5 shows approximate solutions obtained with Euler's method for the problem

$$y' = t^2 - y, \quad y(0) = 1,$$

using stepsizes of $h = 0.05$, 0.5, and 1.0, respectively. The graphs are linear interpolants of the data points generated by the method. The nearly smooth curve corresponds to $h = 0.05$ and follows the exact solution curve fairly closely. The other two give a good graphical illustration of exactly what Euler's method does. It simply follows the direction field at the current point. ■

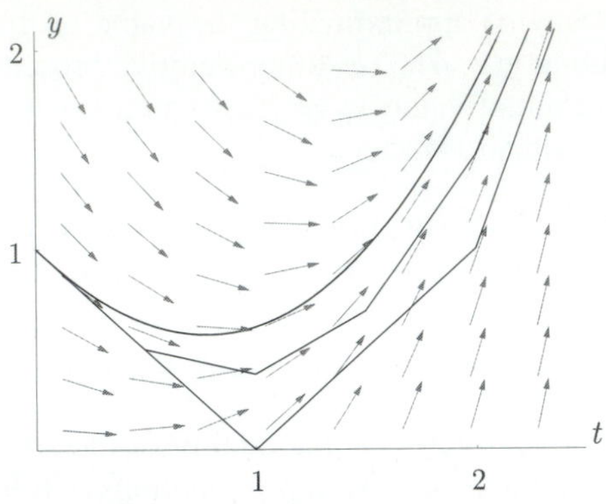

Figure 5

It should be easy to imagine that errors can cause approximations to wander quite far away from the exact solution after many steps. In fact, rather chaotic behavior can arise easily, as illustrated in the next example.

Example 7

Figure 6 shows approximate solutions of

$$y' = t - y^2, \quad y(0) = 1,$$

obtained by Euler's method with stepsizes $h = 0.06$ and 0.6. The approximate solution obtained with $h = 0.06$ follows the exact solution reasonably well, but notice how the approximate solution obtained with $h = 0.6$ starts out poorly and then wanders back close to the correct solution, where it remains for some time before beginning to oscillate wildly. This is not surprising if we look closely at the direction field. ■

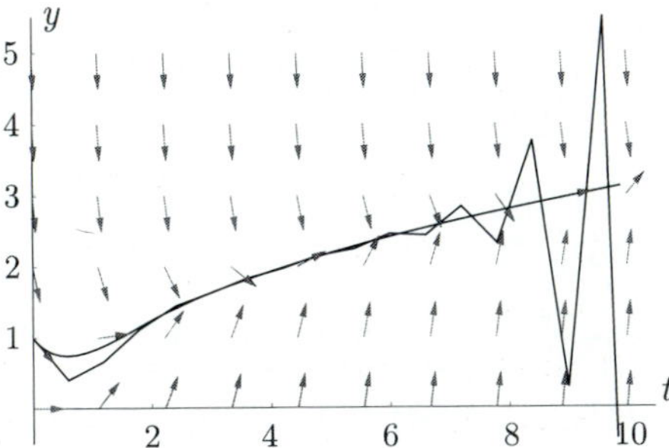

Figure 6

Heun's Method One of many methods that are more accurate than Euler's method at a given stepsize is **Heun's method**:

$$k_1 = f(t_n, y_n),$$
$$k_2 = f(t_n + h, y_n + h\,k_1),$$
$$y_{n+1} = y_n + \frac{h}{2}(k_1 + k_2).$$

Note that k_1 and k_2 are merely "local" calculations that need not be kept track of from step to step. Heun's method is based upon the approximation

$$y(t+h) \approx y(t) + \frac{h}{2}\left(f(t, y(t)) + f\left(t+h, y(t) + h\,f(t, y(t))\right)\right),$$

which arises from a trapezoidal rule approximation to $\int_t^{t+h} f(s, y(s))\,ds$, in which the right-endpoint value is estimated with Euler's method. Heun's method is also known as the *improved Euler's method* and is a simple example of the family of methods known as *Runge-Kutta methods*.

Example 8

Consider again the initial-value problem in Example 1:

$$y' = -ty, \quad y(0) = 1.$$

For this particular problem, Heun's method assumes the form

$$k_1 = -t_n y_n, \quad k_2 = -(t_n + h)(y_n + hk_1), \quad y_{n+1} = y_n + \frac{h}{2}(k_1 + k_2).$$

Four steps with $h = 0.1$ are computed (rounding to five places) as follows:

$$\begin{aligned}
k_1 &= -(0)(1) = 0\\
k_2 &= -(0.1)(1 + (0.1)(0)) = -0.1\\
y(0.1) \approx y_1 &= 1 + (0.05)(0 + (-0.1)) = 0.995\\
k_1 &= -(0.1)(0.995) = -0.0995\\
k_2 &= -(0.2)(0.995 + (0.1)(-0.0995)) = -0.19701\\
y(0.2) \approx y_2 &= 0.99500 + (0.05)(-0.09950 + (-0.19701)) = 0.98018\\
k_1 &= -(0.2)(0.98018) = -0.19604\\
k_2 &= -(0.3)(0.98018 + (0.1)(-0.19604)) = -0.28817\\
y(0.3) \approx y_3 &= 0.98018 + (0.05)(-0.19603 + (-0.28817)) = 0.95596\\
k_1 &= -(0.3)(0.95596) = -0.28679\\
k_2 &= -(0.4)(0.95596 + (0.1)(-0.28679)) = -0.37091\\
y(0.4) \approx y_4 &= 0.95596 + (0.05)(-0.28679 + (-0.37091)) = 0.92308.
\end{aligned}$$

The exact solution of this initial value problem is $y(t) = e^{-t^2/2}$. (See Problem 1.) Thus we can compare the preceding approximations to the correct values (again rounding to five places):

$$y(0.1) = 0.99501, \quad y(0.2) = 0.98020, \quad y(0.3) = 0.95597, \quad y(0.4) = 0.92312.$$

So we see that while Euler's method is giving roughly one-decimal-place accuracy (cf. Example 1), Heun's method gives four-decimal-place accuracy. ■

PROBLEMS

1. Give explicit formulas for the curves in Figures 3 and 4 by finding a general solution of the differential equation:

(a) $y' = t - y$ (b) $y' = -ty$

Problems 2 through 5: Photocopy this page and sketch several isoclines and solution curves over the given direction field. Identify the direction field with one of the equations

$$y' = t - y^2, \quad y' = y - t^2, \quad y' = y\,(y^2 - 4y + 4), \quad y' = y\,(y^2 - 4\,y + 3).$$

What kind of curves are the isoclines?

2.

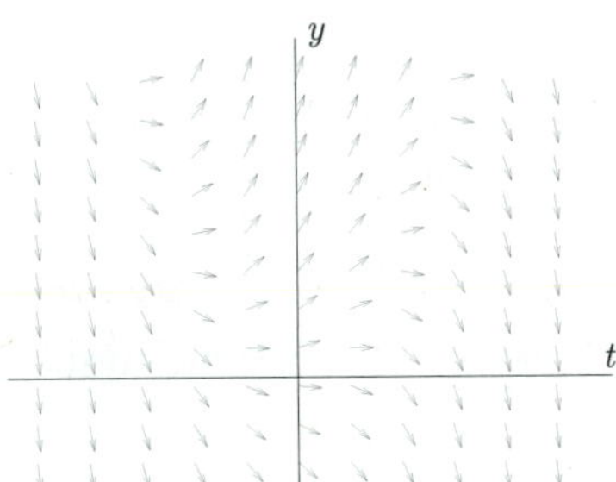

3.

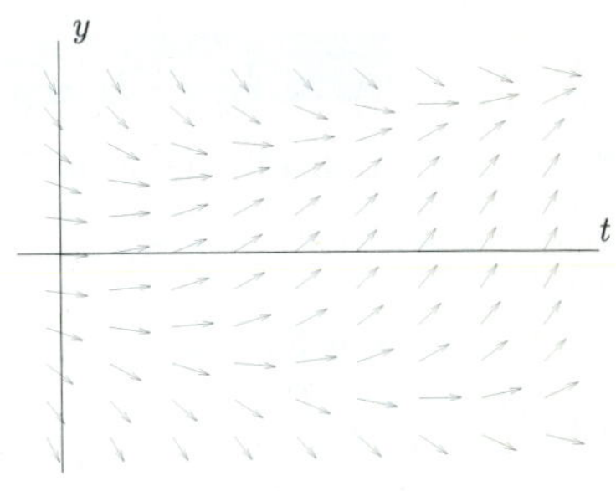

4.

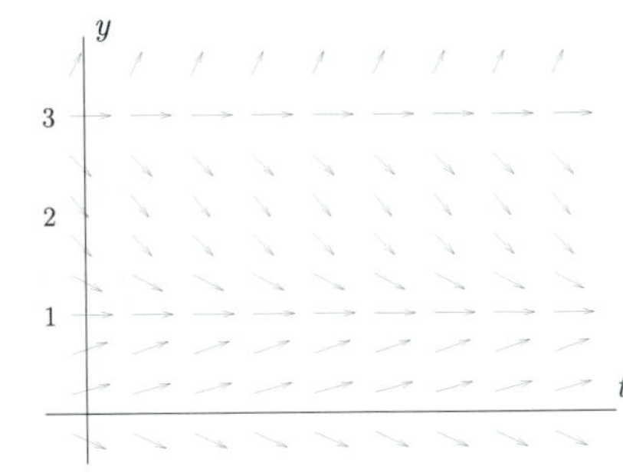

5.

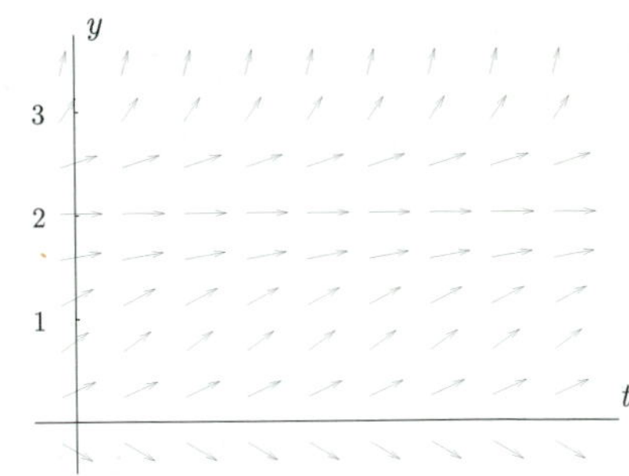

For each of the differential equations in Problems 6 through 9, sketch isoclines corresponding to the indicated slopes. Then sketch the direction field and a few solution curves.

6. $y' = t^2 + y^2, \quad m = 0, 1/4, 1, 4$

7. $y' = e^{1-t^2-y^2}, \quad m = e^{15/16}, e^{3/4}, 1, e^{-5/4}, e^{-3}$

8. $y' = -y + \sin t, \quad m = 0, \pm 1/2, \pm 1$

9. $y' = \tan^{-1}(t - y), \quad m = 0, \pm\pi/6, \pm\pi/4, \pm\pi/3, \pm \tan^{-1} 3$

10. The following figure shows the graph and a contour plot of the function

$$f(t, y) = \frac{4y - t + e^{1-t^2-y^2}}{1 + t^2 + y^2}.$$

The contours are the curves $f(t, y) = m$ for the indicated values of m. (Also notice the visible portions of the t- and y-axes in the surface plot.) Reproduce the contour plot, and sketch over it the direction field associated with f. Then sketch several solution curves of the equation $y' = f(t, y)$.

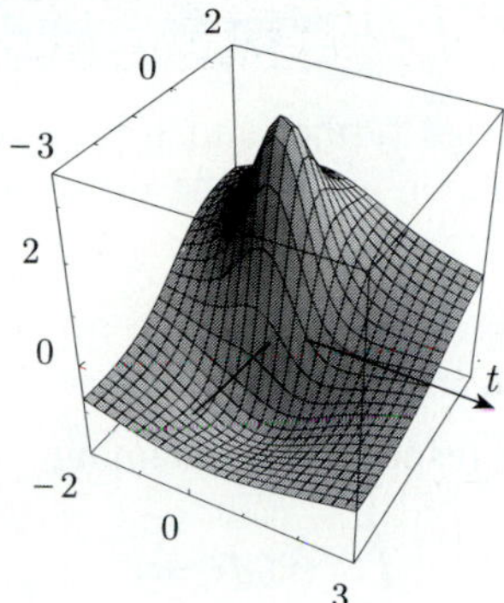

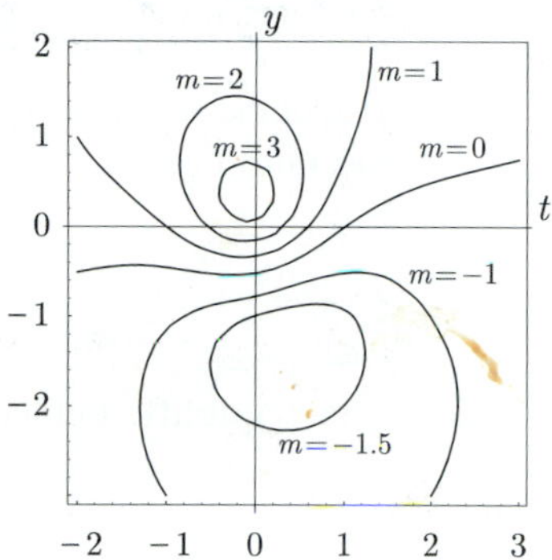

11. Do this problem with a calculator. For the initial-value problem

$$y' = 1/y, \quad y(0) = 1,$$

perform enough steps of Euler's method to reach $t = 0.5$ using stepsize $h = 0.1$. Repeat with $h = 0.05$. Then find the exact solution and compare your approximations of $y(.5)$ to the exact value.

12. Rework Problem 11 using Heun's method.

13. Find, in terms of the stepsize h, the result of one step of Euler's method on

$$y' = y, \quad y(0) = 1.$$

Compute the error as a Taylor series in h.

14. Rework Problem 13 using Heun's method.

15. A method similar to Heun's method, but based on Simpson's rule, is

$$k_1 = f(t_n, y_n), \quad k_2 = f(t_n + \frac{h}{2}, y_n + \frac{h}{2}k_1), \quad k_3 = f(t_n + h, y_n + hk_2),$$
$$y_{n+1} = y_n + \frac{h}{6}(k_1 + 4k_2 + k_3).$$

Rework Problem 11 using this method.

16. Rework Problem 13 using the method in Problem 15.

17. For the problem $y' = y \cos t$, $y(0) = 1$, compute an approximation of $y(0.2)$ with (a) Euler's method and (b) Heun's method, using stepsize $h = 0.1$ in each case. Find the exact value of $y(0.2)$ and compute the error in each approximation.

3.2 Separable Equations

A differential equation is said to be *separable* if it can be written in the form

$$g(y)\, dy = f(t)\, dt.$$

This form suggests the formal procedure of antidifferentiating $g(y)$ and $f(t)$ with respect to y and t, respectively, and equating the results to arrive at an *implicit* general solution of the form

$$\int g(y)\,dy = \int f(t)\,dt + C. \tag{1}$$

(In specific cases, it may be possible to go further and solve explicitly for y.) This procedure can be justified as follows. After writing the differential equation in the form

$$g(y(t))\,\frac{dy}{dt} - f(t) = 0,$$

we can antidifferentiate each term with respect to t, obtaining

$$\int g(y(t))\,\frac{dy}{dt}\,dt - \int f(t)\,dt = C.$$

The "substitution" $y = y(t)$ now gives (1).

Example 1

The equation

$$\frac{dy}{dt} = \frac{1+\cos t}{1+3y^2}$$

is separable, since it can be rewritten in the form

$$(1+3y^2)\,dy = (1+\cos t)\,dt.$$

Antidifferentiation of each side results in an implicit general solution:

$$y + y^3 = t + \sin t + C.$$

A sampling of the resulting solution curves is plotted in Figure 1. ■

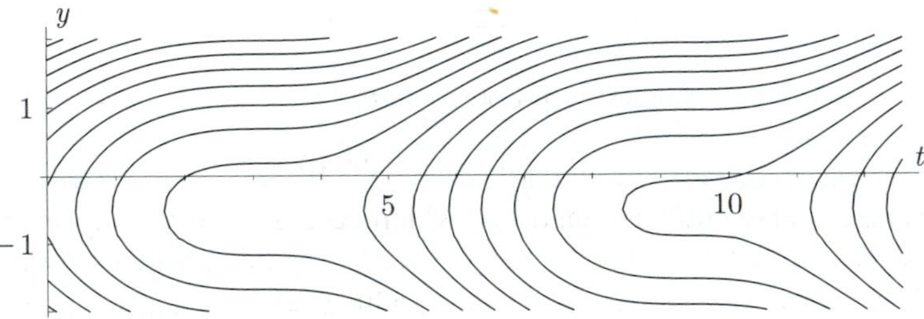

Figure 1

It is often convenient to solve an initial-value problem involving a separable equation and an initial condition $y(t_0) = y_0$ by integrating between limits. After writing the equation in the form $g(y)y' = f(t)$, we integrate each side over $[t_0, t]$:

$$\int_{t_0}^{t} g(y(s))y'(s)\,ds = \int_{t_0}^{t} f(s)\,ds.$$

With the change of variable $u = y(s)$ in the integral on the left side, this becomes

$$\int_{y_0}^{y} g(u)\,du = \int_{t_0}^{t} f(s)\,ds,$$

where the upper limit y on the left is an abbreviation for $y(t)$.

Example 2

Consider the initial-value problem

$$\frac{dy}{dt} = -3t^2y^2, \quad y(2) = 1.$$

Separating variables and preparing to integrate between limits, we obtain

$$\int_1^y \frac{du}{u^2} = -\int_2^t 3s^2\,ds.$$

From this we get $-u^{-1}|_1^y = -s^3|_2^t$, which leads to the implicit solution

$$-\frac{1}{y} + 1 = -t^3 + 8.$$

This is easily solved for y, yielding the explicit solution

$$y = \frac{1}{t^3 - 7}.$$

Note that this solution has the maximal domain $(\sqrt[3]{7}, \infty)$. ■

The Equation $\frac{dy}{dt} = g(y/t)$ Here we consider equations

$$\frac{dy}{dt} = f(t, y)$$

in which the function f has the property that

$$f(xt, xy) = f(t, y)$$

whenever x, t, and y are numbers for which each side of the equation is defined.* For instance, the function $f(t, y) = y/t$ clearly has this property, as does $f(t, y) = y/\sqrt{t^2 + y^2}$. Note that this property of f, with $x = 1/t$, implies that

$$f(1, y/t) = f(t, y),$$

* A function f with this property is said to be *homogeneous of degree* 0. For that reason, many textbooks refer to these as *homogeneous* differential equations. To avoid confusion, we will not use this terminology.

and so we may view $f(t, y)$ as a function of the single variable y/t. Therefore, the differential equation may be expressed in the form

$$\frac{dy}{dt} = g(y/t),$$

where $g(x) = f(1, x)$, which suggests the substitution $z = y/t$, or $y = tz$. It follows by the product rule that, if $y = tz$, then

$$\frac{dy}{dt} = t\frac{dz}{dt} + z.$$

Thus the differential equation becomes a separable one for z:

$$t\frac{dz}{dt} = g(z) - z.$$

Example 3

Consider the equation

$$\frac{dy}{dt} = \frac{y^2 - ty}{t^2 + ty}.$$

The function $f(t, y)$ on the right side has the property that

$$f(xt, xy) = \frac{(xy)^2 - (xt)(xy)}{(xt)^2 + (xt)(xy)} = \frac{y^2 - ty}{t^2 + ty} = f(t, y)$$

whenever x, t, and y are numbers for which the expressions involved are defined. Therefore, $f(t, y)$ can be expressed as a function of y/t. Indeed the differential equation may be rewritten as

$$\frac{dy}{dt} = \frac{(y/t)^2 - y/t}{1 + y/t}.$$

Substituting $y = tz$ as described previously produces

$$t\frac{dz}{dt} = \frac{z^2 - z}{1 + z} - z = -\frac{2z}{1 + z}.$$

By separating variables and integrating,

$$\int \frac{1 + z}{z} dz = -2 \int \frac{1}{t} dt,$$

we arrive at

$$\ln|z| + z = -2\ln|t| + C,$$

which, after some manipulation, becomes

$$t^2 z e^z = C.$$

Finally, replacing z with y/t, we obtain an implicit general solution:

$$t y e^{y/t} = C.$$

A sampling of the resulting solution curves is plotted in Figure 2. ■

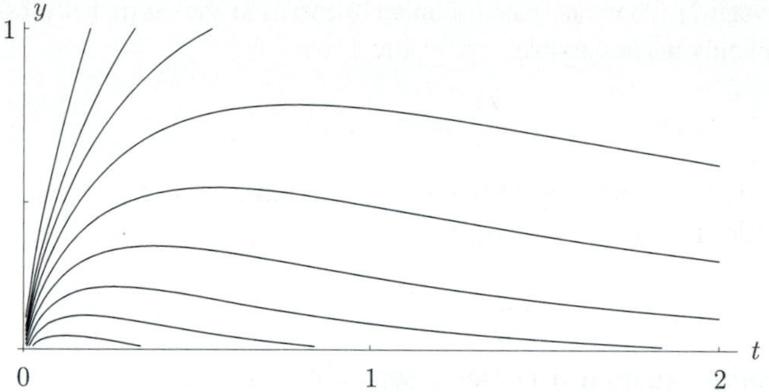

Figure 2

PROBLEMS

In Problems 1 through 6, solve the initial-value problem by separation of variables, and state the maximal domain.

1. $y' = 2ty^2, \quad y(0) = 1/4$

2. $\dfrac{dx}{dt} = 1 - x^2, \quad x(0) = 0$

3. $\dfrac{du}{dt} = e^u, \quad u(0) = 0$

4. $\dfrac{dy}{dx} = \dfrac{1}{2}y^3 \sin x, \quad y(\pi/3) = \sqrt{2}$

5. $y' = y \ln y, \quad y(0) = e$

6. $\dfrac{dz}{ds} = (1 + z^2) \tan^{-1} z, \quad z(0) = 1$

In Problems 7 and 8, obtain an implicit solution by separation of variables:

7. $y' = \cos t \sec y, \quad y(0) = \frac{\pi}{6}$

8. $y' = \sqrt{9 + y^2}, \quad y(0) = 4$

9. Consider the initial-value problem

$$y' = -\sqrt{y}, \quad y(0) = 1.$$

(a) Show that separation of variables produces a function that is defined for all t, yet satisfies the differential equation only on an interval of the form $(-\infty, T)$, where $0 < T < \infty$.

(b) Find a piecewise-defined solution that is valid for all t.

In Problems 10 through 12, obtain an implicit general solution of the equation by means of the substitution $z = y/t$.

10. $y' = \dfrac{t^2 y}{t^3 + y^3}$

11. $y' = \dfrac{t - y}{t + y}$

12. $y' = \dfrac{2t - y}{t - 2y}$

In Problems 13 through 16, solve the initial-value problem and state the maximal domain of the solution.

13. $y' = \dfrac{y^2 + ty}{t^2}, \quad y(1) = 1$

14. $y' = \dfrac{y + \sqrt{t^2 - y^2}}{t}, \quad y(1) = \dfrac{1}{2}$

15. $y' = \dfrac{t^2 + ty + y^2}{t^2}, \quad y(1) = 1$

16. $y' = \dfrac{y + te^{-y/t}}{t}, \quad y(1) = 0$

In Problems 17 through 21, obtain a general solution by means of the suggested substitution. Solve for the solution explicitly when possible.

17. $y' = \dfrac{1}{t+y}$ $(z = t + y)$

18. $y' = \sin^2(t + y + 1)$ $(z = t + y + 1)$

19. $y' = \dfrac{t+y-3}{t-y-1}$ $(y = v + 1, t = u + 2)$

20. $y' = \dfrac{2y}{t} + t^3 \sec^2\left(\dfrac{y}{t^2}\right)$ $(y = t^2 z)$

21. $y' = y^2 \sec^4 t - 2y \tan t$ $(y = z \cos^2 t)$

3.3 Bernoulli and Riccati Equations

This section is devoted to certain types of nonlinear first-order differential equations that become linear after an appropriate substitution. The most basic equations of this type are the so-called **Bernoulli equations**:

$$y' + py = fy^n, \quad n \neq 0, 1, \tag{1}$$

where p and f are continuous on an interval I. (Note that the equation is already linear if n has one of the exceptional values of 0 or 1.) An appropriate substitution for such an equation is

$$y = u^m, \quad y' = mu^{m-1}u',$$

since a judicious choice of m produces a linear equation in u. (See Problem 4.) The process is illustrated in the following example.

Example 1

Consider the equation

$$y' + 2y = y^3.$$

Substituting $y = u^m$, $y' = mu^{m-1}u'$ results in

$$mu^{m-1}u' + 2u^m = u^{3m}.$$

Now dividing by u^{m-1} gives us

$$m\,u' + 2\,u = u^{2\,m+1}.$$

The expression on the right side becomes 1 if we select $m = -1/2$. Then, after multiplying through by -2, we have the simple first-order linear equation

$$u' - 4u = -2,$$

which we solve by inspection, obtaining

$$u = \frac{1}{2} + Ce^{4t} = \frac{1}{2}(1 + Ce^{4t}).$$

Now since $y = u^{-1/2}$, we arrive at

$$y = \frac{\sqrt{2}}{\sqrt{1 + Ce^{4t}}}.$$

Bernoulli Equations with $n = 2$ Bernoulli equations that are quadratic in y (i.e., $n = 2$) are of particular interest, since they arise in numerous applications including population models. For the equation

$$y' + py = fy^2,$$

the substitution $y = u^m$, $y' = mu^{m-1}u'$ results in

$$mu^{m-1}u' + pu^m = fu^{2m},$$

which after division by u^{m-1} becomes

$$mu' + pu = fu^{m+1}.$$

So with $m = -1$, we have

$$u' - pu = -f, \quad \text{and} \quad y = \frac{1}{u}.$$

Example 2

Consider the initial-value problem

$$y' = (1 - e^{-t}y)y, \quad y(t_0) = y_0.$$

By substituting $y = 1/u$ and assuming for the moment that $y_0 \neq 0$, we obtain a linear problem for u:

$$u' + u = e^{-t}, \quad u(t_0) = \frac{1}{y_0}.$$

Next we multiply by the integrating factor e^t, after which the differential equation becomes $(e^t u)' = 1$, and then integrate from t_0 to t:

$$\int_{t_0}^{t} (e^s u(s))' ds = \int_{t_0}^{t} ds.$$

This easily leads to

$$e^t u - \frac{1}{y_0} = t - t_0,$$

and then, after solving for u,

$$u = \frac{1 + y_0(t - t_0)}{y_0 e^t}.$$

Thus the solution of our original initial-value problem, for *any* (t_0, y_0), is

$$y = \frac{y_0 e^t}{1 + y_0(t - t_0)}.$$

Riccati Equations Another class of nonlinear first-order equations that become linear after an appropriate substitution are the **Riccati equations**:

$$y' + py = fy^2 + g, \tag{2}$$

where the functions p, f, g are continuous on an interval I. Note that a Bernoulli equation with $n = 2$ is a special case of a Riccati equation. Indeed, the first step toward finding an appropriate substitution is the "observation" (see Problem 17) that if y and $\tilde{y}$ are any two solutions of (2), then the difference $z = y - \tilde{y}$ satisfies the Bernoulli equation

$$z' + (p - 2\tilde{y}f)z = f z^2.$$

Thus if $\tilde{y}$ is a known particular solution, then z can be found by means of the substitution $z = 1/u$, which produces a linear equation for u:

$$u' - (p - 2\tilde{y}f)u = -f.$$

Note that the successive substitutions that resulted in this linear equation can be combined into one:

$$y = \frac{1}{u} + \tilde{y}.$$

Example 3

Consider the equation

$$y' + y = ty^2 - \frac{1}{t^2}.$$

It is not difficult to see that $\tilde{y} = 1/t$ is a particular solution. So we make the substitution

$$y = \frac{1}{u} + \frac{1}{t}, \quad y' = -\frac{u'}{u^2} - \frac{1}{t^2},$$

obtaining

$$-\frac{u'}{u^2} - \frac{1}{t^2} + \frac{1}{u} + \frac{1}{t} = t\left(\frac{1}{u} + \frac{1}{t}\right)^2 - \frac{1}{t^2},$$

which, after simplification, becomes the linear equation

$$u' + u = -t.$$

With the help of the integrating factor e^t, we find

$$u = 1 - t + Ce^{-t},$$

which then gives

$$y = \frac{1}{1 - t + Ce^{-t}} + \frac{1}{t}. \quad \blacksquare$$

PROBLEMS

In Problems 1 through 3, find a general solution of the equation by substituting $y = u^m$ and determining the appropriate value of m.

1. $y' + y = y^2$

2. $y' + y = \sqrt{y}$

3. $y' + y = \dfrac{1}{y}$

4. Find a general formula for the correct value of m (in terms of n) to use in solving $y' + py = fy^n, n \neq 1$, by means of the substitution $y = u^m$. Observe that the transformed equation always takes the form

$$mu' + pu = f.$$

For the equations in Problems 5 through 13, find a general solution by means of an appropriate substitution of the form $y = u^m$.

5. $y' - \dfrac{1 - e^{-t}}{t + e^{-t}} y = y^2$

6. $y' - y \tan t = y^2$

7. $y' - \dfrac{y}{2t \ln t} = \dfrac{\ln t}{y}$

8. $y' + \dfrac{y}{t + 1} = -y^3$

9. $y' - \dfrac{ty}{t^2 + 1} = \dfrac{2t}{y^2}$

10. $y' - \dfrac{e^t y}{1 + e^t} = -e^t \sqrt[3]{y}$

11. $y' + ty = ty^3$

12. $y' - 3\sqrt{t}y = \sqrt{ty}$

13. $y' - \dfrac{y}{2t^2} = \dfrac{te^{-1/t}}{y}$

In Problems 14 through 16, find (as in Example 2) the solution that satisfies $y(t_0) = y_0$.

14. $y' = (1 - e^t y)y$

15. $y' = (\cot t - y)y$

16. $y' = y\left(\dfrac{e^t}{1 + e^t} - e^{-t} y\right)$

17. Show that if y and $\tilde{y}$ are solutions of the Riccati equation

$$y' + py = fy^2 + g,$$

then the difference $z = y - \tilde{y}$ satisfies the Bernoulli equation

$$z' + (p - 2\tilde{y}f)z = fz^2.$$

For the Riccati equations in Problems 18 through 23,

(a) verify the indicated particular solution $\tilde{y}$;

(b) find a general solution;

(c) find the solution that satisfies the initial condition $y(t_0) = y_0$. (In 20 assume that $\cos t_0 \neq 0$, in 22 assume that $t_0 \neq 0$, and in 23 assume that $\sin t_0 \neq -1$.)

18. $y' + y = y^2 - 2, \quad \tilde{y} = 2$

19. $y' + y = e^t y^2 - e^{-t}, \quad \tilde{y} = e^{-t}$

20. $y' - y \tan t = y^2 \cos t - \sec t, \quad \tilde{y} = \sec t$

21. $y' + (2e^t - 1)y = y^2 + e^{2t}, \quad \tilde{y} = e^t$

22. $y' + \dfrac{1+2t^3}{t} y = ty^2 + t^3 + 2, \quad \tilde{y} = t$

23. $y' + \dfrac{y \cos t}{1+\sin t} = \dfrac{y^2}{1+\sin t} - \sin t, \quad \tilde{y} = \cos t$

24. Let p and f be continuous on an interval I, and let g and g' be continuous on an interval J, with $g'(x) \neq 0$ for all x in J. Show that the substitution $u = g(y)$ transforms the equation

$$\frac{dy}{dt} = \frac{f(t) - p(t)g(y)}{g'(y)},$$

into the linear equation

$$u' + pu = f.$$

Show that the Bernoulli equation (1) is of this form, where $g(y) = y^{1-n}$.

Each of the equations in Problems 25 through 30 can be written in the form of the equation in Problem 24. Find a general solution by means of an appropriate substitution.

25. $y' = \dfrac{e^{-t} - \sin y}{\cos y}$

26. $y' = (e^t + \tan y)\cos^2 y$

27. $y' = e^{-y} - \dfrac{1}{t}$

28. $y' = e^{y-t} - 1$

29. $y' = (1+y^2)(e^t + \tan^{-1} y)$

30. $y' = ty - \dfrac{y \ln y}{t+1}$

3.4 Reduction of Order

Certain second-order equations can be reduced to first order by means of an appropriate substitution. The simplest situation occurs when a second-order equation for y does not explicitly involve y; that is, when the equation has the form

$$y'' = f(t, y').$$

Then the substitution $v = y'$ gives the first-order equation

$$v' = f(t, v).$$

If v can be found explicitly, then y can be obtained by integration.

Example 1

Suppose that we wish to find a solution of

$$y'' = 2t(1 + (y')^2), \quad y(0) = y_0, \quad y'(0) = v_0.$$

Substituting $v = y'$, $v' = y''$ produces the equation

$$v' = 2t(1 + v^2).$$

By separating variables and integrating, we obtain

$$\tan^{-1} v = t^2 + c_1,$$

which leads to

$$v = \tan(t^2 + \tan^{-1} v_0),$$

Now we replace v with y' and integrate each side of this equation from 0 to t, obtaining

$$y = y_0 + \int_0^t \tan(s^2 + \tan^{-1} v_0)\, ds.$$

■

Example 2

Consider the initial-value problem

$$y'' = -ky', \quad y(0) = y_0, \quad y'(0) = v_0.$$

Substituting $v = y'$, $v' = y''$ gives the equation $v' = -kv$. This, together with $y'(0) = v_0$, gives

$$v = v_0 e^{-kt}.$$

Next we integrate and use the initial condition $y(0) = y_0$ to obtain

$$y = \frac{v_0}{k}(1 - e^{-kt}) + y_0.$$

■

A somewhat more complicated situation is presented by equations of the form

$$y'' = f(y, y'),$$

in which f does not explicitly depend on the independent variable. The key here is to treat y as if *it* were the independent variable. Setting $v = y'$ as before, we have by the chain rule that

$$y'' = \frac{dv}{dt} = \frac{dv}{dy}\frac{dy}{dt} = \frac{dv}{dy}v.$$

So, substituting

$$y' = v \quad \text{and} \quad y'' = v\frac{dv}{dy}$$

into the differential equation gives

$$v\frac{dv}{dy} = f(y, v),$$

which we then attempt to solve for v *as a function of* y. The result will be a first-order equation for y.

Example 3

Consider the equation

$$y'' = \frac{y'}{y}(y^2 + y').$$

It is easy to see that this equation is satisfied by any constant. To find nonconstant solutions we substitute $y' = v$ and $y'' = v\,\frac{dv}{dy}$, obtaining a linear first-order equation for v:

$$\frac{dv}{dy} - \frac{1}{y}\,v = y.$$

Using the integrating factor $e^{\int 1/y} = e^{\ln y} = y$ results in

$$\frac{d}{dy}(y\,v) = y^2.$$

So we antidifferentiate, rearrange, and replace v with y', finally arriving at a separable equation for y:

$$y' = \frac{y^3 + C}{3y}.$$ ■

Example 4

Consider the initial-value problem

$$y'' = 2yy', \quad y(0) = 0, \quad y'(0) = 4.$$

Substitution of $y' = v$ and $y'' = v\frac{dv}{dy}$, followed by division by v, gives

$$\frac{dv}{dy} = 2y;$$

thus $v = y^2 + C$. Before going further, let's determine C from the initial conditions. Since y and v simultaneously have values 0 and 4, respectively, we have $4 = 0 + C$. Thus, $v = y^2 + 4$, which is a separable equation for y, since $v = y'$. Separating variables and integrating between limits, we get

$$\int_0^y \frac{du}{u^2 + 4} = \int_0^t ds,$$

which leads to

$$\frac{1}{2}\tan^{-1}\frac{y}{2} = t, \quad \text{or } y = 2\tan 2t.$$

Note that the maximal domain of the solution is $\left(-\frac{\pi}{4}, \frac{\pi}{4}\right)$. ■

PROBLEMS

For each equation in Problems 1 through 6, solve for $v = y'$ in terms of y and t.

1. $y'' = 6t(y')^{2/3}$

2. $y''(y')^2 = \cos t$

3. $y'' + y' = e^t$

4. $y'' = y'\,(e^y - y')$

5. $y''y' = \cos y$

6. $y'' = 3y^2\left(1 + (y')^2\right)$

In Problems 7 through 9, solve the initial-value problem consisting of the equation

$$y'' = 2yy'$$

and the given initial conditions.

7. $y(0) = 0,\ y'(0) = 1$

8. $y(0) = 1,\ y'(0) = 1$

9. $y(0) = 0,\ y'(0) = -1$

In Problems 10 and 11, solve the initial-value problem consisting of the equation

$$y'' = (1 + 2y)\, y'$$

and the given initial conditions.

10. $y(0) = 1,\ y'(0) = 2$

11. $y(0) = 0,\ y'(0) = 1$

In Problems 12 through 14, solve the given initial-value problem.

12. $y'' = 2y(y')^3, \quad y(0) = 1,\ y'(0) = -1$

13. $y'' = (y')^2(y' - \tan y), \quad y(0) = \frac{\pi}{4},\ y'(0) = -1$

14. $y'' = 8y\,(1 + y^2), \quad y(0) = 1,\ y'(0) = -4$

15. If air resistance is ignored, the height $y(t)$ of a mass m falling under constant gravitational acceleration $-g$ satisfies $m\,y'' = -mg$.

(a) Show that the substitution $y' = v,\ y'' = v\frac{dv}{dy}$ leads to the statement that the sum of kinetic and potential energy is constant:

$$\frac{1}{2}mv^2 + mgy = \frac{1}{2}mv_0^2 + mgy_0.$$

(b) Show that an object that falls through a distance y in a vacuum achieves a speed increase of $\sqrt{2gy}$.

(c) Show that the differential equation $m\,y'' = -mg$ can be derived from the conservation of energy statement in part (a).

16. The gravitational attraction of the earth decreases as the inverse square of the distance to the earth's center. If we ignore air resistance, then vertical motion of an object with mass m is governed by

$$m\,r'' = -\frac{GMm}{r^2},$$

where r is the distance to the earth's center, G is the universal gravitation constant, and M is the mass of the earth.

(a) Show that the substitution $r' = v,\ r'' = v\frac{dv}{dr}$ leads to

$$\frac{1}{2}m\,v^2 - \frac{1}{2}m\,v_0^2 = GMm\left(\frac{1}{r} - \frac{1}{r_0}\right).$$

(b) Let r_e denote the radius of the earth, so that $r = r_e + y$, where y is the height above the earth's surface. Also let $y(0) = y_0$, and note that $r_0 = r_e + y_0$. Using the fact that $g = GM/r_e^2$, show that the result of part (a) can be rewritten as

$$v^2 - v_0^2 = -2g\,\frac{r_e^2\,(y - y_0)}{(r_e + y)(r_e + y_0)}.$$

(c) Set $v = 0$ in the result of part (b), and solve for y. Note that this produces a formula for the maximum height of the object.

(d) Find v_0 so that the maximum height is infinite. This is the initial velocity necessary to escape the earth's gravitational pull from a height of y_0. Given $r_e = 6.4$ by 10^6 m and $g = 9.8$ m/s^2, compute a numerical value of the escape velocity (in m/s) for both $y_0 = 0$ and $y_0 = 10^8$ m.

17. Consider the equation of motion of the simple pendulum:

$$\theta'' = -\frac{g}{L} \sin \theta.$$

(a) Show that the angular velocity $\omega = \theta'$ satisfies

$$\frac{d}{d\theta}\left(\omega^2\right) = -\frac{2g}{L} \sin \theta$$

and that the sum of kinetic and potential energy is constant:

$$\frac{1}{2} m L^2 \omega^2 + m g L (1 - \cos \theta) = C.$$

(b) Show that the differential equation $\theta'' = -\frac{g}{L} \sin \theta$ can be derived from the conservation of energy statement in part (a).

18. 18. Let $0 < \theta_0 < \pi$, and suppose that the pendulum in Problem 17 is set in motion at time $t = 0$ with

$$\theta(0) = \theta_0 \text{ and } \omega(0) = 0.$$

(a) Show that

$$\omega = -\sqrt{\frac{2g}{L}} \sqrt{\cos \theta - \cos \theta_0},$$

and therefore

$$-\sqrt{\frac{2g}{L}}\, dt = \frac{d\theta}{\sqrt{\cos \theta - \cos \theta_0}}.$$

(b) Let T be the period of the subsequent motion of the pendulum, noting that $t = T/4$ is then the time at which the pendulum first passes the downward vertical position (i.e., $\theta(T/4) = 0$). By integrating the result of part (a) from $t = 0$ to $t = T/4$, show that the period T is given by

$$T = 2\sqrt{\frac{2L}{g}} \int_0^{\theta_0} \frac{d\theta}{\sqrt{\cos \theta - \cos \theta_0}}.$$

(c) Using a graphing calculator or computer, plot the graph of T versus θ_0 for $0 < \theta_0 < \pi$ with $g = 9.8$ and $L = 1$. Discuss the connection between the graph and the behavior of the pendulum.

19. The purpose of this problem is to show that the period T of the pendulum in Problem 18 can be expressed in terms of a standard, nonelementary, "special function" known as the *elliptic integral of the first kind*:

$$F(\beta, m) = \int_0^{\beta} \frac{d\phi}{\sqrt{1 - m \sin^2 \phi}}.$$

(a) Use trigonometric half-angle formulas to show that

$$T = 2\sqrt{\frac{L}{g}} \int_0^{\theta_0} \frac{d\theta}{\sqrt{\sin^2(\theta_0/2) - \sin^2(\theta/2)}}.$$

(b) Show that the change of variables

$$\sin(\theta/2) = \sin\alpha \, \sin\phi, \quad \text{where } \alpha = \theta_0/2,$$

implies that

$$\frac{1}{2}\cos(\theta/2)\, d\theta = \sin\alpha \, \cos\phi \, d\phi,$$

and so

$$d\theta = \frac{2\sin\alpha \, \cos\phi \, d\phi}{\cos(\theta/2)} = \frac{2\sin\alpha \, \cos\phi \, d\phi}{\sqrt{1 - \sin^2\alpha \, \sin^2\phi}}.$$

(c) Show that

$$T = 4\sqrt{\frac{L}{g}} \int_0^{\pi/2} \frac{d\phi}{\sqrt{1 - \sin^2\alpha \, \sin^2\phi}} = 4\sqrt{\frac{L}{g}}\, F\left(\pi/2, \sin^2\alpha\right).$$

3.5 Nonlinear First-Order Equations in Applications

In this section we'll look at a few applications from physics and engineering that give rise to nonlinear first-order differential equations. With the exception of Torricelli's law, which leads to a differential equation that governs the depth of fluid in a draining tank, the applications here are nonlinear versions of applications that we encountered in Section 2.2.

3.5.1 Motion with Nonlinear Resistance

Recall that in Section 2.2.1, we learned that a model for the vertical motion of a projectile with mass m under constant gravitational force through a resistive medium is

$$m\,\frac{dv}{dt} = -m\,g - k_\ell\, v,$$

where $k_\ell > 0$ is a constant. (Here we are using the subscript ℓ to signify that k_ℓ is the drag coefficient in a *linear* model.)

Quadratic Drag Experiments show that resistance caused by a viscous fluid is, for low velocities, approximately proportional to the velocity, but for higher velocities it is approximately proportional to the square of the velocity.* A quadratic drag force can be written as

$$\rho(v) = k_q v|v|$$

* Whether a velocity is considered "low" or "high" depends upon the size of the object and the viscosity of the fluid. This is related to the concept of the *Reynolds number* in fluid dynamics.

where $k_q > 0$ is a constant. (The expression $v|v|$ is just a compact way of expressing v^2 if $v \geq 0$ and $-v^2$ if $v < 0$.) Now, with constant mass, our equation for the velocity is

$$m\frac{dv}{dt} = -mg - k_q v|v|. \tag{1}$$

This equation is separable. However, its solution requires separate consideration of upward and downward paths. If we restrict our consideration to the *downward* flight of an object, then $v|v| = -v^2$ and so we have the separated equation

$$\frac{m\,dv}{mg - k_q v^2} = -dt.$$

If we set $r = \sqrt{mg/k_q}$, we have

$$\frac{dv}{r^2 - v^2} = -\frac{k_q}{m}\,dt. \tag{2}$$

Now, after performing a partial fraction decomposition on the left side, we integrate and use the initial condition $v(0) = 0$, finally arriving at the implicit solution

$$\ln\left|\frac{v+r}{v-r}\right| = -\frac{2rk_q}{m}\,t. \tag{3}$$

A persistent student should proceed onward to solve for v (see Problem 1); however, we can already observe that the terminal velocity is the quantity $-r = -\sqrt{mg/k_q}$. (Why?)

It is worth noting that the terminal velocity, both here and in the linear case, is simply the *equilibrium solution of the differential equation for* v. Indeed, if we set about finding equilibrium (i.e., constant) solutions $v = v_\infty$ of (1), then we easily find $v_\infty = -\sqrt{mg/k_q}$.

A Hybrid Drag Model A model for drag force that is approximately proportional to v for low velocities and approximately proportional to v^2 for higher velocities is given by

$$\rho(v) = k_\ell\, v + k_q v|v|.$$

The differential equation analogous to (1) is then

$$m\frac{dv}{dt} = -mg - k_\ell\, v - k_q v|v|, \tag{4}$$

which for downward motion ($v < 0$) becomes

$$m\frac{dv}{dt} = -mg - k_\ell\, v + k_q v^2.$$

This is a Riccati equation with a negative equilibrium solution v_∞ (the terminal velocity) that is found by solving

$$0 = -mg - k_\ell\, v_\infty + k_q v_\infty^2.$$

The negative solution of this quadratic equation is

$$v_\infty = \frac{k_\ell - \sqrt{k_\ell^2 + 4gk_q m}}{2k_q}.$$

So in order to solve the Riccati equation, we substitute

$$v = \frac{1}{u} + v_\infty,$$

thus obtaining the linear equation

$$u' - \frac{k_\ell - 2v_\infty k_q}{m}u = -\frac{k_q}{m}.$$

For convenience, let us set

$$\mu = k_\ell - 2v_\infty k_q = \sqrt{k_\ell^2 + 4gk_q m},$$

so that the equation for u becomes

$$u' - \frac{\mu}{m}u = -\frac{k_q}{m}.$$

Since the coefficients here are constants, a general solution is easily found by superposition:

$$u = Ce^{\mu t/m} + \frac{k_q}{\mu}.$$

Since $v = 1/u + v_\infty$, we now have

$$v = \frac{\mu}{Ce^{\mu t/m} + k_q} + v_\infty = \frac{\mu e^{-\mu t/m}}{C + k_q e^{-\mu t/m}} + v_\infty.$$

The initial condition $v(0) = 0$ implies that

$$C = -\frac{\mu}{v_\infty} - k_q.$$

Therefore, the velocity for $t \geq 0$ is

$$v = \frac{v_\infty \mu\, e^{-\mu t/m}}{k_q v_\infty (e^{-\mu t/m} - 1) - \mu} + v_\infty, \tag{5}$$

and the height y can be found by integration. (See Problem 7.)

Variable Mass and Thrust The incorporation of variable mass m and thrust f into the hybrid drag model gives us the equation

$$\frac{d}{dt}(mv) = -mg + f - k_\ell v - k_q v|v|,$$

which upon rearrangement becomes

$$v' + \left(\frac{k_\ell + m'}{m}\right)v = -g + \frac{f - k_q v|v|}{m}. \tag{6}$$

For solutions with constant sign, $v|v|$ becomes either $\pm v^2$, and the equation is a Riccati equation. However, finding a particular solution for constructing the desired substitution promises to be difficult, if possible at all. Thus, numerical approximation becomes our only available method for gaining insight into the solutions of this equation.

Example 1

Consider a small rocket whose velocity is governed by (5) with

$$m = 0.1 + e^{-t} \text{ kg}, \quad f = 500e^{-t} \text{ N},$$
$$k_\ell = 0.01 \text{ N s/m}, \quad k_q = 10^{-4} \text{ N s}^2/\text{m}^2;$$

that is,

$$v' + \left(\frac{0.01 - e^{-t}}{0.1 + e^{-t}}\right) v = -9.8 + \frac{500e^{-t} - 10^{-4}\, v|v|}{0.1 + e^{-t}}.$$

Suppose also that the rocket is launched from rest at time $t = 0$ (i.e., $v(0) = 0$). Figure 1a is a numerically generated plot of the velocity versus time. Notice the horizontal asymptote $v = v_\infty$, where v_∞ is the terminal velocity. Figure 1b is a plot of the rocket's height y versus time, obtained by numerically integrating the velocity approximation. ■

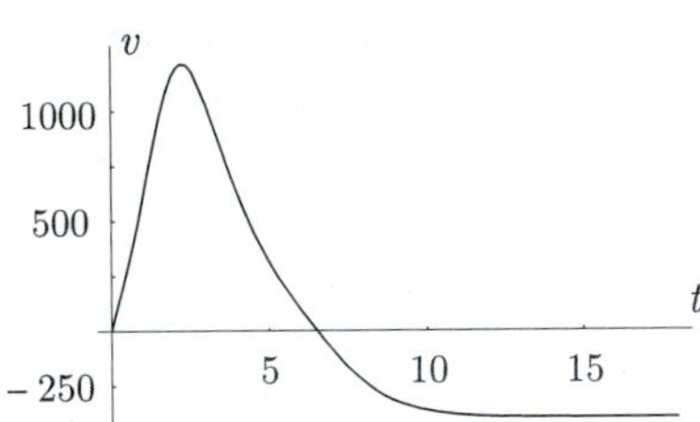

Figure 1a

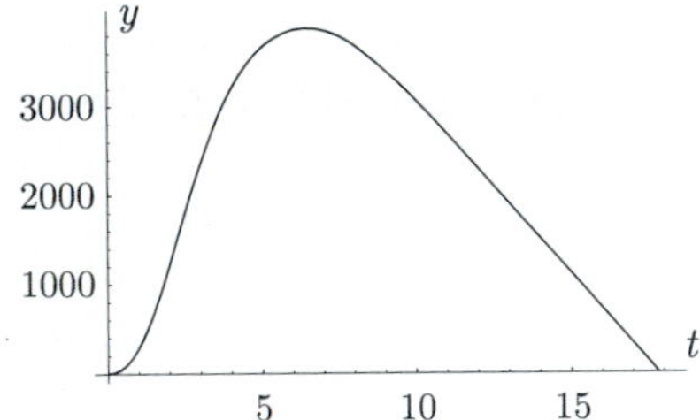

Figure 1b

PROBLEMS

1. Let $\beta = 2rk_q/m = 2\sqrt{k_q g/m}$. Solve (3) for v to obtain

$$v = -r\,\frac{1 - e^{-\beta t}}{1 + e^{-\beta t}}, \quad \text{where} \quad r = \sqrt{\frac{mg}{k_q}}, \quad \beta = \frac{2\,rk_q}{m}.$$

2. The antiderivative of the left side of equation (2) may be expressed in terms of $\tanh^{-1}$, the inverse hyperbolic tangent function. Show that this leads to

$$v = -r \tanh\left(\frac{k_q r}{m}\,t\right)$$

and that this result is equivalent to that of Problem 1.

3. Use the result of Problem 2, along with the formula $\int \tanh u\, du = \ln \cosh u + C$, to find the height y of the projectile on its downward path with zero initial velocity and initial height y_0.
4. Suppose that a ball with a mass of 1 kilogram is dropped from a height of 1000 meters above the ground. Assuming a quadratic drag model with $k_q = 0.001$, how long will it take the ball to hit the ground? What will be the ball's impact velocity?
5. Solve the quadratic drag model for the velocity and height of a projectile during its *upward* flight with initial velocity $v_0 > 0$ and initial position $y_0 = 0$.
6. Consider the general model

$$m\frac{dv}{dt} = -m\,g - \rho(v)$$

for the velocity of a projectile with initial height $y_0 = 0$ and a positive initial velocity v_0. (See Section 2.2.1.) Make the substitution $\frac{dv}{dt} = v\frac{dv}{dy}$ and integrate between limits to obtain the formula

$$H = \int_0^{v_0} \frac{mv\,dv}{mg + \rho(v)}$$

for the projectile's maximum height. To what does this reduce when $\rho(v) = k_q v^2$? To what does it reduce when $\rho(v) = k_\ell v$?

7. Integrate the velocity (5) to obtain the height

$$y = y_0 + v_\infty t - \frac{m}{k_q}\ln\left(\frac{\mu + k_q v_\infty\,(1 - e^{-\mu t/m})}{\mu}\right).$$

8. **(a)** Suppose that, with an initial velocity of $v_0 = 5$ m/s, a certain object with mass $m = 0.1$ kg attains a maximum height of 1.2 m. Assuming that $k_q v^2$ is negligible for velocities this small, use the maximum height formula for the linear drag model to estimate the coefficient k_ℓ in the hybrid model (4).

   (b) Suppose further that the terminal velocity of the same object is observed to be −29 m/s. Estimate the coefficient k_q in the hybrid model (4).

   (c) Use the result of Problem 7 to find the time it takes for the object to hit the ground if it is dropped from a height of 200 meters.

9. Consider the following model for the motion of an object with variable mass in a viscous fluid with no forces present other than drag:

$$\frac{d}{dt}(mv) = -k_\ell v - k_q v|v|.$$

   (a) Show that for motion in the positive direction ($v > 0$) the velocity satisfies the Bernoulli equation

$$mv' + (m' + k_\ell)v = -k_q v^2.$$

   (b) Suppose that the object's mass is given by $m = m_0 + r\,t$ (and that k_ℓ and k_q are positive constants). Assuming that $r + k_\ell \neq 0$, derive the solution

$$v = \frac{(r + k_\ell)\,v_0}{(r + k_\ell + k_q v_0)\left(\frac{m_0 + r\,t}{m_0}\right)^{1 + k_\ell/r} - k_q v_0}.$$

10. If a quadratic drag model is used, the equations for the path of a baseball in two dimensions under constant gravitational force can be obtained as follows. Since $\frac{V(t)}{\|V(t)\|}$ is a unit vector in the same direction as the velocity vector $V(t)$, the quadratic drag expression will be

$$-k_q\|V(t)\|^2\frac{V(t)}{\|V(t)\|} = -k_q V(t)\|V(t)\|,$$

where $\|V(t)\| = \sqrt{x'(t)^2 + y'(t)^2}$ is the length of the velocity vector. The horizontal and vertical components of this drag force are then

$$-k_q x'(t)\sqrt{x'(t)^2 + y'(t)^2} \quad \text{and} \quad -k_q y'(t)\sqrt{x'(t)^2 + y'(t)^2},$$

respectively. Thus the equations become

$$\begin{aligned} mx''(t) &= -k_q x'(t)\sqrt{x'(t)^2 + y'(t)^2}, \\ my''(t) &= -mg - k_q y'(t)\sqrt{x'(t)^2 + y'(t)^2}. \end{aligned}$$

This is a coupled system of nonlinear equations for which a solution in terms of elementary functions is out of the question. Numerical approximation of $(x(t), y(t))$ requires that the system be converted to a first-order system. Show that substitution of $u = x'$ and $v = y'$ results in the following first-order system for (x, u, y, v):

$$x' = u, \quad u' = -ru\sqrt{u^2+v^2}, \quad y' = v, \quad v' = -g - rv\sqrt{u^2+v^2},$$

where $r = k_q/m$. The point is that a numerical method such as Euler or Heun can be adapted to give approximate solutions to such a first-order system.

3.5.2 Torricelli's Law

The situation with which we are concerned here is that of a tank containing a fluid that is draining through a hole in the bottom of the tank. In the 1600s, Evangelista Torricelli* discovered that, if the depth of the fluid is y, then the speed at which the fluid exits the hole is the same as the speed of a stone after free fall through a distance y, which is given by $\sqrt{2gy}$.† So, since the (volume) flow rate through the hole is proportional to the exit speed, it is also proportional to $\sqrt{y}$; that is,

$$\frac{dV}{dt} = -\rho\sqrt{y} \tag{1}$$

where ρ is a positive constant, $V(t)$ is the volume of fluid in the tank, and $y(t)$ is the depth of the fluid. This is Torricelli's law—also known as the Torricelli-Borda law. The constant ρ naturally depends on the size of the hole and the viscosity of the fluid. In fact, if the hole (or exit pipe) has cross-sectional area a, then

$$\rho = ab\sqrt{2g},$$

---

* Torricelli is best known for inventing the barometer in 1643.

† This follows from conservation of energy; see Problem 15 in Section 3.4.

where b, the *Borda constant*, reflects the viscosity of the fluid. For water, $b \approx 0.6$, while for a less viscous fluid, such as alcohol or gasoline, $0.6 < b < 1$.

The first question to address here is how we can turn this law into a differential equation for the depth y. If the tank has a constant horizontal cross-sectional area a (i.e., it's some type of cylinder), then this is easy. The volume is always $V = ay$, and so the differential equation becomes

$$a\frac{dy}{dt} = -\rho\sqrt{y}.$$

However, we may be interested in more general shapes, so we need to look at a variable horizontal cross-sectional area $a(s)$, where we'll agree that s is the distance from the bottom of the tank. Now we need to do an integration to find the volume of fluid in the tank when the depth is y:

$$V = \int_0^y a(s)\,ds.$$

By the Fundamental Theorem of Calculus and the chain rule, this gives us

$$\frac{dV}{dt} = a(y)\,\frac{dy}{dt}.$$

Therefore, Torricelli's law becomes

$$a(y)\frac{dy}{dt} = -\rho\sqrt{y}. \tag{2}$$

Note that this is a separable, nonlinear differential equation for the depth y.

Example 1

Suppose that the tank in question is a paraboloid generated by revolving the arc $y = x^2$, $0 \le x \le 2$, about the y-axis. Suppose further that the tank is initially full and that the fluid in it is draining through a hole located at the vertex. To determine the depth $y(t)$ of the fluid in the tank at time $t \ge 0$, we first note that the horizontal cross-sectional area of the tank is $a(s) = \pi s$ at height s above the vertex. So the initial-value problem of interest is

$$\pi\,y\,\frac{dy}{dt} = -\rho\sqrt{y}, \quad y(0) = 4.$$

Separation of variables leads to

$$\int_4^y \sqrt{u}\,du = \int_0^t -\frac{\rho}{\pi}\,dt,$$

and then

$$\frac{2}{3}\left(y^{3/2} - 8\right) = -\frac{\rho}{\pi}\,t.$$

We solve for y and find that

$$y = \left(8 - \frac{3\rho}{2\pi}\,t\right)^{2/3}$$

up to time $t = \frac{16\pi}{3\rho}$, at which the tank will be completely empty. Thus the full solution of the problem is

$$y = \begin{cases} \left(8 - \dfrac{3\rho}{2\pi}t\right)^{2/3}, & \text{if } 0 \le t \le \dfrac{16\pi}{3\rho}; \\ 0, & \text{if } t > \dfrac{16\pi}{3\rho}. \end{cases}$$

■

Emptying Time Let's look at the specific problem of determining the time at which a tank becomes empty. If the initial depth is y_0 and we agree to call the emptying time T, then we have after separating variables and integrating over $0 \le t \le T$ that

$$\int_{y_0}^{0} \frac{a(y)}{\sqrt{y}}\, dy = -\rho \int_0^T dt.$$

Thus we arrive at the formula

$$T = \frac{1}{\rho} \int_0^{y_0} \frac{a(y)}{\sqrt{y}}\, dy.$$

Example 2

Suppose that a fuel tank is a cylinder with radius 3 feet, length 10 feet, and a horizontal central axis. (The flat ends are vertical.) The tank's horizontal cross-sectional area at height s is

$$a(s) = 2\sqrt{9 - (s-3)^2} \cdot 10 = 20\sqrt{6s - s^2}.$$

Therefore, if the tank is initially full and begins to drain through an exit pipe at the bottom, then the time required for the tank to drain completely is

$$\begin{aligned} T &= \frac{1}{\rho} \int_0^6 \frac{20\sqrt{6y - y^2}}{\sqrt{y}}\, dy \\ &= \frac{20}{\rho} \int_0^6 \sqrt{6 - y}\, dy = \frac{80\sqrt{6}}{\rho}. \end{aligned}$$

Suppose that the Borda constant associated with the fuel is .75. Then, with a 2-inch-diameter exit pipe, the constant ρ would be $.75\pi(1/12)^2\sqrt{(2)(32)} \approx 0.13$, and so $T \approx 1500\,\text{s} = 25$ min. If the diameter of the exit pipe is 4 inches, then $\rho = 0.75\pi(1/6)^2\sqrt{(2)(32)} \approx 0.52$, and so $T \approx 380\,\text{s} = 6.3$ min. ■

Mixing and Draining Let us now suppose that we have a cylindrical tank with cross-sectional area a, so that volume is related to depth by $V = ay$. Also, we suppose that fluid containing a contaminant at a concentration k_{in} is being pumped into the tank at a rate R_{in} and that outflow is due to drainage through a hole in the bottom of the tank. As we saw in Section 2.2.2, the amount of contaminant in the tank satisfies

$$\frac{dA}{dt} = k_{\text{in}} R_{\text{in}} - \frac{A}{V} R_{\text{out}}, \quad A(0) = A_0,$$

and the volume of fluid in the tank satisfies

$$\frac{dV}{dt} = R_{\text{in}} - R_{\text{out}}, \qquad V(0) = V_0.$$

Since the outflow is due to drainage, Torricelli's law tells us that the outflow rate is

$$R_{\text{out}} = \rho\sqrt{y} = \rho\sqrt{V/a},$$

where ρ is a positive constant. Therefore, the initial-value problems for A and V become the coupled system

$$\begin{aligned} \frac{dA}{dt} &= k_{\text{in}} R_{\text{in}} - \frac{\rho A}{\sqrt{aV}}, \qquad A(0) = A_0 = k_0 V_0, \\ \frac{dV}{dt} &= R_{\text{in}} - \rho\sqrt{V/a}, \qquad V(0) = V_0. \end{aligned} \tag{3}$$

In principle, since the V-equation does not involve A, we could attack the system by solving the second equation for $V(t)$, substituting the result into the first equation, and then solving the first equation for $A(t)$. However, this plan leads no further than an implicit solution for $V(t)$.

If the tank has variable cross-sectional area $a(s)$, where s is the distance from the bottom of the tank, and if V is known in terms of y, say $V = V(y)$, then we have the following system for the amount A and the depth y:

$$\begin{aligned} \frac{dA}{dt} &= k_{\text{in}} R_{\text{in}} - \frac{\rho\sqrt{y}}{V(y)} A, \qquad A(0) = A_0 = k_0 V(y_0) \\ \frac{dy}{dt} &= \frac{R_{\text{in}} - \rho\sqrt{y}}{a(y)}, \qquad y(0) = y_0. \end{aligned} \tag{4}$$

Again, this is not a problem in which we expect to make much progress toward finding the exact solution. However, having set up the equations, we can tackle the problem numerically.

Example 3

Suppose that a conical tank, whose height and diameter are each 3 feet, is initially full of pure water and that, at time $t = 0$, brine with salt concentration 0.1 lb/ft^3 begins to flow into the tank at a rate of 1 ft^3/min, and the well-mixed contents of the tank begin to drain through a hole at the vertex of the tank. Suppose further that the size of the hole is such that the constant ρ in Torricelli's law is 1. Since the cross-sectional area at height s above the vertex is $a(s) = \pi s^2$, and since the volume of solution with depth y is $V(y) = \pi y^3/3$, the system (4) becomes

$$\begin{aligned} \frac{dA}{dt} &= 0.1 - \frac{3A}{\pi y^{5/2}}, \qquad A(0) = 0, \\ \frac{dy}{dt} &= \frac{1 - \sqrt{y}}{\pi y^2}, \qquad y(0) = 3. \end{aligned}$$

Numerical approximations of y and A are plotted in Figures 1a and 1b. Note that, as $t \to \infty$, the "solution pair" (y, A) approachs the equilibrium solution of the system, $(y_\infty, A_\infty) = (1, \pi/30)$. ■

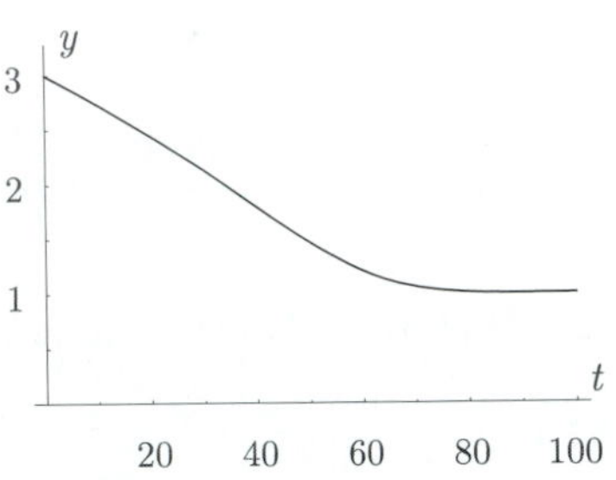

Figure 1a

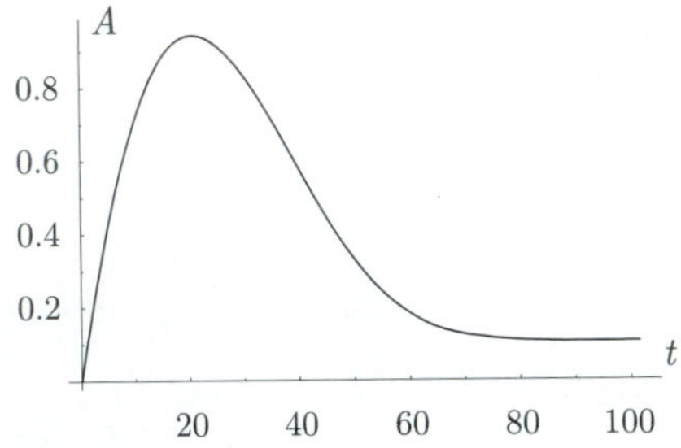

A
0.8
0.6
0.4
0.2
t
20 40 60 80 100

Figure 1b

PROBLEMS

1. **(a)** A tank in the shape of a right circular cone with radius 5 ft and height 10 ft is initially full and drains through a hole at its bottom. The vertex of the tank points downward and the constant ρ is known to be approximately 0.440. Solve the differential equation for the depth y of fluid in the tank. For what time interval is the solution valid? Write the solution in a form that is valid for all $t \geq 0$, and plot its graph. (See Example 1.)

   (b) Consider the same problem with the vertex of the tank pointing upward. Obtain an implicit solution of the differential equation for the depth y, and determine the draining time without solving for y. Plot the graph of t as a function of y; then reflect the graph to obtain the graph of y versus t.

2. Consider a cylindrical tank with height H and radius R. Compute the time required for the tank, initially full, to drain completely through a hole at its bottom if

   (a) the tank is upright (i.e., flat side down); **(b)** the tank is lying on its side.

   Suppose that $H = 2R$, so that the initial depth is the same for each of the two positions (and the contest is fair).

   (c) In which of the two positions does the tank drain faster?

3. A spherical water tank has a radius of 8 feet and is initially full of water, which then flows out of a hole at the bottom of the tank.

   (a) Set up the initial-value problem for the depth of water in the tank, and find an implicit solution.

   (b) Find the time T when the tank becomes empty.

   (c) Compute the tank's draining time T in minutes if the radius of the hole is 1 inch and the Borda constant is $b = 0.6$.

4. A spherical water tank has a diameter of 25 feet and rests atop a 50-foot tower. Water flows out of the bottom of the tank, straight down, through and out of a 48-foot length of pipe. Assume that the tank and the pipe are initially full.

(a) Set up the initial-value problem for the depth of water in the tank, and find an implicit solution.

(b) Find the time T when the tank (not the pipe) becomes empty.

(c) Compute the tank's draining time T in minutes if the radius of the pipe is 2 inches and the Borda constant is $b = 0.6$.

5. Consider the draining tank in Example 2.

(a) How long does it take for the tank to become half empty? (or half full?)

(b) What fraction of the initial contents of the tank remains after one-half of the draining time?

6. Consider an upright cylindrical tank with radius R and a circular hole in its bottom with radius r. Show that, if the tank initially contains fluid with a depth of y_0, then the time it takes for the tank to drain completely is

$$T = \sqrt{\frac{2y_0}{b^2 g}} \left(\frac{R}{r}\right)^2,$$

where b is the Borda constant associated with the fluid.

7. Consider a tank in the shape of a frustrum of a cone with height H. Let R and r be the radii at the top and bottom of the tank, respectively, with $R > r > 0$. Show that, if the tank is initially full of a fluid with Borda constant b, and if the bottom falls out of the tank, then the time it takes for the tank to empty is

$$T = \sqrt{\frac{2H}{b^2 g}} \, \frac{8r^2 + 4rR + 3R^2}{15r^2}.$$

8. A water tank with horizontal cross-sectional area $a(s)$ at height s drains through a hole in the bottom of the tank. At the same time water is being pumped into the tank at a constant rate R. Write the differential equation for the depth y. Find the equilibrium solution of the differential equation. Under what condition will it be meaningful?

9. An upright cylindrical tank with radius R and height H has two identical spigots attached—one at the very bottom and another half-way up the side. Both spigots are opened fully at time $t = 0$. Write down the differential equation governing the depth of fluid in the tank while (a) the depth is greater than $H/2$, and (b) the depth is less than $H/2$.

10. This problem refers to Example 1. Note that if the solution $y = \left(8 - \frac{3\rho}{2\pi} t\right)^{2/3}$ were valid for all $t > 0$, then that would indicate that the tank begins to fill up again after it empties. Check that this function does not satisfy the differential equation for $t > \frac{16\pi}{3\rho}$.

11. Can you design a tank, formed by revolving some arc about the y-axis, such that water draining through a hole at the bottom causes the water level to fall at a *constant* rate? Such a tank could be used as a time-keeping device if a uniform scale were marked down its side. This "water clock" is called a *clepsydra*.

12. Consider the integral formula derived for the emptying time of a tank with horizontal cross-sectional area $a(y)$. Notice, for example, that if $a(y)$ is constant, then the integral is a convergent improper integral. Can you imagine a tank for which there is no finite emptying time? That is, can the integral be a *divergent* improper integral?

13. Consider the initial-value problem

$$\frac{dy}{dt} = -\frac{\rho}{a}\sqrt{y}, \quad y(t_0) = 0$$

for the depth of water in an *empty* cylindrical tank. Clearly, there should be *only* the trivial solution $y(t) = 0$ for $t > t_0$. But what about for $t < t_0$? The trivial solution is *one* solution, but should there be others? What do these "backward solutions" represent? Can you expect the problem to have just one solution?

14. **(ab)** Find equilibrium solutions for the differential equations in the systems (3), (4).

15. This is not a draining tank problem. Suppose a water tank with height H has a horizontal cross-sectional area $a(s)$ at height s. Water evaporates from the tank at a rate that is proportional to the area of the water's surface.

(a) Show that the depth of water in the tank decreases at a constant rate regardless of the shape of the tank.

(b) Suppose that the tank is a right circular cone with height H and radius R. Suppose further that the initial depth of the water is $H/2$ and that, in addition to the evaporation, water is also being pumped into the tank at a constant rate q. Set up the initial-value problem for the depth of water in the tank. Find the equilibrium solution of the differential equation. Under what condition will it be meaningful?

3.5.3 Nonlinear Circuits

In this section we revisit two of the simple electric circuits from Section 2.2.3 and consider nonlinear resistors ("varistors") for which the voltage-current relationship of Ohm's law is replaced by the more general

$$V_R = \rho(I),$$

where ρ is an invertible continuous function. We continue to assume the same linear characteristics of capacitors and inductors. Under these assumptions, the differential equation for the on the capacitor in an RC circuit is

$$Q' = \rho^{-1}\left(-\frac{Q}{C}\right), \tag{1}$$

and the differential equation for the current in an RL circuit with source is

$$L\,I' + \rho(I) = v_s(t). \tag{2}$$

Example 1

Consider a varistor for which

$$V_R = R\sqrt[n]{I},$$

where R is a positive constant and $n \geq 3$ is an odd positive integer. This voltage-current relationship (with $n = 21$) is plotted in Figure 1. The differential equation (2) becomes

$$LI' + R\sqrt[n]{I} = v_s(t).$$

To get an idea of how such a circuit will behave, suppose that R/L is large and the source voltage is a constant v_0. Rewriting the differential equation as

$$I' = \frac{R}{L}\left(\frac{v_0}{R} - \sqrt[n]{I}\right)$$

makes it clear that the current $I(t)$ will rapidly approach the equilibrium current

$$I_\infty = v_0^n / R^n.$$

For small source voltages v_0 (significantly less than R), this equilibrium current is *very small*. However, for large source voltages v_0 (significantly greater than R), the equilibrium current is *very large*. Thus the varistor tends to behave almost like a switch that is open when the source voltage is small and closed when the source voltage is large. (The greater n is, the more dramatic is this effect.) In fact, zinc oxide varistors with similar characteristics are used in surge-protection circuits. ■

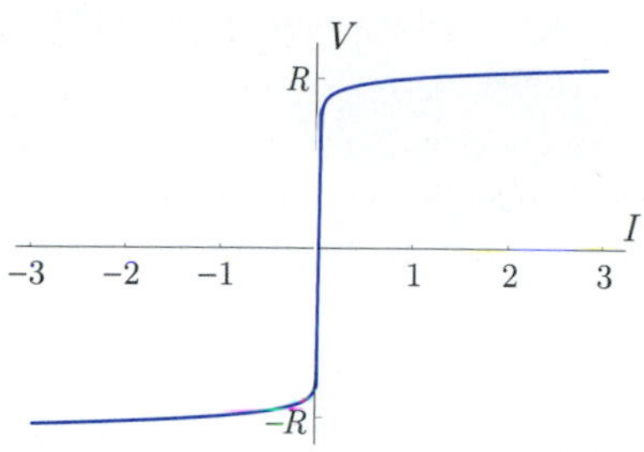

Figure 1

PROBLEMS

1. Solve for the charge Q in an RC circuit in which $C = 1$ and $V_R = \sqrt[3]{I}$. Express the solution in terms of the initial charge Q_0. Describe how the behavior of the circuit is different from that of the linear case in which $V_R = I$.
2. Solve for the current in an RL circuit in which $L = 1$ and $V_R = \sqrt[3]{I}$, if the source voltage is zero and the initial current is I_0. Describe how the behavior of the circuit is different from that of the linear case in which $V_R = I$.
3. Repeat Problem 1 with $V_R = I^3$.
4. Repeat Problem 2 with $V_R = I^3$.

For Problems 5 through 8, consider the initial-value problem

$$I' + 2I - I^2 = v_0, \quad I(0) = 0,$$

for the current in an RL circuit in which $L = 1$, the voltage source is constant ($v_s(t) = v_0$), and the voltage-current relationship of the resistor is $V_R = 2I - I^2$. (Note that the differential equation is a Riccati equation and that the initial current is zero.)

5. Suppose that $v_0 < 0$. Find the negative equilibrium current I_∞, solve the initial-value problem, and show that $\lim_{t\to\infty} I(t) = I_\infty$.
6. Suppose that $0 < v_0 < 1$. Find the least positive equilibrium current I_∞, solve the initial-value problem, and show that $\lim_{t\to\infty} I(t) = I_\infty$.
7. Suppose that $v_0 = 1$. Show that $I_\infty = 1$ is the only equilibrium current, solve the initial-value problem, and show that $\lim_{t\to\infty} I(t) = 1$.
8. Suppose that $v_0 > 1$. Show that no equilibrium current exists and that $I' \geq v_0 - 1 > 0$ for all $t \geq 0$. Conclude that $\lim_{t\to\infty} I(t) = \infty$.
9. Solve the initial-value problem

$$I' + I + I^3 = 0, \quad I(0) = I_0,$$

for the current in an RL circuit in which $L = 1$, the voltage source is zero, and the voltage-current relationship of the resistor is $V_R = I + I^3$. (Note that the differential equation is a Bernoulli equation.)

10. Consider the initial-value problem

$$I' + \rho(I) = \cos t, \quad I(0) = 0,$$

for the current in an RL circuit where

$$\rho(I) = \begin{cases} 10I, & \text{if } I < 0; \\ 0, & \text{if } I \geq 0. \end{cases}$$

Verify and plot the graph of the solution given, for $0 \leq t \leq 11.0952$, by

$$I(t) = \begin{cases} \sin t, & \text{if } 0 \leq t \leq \pi; \\ \dfrac{1}{101}\left(10e^{10(\pi - t)} + 10\cos t + \sin t\right), & \text{if } \pi < t \leq 4.81206; \\ \sin t + 0.99504, & \text{if } 4.81206 < t \leq 11.0952. \end{cases}$$

11. Consider the initial-value problem

$$I' + \rho(I) = \cos t, \quad I(0) = 0,$$

where

$$\rho(I) = \begin{cases} -\dfrac{1}{2}, & \text{if } I < 0; \\ 0, & \text{if } I \geq 0. \end{cases}$$

Find the solution for $0 \leq t \leq t_3$, where t_3 is the third positive zero of the solution. Also plot the graph.

CHAPTER

4

Nonlinear First-Order Equations II

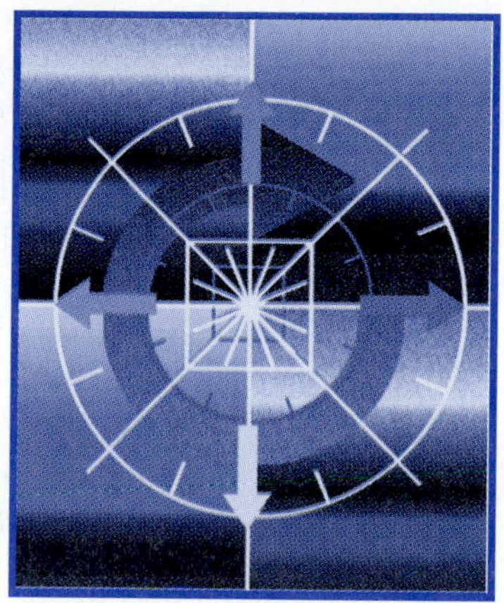

Most of our work so far has dealt with problems for which appropriate procedures allow the derivation of elementary solutions. In this chapter we will discuss some fundamental concepts that provide a basis for studying problems for which elementary solutions are either difficult or impossible to find. The first three sections introduce the fundamental notions of existence, uniqueness, and asymptotic behavior of solutions. The fourth section applies these fundamental concepts to the logistic equation, a basic model from mathematical biology. Section 4.5 focuses on numerical approximation with emphasis on error analysis, and Section 4.6 introduces coupled systems of two equations.

4.1 Construction of Local Solutions

The two most fundamental questions concerning any initial-value problem are as follows: Does the problem have a solution, and if so, does it have more than one solution? The first of these questions is about the *existence* of a solution, and the second is about *uniqueness*. In this section we will consider the existence question for first-order initial-value problems of the form

$$y' = f(t, y), \quad y(t_0) = y_0, \tag{1}$$

where f is a continuous function of two variables that is defined on some subset of the (t, y)-plane containing the point (t_0, y_0). The uniqueness question will be addressed in the next section.

Before going any further, we state a precise definition of what we mean by a solution of (1).

Definition A (local) **solution** of (1) is a continuously differentiable function y, defined on an interval I containing t_0, such that $y(t_0) = y_0$ and $y'(t) = f(t, y(t))$ holds for each t in I. ◆

When the differential equation in (1) is sufficiently simple, the question of existence can be settled by *finding* a solution, as was the case in Chapter 2 for linear equations and nonlinear separable equations. However, finding a solution may be easier said than done—and often literally impossible. Knowing whether (1) *has* a solution is especially crucial when we attempt to compute an approximate solution numerically or otherwise study the problem without having the luxury of an elementary formula for the solution.

The way mathematicians show that a problem such as (1) has a solution for a class of functions f typically involves

i) the description of a well-defined procedure for generating successively improved approximations, and

ii) a proof that the approximations converge in an appropriate sense and that the limit is indeed a solution.

This plan requires a set of conditions on the function f under which the "construction" succeeds and the proof is manageable.

In what follows, we will describe three procedures for the construction of local solutions of (1). All three are based on elementary ideas, but the corresponding proofs are highly technical and therefore omitted. However, the second construction is used in the proof of our main theorem on existence and uniqueness, which is stated in the next section and proved in Appendix III.

Taylor Series Suppose that f has continuous partial derivatives of all orders in the interior of a disk centered at the point (t_0, y_0). Then a potential solution of (1) is the Taylor series

$$y(t) = y_0 + \sum_{k=1}^{\infty} \frac{y^{(k)}(t_0)}{k!}(t - t_0)^k,$$

where the derivative values at t_0 can be computed recursively as follows:

$$\begin{aligned}
y'(t_0) &= f(t, y(t))|_{t=t_0} \\
y''(t_0) &= (f_t + f_y y')|_{t=t_0} \\
y^{(3)}(t_0) &= (f_{tt} + f_{yt} y' + f_y y'' + f_{ty} y' + f_{yy} y' y' + f_y y'')|_{t=t_0} \\
&\vdots
\end{aligned}$$

It can be shown that, if $y(t)$ converges for all t in $I = (t_0 - \delta, t_0 + \delta)$, then y is a solution of (1) on I.

Example 1

Consider the problem

$$y' = t \ln y, \quad y(0) = e.$$

The computation of derivatives proceeds as follows:

$$
\begin{aligned}
&y'(t) = t \ln y, && y'(0) = 0\\
&y''(t) = \ln y + t\tfrac{y'}{y}, && y''(0) = 1\\
&y^{(3)}(t) = 2\frac{y'}{y} + t\frac{y''y - y'y'}{y^2}, && y^{(3)}(0) = 0\\
&y^{(4)}(t) = 3\frac{y''y - y'y'}{y^2} + t\left(\frac{y''y - y'y'}{y^2}\right)', && y^{(4)}(0) = \frac{3}{e}\\
&y^{(5)}(t) = 4\frac{y'''y^3 - 3y''y'y^2 + 2y(y')^3}{y^4} + t\left(\frac{y'''y^3 - 3y''y'y^2 + 2y(y')^3}{y^4}\right)', && y^{(5)}(0) = 0\\
&\qquad\vdots && \vdots
\end{aligned}
$$

This process produces the Taylor series

$$y(t) = e + \frac{1}{2}t^2 + \frac{3}{4!\,e}t^4 + \cdots .$$

We will not pursue this type of construction any further, because requiring f to have continuous partial derivatives of all orders places far more stringent conditions on f than we would like. In particular, we want to assume as little as possible about f beyond continuity. ■

Picard Iteration Suppose that f is continuous in the interior of a disk centered at the point (t_0, y_0). It is easy to verify that y satisfies the initial-value problem (1) on I if and only if it satisfies the *integral equation*

$$y(t) = y_0 + \int_{t_0}^{t} f(s, y(s))\,ds$$

for all t in I. Motivated by this integral equation, we might hope to choose some initial approximation y_1 and then compute a sequence of approximate solutions $y_2, y_3, y_4 \ldots$ of (1) recursively according to

$$y_{n+1}(t) = y_0 + \int_{t_0}^{t} f(s, y_n(s))\,ds, \quad n = 1, 2, 3, \ldots \tag{2}$$

Of course there is no immediate guarantee that such a plan will work, but under appropriate conditions on f, the sequence $y_1, y_2, y_3, \ldots$, as well as the corresponding sequence of derivatives $y_1', y_2', y_3', \ldots$, will converge in an appropriate sense on some interval containing t_0, and the function to which $y_1, y_2, y_3, \ldots$ converges will be a solution of (1) on that interval—thus proving that a solution exists.

The process described by (2) is known as *Picard iteration* and has the advantage of requiring *no differentiation of f*! The process is illustrated by the following example.

Example 2

Consider the problem

$$y' = 2t - y^2, \quad y(0) = 1.$$

The equivalent integral equation is

$$y(t) = 1 + \int_0^t (2\,s - y(s)^2)\,ds,$$

and the iteration in (2) becomes

$$y_{n+1}(t) = 1 + \int_0^t (2s - y_n(s)^2)\,ds, \quad n = 1, 2, 3, \ldots .$$

To begin the iteration, let's choose $y_1(t) = 1 - t$, since $y(0) = 1$ and $y'(0) = -1$, and compute (with *CAS* assistance)

$$y_2(t) = 1 + \int_0^t (s - (1 - s)^2)\,ds = 1 - t + 2t^2 - \frac{t^3}{3},$$

and then

$$\begin{aligned} y_3(t) &= 1 + \int_0^t \left(s - \left(1 - s + 2s^2 - \frac{s^3}{3}\right)^2\right) ds \\ &= 1 - t + 2t^2 - \frac{5t^3}{3} + \frac{7t^4}{6} - \frac{14t^5}{15} + \frac{2t^6}{9} - \frac{t^7}{63} . \end{aligned}$$

The next iteration—and for obvious reasons the last one we will show—is

$$\begin{aligned} y_4(t) &= 1 + \int_0^t (s - y_3(s)^2)\,ds \\ &= 1 - t + 2t^2 - \frac{5t^3}{3} + \frac{11t^4}{6} - \frac{29t^5}{15} + \frac{163t^6}{90} - \frac{439t^7}{315} + \frac{2551t^8}{2520} \\ &\quad - \frac{151t^9}{252} + \frac{1409t^{10}}{4725} - \frac{2272t^{11}}{17325} + \frac{61t^{12}}{1620} - \frac{32t^{13}}{5265} + \frac{2t^{14}}{3969} - \frac{t^{15}}{59535} . \end{aligned}$$

Figure 1a shows the graphs of $y_1, \ldots, y_4$, as well as subsequent iterates y_5 and y_6. Figure 1b magnifies the plot near the initial point (0, 1).

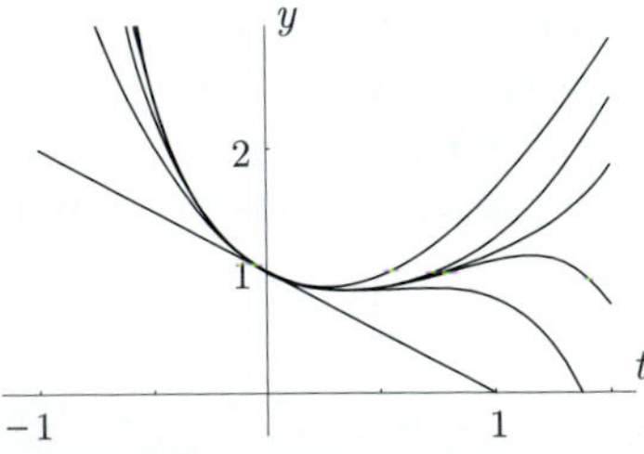

Figure 1a

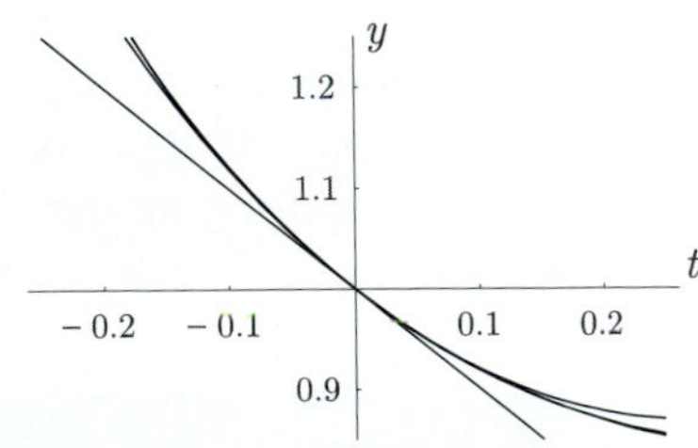

Figure 1b

It should be clear that in this particular example the iteration (2) generates a power series about $t = 0$. The key questions then are whether that power series has a positive radius of convergence and, if so, whether the power series satisfies the differential equation on its interval of convergence. ■

In Appendix III, we prove an existence (and uniqueness) theorem by means of Picard iteration. While the construction itself requires only continuity of f, the proof that it produces a solution requires an additional condition on f, which turns out to imply uniqueness as well. We will formally state this theorem in the next section.

The Cauchy-Euler Construction The *Cauchy-Euler construction* is based upon Euler's method. We assume that f is continuous in an open region G containing (t_0, y_0) in which

$$|f(t, y)| \leq M \quad \text{for all} \quad (t, y) \text{ in } G,$$

where M is a constant. Let $r > 0$ be sufficiently small so that the rectangle

$$R = [t_0 - r, t_0 + r] \times [y_0 - rM, y_0 + rM]$$

is contained in G. Then, for each $n = 1, 2, 3 \ldots$, let y_n be the linear interpolation of the points

$$(t_{-n}, \hat{y}_{-n}), \ldots, (t_{-1}, \hat{y}_{-1}), (t_0, \hat{y}_0), (t_1, \hat{y}_1), \ldots, (t_n, \hat{y}_n)$$

generated by

$$\hat{y}_{k+1} = \hat{y}_k + \frac{r}{n} f\left(t_0 + \frac{kr}{n}, \hat{y}_k\right), \qquad k = 0, 1, \ldots, n-1,$$

$$\hat{y}_{k-1} = \hat{y}_k - \frac{r}{n} f\left(t_0 + \frac{kr}{n}, \hat{y}_k\right), \qquad k = 0, -1, \ldots, -n+1,$$

with $\hat{y}_0 = y_0$. Note that these are forward and backward Euler's method formulas, respectively, each starting at (t_0, y_0). (Figure 2 shows a typical y_3. Note that there

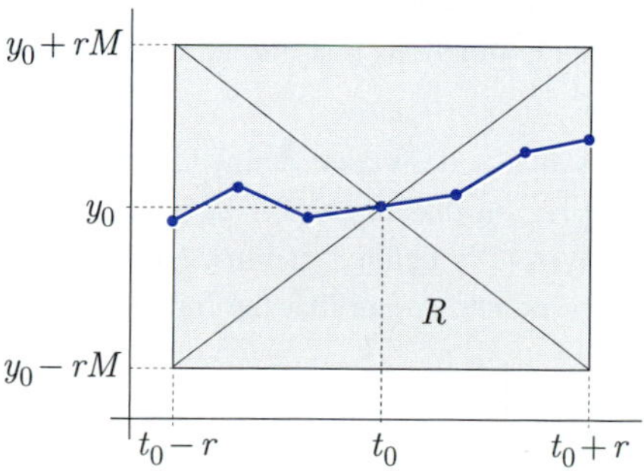

Figure 2

are six segments.) The graph of each y_n lies in R. (Why?) It turns out that the sequence $y_1, y_2, y_3, \ldots$ must have a subsequence that converges in an appropriate sense to a continuously differentiable function y that satisfies (1) on $(t_0 - r, t_0 + r)$. The proof of this fact is difficult, and we will not even attempt to describe it here. Yet this construction of a local solution is important, because the proof that it succeeds requires *only* that f be continuous and bounded near (t_0, y_0).

PROBLEMS

In Problems 1 through 4, find the first five terms of the solution's Taylor series about $t = 0$.

1. $y' = t\,y + y^2, \quad y(0) = 1$

2. $y' = e^y + t, \quad y(0) = 0$

3. $y' = t^2 - \ln y, \quad y(0) = e$

4. $y' = \sin(\frac{\pi}{2} \sin y), \quad y(0) = \frac{\pi}{2}$

In Problems 5 through 7, let y_1 be the constant function that satisfies the initial condition, and compute Picard iterates y_2, y_3, and y_4. Graph each iterate with the help of a calculator or computer.

5. $y' = y, \quad y(0) = 1$

6. $y' = y/t, \quad y(1) = 1$

7. $y' = y \sin t, \quad y(0) = 1$

8. Suppose that y is a solution of $y' = f(t, y)$, $y(t_0) = y_0$, on an interval (a, b) and takes on values in the interval (c, d). Let

$$R = (a, b) \times (c, d) = \{(t, y) |\ a < t < b,\ c < y < d\}.$$

(a) Show that if f_t and f_y are continuous on R, then y'' is continuous on (a, b).

(b) Show that if f has continuous second-order partial derivatives on R, then y''' is continuous on (a, b).

(c) Prove by induction that, for any positive integer n, if f has continuous n^{th}-order partial derivatives on R, then $y^{(n+1)}$ is continuous on (a, b).

9. Concoct a simple example of an initial-value problem $y' = f(t, y)$, $y(0) = y_0$, that has a solution with continuous derivatives of all orders, yet neither f_t nor f_y exists at (t_0, y_0). (*Hint*: Think *really* simple: $y(t) = 0$ is a solution if $y_0 = 0$ and y is a factor of $f(t, y)$.)

10. This problem is related to why we only consider *continuously* differentiable solutions of (1), thus excluding discontinuous functions f and solutions y that might be differentiable with y' discontinuous. It turns out that differentiable functions that are not continuously differentiable can behave quite strangely.

(a) A theorem from calculus states that a derivative defined on an open interval must possess the *intermediate-value property* on that interval; that is, between any two points in an open interval on which a derivative is defined, it must take on every intermediate value between its values at those two points. Argue that the function

$$f(x) = \begin{cases} 1, & \text{if } x \geq 0 \\ -1, & \text{if } x < 0 \end{cases}$$

does not have the intermediate-value property on any open interval containing $x = 0$ and therefore cannot be the derivative of any function.

(b) A function is said to have a "jump" discontinuity at a point $x = a$ if both one-sided limits exist at $x = a$ but do not have the same value. Argue that a function with a jump discontinuity at $x = a$ cannot have the intermediate-value property in any open interval containing $x = a$ and therefore cannot be the derivative of any function.

(c) The derivative of the absolute value function $f(x) = |x|$ is

$$f'(x) = \begin{cases} -1, & \text{if } x < 0; \\ 1, & \text{if } x > 0. \end{cases}$$

Why doesn't this contradict the fact that derivatives must have the intermediate-value property?

(d) Since no derivative can have a jump discontinuity in any open interval where it is defined, it may be tempting to believe that a derivative must be continuous at any point where it is defined. The purpose of what follows is to demonstrate that a differentiable function can indeed have a discontinuous derivative. Let

$$f(x) = \begin{cases} x^2 \sin \frac{1}{x}, & \text{if } x \neq 0 \\ 0, & \text{if } x = 0. \end{cases}$$

Show that

$$f'(x) = \begin{cases} 2x \sin \frac{1}{x} - \cos \frac{1}{x}, & \text{if } x \neq 0 \\ 0, & \text{if } x = 0, \end{cases}$$

which is defined but discontinuous at $x = 0$. (*Hint*: Note that the challenging part here is in showing that $f'(0) = 0$.) Then sketch the graphs of f and f', and convince yourself that f' does have the intermediate-value property on every interval $[-a, a]$, $a > 0$. Note that f' is bounded near $x = 0$ and that the discontinuity of f' at $x = 0$ is not a "jump" discontinuity; it is a discontinuity at which the one-sided limits of f' fail to exist. It is this type of nastiness that is avoided by assuming continuous differentiability.

4.2 Existence and Uniqueness

We begin this section with an example that illustrates *nonuniqueness*. It is an especially relevant example, since it concerns a concrete physical problem that has many distinct—and meaningful—solutions.

Example 1

Consider the draining tank problem of Section 3.5.2. Assume that the tank is cylindrical with cross-sectional area A and that the fluid and the size of the hole are such that $\rho/A = 1$. By Torricelli's law, the depth $y(t)$ of the fluid satisfies $y' = -\sqrt{y}$. Now consider the initial-value problem

$$y' = -\sqrt{y}, \quad y(T) = 0,$$

where T is some fixed positive time at which the tank is known to be empty. Certainly the tank will remain empty for all times $t > T$. But should it be possible to determine the *past history* of the depth of water in the tank? Surely not! If we only know that the tank is now empty, we can't know whether it has only just finished draining or has been empty for a long time—or perhaps it was always empty. Let's see if the solutions of the differential equation reflect this observation. One solution,

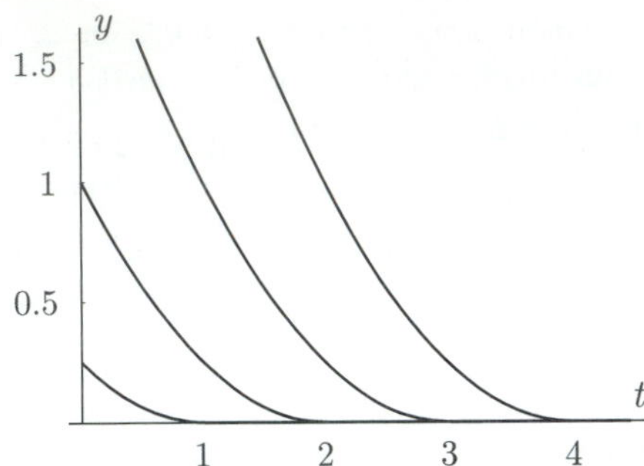

Figure 1

which we can obtain by inspection, is $y = 0$ for all t. This is the situation in which the tank has always been empty. Now let's assume that the tank did once contain water, and let $t_0 \leq T$ denote the precise time at which the tank first became empty. Solving by separation of variables with the condition $y(t_0) = 0$, we find that

$$y = \begin{cases} (t_0 - t)^2/4, & \text{if } t \leq t_0 \\ 0, & \text{if } t > t_0 \end{cases}$$

gives a family of solutions in which t_0 is any time at or before T. Several of these solutions appear in Figure 1. The important point here is that the problem has infinitely many solutions and that each of these solutions is physically meaningful. ■

Let's consider again initial-value problems of the form

$$y' = f(t, y), \quad y(t_0) = y_0, \tag{1}$$

where f is a continuous function of two variables that is defined on some subset of the (t, y)-plane containing the point (t_0, y_0). The following three definitions establish terminology related to the questions of existence and uniqueness.

Definition We say that (1) has a **unique local solution** if there is some interval I containing t_0 on which there is exactly one solution y; that is, any solution defined on I or some larger interval must agree with y on I. ◆

Clearly, if there were a unique solution on I, then there would also be a unique solution on any subinterval of I. We must point out that (1) may actually have many different unique local solutions, but only in the sense that each is defined on a different interval. This emphasizes the fact that this concept of uniqueness is closely tied to the local nature of the solution.

Definition Given a solution y (unique or not) defined on an interval I containing t_0, we say that a solution $\tilde{y}$, defined on an interval J containing I, is a **continuation** of y if $y(t) = \tilde{y}(t)$ for all t in I. ◆

The graph of y is actually a subset of the graph of any continuation $\tilde{y}$. Even if a solution y is unique on I, it may have more than one continuation on a given interval J containing I. (See Example 2 that follows.)

Definition A **maximal solution** is a solution whose only continuation is itself. A solution that is defined for $-\infty < t < \infty$ is a **global solution.** ◆

A global solution is necessarily maximal, but as we will see shortly, a maximal solution need not be global. We will also see in the following example that a local solution y may have more than one maximal continuation, even when y itself is unique.

Example 2

Consider the initial-value problem:

$$y' = 3y^{2/3}, \quad y(-1) = -1.$$

It is easy to find, by separating variables, that $y(t) = t^3$ is a global solution. However, it is easy to check that

$$y = \begin{cases} t^3, & \text{if } t < 0 \\ 0, & \text{if } 0 \le t \le T \\ (t-T)^3, & \text{if } t > T \end{cases}$$

is also a global solution for any $T \ge 0$. A few of these are shown in Figure 2. This problem actually has a unique local solution $y(t) = t^3$ on any interval (a, b) with $a < -1 < b < 0$. Given any one of these unique local solutions, each of the global solutions previously described is a maximal continuation of it. ■

If a solution y has a unique maximal continuation $\tilde{y}$, then there is usually little point in making any distinction between the two. We can more or less equate y with $\tilde{y}$ and call the interval on which it is defined the **maximal domain** of the solution y (which was discussed briefly in Section 1.3. So, when we talk about *the solution* of (1), we normally mean the unique maximal solution, provided there is one.

There are two questions that we would now like to address:

$Q_{1:}$: *What conditions on f guarantee that (1) has a unique local solution?*
$Q_{2:}$: *What further conditions on f guarantee that a unique local solution of (1) has a unique maximal continuation?*

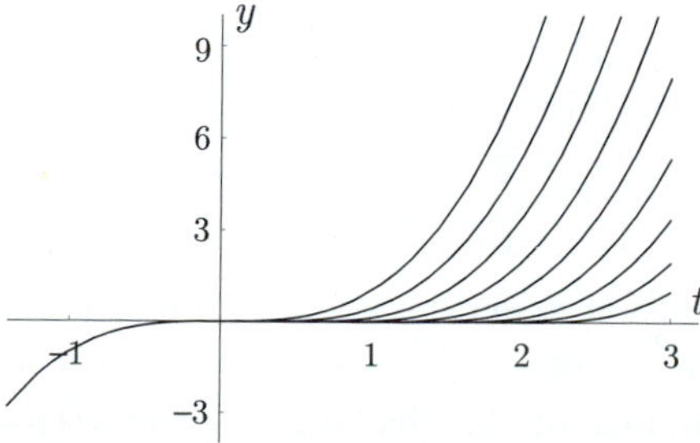

Figure 2

The following is our **local existence and uniqueness theorem** for first-order initial-value problems. It provides an answer to Q_1.

Theorem 1

Suppose that f and $\frac{\partial f}{\partial y}$ are continuous on some closed rectangle

$$R = [a, b] \times [c, d] = \{(t, y) | a \le t \le b,\ c \le y \le d\}$$

that contains (t_0, y_0) in its interior; that is, $a < t_0 < b$ and $c < y_0 < d$. Then the initial-value problem

$$y' = f(t, y), \quad y(t_0) = y_0,$$

has a unique local solution, which can be uniquely continued for $t < t_0$ and $t > t_0$ at least until $(t, y(t))$ reaches a boundary point of R. ▲

Before we look at how Theorem 1 applies to some specific examples, a couple of remarks about the hypotheses are in order. The region R is taken to be a rectangle merely for simplicity; the theorem remains true if R is any bounded, closed region containing (t_0, y_0) in its interior. Also, we emphasize that the boundedness and closedness of R are crucial, because, together with the continuity of f and $\frac{\partial f}{\partial y}$, they guarantee that f and $\frac{\partial f}{\partial y}$ are each *bounded* on R, properties which are prominent in the proof of the theorem. (See Appendix III.)

Let's now consider how Theorem 1 relates to the preceding Examples 1 and 2. In Example 1, we have the function $f(t, y) = -\sqrt{y}$ and the initial point $(T, 0)$. While it is true that $f(t, y)$ is undefined for $y < 0$, this is not really the cause of the nonuniqueness exhibited by this problem. In fact we could replace f with $f(t, y) = \sqrt{|y|}$, so that it would be defined and continuous in the entire (t, y)-plane, and we would still find multiple solutions through any point on the t-axis. The trouble here is with

$$\frac{\partial f}{\partial y} = -\frac{1}{2\sqrt{y}},$$

which is undefined on (and unbounded near) the t-axis. Thus, since the initial point $(T, 0)$ is on the t-axis, there is no rectangle R for which the conditions of the theorem can be met. However, the theorem *does* guarantee the existence of a unique local solution through any point (t_0, y_0) with $y_0 \neq 0$. In this case, we can take R to be any closed rectangle that contains (t_0, y_0) in its interior and does not touch the t-axis.

In Example 2, our function is $f(t, y) = 3y^{2/3}$, and the initial point is $(-1, -1)$. The function f is defined and continuous everywhere, but

$$\frac{\partial f}{\partial y} = \frac{2}{\sqrt[3]{y}},$$

which is undefined on (and unbounded near) the t-axis. If we take R to be any closed rectangle $[a, b] \times [c, d]$ that contains $(-1, -1)$ and does not touch the t-axis (i.e., $d < 0$), then $\frac{\partial f}{\partial y}$ will be continuous on R. Thus the solution $y(t) = t^3$ through $(-1, -1)$ can be continued uniquely for $t > 0$ until it reaches the line $y = d$. But since d can be as close to 0 as we like, the solution can actually be continued uniquely for $t > 0$ until it reaches the the t-axis (where uniqueness can no longer be guaranteed). This happens at the origin, where the solution may follow any of several different "branches" out of that point.

Example 3

Let's determine all initial points (t_0, y_0) for which Theorem 1 guarantees the existence of a unique local solution of the initial-value problem

$$y' = \frac{(t-y)^{1/3}}{1-t}, \qquad y(t_0) = y_0.$$

Here we have

$$f(t, y) = \frac{(t-y)^{1/3}}{1-t}, \qquad \text{and thus} \qquad \frac{\partial f}{\partial y} = -\frac{1}{3(t-y)^{2/3}(1-t)}.$$

Therefore, so long as $t_0 \neq 1$ and $y_0 \neq t_0$, there is a closed rectangle R for which the conditions of Theorem 1 are met. In fact, any closed rectangle R containing (t_0, y_0) in its interior will do, provided that R does not touch either of the lines $t = 1$ or $y = t$. ■

REMARK It is important to realize that Theorem 1 only tells us when we can expect an initial-value problem to have a unique local solution. It tells us nothing about when a problem will not have a unique local solution. That is, it only gives *sufficient* (but not necessary) conditions for existence and uniqueness. It may well happen that a problem for which one or more of the hypotheses fails will still have a unique local solution. However, if we know that existence or uniqueness fails for a certain problem, *then* at least one of the hypotheses must have been violated.

Consequences of Uniqueness It turns out that local uniqueness of solutions is a simple but powerful tool for proving some very important properties of maximal solutions. We have the following corollaries to Theorem 1, the second of which tells us that when there is a unique local solution through *every* point, there is exactly one maximal solution through *any* point. This corollary gives an answer to Q_2.

COROLLARY 1 *If f and $\frac{\partial f}{\partial y}$ are continuous on a closed rectangle R, then distinct solution curves of $y' = f(t, y)$ never intersect in the interior of R.*

Proof Suppose for the sake of contradiction that y and $\tilde{y}$ are two distinct solutions whose graphs intersect in the interior of R. Then there is some t^* such that $y(t^*) = \tilde{y}(t^*) = y^*$, (t^*, y^*) is in the interior of R, and $y(t) \neq \tilde{y}(t)$ on one (or both) of the intervals $t^* - \delta < t < t^*$ or $t^* < t < t^* + \delta$, where δ is some small positive number. This implies that there are two distinct local solutions through the point (t^*, y^*), which contradicts Theorem 1. ●

COROLLARY 2 *If f and $\frac{\partial f}{\partial y}$ are continuous on $\mathbb{R}^2$, then for any (t_0, y_0) the initial-value problem (1) has a unique maximal solution.*

Proof Suppose that y and $\tilde{y}$ are distinct maximal solutions of (1) through the initial point (t_0, y_0). Then there is some t^* (as in the proof of Corollary 1) such that $y(t^*) = \tilde{y}(t^*) = y^*$, and $y(t) \neq \tilde{y}(t)$ on one (or both) of the intervals $t^* - \delta < t < t^*$ or $t^* < t < t^* + \delta$, where δ is some small positive number. Thus there is no unique local solution through (t^*, y^*). But f satisfies the hypotheses of Theorem 1 on any closed rectangle whose interior contains (t^*, y^*), which ensures that there is a unique local solution through (t^*, y^*). This contradiction leads us to conclude that there is a unique maximal solution through (t_0, y_0). ●

In many applications, such as models of populations or chemical reactions, the quantity of interest is known to be nonnegative. Thus it is important for the model to reflect this basic property—that is, the property that a nonnegative initial value results in a nonnegative maximal solution. The following corollary gives simple criteria for an initial-value problem to have this property.

COROLLARY 3 *Suppose that f and $\frac{\partial f}{\partial y}$ are continuous on some closed "infinite strip" containing the t-axis:*

$$S = \{(t, y) | -\delta \leq y \leq \delta\}, \quad \delta > 0.$$

If $f(t, 0) = 0$ for all t, then no solution of $y' = f(t, y)$ can change sign. Moreover, no nontrivial solution can ever have a zero value.

Proof We need only prove the second assertion. Since the equation has the trivial solution $y(t) = 0$, and since Corollary 1 tells us that distinct solution curves cannot intersect in the interior of any closed rectangle inside S, we conclude that any solution with a zero value at some t must be zero for all t. ●

Example 4

Corollary 3 guarantees that every nontrivial solution of each of the following equations is either always positive or always negative.

$$y' = -y^3 + y \qquad y' = y\cos t \qquad y' = 1 - e^y \qquad y' = t\sin y$$ ■

PROBLEMS

1. Find all solutions of $ty' = y$, $y(0) = 0$. Why doesn't Theorem 1 guarantee a unique solution?
2. Following Examples 1 and 2, find all solutions of $y' = 3y^{1/3}$, $y(0) = 0$. Make a sketch of several of solutions. Why doesn't Theorem 1 guarantee a unique solution?
3. Consider the initial-value problem: $y' = 3y^{1/3}$, $y(1) = 1$.

   (a) Find a unique local solution.

   (b) Find a global continuation of the local solution in (a). Is this global solution unique? (*Hint*: Consider the direction field for the differential equation.)
4. Consider the initial-value problem: $y' = 6ty^{1/3}$, $y(1) = 1$.

   (a) Find a unique local solution.

   (b) Find a global continuation of the local solution in (a). Is this global solution unique?

For each function f in Problems 5 through 11, describe all points (t_0, y_0) for which (1) is guaranteed a unique solution by Theorem 1.

5. $f(t, y) = -t^2 y^3$

6. $f(t, y) = \dfrac{y}{t^2 - 1}$

7. $f(t, y) = (y - 1)^{1/3}$

8. $f(t, y) = (1 - t^2 - y^2)^{2/3}$

9. $f(t, y) = \tan(t^2 + y^2)$

10. $f(t, y) = e^{-t} |\sin y|$

11. For which of the following functions f will the maximal solution of

$$y' = f(t, y), \quad y(0) = y_0,$$

be nonnegative whenever y_0 is nonnegative?

(a) $f(t, y) = -t^2 y^3$

(b) $f(t, y) = \sin y - y$

(c) $f(t, y) = y^2 - t$

(d) $f(t, y) = -y e^{-y}$

12. According to Corollary 2, the initial-value problem $y' = y^2$, $y(t_0) = y_0$, has a unique maximal solution for any (t_0, y_0). Show that, if $y_0 \neq 0$, the maximal solution is not global. Describe the maximal domain.

13. Let $p > 1$. Find the domain of the maximal solution of $y' = y^p$, $y(0) = 1$.

14. Solve $y' = e^y$, $y(0) = y_0$. Show that the maximal solution is not global for any y_0.

4.3 Qualitative and Asymptotic Behavior

In this section we will concentrate on initial-value problems of the form

$$y' = f(y), \quad y(0) = y_0. \tag{1}$$

Notice that the function f depends only on y and not on t. Such an equation is called *autonomous*, which means "self-governing." We will assume that f and f' are continuous on all of $\mathbb{R}$, so that (1) will have a unique maximal solution for any y_0 by Corollary 1 of the preceding section. Our goal here is to gain qualitative information about solutions of (1) and, in particular, to say as much as we can about what happens to solutions as $t \to \infty$, all without actually finding the solutions. As you might have guessed, *asymptotic behavior* is a term used for the behavior of a solution as $t \to \infty$.

Example 1

Consider the initial-value problem

$$y' = y(1 - y)^3, \quad y(0) = y_0.$$

Although it could be done, finding the solution y would be a rather tedious task and would ultimately give us only an implicit description of the solution curves. So let's see what we can learn about the solution from the differential equation itself. First, it is easy to see that the differential equation has two equilibrium solutions: $y = 0$ and $y = 1$. Thus, because of uniqueness, if $y_0 = 0$, then $y(t) = 0$ for all t, and if $y_0 = 1$, then $y(t) = 1$ for all t. But what happens for other values of y_0? Notice that

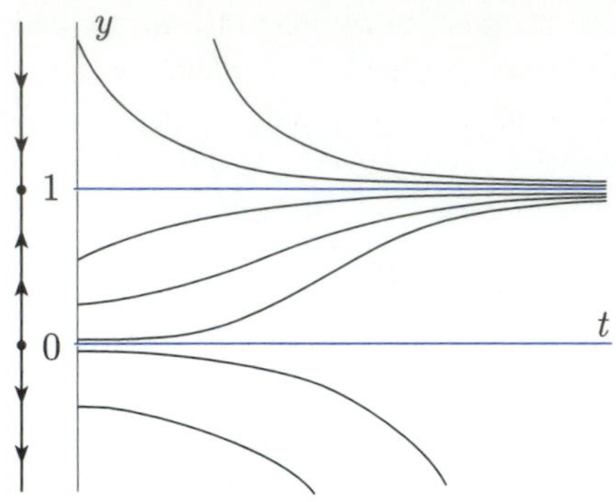

Figure 1

$f(y) = y(1-y)^3$ has the following sign pattern:

$$f(y) < 0 \text{ if } y < 0, \quad f(y) > 0 \text{ if } 0 < y < 1, \quad f(y) < 0 \text{ if } y > 1.$$

Therefore, y is decreasing whenever $y < 0$, increasing whenever $0 < y < 1$, and decreasing whenever $y > 1$. Moreover, because of uniqueness, no solution curve can cross over either of the lines $y = 0$ and $y = 1$. All this leads us to Figure1, which shows a sketch of several solution curves alongside an extra copy of the y-axis, upon which arrows are drawn to indicate the increasing/decreasing behavior of solutions. Assuming that our sketch of solution curves is correct, it appears that every positive solution approaches the equilibrium solution $y = 1$ as $t \to +\infty$, while negative solutions do not approach any limit as $t \to +\infty$. In fact, negative solutions are not even global—they blow up after some finite positive time. ■

A critical ingredient in all of this is uniqueness of solutions; in particular, Corollary 1 of the preceding section, which states that distinct solution curves cannot intersect. But how can we know that the nonequilibrium solution curves actually approach the equilibrium solutions in the way that is shown in the picture? The following two theorems allow us to justify this.

Theorem 1

Suppose that $g(t)$ is a continuous function defined on $[t_0, \infty)$. If $g(t)$ is decreasing and bounded below on $[t_0, \infty)$ or if $g(t)$ is increasing and bounded above on $[t_0, \infty)$, then $g(t)$ has a finite limit as $t \to \infty$. An analogous statement about the limit as $t \to -\infty$ of a function defined on $(-\infty, t_0]$ is true. ▲

Theorem 2

Suppose that y is the solution of (1). If either

$$\lim_{t\to-\infty} y(t) = b \quad \textit{or} \quad \lim_{t\to\infty} y(t) = b,$$

then $f(b) = 0$; that is, $y = b$ is an equilibrium solution of $y' = f(y)$. ▲

Theorem 1 is a theorem from advanced calculus. The interested reader is encouraged to seek out its proof in an advanced calculus or introductory real analysis textbook. An informal argument for Theorem 2 is outlined as follows. If y is defined on $[0, \infty)$ and $\lim_{t\to\infty} y(t) = b$, then $\lim_{t\to\infty} f(y(t)) = f(b)$, since f is continuous. Thus, $\lim_{t\to\infty} y'(t) = f(b)$; i.e., for large t the graph of $y(t)$ is approximately parallel to the line $y = f(b)t$. This implies that $\lim_{t\to\infty} y(t) = \infty$ if $f(b) > 0$, and it implies that $\lim_{t\to\infty} y(t) = -\infty$ if $f(b) < 0$. Consequently, we must have $f(b) = 0$. A similar argument can be made if $\lim_{t\to-\infty} y(t) = b$.

Now, how do these theorems justify our analysis in Example 1? When $y_0 > 1$, Theorem 1 tells us that $y(t)$ has a limit as $t \to \infty$, since $y(t)$ is decreasing and bounded below by the equilibrium solution $y = 1$. By Theorem 2, that limit must be one of the equilibrium solutions; therefore, $y(t) \to 1$ as $t \to \infty$. Likewise, if $0 < y_0 < 1$, then $y(t)$ has limits as $t \to \pm\infty$ by Theorem 1, since $y(t)$ is increasing and bounded above and below by $y = 1$ and $y = 0$. By Theorem 2, these limits must be the equilibrium solutions $y = 1$ and $y = 0$. Finally, $y(t)$ is decreasing when $y_0 < 0$, and so Theorems 1 and 2 together imply that there can be no lower bound, since there are no negative equilibrium solutions.

Stability Because of the autonomous nature of (1), the asymptotic behavior of its solution is limited to only a few possibilities. Specifically, all possibilities are summarized in the following theorem.

Theorem 3

For any y_0, the solution of (1) has one of the following properties:

i) $\lim_{t\to\infty} y(t) = \infty;$

ii) $\lim_{t\to\infty} y(t) = -\infty;$

iii) $\lim_{t\to\infty} y(t) = b$, *where* $f(b) = 0$.

The same statement is true if "$t \to \infty$" is replaced with "$t \to -\infty$." ▲

The proof of this theorem, based upon Theorems 1 and 2, is not difficult. Yet it will perhaps be more instructive simply to illustrate it with a few examples, as we will do after introducing the following terminology.

An equilibrium solution $y = b$ of (1) is **isolated** if there is an open interval containing b in which b is the only equilibrium solution $y = b$ of (1). This is true if b is an isolated zero of f.

The **basin of attraction** of an equilibrium solution $y = b$ is the largest set $\mathcal{B}$ for which $\lim_{t\to\infty} y(t) = b$ when y_0 is in $\mathcal{B}$. In Example 1, the basin of attraction of $y = 1$ is the interval $(0, \infty)$, since $\lim_{t\to\infty} y(t) = 1$ whenever $0 < y_0 < \infty$. For the problems under consideration here, the basin of attraction of an equilibrium solution $y = b$ will be either $\{b\}$ or an interval containing b. When a basin of attraction is an interval, we will call it an **interval of attraction**.

An equilibrium solution $y = b$ of (1) is **asymptotically stable** if its basin of attraction is an open interval. In this case, all solutions of (1) with y_0 sufficiently close to b approach b as $t \to \infty$. An asymptotically stable equilibrium solution is necessarily isolated and is sometimes called an **attractor**, since it attracts all solutions that begin sufficiently close to it. In Example 1, $y = 1$ is asymptotically stable with interval of attraction $(0, \infty)$.

An equilibrium solution $y = b$ of (1) with a half-open interval of attraction in the form of either $(c, b]$ or $[b, d)$ is **semistable**.

An *isolated* equilibrium solution $y = b$ of (1) is **unstable** if its basin of attraction is $\{b\}$. If $y = b$ is isolated and unstable, then all solutions of (1) with $y_0 \neq b$ either approach another equilibrium solution or tend to $\pm\infty$ as $t \to +\infty$. An isolated, unstable equilibrium solution is sometimes called a **repellor**. In Example 1, $y = 0$ is an isolated and unstable equilibrium solution.

Every isolated equilibrium solution of (1) is either asymptotically stable, semistable, or unstable.*

Example 2

The differential equation

$$y' = y(1 - y)^2(2 - y)$$

has three equilibrium solutions: $y = 0$, $y = 1$, and $y = 2$. Considerations similar to those in Example 1 lead to the picture in Figure 2. It is evident that $y = 0$ is unstable, $y = 1$ is asymptotically stable with interval of attraction $(0, 2)$, and $y = 2$ is semistable with interval of attraction $(2, \infty)$. ■

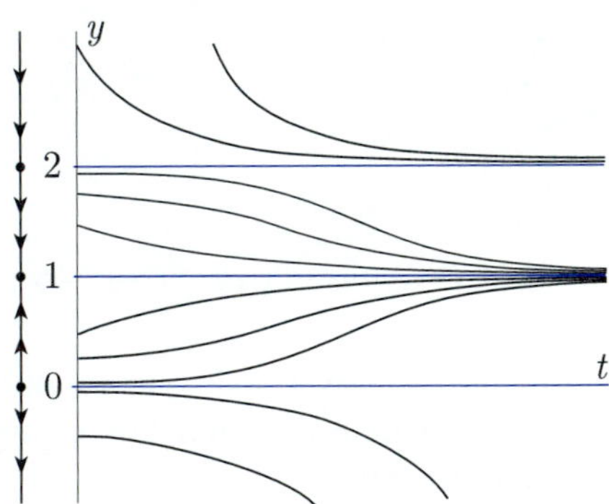

Figure 2

The following theorem states simple criteria for determining the stability of an equilibrium solution of $y' = f(y)$ corresponding to a simple root of f.

Theorem 4

Suppose that $f(b) = 0$, so that $y = b$ is an equilibrium solution of $y' = f(y)$. If $f'(b) < 0$, then $y = b$ is asymptotically stable. If $f'(b) > 0$, then $y = b$ is unstable. ▲

* A nonisolated equilibrium solution of (1) may be semistable, unstable, or *neutrally stable*. See Problems 14–17.

Proof If $f(b) = 0$ and $f'(b) < 0$, then there is an interval $(b - \varepsilon, b)$ on which $f(y) > 0$ and an interval $(b, b + \varepsilon)$ on which $f(y) < 0$. Thus $y(t)$ is increasing when $b - \varepsilon < y(t) < b$ and decreasing when $b < y(t) < b + \varepsilon$. Therefore, $\lim_{t\to\infty} y(t) = b$ whenever $b - \varepsilon < y < b + \varepsilon$, and so $y = b$ is asymptotically stable. If $f(b) = 0$ and $f'(b) > 0$, then there is an interval $(b - \varepsilon, b)$ on which $f(y) < 0$ and an interval $(b, b + \varepsilon)$ on which $f(y) > 0$. Thus $y(t)$ is decreasing when $b - \varepsilon < y(t) < b$ and increasing when $b < y(t) < b + \varepsilon$. Therefore, $y(t)$ cannot approach b as $t \to \infty$ unless $y_0 = b$. Thus $y = b$ is unstable. ●

Example 3

Let f be the function whose graph appears in Figure 3, and consider the equation $y' = f(y)$. The equilibrium solutions are $y = 0$, 1, and 2. From the graph we can see that

$$f'(0) > 0, \quad f'(1) < 0, \quad f'(2) > 0.$$

Therefore, $y = 0$ and $y = 2$ are unstable, while $y = 1$ is asymptotically stable. ■

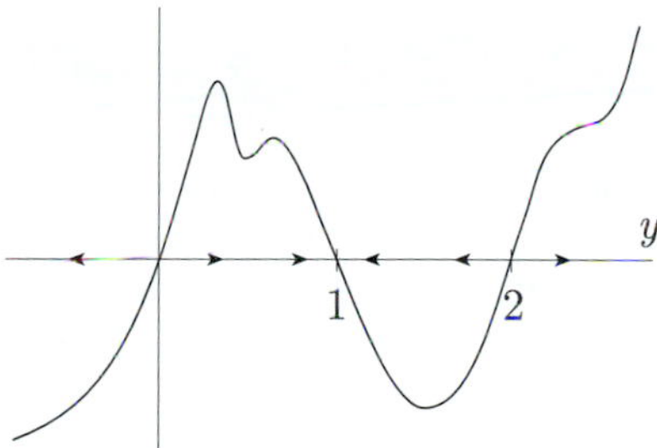

Figure 3

Notice that Theorem 4 only applies to simple roots of f and does not give any indication of when an equilibrium solution is semistable. It is not difficult to generalize the theorem to completely characterize isolated equilibrium solutions in terms of how $f(y)$ changes sign.

Theorem 5

Suppose that $y = b$ is an isolated equilibrium solution of $y' = f(y)$.

i) If $f(y)$ changes from positive to negative at $y = b$, then $y = b$ is asymptotically stable.

ii) If $f(y)$ changes from negative to positive at $y = b$, then $y = b$ is unstable.

iii) If $f(y)$ does not change sign at $y = b$, then $y = b$ is semistable. ▲

The proof of Theorem 5 follows along the same lines as that of Theorem 4.

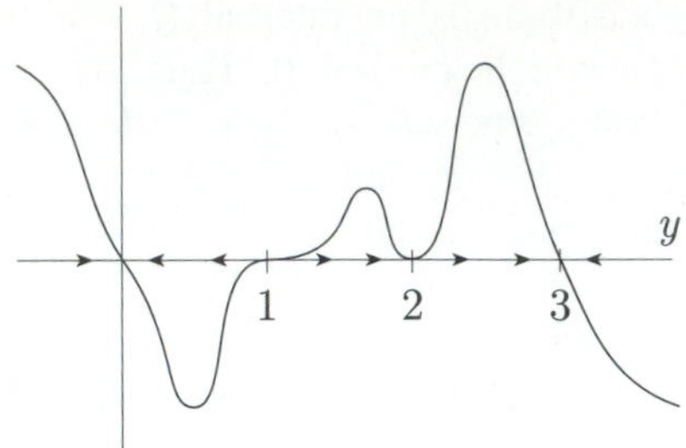

Figure 4

Example 4

Let f be the function whose graph appears in Figure 4, and consider the equation $y' = f(y)$. The equilibrium solutions are $y = 0, 1, 2,$ and 3. From the graph and Theorem 4, we see that $y = 0$ and $y = 3$ are asymptotically stable, $y = 1$ is unstable, and $y = 2$ is semistable (with $(1, 2]$ as its interval of attraction). ■

PROBLEMS

For each equation in Problems 1 through 9, find all equilibrium solutions and classify each of them as asymptotically stable, unstable, or semistable. When an equilibrium solution is asymptotically stable or semistable, state its interval of attraction. Make a sketch by hand showing the equilibrium solutions and a sufficient number of other solution curves to indicate the behavior of all solutions for $t \geq 0$.

1. $y' = -y(3 - y)$

2. $y' = y(1 - e^{y-1})$

3. $y' = y^2(3 - y)^2$

4. $y' = y(2 - y)^3$

5. $y' = -y^2 + 8\,y - 15$

6. $y' = y^3 - y$

7. $y' = \sin y$

8. $y' = \sin^2 y$

9. $y' = \ln(\frac{1}{2}(y^2 + 1))$

10. Consider the equation $y' = y(30 - y) - k$.

(a) Set $k = 0$ and make a simple sketch that indicates the behavior of all solutions for $t \geq 0$.

(b–f) Repeat (a) with $k = 29, 125, 200, 225, 300$. In each case, also sketch a separate graph of the function $f(y) = y(30 - y) - k$.

11. Consider the equation $y' = y(100 - y) - k$.

(a) Find k so that $y = 75$ is an asymptotically stable equilibrium solution. Then make a simple sketch that indicates the behavior of all solutions for $t \geq 0$.

(b) For what values of k will $y' = y(100-y)-k$ have an asymptotically stable equilibrium solution? What are all the possible equilibrium solutions?

12. For each of the following values of k, sketch the graph of the function

$$f(y) = -y(y^2 - 8y + k)$$

and make a separate sketch that indicates the behavior of all solutions of $y' = f(y)$ for $t \geq 0$.

(a) $k = 0$ (b) $k = 7$ (c) $k = 15$ (d) $k = 16$ (e) $k = 20$

13. Without solving the equation, show that a solution curve of $y' = y(M - y)$, where M is a constant, has an inflection point only where the solution has the value $M/2$. Emphasize this fact on a sketch of several solution curves.

Problems 14 through 17 concern nonisolated equilibrium solutions. The definitions of asymptotic stability and semistability stated in text apply to nonisolated equilibrium solutions as well as isolated ones. No nonisolated equilibrium solution $y = b$ can be asymptotically stable, since every open interval containing b must also contain another equilibrium solution. However, a nonisolated equilibrium solution $y = b$ can be semistable. A definition of instability that applies to both isolated and nonisolated equilibrium solutions is as follows: *An equilibrium solution $y = b$ is unstable if every open interval containing b also contains a number y_0 for which the solution of (1) with $y(0) = y_0$ is (i) not an equilibrium solution and (ii) approaches either $\pm\infty$ or some equilibrium solution other than $y = b$ as $t \to \infty$.* Nonisolated equilibrium solutions that are neither semistable nor unstable are **neutrally stable**.

14. Consider the equation $y' = f(y)$, where $f(y) = \frac{1}{2}(|y - 1| + |y - 2| - 1)$.

(a) Sketch the graph of $f(y)$ versus y and then make a simple sketch that indicates the behavior of all solutions of $y' = f(y)$ for $t \geq 0$.

(b) Classify each equilibrium solution as unstable, asymptotically stable, semistable, or neutrally stable.

15. Repeat Problem 14 for $f(y) = \frac{1}{2}(|y - 2| - |y - 1| + 2y - 3)$.

16. Repeat Problem 14 for $f(y) = \frac{1}{2}(|y - 1| - |y - 2| - 2y + 3)$.

17. Repeat Problem 14 for $f(y) = y(3 - y)(1 - |y - 1| - |y - 2|)$.

4.4 The Logistic Population Model

We are concerned here with modeling the growth of a single population in a closed, static environment. Let $P(t)$ denote the size of the population at time t. In reality, the population size would take on only integer values, but we will assume that $P(t)$ is a smooth function that reasonably approximates the discrete population size. An important quantity in modeling is the **per capita growth rate*** of the population, which is defined as $\frac{dP}{dt}/P$. The most basic model assumes that the per capita growth rate is a constant k. Thus the rate of growth of the population is proportional to the size of the population at any given time. This gives us the initial-value problem

$$\frac{dP}{dt} = kP, \quad P(0) = P_0,$$

* The per capita growth rate may be thought of as the percent increase in the population per unit time.

which has the solution $P(t) = P_0 e^{kt}$. This model, which predicts unlimited exponential growth (if $k > 0$), is known as the *Malthusian* model, so named after the English economist Thomas Malthus of the late eighteenth century. The Malthusian model is clearly not realistic for a population that is naturally limited by environmental factors such as finite space and finite food supply.

The constant k can be viewed as the difference in the per capita birth and death rates of the population. Thus k will be negative if the death rate exceeds the birth rate. In this case, the model predicts that $P(t) \to 0$ as $t \to \infty$.

The key to making the model more realistic is to allow the per capita growth rate to depend on the population size itself. It makes sense that this growth rate should decrease as the population increases and to be negative if the population size is too large for the environment to sustain. A simple model that takes these considerations into account comes from assuming that the per capita growth rate depends linearly on P. We replace the constant k with the linear expression $k(1 - P/M)$, where k and M are positive constants. Thus the model becomes

$$\frac{dP}{dt} = kP(1 - P/M), \quad P(0) = P_0. \tag{1}$$

This is called the *logistic* model and is credited to Pierre Verhulst, a Belgian mathematician of the mid-nineteenth century.

Note that the constant M in (1) is the population level at which there is no growth; that is, the population level at which birth and death rates are the same. It is called the **carrying capacity** of the environment, because it represents the maximum population that the environment can sustain. We will see shortly that M also represents the size that the population will approach after a long period of time. The coefficient k is called the **intrinsic growth rate**, because if P is very small relative to M, then the model becomes approximately one of simple exponential growth; that is, $P' \approx kP$. In other words, k would be the exponential growth rate if the carrying capacity were infinite. Thus, the constant k is influenced by natural tendencies of the population and not by environmental factors.

The differential equation in (1) may be explicitly solved by either of two elementary techniques. It is both a separable equation and a Bernoulli equation. You will be asked to solve it both ways in Problem 1 at the end of this section. The result is

$$P(t) = \frac{MP_0}{P_0 + (M - P_0)e^{-kt}}. \tag{2}$$

Our purpose here, however, is to analyze the behavior of solutions using the ideas of the preceding section.

First observe that the differential equation in (1) has two equilibrium solutions $y = 0$ and $y = M$. Observe further that with $f(P) = kP(1 - P/M)$ we have

$$f(P) < 0 \text{ if } P < 0, \quad f(P) > 0 \text{ if } 0 < P < M, \quad f(P) < 0 \text{ if } P > M.$$

These observations lead us to sketch solution curves as in Figure 1. In fact, we also see that $y = 0$ is an unstable equilibrium solution and that $y = M$ is a stable equilibrium solution with interval of attraction $(0, \infty)$. Thus *every* positive solution approaches the steady value of $P = M$ in the long run. (Negative solutions are of no interest here.)

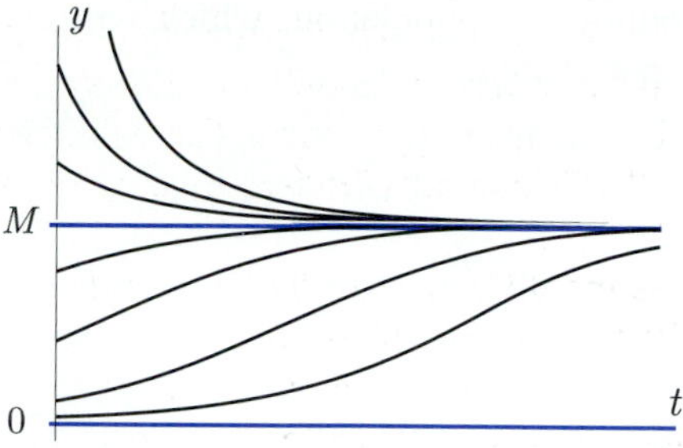

Figure 1

Harvesting Consider a population that is governed by (1) when it is undisturbed by environmental changes or forces from outside the environment. We wish to consider the effect of harvesting on such a population. By harvesting, we mean continuous, deliberate removal of members of the population at some specified rate. We will consider two types of harvesting, namely *proportional harvesting* and *constant harvesting*.

With proportional harvesting, the rate of harvesting is proportional to the size of the population. This has the advantage that a large harvest will be taken from a large population, while a small population perhaps won't be wiped out by harvesting. The disadvantage is that this requires knowledge of the population size on the part of the harvester. Nevertheless, if the harvesting rate is given by βP, where $\beta > 0$, then the problem becomes

$$\frac{dP}{dt} = kP(1 - P/M) - \beta P, \qquad P(0) = P_0. \tag{3}$$

We would like to examine the effect of this additional term on the solutions of the differential equation. Notice that a little rearranging results in

$$\frac{dP}{dt} = (k - \beta)P\left(1 - \frac{P}{M(1 - \beta/k)}\right),$$

which has exactly the same form as the original logistic equation with k replaced by $k - \beta$ and M replaced by $M(1 - \beta/k)$. (Thus the solution can be written easily by modifying (2).) In particular, the equilibrium solutions are $P = 0$ and $P = M(1 - \beta/k)$.

If $1 - \beta/k > 0$, then the equilibrium solution $P = M(1 - \beta/k)$ is asymptotically stable, and so the effect of proportional harvesting is simply to decrease the carrying capacity of the environment by the amount $M\beta/k$. With this modification, the solution curves look the same as those of the original logistic equation.

If $1-\beta/k = 0$, then $P = 0$ is the only equilibrium solution. It is semistable and attracts all positive solutions. If $1-\beta/k < 0$, then $P = 0$ is the asymptotically stable equilibrium solution, while $P = M(1 - \beta/k)$ is unstable. Still, $P = 0$ attracts all positive solutions. (See Figure 2.) The upshot of all of this is that if $\beta < k$, then any nonzero population will approach $P = M(1 - \beta/k)$, but if $\beta \ge k$, then any nonzero population will approach zero as $t \to \infty$. Thus, excessive proportional harvesting can result in an ever-dwindling (continuous) population, which will eventually lead to extinction of the actual discrete population.

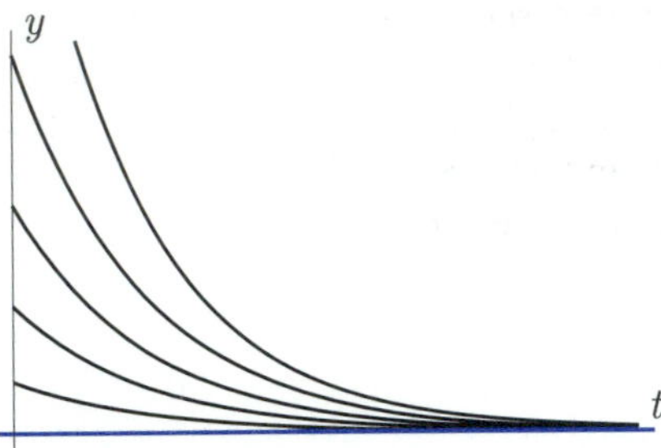

Figure 2

The harvesting rate βP is also called the *yield*. When $1-\beta/k > 0$, the population will approach the equilibrium solution $P = M(1 - \beta/k)$ as $t \to \infty$. Thus the yield will approach a steady value of $\beta M(1 - \beta/k)$ as $t \to \infty$. This steady yield is maximized by $\beta = k/2$ and has a maximum value $kM/4$. This quantity is called the *maximum sustainable yield*.

The second harvesting method we consider is constant harvesting. Here the rate of harvesting is constant, and thus will be the same regardless of the size of the population. Certainly, this is a much easier method for the harvester, since it does not require knowledge of the population size. However, it should be obvious that a small population could easily be killed off completely by this method.

We assume that the harvesting rate is a positive constant, say R. Now the problem becomes

$$\frac{dP}{dt} = kP(1 - P/M) - R, \quad P(0) = P_0. \tag{4}$$

Equilibrium solutions are given by the roots of the quadratic equation

$$-kP^2 + kMP - RM = 0,$$

which are

$$P_\pm = \frac{1}{2}\left(M \pm \sqrt{M(M - 4R/k)}\right). \tag{5}$$

The quantity $M - 4R/k$ determines whether these roots are complex, real and repeated, or real and distinct. So we have the following three cases.

Case 1: $M - 4R/k < 0$ (i.e., $R > kM/4$). Here there are no equilibrium solutions. Moreover, the quantity $kP(1 - P/M) - R$ is negative for all P. Thus solutions are always decreasing and reach zero in a finite amount of time, as shown in Figure 3. This situation is called *overharvesting* and always results in the population being driven to extinction.

Case 2: $M - 4R/k = 0$ (i.e., $R = k\,M/4$). Here we have just one equilibrium solution, $P = M/2$. Notice also that the quantity $kP(1 - P/M) - R$ is negative for all $P \neq M/2$. Thus, the equilibrium solution is semistable. When $P_0 > M/2$, the solution of (4) approaches $M/2$ as $t \to \infty$, but when $P_0 < M/2$, the solution is driven to zero in a finite amount of time, as indicated in Figure 4. This situation is called critical harvesting. Now imagine a scenario in which *critical harvesting* has been done for some time, and the population size has settled down to a nearly steady value just above $M/2$. If the population is disturbed in some way (perhaps by some brief period of environmental stress), and as a result the population size drops below $M/2$, then the population will be driven to extinction if harvesting continues.

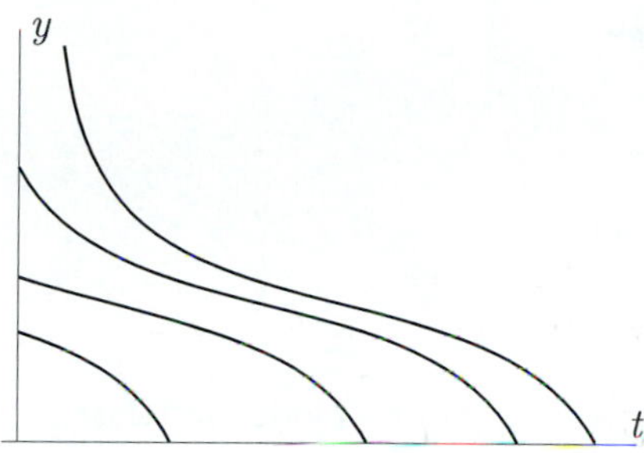

Figure 3

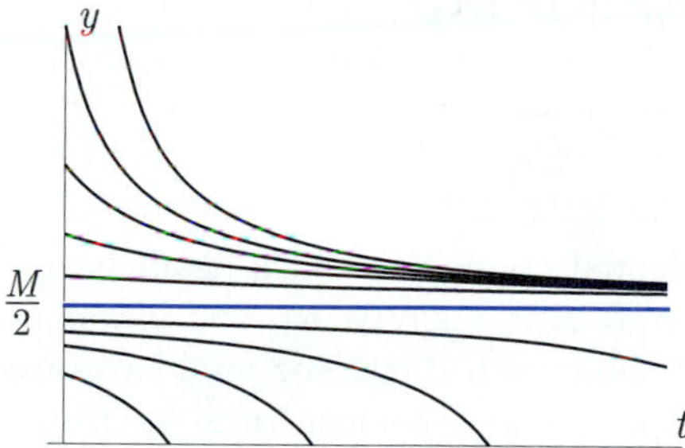

Figure 4

Case 3: $M - 4R/k > 0$ (i.e., $R < k\,M/4$). This is *moderate harvesting*. There are two positive equilibrium solutions given by (5). The smaller of these,

$$P_- = \frac{1}{2}\left(M - \sqrt{M(M - 4R/k)}\right),$$

is unstable, and the larger,

$$P_+ = \frac{1}{2}\left(M + \sqrt{M(M - 4R/k)}\right),$$

is stable. Several solution curves are seen in Figure 5. The interval between these equilibrium solutions can be thought of as a sort of "safety zone." Imagine a scenario in which moderate harvesting has been done for some time, and the population size has settled down near the stable equilibrium solution P_+. If the population is disturbed in some way (perhaps by some brief period of environmental stress), and as a result the population size drops below P_+, then the population will recover and *not* be driven to extinction by continued harvesting, unless it also drops below P_-.

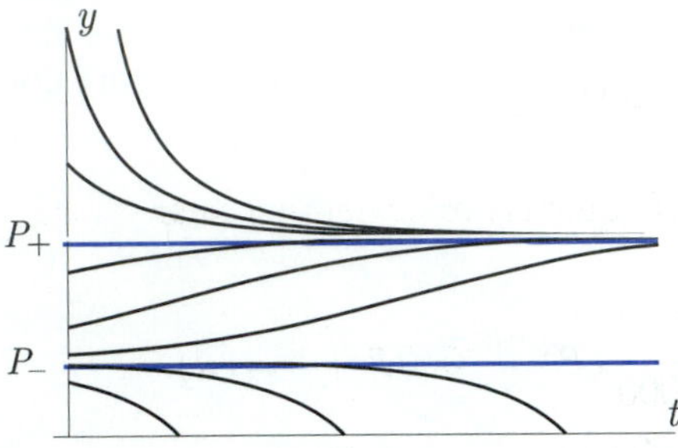

Figure 5

We should point out that the harvesting will be moderate (Case 3) if and only if there is a stable equilibrium greater than $M/2$. Furthermore, the two equilibria are related by

$$P_- + P_+ = M.$$

Example 1

The squirrel population in a residential community on a small island had reached a steady level of about 2000 squirrels after many years, and it was decided by the residents that the large number of squirrels had become an annoyance. A one-time relocation project relocated 500 squirrels to a wooded rural area on the mainland. During the next two years, the squirrel population on the island recovered to about 1800 squirrels. *Question*: If we wanted to maintain in the long run a steady population of 1500 squirrels, at what rate should squirrels be relocated? What if we wanted a steady population of 1000 squirrels? ■

SOLUTION: We assume that when undisturbed the population obeys a logistic equation $P' = kP(1 - P/M)$. The value of M is 2000, and taking $t = 0$ immediately after the relocation project, we have $P(0) = 1500$ and $P(2) = 1800$. We can determine k from

$$\int_{1500}^{1800} \frac{2000\,dP}{(2000 - P)P} = \int_0^2 k\,dt.$$

By a partial fraction decomposition,

$$\int_{1500}^{1800} \left(\frac{1}{2000 - P} + \frac{1}{P}\right) dP = \int_0^2 k\,dt.$$

Therefore,

$$2k = \ln\left(\frac{P}{2000 - P}\right)\Bigg|_{1500}^{1800} = \ln 9 - \ln 3 = \ln 3.$$

So $k \approx 0.5493$. Now when we include a constant harvesting term R, the differential equation becomes

$$\frac{dP}{dt} = 0.5493P(1 - P/2000) - R$$

where R represents the number of squirrels relocated per year. Since $\frac{2000}{.5493} \approx 3641$, we rewrite the equation as

$$\frac{dP}{dt} = -\frac{.5493}{2000}(P^2 - 2000P + 3641R).$$

So there will be an equilibrium solution at $P = 1500$ if

$$1500^2 - 2000(1500) + 3641R = 0,$$

which results in $R \approx 206$ squirrels per year. Since $1500 > M/2$, this will be the stable equilibrium P_+, and the unstable equilibrium will be $P_- = M - P_+ = 500$. (To be fairly consistent with the continuous nature of our model, this solution should be understood to mean relocating perhaps 17 per month or 4 per week.)

Since $M/2 = 1000$, the only way to have an equilibrium at $P = 1000$ is to have $P_+ = P_- = 1000$ (i.e., critical harvesting). The harvesting rate R satisfies

$$1000^2 - 2000(1000) + 3641R = 0,$$

which results in $R = 10^6/3641 \approx 275$ squirrels per year. Notice that with $P_+ = 1500$ we have a type of "safety zone" (or margin for error) in the sense that if the population accidentally drops below 1500, but remains above 500, it will recover, even with continued relocation. However, with $P_+ = P_- = 1000$, there is no safety zone. If the population ever drops below 1000, it will be driven to zero by continued relocation.

Curve-fitting Issues As in the preceding example, a matter of practical concern is the determination of the parameters in the logistic model from a number of observations of the population size. In the preceding example, we were fortunate that one of the observations was the carrying capacity M itself. In Problems 8 and 9 at the end of this section, we will see that under certain circumstances, the calculation of k and M can be quite sensitive to small changes in the observations, which in practice are subject to error. The point is that a small error in an observation can lead to a large error in the computation of k and M.

Suppose that the parameters in the logistic model are computed from inexact observations of the population for the purpose of determining a reasonable harvesting rate that will not cause extinction. If, for example, a small error in an observation leads to a large overestimate of M, then we would be lead to recommend a harvesting rate that is too large and consequently drive the population to extinction.

Fitting a logistic curve through inexact data points is an inherently difficult problem. The reason is that it is possible for two logistic curves to follow each other quite closely for some time and then approach two very different limits as $t \to \infty$.

This is illustrated in Figure 6. Thus, inexact observations can lead to very poor predictions of long term behavior. This sensitivity to errors in the data is especially pronounced when the time interval between observations is relatively short.

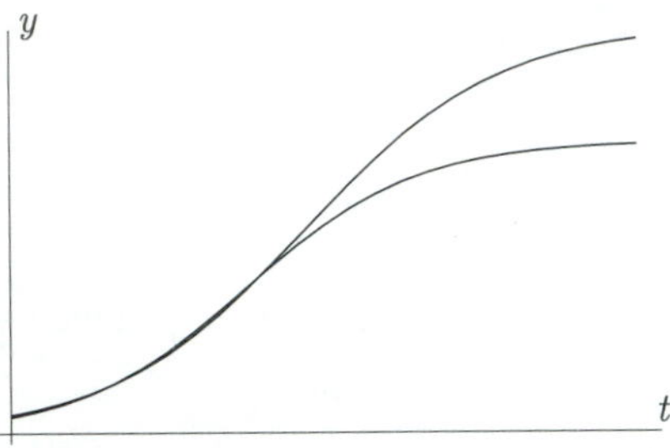

Figure 6

Variable Coefficients In more realistic models, the coefficients k and M would be allowed to depend on t in order to take into account variations in environmental conditions and reproductive habits. Changes of seasons might be responsible for such variation. Recall that the constant M is the carrying capacity of the environment, which may vary with the food supply, for example. Recall also that the intrinsic growth rate k would be the exponential growth rate if the carrying capacity were infinite. So k may vary according to reproductive habits, for example.

Example 2

If an animal species reproduces primarily in the spring of the year, then k might be modeled by some periodic function that attains a maximum in the spring of each year, such as

$$k(t) = k_0 + \alpha \cos(2\pi(t - .25)),$$

where $0 < \alpha \leq k_0$ are constants and t measures time in years. Also, if the food supply peaks each year in the autumn, then M might look something like

$$M(t) = M_0 + \beta \cos(2\pi(t - .75)),$$

where $0 < \beta \leq M_0$ are constants. Note that when M is not constant, there is no positive equilibrium solution. However, when M and k are periodic in t, one might expect that solutions would asymptotically approach some periodic solution. Figure 7 shows several solution curves in the case where

$$k(t) = 1. + .75 \cos(2\pi(t - .25))$$

and

$$M(t) = 10 + 5 \cos(4\pi(t - .25)).$$

These correspond to a population whose reproductive rate peaks in the spring, while the food supply peaks in both the spring and autumn. Notice that all the curves appear to approach the same periodic solution as $t \to \infty$. ■

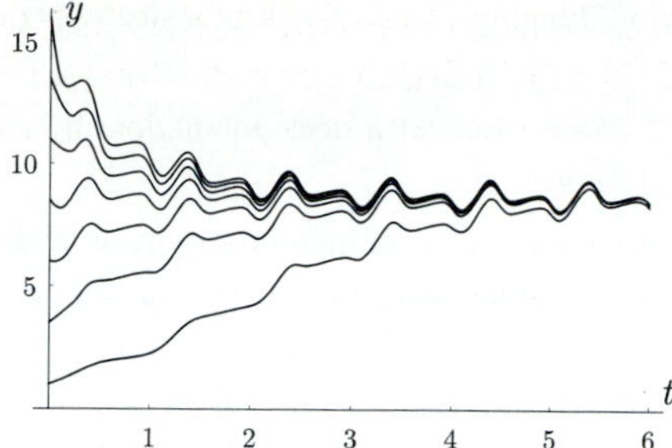

Figure 7

PROBLEMS

1. Solve (1) to obtain the solution (2) by

   (a) separation of variables and partial fraction decomposition;

   (b) treating the differential equation as a Bernoulli equation.

2. Suppose that a lake was initially stocked many years ago with 100 trout. After two years the population had grown to about 300. In recent years the population has had a steady value of about 1000 trout. No fishing is allowed in the lake, and fairly constant environmental conditions have persisted since the lake was first stocked.

   (a) Assuming a logistic model, write down the solution of the appropriate initial-value problem in terms of the unknown coefficient k.

   (b) The observation $P(2) = 300$ leads to an equation for k. Solve for k.

3. Suppose that the number $P(t)$ of mature trees per acre in a large hardwood forest obeys the logistic equation (1). For years, $P(t)$ had maintained a steady value of about 400 trees. Ten years ago, a logging company began harvesting at a rate corresponding to one out of twenty mature trees per year. The population of mature trees is now estimated to be about 280 trees per acre.

   (a) Write down, in terms of the coefficient k, the solution of the appropriate initial-value problem.

   (b) The observation $P(10) = 280$ leads to a transcendental equation for k. Estimate k by approximating the solution of this equation.

   (c) Predict the number of mature trees per acre after many years of such harvesting.

4. Records show that the average number of mature trees per acre in a large forest 150 years ago was 200, and 50 years later the number had reached 280. For the past 20 years the number has remained about 300. Based on this information, what fraction of the trees in this forest can be harvested each year and still maintain a steady stand of 200 trees per acre in the long run?

5. Suppose that 20 years ago a deer population in a wildlife refuge was estimated to be 500 deer, and 10 years ago the population had reached 800 deer. The population has remained nearly constant at about 1000 deer for the past several years.

   (a) Based on the given data, find the differential equation of the form (1) that governs the population.

(b) What rate of constant harvesting (hunting) would result in a steady population of 700 deer?

6. Suppose that three consecutive annual observations of a deer population in a wildlife refuge showed populations of 100, 133, and 170 deer.

(a) Predict the long-term steady population if the community remains undisturbed.

(b) What rate of constant harvesting (hunting) would result in a steady population of 300 deer?

(c) Redo part (a) if the third observation were 173 instead of 170. Comment on the reliability of predicting the long-term steady population in this manner. If a hunting policy were implemented based upon a bad prediction, what impact might it have upon the population?

7. Suppose that P satisfies (1). Show that $\frac{dP}{dt}$ is a maximum when $P = M/2$ and the maximum value of $\frac{dP}{dt}$ is $kM/4$. Make sense of the fact that this number is also the maximum sustainable yield.

8. Consider the problem of determining the carrying capacity M, where observations of the population P_0, $P_1 = P(t_1)$, and $P_2 = (2t_1)$ are known. We assume that either $0 < P_0 < P_1 < P_2 < M$ or $P_0 > P_1 > P_2 > M \geq 0$.

(a) Show that

$$e^{-kt_1} = \frac{P_0(M - P_1)}{P_1(M - P_0)} \quad \text{and} \quad e^{-2kt_1} = \frac{P_0(M - P_2)}{P_2(M - P_0)},$$

and therefore

$$P_0P_2(M - P_1)^2 = P_1^2(M - P_2)(M - P_0).$$

Now solve for M to obtain

$$M = \frac{P_1(2P_0P_2 - P_1P_0 - P_1P_2)}{P_0P_2 - P_1^2}.$$

(b) Suppose that $P_0 < P_1 < P_2$. Show that the formula for M yields $M > P_2$ if and only if $P_0P_2 - P_1^2 < 0$.

(c) Suppose that $P_0 > P_1 > P_2$. Show that the formula for M yields $M < P_2$ if and only if $P_0P_2 - P_1^2 > 0$.

(d) Explain why a small change in one of the observations can result in a large change in the calculated M if $P_0P_2 - P_1^2 \approx 0$. Relate this to Problem 6c.

(e) For what type of functions f is it true that $f(t)f(t + 2h) - f(t + h)^2 = 0$? Describe the circumstance under which one might expect $P_0P_2 - P_1^2 \approx 0$ in part (d), even if the observations were exact.

9. Consider the problem of determining the intrinsic growth rate k, where M, P_0, and $P_1 = P(t_1)$ are known with either $0 < P_0 < P_1 < M$ or $0 \leq M < P_1 < P_0$.

(a) Show that

$$k = \frac{1}{t_1} \ln\left(\frac{P_1(M - P_0)}{P_0(M - P_1)}\right).$$

(b) Consider the case $0 < P_0 < P_1 < M$. Suppose that $M > 1$ and units are chosen such that $P_0 = 1$ and $t_1 = 1$. Also set $P_1 = rM$ where $0 < r < 1$. Show that

$$k = \left(\ln\left(\frac{r}{1 - r}\right) + \ln(M - 1)\right).$$

(c) Graph the function $f(r) = \ln(\frac{r}{1-r})$ for $0 < r < 1$. Explain why a small change in the observation P_1 can lead to a very large change in the calculation of k if $r \approx 1$ or $r \approx 0$; that is, if $P_1 \approx M$ or $P_1 \approx P_0 \approx 0$.

Problems 10 through 12 explore properties of solutions of the logistic equation (also a Bernoulli equation) with time-dependent coefficients:

$$y' = k(t)y - f(t)y^2, \quad \text{where } k(t)/f(t) = M(t). \tag{$*$}$$

We assume that k and f are continuous and positive.

10. **(a)** Show that the solution of $(*)$ with $y(0) = y_0$ is given by

$$y = \frac{y_0 e^{K(t)}}{1 + y_0 \int_0^t e^{K(s)} f(s)\, ds}, \quad \text{where } K(t) = \int_0^t k(r)\, dr.$$

(b) Using l'Hôpital's rule, show that if $\lim_{t\to\infty} K(t) = \infty$ and $\lim_{t\to\infty} M(t)$ exists, then $\lim_{t\to\infty} y(t) = \lim_{t\to\infty} M(t)$. What condition on k (other than it being constant) will guarantee that $\lim_{t\to\infty} K(t) = \infty$?

11. Suppose that k and M are periodic with common period T; that is, $k(t+T) = k(t)$ and $M(t+T) = M(t)$ for all $t \ge 0$. (Hence, $f(t+T) = f(t)$ for all $t \ge 0$ as well.)

(a) Use 10a to show that there is a unique positive y_0 such that $y(T) = y_0$.

(b) Let y be the solution of $(*)$ with $y(T) = y(0) = y_0$ resulting from part (a), and define $\tilde{y}(t) = y(t+T)$ for all $t \ge 0$. Show that y and $\tilde{y}$ satisfy the same initial-value problem. Consequently, by uniqueness, $y(t) = \tilde{y}(t)$ for all $t \ge 0$, which proves that y is periodic with period T.

12. Suppose that there are positive constants $\alpha, \beta, \gamma, \delta$ such that

$$\alpha < k(t) < \beta \text{ and } \gamma < M(t) < \delta \text{ for } -\infty < t < \infty.$$

(a) Show that $\delta e^{K(t)} < \int_{-\infty}^t f(s)e^{K(s)}\, ds < \gamma e^{K(t)}$ for all t.

(b) Show that the equation $u' + k\,u = f$, which is satisfied by $u = 1/y$, has a particular solution given by

$$\tilde{u} = e^{-K(t)} \int_{-\infty}^t f(s)e^{K(s)} ds.$$

(c) Show that the solution $\tilde{u}$ in part (b) is the only positive solution of $u' + ku = f$ that approaches neither 0 nor ∞ as $t \to \pm\infty$. Conclude that $\tilde{y} = 1/\tilde{u}$ is the only positive solution of $(*)$ that approaches neither 0 nor ∞ as $t \to \pm\infty$.

(d) Show that every positive solution y of $(*)$ satisfies

$$\lim_{t\to\infty} (y(t) - \tilde{y}(t)) = 0.$$

4.5 Numerical Methods

In this section we will undertake a somewhat more detailed study of numerical methods than was attempted in Section 3.1. Whenever a numerical method is used to compute an approximate solution, it is important to have some knowledge of how large the error will be. This knowledge typically takes the form of an upper bound on the absolute value of the error, which we call an *error estimate*.

In addition to round-off error, which is inherent in all finite-precision numerical calculation, another kind of error, known as *truncation error*, is introduced by the method itself. Roughly speaking, truncation error is the difference between the solution of the original problem and the approximate solution generated by the method in the absence of round-off error.

As we have seen, numerical methods for initial-value problems approximate solution values recursively at a sequence of values of t. This complicates the notion of truncation error, since one must consider not only the "local" truncation error of each individual step, but also the accumulation of truncation errors as the recursive calculation proceeds.

We will try to make these vaguely stated ideas more clear as we develop estimates for the truncation error of Euler's method.

Error Estimates for Euler's method Consider the usual initial-value problem

$$y' = f(t, y), \quad y(0) = y_0, \tag{1}$$

and assume that f has continuous second-order partial derivatives. Suppose that $\hat{y}(t+h)$ is the Euler's method approximation to $y(t+h)$ computed from an *exact* solution value $y(t)$; that is, let

$$\hat{y}(t+h) = y(t) + h\, f(t,\, y(t)). \tag{2}$$

By Taylor's theorem (consult your calculus book),

$$y(t+h) = y(t) + h\, f(t,\, y(t)) + \frac{1}{2} h^2 y''(\tau) \tag{3}$$

for some τ between t and $t+h$. Thus, by subtracting (2) from (3) and taking absolute value, we have

$$|\hat{y}(t+h) - y(t+h)| = \frac{1}{2} h^2 |y''(\tau)| \le \frac{1}{2} M\, h^2, \tag{4}$$

where M is any upper bound on $|y''(\tau)|$ for $t \le \tau \le t+h$; its maximum value for instance. The quantity $|\hat{y}(t+h) - y(t+h)|$ is called the **local truncation error**. Because of (4), we say that the local truncation error for Euler's method is *of order* h^2, or simply $\mathcal{O}(h^2)$—read "big-oh of h^2."

The local truncation error tells only part of the story. Suppose that we want to use Euler's method to approximate the solution of (1) on an interval $0 \le t \le T$, beginning with the given initial point $(0, y_0)$. We will compute approximations $y_1, y_2, \ldots, y_n$ to solution values $y(t_1), y(t_2), \ldots, y(t_n)$, respectively, where $t_n = T$ and $h = T/n$. By the time we have computed y_n, our approximation to $y(T)$, local truncation errors have been made n times. Estimates of these local truncation errors, as in (4), contribute to an estimate of the **accumulated truncation error** at T:

$$|y_n - y(T)| \le \frac{1}{2} \sum_{k=0}^{n-1} M_k\, h^2,$$

where M_k is an upper bound on $|y''(\tau)|$ for $t_k \le \tau \le t_{k+1}$. In terms of the average $\overline{M} = \frac{1}{n}\sum_{k=0}^{n-1} M_k$ of the M_k's, this estimate becomes

$$|y_n - y(T)| \le \frac{n}{2}\,\overline{M}\,h^2.$$

Now, since $nh = T$, we have

$$|y_n - y(T)| \le \frac{1}{2}\overline{M}Th. \tag{5}$$

Thus the accumulated truncation error at T is of order h; i.e., $\mathcal{O}(h)$.

Since the "final" time T need not be viewed as fixed, (5) may be interpreted as follows: *The accumulated truncation error in the Euler's method approximation of $y(t)$ using stepsize h is $\mathcal{O}(th)$.* In other words, for fixed h the accumulated truncation error is roughly proportional to t, while for fixed t it is roughly proportional to h.

In summary, because of (4) and (5), the truncation error for Euler's method is said to be *locally* $\mathcal{O}(h^2)$ and *globally* $\mathcal{O}(h)$. It should not be difficult to see that, in general, if the truncation error a method is locally $\mathcal{O}(h^n)$, then it will be globally $\mathcal{O}(h^{n-1})$.

The role of the constant $\overline{M}$ in (5) should not be understated. It depends upon the solution itself and may be quite large—or small. However, due to its definition, $\overline{M}$ will be roughly independent of the number of steps needed to reach T. This will be demonstrated in the following example by computation of the ratio $\frac{|y_n - y(t_k)|}{ht_k}$ at each step.

Example 1

To illustrate the effect of the stepsize on Euler's method, let us consider the "toy problem"

$$y' = t^2 - y, \quad y(0) = 1,$$

of which the exact solution is

$$y = t^2 - 2t + 2 - e^{-t}.$$

We will approximate the solution with Euler's method on the interval $0 \le t \le 1$, first with stepsize $h = 0.1$ and then with $h = 0.01$ and $h = 0.001$. Table 1 shows the result with $h = 0.1$. The Euler's method approximations are in the third column, and exact solution values are in the fourth. The fifth column lists the error at each point, and the final column lists the ratio of the error and the product ht_k, which according to (5) should be bounded by a constant that depends upon the second derivative of the solution.

Notice that the estimate of the accumulated truncation error given by (5) is demonstrated by the numbers in the last column. However, the most important demonstration of (5) is seen by comparing the last column of Table 1 with the last columns of the following Tables 2 and 3, which are constructed with $h = 0.01$ and $h = 0.001$, respectively.

Table 1. Euler, $h = 0.1$

k	t_k	y_k	$y(t_k)$	$\lvert y_k - y(t_k)\rvert$	$\dfrac{\lvert y_k - y(t_k)\rvert}{h\,t_k}$
0	0	1	1	0	—
1	0.1	0.9	0.905163	0.005163	0.516258
2	0.2	0.811	0.821269	0.010270	0.513462
3	0.3	0.7339	0.749182	0.015282	0.509393
4	0.4	0.66951	0.689680	0.020170	0.504249
5	0.5	0.618559	0.643469	0.024910	0.498207
6	0.6	0.581703	0.611188	0.029485	0.491421
7	0.7	0.559533	0.593415	0.033882	0.484027
8	0.8	0.552580	0.590671	0.038092	0.476144
9	0.9	0.561322	0.603430	0.042109	0.467875
10	1.	0.586189	0.632121	0.045931	0.459312

Table 2. Euler, $h = 0.01$

k	t_k	y_k	$y(t_k)$	$\lvert y_k - y(t_k)\rvert$	$\dfrac{\lvert y_k - y(t_k)\rvert}{h\,t_k}$
0	0	1	1	0	—
10	0.1	0.904662	0.905163	0.000501	0.500836
20	0.2	0.820272	0.821269	0.000997	0.498558
30	0.3	0.747697	0.749182	0.001485	0.495050
40	0.4	0.687718	0.689680	0.001962	0.490499
50	0.5	0.641044	0.643469	0.002425	0.485069
60	0.6	0.608315	0.611188	0.002873	0.478907
70	0.7	0.590110	0.593415	0.003305	0.472138
80	0.8	0.586952	0.590671	0.003719	0.464877
90	0.9	0.599315	0.603430	0.004115	0.457221
100	1.	0.627628	0.632121	0.004493	0.449258

Notice that the final columns of Tables 1 through 3 are roughly the same. This is indicative of the fact mentioned earlier that the constant $\overline{M}$ in (5) is roughly independent of the number of steps needed to reach a particular time t. ■

Runge-Kutta Methods A family of now-classical numerical methods arose from the work of German mathematicians C. D. Runge and W. Kutta around the turn of the twentieth century. Each of these methods uses a weighted average of intermediate derivative approximations to compute highly accurate numerical solutions.

Table 3. **Euler, $h = 0.001$**

k	t_k	y_k	$y(t_k)$	$\lvert y_k - y(t_k)\rvert$	$\frac{\lvert y_k - y(t_k)\rvert}{h\,t_k}$
0	0	1	1	0	—
100	0.1	0.905113	0.905163	0.0000499	0.499369
200	0.2	0.821170	0.821269	0.0000994	0.497138
300	0.3	0.749034	0.749182	0.0001481	0.493681
400	0.4	0.689484	0.689680	0.0001957	0.489185
500	0.5	0.643227	0.643469	0.0002419	0.483812
600	0.6	0.610902	0.611188	0.0002866	0.477707
700	0.7	0.593085	0.593415	0.0003297	0.470997
800	0.8	0.590300	0.590671	0.0003710	0.463794
900	0.9	0.603020	0.603430	0.0004106	0.456196
1000	1.	0.631672	0.632121	0.0004483	0.448288

Recall that Heun's method was introduced in Section 2.1 and described there as a variation on the Trapezoidal rule. Heun's method is also a simple example of a *two-stage* Runge-Kutta method. A derivation of it in the Runge-Kutta context—providing a truncation error estimate as well—proceeds as follows.

For some θ in $[0, 1]$ to be determined, and given (t_n, y_n), we will compute derivative approximations k_1 at t_n and k_2 at $t_n + \theta h$ according to

$$k_1 = f(t_n, y_n), \quad k_2 = f(t_n + \theta h, \; y_n + \theta h\, k_1),$$

and then use a weighted average of k_1 and k_2 to compute y_{n+1}:

$$y_{n+1} = y_n + h(w_1 k_1 + w_2 k_2),$$

where $w_1 + w_2 = 1$. The goal is to determine θ, w_1, and w_2 so that the local truncation error is $\mathcal{O}(h^3)$, which will in turn give an accumulated truncation error that is $\mathcal{O}(h^2)$.

So we begin by noting that from Taylor's theorem, the equation $y' = f(t, y)$, and the chain rule, we have

$$\begin{aligned} y(t+h) &= y(t) + hf(t, y(t)) + \frac{h^2}{2}\frac{d}{dt} f(t, y(t)) + \mathcal{O}(h^3) \\ &= y(t) + hf(t, y(t)) + \frac{h^2}{2}(f_t(t, y(t)) + f_y(t, y(t)) f(t, y(t))) + \mathcal{O}(h^3). \end{aligned} \tag{6}$$

Letting $\hat{y}(t+h)$ denote our approximation to $y(t+h)$ based upon the exact value of $y(t)$, we compute

$$k_1 = f(t, \; y(t)), \quad k_2 = f(t + \theta h, \quad y(t) + \theta h\, k_1),$$
$$\hat{y}(t+h) = y(t) + h\,(w_1 k_1 + w_2 k_2),$$

which can be condensed to

$$\hat{y}(t+h) = y(t) + h(w_1 f(t, y(t)) + w_2 f(t+\theta h, y(t)+\theta h k_1)). \tag{7}$$

By Taylor's theorem for functions of two variables we have

$$f(t+\theta h,\ y(t)+\theta h\, k_1) = f(t, y(t)) + f_t(t, y(t))\,\theta h + f_y(t, y(t))\theta h\, k_1 + \mathcal{O}(h^2).$$

Substituting this into (7) results in

$$\begin{aligned} \hat{y}(t+h) &= y(t) + h(w_1+w_2) f(t, y(t)) \\ &\quad + w_2\theta h^2 (f_t(t, y(t)) + f_y(t, y(t)) f(t, y(t))) + \mathcal{O}(h^3). \end{aligned} \tag{8}$$

Now, comparing (6) and (8), we see that $y(t+h) - \hat{y}(t+h) = \mathcal{O}(h^3)$ if

$$w_1 + w_2 = 1 \quad \text{and} \quad w_2\theta = \frac{1}{2}.$$

One simple solution of these is $w_1 = w_2 = 1/2$ and $\theta = 1$, from which we obtain Heun's method:

$$\begin{gathered} k_1 = f(t_n, y_n), \quad k_2 = f(t_n + h, y_n + h\, k_1), \\ y_{n+1} = y_n + \frac{1}{2} h(k_1 + k_2), \end{gathered}$$

with truncation error that is locally $\mathcal{O}(h^3)$ and globally $\mathcal{O}(h^2)$.

Notice that the derivation of the method essentially amounted to choosing coefficients so that $\hat{y}(t+h)$ would agree with the second-degree Taylor polynomial approximation of $y(t+h)$ centered at t. For this reason, Heun's method is said to be a *second-order* method. In this terminology, Euler's method is a *first-order* method.

Example 2

Table 4 shows the result of Heun's method applied to the same problem as in Example 1,

$$y' = t^2 - y, \quad y(0) = 1,$$

with stepsize $h = 0.1$. Note that the last column lists the error at each step divided by $h^2 t_k$. While the errors increase gradually, those entries do not. ■

The most widely used Runge-Kutta method—perhaps the most widely used numerical method of any kind for ordinary differential equations—is a particular four-stage method. The general form of any four-stage Runge-Kutta method is

$$\begin{gathered} k_1 = f(t_n, y_n), \quad k_i = f(t_n + \theta_{i-1} h,\ y_n + \theta_{i-1} h\, k_{i-1}), \ i = 2, 3, 4, \\ y_{n+1} = y_n + h(w_1 k_1 + w_2 k_2 + w_3 k_3 + w_4 k_4). \end{gathered}$$

With an appropriate choice of $\theta_1, \ldots, \theta_3$ and $w_1, \ldots, w_4$, the local truncation error of the method will be $\mathcal{O}(h^5)$, thus giving a *fourth-order* method.

As you might imagine from the preceding derivation of Heun's method, the derivation of a fourth-order method will likely be a very tedious exercise. However,

Table 4. **Heun, $h = 0.1$**

k	t_k	y_k	$y(t_k)$	$\lvert y_k - y(t_k)\rvert$	$\frac{\lvert y_k - y(t_k)\rvert}{h^2 t_k}$
1	0.1	0.9055	0.905163	0.000337	0.337418
2	0.2	0.821928	0.821269	0.000658	0.329127
3	0.3	0.750144	0.749182	0.000963	0.320869
4	0.4	0.690931	0.689680	0.001251	0.312679
5	0.5	0.644992	0.643469	0.001523	0.304583
6	0.6	0.612968	0.611188	0.001780	0.296605
7	0.7	0.595436	0.593415	0.002021	0.288762
8	0.8	0.592920	0.590671	0.002249	0.281072
9	0.9	0.605892	0.603430	0.002462	0.273545
10	1.	0.634782	0.632121	0.002662	0.266192

a useful insight is provided by the formula

$$y(t+h) = y(t) + \frac{1}{6}h\left(y'(t) + 4y'(t+\frac{1}{2}h) + y'(t+h)\right) + \mathcal{O}(h^5), \tag{9}$$

which can be derived by a fairly simple manipulation of the cubic Taylor polynomial for y'. (See Problem 10.) This formula suggests that

$$\begin{aligned}
k_1 &= f(t_n,\ y_n),\\
k_2 &= f\left(t_n + \frac{1}{2}h,\ y_n + \frac{1}{2}h\,k_1\right),\\
k_3 &= f\left(t_n + \frac{1}{2}h,\ y_n + \frac{1}{2}h\,k_2\right),\\
k_4 &= f(t_n + h,\ y_n + h\,k_3),\\
y_{n+1} &= y_n + \frac{1}{6}h(k_1 + 2k_2 + 2k_3 + k_4).
\end{aligned} \tag{10}$$

will perhaps be a fourth-order Runge-Kutta method. This is indeed the case, and (10) is commonly known as *the* **fourth-order Runge-Kutta method**. The proof that the local truncation error of (10) is $\mathcal{O}(h^5)$ is too tedious and technical for this setting. We will simply demonstrate the fact in the context of a simple problem.

Example 3

Consider the problem

$$y' = y, \quad y(0) = 1,$$

for which we know that the value of the solution at $t + h$ is

$$y(t+h) = e^{t+h} = e^t\,e^h = e^t\left(1 + h + \frac{1}{2}h^2 + \frac{1}{6}h^3 + \frac{1}{24}h^4\right) + \mathcal{O}(h^5). \tag{11}$$

One step of the method described above, starting at (t, e^t), results in

$$\begin{aligned}
k_1 &= e^t, \quad k_2 = e^t + \frac{1}{2} h\, e^t = e^t \left(1 + \frac{1}{2} h\right), \\
k_3 &= e^t + \frac{1}{2} h\, e^t \left(1 + \frac{1}{2} h\right) = e^t \left(1 + \frac{1}{2} h + \frac{1}{4} h^2\right), \\
k_4 &= e^t + h\, e^t \left(1 + \frac{1}{2} h + \frac{1}{4} h^2\right) = e^t \left(1 + h + \frac{1}{2} h^2 + \frac{1}{4} h^3\right), \\
\hat{y}(t+h) &= e^t + \frac{h\, e^t}{6} \left(1 + 2\left(1 + \frac{1}{2} h\right)\right. \\
&\qquad \left. + 2\left(1 + \frac{1}{2} h + \frac{1}{4} h^2\right) + 1 + h + \frac{1}{2} h^2 + \frac{1}{4} h^3\right) \\
&= e^t \left(1 + h + \frac{1}{2} h^2 + \frac{1}{6} h^3 + \frac{1}{24} h^4\right).
\end{aligned}$$

Comparing this with (11) we see that $y(t+h) - \hat{y}(t+h) = \mathcal{O}(h^5)$. ■

Example 4

Table 5 shows the result of the fourth-order Runge-Kutta method applied to the same problem as in Examples 1 and 2,

$$y' = t^2 - y, \quad y(0) = 1,$$

with stepsize $h = 0.1$. Note that the last column lists the error at each step divided by $h^4\, t_k$.

Table 5. Fourth-order Runge-Kutta, $h = 0.1$

k	t_k	y_k	$y(t_k)$	$\lvert y_k - y(t_k)\rvert$	$\frac{\lvert y_k - y(t_k)\rvert}{h^4\, t_k}$
0	0	1	1	0	—
1	0.1	0.905163	0.905163	1.26×10^{-7}	0.012637
2	0.2	0.821269	0.821269	2.49×10^{-7}	0.012426
3	0.3	0.749182	0.749182	3.66×10^{-7}	0.012203
4	0.4	0.689680	0.689680	4.79×10^{-7}	0.011972
5	0.5	0.643470	0.643469	5.87×10^{-7}	0.011734
6	0.6	0.611189	0.611188	6.89×10^{-7}	0.011491
7	0.7	0.593415	0.593415	7.87×10^{-7}	0.011246
8	0.8	0.590672	0.590671	8.80×10^{-7}	0.010999
9	0.9	0.603431	0.603430	9.68×10^{-7}	0.010752
10	1.	0.632122	0.632121	1.05×10^{-6}	0.010506

Table 6. Fourth-order Runge-Kutta, $h = 0.2$

k	t_k	y_k	$y(t_k)$	$\lvert y_k - y(t_k)\rvert$	$\frac{\lvert y_k - y(t_k)\rvert}{h^4 t_k}$
0	0	1	1	0	—
1	0.2	0.821273	0.821269	0.0000041	0.012770
2	0.4	0.689688	0.689680	0.0000079	0.012344
3	0.6	0.611200	0.611188	0.0000114	0.011880
4	0.8	0.590686	0.590671	0.0000146	0.011397
5	1.	0.632138	0.632121	0.0000175	0.010907

For the sake of comparison, Table 6 shows the result of the fourth-order Runge-Kutta method applied to the same problem, but with an even larger stepsize, $h = 0.2$. Notice that the numbers in the last column are quite close to their counterparts in Table 5, indicating that the error in the approximation of $y(t)$ is roughly proportional to $h^4 t$ with a proportionality constant independent of the number of steps taken to reach t. ■

PROBLEMS

1. Verify the calculations in the second and third rows of Table 1.

2. Verify the calculations in the second and third rows of Table 4.

3. Verify the calculations in the second and third rows of Table 5.

4. For the initial-value problem $y' = -t\,y^2$, $y(0) = 1$, and stepsize $h = 0.1$, compute *by hand* (and calculator) one step of

a) Euler's method;

b) Heun's method;

c) fourth-order Runge-Kutta.

d) Find the exact value of $y(0.1)$ and compute the error in each of the results.

5. In this problem, automate the calculations as you desire. For the initial-value problem in Problem 4, compute an approximation to $y(1)$ using stepsize $h = 0.1$ and

a) Euler's method;

b) Heun's method;

c) fourth-order Runge-Kutta.

d) Find the exact value of $y(1)$ and compute the error in each of the results.

6. Suppose that Euler's method is applied to $y' = -10y + 10t + 1$, $y(0) = 1$.

(a) Use the error estimate in (4), together with the knowledge of the exact solution, to find a stepsize h for which the error after one step will be no more than 0.0005. Check the result numerically.

(b) The constant $\overline{M}$ in (5) may be approximated by $\frac{1}{T}\int_0^T y''(t)dt$. Use this fact and (5) to find a stepsize h for which $y(1)$ will be approximated to within 0.0005.

7. Show that Euler's method with stepsize h, applied to $y' = ky$, $y(0) = y_0$, produces exact values of the solution of $y' = k_h y$, $y(0) = y_0$, where $k_h = \frac{1}{h}\ln(1+hk)$. Then show that $\lim_{h\to 0^+} k_h = k$.

8. Show that Heun's method with stepsize h, applied to $y' = ky$, $y(0) = y_0$, produces exact values of the solution of $y' = k_h y$, $y(0) = y_0$, where

$$k_h = \frac{1}{h}\ln(1 + hk + h^2k^2/2).$$

Then show that $\lim_{h\to 0^+} k_h = k$.

9. Consider the initial-value problem $y' = y^2$, $y(0) = 1$; and let y_1 be the approximation to $y(h)$ given by one step of fourth-order Runge-Kutta with stepsize h. Show that $y(h) - y_1 = \mathcal{O}(h^5)$.

10. Derive formula (9) as follows.

(a) Express $y'(t+x)$ as a cubic Taylor polynomial in x, centered at t, plus a $\mathcal{O}(x^4)$ error term.

(b) Obtain expressions for $y'(t+h/2)$ and $y'(t+h)$ by evaluating the expression from part (a) at $x = h/2$ and $x = h$.

(c) Use the expressions from part (b) to compute $h(y'(t) + 4y'(t+h/2) + y'(t+h))/6$ and compare the result with $y(t+h) - y(t)$.

11. Show that formula (9) applied to the equation $y' = f(t)$ leads to Simpson's rule:

$$\int_t^{t+h} f(x)\,dx = \frac{1}{6}h\left(f(t) + 4f\left(t + \frac{1}{2}h\right) + f(t+h)\right) + \mathcal{O}(h^5).$$

12. The equations that lead to Heun's method, $w_1 + w_2 = 1$ and $w_2\theta = 1/2$, also lead to other second-order Runge-Kutta methods. Formulate the method corresponding to $\theta = 2/3$, $w_1 = 1/4$, and $w_2 = 3/4$, and compare its performance to that of Heun's method in Problem 5.

In Problems 13 through 16, we investigate "three-stage" Runge-Kutta methods:

$$\begin{aligned}
k_1 &= f(t_n, y_n),\\
k_2 &= f(t_n + \theta_1 h,\ y_n + \theta_1 h\, k_1),\\
k_3 &= f(t_n + \theta_2 h,\ y_n + \theta_2 h\, k_2),\\
y_{n+1} &= y_n + h\left(w_1k_1 + w_2k_2 + w_3k_3\right).
\end{aligned}$$

13. Based on the requirement that one step of the method applied to $y' = y$, $y(0) = 1$, must compute e^h to within $\mathcal{O}(h^4)$ (cf. Example 3), derive a system of three algebraic equations in the variables θ_1, θ_2, w_1, w_2, and w_3.

For each of the choices of θ_1 and θ_2 in Problems 14 through 16,

(a) solve for the weights w_1, w_2, w_3 and write down the resulting three-stage Runge-Kutta method;

(b) using $h = 0.1$, compute an approximation to $y(1)$ where $y' = -t\,y^2$, $y(0) = 1$. Compare the performance of the method with that of Heun's method and fourth-order Runge-Kutta (cf. Problems 4 and 5).

14. $\theta_1 = \frac{1}{2}$, $\theta_2 = 1$

15. $\theta_1 = \frac{1}{2}$, $\theta_2 = \frac{3}{4}$

16. $\theta_1 = \frac{1}{3}$, $\theta_2 = \frac{2}{3}$

4.6 A First Look at Systems

This section is a cursory first look at coupled systems of differential equations, with the discussion limited to systems of two equations. A far more thorough treatment of systems comprises Chapters 8 through 10. Here we will mainly build upon ideas that are already familiar in the context of a single equation, namely direction fields and numerical approximation.

Applications often give rise to systems of first-order differential equations for two or more quantities. The rate of change in any one of the quantities involved may depend on any or all of the others. A *system* of two first-order differential equations takes the form

$$\frac{dx}{dt} = f(x, y), \quad \frac{dy}{dt} = g(x, y), \tag{1}$$

provided that the rates of change in x and y depend upon the values of x and y in a way that is independent of the time at which those values occur. Such a system is called *autonomous*. A solution of this system consists of a pair of functions $x(t)$, $y(t)$. An initial-value problem involving such a pair of differential equations would specify initial values for both x and y.

Example 1

It is easy to verify (and you should) that the pair of functions

$$x(t) = 2\sin t + \cos t \quad \text{and} \quad y(t) = \sin t$$

satisfies the initial-value problem

$$\begin{aligned} \frac{dx}{dt} &= 2x - 5y, \quad x(0) = 1 \\ \frac{dy}{dt} &= x - 2y, \quad y(0) = 0. \end{aligned}$$

(A good strategy here would be to check that $x' - 2x = -5y$ and $y' + 2y = x$.) ■

The system (1) is said to be **linear** if f and g are linear functions of x and y and **homogeneous** if (cx, cy) is a solution for any constant c whenever (x, y) is a solution. If (1) is linear, then it is homogeneous if and only if $f(0, 0) = g(0, 0) = 0$. The system in Example 1 is both linear and homogeneous.

Equilibrium Solutions As we shall see in Chapters 8 through 10, solutions of (1) can exhibit far more complicated behavior than solutions of a single autonomous equation, which we studied in Section 4.3. Indeed, Example 1 shows that a solution can be bounded for all t, yet not approach a constant equilibrium solution as $t \to \infty$. This is impossible in the single-equation case. So the notion of an equilibrium solution is far more subtle for systems. A periodic solution, for instance, could be viewed as a type of equilibrium solution.

As usual, however, constant solutions are equilibrium solutions. In the context of (1), this means a pair of numbers (a, b) such that

$$f(a, b) = g(a, b) = 0,$$

which we call an **equilibrium point**.

Example 2

Consider the (nonlinear) system

$$\frac{dx}{dt} = x(5 - x - y), \quad \frac{dy}{dt} = -y(1 - x + y).$$

Equilibrium points are solutions of the algebraic system

$$x(5 - x - y) = 0, \quad -y(1 + y - x) = 0.$$

It is easy to observe three solutions in which $x = 0$ or $y = 0$:

$$(0, 0), \quad (0, -1), \quad \text{and } (5, 0).$$

A fourth solution comes from the linear system

$$x + y = 5, \quad x - y = 1$$

and turns out to be $(3, 2)$. ■

The Phase Plane Solutions of (1) may be visualized as parametric curves in the (x, y)-plane. These curves are called **phase-plane orbits** and collectively comprise the **phase portrait** of the system. The direction field associated with these curves is visualized as an array of arrows of uniform length, with slopes given by

$$\frac{dy}{dx} = \frac{dy/dt}{dx/dt} = \frac{g(x, y)}{f(x, y)}.$$

Isoclines are the curves

$$\frac{g(x, y)}{f(x, y)} = m, \quad -\infty < m < \infty,$$

together with $f(x, y) = 0$, which corresponds to $m = \pm\infty$ (i.e., an up or down direction). Furthermore, the direction field has the same direction as the *vector* field $(f(x, y), g(x, y))$ at each point (x, y). Thus directions are determined by slopes and the individual signs of $f(x, y)$ and $g(x, y)$. Regions of constant sign are separated by **nullclines**, which are the isoclines along which directions are either vertical (where $f(x, y) = 0$) or horizontal (where $g(x, y) = 0$). Nullclines intersect at equilibrium points.

Example 3

Consider the simple system

$$\frac{dx}{dt} = y, \quad \frac{dy}{dt} = -x - y.$$

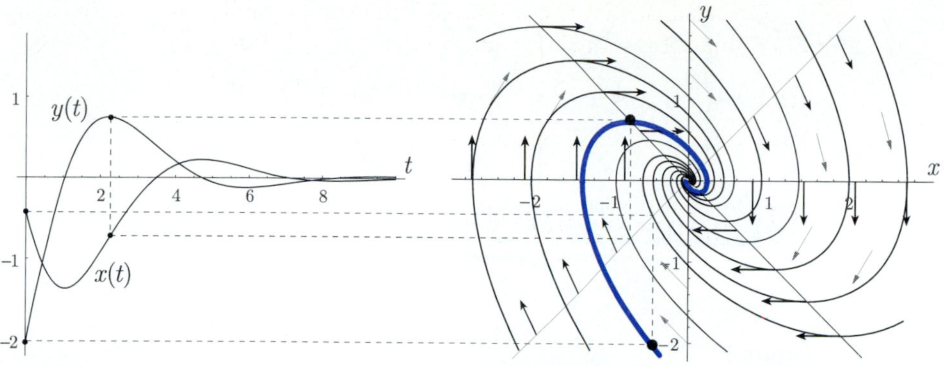

Figure 1

The plot on the left in Figure 1 shows the graphs of x and y versus t for the particular pair of initial values $x(0) = -.4$ and $y(0) = -2$. The plot on the right shows the phase portrait for the system. The thickened curve represents the particular orbit corresponding to the graphs on the left. (Each of the dashed rectangles indicates the correspondence between a pair of points on the graphs of $x(t)$ and $y(t)$ and a single point on the phase-plane orbit.) Notice that all of the phase-plane orbits spiral in toward the origin, corresponding to the fact that all (nontrivial) solutions oscillate while approaching 0 as $t \to \infty$. The nullcline on which $\frac{dx}{dt} = 0$ is the line $y = 0$ (i.e., the x-axis). On the positive half of the x-axis, the direction field points downward, because $\frac{dy}{dt} = -x < 0$ indicates that y is decreasing there. On the negative half of the x-axis, the direction field points upward, because $\frac{dy}{dt} = -x > 0$ indicates that y is increasing there. The nullcline on which $\frac{dy}{dt} = 0$ is the line $-x - y = 0$ (i.e., $y = -x$). On the second-quadrant portion of this line, the direction field points to the right, because $\frac{dx}{dt} = y > 0$ indicates that x is increasing there. On the fourth-quadrant portion of the line $y = -x$, the direction field points to the left, because $\frac{dx}{dt} = y < 0$ indicates that x is decreasing there. The other isoclines for this system are also lines through the origin:

$$\frac{-x - y}{y} = m, \quad \text{i.e., } x = -(m + 1)y.$$

For instance, the direction field has slope $m = -1$ along the y-axis and slope $m = -2$ along the line $y = x$. Notice these facts in Figure 1. ■

Often a rough approximation of a direction field can be obtained by simply considering nullclines and the signs of $f(x, y)$ and $g(x, y)$. Figure 2 indicates the eight directional possibilities determined by the signs of $f(x, y)$ and $g(x, y)$. Any point where $f(x, y) = g(x, y) = 0$ is an equilibrium point. In the vicinity of such a point, the direction field typically points in many different directions.

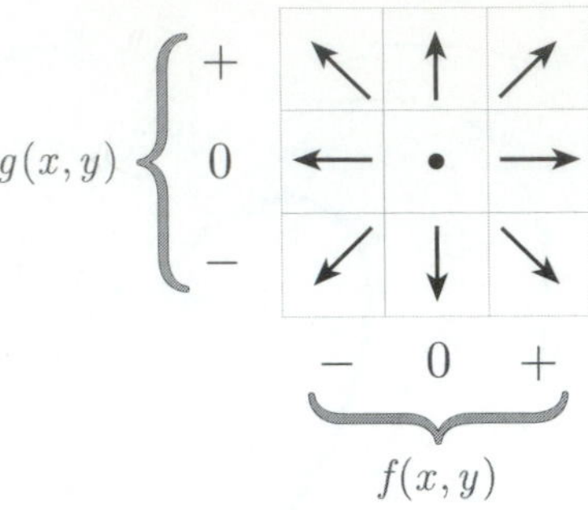

Figure 2

Example 4

Consider the system

$$\frac{dx}{dt} = x + y, \qquad \frac{dy}{dt} = 1 - x^2 - y^2.$$

Nullclines for this system are given by

$$x + y = 0 \quad \text{and } 1 - x^2 - y^2 = 0.$$

On the line $x + y = 0$, directions are vertical, and on the unit circle $x^2 + y^2 = 1$, directions are horizontal. Together, these nullclines separate the plane into four distinct regions corresponding to the four sign combinations of $\frac{dx}{dt}$ and $\frac{dy}{dt}$. These regions are shown in Figure 3a, where the sign combinations of $\frac{dx}{dt}$ and $\frac{dy}{dt}$ are indicated by arrows as in Figure 2. Note that $\frac{dx}{dt}$ is positive above the line $x + y = 0$ and negative below it, while $\frac{dy}{dt}$ is positive inside the circle $x^2 + y^2 = 1$ and negative outside of that circle. This information tells how to direct the horizontal and vertical arrows on the nullclines, as seen in Figure 3b. We also note that the system has constant solutions corresponding to the nullclines' two points of intersection. All of this provides enough information to sketch several crude phase-plane orbits as indicated in Figure 3c. Even though the sketch is quite crude, it does provide a great deal of qualitative information about the behavior of the system's solutions. ■

Example 5

Figure 4 shows direction fields and phase portraits of the following systems:

$$\begin{cases} \dfrac{dx}{dt} = x + y \\ \dfrac{dy}{dt} = x - y \end{cases} \quad \begin{cases} \dfrac{dx}{dt} = x - y \\ \dfrac{dy}{dt} = 2x - y \end{cases} \quad \begin{cases} \dfrac{dx}{dt} = -x - y \\ \dfrac{dy}{dt} = 2x - y \end{cases} \quad \begin{cases} \dfrac{dx}{dt} = y \\ \dfrac{dy}{dt} = -\sin x \end{cases}.$$

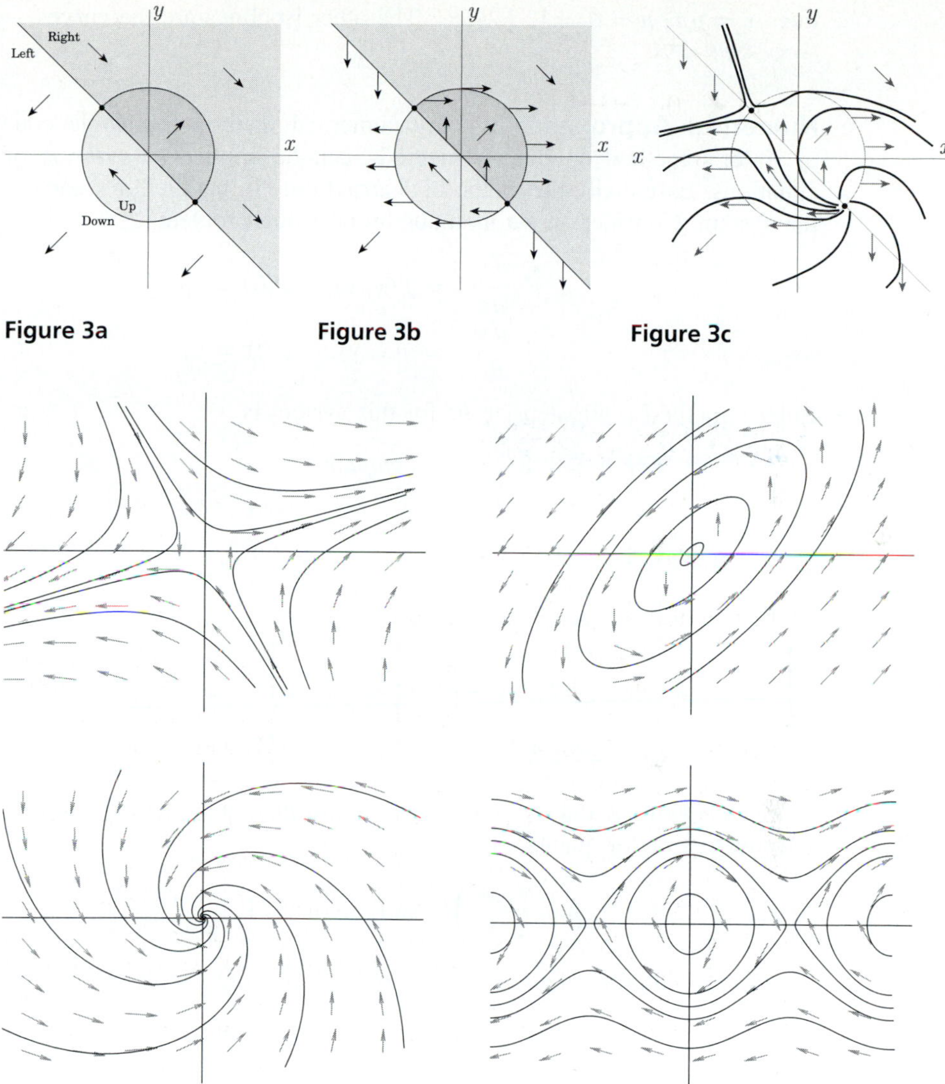

Figure 3a **Figure 3b** **Figure 3c**

Figure 4

We will learn in Chapter 8 that the first three of these phase portraits are typical of *linear* autonomous systems. Notice that for each of the first three systems, nullclines are lines through the origin. The fourth system describes the motion of a pendulum. There x represents the angle that the arm of the pendulum makes with the downward vertical, and y represents the angular velocity. (See Section 1.2.) The closed orbits describe periodic, back-and-forth motion, while the wavy orbits correspond to the pendulum spinning around the pivot. The nullclines are the x-axis and the vertical lines $x = n\pi$, $n = 0, \pm 1, \pm 2, \ldots$. The other isoclines are the curves $-\sin x = my$, $-\infty < m < \infty$. ■

Numerical Approximation Any numerical method for single equations—such as Euler's method or Heun's method—can be applied to systems of differential equations. Essentially the method is carried out "in parallel" for each component of the system. Consider an autonomous initial-value problem

$$\begin{aligned} \frac{dx}{dt} &= f(x, y), \quad x(0) = x_0 \\ \frac{dy}{dt} &= g(x, y), \quad y(0) = y_0. \end{aligned} \tag{2}$$

Euler's method (with stepsize h) for this system is

For $n = 0, 1, 2, \ldots,$ *compute*

$$x_{n+1} = x_n + h\, f(x_n, y_n) \quad \text{and } y_{n+1} = y_n + h\, g(x_n, y_n).$$

Heun's method is

For $n = 0, 1, 2, \ldots,$ *compute*

$$\left\{ \begin{aligned} k_1 &= f(x_n, y_n) \\ k_2 &= f(x_n + h\,k_1, y_n + h\ell_1) \\ x_{n+1} &= x_n + \frac{h}{2}(k_1 + k_2) \end{aligned} \right\} \quad \text{and} \quad \left\{ \begin{aligned} \ell_1 &= g(x_n, y_n) \\ \ell_2 &= g(x_n + hk_1, y_n + h\ell_1) \\ y_{n+1} &= y_n + \frac{h}{2}(\ell_1 + \ell_2) \end{aligned} \right\}.$$

The statement and the implementation of any such method can be simplified by introducing vector quantities

$$\mathbf{u} = \begin{pmatrix} x \\ y \end{pmatrix} \quad \text{and} \quad F(\mathbf{u}) = \begin{pmatrix} f(x, y) \\ g(x, y) \end{pmatrix}.$$

The initial-value problem (2) now may be expressed simply as

$$\mathbf{u}' = F(\mathbf{u}), \quad \mathbf{u}(t_0) = \mathbf{u}_0. \tag{3}$$

With the notation

$$\mathbf{u}_k = \begin{pmatrix} x_k \\ y_k \end{pmatrix},$$

where (x_k, y_k) is the approximation to $(x(kh), y(kh))$, Euler's method for calculating $\mathbf{u}_{k+1}$ from $\mathbf{u}_k$ with stepsize h takes the form

$$\mathbf{u}_{k+1} = \mathbf{u}_k + h\, F(\mathbf{u}_k), \quad k = 0, 1, 2, \ldots,$$

while Heun's method becomes

$$\begin{aligned} \mathbf{k}_1 &= F(\mathbf{u}_k) \\ \mathbf{k}_2 &= F(\mathbf{u}_k + h\,\mathbf{k}_1) \qquad k = 0, 1, 2, \ldots. \\ \mathbf{u}_{k+1} &= \mathbf{u}_k + \frac{h}{2}(\mathbf{k}_1 + \mathbf{k}_2) \end{aligned}$$

The point here is that when we view the system in the vector form of (3), a numerical method is written and implemented in basically the same way as in the case of a single equation.

PROBLEMS

1. **(a)** Check that $x(t) = 2e^t - e^{-t}$ and $y(t) = e^t - e^{-t}$ together satisfy the system

$$x' = 3x - 4y, \quad y' = 2x - 3y.$$

(b) Check that the same system is satisfied by $x(t) = 2c_1e^t + c_2e^{-t}$ and $y(t) = c_1e^t + c_2e^{-t}$ for any constants c_1 and c_2.

2. **(a)** Check that $x(t) = 2\cos t - \sin t$ and $y(t) = 3\cos t + \sin t$ together satisfy

$$x' = x - y, \quad y' = 2x - y.$$

(b) Check that the same system is satisfied by $x(t) = c_1(\cos t - \sin t) + c_2(\cos t + \sin t)$ and $y(t) = 2c_1 \cos t + 2c_2 \sin t$ for any constants c_1 and c_2.

In Problems 3 through 6, find all the equilibrium points of the system.

3. $\dfrac{dx}{dt} = 2x - y,$ $\quad \dfrac{dy}{dt} = x - y^2$

4. $\dfrac{dx}{dt} = x + y - 1,$ $\quad \dfrac{dy}{dt} = 1 - x - y$

5. $\dfrac{dx}{dt} = 6 - xy,$ $\quad \dfrac{dy}{dt} = 5 - x - y$

6. $\dfrac{dx}{dt} = (x + y)(3 - x - y),$ $\quad \dfrac{dy}{dt} = x^2 - y^2$

The systems in Problems 7 through 10 are those of Example 4.6. For each of them make a sketch showing

(a) the nullclines and several arrows indicating the direction field along each of them;

(b) several arrows indicating the direction field along the x- and y-axes (if not already covered by part (a));

(c) several phase-plane orbits (as shown in Figure 4) consistent with the arrows on the nullclines and axes.

7. $\begin{cases} \dfrac{dx}{dt} = x + y \\ \dfrac{dy}{dt} = x - y \end{cases}$

8. $\begin{cases} \dfrac{dx}{dt} = x - y \\ \dfrac{dy}{dt} = 2x - y \end{cases}$

9. $\begin{cases} \dfrac{dx}{dt} = -x - y \\ \dfrac{dy}{dt} = 2x - y \end{cases}$

10. $\begin{cases} \dfrac{dx}{dt} = y \\ \dfrac{dy}{dt} = -\sin x \end{cases}$

In Problems 11 through 16, sketch the direction field and a few phase-plane orbits for the given system. Begin by sketching the nullclines.

11. $\dfrac{dx}{dt} = y,$ $\quad \dfrac{dy}{dt} = -x$

12. $\dfrac{dx}{dt} = x - y,$ $\quad \dfrac{dy}{dt} = 1 - x^2 - y^2$

13. $\dfrac{dx}{dt} = 2 - x^2 - y,$ $\quad \dfrac{dy}{dt} = x^2 - y$

14. $\dfrac{dx}{dt} = 2 - x^2 - y,$ $\quad \dfrac{dy}{dt} = y - x^2$

15. $\dfrac{dx}{dt} = xy,$ $\quad \dfrac{dy}{dt} = -xy$

16. $\dfrac{dx}{dt} = -xy,$ $\quad \dfrac{dy}{dt} = xy - y$

17. Use Euler's method with stepsize $h = 0.1$ to compute approximations of $x(0.3)$ and $y(0.3)$, where

$$\frac{dx}{dt} = xy, \ x(0) = 1, \ \text{and} \ \frac{dy}{dt} = x, \ y(0) = 0.$$

18. Rework Problem 17 with Heun's method.

19. Consider the initial-value problem

$$\frac{dx}{dt} = -y, \ x(0) = 1,$$
$$\frac{dy}{dt} = x, \ y(0) = 0.$$

(a) Show that the solution is $x = \cos t$, $y = \sin t$.

(b) Show that one step of Euler's method with stepsize h produces the Taylor series of the solution at $t = h$ up through terms of degree 1.

(c) Show that one step of Heun's method with stepsize h produces the Taylor series of the solution at $t = h$ up through terms of degree 2.

(d) Show that one step of fourth-order Runge-Kutta with stepsize h (see Section 4.5) produces the Taylor series of the solution at $t = h$ up through terms of degree 4.

20. Let $(x(t), y(t))$ be any solution of the system

$$\frac{dx}{dt} = -x - y, \quad \frac{dy}{dt} = x.$$

(a) Show that $x'' + x' + x = 0$.

(b) Show that $y'' + y' + y = 0$.

21. Let $(x(t), y(t))$ be any solution of the system

$$\frac{dx}{dt} = yp(x, y), \quad \frac{dy}{dt} = -xp(x, y),$$

where p is a continuous function. Show that

$$\frac{d}{dt}(x^2 + y^2) = 0.$$

Conclude that $x^2 + y^2$ is constant. What does this say about the phase portrait of the system?

22. Let $(x(t), y(t))$ be any solution of the system

$$\frac{dx}{dt} = x - y, \quad \frac{dy}{dt} = 2x - y.$$

Show that

$$\frac{d}{dt}(2x^2 - 2xy + y^2) = 0.$$

Conclude that $2x^2 - 2xy + y^2$ is constant. How does this help explain the second phase portrait in Figure 4?

23. Let $(x(t), y(t))$ be any solution of the system

$$\frac{dx}{dt} = -x - y, \quad \frac{dy}{dt} = 2x - y.$$

Show that

$$\frac{d}{dt}(2x^2 + y^2) = -2(2x^2 + y^2).$$

Conclude that $2x^2 + y^2 = Ce^{-2t}$ for some constant C. How does this help explain the third phase portrait in Figure 4?

CHAPTER

5

SECOND-ORDER LINEAR EQUATIONS I

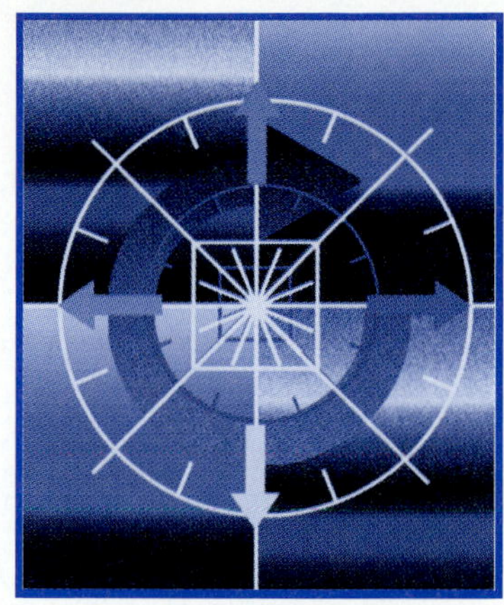

In this chapter we undertake a study of the basic structure of solutions of second-order linear differential equations. However, we will first look at some models of simple vibrations that lead to second-order linear differential equations and then, in the second section, become acquainted with methods for visualizing solutions and numerical approximation. The remainder of the chapter is devoted to general ideas related to the construction and representation of solutions.

5.1 Introduction: Modeling Vibrations

Many mechanical systems are capable of oscillatory motion, or vibrations. The most basic model of such a system is a mass attached to a spring. As you might imagine, whether—and in what manner—such a system vibrates depends upon a combination of factors: the stiffness of the spring, the size of the mass, the amount of friction present, and the nature of any external forces present.

The Basic Spring-mass System Consider a mass m attached to the end of a spring that hangs vertically from a fixed platform as in Figure 1. The first part of the figure shows the spring hanging under its own weight. The middle part of the figure shows the spring with the attached mass at rest (or equilibrium). The elongation of the spring at this equilibrium position is denoted by s. The last part of the figure is a snapshot of the system in motion. The displacement of the mass from equilibrium is denoted by $y(t)$, with the downward direction taken to be positive. To describe the motion of the mass, we intend to set up and solve a differential equation for the displacement $y(t)$. For the time being we will assume that the only forces acting on the mass are its weight and the force exerted by the spring.

First consider the system at equilibrium. The force on the mass due to gravity is its weight mg, where g is the usual gravitational acceleration. This force is positive because we're assuming that downward is the positive direction. Precisely balancing this downward force is the upward force that the spring exerts on the mass. We assume that Hooke's law applies so that the *restorative* force exerted by the spring

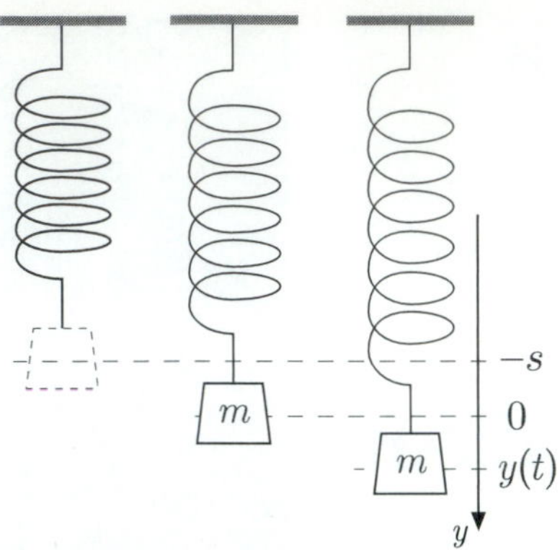

Figure 1

on the mass is given by $-k\,s$, where k is a positive constant representing the stiffness of the spring. This force is negative because it is in the upward direction. Now, because the mass is at equilibrium, the sum of these two forces must be zero; that is,

$$k\,s = m\,g. \tag{1}$$

Now consider the mass at time t, when its displacement from equilibrium is $y(t)$. The spring is stretched an amount $y(t) + s$, and so the restorative force of the spring is $-k\,(y(t) + s)$. The sum of the forces on the mass is therefore $m\,g - k\,(y(t) + s)$. Thus Newton's second law gives

$$m\,y''(t) = m\,g - k\,(y(t) + s).$$

The relationship $k\,s = m\,g$ and a little rearrangement result in

$$m\,y'' + k\,y = 0,$$

in which each term in the equation has units of force. If we divide through by m and set $\omega = \sqrt{k/m}$, then the equation becomes

$$y'' + \omega^2 y = 0. \tag{2}$$

It is easy to see that $\cos\omega\,t$ and $\sin\omega\,t$ are particular solutions. Therefore, because of the linearity and homogeneity of (2), it follows that

$$y(t) = c_1 \cos\omega\,t + c_2 \sin\omega\,t$$

is a solution for any constants c_1 and c_2. Also, the constants c_1 and c_2 can be determined from the initial displacement and the initial velocity of the mass. Thus the motion of the mass is *simple harmonic motion*.

It is interesting to note that (2) also governs the motion of other simple spring-mass systems such as the system depicted in Figure 2. There, $y(t)$ denotes the horizontal displacement from equilibrium, with positive displacement corresponding to stretching of the spring. We assume for the present that there is no friction or any other force acting on the mass other than that provided by the spring. The equation for $y(t)$ is actually simpler to derive in this case, since the weight of the mass does

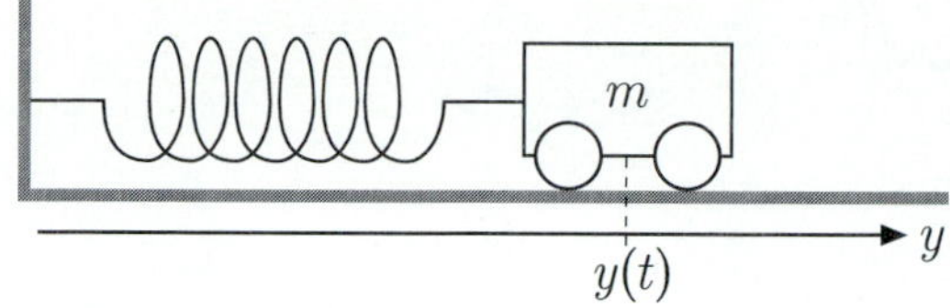

Figure 2

not play a role. We leave it to the reader to set up the differential equation by applying Newton's second law.

Another interesting spring-mass system that leads to equation (2) is one in which the arrangement of Figure 1 is inverted; that is, the mass sits atop the spring rather than hanging from it. (See Problem 9.)

Damping We now consider the effect of including a friction or damping force in either of the systems in Figures 1 and 2. Our assumption will be that such a force is proportional to the velocity of the mass and in a direction opposite that of the motion. Thus the force takes the form of $-r\,y'$, where r is a positive constant. So Newton's second law applied to the systems in Figures 1 and 2 results in

$$m\,y'' = m\,g - k(y+s) - r\,y' \quad \text{and} \quad m\,y'' = k\,y - r\,y',$$

respectively. Simplification and rearrangement of each of these results in

$$m\,y'' + r\,y' + k\,y = 0, \tag{3}$$

in which each term has units of force. Thus, the equation takes the form of a *homogeneous, second-order, linear equation* with *constant coefficients*.

External Forces One way in which an external force may be applied to the mass in a spring-mass system is by moving the platform to which the opposite end of the spring is attached. Imagine that you are holding the top end of a spring and that a mass is attached to the other end. Moving your hand up and down in some way will cause the mass to move as well, but in a way that doesn't necessarily follow closely the motion of your hand.

So suppose that a function $f(t)$ describes the motion of the platform (or your hand) to which the top end of the spring is attached; see Figure 3. Assume that the system is at its equilibrium position when $f(t) = y(t) = 0$ and that downward is the positive direction for $f(t)$ as well as for $y(t)$. The only difference that this makes in the derivation of equation (2) is that at time t the spring is stretched by an amount $y(t)+s-f(t)$, and so the Hooke's law force becomes $-k(y(t)+s-f(t))$. Ignoring damping, Newton's second law now gives

$$m\,y''(t) = m\,g - k(y(t) + s - f(t)).$$

Since $k\,s = m\,g$, the resulting equation for undamped, forced motion is

$$m\,y'' + k\,y = k\,f(t). \tag{4}$$

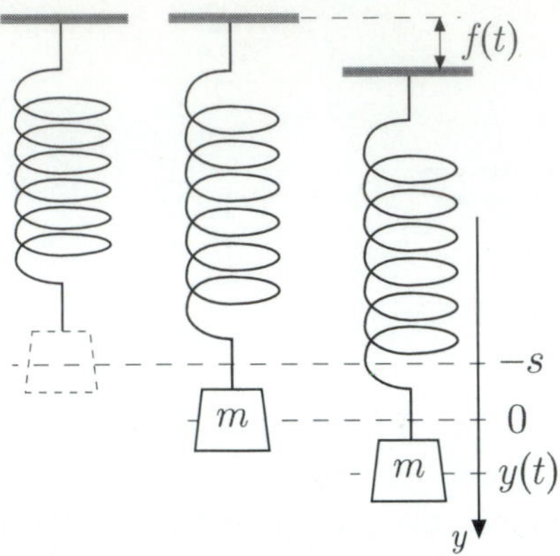

Figure 3

Note that the external *force* resulting from the motion $f(t)$ of the platform is $k\,f(t)$, where k is the spring stiffness constant. If damping is included, then the equation is

$$m\,y'' + ry' + k\,y = kf(t). \tag{5}$$

In either case, the equation takes the form of a *nonhomogeneous*, second-order, linear equation with constant coefficients.

Of course, an external force may be applied to the mass in other ways; for example, by some varying magnetic field. If the external force is described by a function $F(t)$, then the differential equation is

$$m\,y'' + r\,y' + k\,y = F(t).$$

What is important to notice here is that the nonhomogeneous term on the right side of the equation is precisely the applied force, but *only if* the terms on the left side each have units of force as well.

We remark, finally, that the equations discussed so far represent a quite idealized model. In a more realistic setting, it may be that

i. the damping force is some nonlinear function ρ of y and/or y',
ii. the restorative force of the spring is some nonlinear function σ of y and/or y', and
iii. the external force F is a function of both t and y.

Moreover, the mass m need not be constant. Under any of these circumstances, the differential equation of the motion has the general form

$$(m\,y')' = \rho(y, y') + \sigma(y, y') + F(t, y),$$

where the functions ρ and σ represent damping and restorative forces, respectively.

Circuits We saw in Section 2.2.3 that the current in an RC or an RL circuit obeys a first-order linear differential equation. The reader may wish to review that section

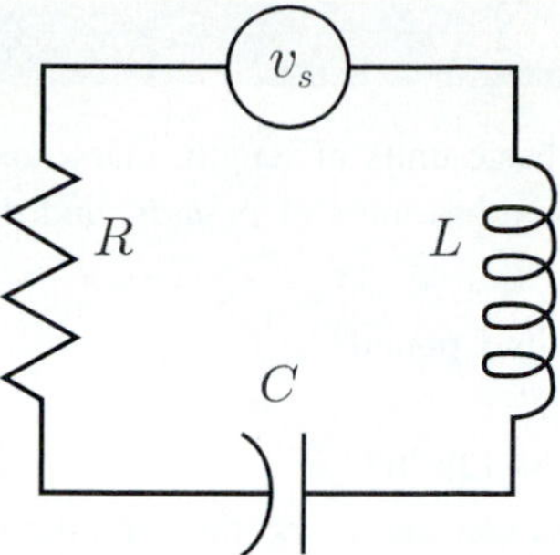

Figure 4

for descriptions of the basic circuit components and statement of Kirchhoff's voltage law.

Interestingly, the current in a simple RLC electrical circuit exhibits behavior that is completely analogous to that of the mechanical spring-mass model. Here, we will look at an RLC circuit with a voltage source, as shown schematically in Figure 4. Application of Kirchhoff's voltage law results in

$$V_L + V_R + V_C - v_s(t) = 0.$$

Using $V_R = IR$ and $V_L = L\,I'$, this becomes

$$L\,I' + R\,I + V_C = v_s(t).$$

Now, since $I = C\,V_C'$, this becomes

$$LV'' + RV' + \frac{1}{C}V = \frac{1}{C}v_s(t), \tag{6}$$

where $V = V_C$ and we have divided through by C. This is a second-order linear differential equation with constant coefficients for the voltage across the capacitor.

If we differentiate with respect to t, multiply by C, and use $CV_C' = I$, then we get

$$L\,I'' + R\,I' + \frac{1}{C}I = v_s'(t), \tag{7}$$

a second-order linear equation with constant coefficients for the current $I(t)$.

Notice that equations (6) and (7) are exactly the same sort of equation as (5), which governs the motion of a damped spring-mass system. Thus both problems have the same type of solutions. In fact, direct analogies can be made between

- the inductance L and the mass m;
- the resistance R and the damping coefficient r;
- the "elastance" $\frac{1}{C}$ and the spring stiffness k.

As in the spring-mass problem, what we have here is an idealized model of a circuit, resulting from the assumption of *linear* device characteristics.

PROBLEMS

1. In the British engineering system of units, the basic units of length, mass, and time are the *foot*, the *slug*, and the *second*, respectively. Force has units of *pounds*, and the acceleration due to gravity is approximately $g = 32.1$ ft/s^2.

   (a) What is the mass of an object that weighs 1 pound?

   (b) How much does a mass of 1 slug weigh?

   (c) What is the mass of a person who weighs 128 lb?

   (d) What are the units of the spring stiffness constant k and the damping coefficient r in this system? (*Hint*: ky and ry' have units of force.)

2. In the *mks* (or SI) system of units, the basic units of length, mass, and time are the *meter*, the *kilogram*, and the *second*, respectively. Force has units of *Newtons*, and the acceleration due to gravity is approximately $g = 9.8$ m/s^2.

   (a) What is the mass of an object that weighs 1 Newton?

   (b) What is the weight (in Newtons) of a 1-kilogram mass?

   (c) What is the mass of an object that weighs 49 Newtons?

   (d) What are the units of the spring stiffness constant k and the damping coefficient r in this system?

3. In the *cgs* system of units, the basic units of length, mass, and time are the *centimeter*, the *gram*, and the *second*, respectively. Force has units of *dynes*, and the acceleration due to gravity is approximately $g = 980$ cm/s^2.

   (a) What is the mass of an object that weighs 1 dyne?

   (b) What is the weight (in dynes) of a 1-gram mass?

   (c) What is the mass of an object that weighs 2940 dynes?

   (d) What are the units of the spring stiffness constant k and the damping coefficient r in this system?

4. **(a)** A 10-lb weight stretches a spring 1 inch. What is the spring stiffness k?

   (b) A 0.3-kg mass stretches a spring 0.06 meters. What is the spring stiffness k?

   (c) An object is attached to a spring with known stiffness $k = 147$ dynes/cm, stretching it 4 cm. What are the weight and mass of the object (in cgs units)?

5. Suppose that in Figure 1, the mass is 2 kg, and it stretches the spring 0.1 m.

   (a) Write down the differential equation governing the undamped, unforced motion.

   (b) If the mass is pulled down an additional 0.2 meters and gently released, what is the appropriate initial value problem for the subsequent motion?

6. An unknown mass stretches the spring in Figure 1 by an amount $s = 0.04$ m.

   (a) Write down the differential equation governing the undamped, unforced motion.

   (b) If the mass is pushed up 0.1 meters above equilibrium and projected downward with an initial velocity of 1 m/s, what is the appropriate initial value problem for the subsequent motion of the mass?

7. Suppose that in Figure 1, the mass is 2 kg, and it stretches the spring 0.1 m. In addition, a damping force of 4 N per unit velocity is present. Write down the differential equation governing the damped, unforced motion.

8. Suppose that for the same system as in Problem 7, the platform moves according to $f(t) = 0.1 \sin \pi t$ for $t \geq 0$.

(a) Write down the differential equation governing the mass's damped, forced motion.

(b) If the mass is at rest at time $t = 0$, what is the appropriate initial value problem for the subsequent motion of the mass?

9. Suppose that the spring-mass system of Figure 1 is upside down, so that the weight of the mass compresses the spring rather than stretching it.

(a) Taking the upward direction to be positive, set up the differential equation for the undamped motion of the mass.

(b) Include into the equation a damping term and a force term caused by motion of the platform.

10. A simple spring-mass system can be thought of as a crude model of your car's suspension system. Suppose that the car has mass m, the stiffness of the suspension is k, and the shock absorbers provide a damping force with coefficient r; in other words, unforced vertical motion of the body of the car obeys $m\,y'' + r\,y' + k\,y = 0$. Now suppose that you're driving on a "washboarded" road whose surface is sinusoidal with amplitude a ft and high spots that are δ feet apart. If your speed is S ft/s, what is an appropriate differential equation for the forced vertical motion of the body of the car?

11. Suppose that for a damped system as in Figure 1, the spring stiffness is $k = 1$ N/m and a damping coefficient is $r = 2$ N·s/m. Suppose also that the mass is a bucket of sand with a hole in the bottom, through which sand runs out at a constant rate of .01 kg/s. The bucket itself has a mass of 1 kg and initially contains 3 kg of sand. Write down the differential equation that governs the damped, unforced motion of the spring up to the time when all the sand has run out. (*Hint*: The term $m\,y''$ is replaced by $(my')'$.)

12. Consider an undamped spring-mass system as in Figure 2 with mass $m = 10$ kg and spring stiffness $k = 50$ N/m. The mass is composed of iron, and the other end of the spring is attached to a fixed magnet that exerts a force on the mass that is inversely proportional to the square of the distance between it and the mass. The natural length of the spring is 1.2 m, and the magnetic force on the mass compresses the spring by .2 m. Letting $y(t)$ denote the displacement from equilibrium, find the differential equation governing the undamped, forced motion of the spring.

13. Consider the arrangement of Figure 2, with a cart attached to a spring with stiffness k. Suppose that the cart has an engine that burns fuel at a constant rate of ρ kg/s and provides a constant thrust of T Newtons in the positive direction. Let the total initial mass of the cart, including fuel, be m_0 kg. With $y(t)$ denoting the rest position of the cart before the engine is started, set up the differential equation for the undamped motion of the cart up until the time when the fuel runs out.

14. **(a)** Set up the differential equation for the undamped motion of the mass in Figure 1, given that the force exerted by the spring is proportional to the cube of the amount that the spring is stretched.

(b) Repeat for the mass in Figure 2.

5.2 State Variables and Numerical Approximation

Consider the second-order linear initial-value problem

$$y'' + p\,y' + q\,y = f, \quad y(t_0) = y_0, \quad y'(t_0) = v_0. \tag{1}$$

As we saw in Section 5.1, the behavior of certain physical systems such as simple spring-mass systems and electrical circuits may be modeled by (1). For any system governed by the differential equation in (1), the values of y and y' at any time t_0 completely describe the state of the system at that time in the sense that those values completely determine the subsequent state of the system for all times t (so long as a unique solution exists). For this reason, we refer to y and y' as *state variables* for the differential equation and to the pair $(y(t), y'(t))$ as the *state* of the solution at time t. For a spring-mass system the position and velocity of the mass determine the state of the system at any time. The state of an RLC circuit is determined at any time t by, for example, the voltage across the resistor and the current through the resistor at that instant.

With this notion of state in mind, it makes sense to visualize the solution of (1) as a curve in $\mathbb{R}^3$ with parametric description $(t, y(t), y'(t))$. We will refer to such a curve as a **time-state trajectory**. We can also visualize the solution of (1) as a curve in the plane with parametric description $(y(t), y'(t))$, which we call a **phase-plane orbit**.

Example 1

The initial-value problem

$$y'' + 2\,y' + 26\,y = 0, \quad y(0) = 5, \quad y'(0) = 0$$

could arise in the case of a damped, unforced spring-mass system. Figure 1 shows the three-dimensional time-state trajectory of the solution in *time-state space* (i.e., (t, y, y')-space) along with projections onto each of the (t, y)-, (t, y')-, and (y, y')-planes. The curve in the (y, y')-plane is the phase-plane orbit. ■

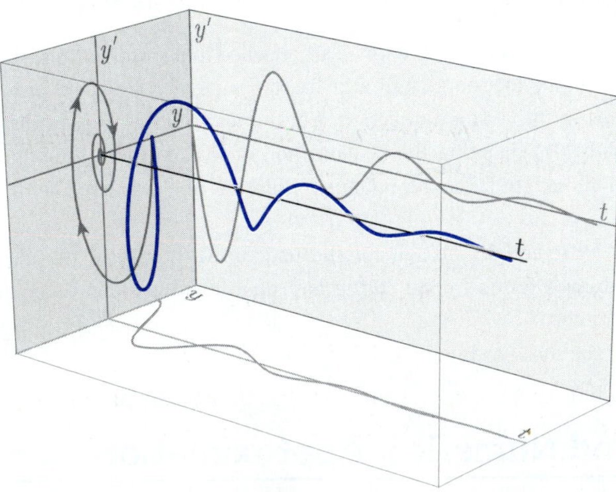

Figure 1

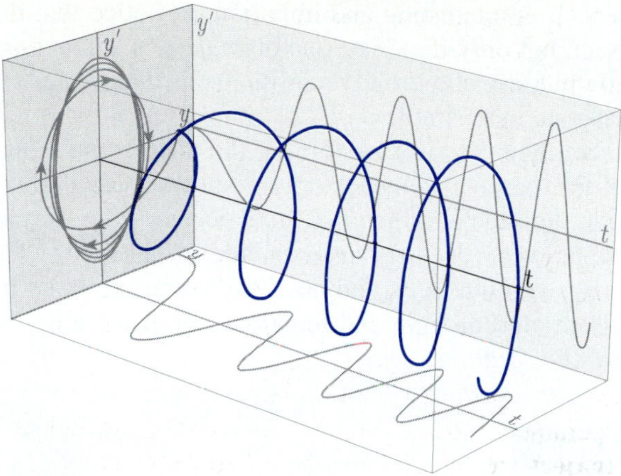

Figure 2

Example 2

Figure 2, which shows the solution of

$$y'' + t\,y = 0, \quad y(0) = 10, \quad y'(0) = 0,$$

is analogous to the corresponding figure in Example 1. Notice here that the phase-plane orbit (in the (y, y')-plane) crosses itself many times, while the time-state trajectory passes through a given point at most once. ■

Existence and Uniqueness The following theorem is concerned with the existence and uniqueness of solutions to initial value problems of the form in (1). It tells us that the state at any time t_0 in I uniquely determines the state at all times t in I.

Theorem 1

Suppose that the functions p, q, and f are each continuous on an open interval I containing the initial point t_0. Then for any y_0 and m_0, (1) has a unique solution on I; that is, there is a unique time-state trajectory through any point (t_0, y_0, v_0) with t_0 in I. ▲

This theorem is really just an application of a two-equation version of Theorem 1 from Section 4.2 to the equivalent first-order system

$$y' = v, \quad v' = -p\,v - q\,y + f.$$

Notice that on the right side of each of these equations the partial derivatives with respect to y are 0 and $-q$, and those with respect to v are 1 and $-p$. These are continuous by assumption.

Note that uniqueness implies that *distinct time-state trajectories of (1) cannot intersect at any time t in I*. However, uniqueness does not imply that phase-plane orbits must not intersect. It is interesting and important to notice that distinct phase-plane orbits *can* intersect, but only if at least one of p, q, f is a nonconstant function of t. Moreover, a single phase-plane orbit can cross itself; that is, the same state can occur at different times—as in Example 2.

When p, q, f are constants (as in Example 1), the differential equation in (1) is called *autonomous*. In this case, y'' depends only upon the state $(y(t), y'(t))$ at any given time and not upon the time at which that state occurs. Thus, any two solutions with the same initial state $(y(t_0), y'(t_0))$ must coincide for all t in I. That is, no two distinct phase-plane orbits can intersect, and no phase-plane orbit can intersect itself without forming a smooth closed curve and remaining on it for all t. Such a closed curve represents a *periodic solution*.

Example 3

Figure 3 shows the helical time-state trajectory of the solution

$$y = \cos 2t + \tfrac{1}{2}\sin 2t$$

of the problem

$$\begin{aligned} y'' + 4y &= 0, \\ y(0) &= 1, \\ y'(0) &= 1. \end{aligned}$$

Notice that the phase-plane orbit (in the (y, y')-plane) is a closed curve, reflecting the fact that the solution is periodic. ■

Numerical Approximation Recall from Sections 2.2 and 4.5 that there are various numerical procedures such as Euler's method, Heun's method, and fourth-order Runge-Kutta for computing approximate solutions to first-order equations

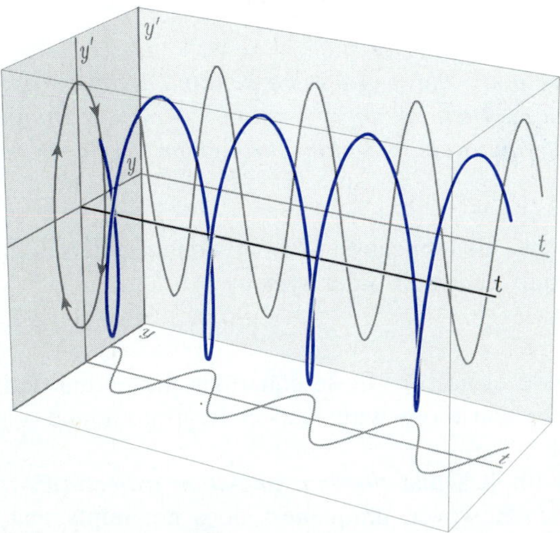

Figure 3

$y' = f(t, y)$. Any such method can also be applied to the second-order problem (1). The key to this is that (as mentioned after Theorem 1) the second-order differential equation in (1) may be reformulated as a pair of coupled first-order equations for the state variables y and y'. That is, (1) may be converted to a *first-order system*.

To do this, let's first set $v = y'$. The equation for y then is simply $y' = v$, and the equation for v comes from noting that

$$v' = y'' = f - p\,y' - q\,y = f - p\,v - q\,y.$$

Thus the resulting system may be written in vector form as

$$\begin{pmatrix} y \\ v \end{pmatrix}' = \begin{pmatrix} v \\ f - p\,v - q\,y \end{pmatrix}.$$

To simplify notation it is advantageous to introduce the vector quantities

$$\mathbf{u} = \begin{pmatrix} y \\ v \end{pmatrix} \quad \text{and} \quad F(t, \mathbf{u}) = \begin{pmatrix} v \\ f(t) - p(t)v - q(t)y \end{pmatrix}.$$

The initial-value problem (1) now may be expressed simply as

$$\mathbf{u}' = F(t, \mathbf{u}), \quad \mathbf{u}(t_0) = \begin{pmatrix} y_0 \\ v_0 \end{pmatrix}.$$

Note that we could refer to $\mathbf{u}$ as a *state vector*, since it contains the two state variables y and $v = y'$.

Euler's Method With the notation

$$\mathbf{u}_k = \begin{pmatrix} y_k \\ v_k \end{pmatrix},$$

where $t_k = t_0 + kh$ and y_k, v_k are approximations to $y(t_k)$, $v(t_k)$, Euler's method for calculating $\mathbf{u}_{k+1}$ from $\mathbf{u}_k$ with stepsize h takes the form

$$\mathbf{u}_{k+1} = \mathbf{u}_k + h\,F(t_k, \mathbf{u}_k), \quad k = 0, 1, 2, \ldots. \tag{2}$$

Example 4

Consider the initial-value problem

$$y'' + y' + t\,y = \cos t, \quad y(0) = 1, y'(0) = 0.$$

Setting $v = y'$, we find that $v' = y'' = \cos t - v - t\,y$, and so the equation may be expressed as the first-order system

$$\begin{pmatrix} y \\ v \end{pmatrix}' = \begin{pmatrix} v \\ \cos t - v - ty \end{pmatrix}.$$

This is equivalent to

$$\mathbf{u}' = F(t, \mathbf{u}), \quad \mathbf{u}(0) = \begin{pmatrix} 1 \\ 0 \end{pmatrix},$$

where

$$\mathbf{u} = \begin{pmatrix} y \\ v \end{pmatrix} \quad \text{and} \quad F(t, \mathbf{u}) = \begin{pmatrix} v \\ \cos t - v - t\,y \end{pmatrix}.$$

Two steps of Euler's method (2) with $h = 0.1$ are

$$\mathbf{u}(0.1) \approx \mathbf{u}_1 = \begin{pmatrix} 1 \\ 0 \end{pmatrix} + 0.1 \begin{pmatrix} 0 \\ \cos 0 - 0 - (0)(1) \end{pmatrix} = \begin{pmatrix} 1 \\ 0.1 \end{pmatrix},$$

$$\mathbf{u}(0.2) \approx \mathbf{u}_2 = \begin{pmatrix} 1 \\ 0.1 \end{pmatrix} + 0.1 \begin{pmatrix} 0.1 \\ \cos 0.1 - 0.1 - (0.1)(1) \end{pmatrix} = \begin{pmatrix} 1.01 \\ 0.1795 \end{pmatrix}. \quad \blacksquare$$

Heun's Method Heun's method for calculating $\mathbf{u}_{k+1}$ from $\mathbf{u}_k$ with stepsize h takes the form

$$\begin{aligned} \mathbf{k}_1 &= F(t_k, \mathbf{u}_k), \\ \mathbf{k}_2 &= F(t_{k+1}, \mathbf{u}_k + h\mathbf{k}_1) \quad k = 0, 1, 2, \ldots . \\ \mathbf{u}_{k+1} &= \mathbf{u}_k + \frac{h}{2}(\mathbf{k}_1 + \mathbf{k}_2) \end{aligned} \tag{3}$$

Example 5

For the same initial-value problem as in Example 4, one step of Heun's method goes as follows:

$$\begin{aligned} \mathbf{k}_1 &= F\left(0, \begin{pmatrix} 1 \\ 0 \end{pmatrix}\right) = \begin{pmatrix} 0 \\ \cos 0 - 0 - (0)(1) \end{pmatrix} = \begin{pmatrix} 0 \\ 1 \end{pmatrix} \\ \mathbf{k}_2 &= F\left(0.1, \begin{pmatrix} 1 \\ 0 \end{pmatrix} + 0.1 \begin{pmatrix} 0 \\ 1 \end{pmatrix}\right) = F\left(0.1, \begin{pmatrix} 1 \\ 0.1 \end{pmatrix}\right) \\ &= \begin{pmatrix} 0.1 \\ \cos 0.1 - 0.1 - (0.1)(1) \end{pmatrix} = \begin{pmatrix} 0.1 \\ 0.7950 \end{pmatrix} \\ \mathbf{u}(0.1) \approx \mathbf{u}_1 &= \begin{pmatrix} 1 \\ 0 \end{pmatrix} + 0.05 \left(\begin{pmatrix} 0 \\ 1 \end{pmatrix} + \begin{pmatrix} 0.1 \\ 0.7950 \end{pmatrix} \right) = \begin{pmatrix} 1.005 \\ 0.08975 \end{pmatrix} \quad \blacksquare \end{aligned}$$

The main point here is that numerical approximation procedures for first-order equations can be applied to second-order equations by converting the equation to a first-order system and viewing the system as a single equation involving vector quantities.

Nonlinear Problems All of the ideas presented so far in this section apply as well to nonlinear second-order initial-value problems

$$y'' = g(t, y, y'), \quad y(t_0) = y_0,\ y'(t_0) = v_0. \tag{4}$$

Such a problem can be rewritten as a first-order problem

$$\mathbf{u}' = F(t, \mathbf{u}), \quad \mathbf{u}(t_0) = \begin{pmatrix} y_0 \\ v_0 \end{pmatrix},$$

where

$$\mathbf{u} = \begin{pmatrix} y \\ v \end{pmatrix} \text{ and } F(t, \mathbf{u}) = \begin{pmatrix} v \\ g(t, y, v) \end{pmatrix}.$$

Euler's method and Heun's method are then exactly the same as in (2) and (3), respectively.

Time-state trajectories and phase-plane orbits for solutions of (4) may be considered in exactly the same way as for linear problems. The only modification that needs to be made in the preceding remarks concerns the conditions that guarantee existence and uniqueness of solutions. An existence and uniqueness theorem in terms of time-state trajectories is the following.

Theorem 2

Suppose that g, $\frac{\partial g}{\partial y}$, and $\frac{\partial g}{\partial v}$ are each continuous on a closed box

$$G = \{(t, y, v) | \quad a \le t \le b, c \le y \le d, h \le v \le k\}.$$

If (t_0, y_0, v_0) is contained in the interior of G, then there is a unique time-state trajectory of (4) through (t_0, y_0, v_0), which can be uniquely continued for $t < t_0$ and $t > t_0$ at least until $(t, y(t), y'(t))$ reaches a boundary point of G. ▲

This theorem is an extension of Theorem 1 in Section 5.2. The reader is encouraged to review the discussion surrounding that theorem. Note also that Theorem 1 of this section follows as a special case of Theorem 2.

Example 6.

Consider the second-order equation

$$y'' + y^3 y' = \sin\, t.$$

In the preceding theorem we have $g(t, y, v) = \sin t - y^3 v$, which clearly satisfies the hypotheses for any open and connected G in $\mathbb{R}^3$. Thus the theorem guarantees that the equation has a unique time-state trajectory through any point (t_0, y_0, v_0). To compute a numerical approximation we would first set $v = y'$ and $v' = y''$ and convert to the first-order system

$$\begin{pmatrix} y \\ v \end{pmatrix}' = \begin{pmatrix} v \\ \sin t - y^3 v \end{pmatrix}.$$

Then Euler's method (2) or Heun's method (3) may be carried out with

$$\mathbf{u} = \begin{pmatrix} y \\ v \end{pmatrix} \quad \text{and} \quad F(t, \mathbf{u}) = \begin{pmatrix} v \\ \sin t - y^3 v \end{pmatrix}.$$ ■

PROBLEMS

1. **(a)** Verify that the solution of the initial-value problem in Example 1 is

$$y = e^{-t}(5\cos(5t) + \sin(5t)).$$

(b) Show that the phase-plane orbit in Figure 1 satisfies

$$\left(\frac{26y + y'}{130}\right)^2 + \left(\frac{y'}{26}\right)^2 = e^{-2t}.$$

How does this explain the type of trajectory we see?

Convert each of the linear equations in Problems 2 through 4 to a first-order system.

2. $y'' + e^{-t}y = \sin t$

3. $y'' + ty' + t^2y = 0$

4. $y''' + y'' + y' + y = 0$

Convert each of the nonlinear equations in Problems 5 through 7 to a first-order system.

5. $y'' = \sin y$

6. $y'' + yy' = \cos t$

7. $y'' + (y')^3 = e^{-y}$

The first-order system that arises from a linear second-order equation can always be expressed in the form

$$\mathbf{u}' = A\mathbf{u} + \mathbf{b},$$

where $\mathbf{u} = \begin{pmatrix} y \\ v \end{pmatrix}$ with $v = y'$, A is a 2×2 matrix, and $\mathbf{b}$ is a column vector. If the equation is not autonomous, then the entries in A and $\mathbf{b}$ may depend on t. Express each of the equations in Problems 8 through 11 in this form.

8. $y'' + y' + y = 0$

9. $y'' + ty' + 2y = t$

10. $y'' + ty' + t^2y = t^3$

11. **(a)** Consider a general, linear, second-order equation: $y'' + py' + qy = f$. Write the equation as a first-order system $\mathbf{u}' = A\mathbf{u} + \mathbf{b}$.
(b) Do the same for the general third-order equation: $y''' + py'' + qy' + ry = f$.
(c) Comment on the general form of the matrix A and the vector $\mathbf{b}$ obtained in this way from a linear nth-order equation.

Use a calculator to do the following.

12. Perform three steps of Euler's method, with $h = 0.1$, on the problem

$$y'' + (t + 1)y = \cos t, \quad y(0) = 1, \, y'(0) = 0.$$

13. Perform two steps of Heun's method, with $h = 0.1$, on the problem

$$y'' + y' + y = 0, \quad y(0) = 1, \, y'(0) = 1.$$

14. The initial-value problem in Example 3 is of the general form

$$my'' + ky = 0, \quad y(0) = y_0, \, y'(0) = v_0,$$

which arises in the case of an undamped, unforced spring-mass system. Use the integrating factor y' to show that

$$\frac{1}{2}mv^2 + \frac{1}{2}ky^2 = C,$$

where $v = y'$. (Thus the phase-plane orbits are concentric ellipses centered at the origin.) Express the constant of integration C in terms of initial values to obtain

$$\frac{1}{2}mv^2 + \frac{1}{2}ky^2 = \frac{1}{2}mv_0^2 + \frac{1}{2}ky_0^2,$$

which may be interpreted as a conservation of energy statement: *The sum of kinetic and potential energy is constant.*

5.3 Operators and Linearity

The terms *function, transformation, mapping*, and *operator* are all more or less synonymous in mathematics. Each simply means a rule that "maps" elements of one set to unique elements of another set. Sometimes tradition calls for the use of one of these terms over the others within specific areas of mathematics. For example, in linear algebra we study linear *transformations* from $\mathbb{R}^n$ to $\mathbb{R}^m$ and learn that these may always be represented by matrix multiplication. In differential equations we typically use the term *operator* to signify a function that acts on functions. A couple of familiar examples are differentiation and definite integration; that is, we can define an operator $\mathcal{D}$ by

$$\mathcal{D}u = u',$$

acting on the differentiable functions on some interval, and an operator $\mathcal{R}$ by

$$\mathcal{R}u = \int_a^b u(x)dx,$$

acting on the $\mathcal{R}$iemann integrable functions on the interval $[a, b]$. Note that for each differentiable u, $\mathcal{D}u$ is another function, while for each integrable u, $\mathcal{R}u$ is a real number. That is, $\mathcal{D}$ maps functions to functions, and $\mathcal{R}$ maps functions to numbers.

Another important operator is the identity operator $\mathcal{I}$, defined by

$$\mathcal{I}u = u.$$

For the most part, we are concerned in differential equations with operators that map functions to functions, and in particular, operators that involve differentiation.

Example 1

Define the operator $\mathcal{T}$ by

$$\mathcal{T}u = u'' - 3u' + u,$$

acting on the twice differentiable functions on $(-\infty, \infty)$. Note the following actions of $\mathcal{T}$ applied to some simple functions:

$$\mathcal{T}\sin t = -3\cos t, \quad \mathcal{T}(t^3 + t) = t^3 - 9t^2 + 7t - 3, \quad \mathcal{T}e^{2t} = -e^{2t}. \quad \blacksquare$$

Example 2

Define $\mathcal{T}$ by

$$\mathcal{T}u = uu',$$

acting on the differentiable functions on $(-\infty, \infty)$. Note the following actions of $\mathcal{T}$ applied to some simple functions:

$$\mathcal{T}\sin t = \sin t \cos t, \quad \mathcal{T}(t^3 + t) = (t^3 + t)(3t^2 + 1), \quad \mathcal{T}e^{2t} = 2e^{4t}.$$ ■

We define addition, scalar multiplication, and composition of operators in much the same way as is done for ordinary functions. Given an operator $\mathcal{T}$ and a real number (scalar) k, we define the operator $k\mathcal{T}$ by

$$(k\mathcal{T})u = k(\mathcal{T}u).$$

Given two operators $\mathcal{S}$ and $\mathcal{T}$ (acting on the same set of functions), we define $\mathcal{S} \pm \mathcal{T}$ by

$$(\mathcal{S} + \mathcal{T})u = \mathcal{S}u + \mathcal{T}u \text{ and } (\mathcal{S} - \mathcal{T})u = \mathcal{S}u + (-1)\mathcal{T}u,$$

and the composition (or "product") $\mathcal{S}\mathcal{T}$ by

$$\mathcal{S}\mathcal{T}u = \mathcal{S}(\mathcal{T}u).$$

We also define powers $\mathcal{T}^2$, $\mathcal{T}^3, \ldots$ to denote repeated application of $\mathcal{T}$:

$$\mathcal{T}^2u = \mathcal{T}(\mathcal{T}u), \quad \mathcal{T}^3u = \mathcal{T}(\mathcal{T}(\mathcal{T}u)), \text{ and so on.}$$

In the case of the operator $\mathcal{D}$ defined by $\mathcal{D}u = u'$, these powers are the familiar higher-order derivatives:

$$\mathcal{D}^2u = u'', \quad \mathcal{D}^3u = u''', \quad \text{and so on.}$$

The zeroth power of any operator is defined by $\mathcal{T}^0 = \mathcal{I}$.

We will be particularly interested in polynomial expressions in $\mathcal{D}$ with coefficients that are constants or other functions of the independent variable; for example,

$$(\mathcal{D}^2 + 3t^2\mathcal{D})u = u'' + 3t^2u', \quad (\mathcal{D}^2 - \mathcal{D} - 2\mathcal{I})u = u'' - u' - 2u.$$

When coefficients are constants, it makes sense to factor polynomial expressions in $\mathcal{D}$ in the natural way. For example, literal application of $\mathcal{D}(\mathcal{D} + 3\mathcal{I})$ to u produces

$$\mathcal{D}(\mathcal{D} + 3\mathcal{I})u = \mathcal{D}(u' + 3u) = u'' + 3u',$$

and this is precisely the same as $(\mathcal{D}^2 + 3\mathcal{D})u$. Similarly, you should convince yourself that $(\mathcal{D}+\mathcal{I})(\mathcal{D}-2\mathcal{I})u$ gives precisely the same expression as $(\mathcal{D}^2-\mathcal{D}-2\mathcal{I})u$. Thus, statements about operators such as

$$\mathcal{D}(\mathcal{D} + 3\mathcal{I}) = \mathcal{D}^2 + 3\mathcal{D}, \quad (\mathcal{D} + \mathcal{I})(\mathcal{D} - 2\mathcal{I}) = \mathcal{D}^2 - \mathcal{D} - 2\mathcal{I}$$

are meaningful in the sense that *two operators are equal if and only if they do the same thing to any given function*. The gist of this is that polynomials in $\mathcal{D}$ can be manipulated in much the same way as if $\mathcal{D}$ represented a real number. However, one must be careful when "multiplying" expressions with nonconstant coefficients. To emphasize this point, we leave it to the reader to verify that

$$(\mathcal{D}-2\mathcal{I})(\mathcal{D}+2\mathcal{I})=\mathcal{D}^2-4\mathcal{I} \quad \text{but} \quad (\mathcal{D}-t\mathcal{I})(\mathcal{D}+t\mathcal{I})=\mathcal{D}^2-(1+t^2)\mathcal{I}.$$

Linearity The concept of linearity is fundamental in many areas of mathematics. The following definition applies to all types of operators, including functions from $\mathbb{R}$ to $\mathbb{R}$ or from $\mathbb{R}^n$ to $\mathbb{R}^m$, as well as the differential and integral operators discussed previously.

Definition An operator $\mathcal{T}$ is said to be a **linear operator** if

$$\text{(i) } \mathcal{T}(cu)=c\,\mathcal{T}u \quad \text{and} \quad \text{(ii) } \mathcal{T}(u+v)=\mathcal{T}u+\mathcal{T}v$$

for any real number c and any u and v for which $\mathcal{T}u$ and $\mathcal{T}v$ are defined. (Implied here is that $\mathcal{T}(cu)$ and $\mathcal{T}(u+v)$ must be defined whenever $\mathcal{T}u$ and $\mathcal{T}v$ are defined; that is, the domain of $\mathcal{T}$ is a *vector space*.) ◆

Basic facts from calculus tell us that the differentiation and integration operators $\mathcal{D}$ and $\mathcal{R}$ defined previously are linear. (Why?)

Example 3

To show that the operator $\mathcal{T}$ defined by $\mathcal{T}u=uu'$ is *not* linear, it suffices to observe that

$$\mathcal{T}(cu)=(cu)(cu)'=c^2uu'=c^2\mathcal{T}u\neq c\,\mathcal{T}u,$$

since linearity requires that *both* conditions (i) and (ii) be satisfied. It happens that (ii) also fails for this operator. ■

Example 4

To show that the operator $\mathcal{T}=\mathcal{D}^2-t^3\mathcal{D}+3\mathcal{I}$ is linear, we check each of the requirements as follows:

$$\begin{aligned}
\mathcal{T}(cu) &= (cu)''-t^3(cu)'+3(cu)\\
&= cu''-t^3cu'+3cu\\
&= c(u''-t^3u'+3u)\\
&= c\,\mathcal{T}u;\\
\mathcal{T}(u+v) &= (u+v)''-t^3(u+v)'+3(u+v)\\
&= u''-t^3u'+3u+v''-t^3v'+3v\\
&= \mathcal{T}u+\mathcal{T}v.
\end{aligned}$$

Therefore, $\mathcal{T}$ is linear. Note that the nonlinear coefficient t^3 does not prevent the operator from being linear. ■

In the same manner as in Example 4, the following theorem can be proven.

Theorem 1

The operator $P(\mathcal{D}) = p_n\mathcal{D}^n + p_{n-1}\mathcal{D}^{n-1} + \cdots + p_1\mathcal{D} + p_0\mathcal{I}$, *where* $p_n, p_{n-1}, \ldots, p_0$ *are functions of the independent variable* t, *is a linear operator.* ▲

Linear Differential Equations Combining our definition of a linear differential equation in Section 1.5 with the notations of the current section, we can say that linear differential equations are of the form

$$\mathcal{T}y = f, \tag{1}$$

where $\mathcal{T}$ is a polynomial in $\mathcal{D}$ (with coefficients that may depend on the independent variable); that is,

$$\mathcal{T} = P(\mathcal{D}) = p_n\mathcal{D}^n + p_{n-1}\mathcal{D}^{n-1} + \cdots + p_1\mathcal{D} + p_0\mathcal{I}.$$

Note that such an equation is *homogeneous* if $f = 0$.

A major consequence of the linearity of an operator $\mathcal{T}$ is the following **superposition principle**:

Theorem 2

Let $\mathcal{T}$ *be a linear operator. If* $\mathcal{T}u = f$ *and* $\mathcal{T}v = g$, *then* $w = c_1u + c_2v$ *satisfies* $\mathcal{T}w = c_1 f + c_2 g$ *for any constants* c_1 *and* c_2. ▲

The proof of this theorem is simple exploitation of the definition of linearity. If $\mathcal{T}$ is linear and $\mathcal{T}u = f$, $\mathcal{T}v = g$, and $w = c_1u + c_2v$, then

$$\mathcal{T}w = \mathcal{T}(c_1u + c_2v) = \mathcal{T}(c_1u) + \mathcal{T}(c_2v) = c_1\mathcal{T}u + c_2\mathcal{T}v = c_1 f + c_2 g.$$

Example 5

Suppose we wish to find a particular solution of the equation

$$y' + y = 3e^t - 5\cos t.$$

Theorem 2 tells us that we can accomplish this by first finding u and v so that

$$u' + u = e^t \quad \text{and} \quad v' + v = \cos\, t,$$

and then writing $y = 3u - 5v$. Substituting $u = ae^t$ and $v = b\cos t + c\sin t$ leads us to $u = \frac{1}{2}e^t$ and $v = \frac{1}{2}(\cos t + \sin t)$, and thus $y = \frac{1}{2}(3e^t - 5\cos t - 5\sin t)$. ■

We will list now a few corollaries to Theorem 2 applied to the problem of solving equation (1) where $\mathcal{T}$ is a linear operator. The proofs are left as exercises.

COROLLARY 1 *If* u *is a solution of the homogeneous equation* $\mathcal{T}y = 0$, *then so is* cu *for any constant* c.

COROLLARY 2 *If* $u_1, u_2, \ldots, u_n$ *are solutions of the homogeneous equation* $\mathcal{T}y = 0$, *then so is* $c_1u_1 + c_2u_2 + \cdots + c_nu_n$ *for any constants* $c_1, c_2, \ldots, c_n$.

COROLLARY 3 *If $u_1, u_2, \ldots, u_n$ are solutions of the homogeneous equation $\mathcal{T}y = 0$ and v is any solution of $\mathcal{T}y = f$, then $c_1u_1 + c_2u_2 + \cdots + c_nu_n + v$ is a solution of $\mathcal{T}y = f$ for any constants $c_1, c_2, \ldots, c_n$.*

COROLLARY 4 *If u and v are any two solutions of $\mathcal{T}y = f$, then $u - v$ is a solution of $\mathcal{T}y = 0$.*

The expression $c_1u_1 + c_2u_2 + \cdots + c_nu_n$ that appears in Corollaries 2 and 3 is called a **linear combination** of the functions $u_1, u_2, \ldots, u_n$. Thus Corollary 2 says that *any linear combination of solutions of a homogeneous linear equation is also a solution of that equation.* (Students who have studied linear algebra should recognize that this essentially says that the solutions of a homogeneous linear equation form a *vector space.*)

PROBLEMS

In Problems 1 through 4,

(a) prove whether or not the operator $\mathcal{T}$ is linear;

(b) compute $\mathcal{T}u$ where $u(t) = e^{-t}$;

(c) compute $\mathcal{T}u$ where $u(t) = 2 - t^2$.

1. $\mathcal{T}u = u' + 1$

2. $\mathcal{T}u = u'' + 3u'$

3. $\mathcal{T}u = u' - t^2u$

4. $\mathcal{T}u = u' + tu^2$

Each operator $\mathcal{T}$ in Problems 5 through 8 acts on functions defined on the interval $[0, 1]$. Prove whether or not each operator $\mathcal{T}$ is linear. Also state whether $\mathcal{T}$ maps functions to functions or maps functions to numbers.

5. $\mathcal{T}u = \int_0^1 t^2u(t)dt$

6. $\mathcal{T}u = u - \int_0^1 u(t)dt$

7. $\mathcal{T}u = \dfrac{u(0) + u(1)}{2}$

8. $\mathcal{T}u = u(0)t$

9. Show that $(\mathcal{D} + 1)(\mathcal{D} + 2)u = (\mathcal{D}^2 + 3\mathcal{D} + 2)u$ for any twice differentiable function u.

10. Define $\mathcal{T}$ by $\mathcal{T}u = tu'$ and $\mathcal{S}$ by $\mathcal{S}u = u + u'$. Show that $\mathcal{S}\mathcal{T} \neq \mathcal{T}\mathcal{S}$.

11. **(a)** Show that if a and b are constants, then $(\mathcal{D} + a\mathcal{I})(\mathcal{D} + b\mathcal{I}) = (\mathcal{D} + b\mathcal{I})(\mathcal{D} + a\mathcal{I})$.

(b) Show that $(\mathcal{D} + a\mathcal{I})(\mathcal{D} + b\mathcal{I})$ may not be equal to $(\mathcal{D} + b\mathcal{I})(\mathcal{D} + a\mathcal{I})$ when a and b are not constants.

(c) What is the correct "expansion" of $(\mathcal{D} + a\mathcal{I})(\mathcal{D} + b\mathcal{I})$ if a and b are (nonconstant) functions of t?

12. In this problem assume that k, a, and b are constants.

(a) Show that $\mathcal{D}(e^{kt}u) = e^{kt}(\mathcal{D} + k\mathcal{I})u$.

(b) Show that $(\mathcal{D} + a\mathcal{I})(e^{kt}u) = e^{kt}(\mathcal{D} + (k + a)\mathcal{I})u$.

(c) Show that $(\mathcal{D} + a\mathcal{I})(\mathcal{D} + b\mathcal{I})(e^{kt}u) = e^{kt}(\mathcal{D} + (k + a)\mathcal{I})(\mathcal{D} + (k + b)\mathcal{I})u$.

13. **(a)** Show by mathematical induction that $\mathcal{D}^n(e^{kt}u) = e^{kt}(\mathcal{D} + k\mathcal{I})^nu$ for all $n = 1, 2, 3, \ldots$.

(b) Show that $P(\mathcal{D})(e^{kt}u) = e^{kt}P(\mathcal{D}+k\mathcal{I})u$, where $P(\mathcal{D})$ is any polynomial in $\mathcal{D}$ (with coefficients that can depend on t).

14. **(a–d)** Prove Corollaries 1–4.

15. Let $\mathcal{T} = \mathcal{D}^2 - a^2\mathcal{I}$, where a is constant.

(a) Check that e^{at} and e^{-at} are particular solutions of $\mathcal{T}y = 0$.

(b) Check that $y = c_1e^{at} + c_2e^{-at}$ satisfies $\mathcal{T}y = 0$ for any constants c_1, c_2.

(c) Check that $y = c_1e^{at} + c_2e^{-at} + \frac{e^{bt}}{b^2-a^2}$ satisfies $\mathcal{T}y = e^{bt}$ for any constants c_1, c_2, provided that $b \neq \pm a$.

16. Let $\mathcal{T} = \mathcal{D}^2 + a^2\mathcal{I}$, where a is constant.

(a) Check that $\sin at$ and $\cos at$ are particular solutions of $\mathcal{T}y = 0$.

(b) Check that $y = c_1 \sin at + c_2 \cos at$ satisfies $\mathcal{T}y = 0$ for any constants c_1, c_2.

(c) Check that $y = c_1 \sin at + c_2 \cos at + \frac{\sin bt}{a^2-b^2}$ satisfies $\mathcal{T}y = \sin bt$ for any constants c_1, c_2, provided that $b \neq \pm a$.

17. What basic theorem from calculus does Corollary 4 state when $\mathcal{T} = \mathcal{D}$?

18. Suppose that u and v are any two solutions of $y' - y\tan t = \frac{1}{t^2+1}$. Show that $u(t) = v(t) + C\sec t$ for some constant C.

19. Suppose that u and v are the solutions of the initial-value problems

$$y' - \frac{2}{2+t}y = \cos t, \quad y(0) = 1, \quad \text{and} \quad y' - \frac{2}{2+t}y = \cos t, \quad y(0) = 0,$$

respectively. Show that $u(t) = v(t) + \frac{1}{4}(2+t)^2$.

5.4 Solutions and Linear Independence

Recall from Section 1.3 that a second-order linear differential equation is one that may be expressed in the form

$$p_2y'' + p_1y' + p_0y = f,$$

where p_2, p_1, p_0, and f are given functions of the independent variable, which we will assume is t. We assume that p_2, p_1, and p_0 are continuous on some interval I and that $p_2(t) \neq 0$ for all t in I, so that we may divide by p_2 to obtain an equation of the form

$$y'' + py' + qy = f, \tag{1}$$

where p and q are continuous on I. Note that this equation may be written in operator form as

$$\mathcal{T}y = f,$$

where the operator $\mathcal{T}$ is defined by $\mathcal{T}u = u'' + pu' + qu$ for twice differentiable functions u. Note also that T may be expressed as

$$\mathcal{T} = \mathcal{D}^2 + p\mathcal{D} + q\mathcal{I}$$

and that $\mathcal{T}$ is a linear operator by Theorem 1 in Section 5.3. Thus superposition tells us the following important facts:

i. Let $\tilde{y}$ be any particular solution of (1). Then $y + \tilde{y}$ is also a solution of (1) if y is any solution of the corresponding homogeneous equation

$$y'' + py' + qy = 0. \tag{2}$$

ii. If u and v are any two solutions of (2), then $c_1u + c_2v$ is also a solution for any pair of constants c_1 and c_2.

Linear Independence We saw in Chapter 2 that for homogeneous first-order linear equations, every solution could be expressed as a constant multiple of a single particular solution. For instance, a particular solution of $y' + y \sin t = 0$ is $y = e^{\cos t}$, and the general solution is given by $y = Ce^{\cos t}$. So it may seem plausible that for homogeneous second-order linear equations, every solution should be expressible using two particular solutions, say u and v, as $y = c_1u + c_2v$, where c_1 and c_2 are constants; that is, as a *linear combination* of u and v. This is indeed true, but we must impose a fairly simple condition upon u and v. This condition is *linear independence.*

Definition Two functions u and v are **linearly dependent** on an interval I if there are constants c_1 and c_2, not both zero, so that $c_1u(t) + c_2v(t) = 0$ for all t in I. If u and v are not linearly dependent on I, then we say that they are **linearly independent**. ◆

An alternative way of defining linear independence is to say that two functions u and v are linearly independent when

$$c_1u + c_2v = 0 \text{ on } I \text{ only if } c_1 = c_2 = 0.$$

In other words, the zero function cannot be obtained as a linear combination of u and v except in the trivial manner.

Note that two functions are linearly dependent if and only if one of them is a constant multiple of the other. So we can say that two functions are linearly independent if neither is a constant multiple of the other. (Note that the zero function is a constant multiple of *any* other function.) For sets of three or more functions, the analogous interpretation of linear independence becomes complicated, while the definition just given has a rather obvious and simple extension. This is why the definition is stated the way that it is.

Example 1

Let's show that the functions $u(t) = t^m$ and $v(t) = t^n$, where $m, n \geq 0$, are linearly independent on any interval I unless $m = n$. To do this, suppose that $c_1t^m + c_2t^n = 0$ for all t in I with at least one of c_1, c_2 not zero, say $c_1 \neq 0$. If $m \neq n$, this contradicts the fact that a nonzero polynomial cannot have infinitely many zeros, which follows from the *fundamental theorem of algebra*. ■

Example 2

Let's show that the functions $u(t) = e^{at}$ and $v(t) = e^{bt}$ are linearly independent on any interval I unless $a = b$. To do this, suppose that $c_1e^{at} + c_2e^{bt} = 0$ for all t in

I with at least one of c_1, c_2 not zero, say $c_1 \neq 0$. Now, dividing by e^{bt}, we have $c_1 e^{(a-b)t} + c_2 = 0$ for all t in I. Since $c_1 \neq 0$, this says that $e^{(a-b)t}$ is constant for all t in I, which we know is certainly not true unless $a = b$. Therefore, it must be the case that $c_1 = c_2 = 0$ if $a \neq b$. ■

Example 3

Consider the functions $u(t) = t$ and $v(t) = |t|$. On any interval I contained in $[0, \infty)$, these functions are identical and thus linearly dependent on I. Also, if I is contained in $(-\infty, 0]$, then $u = -v$, and again u and v are linearly dependent on I. However, if I intersects both $(-\infty, 0)$ and $(0, \infty)$, then u and v are linearly independent on I. For example, if $I = (-1, 1)$ and $c_1 t + c_2 |t| = 0$, then for each $t < 0$ in I, we can divide by t and see that $c_1 - c_2 = 0$. Doing the same thing with $t > 0$ shows that $c_1 + c_2 = 0$. Solving these two equations for c_1 and c_2 results in $c_1 = c_2 = 0$. ■

The Structure of General Solutions The main result in this section is the following important theorem.

Theorem 1

There exists a pair of linearly independent solutions of (2) on I. Moreover, given any such pair u and v,

i) every solution of (2) on I is given by $y = c_1 u + c_2 v$ for some pair of constants c_1 and c_2.

If, in addition, $\tilde{y}$ is any particular solution of (1), then

ii) every solution of (1) on I is given by $y = c_1 u + c_2 v + \tilde{y}$ for some pair of constants c_1 and c_2. Furthermore, c_1 and c_2 can be chosen so that y satisfies arbitrary initial values $y(t_0) = y_0$ and $y'(t_0) = m_0$ with t_0 any point in I. ▲

Let us reconsider the notion of general solution for (1) and (2) in light of Theorem 1. According to the theorem, if u and v are linearly independent solutions of (1), then the general solution of (1) that consists of all linear combinations $c_1 u + c_2 v$ must contain *all* solutions of (1). Thus all such general solutions of (1) are equivalent. Similarly, a general solution of (2) given by $c_1 u + c_2 v + \tilde{y}$, where $\tilde{y}$ is a particular solution of (2), must contain all solutions of (2). Thus all such general solutions of (2) are equivalent. So henceforth we will refer to any such general solution as *the* general solution.

Example 4

Consider the equation

$$y'' + y = 5e^{-2t}.$$

The homogeneous equation $y'' + y = 0$ is easily seen to have linearly independent solutions $\cos t$ and $\sin t$ on $(-\infty, \infty)$. Thus, part (i) of Theorem 1 guarantees that *every* solution of $y''+y=0$ is of the form $y=c_1\cos t+c_2\sin t$. A particular solution of $y''+y=5e^{-2t}$ is $\tilde{y}=e^{-2t}$, and so part (ii) of Theorem 1 tells us that *every* solution of $y''+y=5e^{-2t}$ is of the form $y=c_1\cos t+c_2\sin t+e^{-2t}$. Therefore, we say that $y=c_1\cos t+c_2\sin t+e^{-2t}$ is *the* general solution. Furthermore, the second assertion in part (ii) of Theorem 1 guarantees that c_1 and c_2 can be found so that any initial conditions $y(t_0)=y_0$ and $y'(t_0)=m_0$ are satisfied. ■

The remainder of this section is devoted to the proof of Theorem 1.

The Wronskian Suppose that we are interested in solving an initial value problem of the form

$$y''+py'+qy=f,\quad y(t_0)=y_0,\quad y'(t_0)=m_0.$$

Further suppose that we have two solutions u and v of $y''+py'+qy=0$ and a particular solution $\tilde{y}$ of $y''+py'+qy=f$. We would then use superposition to write $y=c_1u+c_2v+\tilde{y}$ and attempt to solve for the constants c_1 and c_2. These constants are the solution of the following pair of linear algebraic equations:

$$\begin{aligned}c_1u(t_0)+c_2v(t_0)&=y_0-\tilde{y}(t_0)\\ c_1u'(t_0)+c_2v'(t_0)&=m_0-\tilde{y}'(t_0).\end{aligned}$$

This system can be solved for a unique pair of constants c_1 and c_2 precisely when the matrix of coefficients

$$\begin{pmatrix} u(t_0) & v(t_0)\\ u'(t_0) & v'(t_0)\end{pmatrix}$$

has a nonzero determinant; that is, when

$$u(t_0)v'(t_0)-u'(t_0)v(t_0)\neq 0.$$

This proves that (3) has a unique solution of the form $y=c_1u+c_2v+\tilde{y}$, for any t_0 in I and any initial values y_0 and m_0, precisely when the function $uv'-u'v$ is never zero in I.

The **Wronskian** of two differentiable functions u and v is defined by

$$W(u,v)=uv'-u'v.$$

Note that this may be expressed as the determinant of a 2×2 matrix:

$$W(u,v)=\begin{vmatrix} u & v\\ u' & v'\end{vmatrix}.$$

Here are a few simple examples of Wronskians:

$$\begin{aligned}W(e^{-t},e^t)&=e^{-t}e^t+e^{-t}e^t=2,\\ W(t,t^2)&=2t^2-t^2=t^2,\\ W(\cos t,\sin t)&=\cos^2 t+\sin^2 t=1.\end{aligned}$$

Notice that $W(u, v)$ is a function of the independent variable. We sometimes even write $W(t)$ to emphasize this when u and v are understood from context. For example,

$$\text{if } u(t) = \frac{1}{t}, \text{ and } v(t) = t, \text{ then } W(t) = \frac{1}{t}(1) - \left(-\frac{1}{t^2}\right)t = \frac{2}{t}.$$

There is an intimate connection between the Wronskian and the concept of linear independence. Suppose that u and v are linearly dependent, differentiable functions on an interval I, with $u(t) = cv(t)$ for all t in I. Then

$$W(u, v) = cvv' - cv'v = 0 \quad \text{on} \quad I.$$

That is, if u and v are differentiable and linearly dependent on I, then $W(u, v)$ is identically zero on I. Now suppose that $W(u, v) = 0$ on I (i.e., for *all* t in I). If we also suppose that $u \neq 0$ on I, then division by u^2 shows that

$$\frac{uv' - u'v}{u^2} = 0.$$

This implies that

$$\frac{v}{u} = c, \quad \text{and so} \quad v = cu \quad \text{on} \quad I.$$

Thus u and v are linearly dependent. We have essentially proven the following:

Theorem 2

Let u and v be differentiable on an interval I.

a) If u and v are linearly dependent on I, then $W(u, v) = 0$ on I.

b) If $W(u, v) = 0$ on I and either $u \neq 0$ or $v \neq 0$ on I, then u and v are linearly dependent on I. ▲

We point out that if u and v are differentiable on I and $W(u, v) \neq 0$ at *any* t in I, then Theorem 2 tells us that u and v are linearly *independent* on I.

Example 5

According to Theorem 2, $\cos t$ and $\sin t$ are linearly independent on any interval I, since $W(\cos t, \sin t) = 1$ for all t. Also, t and t^2 are linearly independent on any interval I since $W(t, t^2) = t^2$. ■

Example 6

In order to cast some light on the extra assumption that either $u \neq 0$ or $v \neq 0$ on I in part (b) of Theorem 2, let's consider the functions $u(t) = t^2$ and $v(t) = t|t|$ on the interval $I = (-1, 1)$. Note that each of these functions has a root at $t = 0$. The derivatives are $u'(t) = 2t$ and $v'(t) = 2|t|$. (Why?) Thus

$$W(t) = 2t^2|t| - 2t^2|t| = 0 \quad \text{for all } t \text{ in } I.$$

However, u and v are actually linearly independent on I, because if $c_1t^2 + c_2t|t| = 0$ on I, then dividing through by t^2 with $t > 0$ shows that $c_1 + c_2 = 0$, while dividing through by t^2 with $t < 0$ shows that $c_1 - c_2 = 0$. Therefore, we conclude that $c_1 = c_2 = 0$, which proves that u and v are linearly independent on I. This shows that part (b) of Theorem 2 is not true in general without the extra assumption that either $u \neq 0$ or $v \neq 0$ on I. ■

If u and v are solutions of a homogeneous second-order linear differential equation, then we can say even more about the Wronskian of u and v. Suppose that u and v are any two solutions of $y'' + py' + qy = 0$ on an interval I and that there is some point t_0 in I where $W(t_0) = 0$. Since $W(t_0)$ is the determinant of the matrix

$$M = \begin{pmatrix} u(t_0) & v(t_0) \\ u'(t_0) & v'(t_0) \end{pmatrix},$$

linear algebra tells us that the matrix equation

$$\begin{pmatrix} u(t_0) & v(t_0) \\ u'(t_0) & v'(t_0) \end{pmatrix} \begin{pmatrix} c_1 \\ c_2 \end{pmatrix} = \begin{pmatrix} 0 \\ 0 \end{pmatrix}$$

has a nontrivial solution. That is, there are constants c_1 and c_2, not both zero, such that

$$c_1u(t_0) + c_2v(t_0) = 0,$$
$$c_1u'(t_0) + c_2v'(t_0) = 0.$$

Now, with *this* choice of c_1 and c_2, consider the function $y = c_1u + c_2v$, and note that y satisfies the initial-value problem

$$y'' + py' + qy = 0, \quad y(t_0) = y'(t_0) = 0.$$

Theorem 1 in Section 5.2 tells us that $y(t) = 0$ for all t in I, because the initial-value problem can have only one solution. Now, since $y(t) = c_1u(t) + c_2v(t) = 0$ for all t in I, we see that u and v are linearly dependent on I, and consequently $W(t) = 0$ for *all* t in I because of Theorem 2. All of this is summarized in the following theorem.

Theorem 3

If u and v are any two solutions of $y'' + py' + qy = 0$ on an interval I and there is some point t_0 in I where $W(t_0) = 0$, then u and v are linearly dependent on I, and consequently $W(t) = 0$ for all t in I. ▲

The following is an easy corollary to Theorem 3.

Corollary 1 *If u and v are any two solutions of $y'' + py' + qy = 0$ on an interval I, then either $W(t) = 0$ for all t in I, or else $W(t) \neq 0$ for all t in I.*

Example 7

Note that $u = \cos t$ and $v = \sin t$ satisfy $y'' + y = 0$ (i.e., (2) with $p = 0$ and $q = 1$). We have seen already that $W(t) = 1 \neq 0$ for all t. Also, note that $u = e^{-t}$ and $v = e^t$ satisfy $y'' - y = 0$. In this case, $W(t) = 2 \neq 0$ for all t. ■

Example 8

Consider the functions $u = t$ and $v = t^2$. Note that $W(t) = t^2$. Since $W(0) = 0$ and $W(t) \neq 0$ if $t \neq 0$, Corollary 1 tells us that t and t^2 cannot both be solutions of $y'' + py' + qy = 0$ on an interval I containing $t = 0$ with p and q continuous on I. In other words, if t and t^2 are solutions of $y'' + py' + qy = 0$ on I, then either I does not contain 0 or else at least one of p, q is not continuous on I. ■

The Proof of Theorem 1 Finally, we are in a position to prove Theorem 1. The first step is to show that (2) has a pair of linearly independent solutions. So let t_0 be any number in I, and let u and v be the "fundamental" solutions of (2) that satisfy the initial conditions

$$u(t_0) = 1, \quad u'(t_0) = 0 \quad \text{and} \quad v(t_0) = 0, \quad v'(t_0) = 1.$$

These solutions are guaranteed to exist by Theorem 1 in Section 5.2. Computing the Wronskian of u and v at $t = t_0$,

$$W(t_0) = u(t_0)v'(t_0) - u'(t_0)v(t_0) = (1)(1) - (0)(0) = 1,$$

shows that u and v are linearly independent. (This is just one pair of linearly independent solutions; there are plenty more. Indeed, any solutions u and v whose initial values at t_0 satisfy $u(t_0)v'(t_0) - u'(t_0)v(t_0) \neq 0$ are linearly independent.)

Now suppose that u and v are linearly independent solutions of $y''+py'+qy = 0$ on an interval I, and let $W(t)$ be the Wronskian of u and v. We want to prove that every solution of $y'' + py' + qy = 0$ on I is given by $y = c_1u + c_2v$ for some constants c_1 and c_2.

So suppose that w is any solution whatsoever, and pick any point t_0 in I. Because $W(t_0) \neq 0$, we can pick c_1 and c_2 so that

$$\begin{pmatrix} u(t_0) & v(t_0) \\ u'(t_0) & v'(t_0) \end{pmatrix} \begin{pmatrix} c_1 \\ c_2 \end{pmatrix} = \begin{pmatrix} w(t_0) \\ w'(t_0) \end{pmatrix};$$

that is, so that

$$c_1u(t_0) + c_2v(t_0) = w(t_0), \quad c_1u'(t_0) + c_2v'(t_0) = w'(t_0).$$

This means that we can pick c_1 and c_2 so that both w and $c_1u + c_2v$ satisfy the same initial conditions at t_0. Therefore, by uniqueness, we conclude that w and c_1u+c_2v must be identical on I. This proves that *every* solution on I can be written as $c_1u + c_2v$.

To prove the second assertion in Theorem 1, suppose again that u and v are linearly independent solutions of $y'' + py' + qy = 0$ on an interval I and suppose further that $\tilde{y}$ is a particular solution of $y'' + py' + qy = f$ on I. Let w be *any* solution of (1) on I. Note that $w - \tilde{y}$ satisfies $y'' + py' + qy = f$. Therefore, the previous argument shows that we can pick c_1 and c_2 so that $w - \tilde{y} = c_1u + c_2v$ on I; that is, $w = c_1u + c_2v + \tilde{y}$ on I. Thus *every* solution of $y'' + py' + qy = f$ on I can be written as $c_1u + c_2v + \tilde{y}$. The preceding argument that motivated the definition of the Wronskian shows that if $W(t) \neq 0$ on I, then c_1 and c_2 can be chosen so that $y = c_1u + c_2v + \tilde{y}$ satisfies any given initial conditions at any t_0 in I. This completes the proof.

PROBLEMS

1. Suppose that u and v are defined, but not necessarily differentiable, on the interval [0, 1] with $u(0) = 0$, $v(0) = 1$, $u(1) = 1$, $v(1) = -1$. Show that u and v are linearly independent on [0,1].

2. Suppose that u and v are linearly *dependent* solutions of (2). Show that the two-parameter family of solutions $y = c_1u + c_2v$ can be described as a one-parameter family and therefore is not a general solution of (2). (See Section 1.3.)

3. **(a)** Show directly that the system of equations

$$ax + by = 0, \quad cx + dy = 0$$

has only the trivial solution $x = y = 0$ if $ad - bc \neq 0$.

(b) Suppose that u and v are defined, but not necessarily differentiable, on an interval I. Show that if there are points t_1, t_2 in I such that $u(t_1)v(t_2) \neq u(t_2)v(t_1)$, then u and v are linearly independent on I.

4. Suppose that u and v are defined, but not necessarily differentiable, on an interval I. Show that if there are points t_1, t_2 in I such that $u(t_1) = v(t_1)$ and $u(t_2) \neq v(t_2)$, then u and v are linearly independent on I.

Use the Wronskian and Theorem 2 to show that each of the pairs of functions in Problems 5 through 9 are linearly independent on $I = (-\infty, \infty)$.

5. e^{at} and e^{bt}, if $a \neq b$

6. $\cos t$ and $\cos 2t$

7. $\sin at$ and $\sin bt$, if $a^2 \neq b^2$

8. $e^{at} \cos bt$ and $e^{at} \sin bt$, if $b \neq 0$

9. $u(t)$, $u(t) + c$ where $c \neq 0$ and u is differentiable and nonconstant on $(-\infty, \infty)$

10. Give examples to show that these statements are *false*:

(a) If $c \neq 1$ and u is not constant, then $u(t)$ and $u(ct)$ are linearly independent on any interval I on which they are both defined.

(b) If $c \neq 0$ and u is not constant, then $u(t)$ and $u(t + c)$ are linearly independent on any interval I on which they are both defined.

Write down the general solution of each equation in Problems 11 through 13.

11. $y'' + y = 1$

12. $y'' - y = 5 - t$

13. $y'' = 6t$

Solve the initial-value problem in each of Problems 14 through 16.

14. $y'' + y = 1,\ y(0) = y'(0) = 0$

15. $y'' - y = 0,\ y(0) = 1,\ y'(0) = 0$

16. $y'' + y = 1,\ y(0) = 1,\ y'(0) = -1$

Solve Problems 17 through 20 by inspection.

17. $y'' + t^3y' + 2y = 4,\ y(0) = 2,\ y'(0) = 0$
18. $y'' + t^3y' + 2y = t^3 + 2t,\ y(0) = 0,\ y'(0) = 1$
19. $y'' + y' + y = 2 + 2t + t^2,\ y(0) = 0,\ y'(0) = 0$
20. $y'' + y' + y = 2 + t,\ y(0) = 1,\ y'(0) = 1$

In Problems 21 and 22, let p and q be continuous on an interval I.

21. Show that e^t and $\cos t$ cannot both be solutions of $y'' + py' + qy = 0$ on I if I has length π or greater.

22. Show that t^m and e^{at} cannot both be solutions of $y'' + py' + qy = 0$ on I if I contains $t = \frac{m}{a}$.

23. Suppose that u and v satisfy, respectively,

$$\begin{cases} u'' + pu' + qu &= 0 \\ u(0) = 1, u'(0) &= 0 \end{cases} \quad \text{and} \quad \begin{cases} v'' + pv' + qv &= 0 \\ v(0) = 0, v'(0) &= 1. \end{cases}$$

Check that $y = y_0u + m_0v$ solves

$$y'' + py' + qy = 0, \quad y(0) = y_0, y'(0) = m_0.$$

Such functions u and v are called **fundamental solutions** of $y'' + py' + qy = 0$.

24. Suppose that u and v satisfy, respectively,

$$\begin{cases} u'' + pu' + qu &= 0 \\ u(0) = y_0, u'(0) &= m_0 \end{cases} \quad \text{and} \quad \begin{cases} v'' + pv' + qv &= f(t) \\ v(0) = 0, v'(0) &= 0. \end{cases}$$

Check that $y = u + v$ solves

$$y'' + py' + qy = f(t), \quad y(0) = y_0, \quad y'(0) = m_0.$$

The function v here, which satisfies zero initial conditions, is called the **rest solution** of $y'' + py' + qy = f$ (with respect to the initial point $t_0 = 0$).

25. Consider the equation $t^2y'' - 2mty' + (m(m+1) + \omega^2t^2)y = 0$.

(a) Show that the substitution $y(t) = t^m u(t)$ produces the equation $u'' + \omega^2 u = 0$.

(b) Conclude that the general solution is $y = t^m(c_1 \cos \omega t + c_2 \sin \omega t)$.

26. Consider the equation $t^2y'' - 2mty' + (m(m+1) - \omega^2t^2)y = 0$.

(a) Show that the substitution $y(t) = t^m u(t)$ produces the equation $u'' - \omega^2 u = 0$.

(b) Conclude that the general solution is $y = t^m(c_1 \cosh \omega t + c_2 \sinh \omega t)$.

27. Let u and v be any two solutions of $y'' + py' + qy = 0$ on an interval I, and let $W(t)$ be the Wronskian of u and v.

(a) Show that $W' = uv'' - u''v$ and consequently

$$W' + pW = 0 \quad \text{on } I.$$

This is known as **Abel's identity**.

(b) Conclude from Abel's identity that $W(t) = Ce^{-\int p}$ for some constant C. Now give an alternative proof of Corollary 1.

28. Finding a second solution. Suppose that u is a known, nonzero solution of $y''+py'+qy = 0$ on an interval I. We will use u to "generate" a second solution v such that u and v are linearly independent.

(a) For any twice differentiable function v on I, show that $W(u, v)$ satisfies

$$W' + pW = u(v'' + pv' + qv) \quad \text{on } I.$$

(b) Conclude that v is a solution of $y'' + py' + qy = 0$ on I, if and only if $W(u, v)$ satisfies $W' + pW = 0$ on I.

(c) Conclude that v is a second *linearly independent* solution of $y'' + py' + qy = 0$ on I, if and only if

$$uv' - u'v = Ce^{-\int p} \quad \text{on } I$$

for some nonzero constant C. (cf. Problem 27)

(d) Using the integrating factor $1/u^2$, show from the identity in part (c) that a second linearly independent solution v of $y'' + p\,y' + q\,y = 0$ on I is given by

$$\left(\frac{v}{u}\right)' = \frac{Ce^{-\int p}}{u^2}, \quad C \neq 0.$$

The constant C is typically chosen to be 1.

29. Given that $u = e^t$ satisfies $y'' - 5y' + 4y = 0$ on $(-\infty, \infty)$, use the result of Problem 28d to find a second linearly independent solution v.

30. Given that $u = t^2$ satisfies $y'' + \frac{2}{t^2}y = 0$ on $(0, \infty)$, use the result of Problem 28d to find a second linearly independent solution v.

31. Elimination of the first-order term. Show that y satisfies $y'' + py' + qy = 0$, if and only if

$$u = e^{\frac{1}{2}\int p} y \ \text{ satisfies } \ u'' + \left(q - \frac{p'}{2} - \frac{p^2}{4}\right)u = 0.$$

For the equations in Problems 32 through 35, find an equivalent equation that lacks a first-order term.

32. $y'' + 2ty' + t^2y = 0$

33. $t^2y'' + ty' + y = 0$

34. $y'' + 2\sin ty' - \cos^2 ty = 0$

35. $y'' + 2t^2y' + 2ty = 0$

36. Constant coefficients. Suppose that p and q are constants, and let $\omega = -q + p^2/4$. Given the general solutions

$$\begin{aligned} u'' - a^2u = 0 \quad & u = c_1e^{at} + c_2e^{-at} \\ u'' = 0 \quad & u = c_1 + c_2t \\ u'' + a^2u = 0 \quad & u = c_1\cos at + c_2\sin at, \end{aligned}$$

where a is a positive constant, eliminate the first-order term (as in Problem 31) in order to find the general solution of $y'' + py' + qy = 0$ in each of the following cases:

(a) $\omega < 0$ (b) $\omega = 0$ (c) $\omega > 0$

37. Use the results of Problem 36 to write down the general solution of

(a) $y'' + 4y' + 3y = 0$ (b) $y'' + 4y' + 4y = 0$ (c) $y'' + 4y' + 5y = 0$

5.5 Variation of Constants and Green's Functions

Suppose that we wish to solve the nonhomogeneous initial-value problem

$$\mathcal{T}y = f, \quad y(0) = 0, \quad y'(0) = 0, \tag{1}$$

where $\mathcal{T}$ is defined by

$$\mathcal{T}y = y'' + py' + qy$$

and p, q, and f are given continuous functions of t. Because of the zero initial conditions, the solution describes the behavior of a system that is initially at rest; therefore, we will refer to this solution as the **rest solution** of the equation $\mathcal{T}y = f$ (not to be confused with an *equilibrium* solution).

If we think of the equation as a model of some physical system such as a spring-mass system or electrical circuit, it is natural to view the functions p and q as characteristic of the system itself and to view f as the *input* to the system. The rest solution y can be viewed as the output, or *response*, of the system to the input f. In other words, y is solely and uniquely determined by f. This correspondence that maps f to y is the *solution operator* for the problem (1) and is, in fact, a linear operator.

A technique known as **variation of constants** (often called *variation of parameters*) allows us to derive a *representation* of the rest solution y in terms of f and thus describe the solution operator for the problem (1). We begin by assuming that u and v are linearly independent solutions of the corresponding homogeneous equation

$$\mathcal{T}y = 0. \tag{2}$$

Thus, by the main result of Section 5.4, every solution of (2) is of the form $y = c_1u + c_2v$. This motivates us to look for a solution of (1) in the form

$$y = au + bv,$$

where a and b are functions of t to be determined. We first find that

$$y' = au' + bv' + a'u + b'v,$$

but before computing y'', let's notice that we can avoid dealing with a'' and b''—and perhaps obtain *first-order* equations for a and b—if we decide at this point to require that

$$a'u + b'v = 0.$$

Having done that, we have

$$y = au + bv, \quad y' = au' + bv', \quad y'' = a'u' + au'' + b'v' + bv''. \tag{3}$$

Now we find that

$$\begin{aligned} y'' + py' + qy &= a(u'' + pu' + qu) + b(v'' + pv' + qv) + a'u' + b'v' \\ &= a'u' + b'v'. \end{aligned}$$

Thus y satisfies $\mathcal{T}y = f$ if

$$a'u + b'v = 0 \quad \text{and} \quad a'u' + b'v' = f.$$

Solving these two linear equations for a' and b' shows that

$$a' = -\frac{vf}{W(u, v)} \quad \text{and} \quad b' = \frac{uf}{W(u, v)},$$

where $W(u, v) = uv' - u'v$ is the Wronskian of u and v. (Recall that since u and v are linearly independent solutions of (2), Theorem 3 of Section 5.4 guarantees that $W(u, v)$ is never zero.) Because of the first two equations in (3), the initial conditions $y(0) = y'(0) = 0$ will be satisfied if we choose a and b such that $a(0) = b(0) = 0$. Thus we simply integrate from 0 to t to obtain

$$a(t) = -\int_0^t \frac{v(s)f(s)}{W(s)}\,ds \quad \text{and} \quad b(t) = \int_0^t \frac{u(s)f(s)}{W(s)}\,ds$$

and the formula

$$y(t) = v(t)\int_0^t \frac{u(s)f(s)}{W(s)}\,ds - u(t)\int_0^t \frac{v(s)f(s)}{W(s)}\,ds,$$

which may be written with one integral as

$$y(t) = \int_0^t \frac{u(s)v(t) - u(t)v(s)}{W(s)} f(s)\,ds.$$

This gives rise to the compact formula

$$y(t) = \int_0^t G_0(t, s) f(s)\,ds \quad \text{where} \quad G_0(t, s) = \frac{u(s)v(t) - u(t)v(s)}{W(s)}. \tag{4}$$

We refer to the function G_0 as a *Green's function*, after George Green, an English physicist who introduced formulas analogous to (4) for solutions of certain partial differential equations in a paper published in 1828. Notice that G_0 determines a linear integral operator $\mathcal{G}$ acting on the function f, given by

$$[\mathcal{G}f](t) = \int_0^t G_0(t, s) f(s)\,ds. \tag{5}$$

This linear operator $\mathcal{G}$ is precisely the *solution operator* for problem (1) that we wanted to determine. Our derivation shows that

$$\left.\begin{array}{r} \mathcal{T}y = f \\ y(t_0) = y'(t_0) = 0 \end{array}\right\} \quad \text{if and only if} \quad y = \mathcal{G}f.$$

Moreover, if we define an operator $\mathcal{T}_0$ to be the restriction of $\mathcal{T}$ to a domain consisting of twice differentiable functions y satisfying $y(t_0) = y'(t_0) = 0$, then it makes sense to consider $\mathcal{G}$ to be precisely the *inverse* of $\mathcal{T}_0$; that is,

$$\mathcal{G} = \mathcal{T}_0^{-1}, \quad \text{since} \quad \mathcal{T}_0 y = f \quad \text{if and only if} \quad y = \mathcal{G}f.$$

Example 1

Consider the equation

$$y'' + y = f.$$

Solutions of the homogeneous equation $y'' + y = 0$ are $u = \cos t$ and $v = \sin t$. The Wronskian is therefore $W = \cos^2 t + \sin^2 t = 1$. Thus we have the Green's function

$$G_0(t, s) = \cos s \sin t - \cos t \sin s,$$

and the rest solution

$$y = \int_0^t (\cos s \sin t - \cos t \sin s) f(s)\, ds.$$

If, for example, $f(t) = \cos t$, then

$$\begin{aligned} y &= \sin t \int_0^t \cos^2 s\, ds - \cos t \int_0^t \sin s \cos s\, ds \\ &= \frac{1}{2} \sin t \left(t + \frac{1}{2} \sin 2t \right) - \frac{1}{2} \cos t \sin^2 t \\ &= \frac{1}{2} \sin t \left(t + \frac{1}{2} \sin 2t - \cos t \sin t \right) = \frac{t}{2} \sin t. \end{aligned}$$

■

Example 2

Consider the equation

$$y'' - \frac{2}{t} y' + \frac{2}{t^2} y = f.$$

If we look for solutions of the homogeneous equation $y'' - \frac{2}{t}y' + \frac{2}{t^2}y = 0$ in the form $y = t^m$, we find that m should satisfy $m^2 - 3m + 2 = 0$, which gives $m = 1, 2$. So we have $u = t$ and $v = t^2$, for which we find $W = (t)(2t) - (1)(t^2) = t^2$. Thus u and v are linearly independent solutions on any interval not containing $t = 0$. Using the Green's function

$$G_0(t, s) = \frac{st^2 - ts^2}{s^2} = \frac{t^2 - s, t}{s},$$

we write

$$y = \int_{t_0}^t \frac{t^2 - st}{s} f(s)\, ds$$

as the rest solution for $t_0 \neq 0$. The solution is valid on $(-\infty, 0)$ if $t_0 < 0$ and on $(0, \infty)$ if $t_0 > 0$. If, for example, $f(t) = t^{-1}$, then

$$\begin{aligned} y &= \int_{t_0}^t \frac{t^2 - st}{s^2}\, ds = t^2 \left(\frac{1}{t} - \frac{1}{t_0} \right) - t(\ln t - \ln t_0) \\ &= t \left(1 - \frac{t}{t_0} - \ln \frac{t}{t_0} \right). \end{aligned}$$

■

PROBLEMS

1. Follow Example 1 to find the rest solution of $y'' + y = \sin t$.

In Problems 2 through 4, let $\mathcal{T}$ be the operator defined by $\mathcal{T}y = y'' - y$.

2. **(a)** Find, by inspection or any other means, two linearly independent solutions of $\mathcal{T}y = 0$.

(b) Construct the Green's function $G_0(t, s)$ corresponding to the initial point $t_0 = 0$.

(c) Use (4) to express the rest solution of $\mathcal{T}y = f$ in terms of G_0 and f.

3. Compute the rest solution of $\mathcal{T}y = e^{-2t}$.

4. Compute the rest solution of $\mathcal{T}y = e^{-t}$.

In Problems 5 through 7, let $\mathcal{T}$ be the operator defined by $\mathcal{T}y = y'' + 4y' + 3y$.

5. **(a)** Check that $u = e^{-3t}$ and $v = e^{-t}$ are solutions of $\mathcal{T}y = 0$.

(b) Construct the Green's function $G_0(t, s)$ corresponding to the initial point $t_0 = 0$.

(c) Use (4) to express the rest solution of $\mathcal{T}y = f$ in terms of G_0 and f.

6. Compute the rest solution of $\mathcal{T}y = e^{-2t}$.

7. Compute the rest solution of $\mathcal{T}y = e^{-t}$.

In Problems 8 and 9, let $\mathcal{T}$ be the operator defined by $\mathcal{T}y = y'' + \frac{1}{t}y' - \frac{1}{t^2}y$.

8. **(a)** Find two linearly independent solutions of $\mathcal{T}y = 0$ in the form t^m.

(b) Construct the Green's function $G_0(t, s)$ corresponding to an initial point $t_0 \neq 0$.

(c) Use (4) to express the rest solution of $\mathcal{T}y = f$, with respect to an $t_0 \neq 0$, in terms of G_0 and f.

9. Compute the rest solution of $\mathcal{T}y = t$, with respect to an initial point $t_0 \neq 0$.

10. Let $\mathcal{T}$ be the linear operator defined by $\mathcal{T}y = y'' + py' + qy$. Show that the solution of the initial-value problem

$$\mathcal{T}y = f, \quad y(t_0) = y_0, \quad y'(t_0) = m_0$$

can be constructed by adding the rest solution of $\mathcal{T}y = f$ to the solution of

$$\mathcal{T}y = 0, \quad y(t_0) = y_0, \quad y'(t_0) = m_0.$$

11. Verify that the operator $\mathcal{G}$ defined by (5) is a linear operator.

12. Show that the Green's functions in Example 1 and Problem 2 are of the form $G_0(t - s)$; that is, they can be expressed in terms of a function of one variable evaluated at $t - s$. Is this possible for the Green's functions in Example 2 and Problem 8?

13. Prove that the Green's function G_0 is independent of the choice of u and v, provided they satisfy $\mathcal{T}y = 0$ and are linearly independent. (*Hint*: Given any particular choice of u and v, any other choice, say $\tilde{u}$ and $\tilde{v}$, can be expressed as linear combinations of u and v.)

14. Use a variation of constants approach to derive a representation for the solution of the first-order initial-value problem $y' + py = f$, $y(0) = 0$. Express the result in the form $y = \int_0^t G_0(t, s) f(s)\, ds$, and so determine the Green's function for this problem.

5.6 Power-Series Solutions

We have seen on numerous occasions that differential equations may have solutions that are not expressible in terms of elementary functions. Solutions of such equations can often be expressed as power series. In this section we will present an elementary technique for finding solutions in the form of power series.

A function φ defined on an open interval I containing 0 is said to have a power-series expansion

$$\varphi(t) = \sum_{k=0}^{\infty} a_k (t - t_0)^k$$

about $t = t_0$ provided that the series converges to $\varphi(t)$ for all t in some open interval $(t_0 - r, t_0 + r)$, $r > 0$. The radius r of the largest such interval is called the *radius of convergence* of the series. In what follows, we will consider only power-series expansions about $t = 0$; so when we refer to a power-series expansion, we mean a power-series expansion about $t = 0$:

$$\varphi(t) = a_0 + a_1 t + a_2 t^2 + a_3 t^3 + \cdots = \sum_{k=0}^{\infty} a_k t^k . \tag{1}$$

Basic results from calculus give the following facts.

i. If two functions φ and ψ each have power-series expansions

$$\varphi(t) = \sum_{k=0}^{\infty} a_k t^k \quad \textit{and} \quad \psi(t) = \sum_{k=0}^{\infty} b_k t^k,$$

convergent on an interval I, then $\varphi(t) = \psi(t)$ for all t in I if and only if $a_k = b_k$ for all $k = 0, 1, 2, \ldots$. In particular, $\varphi(t) = 0$ for all t in I, if and only if $a_k = 0$ for all $k = 0, 1, 2, \ldots$.

ii. If φ has a power-series expansion, then so do $\varphi', \varphi'', \ldots$, and furthermore φ can be differentiated term by term:

$$\begin{aligned} \varphi'(t) &= a_1 + 2a_2 t + 3a_3 t^2 + 4a_4 t^3 + 5a_5 t^4 + \cdots \\ &= \sum_{k=1}^{\infty} k a_k t^{k-1} = \sum_{n=0}^{\infty} (n+1) a_{n+1} t^n, \end{aligned} \tag{2}$$

$$\begin{aligned} \varphi''(t) &= 2a_2 + 6a_3 t + 12a_4 t^2 + 20a_5 t^3 + \cdots \\ &= \sum_{k=2}^{\infty} (k-1) k a_k t^{k-2} = \sum_{n=0}^{\infty} (n+1)(n+2) a_{n+2} t^n, \end{aligned} \tag{3}$$

and so on, where each series has the same radius of convergence as the expansion of φ.

iii. The coefficients $a_0, a_1, a_2, \ldots$ are related to the derivatives of φ at 0 as follows:

$$a_0 = \varphi(0), \;\; a_1 = \varphi'(0), \;\; a_2 = \frac{1}{2}\varphi''(0), \ldots, \;\; a_n = \frac{1}{n!}\varphi^{(n)}(0), \ldots .$$

For reasons that will soon become clear, it is crucial to understand how the final forms of the series for φ' and φ'' in (2) and (3) come about, since they reveal precisely the general form of the coefficients of t^n in terms of n. Each is the result of *index shifting*. For instance, in the series

$$\varphi''(t) = \sum_{k=2}^{\infty}(k-1)ka_k t^{k-2},$$

we shift the index by substituting $n = k - 2$, or $k = n + 2$, to obtain

$$\varphi''(t) = \sum_{n=0}^{\infty}(n+1)(n+2)a_{n+2}t^n.$$

To illustrate the idea further, let's look at the product

$$t^2\varphi'(t) = t^2\sum_{k=1}^{\infty}ka_k t^{k-1} = \sum_{k=1}^{\infty}ka_k t^{k+1}.$$

Here we substitute $n = k + 1$, or $k = n - 1$, to obtain

$$t^2\varphi'(t) = \sum_{n=2}^{\infty}(n-1)a_{n-1}t^n.$$

Note that the starting value of the new index is $n = 2$, since t^2 clearly is the lowest power of t in $t^2\varphi''$. However, note that it is valid to write

$$t^2\varphi'(t) = \sum_{n=0}^{\infty}(n-1)a_{n-1}t^n,$$

if we simply *define* $a_{-1} = 0$.

We now proceed with several examples that illustrate a method for finding power-series solutions of linear second-order differential equations. As we will see, the crux of the method is the derivation of a *recurrence formula* that can be used to compute the power-series coefficients.

Example 1

Consider *Airy's equation*:

$$y'' - ty = 0.$$

We will find the general solution in the form $y = c_1u + c_2v$, where u and v are "fundamental" power-series solutions satisfying

$$u(0) = 1,\ u'(0) = 0 \text{ and } v(0) = 0,\ v'(0) = 1.$$

Toward that end, we first consider any solution y in the form of a power series

$$y = \sum_{k=0}^{\infty}a_k t^k.$$

The series expansions of the terms in the differential equation are then

$$y'' = \sum_{k=2}^{\infty}(k-1)ka_k t^{k-2} = \sum_{n=0}^{\infty}(n+1)(n+2)a_{n+2}t^n$$

and

$$-ty = -\sum_{k=0}^{\infty}a_k t^{k+1} = -\sum_{n=0}^{\infty}a_{n-1}t^n,$$

where we define $a_{-1} = 0$. So in order for the differential equation to be satisfied, it must be the case that

$$\sum_{n=0}^{\infty}((n+1)(n+2)a_{n+2} - a_{n-1})t^n = 0,$$

which is equivalent to

$$(n+1)(n+2)a_{n+2} - a_{n-1} = 0, \quad n = 0, 1, 2, \ldots .$$

This suggests that the coefficients can be generated by the *recurrence formula*

$$a_{n+2} = \frac{a_{n-1}}{(n+2)(n+1)}, \quad n = 0, 1, 2, \ldots . \tag{4}$$

Since we defined $a_{-1} = 0$, the recurrence formula (4) implies that

$$a_2 = a_5 = a_8 = a_{11} = \cdots = 0$$

in any power-series solution. This is consistent with the fact that the differential equation implies that $y''(0) = 0$, which in turn implies that $a_2 = 0$. We also observe from (4) that a_0 determines $a_3, a_6, a_9, \ldots$, while a_1 determines $a_4, a_7, a_{10}, \ldots$.

Now we return our attention specifically to the fundamental solutions u and v. Since $u(0) = 1$ and $u'(0) = 0$, we will find u by setting

$$a_0 = 1 \quad \textit{and} \quad a_1 = 0$$

and then with (4) computing

$$a_3 = \frac{1}{3 \cdot 2} = \frac{1}{3!}, \; a_6 = \frac{1}{6 \cdot 5 \cdot 3 \cdot 2} = \frac{4}{6!} \; a_9 = \frac{1}{9 \cdot 8 \cdot 6 \cdot 5 \cdot 3 \cdot 2} = \frac{7 \cdot 4}{9!}, \ldots$$

and

$$a_4 = 0, \; a_7 = 0, \; a_{10} = 0, \ldots ,$$

thus arriving at

$$u = 1 + \frac{t^3}{3!} + \frac{4t^6}{6!} + \frac{7 \cdot 4 \cdot t^9}{9!} + \frac{10 \cdot 7 \cdot 4 \cdot t^{12}}{12!} + \cdots .$$

Since $v(0) = 0$ and $v'(0) = 1$, we similarly find v by setting

$$a_0 = 0 \quad \text{and} \quad a_1 = 1$$

and then with (4) computing

$$a_4 = \frac{1}{4 \cdot 3} = \frac{2}{4!}, a_7 = \frac{1}{7 \cdot 6 \cdot 4 \cdot 3} = \frac{5 \cdot 2}{7!}, \; a_{10} = \frac{1}{10 \cdot 9 \cdot 7 \cdot 6 \cdot 4 \cdot 3} = \frac{8 \cdot 5 \cdot 2}{10!}, \ldots$$

and

$$a_5 = 0,\ a_8 = 0,\ a_{11} = 0,\ \ldots,$$

thus arriving at

$$v = t + \frac{2t^4}{4!} + \frac{5 \cdot 2 \cdot t^7}{7!} + \frac{8 \cdot 5 \cdot 2 \cdot t^{10}}{10!} + \frac{11 \cdot 8 \cdot 5 \cdot 2 \cdot t^{13}}{13!} + \cdots.$$

The general solution is therefore

$$y(t) = c_1 \left(1 + \frac{t^3}{3!} + \frac{4t^6}{6!} + \frac{7 \cdot 4 \cdot t^9}{9!} + \cdots\right) + c_2 \left(t + \frac{2t^4}{4!} + \frac{5 \cdot 2 \cdot t^7}{7!} + \cdots\right),$$

where the parameters c_1 and c_2 represent the initial values $y(0)$ and $y'(0)$, respectively. Figures 1ab illustrate the convergence of the series defining u and v. Figure 1a shows the graphs of the partial sums of $u(t)$ with degrees 3, 6, 9, 12, and 15. Figure 1b shows graphs of the partial sums of $v(t)$ with degrees 4, 7, 10, 13, and 16.

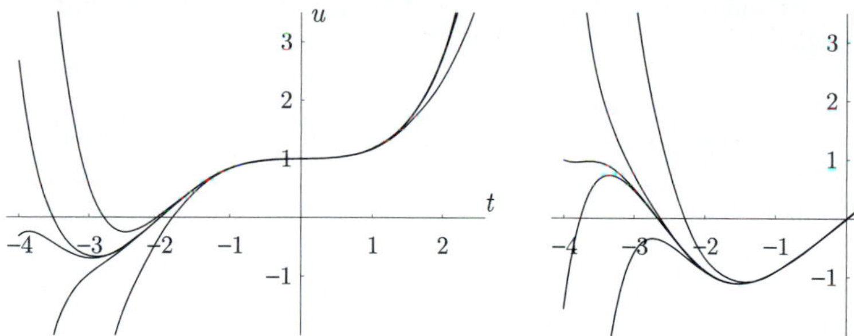

Figure 1a

Figure 1b

The (approximate) graphs of u and v in Figures 2ab were obtained by plotting partial sums of degree 90 and 91, respectively, which approximate the true solutions reasonably well on the interval shown. ■

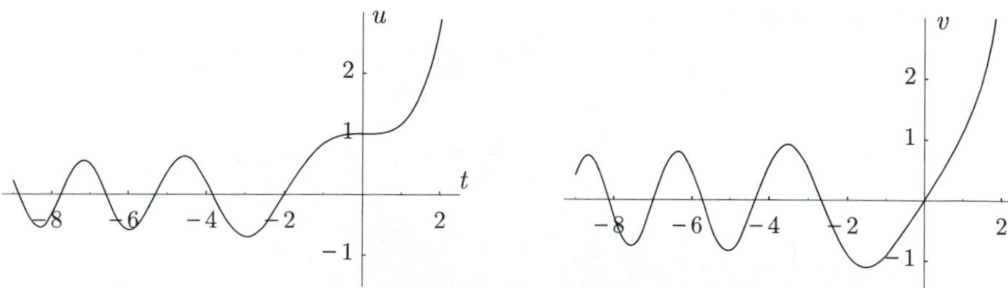

Figure 2a

Figure 2b

Example 2

Consider the equation

$$y'' + y' + ty = 0,$$

and suppose that y is a solution in the form of a power series

$$y = \sum_{k=0}^{\infty} a_k t^k.$$

The series expansions of the terms in the differential equation are

$$y'' = \sum_{k=2}^{\infty}(k-1)ka_k t^{k-2} = \sum_{n=0}^{\infty}(n+1)(n+2)a_{n+2}t^n,$$

$$y'(t) = \sum_{k=1}^{\infty} ka_k t^{k-1} = \sum_{n=0}^{\infty}(n+1)a_{n+1}t^n,$$

$$ty = \sum_{k=0}^{\infty} a_k t^{k+1} = \sum_{n=0}^{\infty} a_{n-1}t^n,$$

where we define $a_{-1} = 0$. So in order for the differential equation to be satisfied, we must have

$$\sum_{n=0}^{\infty}((n+1)(n+2)a_{n+2} + (n+1)a_{n+1} + a_{n-1})t^n = 0,$$

which is equivalent to

$$(n+1)(n+2)a_{n+2} + (n+1)a_{n+1} + a_{n-1} = 0, \quad n = 0, 1, 2, \ldots.$$

So we arrive at the *three-term* recurrence formula:

$$a_{n+2} = -\frac{a_{n+1}}{n+2} - \frac{a_{n-1}}{(n+1)(n+2)}, \quad n = 0, 1, 2, \ldots.$$

Now, as in Example 1, we will find the general solution in the form

$$y = c_1 u + c_2 v,$$

where u and v are "fundamental" power-series solutions satisfying $u(0) = 1$, $u'(0) = 0$, $v(0) = 0$, and $v'(0) = 1$. To find u, we set $a_0 = 1$ and $a_1 = 0$ and then use the recurrence formula to find

$$a_2 = -\frac{a_1}{2} - \frac{a_{-1}}{2} = 0,$$

$$a_3 = -\frac{a_2}{3} - \frac{a_0}{2 \cdot 3} = -\frac{1}{2 \cdot 3} = -\frac{1}{3!},$$

$$a_4 = -\frac{a_3}{4} - \frac{a_1}{3 \cdot 4} = \frac{1}{2 \cdot 3 \cdot 4} = \frac{1}{4!},$$

$$a_5 = -\frac{a_4}{5} - \frac{a_2}{4 \cdot 5} = -\frac{1}{2 \cdot 3 \cdot 4 \cdot 5} = -\frac{1}{5!},$$

$$a_6 = -\frac{a_5}{6} - \frac{a_3}{5\cdot 6} = \frac{1}{5!\cdot 6} + \frac{1}{3!\cdot 5\cdot 6} = \frac{5}{6!},$$
$$a_7 = -\frac{a_6}{7} - \frac{a_4}{6\cdot 7} = -\frac{5}{6!\cdot 7} - \frac{1}{4!\cdot 6\cdot 7} = -\frac{10}{7!},$$
$$a_8 = -\frac{a_7}{8} - \frac{a_5}{7\cdot 8} = \frac{10}{7!\cdot 8} + \frac{1}{5!\cdot 7\cdot 8} = \frac{16}{8!},$$
$$a_9 = -\frac{a_8}{9} - \frac{a_6}{8\cdot 9} = -\frac{16}{8!\cdot 9} - \frac{5}{6!\cdot 8\cdot 9} = -\frac{51}{9!},$$

and so on. Thus we have

$$u = 1 - \frac{t^3}{3!} + \frac{t^4}{4!} - \frac{t^5}{5!} + \frac{5t^6}{6!} - \frac{10t^7}{7!} + \frac{16t^8}{8!} - \frac{51t^9}{9!} + \cdots.$$

Similarly, to find v, we set $a_0 = 0$ and $a_1 = 1$ and then use the recurrence formula to find (with details omitted)

$$a_2 = -\frac{1}{2},\ a_3 = \frac{1}{3!},\ a_4 = -\frac{3}{4!},\ a_5 = \frac{6}{5!},\ a_6 = -\frac{10}{6!},\ a_7 = \frac{25}{7!},\ a_8 = -\frac{61}{8!}, \ldots.$$

So we have

$$v = t - \frac{t^2}{2} + \frac{t^3}{3!} - \frac{3t^4}{4!} + \frac{6t^5}{5!} - \frac{10t^6}{6!} + \frac{25t^7}{7!} - \frac{61t^8}{8!} + \cdots.$$

The general solution may now be written as $y = c_1 u + c_2 v$. ■

Example 3

Consider the initial-value problem

$$y'' + ty = 1, \quad y(0) = y'(0) = 0,$$

in which we seek the *rest solution* of the differential equation. We will look for the solution in the form of a power series. Toward this end, we set

$$y = \sum_{k=0}^{\infty} a_k t^k$$

and borrow from the previous examples the expansions

$$y'' = \sum_{n=0}^{\infty}(n+1)(n+2)a_{n+2}t^n \text{ and } ty = \sum_{n=0}^{\infty} a_{n-1}t^n,$$

where $a_{-1} = 0$ by definition. The differential equation now requires that

$$\sum_{n=0}^{\infty}((n+1)(n+2)a_{n+2} + a_{n-1})t^n = 1. \tag{5}$$

Now, because of the initial conditions, we must have $a_0 = a_1 = 0$. Also, by equating the constant terms on each side of (5), we find that $2a_2 = 1$; so $a_2 = 1/2$. (Notice that the differential equation requires that $y''(0) = 2a_1 = 1$ for *any* solution.) Furthermore, the remaining coefficients must satisfy

$$(n+2)(n+1)a_{n+2} + a_{n-1} = 0 \quad \text{for } n \geq 3,$$

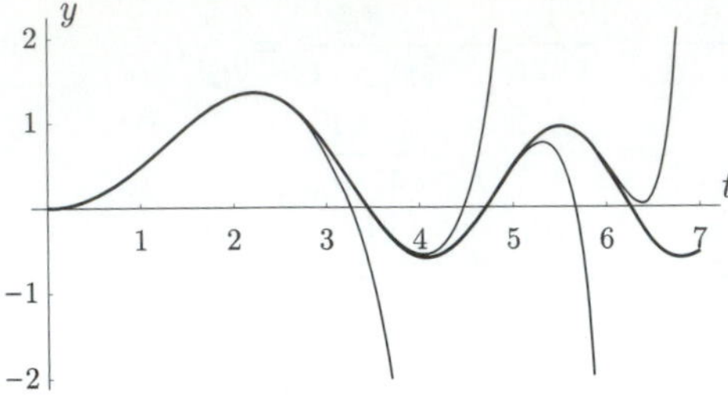

Figure 3

which amounts to the recurrence formula

$$a_{n+2} = \frac{-a_{n-1}}{(n+2)(n+1)} \quad \text{for } n \geq 3.$$

Thus we find that

$$a_0 = a_3 = a_6 = \cdots = 0, \;\; a_1 = a_4 = a_7 = \cdots = 0,$$
$$a_2 = \frac{1}{2}, \;\; a_5 = \frac{-1}{5 \cdot 4 \cdot 2}, \;\; a_8 = \frac{1}{8 \cdot 7 \cdot 5 \cdot 4 \cdot 2}, \;\; a_{11} = -\frac{9 \cdot 6 \cdot 3}{11!}, \;\; \ldots .$$

Therefore, the solution is

$$y = \frac{t^2}{2} - \frac{t^5}{40} + \frac{t^8}{2240} - \frac{t^{11}}{246400} + \cdots .$$

Partial sums of degrees 11, 20, 29, 38, and 47 are plotted for $t \geq 0$ in Figure 3. ■

Convergence We begin our brief discussion of convergence with a theorem that provides an estimate of the radius of convergence for power-series solutions (about $t = 0$) of

$$y'' + py' + qy = 0, \tag{6}$$

in the case where p and q have power-series expansions about $t = 0$; that is, where $t = 0$ is a so-called *ordinary point*. We refer the reader to Coddington [7] or Redheffer [9] for the proof, which is technical and beyond the scope of this book.

Theorem 1

Suppose that p and q have power-series expansions that converge for all t in $(-r, r)$. Then the radius of convergence of any power-series solution of (6) is at least r. In particular, if p and q are polynomials, then the radius of convergence of any power-series solution of (6) is ∞. ▲

Example 4

Consider the equation

$$y'' + y' \sin t + \frac{y}{25 + t^2} = 0.$$

The Maclaurin series

$$\sin t = \sum_{n=0}^{\infty}(-1)^n t^{2n+1}/(2n+1)!$$

converges for all t, while the Maclaurin series

$$\frac{1}{25+t^2} = \frac{1}{25}\sum_{n=0}^{\infty}(-1)^n\left(\frac{t}{5}\right)^{2n},$$

which is similar to a geometric series, converges for $-5 < x < 5$. Therefore, both series converge for $-5 < x < 5$, and so, by Theorem 1, the radius of convergence of any power-series solution of the differential equation is at least 5. ■

Let us consider now the equation

$$p_2 y'' + p_1 y' + p_0 y = 0, \tag{7}$$

where $p_2(0) \neq 0$ and p_0, p_1, and p_2 are polynomials sharing no common factor. Division by p_2 results in

$$y'' + \frac{p_1}{p_2}y' + \frac{p_0}{p_2}, y = 0,$$

which is the same as (7) with $p = p_1/p_2$ and $q = p_0/p_2$; that is, with p and q rational functions that are defined at $t = 0$. It is possible to show (again, see Redheffer [9]) that any rational function f/g, where $g(0) \neq 0$ and f and g share no common factors, has a power-series expansion, whose radius of convergence is precisely equal to the distance from 0 to the nearest complex zero of g; that is, the distance from 0 to the nearest *singularity* of f/g.

Example 5

Consider the rational function

$$q(t) = \frac{1}{t^2+2t+2}$$

This function has a power-series expansion,* which turns out to be

$$\frac{1}{t^2+2t+2} = \frac{1}{2}\left(1 - t + \frac{t^2}{2} - \frac{t^4}{4} + \frac{t^5}{4} - \frac{t^6}{8} + \frac{t^8}{16} - \frac{t^9}{16} + \frac{t^{10}}{32} - \frac{t^{12}}{64} + \cdots\right).$$

Since the singularities of q are $t = -1 \pm i$, and each of these have magnitude

$$|-1 \pm i| = \sqrt{(-1)^2 + 1^2} = \sqrt{2}$$

it follows that the radius of convergence of the series expansion is $\sqrt{2}$; that is, the series converges for $|t| < \sqrt{2}$ and diverges for $|t| > \sqrt{2}$. ■

* This may be obtained from the geometric series $\frac{1}{1-x} = \sum_{n=0}^{\infty} x^n$, since $q(t) = \frac{1}{1+(t+1)^2}$, or, of course, with a CAS.

Since the singularities of the rational coefficients p_1/p_2 and p_0/p_2 are precisely the zeros of p_2, we have the following corollary to Theorem 1.

COROLLARY 1 *Let p_0, p_1, and p_2 be polynomials. If p_2 is constant, then the radius of convergence of any power-series solution of (7) is ∞. Otherwise, the radius of convergence of any power-series solution of (7) is at least $|z|$, where z is the (complex) zero of p_2 that is closest to 0.*

Example 6

Consider the equation

$$(t^2+4)y''+ty'+t^2y=0.$$

Since the zeros of t^2+4 are $\pm 2i$, and these each have magnitude 2, it follows that any power-series solution (about $t=0$) converges at least for $-2<t<2$. ■

The Ratio Test It is often practical to determine the exact radius of convergence of a power-series solution by means of the **ratio test**—primarily when the recurrence formula for the coefficients is a *two-term* recurrence formula. Recall (consult your calculus book) that a series $\sum_{k=1}^{\infty} x_n$ converges (absolutely) if

$$\lim_{k\to\infty} |x_{k+1}/x_k| < 1,$$

and diverges if this limit is greater than 1. (If the limit equals 1, then no conclusion about convergence is reached.)

Example 7

Consider the equation

$$(t^2+2)y''+ty'+y=0.$$

The associated recurrence formula turns out to be (see Problem 8)

$$a_{n+2}=-\frac{(n^2+1)a_n}{2(n^2+3n+2)}.$$

In each of the series expansions of fundamental solutions u and v, the ratio of consecutive nonzero terms is

$$\frac{a_{n+2}t^{n+2}}{a_nt^n}=-\frac{(n^2+1)t^2}{2(n^2+3n+2)}.$$

Therefore, by the ratio test, the series expansions of u and v converge if

$$\lim_{n\to\infty}\frac{(n^2+1)t^2}{2(n^2+3n+2)}<1$$

and diverge if that limit is greater than 1. Computing the limit on the left side, we find that

$$\lim_{n\to\infty} \frac{(n^2+1)t^2}{2(n^2+3n+2)} = \frac{1}{2}t^2.$$

Therefore, the series expansions of u and v converge when $t^2 < 2$ and diverge when $t^2 > 2$; that is, they converge when $|t| < \sqrt{2}$ and diverge when $|t| > \sqrt{2}$. Since every power-series solution can be constructed from u and v, we conclude that the radius of convergence of every power-series solution is $\sqrt{2}$. Note that Corollary 1 tells us that the radius of convergence is *at least* $\sqrt{2}$. We now know that it is exactly $\sqrt{2}$. (Though it often happens that the actual radius of convergence is the same as the lower bound provided by Corollary 1, it is not always the case. Problem 7 provides an example.)

Let's now look at the general case involving a two-term recurrence formula. Suppose that the coefficients in a particular series solution of a differential equation satisfy a two-term recurrence formula of the form

$$a_{n+k} = f(n)a_n,$$

beginning with coefficients $a_0, a_1, \ldots, a_{k-1}$, of which only one is not zero. The ratio of consecutive nonzero terms is then

$$\frac{a_{n+k}t^{n+k}}{a_n t^n} = f(n)t^k.$$

Thus, by the ratio test, the series converges absolutely for all t such that

$$|t|^k \lim_{n\to\infty} |f(n)| < 1$$

and diverges for all t for which that limit is greater than 1. Consequently, the radius of convergence ρ for any such series solution is given by

$$\rho = \sqrt[k]{\lim_{n\to\infty} \frac{1}{|f(n)|}}. \tag{8}$$

In particular, we conclude that $\rho = \infty$ whenever $\lim_{n\to\infty} f(n) = 0$. ■

Example 8

Consider the equation

$$(t^3 + \beta)y'' + ty = 0.$$

The associated recurrence formula turns out to be (see Problem 9)

$$a_{n+3} = -\frac{n(n-1)+1}{\beta(n+3)(n+2)}a_n.$$

Using (8), we find that the radius of convergence of fundamental solutions u and v in the form of power series is

$$\rho = \sqrt[3]{\lim_{n\to\infty} \frac{|\beta|(n+3)(n+2)}{n(n-1)+1}} = \sqrt[3]{|\beta|}.$$

Therefore, every power-series solution has radius of convergence $\rho = \sqrt[3]{|\beta|}$. ■

PROBLEMS

1. Using the known power-series expansion

$$e^t = \sum_{k=0}^{\infty} \frac{t^k}{k!},$$

describe the coefficient a_n of t^n, $n = 0, 1, 2, \ldots$, in the power-series expansion of each of the following.

(a) $e^t + te^{-t}$ **(b)** $\dfrac{e^t - 1}{t}$ **(c)** $(t^2 - t + 1)e^t$

2. Using the known power-series expansions

$$\cos t = \sum_{k=0}^{\infty} \frac{(-1)^k t^{2k}}{(2k)!} \quad \text{and} \quad \sin t = \sum_{k=0}^{\infty} \frac{(-1)^k t^{2k+1}}{(2k+1)!},$$

describe the coefficient a_n of t^n, $n = 0, 1, 2, \ldots$, in the power-series expansion of each of the following.

(a) $(1+t)\cos t$ **(b)** $\cos t - \dfrac{\sin t}{t}$ **(c)** $\dfrac{\cos t - 1}{t}$

3. Using the known power-series expansion

$$\frac{1}{1-t} = \sum_{k=0}^{\infty} t^k,$$

describe the coefficient a_n of t^n, $n = 0, 1, 2, \ldots$, in the power-series expansion of each of the following.

(a) $\dfrac{1}{1+t} + \dfrac{t}{1-2t}$, **(b)** $\dfrac{t^3}{1+t^2}$ **(c)** $\dfrac{1+t}{1-t}$

4. **(a)** Find the recurrence formula for the coefficients in any power-series solution of $y'' + y = 0$.

(b) Derive the power-series expansions of $\cos t$ and $\sin t$ by finding a pair of fundamental solutions in the form of power series for the equation $y'' + y = 0$.

5. The solution of

$$y'' + 2y' + 2y = 0, \quad y(0) = 1, \quad y'(0) = -1,$$

is $y = e^{-t}\cos t$. Verify that fact, and then find the series expansion of $e^{-t}\cos t$ by finding a power-series solution of the initial-value problem.

6. The solution of

$$y'' + 2y' + 2y = 0, \quad y(0) = 0, \quad y'(0) = 1,$$

is $y = e^{-t}\sin t$. Verify that fact, and then find the series expansion of $e^{-t}\sin t$ by finding a power-series solution of the initial-value problem.

7. For the equation

$$(t-1)y'' - ty' + y = 0,$$

(a) use Corollary 1 to estimate the radius of convergence of every nonterminating power-series solution;

(b) find the recurrence formula for the coefficients in any power-series solution;

(c) find the fundamental power-series solutions and identify them as familiar elementary functions. What is the actual radius of convergence of every power-series solution?

The equations in Problems 8 and 9 are from Examples 7 and 8, respectively. For each of them, derive the recurrence formula stated in the corresponding example, and then find fundamental power-series solutions u and v.

8. $(t^2+2)y'' + ty' + y = 0$

9. $(t^3+\beta)y'' + ty = 0$

For the equation in each of Problems 10 through 17,

(a) find the recurrence formula for the coefficients in any power-series solution;

(b) find fundamental power-series solutions u and v;

(c) use Corollary 1 to estimate the radius of convergence for any power-series solution. If that result is not ∞, then use the recurrence formula and the ratio test to compute the exact radius of convergence.

10. $y'' - ty' + y = 0$

11. $y'' - ty' - y = 0$

12. $y'' - t^2y = 0$

13. $(4-t^2)y'' - y = 0$

14. $(8+t^3)y'' + ty = 0$

15. $(1+t^2)y'' + ty' - y = 0$

16. $y'' - 2ty' + 4y = 0$

17. $(1-t^2)y'' - 2ty' + 6y = 0$

For each of Problems 18 through 23, find the rest solution of the equation in the form of a power series.

18. $y'' + ty = 1$

19. $y'' + ty = t$

20. $y'' + t^2y = 1$

21. $y'' - ty' + y = 1$

22. $y'' - ty' - y = t$

23. $y'' - t^2y = t^2$

24. Find the value of the numerical sum $\sum_{n=1}^{\infty} \frac{n2^n}{(n+1)!}$ as follows.

(a) Note that the sum is equal to $y(2)$, where

$$y(t) = \sum_{n=1}^{\infty} \frac{nt^n}{(n+1)!},$$

and show that $ty'(t) + y(t)$ is the power-series expansion of a familiar function.

(b) Find the solution of the resulting linear, first-order differential equation subject to the initial condition $\lim_{t\to 0} y(t) = 0$. Then compute $y(2)$.

25. Find a "closed form" for the function

$$y(t) = \sum_{n=0}^{\infty} \frac{n^2 - n + 4}{4n!} t^{2n}$$

by recognizing a closed form of $y'(t) - 2ty(t)$ and then solving an initial-value problem.

26. Consider the equation $t^2y'' - 2mty' + (m(m+1) - t^3)y = 0$.

(a) Show that the substitution $y(t) = t^m u(t)$ produces Airy's equation $u'' - tu = 0$.

(b) Find the general solution of $t^2y'' - 2ty' + (2 - t^3)y = 0$ in terms of fundamental power-series solutions of Airy's equation.

(c) Find the general solution of $t^2y'' - ty' + (\frac{3}{4} - t^3)y = 0$ in terms of fundamental power-series solutions of Airy's equation.

27. Consider the equation $4ty'' + 2y' + y = 0$.

(a) Show that the method of this section does not produce two linearly independent solutions.

Suppose that ν is a constant such that $t^{-\nu}y$ has a convergent power-series expansion on an interval $I = [0, r)$. Then

$$y = t^{\nu} \sum_{k=0}^{\infty} a_k t^k = \sum_{k=0}^{\infty} a_k t^{k+\nu} \quad \text{for } 0 \le t < r.$$

Such an expansion is called a **Frobenius expansion**. Note that we may as well assume that $a_0 \neq 0$, for if $a_0 = a_1 = \cdots = a_{\ell-1} = 0$, then we can simply replace ν with $\nu + \ell$ and shift the summation index by ℓ.

(b) Show that the resulting form of $4ty'' + 2y' + y = 0$ is

$$2\nu(2\nu - 1)a_0 t^{\nu-1} + \sum_{n=0}^{\infty}((2n + 2\nu + 2)(2n + 2\nu + 1)a_{n+1} + a_n)t^{n+\nu} = 0.$$

(c) Observe that $4ty'' + 2y' + y = 0$ is true for all t in I, if and only if

(i) $2\nu(2\nu - 1) = 0$,

and

(ii) $a_{n+1} = \dfrac{-a_n}{(2n + 2\nu + 2)(2n + 2\nu + 1)}, \quad n = 0, 1, 2, 3, \ldots.$

(d) For each solution of $2\nu(2\nu - 1) = 0$, take $a_0 = 1$ and find a nontrivial solution in the form of a Frobenius expansion.

(For theory related to the *Method of Frobenius*, see Redheffer [9] or Simmons [10].)

For each equation in Problems 28 through 31, find a pair of linearly independent solutions by the method of Problem 27.

28. $2ty'' + y' - 2y = 0$

29. $t^2y'' + ty' + \left(t^2 - \frac{1}{9}\right)y = 0$

30. $t(1 - t)y'' + \left(\frac{1}{2} - 3t\right)y' - y = 0$

31. $9t^2y'' + t(2 + 3t^2)y' + (3t^2 - 2)y = 0$

5.7 Polynomial Solutions

In this very brief section, we look at an important example of a differential equation that, for certain values of a parameter in the equation, has a terminating power-series solution—in other words, a *polynomial* solution. The result is a family of polynomials with numerous interesting and useful properties. Other examples are explored in the problems at the end of the section.

Example 1

Chebyshev's equation is

$$(1-t^2)y''-ty'+m^2y=0, \tag{1}$$

where m is a constant. We will look for power-series solutions

$$y=\sum_{k=0}^{\infty}a_kt^k.$$

The necessary power-series expansions are

$$y''=\sum_{n=0}^{\infty}(n+2)(n+1)a_{n+2}t^n, \quad t^2y''=\sum_{n=0}^{\infty}n(n-1)a_nt^n, \quad ty'=\sum_{n=0}^{\infty}na_nt^n,$$

from which we see that in order for the differential equation to be satisfied, we must have

$$(n+2)(n+1)a_{n+2}-n(n-1)a_n-na_n+m^2a_n=0, \quad n=0,1,2,\dots.$$

From this we obtain the recurrence formula

$$a_{n+2}=-\frac{(m^2-n^2)a_n}{(n+2)(n+1)}, \quad n=0,1,2,\dots.$$

As in Examples 1 and 2 of Section 5.6, we will find fundamental solutions u and v satisfying $u(0)=1$, $u'(0)=0$ and $v(0)=0$, $v'(0)=1$. In the expansion for u we have $a_0=1$ and $a_1=0$ and therefore

$$a_2=-\frac{m^2}{2}, \quad a_4=\frac{m^2(m^2-2^2)}{4\cdot3\cdot2}, \quad a_6=-\frac{m^2(m^2-2^2)(m^2-4^2)}{6!}, \dots,$$
$$a_3=a_5=a_7=\cdots=0.$$

In the expansion for v we have $a_0=0$ and $a_1=1$ and therefore

$$a_2=a_4=a_6=\cdots=0,$$
$$a_3=-\frac{m^2-1}{3\cdot2}, \quad a_5=\frac{(m^2-1)(m^2-3^2)}{5!}, \quad a_7=-\frac{(m^2-1)(m^2-3^2)(m^2-5^2)}{7!}, \dots.$$

The ratio test shows that any power-series solution converges for $|t| < 1$ and diverges for $|t| > 1$. It turns out that power-series solutions converge for $t = \pm 1$. (See Problem 8.) Notice that when m is a nonnegative integer, either u or v will have a terminating series expansion. If m is even, then all coefficients beyond a_m in the expansion of u will be zero. If m is odd, then all coefficients beyond a_m in the expansion of v will be zero. Therefore, whenever m is a nonnegative integer, the differential equation has a polynomial solution of degree m. For even m, the polynomial solution is

$$\begin{aligned} u_m = 1 - \frac{m^2}{2}t^2 + \frac{m^2(m^2-2^2)}{4!}t^4 - \cdots \\ +(-1)^{\frac{m}{2}} \frac{m^2(m^2-2^2)\cdots(m^2-(m-2)^2)}{m!}t^m. \end{aligned}$$

For odd m, the polynomial solution is

$$\begin{aligned} v_m = t - \frac{m^2-1}{3!}t^3 + \frac{(m^2-1)(m^2-3^2)}{5!}t^5 - \cdots \\ +(-1)^{\frac{m+1}{2}} \frac{(m^2-1)\cdots(m^2-(m-2)^2)}{m!}t^m. \end{aligned}$$

The first few of these are

$$1,\ t,\ 1-2t^2,\ t-\frac{4}{3}t^3,\ 1-8t^2+8t^4,\ t-4t^3+\frac{16}{5}t^5, \ldots .$$

Scaling to obtain integer coefficients produces the **Chebyshev polynomials*** $T_m(t)$, given for $m = 0, 1, 2, 3, \ldots$ by

$$1,\ t,\ 2t^2-1,\ 4t^3-3t,\ 8t^4-8t^2+1,\ 16t^5-20t^3+5t, \ldots .$$

The first five Chebyshev polynomials are plotted on $[-1, 1]$ in Figure 1.

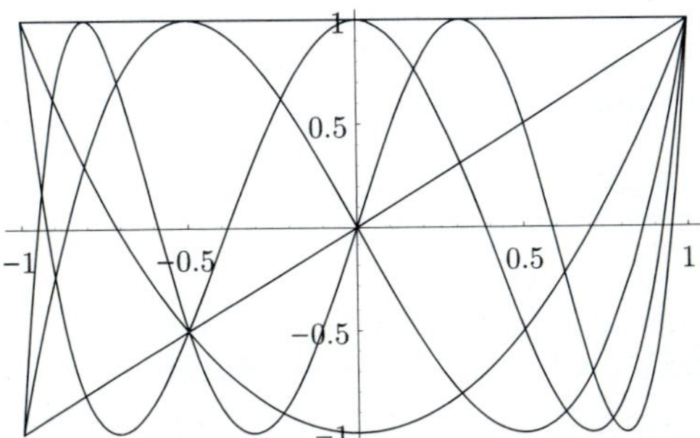

Figure 1

---

* An alternative spelling of Chebyshev is *Tschebychev*; thus the letter "T" for the Chebyshev polynomials.

The Chebyshev polynomials have many interesting properties that make them especially important in the field of numerical analysis (see, e.g., Kincaid and Cheney [21]). One such property is that each T_m has exactly m simple zeros, all in the interval $[-1, 1]$. (See Problem 9.) This illustrated by Figure 1. Another property of the Chebyshev polynomials T_m is that they are *mutually orthogonal* on the interval $[-1, 1]$ with respect to the *weight function* $w(t) = (1 - t^2)^{-1/2}$; that is, if $m \neq n$, then

$$\int_{-1}^{1} T_m(t)T_n(t)\frac{dt}{\sqrt{1-t^2}} = 0. \tag{2}$$

(Note that the integral is improper.) This property can be derived directly from Chebyshev's equation as follows. First, to simplify notation, define the operator $\mathcal{L}$ by

$$\mathcal{L}y = -\sqrt{1-t^2}\left(\sqrt{1-t^2}y'\right)' = -(1-t^2)y'' + ty',$$

and note that Chebyshev's equation can then be written as

$$\mathcal{L}y = m^2 y.$$

The crucial property of $\mathcal{L}$ is a type of symmetry; namely, that

$$\int_{-1}^{1} f\mathcal{L}g\frac{dt}{\sqrt{1-t^2}} = \int_{-1}^{1} g\mathcal{L}f\frac{dt}{\sqrt{1-t^2}} \tag{3}$$

for any twice differentiable functions f and g on $[-1, 1]$ for which the improper integrals involved are finite. To see this, first observe that

$$\begin{aligned}\int_{-1}^{1} f\mathcal{L}g\frac{dt}{\sqrt{1-t^2}} &= -\int_{-1}^{1} f\sqrt{1-t^2}\left(\sqrt{1-t^2}g'\right)'\frac{dt}{\sqrt{1-t^2}} \\ &= -\int_{-1}^{1} f(\sqrt{1-t^2}g')'\,dt,\end{aligned}$$

and so integrating by parts gives

$$\begin{aligned}\int_{-1}^{1} f\mathcal{L}g\frac{dt}{\sqrt{1-t^2}} &= -fg'\sqrt{1-t^2}\Big|_{-1}^{1} + \int_{-1}^{1} f'g'\sqrt{1-t^2}\,dt \\ &= \int_{-1}^{1} f'g'\sqrt{1-t^2}\,dt.\end{aligned}$$

Similarly, or by simply interchanging f and g, we find that

$$\int_{-1}^{1} g\mathcal{L}f\frac{dt}{\sqrt{1-t^2}} = \int_{-1}^{1} f'g'\sqrt{1-t^2}\,dt,$$

which establishes the truth of (3). Now let $f = T_m$ and $g = T_n$, where m and n are nonnegative integers. Then

$$\mathcal{L}T_m = m^2 T_m \quad \text{and} \quad \mathcal{L}T_n = n^2 T_n,$$

and (3) becomes

$$n^2 \int_{-1}^{1} T_m T_n \frac{dt}{\sqrt{1-t^2}} = m^2 \int_{-1}^{1} T_m T_n \frac{dt}{\sqrt{1-t^2}};$$

that is,

$$(n^2 - m^2) \int_{-1}^{1} T_m T_n \frac{dt}{\sqrt{1-t^2}} = 0.$$

Therefore, if $m \neq n$, then the integral must be 0. That proves (2). (Note that (2) is true not just for distinct pairs of Chebyshev polynomials, but for any two solutions of (1) corresponding to different values of the parameter m.) ■

PROBLEMS

1. *Legendre's equation* is

$$(1-t^2)y'' - 2ty' + m(m+1)y = 0,$$

where m is a nonnegative constant.

(a) Find the recurrence formula for the coefficients in any power-series solution.

(b) Find a pair of fundamental solutions in the form of power series.

(c) If m is a nonnegative integer, then Legendre's equation has a polynomial solution P_m of degree m. Scaled so that $P_m(1) = 1$, these are the *Legendre polynomials.* Find $P_0, P_1, \ldots, P_4$.

(d) Find the radius of convergence of any nonterminating power-series solution.

2. An important property of the Legendre polynomials P_m (see Problem 1) is that, just like the Chebyshev polynomials, they are mutually orthogonal on the interval $[-1, 1]$; that is, if $m \neq n$, then

$$\int_{-1}^{1} P_m(t) P_n(t)\, dt = 0.$$

This property can be derived from Legendre's equation as follows. First, to simplify notation, define the operator $\mathcal{L}$ by

$$\mathcal{L}y = -((1-t^2)y')',$$

noting that Legendre's equation can then be written as $\mathcal{L}y = m(m+1)y$.

(a) Using integration by parts, show that

$$\int_{-1}^{1} f \mathcal{L} g\, dt = \int_{-1}^{1} g \mathcal{L} f\, dt$$

for any twice differentiable functions f and g on $[-1, 1]$.

(b) Suppose that f and g satisfy $\mathcal{L}f = m(m+1)f$ and $\mathcal{L}g = n(n+1)g$ on $[-1, 1]$. Using the result of part (a), show that if $m \neq n$, then $\int_{-1}^{1} f(t)g(t)\, dt = 0$.

3. *Hermite's equation* is

$$y'' - 2ty' + 2my = 0,$$

where m is a nonnegative constant.

(a) Find the recurrence formula for the coefficients in any power-series solution.

(b) Find a pair of fundamental solutions in the form of power series.

(c) If m is a nonnegative integer, then Hermite's equation has a polynomial solution H_m of degree m. Scaled so that the leading coefficient of H_m is 2^m, these are the *Hermite polynomials.* Find $H_0, H_1, \ldots, H_4$.

(d) Find the radius of convergence of any nonterminating power-series solution.

4. An important property of the Hermite polynomials H_m (see Problem 3) is that they are mutually orthogonal on $(-\infty, \infty)$ with respect to the weight function $w(t) = e^{-t^2}$; that is, if $m \neq n$, then

$$\int_{-\infty}^{\infty} H_m(t) H_n(t) e^{-t^2}\, dt = 0.$$

This property can be derived from Hermite's equation as follows. First, to simplify notation, define the operator $\mathcal{L}$ by

$$\mathcal{L}y = -e^{t^2}(e^{-t^2}y')',$$

and note that Hermite's equation can then be written as $\mathcal{L}y = 2my$.

(a) Using integration by parts, show that

$$\int_{-\infty}^{\infty} f\mathcal{L}ge^{-t^2}\,dt = \int_{-\infty}^{\infty} g\mathcal{L}fe^{-t^2}\,dt$$

for any twice differentiable functions f and g on $(-\infty, \infty)$ such that the improper integrals involved are finite.

(b) Suppose that f and g satisfy $\mathcal{L}f = 2mf$ and $\mathcal{L}g = 2ng$ on $(-\infty, \infty)$. Using the result of part (a), show that if $m \neq n$, then $\int_{-\infty}^{\infty} f(t)g(t)e^{-t^2}\,dt = 0$. Assume that all improper integrals involved are finite.

5. If $\mathcal{L}$ has the property that

$$\int_a^b f\mathcal{L}gw\,dt = \int_a^b g\mathcal{L}fw\,dt$$

for all f,g in $\mathcal{D}$, where w is some positive weight function on $[a, b]$, then we say that the operator $\mathcal{L}$ is *symmetric* with respect to w.

(a) Show that if $\mathcal{L}$ is defined by $\mathcal{L}y = (qy')'/w$, where q is a differentiable function on $[a, b]$ and $q(a) = q(b) = 0$, then $\mathcal{L}$ is symmetric with respect to w.

Suppose that $\mathcal{L}$ is a linear operator that acts upon a set of functions $\mathcal{D}$ defined on an interval $[a, b]$. If there is a real number λ and a nonzero function f in $\mathcal{D}$ such that $\mathcal{L}f = \lambda f$ on $[a, b]$, then we say that λ is an *eigenvalue* of $\mathcal{L}$ with corresponding *eigenfunction* f.

(b) Show that if $\mathcal{L}$ is symmetric with respect to w, then any two eigenfunctions of $\mathcal{L}$, corresponding to different eigenvalues λ_1 and λ_2, are orthogonal on the interval $[a, b]$ with respect to the weight function w; that is,

$$\text{if } \mathcal{L}f = \lambda_1 f,\ \mathcal{L}g = \lambda_2 g, \text{ and } \lambda_1 \neq \lambda_2,$$
$$\text{then } \int_a^b f(t)g(t)w(t)\,dt = 0.$$

Note that the orthogonality properties of the Chebyshev, Legendre, and Hermite polynomials all follow as special cases.

6. This is a finite-dimensional analogue of Problem 5. Suppose that A is an $n \times n$ matrix. If there is a real number λ and a nonzero vector x in $\mathbb{R}^n$ such that $Ax = \lambda x$, then we say that λ is an *eigenvalue* of A with corresponding *eigenvector* x.

(a) Show that if A is symmetric (i.e., $A^T = A$), then $x^T Ay = y^T Ax$ for all x, y in $\mathbb{R}^n$. (*Hint*: $x^T Ay = (Ay)^T x$.)

(b) Show that if A is symmetric, then any two eigenvectors of A, corresponding to different eigenvalues λ_1 and λ_2, are orthogonal; that is,

$$\text{if } Ax = \lambda_1 x,\ Ay = \lambda_2 y, \text{ and}$$
$$\lambda_1 \neq \lambda_2, \text{ then } x^T y = 0.$$

7. Let $\eta_m(t) = \cos(m \cos^{-1} t)$ and $\sigma_m(t) = \sin(m \cos^{-1} t)$ for $-1 \leq t \leq 1$.

(a) Show that η_m and σ_m are solutions of Chebyshev's equation on $[-1, 1]$.

(b) Compute the Wronskian $W(\eta_m, \sigma_m)$, and use it to show that η_m and σ_m are linearly independent on $(-1, 1)$, and therefore on $[-1, 1]$. Conclude that the general solution of Chebyshev's equation on $[-1, 1]$ is

$$y = c_1\eta_m + c_2\sigma_m.$$

8. Let m be a nonnegative integer, and let η_m and σ_m be as in Problem 8. Also, let u_m and v_m be the fundamental solutions of Chebyshev's equation (i.e., $u_m(0) = v'_m(0) = 1$ and $u'_m(0) = v_m(0) = 0$). Recall that u_m is a polynomial if m is even, and v_m is a polynomial if m is odd.

(a) Compute $\eta_m(0)$, $\eta'_m(0)$, $\sigma_m(0)$, and $\sigma'_m(0)$.

(b) Suppose that m is even. Find c_1 and c_2 so that $u_m = c_1\eta_m + c_2\sigma_m$.

(c) Suppose that m is odd. Find c_1 and c_2 so that $v_m = c_1\eta_m + c_2\sigma_m$.

(d) What (perhaps surprising) fact can you conclude about η_m when m is any nonnegative integer?

9. The traditional *definition* of the Chebyshev polynomials, for $-1 \leq t \leq 1$, is

$$T_m(t) = \cos(m \cos^{-1} t), \quad m = 1, 2, 3, \ldots .$$

(This is the function η_m from Problems 7 and 8.)

(a) Apply the sum and difference formulas for cosine to T_{m+1} and T_{m-1}, respectively, in order to show that

$$T_{m+1} = 2tT_m - T_{m-1}, \quad m = 1, 2, 3, \ldots .$$

This formula, starting with $T_0(t) = 1$ and $T_1(t) = t$, provides an easy way to calculate the Chebyshev polynomials.

(b) Use the formula from part (a) to argue that every Chebyshev polynomial has integer coefficients and that the leading coefficient is 2^{m-1}.

10. Using the definition in Problem 9, show that the Chebyshev polynomials have the following properties.

(a) $|T_m(t)| \leq 1$ for $-1 \leq t \leq 1$.

(b) $T_m(t) = 0$ for each of

$$t = \cos\left(\frac{(2k-1)\pi}{2m}\right), \quad k = 1, 2, 3, \ldots, m.$$

Consequently, T_m has m distinct (and therefore simple) zeros in $[-1, 1]$.

CHAPTER

6

SECOND-ORDER LINEAR EQUATIONS II

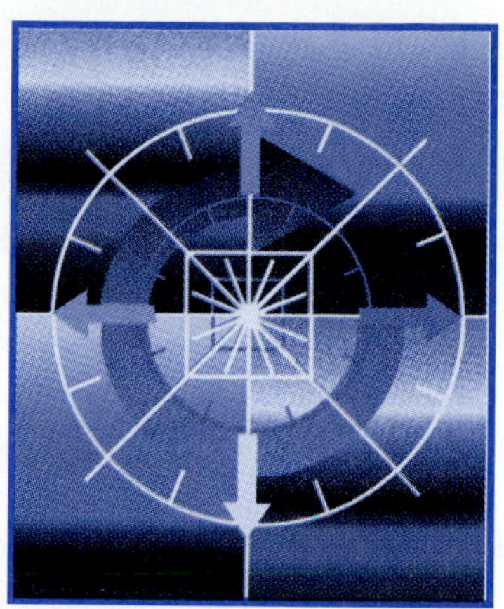

In Chapter 5 we learned about general, second-order, linear differential equations and the structure of their solutions. We also looked at some models of simple vibrations that give rise to such equations, often with constant coefficients. In this chapter we are going to learn how to find explicit solutions to second-order linear equations with constant coefficients, after which we will study their behavior in connection with simple vibration models.

6.1 Homogeneous Equations with Constant Coefficients

We first consider homogeneous second-order linear equations in the form

$$y'' + py' + qy = 0, \tag{1}$$

where p and q are constants, and we assume that $q \neq 0$. (If $q = 0$, the equation is easily reduced to one of first order.) It is sensible to look for solutions of the form e^{rt} by virtue of the fact that differentiation of e^{rt} produces constant multiples of itself. Putting $y = e^{rt}$ into the equation yields

$$r^2e^{rt} + pre^{rt} + qe^{rt} = 0.$$

Now dividing through by e^{rt} produces a quadratic equation for r:

$$r^2 + pr + q = 0. \tag{2}$$

This quadratic equation is called the **characteristic equation** corresponding to the differential equation (1).

The point is that if r satisfies the characteristic equation (2), then e^{rt} satisfies the differential equation (1). The characteristic equation will, of course, have different types of solutions depending on the coefficients p and q. In fact, the quadratic formula gives the roots:

$$r_1, r_2 = \frac{1}{2}\left(-p \pm \sqrt{p^2 - 4q}\right).$$

Thus there are three cases:

i. If $p^2 - 4q > 0$, then there are two distinct real roots r_1, r_2.
ii. If $p^2 - 4q = 0$, then there is only one (repeated) root: $r_1 = r_2 = -p/2$.
iii. If $p^2 - 4q < 0$, then there is a pair of (nonreal) complex conjugate roots.

Distinct Real Roots In the first case, in which $p^2 - 4q > 0$, we have completely solved our problem. That is, since r_1 and r_2 are two distinct real roots of the characteristic equation, $e^{r_1 t}$ and $e^{r_2 t}$ are two linearly independent solutions of the differential equation, and so Theorem 1 of Section 6.4 tells us that the general solution of (1) is given by

$$y = c_1 e^{r_1 t} + c_2 e^{r_2 t},$$

where c_1, c_2 are arbitrary constants.

So let us suppose for the time being that $p^2 - 4q > 0$, so that r_1 and r_2 are real and distinct. Since we assume that $q \neq 0$, we are guaranteed that zero is not a root, and therefore guaranteed that the roots are either both positive, both negative, or of opposite sign. The general behavior of solutions is determined by which of these is true. Examples 1 through 3 illustrate this.

Example 1

Consider the equation

$$y'' - 5y' + 6y = 0.$$

The corresponding characteristic equation is

$$r^2 - 5r + 6 = 0,$$

which upon factoring becomes $(r-2)(r-3) = 0$ and hence has solutions $r_1 = 2$ and $r_2 = 3$. Therefore, e^{2t} and e^{3t} are solutions of the differential equation. Since these are linearly independent, the general solution is given by

$$y = c_1 e^{2t} + c_2 e^{3t}.$$

Consequently, every nonzero solution approaches either $\pm\infty$ as $t \to \infty$. Figure 1a shows a typical solution, with initial values $y(0) = 1$ and $y'(0) = 1$, along with its derivative (dashed). Figure 1b shows several phase-plane orbits, including one corresponding to the solution in Figure 1a. Note that the direction of flow along phase-plane orbits is away from the origin. Phase-plane orbits approach the origin as $t \to -\infty$. ■

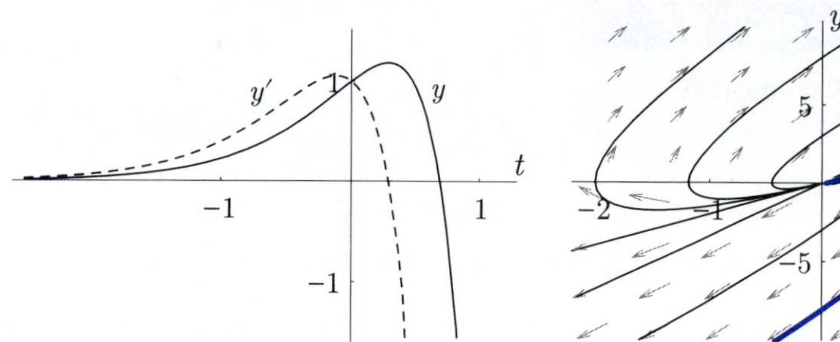

Figure 1a

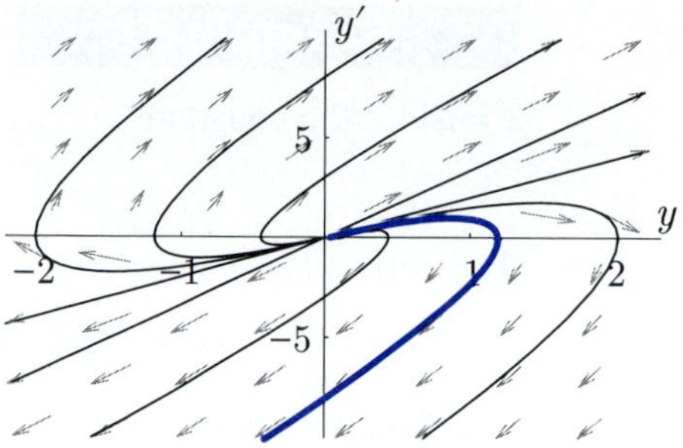

Figure 1b

Example 2

Consider the equation

$$y'' + 5y' + 6y = 0.$$

The corresponding characteristic equation is

$$r^2 + 5r + 6 = 0,$$

which upon factoring becomes $(r+2)(r+3) = 0$ and hence has solutions $r_1 = -2$ and $r_2 = -3$. Therefore, e^{-2t} and e^{-3t} are solutions of the differential equation. Since these are linearly independent, the general solution is given by

$$y = c_1 e^{-2t} + c_2 e^{-3t}.$$

Consequently, every solution approaches zero as $t \to \infty$. Figure 2a shows a typical solution, with initial values $y(0) = 1$ and $y'(0) = 1$, along with its derivative (dashed). Figure 2b shows several phase-plane orbits, including one corresponding to the solution in Figure 2a. Note that the direction of flow along phase-plane orbits is always toward the origin. ■

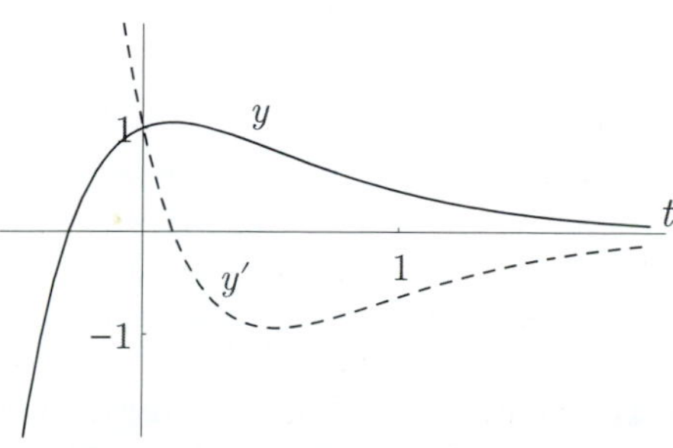

Figure 2a

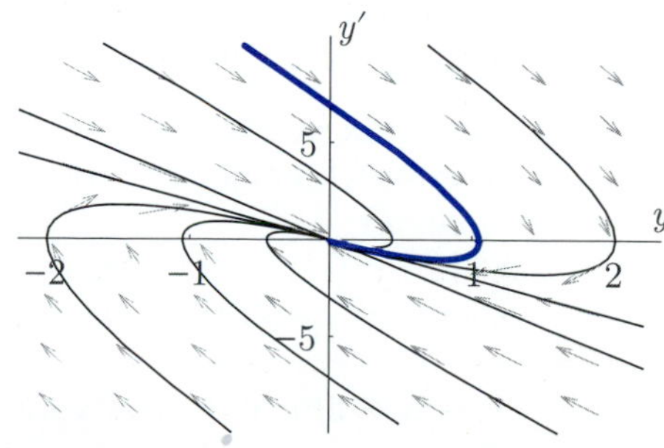

Figure 2b

Example 3

Consider the equation

$$y'' + y' - 6y = 0.$$

The corresponding characteristic equation is

$$r^2 + r - 6 = 0,$$

which upon factoring becomes $(r - 2)(r + 3) = 0$ and hence has solutions $r_1 = 2$ and $r_2 = -3$. Therefore, e^{2t} and e^{-3t} are solutions of the differential equation. Since these are linearly independent, the general solution is given by

$$y = c_1 e^{2t} + c_2 e^{-3t}.$$

Every nonzero solution in which $c_1 \neq 0$ approaches either $\pm\infty$ as $t \to \infty$. If $c_1 = 0$, then the solution approaches zero as $t \to \infty$. Figure 3a shows a typical solution, with initial values $y(0) = 0$ and $y'(0) = 1$, along with its derivative (dashed). Figure 3b shows several phase-plane orbits, including one corresponding to the solution in Figure 3a. ■

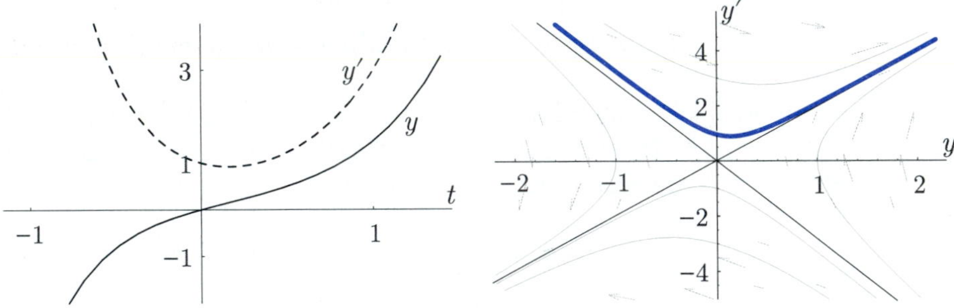

Figure 3a

Figure 3b

Repeated Real Root In the second case, in which the characteristic polynomial has only one (repeated) root given by $r_1 = -p/2$, we find only a one-parameter family of exponential solutions. Since two linearly independent solutions are necessary to describe the general solution, we need to find a second linearly independent solution. This problem will be addressed in Section 6.2.

Complex Roots In the third case, in which the characteristic polynomial has a pair of complex conjugate roots, the preceding method does produce a pair of *complex solutions*:

$$e^{(\alpha \pm \beta i)t}, \quad \text{where} \quad \alpha = -p/2 \quad \text{and} \quad \beta = \sqrt{4q - p^2}/2.$$

It is not difficult to verify this if we simply assume that the differentiation formula $\frac{d}{dt}e^{ct} = ce^{ct}$ is valid when c is a complex constant. However, we have not yet given a meaning to complex powers of e, and, moreover, our goal is to find *real* solutions.

In Section 6.3 we will address these issues and find out how to produce a pair of linearly independent, real solutions from a single complex solution.

Operator Form and the Characteristic Polynomial Notice that we can rewrite equation (1) in operator form as

$$(\mathcal{D}^2 + p\mathcal{D} + q\mathcal{I})y = 0$$

or even more succinctly as

$$P(\mathcal{D})y = 0,$$

where P is the quadratic polynomial

$$P(r) = r^2 + pr + q.$$

This is precisely the quadratic polynomial that appears in the characteristic equation. This polynomial is called the **characteristic polynomial** of the operator $P(\mathcal{D})$. The characteristic polynomial P describes both the differential equation and the characteristic equation; that is, the differential equation is $P(\mathcal{D})y = 0$ and the characteristic equation is $P(r) = 0$. The main point of this section can now be stated as follows:

If r is a real number satisfying $P(r) = 0$,
then $y = e^{rt}$ satisfies the differential equation $P(\mathcal{D})y = 0$.

Now suppose that the characteristic polynomial $P(r) = r^2 + pr + q$ has roots r_1 and r_2. This means that $P(r)$ factors into

$$P(r) = (r - r_1)(r - r_2),$$

which in turn means that the operator $P(\mathcal{D}) = \mathcal{D}^2 + p\mathcal{D} + q\mathcal{I}$ factors into

$$P(\mathcal{D}) = (\mathcal{D} - r_1\mathcal{I})(\mathcal{D} - r_2\mathcal{I}).$$

Thus, when the operator is written in factored form, the roots of the characteristic polynomial are readily seen by inspection. So $e^{r_1 t}$ and $e^{r_2 t}$ satisfy

$$(\mathcal{D} - r_1\mathcal{I})(\mathcal{D} - r_2\mathcal{I})y = 0,$$

and, if r_1 and r_2 are real and distinct, the general solution of this equation may be written down by inspection:

$$y = c_1 e^{r_1 t} + c_2 e^{r_2 t}.$$

More generally, when P is a polynomial of any degree, the following four statements are equivalent:

i. r_0 is a solution of $P(r) = 0$;
ii. $r - r_0$ is a factor of $P(r)$;
iii. $e^{r_0 t}$ is a particular solution of $P(\mathcal{D})y = 0$;
iv. $\mathcal{D} - r_0\mathcal{I}$ is a factor of the operator $P(\mathcal{D})$.

Example 4

Consider the equation

$$(\mathcal{D} - 2\mathcal{I})(\mathcal{D} + 5\mathcal{I})y = 0,$$

which is equivalent to $y'' + 3y' - 10y = 0$. The factored form of the characteristic polynomial is $P(r) = (r - 2)(r + 5)$, which has roots $r_1 = 2$ and $r_2 = -5$. Thus, the general solution of the differential equation is

$$y = c_1 e^{2t} + c_2 e^{-5t}.$$

■

Example 5

The differential equation

$$(\mathcal{D} + 5\mathcal{I})^2 y = 0$$

is equivalent to $y'' + 10y' + 25y = 0$. The characteristic polynomial is $P(r) = (r+5)^2$, which has only one (repeated) real root $r_1 = -5$. Thus, e^{-5t} is a particular solution of the differential equation. A second linearly independent solution must be found by some other means before we can write down the general solution. This issue will be addressed in the next section. ■

PROBLEMS

In Problems 1 through 6, solve the appropriate characteristic equation; then write down the general solution of the differential equation.

1. $y'' + 5y' + 4y = 0$

2. $y'' + 4y' - 5y = 0$

3. $y'' - 2y' - 8y = 0$

4. $y'' - 4y = 0$

5. $y'' + 5y' + 6y = 0$

6. $y'' - 4y' = 0$

In Problems 7 through 10, solve the appropriate characteristic equation; then write down a particular solution of the differential equation.

7. $y'' + 2y' + y = 0$

8. $y'' + 6y' + 9y = 0$

9. $y'' - 4y' + 4y = 0$

10. $y'' = 0$

For each operator $\mathcal{T}$ defined in Problems 11 through 16, express $\mathcal{T}$ in the form $P(\mathcal{D})$, where P is the characteristic polynomial of $\mathcal{T}$. Then find the roots of the characteristic polynomial.

11. $\mathcal{T}y = y'' + 3y' + 2y$

12. $\mathcal{T}y = y'' - 2y' + 5y$

13. $\mathcal{T}y = y'' + 2y' + y$

14. $\mathcal{T}y = y'' + y$

15. $\mathcal{T}y = y'' + 2y' + 2y$

In Problems 16 through 18, write down the general solution by inspection.

16. $(\mathcal{D} - 2\mathcal{I})(\mathcal{D} - 3\mathcal{I})y = 0$

17. $(\mathcal{D} - \mathcal{I})\mathcal{D}y = 0$

18. $(\mathcal{D} + \mathcal{I})(\mathcal{D} + 5\mathcal{I})y = 0$

19. Suppose that p is a constant and consider the *first-order* equation $y' + py = 0$.

(a) Find the characteristic equation by substituting $y = e^{rt}$ into the equation.

(b) Find the characteristic polynomial by writing the equation in operator form.

(c) Solve the characteristic equation and write down the general solution of the differential equation.

20. **(a)** Find three linearly independent solutions of $y''' + y'' - 4y' - 4y = 0$.

(b) Find four linearly independent solutions of $y^{(iv)} - 16y = 0$.

21. **(a)** For any solution $y = c_1e^{2t} + c_2e^{3t}$ of the equation in Example 1, show that

$$y' - 2y = c_2e^{3t} \quad \text{and} \quad y' - 3y = -c_1e^{2t}.$$

Conclude that points along any phase-plane orbit satisfy

$$c_1^3(y' - 2y)^2 + c_2^2(y' - 3y)^3 = 0.$$

(b) For any solution $y = c_1e^{-2t} + c_2e^{-3t}$ of the equation in Example 2, show that

$$y' + 2y = -c_2e^{-3t} \quad \text{and} \quad y' + 3y = c_1e^{-2t}.$$

Conclude that points along any phase-plane orbit satisfy

$$-c_1^3(y' + 2y)^2 + c_2^2(y' + 3y)^3 = 0.$$

(c) For any solution $y = c_1e^{2t} + c_2e^{-3t}$ of the equation in Example 3, show that

$$y' - 2y = -5c_2e^{-3t} \quad \text{and} \quad y' + 3y = 5c_1e^{2t}.$$

Conclude that points along any phase-plane orbit satisfy

$$(y' - 2y)^2(y' + 3y)^3 = c_1^3c_2^2.$$

22. **(a–c)** Using the results of Problem 21, describe all straight-line phase-plane orbits in Examples 1 through 3. In addition, for Examples 1 and 2, investigate both

$$\lim_{t\to\infty} \frac{y'(t)}{y(t)} \quad \text{and} \quad \lim_{t\to-\infty} \frac{y'(t)}{y(t)}$$

to determine the line to which all phase-plane orbits are tangent at the origin and the line through the origin to which all phase-plane orbits become parallel as $t \to \infty$.

6.2 Exponential Shift

Consider an operator $P(\mathcal{D}) = \mathcal{D}^2 + p\mathcal{D} + q\mathcal{I}$ applied to a product of the form $e^{kt}u$ where k is a constant and u is a function of t. First observe that

$$\mathcal{D}e^{kt}u = e^{kt}u' + ke^{kt}u = e^{kt}(u' + ku) = e^{kt}(\mathcal{D} + k\mathcal{I})u,$$

and thus

$$\mathcal{D}^2e^{kt}u = \mathcal{D}e^{kt}(\mathcal{D} + k\mathcal{I})u = e^{kt}(\mathcal{D} + k\mathcal{I})(\mathcal{D} + k\mathcal{I})u = e^{kt}(\mathcal{D} + k\mathcal{I})^2u.$$

Consequently,

$$\begin{aligned} P(\mathcal{D})e^{kt}u &= e^{kt}(\mathcal{D}+k\mathcal{I})^2 u + pe^{kt}(\mathcal{D}+k\mathcal{I})u + qe^{kt}u \\ &= e^{kt}((\mathcal{D}+k\mathcal{I})^2 + p(\mathcal{D}+k\mathcal{I}) + q\mathcal{I})u \\ &= e^{kt}P(\mathcal{D}+k\mathcal{I})u. \end{aligned}$$

The point is that e^{kt} may be "factored" out of $P(\mathcal{D})e^{kt}u$ by replacing $\mathcal{D}$ with $\mathcal{D}+k\mathcal{I}$. This result, which is stated in Theorem 1 for general linear operators, can greatly simply many tedious calculations. The technique it provides is called **exponential shift**.

Theorem 1

Let $P(\mathcal{D}) = p_n\mathcal{D}^n + p_{n-1}\mathcal{D}^{n-1} + \cdots + p_1\mathcal{D} + p_0\mathcal{I}$, where $p_n, \ldots, p_0$ may be constants or functions of the independent variable. Then for any n-times differentiable function u and any constant k,

$$P(\mathcal{D})e^{kt}u = e^{kt}P(\mathcal{D}+k\mathcal{I})u. \quad \blacktriangle$$

REMARK Exponential shift is often applied when $P(\mathcal{D})$ is in factored form; for example,

$$(\mathcal{D}-r_1\mathcal{I})(\mathcal{D}-r_2\mathcal{I})e^{kt}u = e^{kt}(\mathcal{D}+(k-r_1)\mathcal{I})(\mathcal{D}+(k-r_2)\mathcal{I})u.$$

Example 1

As a very simple application of exponential shift, consider the problem of finding the second derivative of the function $e^{-t}\cos 3t$. Normally this would require a few applications of the product rule. However, we can simplify the calculation by using exponential shift as follows:

$$\begin{aligned} \mathcal{D}^2 e^{-t}\cos 3t &= e^{-t}(\mathcal{D}-\mathcal{I})^2\cos 3t \\ &= e^{-t}(\mathcal{D}^2 - 2\mathcal{D} + \mathcal{I})\cos 3t \\ &= e^{-t}(-9\cos 3t + 6\sin 3t + \cos 3t) \\ &= e^{-t}(6\sin 3t - 8\cos 3t). \end{aligned}$$

The reader is invited to check this by performing the calculation with the product rule instead of exponential shift. ■

Repeated Roots It turns out that exponential shift is very useful in finding a second linearly independent solution to a homogeneous, second-order, linear differential equation with constant coefficients in the case where the characteristic equation has a single repeated root $r_1 = -p/2$. In this case, the characteristic polynomial is a perfect square, and thus the differential equation may be written in the form

$$(\mathcal{D}-r_1\mathcal{I})^2 y = 0.$$

Since $Ce^{r_1 t}$ is a particular solution of the equation for any constant C, we will look for a second linearly independent solution in the form $y = ue^{r_1 t}$, where u is a nonconstant function of t. (This is the "variation of constants" technique that we have seen before.) We now substitute $y = ue^{r_1 t}$ into the equation and try to determine u. By exponential shift,

$$(\mathcal{D} - r_1\mathcal{I})^2 ue^{r_1 t} = e^{r_1 t}(\mathcal{D} + r_1\mathcal{I} - r_1\mathcal{I})^2 u = e^{r_1 t}\mathcal{D}^2 u.$$

Therefore,

$$(\mathcal{D} - r_1\mathcal{I})^2 ue^{r_1 t} = 0 \quad \text{if and only if} \quad \mathcal{D}^2 u = 0.$$

Thus a suitable function u would be $u = t$, which in turn yields the solution $y = te^{r_1 t}$. (Many other such functions u are possible, but remember that we are only seeking a particular solution.) The result we have found for the repeated-root case is stated in the following theorem.

Theorem 2

If the characteristic polynomial $P(r) = r^2 + pr + q$ has a single repeated real root r_1, then the differential equation $y'' + py' + qy = 0$ has two linearly independent solutions given by $e^{r_1 t}$ and $te^{r_1 t}$. Therefore, the general solution is given by $y = c_1 e^{r_1 t} + c_2 te^{r_1 t} = (c_1 + c_2 t)e^{r_1 t}$. ▲

Example 2

Consider the equation

$$y'' + 4y' + 4y = 0.$$

The characteristic equation is $r^2 + 4r + 4 = 0$, which is the same as $(r + 2)^2 = 0$. Thus there is only the repeated root $r_1 = -2$, and so e^{-2t} and te^{-2t} are linearly independent solutions of the differential equation. The general solution is therefore

$$y = (c_1 + c_2 t)e^{-2t}.$$

Figure 1a shows a typical solution, with initial values $y(0) = 1$ and $y'(0) = 1$, along with its derivative (dashed). Figure 1b shows several phase-plane orbits, including

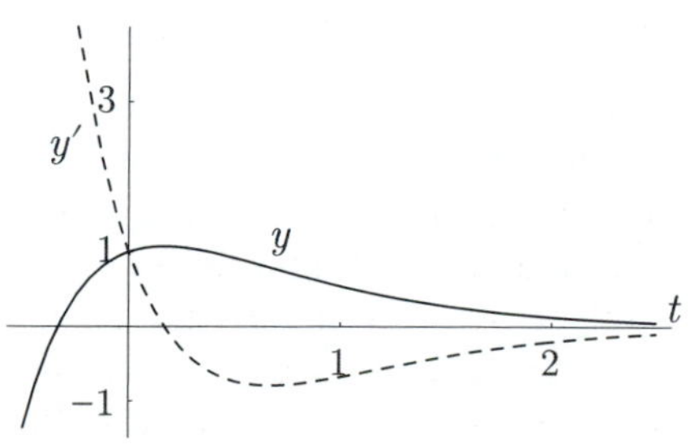

Figure 1a

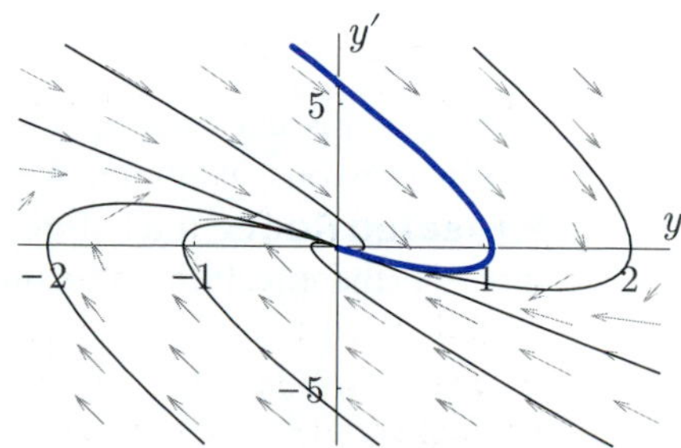

Figure 1b

one corresponding to the solution in Figure 1a. Note that the direction of flow along phase-plane orbits is always toward the origin, consistent with the fact that every solution approaches zero as $t \to \infty$. ■

Particular Solutions of Nonhomogeneous Equations Exponential shift is often useful when looking for a particular solution of a nonhomogeneous equation. We will illustrate its use with a couple of examples.

Example 3

Consider the problem of finding a particular solution of

$$y'' + 3y' + 2y = 3te^t.$$

Since the characteristic polynomial is $r^2 + 3r + 2 = (r + 1)(r + 2)$, we can rewrite the equation in factored operator form as

$$(\mathcal{D} + \mathcal{I})(\mathcal{D} + 2\mathcal{I})y = 3te^t.$$

Now we look for a particular solution of the form $y = ue^t$. Substituting this into the equation and using exponential shift, we find that

$$e^t(\mathcal{D} + 2\mathcal{I})(\mathcal{D} + 3\mathcal{I})u = 3te^t,$$

and thus

$$(\mathcal{D} + 2\mathcal{I})(\mathcal{D} + 3\mathcal{I})u = 3t.$$

Therefore, u must satisfy the equation $u'' + 5u' + 6u = 3t$. To find a particular solution of this, we substitute $u = at + b$ and obtain

$$0 + 5a + 6(at + b) = 3t.$$

Equating coefficients produces $a = 1/2$ and $b = -5/12$. Therefore, we have $u = (6t - 5)/12$ and the resulting particular solution $y = e^t(6t - 5)/12$. ■

Example 4

Here we will find a particular solution of

$$y'' + 3y' + 2y = 3te^{-2t}.$$

Although this does not appear at first to be significantly different from the problem in the previous example, we will see that things work out somewhat differently. Following what we did before, we seek a solution of the form $y = ue^{-2t}$. Substituting this into the equation and using exponential shift, we find that

$$(\mathcal{D} + \mathcal{I})(\mathcal{D} + 2\mathcal{I})ue^{-2t} = 3te^{-2t}$$

becomes

$$e^{-2t}(\mathcal{D} - \mathcal{I})\mathcal{D}u = 3te^{-2t}$$

and then

$$(\mathcal{D} - \mathcal{I})\mathcal{D}u = 3t.$$

We now reduce the order by substituting $v = \mathcal{D}u$ and look for v in the form $v = at + b$. This gives

$$(\mathcal{D} - \mathcal{I})v = (\mathcal{D} - \mathcal{I})(at + b) = a - (at + b) = 3t.$$

Now equating coefficients, we find $a = -3$ and $b = 3$. Thus $v = \mathcal{D}u = -3t + 3$, and so antidifferentiation gives $u = -3t^2/2 + 3t = 3t(2 - t)/2$. Finally, our particular solution is

$$y = \frac{3}{2}t(2 - t)e^{-2t}. \qquad \blacksquare$$

Example 5

Let's look now at the problem of finding a particular solution of

$$y'' + 4y' + 4y = 3te^{-2t},$$

in which the operator of interest is $\mathcal{D}^2 + 4\mathcal{D} + 4\mathcal{I} = (\mathcal{D} + 2\mathcal{I})^2$. As in the last example, we look for a solution of the form $y = ue^{-2t}$. Substitution and exponential shift reveal that

$$(\mathcal{D} + 2\mathcal{I})^2 ue^{-2t} = 3te^{-2t}$$

becomes

$$e^{-2t}\mathcal{D}^2 u = 3te^{-2t}, \quad \text{or,} \quad \mathcal{D}^2 u = 3t.$$

Now integrating twice shows that $u = t^3/2$ is a suitable choice of u. Thus our particular solution is

$$y = \frac{1}{2}t^3 e^{-2t}. \qquad \blacksquare$$

REMARKS Notice the difference between the final forms of the solutions in Examples 3 through 5. In Examples 3 and 4, the operator involved is precisely the same, and the nonhomogeneous term on the right side is of the form $3te^{kt}$. In Example 3, the particular solution ended up being the same *type* of function as the nonhomogeneous term—a linear polynomial times e^t. Yet in Example 4 the particular solution ended up being a *quadratic* polynomial times e^{-2t}, quite different from the nonhomogeneous term. In Example 4, exponential shift caused $\mathcal{D}$ to appear as a factor of the operator, because $\mathcal{D} + 2$ was one of the factors of the original operator—or, equivalently, -2 was one of the roots of the characteristic polynomial. As a result, a factor of t appeared in the solution. In Example 5, -2 was a *double* root of the characteristic polynomial, and so $\mathcal{D}^2$ appeared as a factor of the operator after exponential shift. Then integrating twice introduced a factor of t^2.

What we see here is typical. If the nonhomogeneous term is of the form $p(t)e^{kt}$, where $p(t)$ is a polynomial and k is a root of the characteristic equation, then the differential equation will have a particular solution of the form $t^m \tilde{p}(t)e^{kt}$, where $\tilde{p}(t)$ is a polynomial of the same degree as $p(t)$ and m is the multiplicity of k as a root of the characteristic polynomial.

PROBLEMS

Find the general solution of each equation in Problems 1 through 3.

1. $y'' - 6y' + 9y = 0$

2. $y'' + y' + y/4 = 0$

3. $y'' + 2y' + y = 0$

4. Use exponential shift to compute y'', where $y = t^{-1}e^{2t}$.

5. Use exponential shift to find a solution of $y'' = t^2e^{-t}$ in the form $y = ue^{-t}$.

In Problems 6 through 10, use exponential shift to find a particular solution.

6. $y'' + y = te^{-t}$

7. $y'' - y = te^{-t}$

8. $y'' + 2y' + y = te^{-t}$

9. $y'' + 3y' + 2y = te^{t}$

10. $y'' + 3y' + 2y = te^{-t}$

11. Find the general solution of $y'' + 4y' + 4y = te^{-2t}$.

12. Solve the initial-value problem $y'' - 4y = te^{t}$, $y(0) = y'(0) = 0$.

13. Let $y = (c_1 + c_2t)e^{-2t}$ be a solution of the equation in Example 2.

(a) Show that $y' + 2y = c_2e^{-2t}$. Conclude that any solution in which $c_2 = 0$ leads to the straight-line phase-plane orbit $y' + 2y = 0$.

(b) Investigate both

$$\lim_{t\to\infty} \frac{y'(t)}{y(t)} \quad \text{and} \quad \lim_{t\to-\infty} \frac{y'(t)}{y(t)}$$

to determine the line to which all phase-plane orbits are tangent at the origin and the line through the origin to which all phase-plane orbits become parallel as $t \to -\infty$.

14. **(a)** Find three linearly independent solutions of $y''' - 3y'' + 3y' - y = 0$.

(b) Find n linearly independent solutions of $(\mathcal{D} - k\mathcal{I})^n y = 0$.

6.3 Complex Roots

In this section we will look at the problem of finding a pair of linearly independent, *real* solutions of

$$y'' + pq' + qy = 0 \tag{1}$$

when $p^2 - 4q < 0$ and the roots of the characteristic equation $r^2 + pr + q = 0$ are the complex conjugates

$$r_1, r_2 = \frac{1}{2}\left(-p \pm i\sqrt{4q - p^2}\right),$$

where $i^2 = -1$. Let us first simplify our notation by setting $\alpha = -p/2$ and $\beta = \sqrt{4q - p^2}/2$ so that these roots are $\alpha \pm \beta i$. If we assume that the differentiation rule $\frac{d}{dt}e^{ct} = ce^{ct}$ is valid for any complex constant c, then it is easy to check that $e^{(\alpha+\beta i)t}$ and $e^{(\alpha-\beta i)t}$ are indeed solutions of the differential equation. In addition, these functions can be shown to be linearly independent. However, we are interested only in *real* solutions.

The Euler-DeMoivre Formula The key to "extracting" a pair of linearly independent real solutions from $e^{(\alpha\pm\beta i)t}$ is the **Euler-DeMoivre formula**, which defines imaginary powers of e. It states that, for any real number θ,

$$e^{\theta i} = \cos\theta + i\sin\theta.$$

This definition of $e^{\theta i}$ may be justified by a variety of arguments, including one using Taylor series. Moreover, with this definition we can easily check the differentiation rule $\frac{d}{dt}e^{bit} = bie^{bit}$ for any imaginary constant bi.

Because $\cos x$ is an even function and $\sin x$ is an odd function, the Euler-DeMoivre theorem also gives the identity

$$e^{-\theta i} = \cos\theta - i\sin\theta.$$

Therefore, since $e^{(\alpha\pm\beta i)t} = e^{\alpha t}e^{\pm\beta it}$, we obtain the identity

$$e^{(\alpha\pm\beta i)t} = e^{\alpha t}(\cos\beta t \pm i\sin\beta t).$$

Thus we can write $e^{(\alpha\pm\beta i)t}$ in terms of its real and imaginary parts as

$$e^{(\alpha\pm\beta i)t} = u(t) \pm iv(t),$$

where

$$u(t) = e^{\alpha t}\cos\beta t \quad \text{and} \quad v(t) = e^{\alpha t}\sin\beta t. \tag{2}$$

Real Solutions from Complex Roots Our claim is that u and v from (2), with $\alpha = -p/2$ and $\beta = \sqrt{4q - p^2}/2$, are each real solutions of (1). Note that each of u and v can be expressed as a linear combination of $u(t) \pm iv(t)$:

$$u = \frac{1}{2}((u + iv) + (u - iv)),$$

$$v = \frac{i}{2}((u - iv) - (u + iv)).$$

Therefore, since $u(t) \pm iv(t)$ are solutions, and since the differential equation is homogeneous and linear, it follows that u and v are solutions as well. A straightforward argument shows that u and v are also linearly independent.

We sum up the result of the preceding discussion as a theorem:

Theorem 1

If the characteristic equation $r^2 + pr + q = 0$ has a pair of complex conjugate roots $\alpha \pm \beta i$ with $\beta \neq 0$, then $e^{\alpha t}\cos\beta t$ and $e^{\alpha t}\sin\beta t$ are linearly independent, real solutions of (1). Thus the general solution of (1) is given by

$$y = e^{\alpha t}(c_1\cos\beta t + c_2\sin\beta t). \qquad \blacktriangle$$

Example 1

Consider the equation

$$y'' + 2y' + 10y = 0.$$

The characteristic equation is $r^2 + 2r + 10 = 0$, whose roots are

$$\frac{1}{2}\left(-2 \pm \sqrt{4 - 40}\right) = -1 \pm 3i.$$

These roots have real part $\alpha = -1$ and imaginary part $\pm\beta$ where $\beta = 3$. Thus, $e^{-t}\cos 3t$ and $e^{-t}\sin 3t$ are solutions, and the general solution of the equation is given by

$$y = e^{-t}(c_1 \cos 3t + c_2 \sin 3t).$$

Consequently, every solution approaches zero as $t \to \infty$. Figure 1a shows a typical solution, with initial values $y(0) = 1$ and $y'(0) = 1$, along with its derivative (dashed). Figure 1b shows several phase-plane orbits, including one corresponding to the solution in Figure 1a. Note that the direction of flow along phase-plane orbits is toward the origin. ■

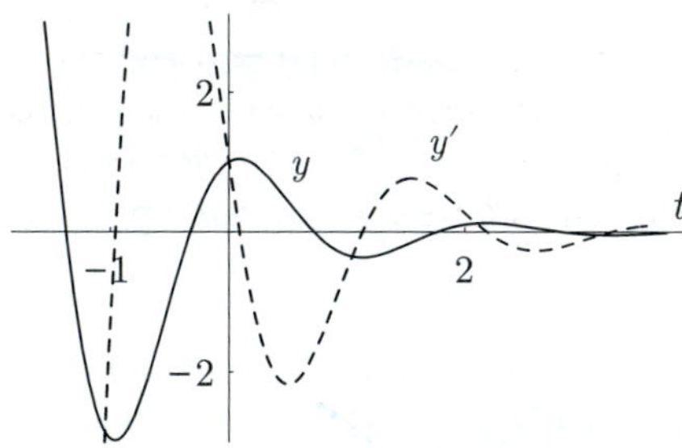

Figure 1a

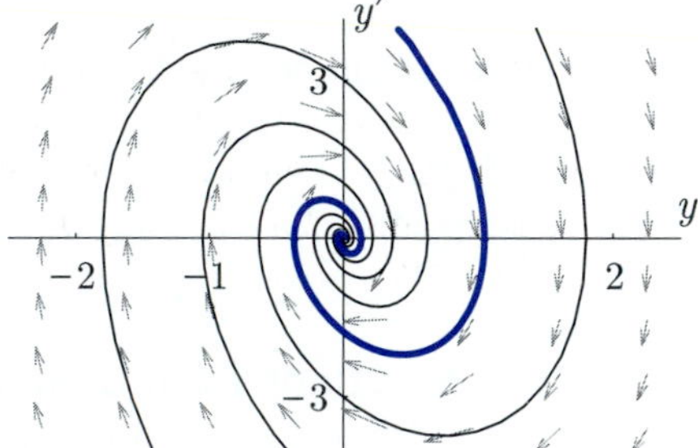

Figure 1b

Example 2

Consider the equation $y'' + 4y = 0$. The characteristic equation $r^2 + 4 = 0$ has purely imaginary roots $\pm 2i$. So the roots have real part $\alpha = 0$ and imaginary part ± 2. Thus, $\cos 2t$ and $\sin 2t$ are solutions, and the general solution of the equation (which we could have written down by inspection) is given by

$$y = c_1 \cos 2t + c_2 \sin 2t.$$

Consequently, every nonzero solution is a simple sinusoid with period π. Figure 2a shows a typical solution, with initial values $y(0) = 1$ and $y'(0) = 1$, along with its derivative (dashed). Figure 2b shows several phase-plane orbits, including one corresponding to the solution in Figure 1a. Note that the direction of flow along phase-plane orbits is clockwise. ■

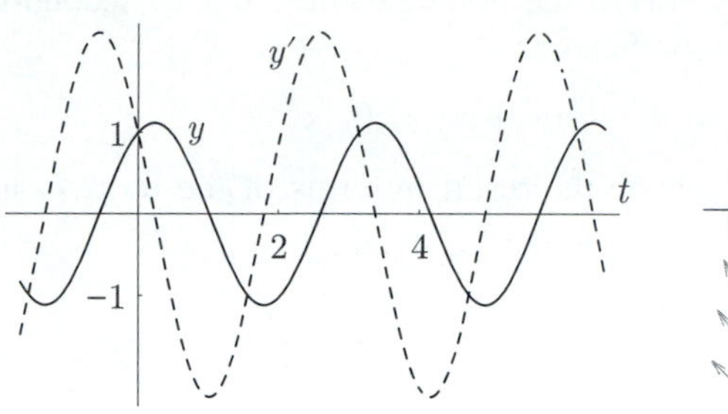

Figure 2a

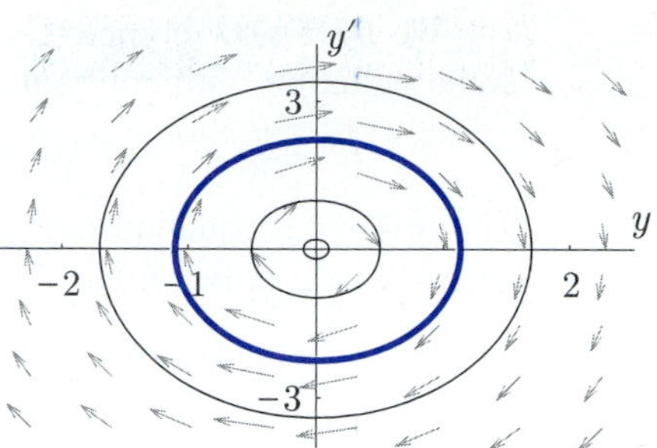

Figure 2b

Example 3

Consider the equation $y'' - y' + 17y/4 = 0$. The associated characteristic equation is $r^2 - r + 17/4 = 0$, whose roots are $\frac{1}{2}\left(1 \pm \sqrt{1 - 17}\right) = 1/2 \pm 2i$. These roots have real part $\alpha = 1/2$ and imaginary part $\pm\beta$ where $\beta = 2$. Thus, $e^{t/2}\cos 2t$ and $e^{t/2}\sin 2t$ are solutions, and the general solution of the equation is given by

$$y = e^{t/2}(c_1 \cos 2t + c_2 \sin 2t).$$

Consequently, the amplitude of every nonzero solution approaches ∞ as $t \to \infty$. Figure 3a shows a typical solution, with initial values $y(0) = 0$ and $y'(0) = 1$, along with its derivative (dashed). Figure 3b shows several phase-plane orbits, including one corresponding to the solution in Figure 3a. Note that the direction of flow along phase-plane orbits is away from the origin. Phase-plane orbits approach the origin as $t \to -\infty$. ■

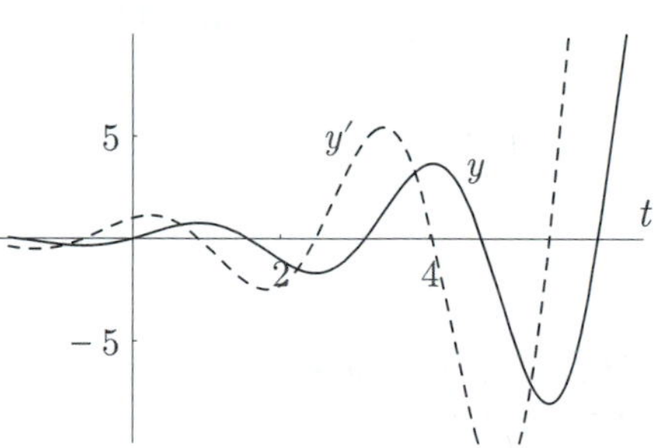

Figure 3a

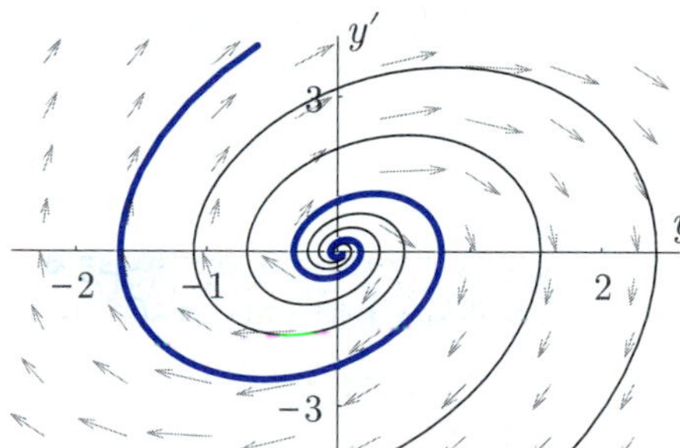

Figure 3b

Summary According to Theorem 1, when the characteristic equation has complex roots, the solution of (1) will be oscillatory in nature. The frequency/period of the oscillation is determined by the imaginary part of the roots. The amplitude of the oscillations decays exponentially if the roots have negative real part, remains constant if the real part of the roots is zero, and grows exponentially if the roots have a positive real part. This is illustrated in the preceding Examples 1–3.

We have now completely solved the homogeneous, linear, second-order differential equation with constant coefficients:

$$y'' + py' + qy = 0.$$

The following equations summarize the result in terms of the roots r_1 and r_2 of the characteristic polynomial $r^2 + pr + q$.

$$r_1, r_2 \text{ real and distinct:} \quad y = c_1 e^{r_1 t} + c_2 e^{r_2 t}$$

$$r_1 = r_2\text{:} \quad y = (c_1 + c_2 t)e^{r_1 t}$$

$$r_1, r_2 = \alpha \pm \beta i, \beta \neq 0\text{:} \quad y = e^{\alpha t}(c_1 \cos \beta t + c_2 \sin \beta t)$$

PROBLEMS

Find the general solution of each equation in Problems 1 through 3.

1. $y'' + 9y = 0$

2. $y'' - 4y' + 5y = 0$

3. $y'' + 6y' + 18y = 0$

In Problems 4 through 6, find the general solution. Where appropriate, use exponential shift to help find a particular solution.

4. $y'' + 4y' + 8y = 8t - 12$

5. $y'' + 2y' + 5y = 4te^{-t}$

6. $y'' + 4y = 16te^{-2t}$

Solve each initial-value problem in Problems 7 through 9.

7. $\begin{cases} y'' + y = 2e^{-t} \\ y(0) = y'(0) = 0 \end{cases}$

8. $\begin{cases} y'' + 2y' + 2y = 0 \\ y(0) = 1,\ y'(0) = 0 \end{cases}$

9. $\begin{cases} y'' + 2y' + 2y = e^{-t} \\ y(0) = y'(0) = 0 \end{cases}$

10. Show that if $p > 0$ and $q \neq 0$, then every solution of $y'' + py' + qy = 0$ approaches zero as $t \to \infty$. What happens in the case where $q = 0$?

11. Let $y = e^{\alpha t}(c_1 \cos \beta t + c_2 \sin \beta t)$. Show that

$$(y' - \alpha y)^2 + \beta^2 y^2 = 2\beta^2 e^{2\alpha t}(c_1^2 + c_2^2).$$

In light of this identity, discuss the phase-plane orbits in Examples 1 through 3.

12. Verify directly that $u = e^{\alpha t} \cos \beta t$ and $v = e^{\alpha t} \sin \beta t$ each satisfy (1) if $\alpha = -p/2$ and $\beta = \sqrt{4q - p^2}/2$.

13. Use exponential shift to show that, if $\alpha = -p/2$ and $\beta = \sqrt{4q - p^2}/2$, then

$$(\mathcal{D}^2 + p\mathcal{D} + q\mathcal{I})e^{\alpha t} \cos \beta t = 0 \quad \text{and} \quad (\mathcal{D}^2 + p\mathcal{D} + q\mathcal{I})e^{\alpha t} \sin \beta t = 0.$$

14. Show that substitution of θi into the Taylor series for e^x suggests the Euler-DeMoivre formula.

15. With $e^{\theta i}$ defined by the Euler-DeMoivre formula and $\frac{d}{dt}(u(t)+iv(t))$ defined to be $u'(t)+iv'(t)$ for any real-valued functions u and v, show that the following are true for any real numbers a and b.

(a) $\dfrac{d}{dt}e^{bit} = bie^{bit}$ (b) $\dfrac{d}{dt}e^{(a+bi)t} = (a+bi)e^{(a+bi)t}$

16. Under the assumptions that $\frac{d}{d\theta}e^{\theta i} = ie^{\theta i}$ and $\frac{d}{dt}(u(t)+iv(t)) = u'(t)+iv'(t)$ for any real-valued functions u and v, justify the Euler-DeMoivre formula by showing that

$$\frac{d}{d\theta}\left(\frac{e^{\theta i}}{\cos\theta + i\sin\theta}\right) = 0 \text{ for all } \theta, \quad \text{and} \quad \frac{e^{\theta i}}{\cos\theta + i\sin\theta} = 1 \text{ when } \theta = 0.$$

17. The **conjugate** of a complex number $z = a+bi$ is defined as $\overline{z} = a-bi$. The **modulus** of a complex number $z = a+bi$ is defined as $|z| = \sqrt{a^2+b^2}$. Show that the following are true for any complex number $z = a+bi$.

(a) $\dfrac{z+\overline{z}}{2} = a$ (b) $\dfrac{z-\overline{z}}{2i} = b$

(c) $z\overline{z} = |z|^2$ (d) $\overline{e^z} = e^{\overline{z}}$

(e) $|e^z| = e^a$

The quotient z_1/z_2 of two complex numbers $z_1 = a_1+b_1i$ and $z_2 = a_2+b_2i \neq 0$ is defined to be the complex number q with the property that $z_1 = qz_2$.

18. Show that $\frac{z_1}{z_2} = \frac{1}{|z_2|^2}z_1\overline{z}_2$.

19. Using the result of Problem 18, compute and simplify: (a) $\dfrac{i}{1+i}$ (b) $\dfrac{1+i}{2-i}$ (c) $\dfrac{1}{i}$

20. Use the Euler-DeMoivre formula and trigonometric identities to show that, for complex numbers $z_1 = a_1+b_1i$ and $z_2 = a_2+b_2i$,

$$e^{z_1}e^{z_2} = e^{z_1+z_2} \quad \text{and} \quad \frac{e^{z_1}}{e^{z_2}} = e^{z_1-z_2}.$$

Any complex number $z = a+bi \neq 0$ can be expressed in **polar form** as

$$z = re^{\theta i} = r(\cos\theta + i\sin\theta),$$

where $r = \sqrt{a^2+b^2}$ and $\tan\theta = b/a$, $-\pi < \theta \leq \pi$. The number r is the **modulus** of z, and θ is called the **argument** of z.

21. Write each number in polar form $re^{\theta i}$:

(a) $z = 1+i$, (b) $z = 2i$

(c) $z = -3$ (d) $z = -i$

22. Write each number in rectangular form $a+bi$:

(a) $z = e^{-\pi i/3}$ (b) $z = e^{\pi i}$

(c) $z = 2e^{5\pi i/6}$ (d) $z = \sqrt{2}e^{-i\pi/4}$

23. Let $z_1 = r_1e^{\theta_1 i}$ and $z_2 = r_2e^{\theta_2 i}$ be two complex numbers in polar form. Use trigonometric identities to show that

$$z_1z_2 = r_1r_2e^{(\theta_1+\theta_2)i} \quad \text{and} \quad \frac{z_1}{z_2} = \frac{r_1}{r_2}e^{(\theta_1-\theta_2)i}.$$

24. The (second-order) **Euler-Cauchy equation** is

$$t^2y'' + aty' + by = 0,$$

where a and b are constants.

(a) Show that the change of independent variable $s = \ln t$ results in an equation with constant coefficients, namely

$$\frac{d^2y}{ds^2} + (a-1)\frac{dy}{ds} + by = 0.$$

$\left(\textit{Hint}: \frac{dy}{dt} = \frac{ds}{dt}\frac{dy}{ds}.\right)$

(b) Let r_1 and r_2 be the roots of the quadratic polynomial $r^2 + (a-1)r + br$. Describe the general solution of the homogeneous Euler-Cauchy equation in each of the three cases: (i) r_1, r_2 real and distinct, (ii) $r_1 = r_2$, and (iii) $r_1, r_2 = \alpha \pm \beta i$.

Find the general solution of the Euler-Cauchy equations in Problems 25 through 30.

25. $t^2y'' - 3ty' + 3y = 0$

26. $t^2y'' + 5ty' + 4y = 0$

27. $t^2y'' + ty' + y = 0$

28. $t^2y'' + 5ty' + 3y = 0$

29. $t^2y'' + 3ty' + 2y = 0$

30. $t^2y'' + ty' - y = 0$

6.4 Real Solutions from Complex Solutions

In the last section we saw how the complex solution $e^{(\alpha+i\beta)t}$ gives rise to two real solutions through its real and imaginary parts, $e^{\alpha t}\cos\beta t$ and $e^{\alpha t}\sin\beta t$. In this section we will use a similar idea to help us find particular solutions of nonhomogeneous equations when the nonhomogeneous term has a trigonometric form. The basis of this is the following theorem.

Theorem 1

Let $\mathcal{T}$ be the operator defined by $\mathcal{T}y = y''+py'+qy$, and suppose that φ is a complex-valued function given by $\varphi(t) = f(t) + ig(t)$, where f and g are real valued for all t. If z satisfies $\mathcal{T}z = \varphi$, where $z(t) = x(t) + iy(t)$ with x and y real valued for all t, then

$$\mathcal{T}x = f \quad \textit{and} \quad \mathcal{T}y = g.$$

The same is true for any linear operator that maps real functions to real functions. ▲

Let's think for a moment about why this is true. Suppose that $z(t) = x(t)+iy(t)$. Then, by definition,

$$z'(t) = x'(t) + iy'(t) \quad \text{and} \quad z''(t) = x''(t) + iy''(t).$$

Thus, a little rearranging shows that

$$z'' + pz' + qz = (x'' + px' + qx) + i(y'' + py' + qy);$$

that is,

$$\mathcal{T}z = \mathcal{T}x + i\,\mathcal{T}y.$$

There is nothing at work here but the linearity of $\mathcal{T}$. Now, because $\mathcal{T}z = \varphi$,

$$\mathcal{T}x + i\,\mathcal{T}y = f + ig.$$

Since two complex numbers are equal if and only if their real parts are equal and their imaginary parts are equal, we now have that $\mathcal{T}x = f$ and $\mathcal{T}y = g$.

Theorem 1 of Section 6.3 is actually a special case of this theorem in which $\varphi = 0$. If $z(t) = x(t) + iy(t)$ and $\mathcal{T}z = 0$, then it follows that $\mathcal{T}x = 0$ and $\mathcal{T}y = 0$, since the real and imaginary parts of 0 are both 0. Thus, the real and imaginary parts of any solution of a linear homogeneous equation are real solutions of that same equation.

Particular Solutions Theorem 1 is particularly useful for finding particular solutions of nonhomogeneous equations with a trigonometric nonhomogeneous term.

Example 1

Consider the problem of finding particular solutions of

$$x'' + x' + 2x = 2\cos t \quad \text{and} \quad y'' + y' + 2y = 2\sin t.$$

Since the nonhomogeneous terms are the real and imaginary parts of $\varphi(t) = 2e^{it}$, respectively, we can obtain *both solutions at once* by finding a particular solution of the complex equation

$$z'' + z' + 2z = 2e^{it}$$

and then extracting real and imaginary parts. Let's look for a solution of the form $z = Ae^{it}$, where A is a (complex) constant. Differentiating z shows that

$$z' = Aie^{it} \quad \text{and} \quad z'' = -Ae^{it}.$$

It follows that $z'' + z' + 2z = -Ae^{it} + iAe^{it} + 2Ae^{it} = (1+i)Ae^{it}$. Substituting into the equation gives us

$$(1+i)Ae^{it} = 2e^{it}.$$

Therefore, the correct value for A is

$$A = \frac{2}{1+i} = \frac{2(1-i)}{2} = 1 - i,$$

and so $z = (1-i)e^{it}$ is a particular solution of our complex equation. Now, we finish the job by finding the real and imaginary parts of z:

$$z = (1-i)e^{it} = (1-i)(\cos t + i\sin t) = \cos t + \sin t + i(\sin t - \cos t).$$

Therefore,

$$x = \cos t + \sin t \quad \text{and} \quad y = \sin t - \cos t$$

are particular solutions of the original real equations. ■

Example 2

Consider the problem of finding a particular solution of

$$y'' + y = e^{-t}\cos t.$$

Since $e^{-t}\cos t$ is the *real* part of the complex function $\varphi = e^{-t}e^{it} = e^{(-1+i)t}$, the particular solution we seek will be the real part of a particular solution of

$$z'' + z = e^{(-1+i)t}.$$

If we substitute $z = Ae^{(-1+i)t}$ into this equation, we find that

$$((-1+i)^2 + 1)Ae^{(-1+i)t} = e^{(-1+i)t},$$

which requires that

$$A = \frac{1}{1-2i} = \frac{1+2i}{5}.$$

Now we have

$$\begin{aligned} z = \frac{1}{5}(1+2i)e^{(-1+i)t} &= \frac{1}{5}(1+2i)e^{-t}(\cos t + i\sin t) \\ &= \frac{1}{5}e^{-t}(\cos t - 2\sin t + i(2\cos t + \sin t)), \end{aligned}$$

from which we can see that the real part is

$$y = \frac{1}{5}e^{-t}(\cos t - 2\sin t).$$

■

Example 3

Let's find a particular solution of the equation

$$y'' + 2y' + 2y = e^{-t}\sin t.$$

Since $e^{-t}\sin t$ is the imaginary part of $\varphi = e^{-t}e^{it} = e^{(-1+i)t}$, the particular solution we seek will be the imaginary part of some particular solution of

$$z'' + 2z' + 2z = e^{(-1+i)t}.$$

Substituting $z = Ae^{(-1+i)t}$ into this equation, we find

$$((-1+i)^2 + 2(-1+i) + 2)Ae^{(-1+i)t} = e^{(-1+i)t}.$$

There is no A that works here because the coefficient of $Ae^{(-1+i)t}$ on the left side is zero! In fact, we should have known this in advance, because $-1+i$ is one of the roots of the characteristic polynomial of the operator $P(\mathcal{D}) = \mathcal{D}^2 + 2\mathcal{D} + 2\mathcal{I}$. So we have to look for a solution in the form $z = ue^{(-1+i)t}$ instead, where u is a function of t. To do this, we'll write the equation in operator form and let exponential shift do most of the work:

$$(\mathcal{D}^2 + 2\mathcal{D} + 2\mathcal{I})(ue^{(-1+i)t})$$
$$= e^{(-1+i)t}((\mathcal{D} + (-1+i)\mathcal{I})^2 + 2(\mathcal{D} + (-1+i)\mathcal{I}) + 2\mathcal{I})u$$
$$= e^{(-1+i)t}(\mathcal{D}^2 + 2i\mathcal{D})u = e^{(-1+i)t}(\mathcal{D} + 2i\,\mathcal{I})\mathcal{D}u.$$

Since we want this expression to equal $e^{(-1+i)t}$, the function u must satisfy

$$(\mathcal{D} + 2i\,\mathcal{I})\mathcal{D}u = 1.$$

So we take $\mathcal{D}u = \frac{1}{2i} = -i/2$ and then integrate to get $u = -it/2$. A particular complex solution is then

$$z = -\frac{t}{2}ie^{(-1+i)t} = -\frac{t}{2}e^{-t}i(\cos t + i\sin t) = -\frac{t}{2}e^{-t}(-\sin t + i\cos t).$$

Finally, we extract the imaginary part and obtain

$$y = \frac{t}{2}e^{-t}\cos t. \qquad \blacksquare$$

PROBLEMS

In Problems 1 through 6, use an appropriate complex equation to find a pair of particular solutions.

1. $\begin{cases} x'' + x = \cos t \\ y'' + y = \sin t \end{cases}$

2. $\begin{cases} x'' + x = e^{-t}\cos t \\ y'' + y = e^{-t}\sin t \end{cases}$

3. $\begin{cases} x'' + x' + 3x = 5\cos t \\ y'' + y' + 3y = 5\sin t \end{cases}$

4. $\begin{cases} x'' + x' + 5x = \cos 2t \\ y'' + y' + 5y = \sin 2t \end{cases}$

5. $\begin{cases} x'' - x = e^{-t}\cos 2t \\ y'' - y = e^{-t}\sin 2t \end{cases}$

6. $\begin{cases} x'' + x' + x = e^{-2t}\cos t \\ y'' + y' + y = e^{-2t}\sin t \end{cases}$

7. Use the results of Problem 1 to write down a particular solution of

(a) $y'' + y = \cos t + \sin t$ (b) $y'' + y = 3\sin t - \cos t$

8. Use the results of Problem 2 to write down a particular solution of

(a) $y'' + y = e^{-t}(\cos t + \sin t)$ (b) $y'' + y = 3e^{-t}(5\sin t - \cos t)$

9. Use the results of Problem 3 to write down a particular solution of

(a) $y'' + y' + 3y = 10\cos t + \sin t$ (b) $y'' + y' + 3y = 15\sin t - 5\cos t$

10. Use the results of Problem 4 to write down a particular solution of

(a) $y'' + y' + 5y = \cos 2t + \sin 2t$ (b) $y'' + y' + 5y = 3\sin 2t - \cos 2t$

11. Use the results of Problem 5 to write down a particular solution of

(a) $y'' - y = 8e^{-t}(\cos 2t + 2\sin 2t)$ (b) $y'' - y = 8e^{-t}(3\sin 2t - 2\cos 2t)$

12. Use the results of Problem 6 to write down a particular solution of

(a) $y'' + y' + y = 13e^{-2t}(\cos t + 2\sin t)$ (b) $y'' + y' + y = 13e^{-2t}(5\cos t - \sin t)$

In Problems 13 through 15, find the general solution.

13. $y'' + 2y' + 2y = 5\cos t$

14. $y'' + 3y' + 2y = e^{-t}\sin t$

15. $y'' + y = \sin t$

In Problems 16 through 18, solve the initial-value problem.

16. $y'' + 3y' + 2y = 10\cos t, \quad y(0) = 1,\ y'(0) = 0$

17. $y'' + 2y' + 5y = 17\cos 2t, \quad y(0) = 0,\ y'(0) = 0$

18. $y'' + 2y' + 2y = 5\cos t, \quad y(0) = 1,\ y'(0) = 0$

19. Let $P(\mathcal{D}) = \mathcal{D}^2 + p\mathcal{D} + q\mathcal{I}$ and consider the equation $P(\mathcal{D})y = e^{\eta t}$, where η is a complex constant.

(a) Suppose that $P(\eta) \neq 0$. Obtain the particular solution $y = e^{\eta t}/P(\eta)$.

(b) Suppose that $P(\eta) = 0$ and $\eta \neq -p/2$. Obtain the particular solution $y = te^{\eta t}/(2\eta + p)$. (Note in particular that this is the case if $P(\eta) = 0$ and η has nonzero imaginary part.)

(c) Suppose that $P(\eta) = 0$ and $\eta = -p/2$. Obtain the particular solution $y = t^2e^{\eta t}/2$. (Note that $P(\eta) = 0$ and $\eta = -p/2$ describe the situation in which η is a double root of the characteristic equation.)

The purpose of Problems 20 through 22 is to look at an alternative approach for finding particular solutions sometimes called the method of **undetermined coefficients**.

20. We seek a particular solution of $y'' + y' + 10y = 10\sin 3t$. Rather than using a complex equation, look for a solution of the form

$$y = A\cos 3t + B\sin 3t.$$

21. We seek a particular solution of $y'' + y' + 10y = 10e^{-t}\sin 3t$. Rather than using a complex equation, look for a solution of the form

$$y = e^{-t}(A\cos 3t + B\sin 3t).$$

(Exponential shift will help here.)

22. We seek a particular solution of $y'' + 2y' + 2y = e^{-t}\sin t$. Rather than using a complex equation, look for a solution of the form

$$y = te^{-t}(A\cos t + B\sin t).$$

(Again, exponential shift will make the job a little easier.) Why do you think that the factor of t is needed in this trial solution?

6.5 Unforced Vibrations

In Section 5.1 we learned that models of simple spring-mass systems and electrical circuits give rise to second-order linear differential equations with constant coefficients. The equations arising from unforced systems are homogeneous. In this section we will look at these equations and their solutions. A brief review of Section 5.1 might be appropriate before proceeding with the rest of this section.

Recall the equation for damped, unforced motion of a spring-mass system:

$$my'' + ry' + ky = 0,$$

where m, r, and k are the mass, damping coefficient, and spring stiffness, respectively. The equation for the current in an unforced RLC circuit is

$$LI'' + RI' + \frac{1}{C}I = 0$$

with L, R, and C being the inductance, resistance, and capacitance, respectively. By dividing through by the leading coefficient in each of these equations, we arrive at an equation of the form

$$y'' + py' + qy = 0.$$

Since $p \geq 0$ and $q > 0$, this a special case of the homogeneous equations we have seen already in this chapter. Note that the coefficient p will be zero only if there is no damping in the system.

Undamped, Unforced Vibrations When no damping is present, we have $p = 0$, and so the equation takes the form $y'' + qy = 0$ with $q > 0$. To simplify the notation in what follows, let $q = \omega^2$, so that the equation becomes

$$y'' + \omega^2 y = 0.$$

The general solution of this equation is easily seen to be

$$y = c_1 \cos \omega t + c_2 \sin \omega t.$$

Thus, ω is the *angular frequency* (in radians per unit time) of the unforced vibration. The *frequency* of the vibration is $\frac{\omega}{2\pi}$ *cycles* per unit time. We call these quantities the *natural* angular frequency and *natural* frequency of the system. An undamped system can vibrate at a different frequency, but only if an external force is applied.

In the case of an undamped spring-mass system, we have $q = k/m$, and so the natural angular frequency is

$$\omega = \sqrt{k/m}.$$

In the case of an LC circuit, we have $q = 1/(LC)$, and so the natural angular frequency is

$$\omega = 1/\sqrt{LC}.$$

We have essentially determined that simple, undamped, unforced vibrations are purely sinusoidal in nature with a natural frequency determined solely by system parameters. All of this is independent of the initial state of the system. The *amplitude* of the vibration, however, is determined by initial conditions.

Consider a solution $y = c_1 \cos \omega t + c_2 \sin \omega t$, where c_1 and c_2 are constants. We would like to find the amplitude of this sinusoidal solution. To do this we need to express this solution as a single sine or cosine term. Recall the trigonometric identity

$$\cos(\alpha - \beta) = \cos \alpha \cos \beta + \sin \alpha \sin \beta.$$

Because of this, we introduce new constants A and ϕ related to c_1 and c_2 by

$$c_1 = A\cos\phi, \quad c_2 = A\sin\phi,$$

and express y as

$$y = A\cos(\omega t - \phi),$$

where

$$A = \sqrt{c_1^2 + c_2^2} \quad \text{and} \quad \tan\phi = \frac{c_2}{c_1}.$$

The quadrant in which the angle ϕ lies is determined by the signs of c_1 and c_2. The quantity A is the amplitude of the solution. The angle ϕ is called the **phase angle** or **angular phase shift**. The actual **phase shift** is given by ϕ/ω, since the graph of $\cos(\omega t - \phi)$ is that of $\cos\omega t$ shifted horizontally by ϕ/ω.

Let us now look at the initial-value problem

$$y'' + \omega^2 y = 0, \quad y(0) = y_0, \quad y'(0) = v_0.$$

It is easy to determine that the solution is

$$y = y_0\cos\omega t + \frac{v_0}{\omega}\sin\omega t.$$

From this and the preceding discussion, we can write the solution as

$$y = A\cos(\omega t - \phi), \quad \text{where} \quad A = \sqrt{y_0^2 + \frac{v_0^2}{\omega^2}} \quad \text{and} \quad \tan\phi = \frac{v_0}{y_0\omega}.$$

Example 1

Consider the initial-value problem

$$y'' + 9y = 0, \quad y(0) = -2\sqrt{3}, \quad y'(0) = 6.$$

The general solution of the differential equation is described by

$$y = c_1\cos 3t + c_2\sin 3t.$$

The initial conditions require $c_1 = -2\sqrt{3}$ and $c_2 = 2$. The amplitude and phase angle are given by

$$A = \sqrt{12 + 4} = 4, \quad \tan\phi = \frac{2}{-2\sqrt{3}} = -\frac{1}{\sqrt{3}}.$$

Since $c_1 < 0$ and $c_2 > 0$, ϕ should be a second quadrant angle. Therefore, we take $\phi = 5\pi/6$, and so the solution may be written as

$$y = 4\cos\left(3t - \frac{5\pi}{6}\right).$$

■

Damped, Unforced Vibrations The motion of a damped, unforced spring-mass system and the current in an unforced RLC circuit are each governed by a differential equation of the form

$$y'' + py' + qy = 0,$$

where $p, q > 0$. The roots of the corresponding characteristic equation are, as usual,

$$r_1, r_2 = \frac{1}{2}\left(-p \pm \sqrt{p^2 - 4q}\right).$$

To examine the solutions of this equation, we need to consider the following three cases.

Case 1: $p^2 - 4q > 0$. Here, we have two distinct real roots r_1 and r_2, which are negative because $p > 0$ and $\sqrt{p^2 - 4q} < p$. Thus the general solution is

$$y = c_1 e^{r_1 t} + c_2 e^{r_2 t},$$

and we can see that in this case there is no vibration and every solution must tend to zero as $t \to \infty$. This is called the *overdamped* case.

Case 2: $p^2 - 4q = 0$. Here, we have the repeated real root $r_1 = r_2 = -p/2$, which is necessarily negative because $p > 0$. Thus the general solution is

$$y = (c_1 + c_2 t)e^{-pt/2},$$

and we can see that again there is no vibration and every solution must tend to zero as $t \to \infty$. This is called the *critically damped* case.

Case 3: $p^2 - 4q < 0$. In this case the roots are the complex conjugates

$$\alpha \pm i\omega = \frac{1}{2}\left(-p \pm i\sqrt{4q - p^2}\right).$$

Thus the general solution is

$$y = e^{-pt/2}(c_1 \cos \omega t + c_2 \sin \omega t),$$

where $\omega = \sqrt{4q - p^2}/2$. Just as in the undamped case we can write this as

$$y = Ae^{-pt/2} \cos(\omega t - \phi),$$

where $A = \sqrt{c_1^2 + c_2^2}$ and $\tan \phi = c_2/c_1$. Here we have damped vibration which permits us to think of the quantity $Ae^{-pt/2}$ as the exponentially decaying amplitude. This is called the *underdamped* case and is the only one of the three cases in which the system actually vibrates. It is important to observe the distinct roles played by the real and imaginary parts of the roots of the characteristic equation. The imaginary part determines the angular frequency of the vibrations and the real part determines the rate of damping. Just as in the first two cases, every solution must tend to zero as $t \to \infty$, because the roots have negative real part.

We have learned that the behavior of a damped, unforced system can be characterized in terms of the roots of the corresponding characteristic equation. The system experiences (damped) vibrations precisely when these roots nonzero imaginary parts. The undamped, unforced case also fits nicely into this picture as the case in which the roots of the characteristic equation are purely imaginary. Note that case 3 reduces to the undamped case when $p = 0$.

Figures 1a–d show the solution of the initial-value problem

$$y'' + py' + 4y = 0, \quad y(0) = \frac{1}{2}, \quad y'(0) = -2$$

with $p = 5, 4, 2$, and 0, respectively. Figures 1a–c illustrate cases 1–3, while Figure 1d illustrates the undamped case. These four pictures essentially illustrate the possible behaviors of an unforced system.

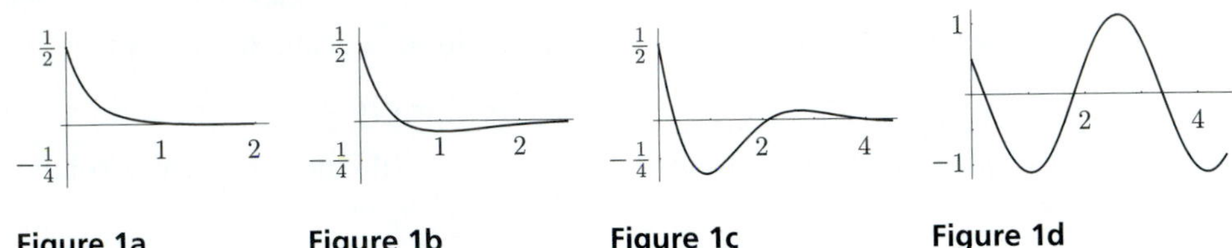

Figure 1a **Figure 1b** **Figure 1c** **Figure 1d**

PROBLEMS

1. Consider an undamped spring-mass system with mass $m = 2$ slugs and stiffness $k = 8$ lb/ft.

(a) Find the natural angular frequency and the natural frequency of the system.

(b) Find the resulting motion if the mass is gently released from an initial position of $y_0 = .75$ ft. What is the amplitude of the motion?

(c) Find the resulting motion if the mass is set in motion from its equilibrium position with an initial velocity of $v_0 = 2$ ft/s. What is the amplitude of the motion?

(d) Find the resulting motion if the mass is set in motion from an initial position of $y_0 = .75$ ft with an initial velocity of $v_0 = 2$ ft/s. What is the amplitude?

2. Consider an LC circuit with inductance $L = 1$ mH and capacitance $C = 10\ \mu$F. Find the natural angular frequency and the natural frequency of the circuit.

3. Consider a damped spring-mass system with mass m, damping coefficient r, and stiffness k. Give conditions on m, r, and k which lead to the (a) overdamped, (b) critically damped, and (c) underdamped cases.

4. Suppose we use a spring-mass system as a simple model of a car's suspension system. We determine that the car has mass $m = 1000$ kg and that the stiffness of the suspension is $k = 16{,}000$ N/m. What amount of damping from shock absorbers would critically damp any unforced motion?

5. Consider an RLC circuit with resistance R, inductance L, and capacitance C. Give conditions on R, L, and C which lead to the (a) overdamped, (b) critically damped, and (c) underdamped cases.

6. An LC circuit with $C = 1\ \mu$F has a natural angular frequency of 10^5 radians per second. What amount of resistance added to the circuit would critically damp unforced oscillations in the current?

7. **(a)** Determine the constants c_1, c_2 in terms of initial values $y(0) = y_0$, $y'(0) = v_0$ for the general solution in the underdamped case.

(b) Write the solution in the form $y = Ae^{-pt/2}\cos(\omega t - \phi)$ with A, ϕ expressed in terms of initial values $y(0) = y_0$, $y'(0) = v_0$.

8. **(a)** Determine the constants c_1, c_2 in terms of initial values $y(0) = y_0$, $y'(0) = v_0$ for the general solution in the overdamped case.

(b) Repeat part (a) for the critically damped case.

9. A damped, unforced spring-mass system is observed to vibrate at an angular frequency of 6 radians per second with an amplitude that decreases by a factor of $1/2$ every 3 seconds.

(a) Find the differential equation that governs the motion.

(b) If the spring stiffness is known to be $k = 1$ lb/ft, find the mass m and the damping coefficient r.

10. Consider a damped spring-mass system with stiffness $k = 8$ dynes/cm and damping coefficient $r = 16$ dynes/(cm/sec). Find the general solution of the differential equation that governs unforced motion of the mass if

(a) $m = 6$ grams

(b) $m = 8$ grams

(c) $m = 10$ grams

11. **(a–c)** For the spring-mass systems in Problem 10, find the motion of the mass if it is gently released from an initial position of $y_0 = 10$ cm.

12. Consider a spring-mass system with mass $m = 2$ kg and stiffness $k = 18$ N/m. Find the general solution of the differential equation that governs unforced motion of the mass if the damping coefficient (in units of N/(m/s)) is

(a) $r = 0$

(b) $r = 4\sqrt{5}$

(c) $r = 12$

(d) $r = 20$

13. **(a–d)** For the spring-mass systems in Problem 12, find the motion of the mass if it is set in motion from its equilibrium position with an initial velocity of $v_0 = 1$ m/s.

14. Solve for the solutions shown in Figures 1a–1d.

6.6 Periodic Force and Response

In the preceding section we investigated the behavior of unforced spring-mass systems and electrical circuits governed by second-order, linear differential equations with constant coefficients. From a physical point of view, an unforced system that is initially at rest should remain at rest for all time. Since the governing differential equation for an unforced system is homogeneous, this principle is reflected in the fact that the "rest" solution (i.e., with zero initial conditions) is in fact the zero solution. Moreover, any nontrivial behavior of an unforced system results solely from being initially perturbed from rest. Such unforced motion is often referred to as the **free response** of the system. One of the lessons of the last section was that the *type* of free response depends only on the coefficients of the differential equation (i.e., system characteristics) and not on the initial conditions. For example, an unforced, underdamped system will behave in an underdamped fashion regardless of how it is set in motion.

In this section we will investigate **forced response**; in particular, response to an external force represented by a linear combination of sinusoidal terms with various frequencies. Because of the form of equations (5) and (6) in Section 5.1, we will write our equation in the form

$$y'' + py' + qy = qf(t). \tag{1}$$

Recall that, for the spring-mass system shown in Figure 3 of Section 5.1, $q = k/m$ and $f(t)$ may be thought of as the position of the platform to which the spring is attached. (The resulting external *force* is actually $kf(t)$.) On the other hand, if y represents the voltage in an *RLC* circuit, then $q = \frac{1}{LC}$ and $f(t)$ is the source voltage that drives the circuit. In any case, we refer to the function $f(t)$ as a *forcing function* or *input* to the system. Our goal is to determine how the system "responds" to a given input.

Recall that any solution of (1) will be composed of a particular solution plus some solution of the corresponding homogeneous equation. Since we already know about solutions of the corresponding homogeneous equation, we will focus here on finding particular solutions. We will also concentrate on simple sinusoidal inputs $f(t)$ in the form

$$f(t) = A_0 \cos \omega_0 t \quad \text{or} \quad A_0 \sin \omega_0 t.$$

Here A_0 and ω_0 represent the amplitude and angular frequency of the input. By linearity, a particular solution corresponding to a sum of such functions can be assembled by summing particular solutions corresponding to these simple sinusoidal terms.

To help us find particular solutions we will use the complex equation

$$z'' + pz' + qz = qA_0 e^{\omega_0 i t}.$$

Once we find a particular complex solution of this equation, we can extract its real and imaginary parts to obtain real particular solutions of (1) corresponding to $f(t) = A_0 \cos \omega_0 t$ and $f(t) = A_0 \sin \omega_0 t$, respectively.

Forced Response, No Damping When there is no damping, we have $p = 0$ and $q = \omega^2$, where ω is the *natural angular frequency* of the system. Our goal here is to find a particular solution of each of the equations

$$x'' + \omega^2 x = \omega^2 A_0 \cos \omega_0 t, \quad y'' + \omega^2 y = \omega^2 A_0 \sin \omega_0 t,$$

by taking real and imaginary parts of a particular solution of

$$z'' + \omega^2 z = \omega^2 A_0 e^{\omega_0 i t}.$$

Case 1: $\omega \neq \omega_0$. Substitution of $z = Ae^{\omega_0 i t}$ into the equation yields

$$\left(-\omega_0^2 + \omega^2\right) A = \omega^2 A_0.$$

When the input frequency is different from the natural frequency of the system (that is, when $\omega \neq \omega_0$), we can solve to find $A = \omega^2 A_0/(\omega^2 - \omega_0^2)$. Consequently,

$$z = \frac{\omega^2}{\omega^2 - \omega_0^2} A_0 e^{\omega_0 i t}$$

is a particular solution. By the Euler-DeMoivre formula, its real and imaginary parts are, respectively,

$$x = \frac{\omega^2}{\omega^2 - \omega_0^2} A_0 \cos \omega_0 t \quad \text{and} \quad y = \frac{\omega^2}{\omega^2 - \omega_0^2} A_0 \sin \omega_0 t. \tag{2}$$

Notice that each of these has amplitude given by

$$A = GA_0, \quad \text{where} \quad G = \frac{\omega^2}{|\omega^2 - \omega_0^2|}.$$

The quantity G is called the ***gain*** of the system. Notice that we can write x and y in terms of G as

$$x = \begin{cases} GA_0 \cos \omega_0 t, & \text{if } \omega_0 < \omega; \\ GA_0 \cos(\omega_0 t - \pi), & \text{if } \omega_0 > \omega; \end{cases} \quad y = \begin{cases} GA_0 \sin \omega_0 t, & \text{if } \omega_0 < \omega; \\ GA_0 \sin(\omega_0 t - \pi), & \text{if } \omega_0 > \omega. \end{cases}$$

Thus there are two important observations to be made here:

- the amplitude of the response is that of the input multiplied by G;
- the response is *in phase* with the input if $\omega_0 < \omega$ and 180° *out of phase* if $\omega_0 > \omega$.

In particular, notice that inputs with frequencies close to the natural frequency are greatly amplified; in fact, it is easy to see that

$$G \to \infty \quad \text{as} \quad \omega - \omega_0 \to 0.$$

Also, if ω is fixed, then $G \to 0$ as $\omega_0 \to \infty$, and if ω_0 is fixed, then $G \to 1$ as $\omega \to \infty$. Therefore, if ω_0 is very large relative to ω, then the amplitude of the response is much smaller than that of the input, and if ω is very large relative to ω_0, then the response is nearly identical to the input.

Example 1

Consider the equation

$$y'' + 16y = 16 \cos \omega_0 t.$$

A particular solution is

$$y = \frac{16}{16 - \omega_0^2} \cos \omega_0 t.$$

In Figure 1a, the solution is plotted for $\omega_0 = 3, 3.55, 3.75, 3.85$ and 3.9. In Figure 1b, the solution is plotted for $\omega_0 = 4.1, 4.15, 4.25, 4.45,$ and 5. Notice that as ω_0 increases, the amplitude of the response also increases while $\omega_0 < \omega = 4$. The amplitude of the response is a decreasing function of ω_0 for $\omega_0 > \omega = 4$. Also, notice that the response is in phase with the input when $\omega_0 < 4$ and 180° out of phase with the input when $\omega_0 > 4$. ■

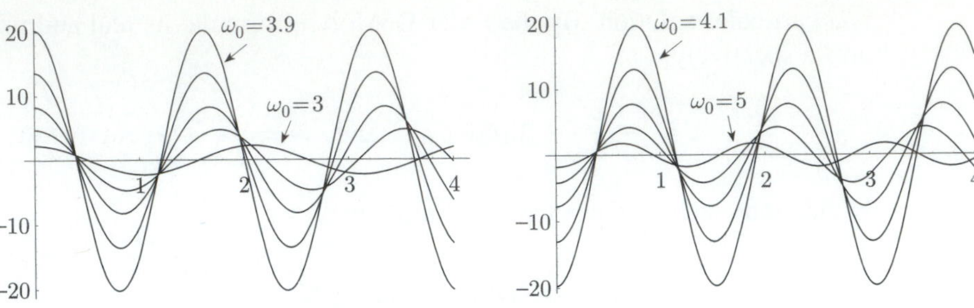

Figure 1a **Figure 1b**

Example 2

Consider the equation

$$y'' + \omega^2 y = \omega^2(\cos t + \cos 3t + \cos 6t).$$

By superposition and (2), a particular solution is

$$y = \omega^2 \left(\frac{\cos t}{\omega^2 - 1} + \frac{\cos 3t}{\omega^2 - 9} + \frac{\cos 6t}{\omega^2 - 36} \right),$$

provided that $\omega \neq 1, 3, 6$. If ω is close to one of these exceptional values, then the corresponding term in the solution will dominate the others. In that sense, the system "selects" any component of the input whose frequency is close to its natural frequency and amplifies it more than the others. So by changing the natural frequency of the system, one can "tune" the system to respond to a particular component frequency of the input. This is illustrated in Figure 2, which shows graphs of the solutions corresponding to $\omega = 1.08, 2, 2.8, 4.5, 6.4$, and 100. The last graph is roughly the same as that of the input, which may be though of as corresponding to the solution with $\omega = \infty$. ■

Case 2: $\omega_0 = \omega$. This is the situation in which the system is forced at its own natural frequency. To find a particular solution of the complex equation

$$z'' + \omega^2 z = \omega^2 A_0 e^{i\omega t},$$

we substitute $z = ue^{i\omega t}$, where u is a (complex) function of t to be determined. We proceed by writing the equation in operator form,

$$(\mathcal{D}^2 + \omega^2 \mathcal{I})ue^{i\omega t} = \omega^2 A_0 e^{i\omega t},$$

and using exponential shift:

$$((\mathcal{D} + i\omega\mathcal{I})^2 + \omega^2\mathcal{I})u = \omega^2 A_0, \quad \text{i.e.,} \quad (\mathcal{D} + 2i\omega\mathcal{I})\mathcal{D}u = \omega^2 A_0.$$

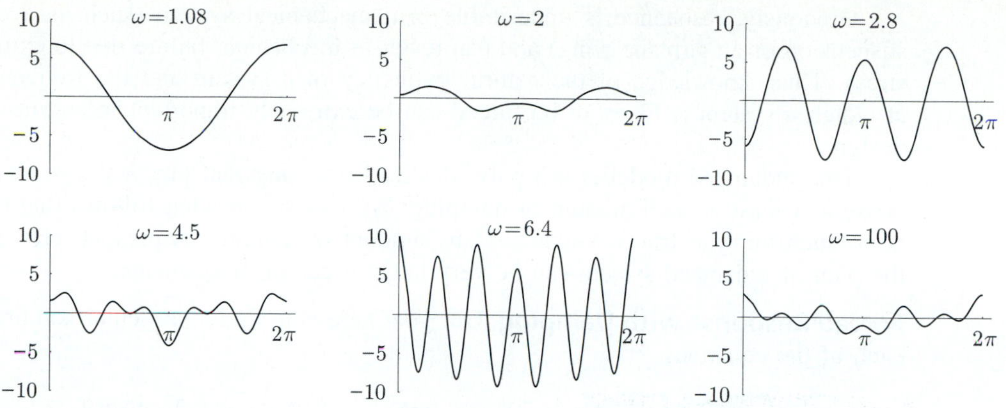

Figure 2

So we can take $\mathcal{D}u$ to be the constant $-i\omega A_0/2$ and then integrate to find $u = -i\omega A_0 t/2$. This gives the complex solution

$$z = -\frac{i}{2}\omega A_0 t e^{i\omega t} = -\frac{i}{2}\omega A_0 t(\cos\omega t + i\sin\omega t)$$
$$= \frac{1}{2}\omega A_0 t(\sin\omega t - i\cos\omega t).$$

Therefore, our real solutions are

$$x = \frac{1}{2}\omega A_0 t \sin\omega t \quad \text{and} \quad y = -\frac{1}{2}\omega A_0 t \cos\omega t.$$

So we see that the response oscillates 90° out of phase with the input. More important, we see that the amplitude of the response grows linearly and without bound as $t \to \infty$. This phenomenon is called ***resonance***. It is what occurs when an undamped system is forced at precisely its own natural frequency. For this reason, the system's natural frequency ω is sometimes called its *resonant frequency*. A resonant response is shown in Figure 3.

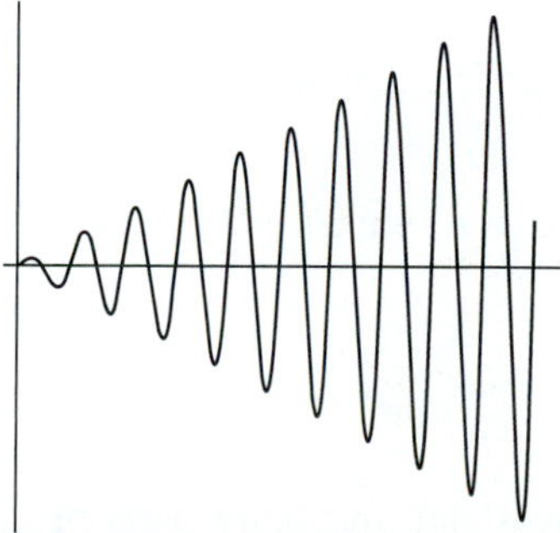

Figure 3

Obviously, resonance is undesirable in a mechanical system (such as a car's suspension or an airplane wing) and can result in mechanical failure due to extreme stress. Thus, knowledge of the natural frequency of a system and the frequencies at which a system is likely to be forced can be extremely important in engineering design.

The undamped model is a highly idealized one; any real physical system will possess at least a small amount of damping. We will see in what follows that there is no such thing as true resonance if any amount of damping is present. However, the gain of a damped system can be very large at certain frequencies.

Forced Response with Damping Our goal here is to find a particular solution of each of the equations

$$x'' + px' + qx = qA_0 \cos \omega_0 t, \quad y'' + py' + qy = qA_0 \sin \omega_0 t,$$

by taking real and imaginary parts of a particular solution of

$$z'' + pz' + qz = qA_0 e^{i\omega_0 t}.$$

We assume for our purposes here that p and q are positive. Recall that the corresponding unforced system can experience damped oscillation at an angular frequency of $\omega = \sqrt{q - p^2/4}$ when $q - p^2/4 > 0$. Substitution of

$$z = Ce^{i\omega_0 t}$$

into the equation gives

$$\left(-\omega_0^2 + ip\omega_0 + q\right) Ce^{i\omega_0 t} = qA_0 e^{i\omega_0 t}.$$

Thus we find that

$$C = \frac{q}{q - \omega_0^2 + p\omega_0 i} A_0, \quad \text{and so} \quad z = \frac{q}{q - \omega_0^2 + p\omega_0} A_0 e^{i\omega_0 t}.$$

Now our job is to find the real and imaginary parts of z. To do this, we first define R and ϕ by

$$R = \sqrt{(q - \omega_0^2)^2 + p^2\omega_0^2}, \quad \text{and} \quad \tan\phi = \frac{p\omega_0}{q - \omega_0^2}, \quad 0 \le \phi \le \pi,$$

so that $q - \omega_0^2 = R\cos\phi$ and $p\omega_0 = R\sin\phi$. The denominator in our expression for z thus becomes

$$q - \omega_0^2 + ip\omega_0 = R(\cos\phi + i\sin\phi) = Re^{i\phi}.$$

With this done, we can write

$$z = \frac{q}{Re^{i\phi}} A_0 e^{i\omega_0 t} = \frac{q}{R} A_0 e^{i(\omega_0 t - \phi)},$$

from which it is easy to observe that the real and imaginary parts of z are

$$x = GA_0 \cos(\omega_0 t - \phi) \quad \text{and} \quad y = GA_0 \sin(\omega_0 t - \phi),$$

where

$$G = \frac{q}{R} = \frac{q}{\sqrt{(q-\omega_0^2)^2 + p^2\omega_0^2}}.$$

We interpret G and ϕ as the **gain** and **phase lag** of the system at the angular input frequency ω_0. The following are several observations concerning the gain G. You will be asked to verify the less obvious of these in the problem set at the end of this section.

i. When $p = 0$ and $q = \omega^2$, G reduces to the gain expression of the undamped case: $G = \omega^2/|\omega^2 - \omega_0^2|$.

ii. When ω_0 is large (relative to p and q), $G \approx q/\omega_0^2$, and so G is small. Thus high-frequency input produces small amplitude response.

iii. For fixed p, $\omega_0 > 0$, the maximum gain is $G = \sqrt{q}/p$, occurring at $q = \omega^2$. (This is where resonance would occur if there were no damping.)

iv. For any fixed p, $q > 0$, G is a *bounded* function of ω_0. Moreover,

- when $0 < q \le p^2/2$, G is a decreasing function of ω_0 with $\lim_{\omega_0 \to 0^+} G = 1$.
- when $q > p^2/2$, a maximum gain of $G = \frac{2q}{p\sqrt{4q-p^2}}$ occurs when

$$\omega_0 = \sqrt{q - p^2/2}.$$

v. $G > 1$ when $\omega_0^2 < 2q - p^2$. This describes the input frequencies that are *amplified* by the system. (This cannot happen if $2q - p^2 \le 0$.) Input frequencies for which $\omega_0^2 > 2q - p^2$ are *attenuated*; that is, $0 < G < 1$.

Figure 4 shows a surface plot of G as a function of q and ω_0 for fixed $p = 1$.

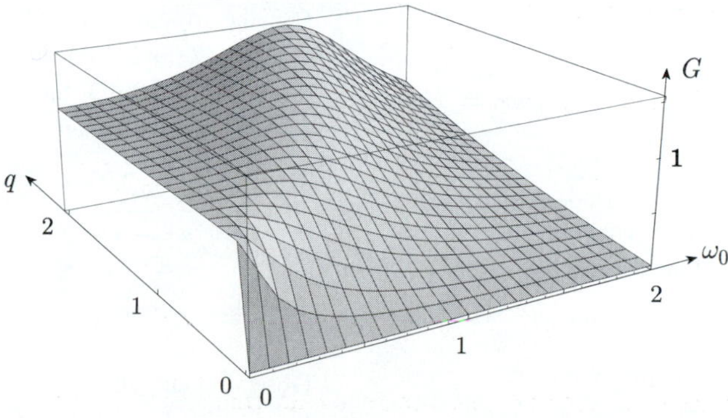

Figure 4

We also make the following observations concerning the phase lag ϕ. Recall that

$$\cos\phi = \frac{q-\omega_0^2}{\sqrt{(q-\omega_0^2)^2 + p^2\omega_0^2}} \quad \text{and} \quad \sin\phi = \frac{p\omega_0}{\sqrt{(q-\omega_0^2)^2 + p^2\omega_0^2}}.$$

i. If there is no damping (i.e., $p = 0$), then either $\phi = 0$ or $\phi = \pi$.
ii. If $\omega_0 \approx 0$, then $\phi \approx 0$. That is, the response is nearly *in phase* with low frequency input.
iii. If $\omega_0 > \sqrt{q}$, then $\pi/2 < \phi < \pi$, and $\phi \to \pi$ as $\omega_0 \to \infty$. That is, the response is nearly 180° out of phase with high frequency input.
iv. If $\omega_0 = \sqrt{q}$, then $\phi = \pi/2$. That is, the response is 90° (or *halfway*) out of phase with input at the natural frequency of the undamped system. (Recall that the resonant response of an undamped system is 90° out of phase with the input.)

PROBLEMS

In Problems 1 through 4, use the formula in (2) to construct a particular solution.

1. $y'' + 4y = 4(3\sin t - 5\cos 3t)$

2. $y'' + y = \cos(\sqrt{0.9}t) + \sin(\sqrt{1.01}t)$

3. $y'' + 9y = 9(\cos 2t + \cos 5t)$

4. $y'' + \frac{\pi^2}{100}y = \cos\left(\frac{\pi}{9}t\right) + \sin\left(\frac{\pi}{12}t\right)$

5. Find a particular solution of

(a) $y'' + 4y = 12\sin 2t$

(b) $y'' + 4y' + 4y = 12\sin 2t$

6. Find a particular solution of

(a) $y'' + 19y = 19\cos 4t$

(b) $y'' + y' + 19y = 19\cos 4t$

In Problems 7 through 10, find a particular solution expressed in the form $A\cos(\omega t - \phi)$ or $A\sin(\omega t - \phi)$. State the phase lag and the gain.

7. $y'' + 4y' + 14y = 13\cos 3t$

8. $y'' + 3y' + 5y = 5\sin t$

9. $y'' + 2y' + y = 5\cos 2t$

10. $y'' + y' + \frac{3}{2}y = 5\cos t$

11. **(a)** Use an appropriate complex equation to find particular real solutions of

$$x'' + x = 8t\cos t, \quad y'' + y = 8t\sin t.$$

(b) Use the results of part (a) to help write down the general solution of

$$y'' + y = 8t(\sin t + \cos t).$$

12. This problem concerns the phenomenon of **beats**, which occurs when an undamped system is forced at a frequency near its natural frequency. A low-frequency oscillation is produced that has a frequency equal to half the difference in the natural frequency and the frequency of the input. Audible beats are often used to tune two guitar or piano strings to the same pitch. When the strings vibrate at nearly the same pitch, beats can be heard. As the frequency of one string is tuned to the other, the beats slow down and disappear.

(a) Show that the rest solution of $y'' + \omega^2 y = \omega^2 A_0 \cos\omega_0 t$ is

$$y = \frac{\omega^2}{\omega^2 - \omega_0^2}(A_0\cos\omega_0 t - \cos\omega t).$$

(b) Use the trigonometric identity $\cos 2\alpha - \cos 2\beta = 2\sin(\beta+\alpha)\sin(\beta-\alpha)$ to show that when $A_0 = 1$ the rest solution in part (a) may be written

$$y = \frac{2\omega^2}{\omega^2 - \omega_0^2}\sin\left(\frac{\omega_0+\omega}{2}t\right) \times \sin\left(\frac{\omega_0-\omega}{2}t\right).$$

Observe that when $\omega_0 \approx \omega$, the low-frequency factor $2\sin\left(\frac{\omega_0-\omega}{2}t\right)$ may be thought of as the slowly oscillating amplitude of the high-frequency factor $\sin\left(\frac{\omega_0+\omega}{2}t\right)$. This slowly oscillating amplitude is sometimes called the *envelope* of the higher-frequency oscillation.

(c) Make a rough sketch of the graph of the rest solution with $A_0 = 1$, $\omega_0 = 7\pi$, and $\omega = 5\pi$ on the interval $[0, 2]$.

13. **(a)** Check that the solution of the initial-value problem

$$y'' + py' + qy = qf(t),$$

$$y(0) = y_0, \quad y'(0) = v_0$$

can be written as $y = u + w$ where u and w satisfy, respectively,

$$u'' + pu' + qu = 0, \quad u(0) = y_0,$$

$$u'(0) = v_0, \quad w'' + pw' + qw = qf(t),$$

$$w(0) = w'(0) = 0.$$

(b) Argue that if $p, q > 0$, then $u(t) \to 0$ as $t \to \infty$, regardless of the initial values, and so $y(t) \approx w(t)$ for large t. Consequently, the long-term behavior of y does not depend on initial values. Such systems are sometimes referred to as *dissipative*. For dissipative systems the effect of initial conditions on a solution is said to be *transient*.

14. **(a–c)** Verify observations iii, iv, and v concerning the gain G of a damped system.

(d) Show also that the maximum gain when $q > p^2/2$ can be written in terms of q and ω_0 as $G = q/\sqrt{q^2 - \omega_0^4}$.

15. Express the gain G of a damped spring-mass system in terms of the mass m, the damping coefficient r, the stiffness k, and the angular input frequency ω_0.

16. Express the gain G of an RLC circuit in terms of the resistance R, the inductance L, the capacitance C, and the angular input frequency ω_0.

17. Find coefficients $p, q > 0$ so that maximum gain occurs at an angular frequency of $\omega_0 = 10$ with a phase lag of $\phi = \pi/4$.

18. Show that a damped system experiences maximum gain at a given angular input frequency ω_0 with a given phase lag ϕ if

$$p = 2\omega_0 \cot\phi \quad \text{and} \quad q = \omega_0^2(1 + 2\cot^2\phi).$$

19. Suppose we wish to have a given maximum gain $G > 1$ occur at a given angular input frequency ω_0. Show that this occurs if

$$p^2 = 2\omega_0^2\left(\sqrt{\frac{G^2}{G^2-1}} - 1\right)$$

$$\text{and} \quad q = \omega_0^2\sqrt{\frac{G^2}{G^2-1}}.$$

20. Suppose that an underdamped system experiences damped, free vibrations with angular frequency ω. Show that maximum gain occurs at an angular input frequency of $\omega_0 = \sqrt{2\omega^2 - q}$.

21. Suppose that the weight of your car compresses its suspension system by 8 inches. Suppose further that when you jump off your car's bumper, you observe damped, free vibration with a frequency of about 1 cycle per second. Estimate the input frequency that will result in forced vibration of maximum amplitude.

22. Suppose that you are driving the same car as in Problem 21 on a washboarded road on which the surface is approximately sinusoidal with successive peaks about 30 feet apart. Estimate the speed (in mph) at which you and the car would experience the most vibration. Also estimate the gain (cf. Problem 14d).

CHAPTER

7

THE LAPLACE TRANSFORM

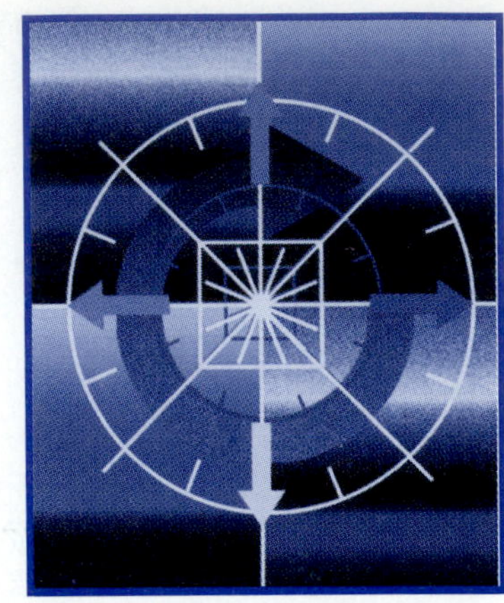

In this chapter we will explore a method for solving linear differential equations with constant coefficients that is widely used in electrical engineering. It involves the transformation of an initial-value problem into an algebraic equation, which is easily solved, and then the inverse transformation back to the solution of the original problem, thereby bypassing the need to solve for arbitrary constants in the general solution. The technique is especially well suited for finding generalized solutions to systems driven by impulses or by discontinuous or periodic forcing functions.

7.1 Definition and Basic Properties

Given a function $f(t)$, defined for $t \geq 0$, we define its **Laplace transform** $F(s)$ by

$$F(s) = \int_0^\infty e^{-st} f(t)\,dt. \tag{1}$$

Notice that the variable s appears as a parameter in an improper integral. We say that the Laplace transform exists if this improper integral converges for all sufficiently large s. The notational convention used here is common. The uppercase version of the function's name denotes its Laplace transform, and s is used for the transform's independent variable. Before we go further, let's illustrate this definition and notation with a couple of simple examples.

Example 1

Consider a constant function $f(t) = c$, $t \geq 0$. The Laplace transform is

$$F(s) = \int_0^\infty e^{-st} c\,dt = \frac{c}{-s} e^{-st}\bigg|_{t=0}^{\infty} = -\frac{c}{s}\left(\lim_{t\to\infty} e^{-st} - e^0\right) = -\frac{c}{s}(0-1) = \frac{c}{s},$$

provided that $s > 0$. Note that the improper integral diverges when $s \leq 0$. ■

Example 2

Let $f(t) = e^{at}$, $t \geq 0$. Then the Laplace transform is

$$\begin{aligned} F(s) &= \int_0^\infty e^{-st} e^{at}\,dt = \int_0^\infty e^{-(s-a)t}\,dt \\ &= -\frac{1}{s-a}\left(\lim_{t\to\infty} e^{-(s-a)t} - e^0\right) = -\frac{1}{s-a}(0-1) = \frac{1}{s-a}, \end{aligned}$$

provided that $s > a$. Note that the improper integral diverges when $s \leq a$. ■

The Laplace transform $F(s)$ of a function $f(t)$ is actually the result of applying a linear operator to f. Denoting this linear operator by $\mathcal{L}$, we can write

$$\mathcal{L}f = F, \quad \text{or} \quad \mathcal{L}[f](s) = F(s).$$

where F is given by (1). Using this notation, the result of Example 2, for instance, is that

$$\mathcal{L}[e^{at}](s) = (s-a)^{-1}, \quad s > a.$$

It is easy to check that the operator $\mathcal{L}$ is linear:

$$\begin{aligned} \mathcal{L}[cf](s) &= \int_0^\infty e^{-st} cf(t)\,dt = c\int_0^\infty e^{-st} f(t)\,dt = cF(s) \\ \mathcal{L}[f+g](s) &= \int_0^\infty e^{-st}(f(t)+g(t))\,dt \\ &= \int_0^\infty e^{-st} f(t)\,dt + \int_0^\infty e^{-st} g(t)\,dt = F(s) + G(s). \end{aligned}$$

Indeed, the linearity of $\mathcal{L}$ is a simple consequence of the linearity properties of integration.

The reason why Laplace transforms are useful in solving differential equations is embodied in the following theorem, which (together with the corollary that follows) we will refer to as the **first differentiation theorem**.

Theorem 1

Suppose that f is a continuous function on $[0, \infty)$ and that the Laplace transforms $F(s) = \mathcal{L}f$ and $\mathcal{L}f'$ each exist. Then

$$\mathcal{L}[f'](s) = sF(s) - f(0). \quad \blacktriangle$$

The proof of Theorem 1 is basically an exercise in integration by parts:

$$\begin{aligned} \mathcal{L}[f'](s) &= \int_0^\infty e^{-st} f'(t)\,dt \\ &= e^{-st} f(t)\Big|_{t=0}^{\infty} - \int_0^\infty (-s)\,e^{-st} f(t)\,dt \\ &= -f(0) + sF(s) \end{aligned}$$

for sufficiently large $s > 0$. Note that in the last step we used the two limits:

$$\lim_{t\to\infty} e^{-st} f(t) = 0 \quad \text{and} \quad \lim_{t\to 0^+} f(t) = f(0).$$

The first of these limits is a property of f that follows from the assumption that the transform F exists; that is, the improper integral defining the transform F converges for sufficiently large s. The second limit follows simply from the continuity assumption on f.

We point out that the *existence* of a Laplace transform—which is *assumed* for f and f' in Theorem 1—is far from automatic. Roughly speaking, the Laplace transform of a function will exist if the function is integrable on bounded intervals $[0, T]$ and does not grow faster than exponentially—which excludes, for instance, a function such as e^{t^2}. This issue is explored further in Problems 41 and 42.

By repeated application of Theorem 1, we arrive at the following corollary.

COROLLARY 1 *Suppose that f and f' are continuous on $[0, \infty)$ and that the Laplace transforms $F(s) = \mathcal{L}f$, $\mathcal{L}f'$ and $\mathcal{L}f''$ each exist. Then*

$$\mathcal{L}[f''](s) = s^2 F(s) - sf(0) - f'(0).$$

Moreover, if f has $n-1$ continuous derivatives on $[0, \infty)$ and if each of these together with $f^{(n)}$ have Laplace transforms, then

$$\mathcal{L}[f^{(n)}](s) = s^n F(s) - s^{n-1} f(0) - s^{n-2} f'(0) - \cdots - f^{(n-1)}(0).$$

Because of these properties, the Laplace transform is particularly well suited for solving linear initial-value problems with constant coefficients.

Example 3

Consider the first-order initial-value problem

$$y' + 2y = 4, \quad y(0) = 1,$$

and let Y denote the transform of the solution y. By applying the operator $\mathcal{L}$ to both sides of the differential equation and using the result of Example 1 on the right side, we find that

$$s\,Y(s) - y(0) + 2Y(s) = \frac{4}{s}.$$

Now we use the given initial value and solve for $Y(s)$:

$$(s + 2)Y(s) = \frac{4}{s} + 1, \quad \text{so} \quad Y(s) = \frac{s + 4}{s(s + 2)}.$$

Now our job is to find the function $y(t)$ that has this transform. A partial fraction decomposition (consult your calculus book) reveals that

$$Y(s) = \frac{2}{s} - \frac{1}{s + 2}.$$

The results of Examples 1 and 2 now tell us that $Y(s)$ is the transform of

$$y(t) = 2 - e^{-2t}.$$

■

Example 4

Consider the initial-value problem

$$y'' + 3y' + 2y = 0, \quad y(0) = 1, \quad y'(0) = 0,$$

and let Y denote the transform of the solution y. By applying the operator $\mathcal{L}$ to both sides of the differential equation, we find that

$$s^2Y(s) - s\ y(0) - y'(0) + 3(sY(s) - y(0)) + 2Y(s) = 0.$$

Using the given initial values and rearranging, this becomes

$$(s^2 + 3s + 2)Y(s) - s - 3 = 0.$$

Now we solve for $Y(s)$ to find

$$Y(s) = \frac{s+3}{s^2 + 3s + 2} = \frac{s+3}{(s+1)(s+2)}.$$

Now we look for the function $y(t)$ that has this transform. A partial fraction decomposition reveals that

$$Y(s) = \frac{2}{s+1} - \frac{1}{s+2}.$$

From the result of Example 2 above, we see that this is the transform of

$$y(t) = 2e^{-t} - e^{-2t}.$$

■

Examples 3 and 4 each illustrate a general procedure for solving initial value problems with the help of Laplace transforms:

(1) Transform each side of the differential equation, using the given initial values.
(2) Solve the resulting *algebraic* equation for the transform $Y(s)$ of the solution $y(t)$.
(3) Find the solution y by identifying the transform $Y(s)$ with known transforms.

The final step amounts to finding the **inverse transform** of $Y(s)$. An important question to ask is whether the operator $\mathcal{L}$ is actually invertible. In other words, given a transform $Y(s)$, is there a unique function $y(t)$, $t \geq 0$, such that $\mathcal{L}y = Y$? The answer to this is no, *unless we require the function y to be continuous*. However, this is not a difficulty in the context of solving differential equations, since solutions will be continuous.

Steps 1 and 2 are easy, provided that the transform of the nonhomogeneous term is easy to get. The challenging part of solving any problem in this way lies in step 3. Extensive tables of known transforms are available to help in this task. As in Examples 3 and 4, partial fraction decomposition is often a useful tool for expressing $Y(s)$ is terms of known transforms. Problems 18 through 33 contain additional discussion of techniques for computing partial fraction decompositions.

Computer algebra systems such as *Mathematica* and *Maple* have the ability to compute partial fraction decompositions as well as all of the "standard" Laplace transforms and inverse Laplace transforms. Thus, even though we present examples with some degree of computational detail, our discussions here are primarily intended to emphasize the conceptual framework of the Laplace transform rather than skill with the often tedious manual computations involved.

To conclude this section, let's now consider what happens in general when we apply the Laplace transform technique to a linear second-order initial-value problem with constant coefficients. Suppose that the differential equation is

$$P(\mathcal{D})y = f, \quad \text{with } P(\mathcal{D}) = \mathcal{D}^2 + p\mathcal{D} + q\mathcal{I},$$

and we have initial conditions

$$y(0) = y_0, \quad y'(0) = v_0.$$

The Laplace transform of the nonhomogeneous term is simply $\mathcal{L}f = F$. Transforming the left side produces

$$\begin{aligned} \mathcal{L}P(\mathcal{D})y &= \mathcal{L}\mathcal{D}^2 y + p\mathcal{L}\mathcal{D}y + q\mathcal{L}y \\ &= s^2 Y(s) - sy_0 - v_0 + p(sY(s) - y_0) + qY(s) \\ &= P(s)Y(s) - (sy_0 + v_0 + py_0). \end{aligned}$$

Thus we find that

$$\mathcal{L}P(\mathcal{D})y = P(s)Y(s) - \gamma(s),$$

where $\gamma(s)$ is a polynomial of degree at most 1 with coefficients determined by the initial values y_0, v_0. The transformed initial-value problem therefore becomes

$$P(s)Y(s) - \gamma(s) = F(s),$$

and solving for $Y(s)$ gives us

$$Y(s) = \frac{F(s)}{P(s)} + \frac{\gamma(s)}{P(s)}. \tag{2}$$

The main point here is that the denominator in the Laplace transform of any solution of $P(\mathcal{D})y = f$ contains $P(s)$ as a factor. Moreover, the term $F(s)/P(s)$ in (2) is the transform of the rest solution of $P(\mathcal{D})y = f$, and the term $\gamma(s)/P(s)$ in (2) is the transform of the solution of the homogeneous equation $P(\mathcal{D})y = 0$ that satisfies the given initial conditions. Moreover, it turns out that the same statements are true for all linear differential equations with constant coefficients, regardless of order.

PROBLEMS

In Problems 1 through 4, use the results of Examples 1 and 2 to write down the Laplace transform of the given function. Express the result as a single quotient.

1. $f(t) = 3 - 5e^t$

2. $f(t) = 2e^{-t} + 3e^{-2t}$

3. $f(t) = \cosh bt$

4. $f(t) = \sinh bt$

In Problems 5 through 10, find the Laplace transform $Y(s)$ of the solution of the given initial-value problem.

5. $y' + 3y = e^{-t}, \ y(0) = 0$

6. $y'' + y = 0, \ y(0) = 1, \ y'(0) = -1$

7. $y'' + y' + y = 0, \ y(0) = 0, \ y'(0) = 1$

8. $y'' + 3y' + 2y = 0, \ y(0) = 1, \ y'(0) = 0$

9. $y'' - 4y = e^{-t}, \ y(0) = 1, \ y'(0) = 1$

10. $y''' - y = 1, \ y(0) = y'(0) = y''(0) = 0$

In Problems 11 through 14, suppose that f has the Laplace transform $F(s)$, and write down *by inspection* the transform $Y(s)$ of the differential equation's rest solution.

11. $y'' + y' + y = f$

12. $y'' + \omega^2 y = f$

13. $y''' + y = f$

14. $y''' - y' + 2y = f$

15. Show that $y = te^{at}$ satisfies

$$y'' - 2ay' + a^2 y = 0, \ y(0) = 0, \ y'(0) = 1,$$

and use that fact to find the Laplace transform of te^{at}.

16. **(a)** Use the fact that $y = \cos \omega t$ is the solution of

$$y'' + \omega^2 y = 0, \ y(0) = 1, \ y'(0) = 0,$$

to find the Laplace transform of $\cos \omega t$.

(b) Similarly find the Laplace transform of $\sin \omega t$.

17. **(a)** Use the fact that $y = t \sin \omega t$ satisfies

$$y'' + \omega^2 y = 2\omega \cos \omega t, \ y(0) = 0, \ y'(0) = 0,$$

to find the Laplace transform of $t \sin \omega t$. Make use of the known transform of $\cos \omega t$ from Problem 16.

(b) Similarly, find the Laplace transform of $t \cos \omega t$.

Partial Fraction Decomposition. A rational function $\frac{N(s)}{P(s)}$ with degree $(N) <$ degree (P) may be decomposed into a sum of rational functions with denominators corresponding to the linear and irreducible quadratic factors of $P(s)$ and with numerators that are either constant or linear. A nonrepeated linear factor $s - a$ of $P(s)$ results in a single simple term of the form $\frac{A}{s-a}$, and repeated linear factors $(s - a)^k$ can give rise to terms

$$\frac{A_1}{s-a} + \frac{A_2}{(s-a)^2} + \cdots + \frac{A_k}{(s-a)^k}.$$

An irreducible quadratic factor $s^2 + bs + c$ of $P(s)$ results in a term of the form $\frac{Bs+C}{s^2+bs+c}$, and repeated quadratic factors $(s^2 + bs + c)^k$ can give rise to terms

$$\frac{B_1 s + C_1}{s^2 + bs + c} + \frac{B_2 s + C_2}{(s^2 + bs + c)^2} + \cdots + \frac{B_k s + C_k}{(s^2 + bs + c)^k}.$$

The number of undetermined constants needed in the decomposition is always equal to the degree of $P(s)$, which is also precisely the number of coefficients needed to specify the numerator $N(s)$ if degree$(N) =$ degree$(P) - 1$. Assuming that degree$(N) <$ degree(P), write down the form of the partial fraction decomposition of each of the rational functions in Problems 18 through 21.

18. $\dfrac{N(s)}{s(s-1)^2}$

19. $\dfrac{N(s)}{s^2(s^2+2s+2)}$

20. $\dfrac{N(s)}{(s^2+1)(s^2+4)^2}$

21. $\dfrac{N(s)}{s^4-1}$

Remark: The coefficients in a partial fraction decomposition can always be found as the solution of a system of n linear equations in n unknowns, where $n = \text{degree}(P)$. However, this is precisely what is involved in determining from initial conditions the constants in the general solution of an nth-order differential equation. Therefore, in order for there to be any advantage to using partial fraction decompositions of Laplace transforms to solve initial-value problems, we must have more efficient ways of finding the constants in partial fraction decompositions. The next several problems illustrate a few such shortcuts.

22. Suppose that $a_1, a_2, \dots, a_n$ are distinct, and consider the partial fraction decomposition

$$\frac{N(s)}{(s-a_1)(s-a_2)\cdots(s-a_n)} = \frac{A_1}{s-a_1} + \frac{A_2}{s-a_2} + \cdots + \frac{A_n}{s-a_n},$$

where $n \geq 2$ and $N(s)$ is a polynomial of degree less than n. Show that A_i may be determined by multiplying both sides by $s - a_i$ and letting $s \to a_i$, resulting, for each $i = 1, \dots, n$, in

$$\begin{aligned} A_i &= \lim_{s\to a_i} \frac{(s-a_i)N(s)}{(s-a_1)(s-a_2)\cdots(s-a_n)} \\ &= \frac{N(a_i)}{(a_i-a_1)\cdots(a_i-a_{i-1})(a_i-a_{i+1})\cdots(a_i-a_n)}. \end{aligned}$$

Use the method of Problem 22 to find the partial fraction decomposition of the rational function in each of Problems 23 through 25.

23. $\dfrac{s}{(s-1)(s-2)}$

24. $\dfrac{12}{s(s+1)(s-3)}$

25. $\dfrac{2s}{(s-1)(s+1)}$

26. Consider the partial fraction decomposition

$$\frac{N(s)}{(s-a)^n} = \frac{A_1}{s-a} + \frac{A_2}{(s-a)^2} + \cdots + \frac{A_n}{(s-a)^n},$$

where $n \geq 2$ and $N(s)$ is a polynomial of degree $k < n$.

(**a**) Show that $A_n = N(a)$ and $A_1 = \lim_{s\to\infty} \frac{N(s)}{(s-a)^{n-1}} = \lim_{s\to\infty} \frac{N(s)}{s^{n-1}}$.

(**b**) Show that if $k < n-1$, then $A_1 = \cdots = A_{n-k-1} = 0$ and

$$A_{n-k} = \lim_{s\to\infty} \frac{N(s)}{(s-a)^k} = \lim_{s\to\infty} \frac{N(s)}{s^k}.$$

Use the method of Problem 26 to help find the partial fraction decomposition of the rational function in each of Problems 27 through 29.

27. $\dfrac{s}{(s-1)^3}$

28. $\dfrac{s^2}{(s+2)^3}$

29. $\dfrac{s^2}{(s-1)^4}$

30. Suppose that $s^2 + bs + c$ is an irreducible quadratic and that $N(s)$ has degree $k < 3$. Consider the partial fraction decomposition

$$\frac{N(s)}{(s-a)(s^2+bs+c)} = \frac{A}{s-a} + \frac{Bs+C}{s^2+bs+c}.$$

Multiply through by $s - a$ and show that

$$A = \frac{N(a)}{a^2 + ba + c} \quad \text{and} \quad A + B = \lim_{s \to \infty} \frac{N(s)}{s^2 + bs + c}.$$

Use the method of Problem 30 to help find the partial fraction decomposition of the rational function in each of Problems 31 through 33.

31. $\dfrac{2s}{(s-1)(s^2+1)}$

32. $\dfrac{2s^2}{(s-1)(s^2+1)}$

33. $\dfrac{s^2+4}{(s-1)(s^2+2s+2)}$

Solve Problems 34 through 39 as in Examples 3 and 4. Use the method of Problem 22 to find the necessary partial fraction decomposition.

33. $y' + y = 1, \ \ y(0) = 0$

35. $y' - y = e^{-t}, \ \ y(0) = 0$

36. $y'' - y = 0, \ \ y(0) = 0, \ \ y'(0) = 1$

37. $y'' - 4y = 8, \ \ y(0) = y'(0) = 0$

38. $y'' + 4y' + 3y = 6, \ \ y(0) = y'(0) = 0$

39. $y''' + 2y'' - y' - 2y = 10,$
$y(0) = y'(0) = y''(0) = 0$

40. Use a comparison test to show that if the improper integral in (1) converges for some $s = s_0$, then it converges for all $s > s_0$.

41. **(a)** Argue that $f(t) = e^{t^2}$ does not have a Laplace transform (i.e., that the defining improper integral is divergent for all s).

(b) Argue that $f(t) = t^t$ does not have a Laplace transform. (*Hint*: $t^t = e^{t \ln t}$.)

42. Existence of the Transform. Suppose that f is integrable on any bounded interval $[0, T]$ and that there are numbers $\sigma \geq 0$ and $M > 0$ such that

$$|f(t)| \leq Me^{\sigma t} \quad \text{for all } t \geq 0.$$

Such functions are said to be *of exponential order*.

(a) Show that, for all $s > \sigma$, the transform $\mathcal{L}[|f|](s)$ exists and

$$\mathcal{L}[|f|](s) \leq \frac{M}{s - \sigma}.$$

(*Hint*: $e^{-st}|f(t)| = e^{-(s-\sigma)t}e^{-\sigma t}|f(t)|$.)

(b) Show that, for all $s > \sigma$, the transform $\mathcal{L}[f](s)$ exists and

$$|\mathcal{L}[f](s)| \leq \frac{M}{s - \sigma}.$$

(*Hint*: $-|f(t)| \leq f(t) \leq |f(t)|$.)

43. Suppose that $\mathcal{L}[f(t)](s) = F(s)$. Prove the following "scaling theorems":

(a) $\mathcal{L}[f(ct)] = \dfrac{1}{c} F\left(\dfrac{s}{c}\right)$

(b) $\mathcal{L}\left[\dfrac{1}{c} f\left(\dfrac{t}{c}\right)\right] = F(cs)$

7.2 More Transforms and Further Properties

In this section we will find Laplace transforms for more of the elementary functions that commonly arise as solutions of linear differential equations with constant coefficients. We will also derive additional properties of Laplace transforms that will help us in finding both transforms and inverse transforms.

First, let's recall the results of Examples 1 and 2 in Section 7.1:

$$\mathcal{L}[c](s) = \frac{c}{s}, \quad s > 0; \tag{1}$$

$$\mathcal{L}[e^{at}](s) = \frac{1}{s-a}, \quad s > a. \tag{2}$$

Note that the first of these is actually a consequence of the second (with $a = 0$) and linearity. Also, the computation that produced the second of these remains valid if a is a complex constant, provided that $s > \text{Re}(a)$. That is,

$$\mathcal{L}[e^{(\alpha+i\beta)t}](s) = \frac{1}{s-\alpha+i\beta} = \frac{s-\alpha-i\beta}{(s-\alpha)^2+\beta^2}, \quad s > \alpha.$$

Now because of the Euler-DeMoivre formula and linearity, we have

$$\mathcal{L}[e^{\alpha t}\cos\beta t](s) = \frac{s-\alpha}{(s-\alpha)^2+\beta^2}, \quad s > \alpha, \tag{3}$$

$$\mathcal{L}[e^{\alpha t}\sin\beta t](s) = \frac{\beta}{(s-\alpha)^2+\beta^2}, \quad s > \alpha, \tag{4}$$

and, in particular,

$$\mathcal{L}[\cos\beta t] = \frac{s}{s^2+\beta^2}, \quad s > 0, \quad \mathcal{L}[\sin\beta t] = \frac{\beta}{s^2+\beta^2}, \quad s > 0. \tag{5, 6}$$

The First Shift Theorem Notice that the transforms of $e^{\alpha t}\cos\beta t$ and $e^{\alpha t}\sin\beta t$ in (3) and (4) can be viewed as *shifted* transforms of $\cos\beta t$ and $\sin\beta t$ obtained by replacing s with $s-\alpha$. It turns out that the same thing happens whenever any function is multiplied by $e^{\alpha t}$. To see this, suppose that $f(t)$ has a known transform $F(s)$ and consider the transform of $e^{\alpha t}f(t)$:

$$\mathcal{L}[e^{\alpha t}f(t)](s) = \int_0^\infty e^{-st}e^{\alpha t}f(t)\,dt = \int_0^\infty e^{-(s-\alpha)t}f(t)\,dt = F(s-\alpha).$$

Thus we have what we will call the **first shift theorem**:

$$\mathcal{L}[e^{\alpha t}f(t)](s) = F(s-\alpha).$$

Example 1

Let's find the inverse transform of

$$F(s) = \frac{s}{s^2+4s+13}.$$

Since the denominator is an irreducible quadratic, the key is to complete the square and then use the first shift theorem. Completing the square shows that the denominator is $(s+2)^2+9$. Now in order take advantage of the shift theorem, we must express the numerator in terms of $s+2$ as well:

$$F(s) = \frac{s+2-2}{(s+2)^2+9} = \frac{s+2}{(s+2)^2+9} - \frac{2}{(s+2)^2+9}.$$

The first term on the right is the transform of $e^{-2t}\cos 3t$ by (3), and the second is the transform of $-\frac{2}{3}e^{-2t}\sin 3t$ by (4). Therefore,

$$f(t) = e^{-2t}\left(\cos 3t - \frac{2}{3}\sin 3t\right). \qquad \blacksquare$$

Example 2

Suppose that we wish to find the rest solution of the equation

$$y'' + 2y' + 10y = 10.$$

The transform of the solution is

$$Y(s) = \frac{10}{s(s^2 + 2s + 10)} = \frac{1}{s} - \frac{s+2}{s^2 + 2s + 10}.$$

The denominator in the second term is $(s+1)^2 + 9$. To take advantage of this, we need the numerator in terms of $s+1$. So we write

$$Y(s) = \frac{1}{s} - \frac{s+1}{(s+1)^2 + 9} - \frac{1}{(s+1)^2 + 9}$$

and observe (by (1), (3), and (4)) that $Y(s)$ is the Laplace transform of

$$y = 1 - e^{-t}\cos 3t - \frac{1}{3}e^{-t}\sin 3t. \qquad \blacksquare$$

The Second Differentiation Theorem The first shift theorem describes the transform of the product $e^{\alpha t}f(t)$ in terms of the transform of f. A similar result describing the transform of the product $t^n f(t)$ would also be useful. To that end, let us consider a function $f(t)$ with known transform $F(s)$:

$$\int_0^\infty e^{-st} f(t)\,dt = F(s), \quad s > s_0.$$

Differentiating with respect to s, we obtain*

$$\int_0^\infty e^{-st}(-tf(t))\,dt = \frac{d}{ds}F(s), \quad s > s_0.$$

Since the left side here is the transform of $-tf(t)$, we have found that

$$\mathcal{L}[tf(t)](s) = -\frac{d}{ds}F(s),$$

a fact which we shall call the **second differentiation theorem**. Repeated application of this result easily produces

$$\mathcal{L}[t^n f(t)](s) = (-1)^n \frac{d^n}{ds^n}F(s), \quad n = 0, 1, 2, 3, \ldots.$$

---

* Here we have made the questionable move of differentiating under the integral sign; however, it can be justified in this situation by a theorem from advanced calculus.

The second differentiation theorem is useful for the derivation of a number of transforms. For instance, with the constant function $f(t) = 1$, $t \geq 0$, we find that $\mathcal{L}[t](s) = -\frac{d}{ds}(s^{-1})$; that is,

$$\mathcal{L}[t](s) = \frac{1}{s^2}, \quad s > 0.$$

Repeating the same procedure produces the transforms of t^2, t^3, and so on:

$$\mathcal{L}[t^n](s) = \frac{n!}{s^{n+1}}, \quad s > 0. \tag{7}$$

Combining (7) with the first shift theorem yields

$$\mathcal{L}[t^n e^{at}](s) = \frac{n!}{(s-a)^{n+1}}, \quad s > a. \tag{8}$$

The second differentiation theorem also produces

$$\mathcal{L}[t\cos\beta t] = \frac{s^2-\beta^2}{(s^2+\beta^2)^2}, \quad s > 0, \tag{9}$$

$$\mathcal{L}[t\sin\beta t] = \frac{2\beta s}{(s^2+\beta^2)^2}, \quad s > 0. \tag{10}$$

To put (9) into a form more useful for finding inverse transforms, we note that

$$\frac{1}{s^2+\beta^2} - \frac{s^2-\beta^2}{(s^2+\beta^2)^2} = \frac{2\beta^2}{(s^2+\beta^2)^2}.$$

Now we combine (6) with (9) to obtain

$$\mathcal{L}\left[\frac{\sin\beta t}{\beta} - t\cos\beta t\right] = \frac{2\beta^2}{(s^2+\beta^2)^2}, \quad s > 0. \tag{11}$$

The following two examples illustrate the use of 8 and 11 in the context of solving initial-value problems.

Example 3

Consider the initial-value problem

$$y'' + 4y' + 4y = 0, \;\; y(0) = 1, \;\; y'(0) = 1.$$

The transform of the solution is easily seen to be

$$Y(s) = \frac{s+5}{s^2+4s+4} = \frac{s+2+3}{(s+2)^2} = \frac{1}{s+2} + \frac{3}{(s+2)^2}.$$

Therefore, because of (2) and (8), we see that Y is the transform of

$$y = e^{-2t} + 3te^{-2t} = (1+3t)e^{-2t},$$

which is easily verified as the solution of the initial-value problem. ■

Example 4

Consider the initial-value problem

$$y'' + 4y = \sin 2t, \;\; y(0) = 1, \;\; y'(0) = 0.$$

The transform of the solution is easily seen to be

$$Y(s) = \frac{2}{(s^2+4)^2} + \frac{s}{s^2+4}.$$

By (5) and (11), we see that

$$y = \frac{1}{4}\left(\frac{1}{2}\sin 2t - t\cos 2t\right) + \cos 2t.$$

■

PROBLEMS

In Problems 1 through 3, find the inverse transform (a) by partial fraction decomposition and (b) by completing the square and using the first shift theorem and the basic transforms

$$\mathcal{L}[\sinh bt] = \frac{b}{s^2-b^2} \quad \text{and} \quad \mathcal{L}[\cosh bt] = \frac{s}{s^2-b^2}.$$

1. $\dfrac{2}{s^2+2s}$

2. $\dfrac{2s}{s^2+4s+3}$

3. $\dfrac{4s}{s^2-6s+5}$

Find the inverse transform of each expression in Problems 4 through 15.

4. $\dfrac{s}{s^2+4s+5}$

5. $\dfrac{s+1}{s^2+6s+13}$

6. $\dfrac{s+8}{s^2+10s+34}$

7. $\dfrac{2s}{s^2+2s+5}$

8. $\dfrac{s}{(s+2)^2}$

9. $\dfrac{2s^2}{(s+1)^3}$

10. $\dfrac{2s}{(s+3)^3}$

11. $\dfrac{s+1}{(s-1)^3}$

12. $\dfrac{12s}{(s^2+4)^2}$

13. $\dfrac{16}{(s^2+4)^2}$

14. $\dfrac{4s^2}{(s^2+4)^2}$

15. $\dfrac{s^3}{(s^2+4)^2}$

Use (10), (11), and the first shift theorem to find the inverse transform of each expression in Problems 16 through 18.

16. $\dfrac{16}{(s^2+2s+5)^2}$

17. $\dfrac{16s}{(s^2+2s+5)^2}$

18. $\dfrac{2s+6}{(s^2+2s+2)^2}$

In Problems 19 through 27, use the Laplace transform to solve the initial-value problem.

19. $y'' + 4y' + 3y = 0,\ y(0) = 1,\ y'(0) = -4$

20. $y'' + 6y' + 13y = 0,\ y(0) = 0,\ y'(0) = 2$

21. $y'' + 4y' + 4y = 0,\ y(0) = 0,\ y'(0) = 1$

22. $y'' + 4y' + 3y = e^{-t},\ y(0) = 0,\ y'(0) = 0$

23. $y'' + 6y' + 13y = t,\ y(0) = 0,\ y'(0) = 0$

24. $y'' + 4y' + 4y = e^{-2t},\ y(0) = 0,\ y'(0) = 0$

25. $y'' + 4y = 6\sin t,\ y(0) = 0,\ y'(0) = 0$

26. $y'' + 9y = \sin 3t,\ y(0) = 0,\ y'(0) = 0$

27. $y'' + y = \cos t,\ y(0) = 0,\ y'(0) = 0$

28. The purpose of this problem is to derive the Laplace transform of $\sqrt{t}$.

(a) Use the definition of the Laplace transform and integration by parts to show that

$$\mathcal{L}\left[\sqrt{t}\right](s) = \frac{1}{2s}\int_0^\infty \frac{e^{-st}}{\sqrt{t}}\,dt.$$

(b) Use the substitution $st = x^2$ to arrive at

$$\mathcal{L}\left[\sqrt{t}\right](s) = \frac{k}{s^{3/2}} \quad \text{where} \quad k = \int_0^\infty e^{-x^2}\,dx.$$

(c) After observing that

$$k^2 = \int_0^\infty e^{-x^2}dx \int_0^\infty e^{-y^2}dy = \int_0^\infty \int_0^\infty e^{-(x^2+y^2)}dx\,dy,$$

use polar coordinates to show that $k^2 = \pi/4$. Conclude that

$$\mathcal{L}\left[\sqrt{t}\right](s) = \frac{\sqrt{\pi}}{2s^{3/2}}.$$

29. Apply the first differentiation theorem to the antiderivative $\int_0^t f(\tau)\,d\tau$ to obtain the **first integration theorem**:

$$\mathcal{L}\left[\int_0^t f(u)\,du\right] = \frac{F(s)}{s}.$$

30. In this problem we will derive the **second integration theorem**:

$$\mathcal{L}\left[\frac{1}{t}f(t)\right] = \int_s^\infty F(\sigma)d\sigma.$$

(a) Suppose that f has the transform $F(s)$ and that $\frac{1}{t}f(t)$ has the transform $\Phi(s)$. Apply the second differentiation theorem to show that $\Phi'(s) = -F(s)$. Hence Φ is *some* antiderivative of $-F$.

(b) By the result of Problem 36, we know that $\Phi(s) \to 0$ as $s \to \infty$. Show that $\int_s^\infty F(\sigma)d\sigma$ has that property, as well as being an antiderivative of $-F(s)$, and therefore must be the desired transform $\Phi(s)$.

In Problems 31 through 34, use the second integration theorem (see Problem 30) to find the Laplace transform of the given function.

31. $\dfrac{1 - e^{at}}{t}$

32. $\dfrac{1 - \cos\beta t}{t}$

33. $\dfrac{\sin\beta t}{t}$

34. $\dfrac{\sinh bt}{t}$

35. Use the second integration theorem (see Problem 30) and the result of Problem 28 to find the Laplace transform of $f(t) = 1/\sqrt{t}$.

36. Asymptotic Behavior of the Laplace Transform. Suppose that $f(t)$ is bounded on every bounded interval $[a, b]$, $a \geq 0$. Suppose further that the Laplace transform $\mathcal{L}[f](s) = F(s)$ exists for all $s > s_0$, where $s_0 \geq 0$. By the following sequence of steps, prove that

$$\lim_{s\to\infty} F(s) = 0.$$

(a) Argue that $e^{-s_1 t} f(t) \to 0$ as $t \to \infty$ for any $s_1 > s_0$.

(b) Argue that because of (a) there is a time T_1 so that $|e^{-s_1 t} f(t)| < 1$ for all $t \geq T_1$, and consequently $|f(t)| < e^{s_1 t}$ for all $t \geq T_1$.

(c) Use the result of (b) to obtain

$$|F(s)| \leq \int_0^{T_1} e^{-st}|f(t)|\,dt + \int_{T_1}^{\infty} e^{-st}e^{s_1 t}\,dt.$$

(d) Use the result of (c) and the fact that f is bounded on $[0, T_1]$ to obtain

$$|F(s)| \leq \frac{M}{s} + \frac{1}{s - s_1} \quad \text{for all } s > s_1,$$

where M is a bound on $|f(t)|$ for all $0 \leq t \leq T_1$.

(e) Finally, conclude that $F(s) \to 0$ as $s \to \infty$. Also observe the stronger fact that $sF(s)$ is bounded for large s.

37. Combine the result of Problem 36 with the first differentiation theorem to show that if f and f' have Laplace transforms and f is continuous, then

$$\lim_{s\to\infty} sF(s) = f(0).$$

38. The *error function* erf and the *complementary error function* erfc are

$$\operatorname{erf}(t) = \frac{2}{\sqrt{\pi}}\int_0^t e^{-x^2}dx \text{ and } \operatorname{erfc}(t) = 1 - \operatorname{erf}(t) = \frac{2}{\sqrt{\pi}}\int_t^{\infty} e^{-x^2}dx$$

(a) Show that $y = e^{-t^2/4}$ satisfies the initial-value problem

$$y' + \frac{1}{2}ty = 0, \quad y(0) = 1.$$

Then show that the transform Y of y satisfies

$$sY - 1 - \frac{1}{2}Y' = 0.$$

(b) Use the appropriate integrating factor (see Section 2.1) to solve for $Y(s)$ subject to the condition $Y(s) \to 0$ as $s \to \infty$ (cf. Problem 36). Conclude that

$$\mathcal{L}[e^{-t^2/4}] = \sqrt{\pi}e^{s^2}\operatorname{erfc}(s).$$

(c) Use the first integration theorem (see Problem 29) and the result of part (b) to show that

$$\mathcal{L}\left[\operatorname{erf}\left(\frac{t}{2}\right)\right] = \frac{1}{s}e^{s^2}\operatorname{erfc}(s).$$

(d) Finally, use the "scaling theorem" $\mathcal{L}[\frac{1}{k}f(\frac{t}{k})] = F(ks)$ (see Problem 43 in Section 7.1) to show that, for $k > 0$,

$$\mathcal{L}\left[e^{-t^2/(4k^2)}\right] = k\sqrt{\pi}e^{k^2s^2}\operatorname{erfc}(ks) \quad \text{and} \quad \mathcal{L}\left[\operatorname{erf}(\frac{t}{2k})\right] = \frac{1}{s}e^{k^2s^2}\operatorname{erfc}(ks).$$

In Problems 39 and 40, use the first integration theorem (see Problem 29) to help in finding the solution of the given *integro-differential* initial-value problem.

39. $y' + \int_0^t y(\tau)\,d\tau = t,\ y(0) = 0$

40. $y'' + \int_0^t y(\tau)\,d\tau = 0,\ y(0) = 3,\ y'(0) = 0$

7.3 Heaviside Functions and Piecewise-Defined Inputs

The Laplace transform is particularly useful for solving initial-value problems in which the differential equation has a piecewise-defined nonhomogeneous term. In order to express such piecewise-defined functions in a convenient manner, we will use the **Heaviside unit step function**, which is defined by

$$h(t) = \begin{cases} 0, & \text{if } t < 0; \\ 1, & \text{if } t \geq 0. \end{cases}$$

This function may be thought of as a "switch" that "turns on" at time $t = 0$. A switch that turns on at time $t = c$ may be obtained by a simple shift:

$$h(t-c) = \begin{cases} 0, & \text{if } t < c; \\ 1, & \text{if } t \geq c. \end{cases}$$

A switch that is "on" during a time interval $a \leq t < b$ can be constructed as

$$h(t-a) - h(t-b) = \begin{cases} 0, & \text{if } t < a; \\ 1, & \text{if } a \leq t < b; \\ 0, & \text{if } b \leq t. \end{cases}$$

Figures 1a–c show the graphs of $h(t)$, $h(t-3)$, and $h(t-3)-h(t-5)$. (Of course, the vertical segments in these pictures simply indicate instantaneous jumps in function values and are not actually parts of the graphs.)

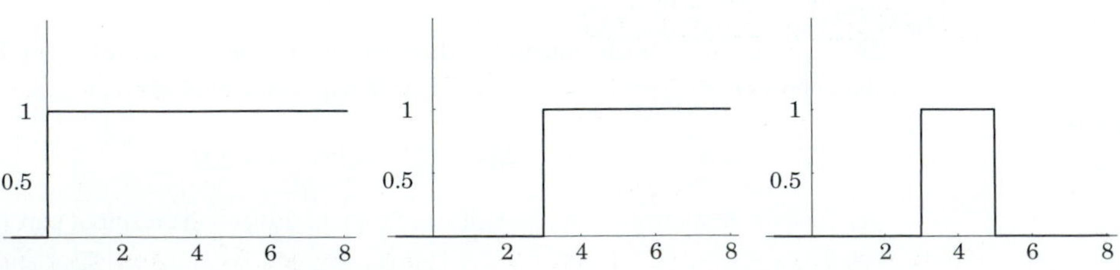

Figure 1a **Figure 1b** **Figure 1c**

For further illustration, Figure 2 shows the graphs of

(a) $h(t) + h(t-2)$,
(b) $h(t) - 2h(t-1) + h(t-2)$,
(c) $h(t) + h(t-1) + h(t-2) - 3h(t-3)$.

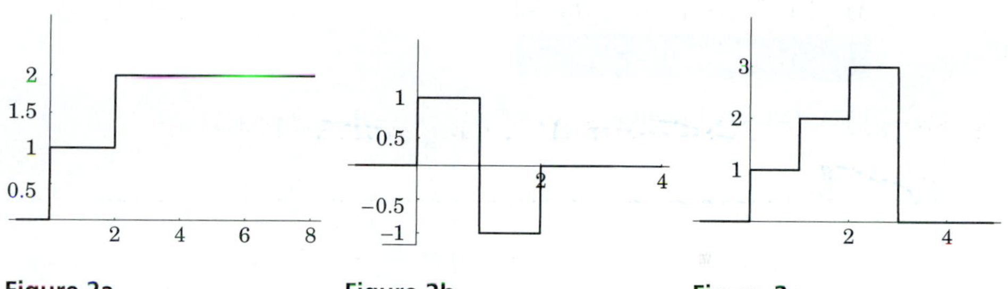

Figure 2a **Figure 2b** **Figure 2c**

All of the preceding examples involve piecewise *constant* functions, or *step functions*. Heaviside functions are also useful for writing simple representations for other types of piecewise-defined functions. Figure 3 shows the graphs of

(a) $th(t) - th(t-2)$,
(b) $th(t) + (2-2t)h(t-1) - (2-t)h(t-2)$,
(c) $t^4h(t) + ((t-2)^4 - t^4)h(t-1) - (t-2)^4h(t-2)$.

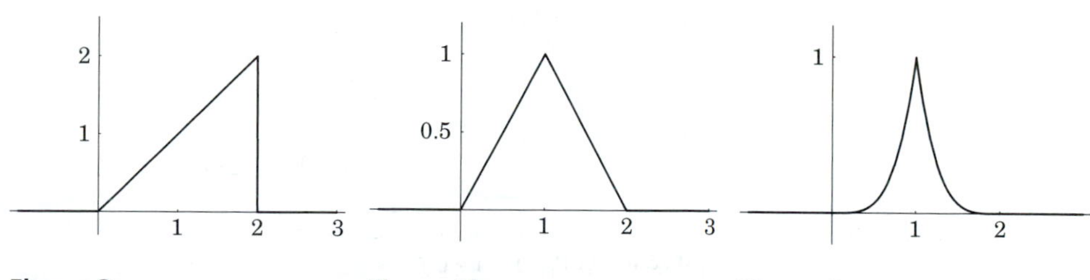

Figure 3a **Figure 3b** **Figure 3c**

Example 1

Let's look closely at the function in Figure 3b with the goal of expressing it in "piecewise form." The function is

$$f(t) = th(t) + (2-2t)h(t-1) + (t-2)h(t-2).$$

When $t < 0$, all three of the Heaviside functions are "off," and so $f(t) = 0$. When $0 \le t < 1$, only the Heaviside function $h(t)$ in the first term is "on," and so $f(t) = t$. When $1 \le t < 2$, both Heaviside functions in the first two terms are "on," and so

$f(t) = t + (2 - 2t) = 2 - t$. When $t \geq 2$, all three Heaviside functions are "on," and so $f(t) = t + (2 - 2t) + (t - 2) = 0$. Thus, f has the piecewise form

$$f(t) = \begin{cases} 0, & t < 0; \\ t, & 0 \leq t < 1; \\ 2 - t, & 1 \leq t < 2; \\ 0, & 2 \leq t. \end{cases}$$ ■

Example 2

Consider the function

$$f(t) = \begin{cases} 0, & t < 0; \\ t^2, & 0 \leq t < 1; \\ (t-1)^2, & 1 \leq t < 2; \\ 0, & 2 \leq t. \end{cases}$$

The graph of f is shown in Figure 4. We can express f in terms of Heaviside functions as follows. First, we note that $t^2 h(t)$ will agree with $f(t)$ for all $t < 1$. To switch from $f(t) = t^2$ to $f(t) = (t-1)^2$ at $t = 1$, we need to add $(t-1)^2 h(t-1)$ *and* subtract $t^2 h(t-1)$. Then to switch from $f(t) = (t-1)^2$ to $f(t) = 0$ at $t = 2$, we need to subtract $(t-1)^2 h(t-2)$. The result is

$$f(t) = t^2 h(t) + ((t-1)^2 - t^2)h(t-1) - (t-1)^2 h(t-2).$$ ■

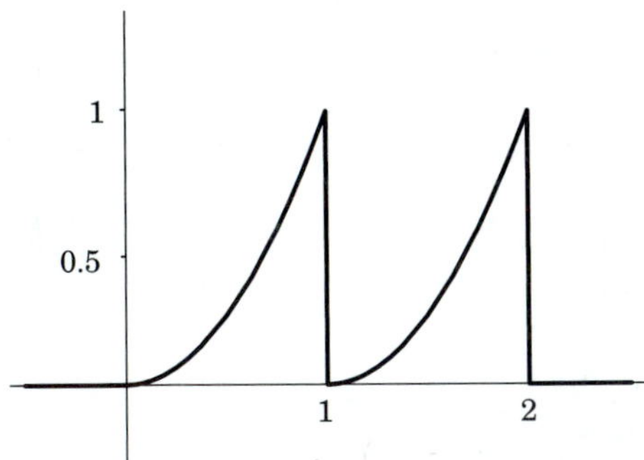

Figure 4

The way in which a general piecewise-defined function can be expressed in terms of Heaviside functions is actually rather easy to describe. Suppose that f is given in piecewise form as

$$f(t) = \begin{cases} 0, & t < 0; \\ f_0(t), & 0 \leq t < t_1; \\ f_1(t), & t_1 \leq t < t_2; \\ f_2(t), & t_2 \leq t < t_3; \\ \vdots & \vdots \end{cases}$$

The representation of f in terms of Heaviside functions is

$$f(t) = f_0(t)h(t) + (f_1(t) - f_0(t))h(t - t_1) + (f_2(t) - f_1(t))h(t - t_2) + \cdots,$$

the form of the typical term being $(f_{\text{new}} - f_{\text{prev}})h(t - c)$. Note that this function f is *right continuous*; that is, $\lim_{t \to a^+} f(t) = f(a)$ for all a.

Given a function $f(t)$ with Laplace transform $F(s)$, consider the function

$$h(t - c)f(t - c),$$

where $c \geq 0$. The graph of this function is the same as that of f shifted to the right by c units and zeroed out for $t < c$. This is illustrated in Figure 5. The Laplace transform of $h(t - c)f(t - c)$ is computed as follows:

$$\begin{aligned} \mathcal{L}[h(t - c)f(t - c)](s) &= \int_0^\infty e^{-st}h(t - c)f(t - c)\,dt \\ &= \int_c^\infty e^{-st}f(t - c)\,dt \\ &= \int_0^\infty e^{-s(t+c)}f(t)\,dt = e^{-cs}F(s). \end{aligned}$$

This computation gives us the **second shift theorem**:

$$\mathcal{L}[h(t - c)f(t - c)](s) = e^{-cs}F(s).$$

As a simple consequence (with $f(t - c) = 1$), we obtain the transform of any simple Heaviside function:

$$\mathcal{L}[h(t - c)](s) = \frac{e^{-cs}}{s}.$$

Note that with $c = 0$ this gives $\mathcal{L}[h(t)](s) = 1/s$, which we have already seen as the transform of the function $f(t) = 1$, $t \geq 0$.

Example 3

Consider the function

$$f(t) = t\,h(t) - t\,h(t - 2).$$

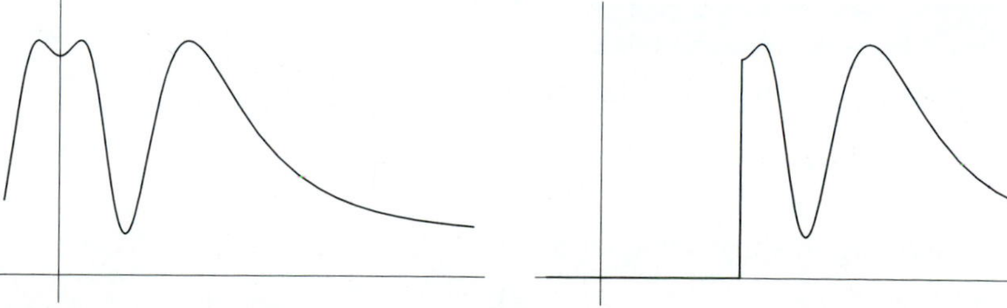

Figure 5

To find the transform $F(s)$, we need to rewrite f as

$$f(t) = th(t) - (t-2)h(t-2) - 2h(t-2).$$

Now we apply the second shift theorem to obtain

$$F(s) = \frac{1}{s^2} - \frac{e^{-2s}}{s^2} - \frac{2}{s} = \frac{1 - e^{-2s} - 2s}{s^2}.$$

Notice that in writing the transform of the second term, we used the transform of the function t, not $t - 2$. ■

Example 4

Consider the function $f(t) = e^t h(t) + (e^{-t} - e^t)h(t-1)$.The function multiplying $h(t-1)$ must be expressed as a function of $t - 1$, so we rewrite f as

$$f(t) = e^t h(t) + (e^{-1}e^{-(t-1)} - e^1 e^{(t-1)})h(t-1).$$

Now we apply the second shift theorem to the transforms of e^t and e^{-t} to obtain

$$F(s) = \frac{1}{s-1} + e^{-s}\left(\frac{e^{-1}}{s+1} - \frac{e}{s-1}\right) = \frac{1 - e^{1-s}}{s-1} + \frac{e^{-1-s}}{s+1}, \quad s > 1.$$ ■

Example 5

Suppose that we want to find the inverse transform of

$$F(s) = \frac{1 - 2e^{-s} + e^{-2s}}{s}.$$

By splitting this into three terms,

$$F(s) = \frac{1}{s} - 2\frac{e^{-s}}{s} + \frac{e^{-2s}}{s},$$

we recognize that

$$f(t) = h(t) - 2h(t-1) + h(t-2).$$

This is the function shown in Figure 2b. ■

Example 6

Suppose we want to find the inverse transform of

$$F(s) = \frac{1 - 2e^{-s} + e^{-2s}}{s^2}.$$

By splitting this into three terms,

$$F(s) = \frac{1}{s^2} - 2e^{-s}\frac{1}{s^2} + e^{-2s}\frac{1}{s^2},$$

we recognize that

$$f(t) = th(t) - 2(t-1)h(t-1) + (t-2)h(t-2).$$

This is the function in Example 1 and Figure 3b. ■

Our main purpose here, of course, is to use all of this to help us solve initial-value problems with piecewise-defined inputs. We will illustrate this with the following example.

Example 7

Consider an undamped spring-mass system with a natural frequency of $\frac{\omega}{2\pi} = 1$ cycle per unit time, so that $\omega = 2\pi$. Suppose that the mass is initially at rest and the platform to which the spring is attached undergoes motion described by the function seen in Figure 2b. Thus we wish to find the rest solution of the equation

$$y'' + 4\pi^2 y = 4\pi^2(h(t) - 2h(t-1) + h(t-2)).$$

Taking Laplace transforms, we find

$$Y(s) = \frac{1 - 2e^{-s} + e^{-2s}}{s} \frac{4\pi^2}{(s^2 + 4\pi^2)}.$$

A partial fraction decomposition shows that

$$Y(s) = (1 - 2e^{-s} + e^{-2s}) \left(\frac{1}{s} - \frac{s}{s^2 + 4\pi^2} \right).$$

The second factor is the transform of $1 - \cos 2\pi t$; therefore

$$\begin{aligned} y(t) = (1 - \cos 2\pi t)h(t) - 2(1 - \cos 2\pi(t-1))h(t-1) \\ +(1 - \cos 2\pi(t-2))h(t-2). \end{aligned}$$

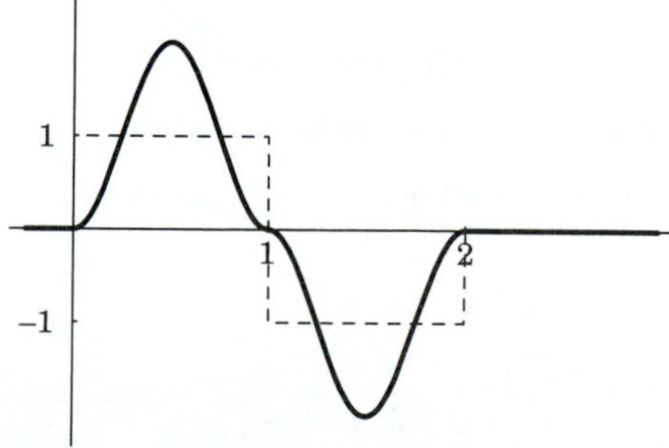

Figure 6

by the second shift theorem. Since $\cos 2\pi t = \cos 2\pi(t-k)$ for all integers k and all t, the solution has the simple piecewise form

$$y(t) = \begin{cases} 0, & t < 0; \\ 1 - \cos 2\pi t, & 0 \le t < 1; \\ \cos 2\pi t - 1, & 1 \le t < 2; \\ 0, & 2 \le t. \end{cases}$$

The graphs of this solution and the forcing function are seen in Figure 6. ■

Note that the solution obtained using the Laplace transform is a *generalized solution* in the same sense as described in Section 2.3. This means that the solution is a continuous function that satisfies the differential equation at each t at which the forcing function is continuous.

PROBLEMS

In Problems 1 through 3, write the given function f in piecewise form. Then sketch the graph.

1. $f(t) = h(t) + h(t-1) + h(t-2) - h(t-3) - h(t-4)$

2. $f(t) = 4t(1-t)(h(t) - h(t-1))$

3. $f(t) = th(t) + (1-t)h(t-1) - h(t-2)$

In Problems 4 through 6, graph the given function *without* writing its piecewise form.

4. $f(t) = 3h(t) - h(t-1) - h(t-2) - h(t-3)$

5. $f(t) = h(t) + (1-t)h(t-1) + (t-2)h(t-2)$

6. $f(t) = h(t) - 2h(t-1) + h(t-2) + h(t-3) - 2h(t-4) + h(t-5)$

In Problems 7 through 9, express the given function in terms of Heaviside functions.

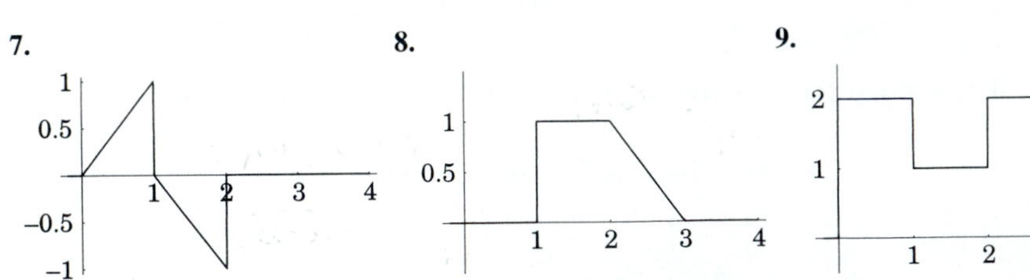

Let floor $\lfloor x \rfloor$ denote the "floor" (or greatest integer) function; that is,

$$\lfloor x \rfloor = n, \text{ where, } n \le x < n + 1 \text{with } n \text{ an integer.}$$

Write each of the functions in Problems 10 through 13 in terms of Heaviside functions. Also sketch the graph.

10. $f(t) = \begin{cases} (-1)^{\lfloor t \rfloor}, & 0 \le t < 3 \\ 0, & \text{elsewhere} \end{cases}$

11. $f(t) = \begin{cases} t - \lfloor t \rfloor, & 0 \le t < 3 \\ 0, & \text{elsewhere} \end{cases}$

12. $f(t) = \begin{cases} t - \lfloor t + .5 \rfloor, & 0 \le t < 2 \\ 0, & \text{elsewhere} \end{cases}$

13. $f(t) = \begin{cases} 3 - \lfloor t \rfloor, & 0 \le t < 3 \\ 0, & \text{elsewhere} \end{cases}$

Problems $n = 14, \ldots, 26$: Find the Laplace transform of the function in Problem $n - 13$.

In Problems 27 through 32, find the inverse Laplace transform and sketch the graph of the result.

27. $\dfrac{1 - 2e^{-2s} + e^{-3s}}{s}$

28. $\dfrac{1 - 2e^{-2s} + e^{-3s}}{s^2}$

29. $\dfrac{s + e^{-2s} - 2se^{-3s}}{s^2}$

30. $\dfrac{1 - 2e^{-s} + e^{-2s}}{s^2 + 4\pi^2}$

31. $\dfrac{4(1 - 2e^{-\pi s} + e^{-2\pi s})}{s(s^2 + 4)}$

32. $\dfrac{2 + 3e^{-\pi s} + e^{-2\pi s}}{s^2 + 1}$

In Problems 33 and 34, solve the initial-value problem; then graph the solution and the nonhomogeneous term on the right side of the differential equation.

33. $y' - y = -eh(t-1),\ \ y(0) = 1$

34. $y' + y = (t+1)(1 - h(t-2)),\ \ y(0) = 0$

In Problems 35 and 36, solve the initial-value problem; then graph the solution and the forcing function.

35. $y'' + \pi^2 y = \pi^2(h(t-1) - h(t-2)),\ \ y(0) = 1,\ \ y'(0) = 0$

36. $y'' + y = th(t) - th(t - 2\pi),\ \ y(0) = 0,\ \ y'(0) = 0$

In Problems 37 through 39, obtain the transform of the given function by direct application of the definition of the transform, and then check by finding the transform with the second shift theorem.

37. $f(t) = h(t-1) - h(t-2)$

38. $f(t) = h(t) - 2h(t-1) + h(t-2)$

39. $f(t) = e^t(h(t) - h(t-1))$

40. A piecewise constant function that is right-continuous for all t, zero on $(-\infty, t_0)$, and discontinuous at $t_0, t_1, t_2, \ldots$ can be expressed as

$$f(t) = j_0 h(t - t_0) + j_1 h(t - t_1) + j_2 h(t - t_2) + \cdots$$

where each j_i is a constant. Show that

$$j_i = \lim_{t \to t_i^+} f(t) - \lim_{t \to t_i^-} f(t).$$

(Thus j_i is the "jump" of $f(t)$ at t_i.) Show further that

$$F(s) = \frac{j_0 e^{t_0 s} + j_1 e^{t_1 s} + j_2 e^{t_2 s} + \cdots}{s}.$$

Using these results, redo Problems 9 and 22.

7.4 Periodic Inputs

We say that a function f defined on $[0, \infty)$ is periodic, with period $p > 0$, if

$$f(t + p) = f(t) \qquad \text{for all } t \geq 0.$$

Such a function may be regarded as the *periodic extension* of the function

$$f_0(t) = f(t)(h(t) - h(t - p)),$$

which agrees with f on $[0, p)$ and is zero elsewhere. We will see that the Laplace transform of the periodic extension f of f_0 can be easily expressed in terms of the transform of f_0.

First note that since $f(t+p) = f(t)$ for all $t \geq 0$, it follows that $f(t) = f(t-p)$ for all $t \geq p$. Thus we can express f_0 as

$$f_0(t) = h(t)f(t) - h(t-p)f(t-p),$$

which allows use of the second shift theorem, producing the transform

$$F_0(s) = F(s) - e^{-ps}F(s),$$

where $F = \mathcal{L}f$ as usual. Now we solve for $F(s)$ to find

$$F(s) = \frac{F_0(s)}{1 - e^{-ps}}. \tag{1}$$

Thus, the transform of the periodic extension of f_0 is simply the transform of f_0 divided by $1 - e^{-ps}$. The transform $F_0 = \mathcal{L}f_0$ in the numerator can be obtained either by expressing f_0 in terms of Heaviside functions and applying the second shift theorem or by evaluating the definite integral

$$F_0(s) = \int_0^p e^{-st} f(s)\, ds.$$

Example 1

The function

$$f(t) = (-1)^{\lfloor t \rfloor}, \quad t \geq 0$$

is the periodic extension, with period $p = 2$, of

$$f_0(t) = h(t) - 2h(t-1) + h(t-2)$$

Graphs of these functions are shown in Figures 1a and 1b. The transform of f_0 is

$$F_0(s) = \frac{1 - 2e^{-s} + e^{-2s}}{s};$$

therefore, the transform of f is

$$F(s) = \frac{1 - 2e^{-s} + e^{-2s}}{s(1 - e^{-2s})} = \frac{1 - e^{-s}}{1 + e^{-s}} \frac{1}{s}.$$ ■

Example 2

The function

$$f(t) = |\sin t|, \quad t \geq 0$$

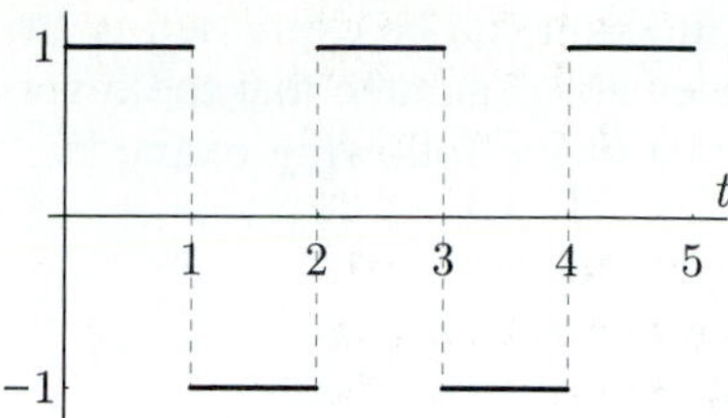

Figure 1a

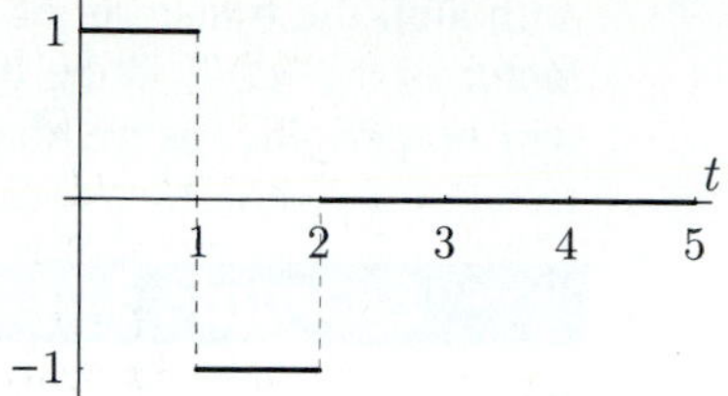

Figure 1b

can be viewed as the periodic extension of

$$\begin{aligned} f_0(t) &= \sin t\,(h(t) - h(t - \pi)) \\ &= h(t)\sin t + h(t - \pi)\sin(t - \pi) \end{aligned}$$

with period $p = \pi$. These functions are plotted in Figures 2a and b. The Laplace transform of f_0 is

$$F_0(s) = (1 + e^{-\pi s})\frac{1}{s^2 + 1};$$

consequently, the transform of f is

$$F(s) = \frac{1 + e^{-\pi s}}{1 - e^{-\pi s}}\frac{1}{s^2 + 1}.$$ ■

Inverse transforms The key to finding the inverse transform of a given transform with $1 - e^{-ps}$ as a factor in its denominator is the geometric series:

$$\frac{1}{1 - x} = 1 + x + x^2 + x^3 + \cdots, \quad \text{if } |x| < 1.$$

Since $0 < e^{-ps} < 1$ for all $s > 0$, we have the expansion

$$\frac{1}{1 - e^{-ps}} = 1 + e^{-ps} + e^{-2ps} + e^{-3ps} + \cdots, \quad s > 0.$$

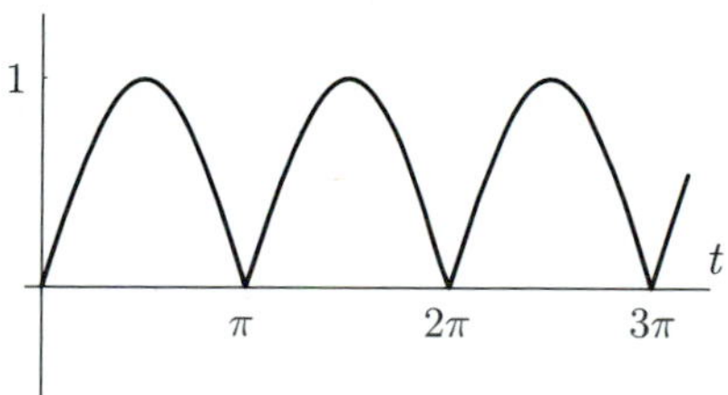

Figure 2a

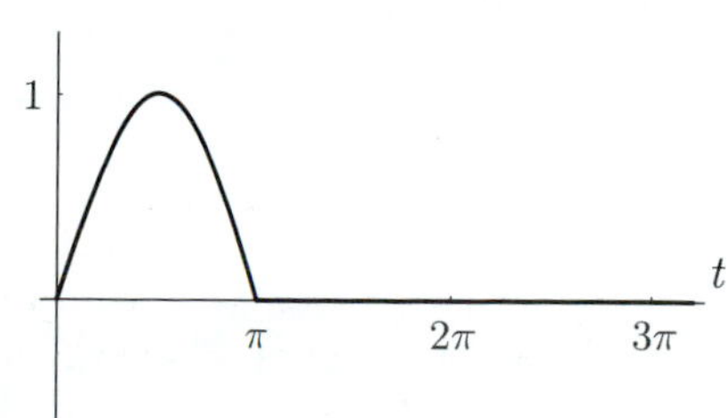

Figure 2b

Although the transform of any periodic function can be expressed in the form (1), a factor of $1 - e^{-ps}$ in the denominator does not guarantee that the inverse transform will be periodic, as we shall see in the first of the following examples.

Example 3

The transform

$$F(s) = \frac{1}{s(1 - e^{-s})}$$

can be written as the series

$$F(s) = \frac{1}{s} + \frac{e^{-s}}{s} + \frac{e^{-2s}}{s} + \frac{e^{-3s}}{s} + \cdots .$$

Taking inverse transforms term by term, we find that

$$f(t) = h(t) + h(t-1) + h(t-2) + h(t-3) + \cdots ,$$

which is easily recognizable as the floor function $f(t) = \lfloor t \rfloor$, $t \geq 0$. ■

Example 4

The transform

$$F(s) = \frac{1 - e^{-s}}{s(1 - e^{-2s})} = \frac{1}{s(1 + e^{-s})}$$

can be written as the series

$$F(s) = \frac{1}{s} - \frac{e^{-s}}{s} + \frac{e^{-2s}}{s} - \frac{e^{-3s}}{s} + \cdots .$$

Taking inverse transforms term by term, we find that

$$f(t) = h(t) - h(t-1) + h(t-2) - h(t-3) + \cdots ,$$

which can be expressed compactly as $f(t) = (1 + (-1)^{\lfloor t \rfloor})/2$, $t \geq 0$. ■

Example 5

The transform

$$F(s) = \frac{1}{(s^2 + 1)(1 - e^{-\pi s})}$$

can be written as

$$F(s) = \frac{1}{s^2 + 1}(1 + e^{-\pi s} + e^{-2\pi s} + e^{-3\pi s} + \cdots).$$

Since the first factor is the transform of $\sin t$, we apply successive shifts to obtain

$$f(t) = h(t) \sin t + h(t - \pi) \sin(t - \pi) + h(t - 2\pi) \sin(t - 2\pi) + \cdots .$$

Since $\sin(t - k\pi) = (-1)^k \sin t$ for any integer k, we can write this function as

$$f(t) = (h(t) - h(t-\pi) + h(t-2\pi) - h(t-3\pi) + \cdots) \sin t$$
$$= \frac{1}{2}(\sin t + |\sin t|), \quad t \geq 0.$$

■

We close this section with a final example in which we will solve a simple initial-value problem with a discontinuous, periodic forcing function.

Example 6

Let f be as in Example 1; that is, f is the periodic extension of

$$f_0(t) = h(t) - 2h(t-1) + h(t-2)$$

with period $p = 2$. Let's find the rest solution of the equation

$$y'' + 4\pi^2 y = 4\pi^2 f(t).$$

The transform of f is

$$F(s) = \frac{1 - 2e^{-s} + e^{-2s}}{s(1 - e^{-2s})};$$

so the transform of the solution is

$$Y(s) = \frac{1 - 2e^{-s} + e^{-2s}}{1 - e^{-2s}} \frac{4\pi^2}{s(s^2 + 4\pi^2)}.$$

We now expand the first factor as a series,

$$\frac{1 - 2e^{-s} + e^{-2s}}{1 - e^{-2s}} = (1 - 2e^{-s} + e^{-2s})(1 + e^{-2s} + e^{-4s} + e^{-6s} + \cdots)$$
$$= 1 - 2e^{-s} + 2e^{-2s} - 2e^{-3s} + \cdots,$$

perform a partial fraction decomposition on the second factor,

$$\frac{4\pi^2}{s(s^2 + 4\pi^2)} = \frac{1}{s} - \frac{s}{s^2 + 4\pi^2},$$

and rewrite the transform as

$$Y(s) = (1 - 2e^{-s} + 2e^{-2s} - 2e^{-3s} + \cdots)\left(\frac{1}{s} - \frac{s}{s^2 + 4\pi^2}\right).$$

Recognizing the second factor as the transform of $1 - \cos 2\pi t$, we apply successive shifts to obtain

$$y(t) = (1 - \cos 2\pi t)h(t) - 2(1 - \cos 2\pi(t-1))h(t-1)$$
$$+ 2(1 - \cos 2\pi(t-2))h(t-2) - 2(1 - \cos 2\pi(t-3))h(t-3) + \cdots.$$

Since $\cos 2\pi(t - k) = \cos 2\pi t$ for any integer k, we finally arrive at the solution

$$y(t) = (1 - \cos 2\pi t)(h(t) - 2h(t-1) + 2h(t-3) - \cdots)$$
$$= (-1)^{\lfloor t \rfloor}(1 - \cos 2\pi t).$$

Notice that this is precisely the periodic extension with period 2 of the solution in Example 7 of Section 3. ■

PROBLEMS

The functions in Problems 1 through 3 are the same as those in Problems 7 through 9 in Section 7.3. For each of them, find the Laplace transform of the periodic extension with period $p = 4$.

1.

2.

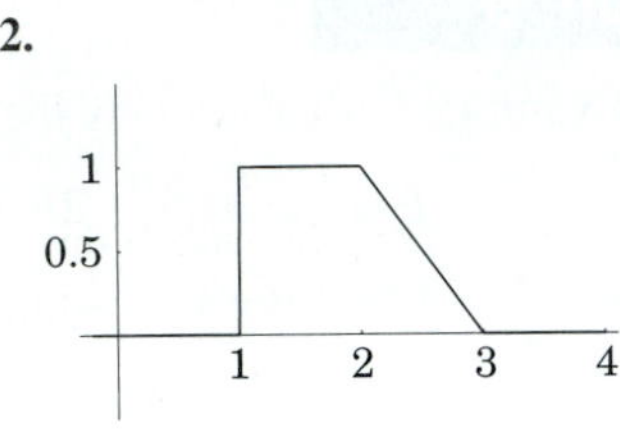

3.

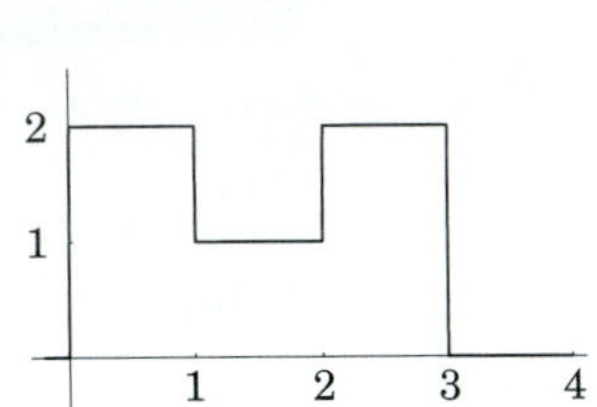

4. Let $f_0(t) = h(t) - h(t-1) - h(t-2) + h(t-3)$.

(a) Sketch the graph of the periodic extension of f_0 with period $p = 3$ and find its Laplace transform.

(b) Sketch the graph of the periodic extension of f_0 with period $p = 4$ and find its Laplace transform.

5. Let $f(t) = t - \lfloor t \rfloor$, $t \geq 0$. Graph f, describe it as a periodic extension of some representative function f_0, and then find its Laplace transform.

6. Let $f(t) = (t - \lfloor t \rfloor)^2$, $t \geq 0$. Graph f, describe it as a periodic extension of some representative function f_0, and then find its Laplace transform.

In Problems 7 through 10, find the inverse Laplace transform and sketch its graph.

7. $\dfrac{s}{(s^2+1)(1-e^{-\pi s})}$

8. $\dfrac{1-e^{-2\pi s}}{(s^2+1)(1-e^{-4\pi s})}$

9. $\dfrac{1-e^{-s}}{s^2(1-e^{-2s})}$

10. $\dfrac{1-e^{-s}}{s^2(1+e^{-2s})}$

In Problems 11 and 12, find the rest solution of the equation and sketch its graph. Note that each equation has a forcing function similar to $t - \lfloor t \rfloor$ from Problem 5.

11. $y'' + 4\pi^2 y = 4\pi^2(2 - (t - \lfloor t \rfloor))$

12. $y'' + \pi^2 y = \pi^3(t - \lfloor t \rfloor - 2)$

13. Let f be the periodic extension of $f_0(t) = h(t) - 2h(t-1) + h(t-2)$ with period $p = 4$. Find the rest solution of the equation

$$y'' + 4\pi^2 y = 4\pi^2 f(t)$$

and sketch its graph along with the graph of f.

A function f defined on $[0, \infty)$ is said to be *antiperiodic* with "antiperiod" k if $f(t+k) = -f(t)$ for all $t \geq 0$. For example, $\sin t$ is antiperiodic with antiperiod π, and $(-1)^{\lfloor t \rfloor}$ is antiperiodic with antiperiod 1. An antiperiodic function with antiperiod k can be described as the antiperiodic extension of

$$f_0(t) = f(t)(h(t) - h(t-k)),$$

which vanishes outside the interval $[0, k]$. Since $f(t+k) = -f(t)$ for all $t \geq 0$, it follows that $f_0(t)$ can be written as

$$f_0(t) = f(t)h(t) + h(t-k)f(t-k).$$

In Problems 14 through 16, assume that f is the antiperiodic extension of f_0 with antiperiod k.

14. Show that the Laplace transform of f can be expressed in terms of the transform of f_0 as follows:

$$F(s) = \frac{F_0(s)}{1+e^{-ks}} = \frac{1}{1+e^{-ks}} \int_0^k e^{-st} f(t)\, dt.$$

15. Let $\phi(t) = |f(t)|$ for all $t \geq 0$. To electrical engineers, $|f(t)|$ is known as the *full-wave rectification* of $f(t)$.

(a) Show that ϕ is periodic with period k and thus

$$\Phi(s) = \frac{1}{1-e^{-ks}} \int_0^k e^{-st} |f(t)|\, dt.$$

(b) Conclude that if $f(t) \geq 0$ for $0 \leq t < k$, then the following relationship holds between the Laplace transforms of $f(t)$ and $\phi(t) = |f(t)|$:

$$\Phi(s) = \frac{1+e^{-ks}}{1-e^{-ks}} F(s) = \coth(ks/2) F(s).$$

16. The function $g(t) = \frac{1}{2}(f(t) + |f(t)|)$ is the *half-wave rectification* of $f(t)$.

(a) Sketch the graph of the half-wave rectification of $\sin 2\pi t$.

(b) Show that if $f(t) \geq 0$ for $0 \leq t < k$, then the following relationship holds between the Laplace transforms of $f(t)$ and $g(t) = \frac{1}{2}(f(t) + |f(t)|)$:

$$G(s) = \frac{1}{1-e^{-2ks}} \int_0^k e^{-st} f(t)\, dt = \frac{1}{1-e^{-ks}} F(s).$$

17. Use the result of Problem 14 to rework Example 1.

18. Use the result of Problem 15 to rework Example 2.

19. Use the result of Problem 15 in a backward fashion to obtain the result of Example 1 from the transform $\mathcal{L}[h(t)] = 1/s$.

In Problems 20 and 21, sketch a graph of the given function, and use the appropriate result from either Problem 15 or 16, along with the first shift theorem, to find its Laplace transform.

20. $g(t) = e^{-t}|\sin t|$

21. $g(t) = \frac{1}{2}e^{-t}(\sin t + |\sin t|)$.

7.5 Impulses and the Dirac Distribution

Consider a spring-mass system being acted upon by some external force. By Newton's second law, force is the rate of change in momentum:

$$F(t) = \frac{d\mu}{dt},$$

where $\mu = mv$. Thus, as illustrated in Figure 1, a force acting on the mass over a time interval $[t_0, t_1]$ results in a change in momentum given by

$$\Delta\mu = \mu(t_1) - \mu(t_0) = \int_{t_0}^{t_1} F(t)\, dt.$$

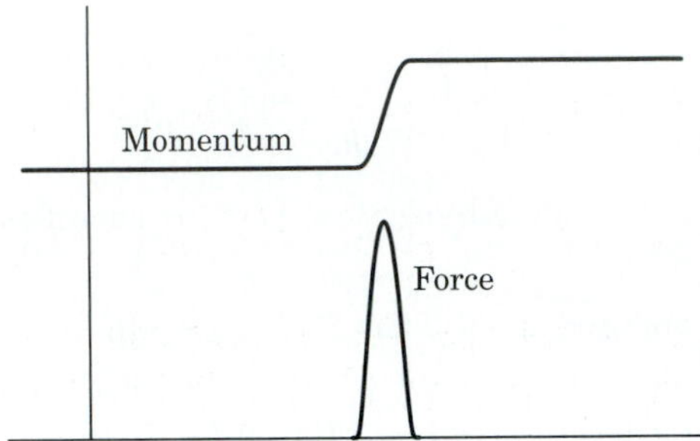

Figure 1

A force acting on the mass over a very small time interval $[t, t + \Delta t]$ might result, for instance, from striking the mass with a hammer. It will be useful to imagine an approximation to such a force in the form of an **impulse**—an idealized force concentrated at a single point t_0 in time and resulting in an instantaneous change in momentum. Our goal here is to develop a model for such a force.

Consider a unit change in momentum caused by a force $F(t)$ whose value is a constant b between time $t = 0$ and $t = \Delta t$ and zero at all other times; that is,

$$F(t) = (h(t) - h(t - \Delta t))b \quad \text{and} \quad \Delta\mu = \int_0^{\Delta t} b\, dt = 1.$$

This arrangement is always possible no matter how small the time interval is. In fact, b and Δt are related by $b\Delta t = 1$; therefore, $b = 1/\Delta t$, and so

$$F(t) = \frac{h(t) - h(t - \Delta t)}{\Delta t}. \tag{1}$$

The limiting behavior of this force as $\Delta t \to 0$ defines a **unit impulse** at time $t = 0$ and may be thought of as a *generalized derivative* of the Heaviside function $h(t)$. Clearly this will not be a function in the usual sense, but we will use notation that seems to suggest that it is. To capture the limiting behavior of F as $\Delta t \to 0$, we define the symbol $\delta(t)$, with units of $time^{-1}$ in the current context, by the following properties:

$$\text{(i) } \delta(t) = 0 \quad \text{for all } t \neq 0; \qquad \text{(ii) } \int_{-\infty}^{\infty} \delta(t)\, dt = 1.$$

We emphasize that $\delta(t)$ should not be thought of as a function of t in the usual sense, because (i) and (ii) are contradictory in the realm of ordinary calculus. Instead, $\delta(t)$ is a peculiar invention that serves as a "limit" for a sequence of functions whose values converge to zero except at a single point, but whose integrals do not converge to zero. It is an example of a class of "generalized function" called *distributions*. This particular distribution $\delta(t)$ is known as the **Dirac distribution**, or the *Dirac delta function*, after the great physicist Paul Dirac.

Let $g(t)$ be a continuous function on $(-\infty, \infty)$. Since $\delta(t)$ represents the "limit" as $\Delta t \to 0$ of the function $F(t)$ in 1, we *define*

$$\int_{-\infty}^{\infty} g(t)\delta(t)\,dt = \lim_{\Delta t\to 0}\int_{-\infty}^{\infty} g(t)\frac{h(t)-h(t-\Delta t)}{\Delta t}\,dt = \lim_{\Delta t\to 0}\frac{1}{\Delta t}\int_{0}^{\Delta t} g(t)\,dt.$$

This last quantity is the limit as $\Delta t \to 0$ of the average value of $g(t)$ over $[0, \Delta t]$. Since g is continuous, this limit must be $g(0)$. Therefore, it is consistent with (i) and (ii) above to *define*

$$\text{(iii)}\quad \int_{-\infty}^{\infty} g(t)\delta(t)\,dt = g(0) \text{ for any continuous } g \text{ on } (-\infty, \infty).$$

A unit impulse at time $t = c$ is represented by $\delta(t - c)$, the unit impulse at $t = 0$ shifted to occur at $t = c$. A simple change of variables applied to (iii) gives

$$\text{(iv)}\quad \int_{-\infty}^{\infty} g(t)\delta(t-c)dt = g(c) \text{ for any continuous } g \text{ on } (-\infty, \infty).$$

These last two properties put us in a position to state the Laplace transform of $\delta(t)$ and $\delta(t - c)$. Applying (iii) and (iv) with $g(t) = e^{-st}$, we find that

$$\mathcal{L}[\delta(t)](s) = 1 \text{ and } \mathcal{L}[\delta(t-c)](s) = e^{-cs}. \tag{2}$$

Initial-Value Problems Impulses can cause discontinuities in the derivatives of the solution of a differential equation, or in the solution itself in the case of a first-order equation. Therefore, initial values should be understood in the sense of left-sided limits. For example, in a second-order initial-value problem, the initial values are understood to mean

$$y(0^-) = \lim_{t\to 0^-} y(t) = y_0 \text{ and } y'(0^-) = \lim_{t\to 0^-} y'(t) = v_0. \tag{3}$$

The *rest solution* of a second-order equation is understood to be the maximal solution y for which $y(t) = 0$ for all $t \le 0$, which implies that each of the statements in (3) holds with $y_0 = v_0 = 0$.

Moreover, the initial values $f(0), f'(0), \ldots$ in the first differentiation theorem on Laplace transforms are understood to be the left-sided limits $f(0^-)$, $f'(0^-)$, and so on.

Example 1

Suppose that an undamped spring-mass system with with mass m, stiffness k, and natural angular frequency $\omega = \sqrt{k/m}$ is initially at rest. At time $t = 0$, the mass is struck with a hammer, imparting an instantaneous transfer of momentum $\Delta\mu$. The subsequent motion may be modeled with the equation

$$my'' + ky = \Delta\mu \cdot \delta(t).$$

(Note that all terms have units of force.) With zero initial values, the transformed equation is

$$ms^2Y(s) + kY(s) = \Delta\mu \cdot 1.$$

Thus the transform of the rest solution is

$$Y(s) = \frac{\Delta\mu}{ms^2 + k} = \frac{\Delta\mu/m}{s^2 + \omega^2},$$

and consequently the solution is

$$y(t) = \frac{\Delta\mu}{m\omega} h(t) \sin \omega t.$$

For $t \geq 0$, this is the same as the solution of $my'' + ky = 0$ with initial position $y(0) = 0$ and initial velocity $y'(0) = \Delta\mu/m$. Also note that, if viewed on $(-\infty, \infty)$, the solution is continuous, but its derivative has a jump discontinuity at $t = 0$. ■

Example 2

Suppose that an undamped spring-mass system with mass m, stiffness k, and natural angular frequency $\omega = \sqrt{k/m}$ is initially in motion with position $y(0) = 0$ and velocity $y'(0) = v_0$. At a later time $t = t_1$, the mass is struck with a hammer, imparting an instantaneous transfer of momentum $\Delta\mu$. The motion of the mass may be modeled by

$$my'' + ky = \Delta\mu \cdot \delta(t - t_1), \quad y(0) = 0, \ y'(0) = v_0.$$

The transformed equation is

$$m(s^2 Y(s) - v_0) + kY(s) = \Delta\mu e^{-t_1 s},$$

so the transform of the solution may be written as

$$Y(s) = \frac{\Delta\mu e^{-t_1 s} + mv_0}{ms^2 + k} = \left(\frac{\Delta\mu e^{-t_1 s}}{m\omega} + \frac{v_0}{\omega}\right)\frac{\omega}{s^2 + \omega^2}.$$

Since the second factor is the transform of $\sin \omega t$, we apply the second shift theorem to obtain

$$y(t) = \frac{\Delta\mu}{m\omega} h(t - t_1) \sin(\omega(t - t_1)) + \frac{v_0}{\omega} h(t) \sin \omega t.$$

Note that if $\Delta\mu = mv_0$ and $t_1 = (2k + 1)\frac{\pi}{\omega}$ for some integer $k \geq 0$, then for $t \geq t_1$ we have

$$\begin{aligned} y(t) &= \frac{v_0}{\omega} \sin(\omega t - (2k + 1)\pi) + \frac{v_0}{\omega} \sin \omega t \\ &= \frac{v_0}{\omega}(\sin(\omega t - \pi) + \sin \omega t) \\ &= 0, \end{aligned}$$

and so the impulse instantaneously stops the motion. ■

Circuits Let $Q(t)$ be the charge on the capacitor in a simple *RC*, *LC*, or *RLC* circuit. (See Sections 2.2.3 and 5.1.) If current flows for some period of time $[t_0, t_1]$, then there is a resulting change in the capacitor's charge given by

$$\Delta Q = Q(t_1) - Q(t_0) = \int_{t_0}^{t_1} Q'(t)\, dt.$$

Since $Q'(t)$ is precisely the current $I(t)$, this becomes

$$\Delta Q = \int_{t_0}^{t_1} I(t)\, dt.$$

So, just as an instantaneous unit change in the momentum of a mass can be described with the Dirac distribution, so can an instantaneous unit change in the charge on a capacitor. A (nearly) instantaneous change in charge might occur due to a quick discharge across a spark gap or a quick charge from a brief spike in source voltage.

Example 3

Consider an RC circuit with $R = C = 1$ and an initial charge of $Q_0 = 0$ on the capacitor. A voltage source produces large spikes at $t = 0$ and $t = 1$ and is zero elsewhere. With y representing the charge, the model is

$$y' + y = a_0\delta(t) + a_1\delta(t-1), \quad y(0) = 0,$$

where a_0 and a_1 are the changes in charge that occur due the voltage spikes. The transform of the solution is

$$Y(s) = (a_0 + a_1 e^{-s})\frac{1}{s+1}.$$

Thus the solution is

$$y = a_0 e^{-t} h(t) + a_1 e^{-(t-1)} h(t-1) = (a_0 h(t) + a_1 e h(t-1))e^{-t}. \quad ■$$

PROBLEMS

Evaluate each integral in Problems 1 through 6.

1. $\int_{-\infty}^{\infty} (t^2+3)\delta(t-2)\, dt$

2. $\int_{-1}^{1} (t^2+3)\delta(t)\, dt$

3. $\int_{-\infty}^{\infty} \delta(t-\pi)\cos t\, dt$

4. $\int_{-\pi}^{\pi} \delta(t-\frac{\pi}{6})\sin t\, dt$

5. $\int_{-1}^{1} \frac{\delta(t)}{1+t^2}\, dt$

6. $\int_{-\infty}^{\infty} \delta(t+1)e^{-t}\, dt$

In Problems 7 through 12, find and graph the rest solution.

7. $y'' = \delta(t) - \delta(t-1)$

8. $y'' = \delta(t) - 2\delta(t-1) + \delta(t-2)$

9. $y'' + y = \delta(t-1)$

10. $y'' + y = \delta(t) + \delta(t-\pi)$

11. $y'' + 3y' + 2y = \delta(t) - \delta(t-\ln 3)$

12. $y'' + 2y' + y = \delta(t-1) - \delta(t-2)$

In Problems 13–15, find and graph the solution satisfying $y(0) = 1$.

13. $y' = \delta(t-1)$

14. $y' + y = \delta(t - \ln 2) - \delta(t - \ln 3)$

15. $y' + y = -\delta(t - \ln 2)$

16. Consider the rest solution of $y' - y = \delta(t) - a\delta(t-1)$. Find a so that the solution is zero for all $t \geq 1$.

17. Show that if p and q are constants, then the rest solution of $y'' + py' + qy = \delta(t)$ and the solution of $y'' + py' + qy = 0$, $y(0) = 0$, $y'(0) = 1$, have the same Laplace transform; hence the solutions agree for all $t \geq 0$.

18. **(a)** Using Laplace transforms, find the solution of $y' = \delta(t)$, $y(0^-) = 0$. What interpretation of $\delta(t)$ does this imply?

(b) Let $a > 0$ and find the solution of $y' = \frac{1}{a}(h(t) - h(t-a))$, $y(0^-) = 0$ in terms of a. Sketch the graph of the forcing function and the solution for $a = 1$, 0.5, and 0.1. What happens as $a \to 0^+$?

19. **(a)** Let $a > 0$, and find the rest solution of $y'' + y = \frac{1}{a}(h(t) - h(t-a))$ in terms of a.

(b) Show that if $t \geq a$, then $y(t) = \frac{1}{a}(\cos(t-a) - \cos t)$.

(c) Argue that if $t > 0$, then $y(t) \to \sin t$ as $a \to 0^+$.

(d) Compare with the rest solution of $y'' + y = \delta(t)$.

20. Find an impulse-driven system whose rest solution is as in the following figure.

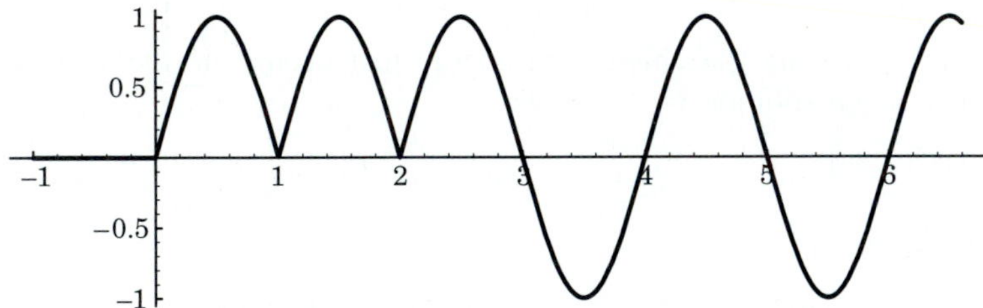

7.6 Convolution

In this section we will study integral equations of the form

$$y(t) + \int_0^t g(t-x)y(x)\,dx = f(t), \tag{1}$$

where f and g are given functions of t. An equation of this form is called a *linear Volterra integral equation of convolution type*. General linear Volterra integral equations are of the form

$$\alpha(t)y(t) + \int_a^t g(t,x)y(x)\,dx = f(t).$$

When $g(t,x)$ can be expressed as a function of $t - x$ as in (1), the integral involved is called a *convolution integral*, which defines the **convolution** of the two functions

g and y, sometimes denoted by $g * y$. That is,

$$(g * y)(t) = \int_0^t g(t-x)y(x)\,dx,$$

and so (1) can be written in the abbreviated form

$$y + g * y = f.$$

Integral equations (and integro-differential equations) arise in applications where behavior is influenced by past history as well as the state at time t. Convolution integrals, in particular, arise when the contribution at time t of the history at time $x < t$ depends upon the elapsed time $t - x$.

Example 1

The convolution of t and t^2 is

$$t * t^2 = \int_0^t (t-x)x^2\,dx = \left(\frac{t}{3}x^3 - \frac{1}{4}x^4\right)\bigg]_{x=0}^{t} = \frac{1}{12}t^4.$$

The convolution of e^t and e^{-t} is

$$e^t * e^{-t} = \int_0^t e^{t-x}e^{-x}\,dx = \int_0^t e^t e^{-2x}\,dx = -\frac{1}{2}e^t e^{-2x}\bigg|_{x=0}^{t}$$
$$= \frac{1}{2}e^t(1 - e^{-2t}) = \sinh t. \quad \blacksquare$$

It is clear from the linearity properties of integration that for a given g the operator $\mathcal{K}$ defined by $\mathcal{K}y = g * y$ is linear; that is,

$$g * (cy) = cg * y \quad \text{and} \quad g * (u + v) = g * u + g * v.$$

It is also true that convolution is commutative; that is, $g * y = y * g$. To see this, simply make a change of variable $\tau = t - x$ in the integral as follows:

$$g * y = \int_0^t g(t-x)y(x)\,dx = \int_t^0 g(\tau)y(t-\tau)(-d\tau)$$
$$= \int_0^t y(t-\tau)g(\tau)\,d\tau = y * g.$$

With an eye toward solving (1), we would like to know how to express the Laplace transform of $g * y$ in terms of the individual transforms of g and y. The answer is provided by the **convolution theorem**:

$$\mathcal{L}[g * y](s) = G(s)Y(s), \tag{2}$$

where the transform exists for all s such that both transforms G and Y exist.

To establish this result, first note that by the definition of the transform,

$$\begin{aligned}\mathcal{L}[g * y](s) &= \int_0^\infty e^{-st}\left(\int_0^t g(t-x)y(x)\,dx\right)dt \\ &= \int_0^\infty \int_0^t e^{-st} g(t-x)y(x)\,dx\,dt.\end{aligned}$$

The region over which we are integrating is $\{(t, x) | 0 \le x \le t < \infty\}$, so interchanging the order of integration produces

$$\begin{aligned}\mathcal{L}[g * y](s) &= \int_0^\infty \int_x^\infty e^{-st} g(t-x)y(x)\,dt\,dx \\ &= \int_0^\infty y(x) \int_x^\infty e^{-st} g(t-x)\,dt\,dx.\end{aligned}$$

This step can be justified by the absolute convergence of the improper integrals involved, which follows from the existence of the Laplace transforms of g and y. Now the change of variable $\tau = t - x$ in the inner integral gives us

$$\begin{aligned}\mathcal{L}[g * y](s) &= \int_0^\infty y(x) \int_0^\infty e^{-s(\tau+x)} g(\tau)d\tau\,dx \\ &= \int_0^\infty e^{-sx} y(x) \int_0^\infty e^{-s\tau} g(\tau)d\tau\,dx \\ &= \int_0^\infty e^{-sx} y(x)G(s)\,dx = G(s)\int_0^\infty e^{-sx} y(x)\,dx = G(s)Y(s).\end{aligned}$$

Example 2

This example indicates how the convolution theorem can be used to help calculate certain convolution integrals. Consider the convolution

$$\cos t * \sin t = \int_0^t \cos(t-x)\sin x\,dx.$$

According to the convolution theorem (2),

$$\mathcal{L}[\cos t * \sin t](s) = \left(\frac{s}{s^2+1}\right)\left(\frac{1}{s^2+1}\right) = \frac{s}{(s^2+1)^2}.$$

By (10) in Section 7.2, we recognize therefore that

$$\int_0^t \cos(t-x)\sin x\,dx = \frac{t}{2}\sin t. \qquad \blacksquare$$

Example 3

This example indicates how the convolution theorem can be used as an alternative to partial fraction decomposition to help find inverse transforms. Consider the transform

$$F(s) = \frac{1}{(s+1)(s+2)}.$$

According to the convolution theorem, this is the transform of

$$f(t) = e^{-t} * e^{-2t} = \int_0^t e^{-(t-x)} e^{-2x}\, dx.$$

A short calculation reveals that $f(t) = e^{-t} - e^{-2t}$. We hope that the reader is not overly impressed with this application of the convolution theorem. ■

Integral Equations We will now look at a few examples to indicate how the convolution theorem can be used to solve linear Volterra integral equations of convolution type.

Example 4

Consider the integral equation

$$y(t) + \int_0^t y(x)\, dx = 1.$$

Viewing the integral as $1 * y$, we see that the transform of the solution satisfies

$$Y(s) + \frac{1}{s} Y(s) = \frac{1}{s}.$$

Solving for $Y(s)$ produces $Y(s) = 1/(s+1)$, and consequently $y(t) = e^{-t}$. Note that this integral equation can be differentiated to obtain $y' + y = 0$ and that the integral equation implies the initial value $y(0) = 1$. Thus it is no surprise that $y = e^{-t}$. ■

Example 5

Consider the integral equation

$$y(t) + \int_0^t (t-x) y(x)\, dx = 1.$$

Viewing the integral as $t * y$, we see that the transform of the solution satisfies

$$Y(s) + \frac{1}{s^2} Y(s) = \frac{1}{s}.$$

Solving for $Y(s)$ produces $Y(s) = s/(s^2+1)$, and consequently $y(t) = \cos t$. ■

Example 6

Consider the integral equation

$$y(t) + 3\int_0^t \sin(t-x)y(x)\,dx = 4.$$

Viewing the integral as $(\sin t) * y$, we see that the transform of the solution satisfies

$$Y(s) + \frac{3}{s^2+1}Y(s) = \frac{4}{s}.$$

Solving for $Y(s)$ produces

$$Y(s) = 4\frac{s^2+1}{s(s^2+4)} = \frac{3}{s} + \frac{3s}{s^2+4}.$$

Therefore, the solution is $y(t) = 1 + 3\cos 2t$. ■

The Green's Function Revisited In Section 5.5, it was shown that, given any two linearly independent solutions u and v of the homogeneous equation $y''+py'+qy=0$, the rest solution of

$$y'' + py' + qy = f$$

could be obtained by variation of constants and expressed as

$$y(t) = \int_0^t G_0(t,x)f(x)\,dx,$$

where $G_0(t,x)$ is the *Green's function*

$$G_0(t,x) = \frac{u(x)v(t)+u(t)v(x)}{u(x)v'(x)-u'(x)v(x)}$$

corresponding to the operator $P(\mathcal{D}) = \mathcal{D}^2+p\mathcal{D}+q\mathcal{I}$. Laplace transform methods and the convolution theorem provide another approach to finding the Green's function for an operator $P(\mathcal{D})$ with constant coefficients.

Suppose that $P(\mathcal{D})$ has constant coefficients and that $f(t)$ is a given function with Laplace transform $F(s)$. Recall that the Laplace transform of the rest solution of $P(\mathcal{D})y = f$ is

$$Y(s) = \frac{1}{P(s)}F(s).$$

Thus, by the convolution theorem,

$$y = \int_0^t G_0(t-x)f(x)\,dx, \quad \text{where} \quad \mathcal{L}G_0 = \frac{1}{P(s)}.$$

An interesting implication is that this type of Green's function G_0—for operators with constant coefficients—is always a function of just one variable.

Example 7

Consider the equation $y'' + y = f$. The transform of the rest solution is

$$Y(s) = \frac{1}{s^2+1}F(s).$$

Thus, $\mathcal{L}G_0 = \frac{1}{s^2+1}$. So $G_0(t) = \sin t$, and by the convergence theorem,

$$y = \int_0^t \sin(t-x)f(x)\,dx.$$ ■

Example 8

Consider the third-order operator $P(\mathcal{D}) = \mathcal{D}^3 + 2\mathcal{D}^2 - \mathcal{D} - 2\mathcal{I}$, which factors as $(\mathcal{D}+\mathcal{I})(\mathcal{D}-\mathcal{I})(\mathcal{D}+2\mathcal{I})$. The corresponding Green's function G_0 is the inverse transform of

$$\frac{1}{P(s)} = \frac{1}{(s+1)(s-1)(s+2)} = \frac{1}{6}\left(\frac{1}{s-1} - \frac{3}{s+1} + \frac{2}{s+2}\right).$$

Therefore,

$$G_0(t) = \frac{1}{6}(e^t - 3e^{-t} + 2e^{-2t}),$$

and the rest solution of $y''' + 2y'' - y' - 2y = f$ can be represented by

$$y = \frac{1}{6}\int_0^t \left(e^{t-x} - 3e^{-(t-x)} + 2e^{-2(t-x)}\right) f(x)\,dx.$$

It is interesting to note that, since the Green's function $G_0(t)$ corresponding to the operator $P(\mathcal{D})$ has the Laplace transform $1/P(s)$, it is actually the rest solution of the equation

$$P(\mathcal{D})y = \delta(t),$$

where $\delta(t)$ is the Dirac distribution. Consequently, for any f with a Laplace transform, *the rest solution of* $P(\mathcal{D})y = f(t)$ *is the convolution of* f *with the rest solution of* $P(\mathcal{D})y = \delta(t)$. Therefore, having found the rest solution of $P(\mathcal{D})y = \delta(t)$, one has, in a sense, also found the rest solution of $P(\mathcal{D})y = f(t)$ for any f. ■

PROBLEMS

In Problems 1 through 4, use the definition (and commutativity) to compute the convolution.

1. $\sqrt{t} * t$ **2.** $t * t$

3. $1 * \cos t$ **4.** $t * e^{-t}$

In Problems 5 through 8, evaluate the convolution by means of the convolution theorem.

5. $\sin t * \sin 2t$ **6.** $e^t * e^t$

7. $e^t * \cos t$

8. $t^m * t^n$, for integers $m, n > 0$

Solve the given integral equation in each of Problems 9 through 14.

9. $y + e^{-t} * y = 1$

10. $e^t * y = e^{-t}$

11. $y + t * y = t$

12. $y + t * y = t^2$

13. $y - t * y = h(t) - h(t-1)$

14. $y - t * y = |1 - t|$

Solve the given integro-differential equation in Problems 15 through 17.

15. $y' + e^{-2t} * y = h(t), \ \ y(0) = 0$

16. $y' + y - 2t * y = 0, y(0) = 5$

17. $y' - (\cos t) * y = h(t), \ \ y(0) = 0$

In Problems 18 and 19, use the results of Problems 15 and 16 in Section 7.4 to help in solving the given integral equation.

18. $(\cos t) * y = |\sin t|$

19. $2(\cos t) * y = \sin t + |\sin t|$

Suppose that v and v have Laplace transforms and that $u(0) = u_0$, $u'(0) = u_1$ $v(0) = v_0$, $v'(0) = v_1$. Use the convolution theorem and the first differentiation theorem to formally verify the identities in Problems 20 through 22.

20. $u * v' - v' * u = u_0 v - v_0 u$

21. $(u * v)' = u * v' + v_0 u = u' * v + u_0 v$

22. $(u*v)'' = u*v'' + v_1 u + v_0 u' = u'' * v + u_1 v + u_0 v'$

23. Conclude from Problem 21 that $y = t * e^{at}$ satisfies $y' = ay + t$. Also, the definition of convolution implies that $y(0) = 0$. Use this to compute $t * e^{at}$.

24. Conclude from Problem 22 that $y = g * \sin at$ satisfies $y'' + a^2 y = ag$. Also conclude from Problem 21 and the definition of convolution that $y'(0) = y(0) = 0$. Use this to compute $\cos bt * \sin at$.

25. Let y be the rest solution of $y'' + py' + qy = f$, where p and q are constants with $q \neq 0$, and let u be the solution of $u'' + pu' = 0$, $u(0) = 0$, $u'(0) = 1$.

(a) Using the results of Problems 20 through 22, show that y satisfies the linear Volterra integral equation of convolution type

$$y + qu * y = u * f.$$

(b) Express the problem $y'' + y' + y = e^t$, $y(0) = y'(0) = 0$, as a linear Volterra integral equation of convolution type.

26. Suppose that f and g have Laplace transforms and that P is a polynomial with constant coefficients. Show that u is the rest solution of $P(\mathcal{D})y = f$, if and only if $g * u$ is the rest solution of $P(\mathcal{D})y = g * f$.

In Problems 27 through 30, find the Green's function $G_0(t)$ of the given operator, and write down the rest solution of $P(\mathcal{D})y = f$.

27. $P(\mathcal{D})y = y'' - y$

28. $P(\mathcal{D})y = y'' + 2y' + 2y$

29. $P(\mathcal{D})y = y'' + 3y' + 2y$

30. $P(\mathcal{D})y = y' + y$

In Problems 31 through 33, find the Green's function $G_0(t)$ of the given operator, and write down the solution of $\mathcal{T}y = f, \ \ y(0) = 0$.

31. $\mathcal{T}y = y' + 2e^{-3t} * y$

32. $\mathcal{T}y = y' + e^{-2t} * y$

33. $\mathcal{T}y = y' + 3\cos t * y$

CHAPTER

8

First-Order Linear Systems

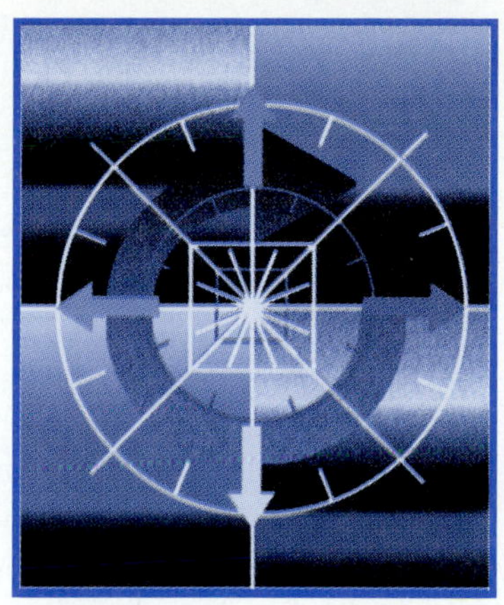

As we have seen on numerous occasions, a differential equation of order two or higher can be converted to a first-order system for the purpose of applying a numerical approximation procedure. First-order systems that arise in this manner have a particular form. First-order systems also arise naturally in various applications that involve two or more quantities. In this chapter, we begin to look at first-order systems in general, concentrating on *linear* systems.

The reader who is not readily familiar with the basic manipulations of linear algebra—such as matrix-vector and matrix-matrix multiplication, transposes, determinants, etc.—is encouraged to study Appendix I before going on here.

8.1 Introduction

Recall that in Chapter 2 we studied linear, first-order (scalar) differential equations, which were of the form

$$y' - ay = f,$$

where a and f are functions of the independent variable.

In this chapter we will study a more general version the same problem. Instead of scalar-valued functions y and f, the problem will involve n-component vector-valued functions $\mathbf{y}$ and $\mathbf{f}$:

$$\mathbf{y} = \begin{pmatrix} y_1 \\ \vdots \\ y_n \end{pmatrix}, \quad \mathbf{f} = \begin{pmatrix} f_1 \\ \vdots \\ f_n \end{pmatrix},$$

and the scalar a will instead be an $n \times n$ matrix A. The system of equations

$$\begin{aligned} y_1' - y_1 + 2ty_2 &= \sin t \\ y_2' - ty_1 + y_2 &= 0, \end{aligned}$$

for example, will be viewed as the single equation

$$\begin{pmatrix} y_1 \\ y_2 \end{pmatrix}' - \begin{pmatrix} 1 & -2t \\ t & -1 \end{pmatrix}\begin{pmatrix} y_1 \\ y_2 \end{pmatrix} = \begin{pmatrix} \sin t \\ 0 \end{pmatrix}$$

for the (column) vector quantity $y = (y_1, y_2)^T$.*

The general form of the systems we shall study is

$$\mathbf{y}' - A\mathbf{y} = \mathbf{f}, \tag{1}$$

where $\mathbf{y}$ and $\mathbf{f}$ are vector-valued functions with n components, A is an $n \times n$ matrix-valued function, and the entries of $\mathbf{f}$ and A are continuous on an interval I. When $\mathbf{f} = \mathbf{0}$, the system is said to be *homogeneous,* and when $\mathbf{f} \neq \mathbf{0}$, the homogeneous equation associated with (1) is

$$\mathbf{y}' - A\mathbf{y} = \mathbf{0}. \tag{2}$$

When A and $\mathbf{f}$ have constant entries, the system is said to be *autonomous.*

Initial-value problems associated with (1) take the form

$$\mathbf{y}' - A\mathbf{y} = \mathbf{f}, \quad \mathbf{y}(t_0) = \mathbf{y}_0, \tag{3}$$

where t_0 in I and $\mathbf{y}_0$ in $\mathbb{R}^n$. Some of our subsequent arguments will rely upon the following existence and uniqueness theorem. Its proof is in Appendix III.

Theorem 1

If the entries of the matrix A and the vector $\mathbf{f}$ are continuous on an interval I, then for any t_0 in I and $\mathbf{y}_0$ in $\mathbb{R}^n$, the initial-value problem (3) has a unique solution on I. ▲

Superposition

The usual consequences of linearity are in force for solutions of (1):

i. If $\mathbf{y}_1$ and $\mathbf{y}_2$ are two solutions of (2), then so is $c_1\mathbf{y}_1 + c_2\mathbf{y}_2$ for any scalar constants c_1 and c_2;
ii. If $\tilde{\mathbf{y}}$ and $\mathbf{y}$ are solutions of (1) and (2), respectively, then $\tilde{\mathbf{y}} + c\mathbf{y}$ is a solution of (1) for any scalar constant c.

These important facts follow easily from the linearity of matrix multiplication.

Example 1

Easy calculations reveal that the vector-valued functions

$$\mathbf{y}_1 = \begin{pmatrix} 3e^t \\ -e^t \end{pmatrix} \quad \text{and} \quad \mathbf{y}_2 = \begin{pmatrix} e^{-t} \\ -e^{-t} \end{pmatrix}$$

* The superscript T signifies the *transpose* of a vector or matrix. (See Appendix I.) The transpose of a row vector is simply the corresponding column vector, and vice versa.

each satisfy the (autonomous) homogeneous equation

$$\mathbf{y}' - \begin{pmatrix} 2 & 3 \\ -1 & -2 \end{pmatrix} \mathbf{y} = \mathbf{0}.$$

Consequently, so does $c_1\mathbf{y}_1 + c_2\mathbf{y}_2$ for any constants c_1, c_2. Also, $\tilde{\mathbf{y}} = (t+1, t-1)^T$ is a particular solution of the nonhomogeneous system

$$\mathbf{y}' - \begin{pmatrix} 2 & 3 \\ -1 & -2 \end{pmatrix} \mathbf{y} = \begin{pmatrix} 2-5t \\ 3t \end{pmatrix};$$

therefore, $c_1\mathbf{y}_1 + c_2\mathbf{y}_2 + \tilde{\mathbf{y}}$ is a solution for any constants c_1, c_2. ■

Example 2

Easy calculations reveal that the vector-valued functions

$$\mathbf{y}_1 = \begin{pmatrix} t \\ -t \end{pmatrix} \quad \text{and} \quad \mathbf{y}_2 = \begin{pmatrix} t^3 \\ t^3 \end{pmatrix}$$

each satisfy the (nonautonomous) homogeneous system

$$\mathbf{y}' - \frac{1}{t}\begin{pmatrix} 2 & 1 \\ 1 & 2 \end{pmatrix} \mathbf{y} = \mathbf{0}.$$

Consequently, so does $c_1\mathbf{y}_1 + c_2\mathbf{y}_2$ for any constants c_1, c_2. Also, $\tilde{\mathbf{y}} = (-t, -t^2)^T$ is a particular solution of the nonhomogeneous system

$$\mathbf{y}' - \frac{1}{t}\begin{pmatrix} 2 & 1 \\ 1 & 2 \end{pmatrix} \mathbf{y} = \begin{pmatrix} t+1 \\ 1 \end{pmatrix};$$

therefore, $c_1\mathbf{y}_1 + c_2\mathbf{y}_2 + \tilde{\mathbf{y}}$ is a solution for any constants c_1, c_2. ■

To motivate the investigation of linear systems of differential equations, we will present here a couple of standard elementary applications—a two-container mixing problem and a two-loop RLC circuit.

Example 3

The following mixing problem involving two well-stirred containers is depicted in Figure 1. (The reader may wish to review Section 2.2.2 at this point.) Suppose that the volume of mixture in each of the two containers is a constant V, and the *amounts* of impurity in the two containers are a_1 and a_2, respectively. Thus the respective impurity *concentrations* are a_1/V and a_2/V. Suppose further that a mixture with an impurity concentration of c_0 is being pumped into the first container at a rate of r_0 units of volume per unit time, and that the well-stirred contents of the second container are being pumped back into the first container at a rate of r_1 units of volume per unit time, while the well-stirred contents of the first container flows into the second container at a rate of $r_0 + r_1$ units of volume per unit time. The

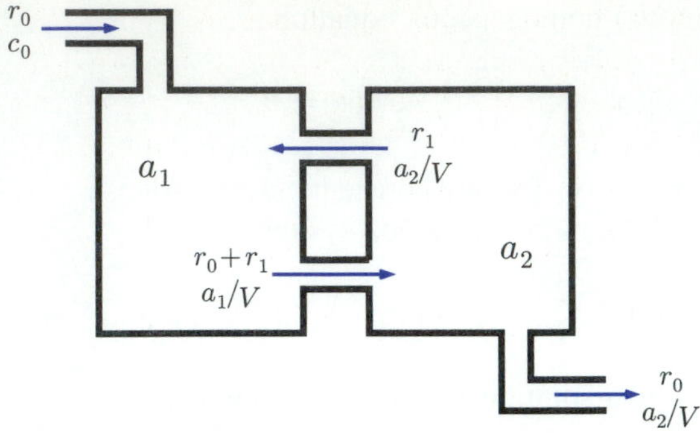

Figure 1

contents of the second container are pumped out at a rate of r_0 units of volume per unit time. The differential equations for the amounts a_1 and a_2 each take the form

$$\frac{da_i}{dt} = \text{rate in} - \text{rate out} .$$

Thus we have the pair of equations

$$\begin{aligned}\frac{da_1}{dt} &= r_0 c_0 + \frac{r_1}{V}a_2 - \frac{r_0+r_1}{V}a_1 \\ \frac{da_2}{dt} &= \frac{r_0+r_1}{V}a_1 - \frac{r_1}{V}a_2 - \frac{r_0}{V}a_2,\end{aligned}$$

or in vector form as in (1), the single equation

$$\frac{d\mathbf{a}}{dt} - \frac{1}{V}\begin{pmatrix} -(r_0+r_1) & r_1 \\ r_0+r_1 & -(r_0+r_1) \end{pmatrix}\mathbf{a} = \begin{pmatrix} r_0 c_0 \\ 0 \end{pmatrix},$$

where $\mathbf{a} = (a_1, a_2)^T$. ■

Example 4

Consider the simple two-loop *RLC* circuit shown in Figure 2, in which a capacitor discharges through a resistor and an inductor connected in parallel. (The reader may wish to review Section 2.2.3 at this point.)

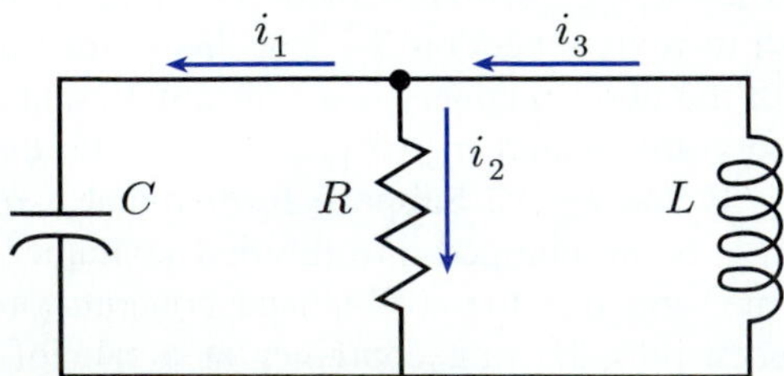

Figure 2

Let the currents i_1, i_2, and i_3 be defined as shown in Figure 2. The system of equations will be obtained by applying Kirchhoff's current law—which says that the sum of the currents entering a node at any instant is zero—and Kirchhoff's voltage law (see Section 2.2.3) to each of the closed loops in the circuit. The current law applied to the top node in this circuit results in

$$-i_1 - i_2 + i_3 = 0, \quad \text{or} \quad i_3 = i_1 + i_2.$$

The voltage law applied to the RL loop on the right results in

$$Li_3' + Ri_2 = 0.$$

The voltage law applied to the RC loop on the left results in $Ri_2 - Q/C = 0$, where Q is the charge on the capacitor. Since $Q' = i_1$, differentiation of that equation produces

$$Ri_2' - \frac{1}{C}i_1 = 0.$$

Combining these three relationships among the currents yields a pair of differential equations for i_1 and i_2:

$$\begin{aligned} i_1' + \frac{1}{RC}i_1 + \frac{R}{L}i_2 &= 0 \\ i_2' - \frac{1}{RC}i_1 &= 0. \end{aligned}$$

The current law then relates i_3 trivially to i_1 and i_2. In vector form as in (1), this pair of differential equations becomes the single equation

$$\frac{d\mathbf{i}}{dt} - \frac{1}{RLC}\begin{pmatrix} -L & -R^2C \\ L & 0 \end{pmatrix}\mathbf{i} = \mathbf{0},$$

where $\mathbf{i} = (i_1, i_2)^T$ and the zero on the right side represents the zero vector $(0, 0)^T$. ■

PROBLEMS

For each of the linear systems in Problems 1 through 3, find the matrix A such that the system can be written as $\mathbf{y}' - A\mathbf{y} = \mathbf{f}$. Identify each system as homogeneous or nonhomogeneous.

1. $y_1' = y_1 - y_2 + \sin t, \quad y_2' - y_2 = ty_1$

2. $y_1' = e^{-t}y_1 - y_2, \quad y_2' = y_1 + 2y_2$

3. $y_1' = y_1 - y_2 + y_3, \quad y_2' + y_2 = -y_3 + y_1, \quad y_3' + y_3 = y_1 + y_2$

In Problems 4 through 6, write down the differential equation for each component of the solution of $\mathbf{y}' - A\mathbf{y} = \mathbf{0}$.

4. $A = \begin{pmatrix} 1 & -3 \\ 2 & -2 \end{pmatrix}$ **5.** $A = \begin{pmatrix} t & -1 \\ 1 & -t^2 \end{pmatrix}$ **6.** $A = \begin{pmatrix} -1 & -3 & 2 \\ 3 & -2 & -2 \\ 1 & 1 & -1 \end{pmatrix}$

7. Find a second-order differential equation for $x(t)$ (that does not involve $y(t)$), given that $x(t)$ and $y(t)$ together satisfy

$$x' - x + y = 0,$$
$$y' + x - 2y = 0.$$

8. Find a 2×2 matrix A such that the functions $y_1 = e^{-t} + e^{-2t}$ and $y_2 = e^{-t} - e^{-2t}$ satisfy $\mathbf{y}' - A\mathbf{y} = \mathbf{0}$. Is the answer unique?

9. Verify each of the solutions $\mathbf{y}_1$, $\mathbf{y}_2$, and $\tilde{\mathbf{y}}$ in Example 1.

10. Verify each of the solutions $\mathbf{y}_1$, $\mathbf{y}_2$, and $\tilde{\mathbf{y}}$ in Example 2.

11. Verify that $y_1 = \cos t$ and $y_2 = \sin t - \cos t$ together satisfy the system

$$\mathbf{y}' - \begin{pmatrix} -1 & -1 \\ 2 & 1 \end{pmatrix} \mathbf{y} = 0.$$

Determine a system of three first-order differential equations that governs each of the systems represented schematically in the following figures. Each is an extension of either Example 3 or Example 4. (Each of the two circuits requires conversion of a second-order equation to a pair of first-order equations.)

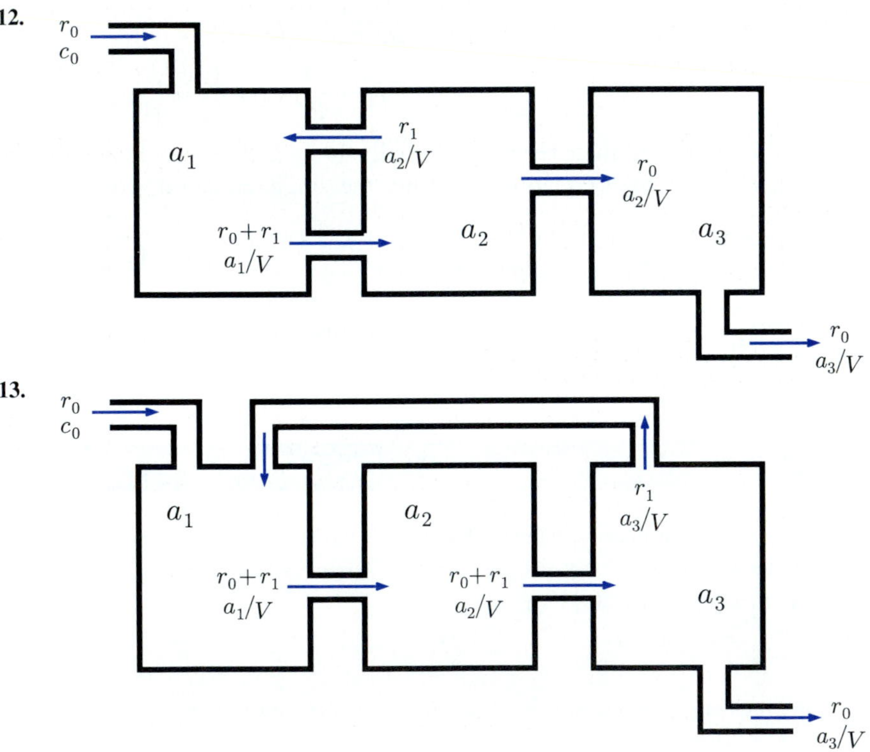

14.

15.

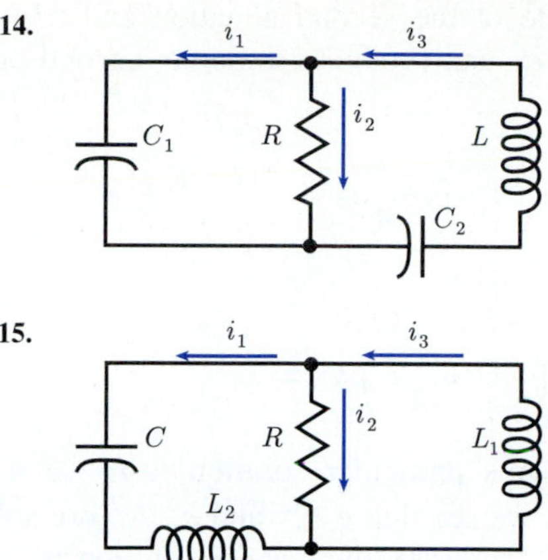

8.2 Two Ad Hoc Methods

In this section we will look at two methods for solving linear systems of differential equations with constant coefficients. These methods are ad hoc in the sense that they are based upon techniques we are already familiar with in the context of a single equation and provide little insight into the general structure of solutions.

The Method of Elimination This method amounts essentially to replacing a linear first-order system with a single higher-order differential equation. For a system of two equations, the result is a single second-order equation, which can be solved by the method of Chapter 6.* We will simply illustrate the method with an example.

Example 1

Consider the system of equations for the two-tank mixing problem in Example 3 of Section 8.1. With parameter values $V = 4\ \text{m}^3$, $r_0 = 3\ \text{m}^3/\text{h}$, $r_1 = 1\ \text{m}^3/\text{h}$, and $c_0 = 1\ \text{kg/m}^3$, and zero initial concentrations in both tanks, we obtain the initial-value problem

$$\mathbf{a}' - \begin{pmatrix} -1 & 1/4 \\ 1 & -1 \end{pmatrix} \mathbf{a} = \begin{pmatrix} 3 \\ 0 \end{pmatrix},$$

$$\mathbf{a}(0) = \begin{pmatrix} 0 \\ 0 \end{pmatrix}.$$

Rewriting the differential equation as a pair of equations in operator form, we have

$$(\mathcal{D}+\mathcal{I})a_1 - \frac{1}{4}a_2 = 3$$
$$-a_1 + (\mathcal{D}+\mathcal{I})a_2 = 0.$$

* Higher-order equations are discussed in Appendix IV.

Now we apply $\mathcal{D}+\mathcal{I}$ to each side of the second equation and add the result to the first equation. This eliminates a_1 and produces a single, second-order equation for a_2:

$$(\mathcal{D}+\mathcal{I})^2 a_2 - \frac{1}{4}a_2 = 3.$$

Simplification results in

$$\left(\mathcal{D}^2 + 2\mathcal{D} + \frac{3}{4}\mathcal{I}\right) a_2 = 3.$$

At this point it is easy to see that a particular solution is $\tilde{a}_2 = 4$. Also, since $D^2 + 2\mathcal{D} + \frac{3}{4}\mathcal{I} = (\mathcal{D}+\frac{1}{2})(\mathcal{D}+\frac{3}{2})$, we see that $e^{-t/2}$ and $e^{-3t/2}$ are solutions of the associated homogeneous equation. Therefore, the general solution is

$$a_2 = c_1 e^{-t/2} + c_2 e^{-3t/2} + 4.$$

Returning now to the second of the two original differential equations, we simply *compute* a_1:

$$\begin{aligned} a_1 &= (\mathcal{D}+\mathcal{I})a_2 \\ &= \frac{c_1}{2}e^{-t/2} - \frac{c_2}{2}e^{-3t/2} + 4. \end{aligned}$$

The zero initial values give us the following system of equations for c_1 and c_2:

$$\begin{aligned} c_1 + c_2 &= -4, \\ c_1 - c_2 &= -8. \end{aligned}$$

Solving these equations gives us $c_1 = -6$ and $c_2 = 2$; thus the solution of the initial-value problem is

$$\mathbf{a} = \begin{pmatrix} -3e^{-t/2} - e^{-3t/2} + 4 \\ -6e^{-t/2} + 2e^{-3t/2} + 4 \end{pmatrix}.$$

The graphs of a_1 and a_2 are seen in Figure 1. Decide for yourself which is which. ■

Laplace Transforms The Laplace transform method extends naturally to systems of linear differential equations. As you might expect, the algebraic step that produces the transform of the solution will involve solving a system of equations in which the coefficients depend on the variable s. In fact, given an initial-value problem of the form

$$\begin{aligned} \mathbf{y}' - A\mathbf{y} &= \mathbf{f}, \\ \mathbf{y}(0) &= \mathbf{y}_0, \end{aligned}$$

the vector-valued Laplace transform $\mathcal{Y}(s)$ of the solution satisfies

$$s\mathcal{Y}(s) - \mathbf{y}_0 - A\mathcal{Y}(s) = \mathbf{F}(s),$$

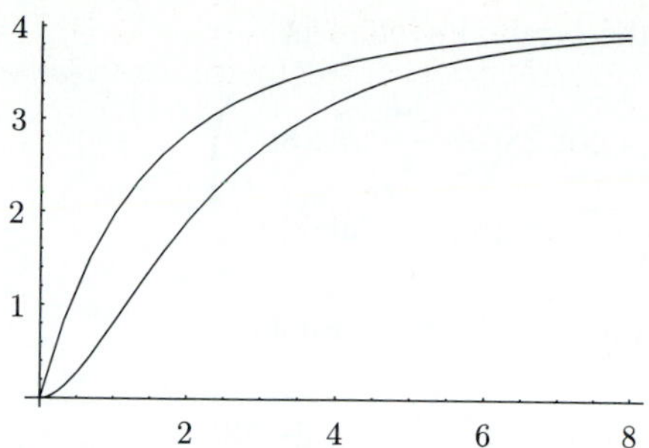

Figure 1

which becomes the linear system

$$(sI - A)\mathcal{Y}(s) = \mathbf{F}(s) + \mathbf{y}_0.$$

Solving this equation produces the transform $\mathcal{Y}(s)$. The inverse transform of each component of $\mathcal{Y}(s)$ is then found by the usual means.

Example 2

Consider the system of equations for the two-loop circuit in Example 4 of Section 8.1. With parameter values $R = 1$ ohm, $L = .01$ henry, and $C = .125$ farad, and initial currents $i_1(0) = 1$ and $i_2(0) = -1$ ampere, we obtain the initial-value problem

$$\mathbf{i}' - \begin{pmatrix} -8 & -100 \\ 8 & 0 \end{pmatrix} \mathbf{i} = \mathbf{0}, \quad \mathbf{i}(0) = \begin{pmatrix} 1 \\ -1 \end{pmatrix}.$$

Using $\mathbf{\Psi}(s) = (\Psi_1(s), \Psi_2(s))^T$ to denote the Laplace transform of $\mathbf{i}(t)$, we have

$$s\mathbf{\Psi}(s) - \begin{pmatrix} 1 \\ -1 \end{pmatrix} - \begin{pmatrix} -8 & -100 \\ 8 & 0 \end{pmatrix} \mathbf{\Psi}(s) = 0,$$

which becomes

$$\begin{pmatrix} s+8 & 100 \\ -8 & s \end{pmatrix} \mathbf{\Psi}(s) = \begin{pmatrix} 1 \\ -1 \end{pmatrix}.$$

The solution of this system is

$$\mathbf{\Psi}(s) = \frac{1}{s^2 + 8s + 800} \begin{pmatrix} s + 100 \\ -s \end{pmatrix}.$$

After a bit of processing, we arrive at

$$\mathbf{\Psi}(s) = \begin{pmatrix} \dfrac{s+4}{(s+4)^2 + 28^2} + \dfrac{24}{7}\,\dfrac{28}{(s+4)^2 + 28^2} \\ -\dfrac{s+4}{(s+4)^2 + 28^2} + \dfrac{1}{7}\,\dfrac{28}{(s+4)^2 + 28^2} \end{pmatrix},$$

which tells us that the solution of the original system is

$$\mathbf{i}(t) = e^{-4t} \begin{pmatrix} \cos 28t + \frac{24}{7} \sin 28t \\ -\cos 28t + \frac{1}{7} \sin 28t \end{pmatrix}.$$

From this we find that the current through the inductor is

$$i_3(t) = i_1(t) + i_2(t) = \frac{25}{7} e^{-4t} \sin 28t.$$

Figure 2 shows the graph of this function. ■

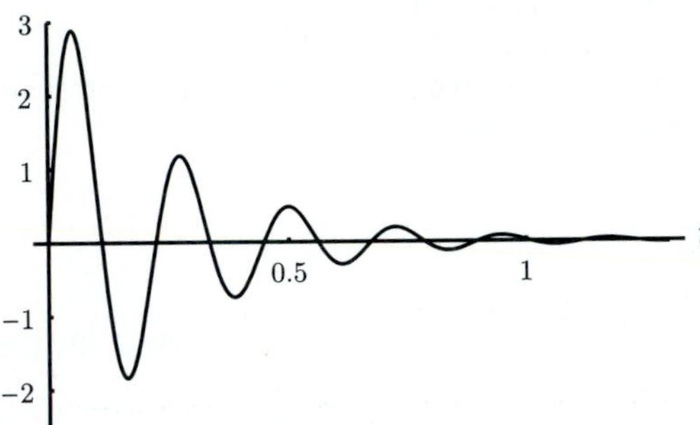

Figure 2

PROBLEMS

1. Rework Example 1 with the parameter values $V = 10\ \text{m}^3$, $r_0 = 5\ \text{m}^3/\text{h}$, $r_1 = 4\ \text{m}^3/\text{h}$, and $c_0 = 1.2\ \text{kg/m}^3$ and zero initial concentrations in both tanks.
2. Rework Example 2 with the parameter values $R = 1$ ohm, $L = 1$ henry, and $C = 2/9$ farad and initial currents $i_1(0) = 1$ and $i_2(0) = -1$ ampere.
3. (ab) Use Laplace transforms to rework Example 1 and Problem 1.
4. (ab) Use the method of elimination to rework Example 2 and Problem 2.

In each of Problems 5 through 12, solve the given initial-value problem

(a) by the method of elimination;

(b) using the Laplace transform.

(Solving the 3 × 3 systems in Problems 9 through 12 by elimination will require the solution of a cubic characteristic equation.)

5. $\mathbf{y}' - \begin{pmatrix} -2 & 1 \\ 1 & -2 \end{pmatrix} \mathbf{y} = \mathbf{0}, \quad \mathbf{y}(0) = \begin{pmatrix} 1 \\ 1 \end{pmatrix}$

6. $\mathbf{y}' - \begin{pmatrix} -2 & -1 \\ 1 & -2 \end{pmatrix} \mathbf{y} = \mathbf{0}, \quad \mathbf{y}(0) = \begin{pmatrix} 1 \\ 0 \end{pmatrix}$

7. $\mathbf{y}' - \begin{pmatrix} -1 & 1 \\ -1 & -1 \end{pmatrix} \mathbf{y} = \mathbf{0}, \quad \mathbf{y}(0) = \begin{pmatrix} 0 \\ 1 \end{pmatrix}$

8. $\mathbf{y}' - \begin{pmatrix} 0 & 4 \\ -1 & 0 \end{pmatrix} \mathbf{y} = \mathbf{0}, \quad \mathbf{y}(0) = \begin{pmatrix} 1 \\ 1 \end{pmatrix}$

9. $\mathbf{y}' - \begin{pmatrix} -3 & 1 & 0 \\ 0 & -3 & 1 \\ -4 & 0 & 0 \end{pmatrix} \mathbf{y} = \mathbf{0}, \quad \mathbf{y}(0) = \begin{pmatrix} 0 \\ 0 \\ 9 \end{pmatrix}$

10. $\mathbf{y}' - \begin{pmatrix} -1 & 5 & 0 \\ -5 & 1 & -5 \\ 0 & 5 & -1 \end{pmatrix} \mathbf{y} = \mathbf{0}, \quad \mathbf{y}(0) = \begin{pmatrix} 1 \\ 0 \\ 1 \end{pmatrix}$

11. $\mathbf{y}' - \begin{pmatrix} -1 & 1 & 0 \\ -1 & 1 & -1 \\ 0 & -1 & -1 \end{pmatrix} \mathbf{y} = \mathbf{0}, \quad \mathbf{y}(0) = \begin{pmatrix} 0 \\ 1 \\ 0 \end{pmatrix}$

12. $\mathbf{y}' - \begin{pmatrix} 0 & 0 & 1 \\ 0 & -2 & 1 \\ -1 & 0 & 0 \end{pmatrix} \mathbf{y} = \mathbf{0}, \quad \mathbf{y}(0) = \begin{pmatrix} 1 \\ 0 \\ 0 \end{pmatrix}$

8.3 Vector-Valued Functions and Linear Independence

Just as in the case of differential equations of order two or higher, the notion of linear independence is crucial to understanding the structure of solutions to linear first-order systems of differential equations. Since we view such solutions as vector-valued functions, certain basic concepts from linear algebra play an important role. We begin with a very brief review of the the notion of linear independence for numeric vectors, after which we will develop the notion of linear independence for vector-valued functions and ultimately describe the general solution of of $\mathbf{y}' - A\,\mathbf{y} = 0$.

Numeric Vectors and Linear Independence A basic definition from linear algebra is that a set of m vectors $\mathbf{x}_1, \mathbf{x}_2, \ldots, \mathbf{x}_m$ in $\mathbb{R}^n$ is **linearly dependent** if there are scalars $c_1, c_2, \ldots, c_m$, with at least one $c_i \neq 0$, such that

$$c_1\mathbf{x}_1 + c_2\mathbf{x}_2 + \cdots + c_m\mathbf{x}_m = \mathbf{0},$$

where $\mathbf{0}$ represents the zero vector in $\mathbb{R}^n$. When the only such scalars that exist are all zero, then we say that $\mathbf{x}_1, \mathbf{x}_2, \ldots, \mathbf{x}_m$ are **linearly independent**.

Let X be the constant $n \times m$ matrix whose j^{th} column is $\mathbf{x}_j$; that is, the ij^{th} entry of X is $x_{i,j}$:

$$X = (\mathbf{x}_1\ \mathbf{x}_2 \cdots \mathbf{x}_m) = \begin{pmatrix} x_{1,1} & x_{1,2} & & x_{1,m} \\ x_{2,1} & x_{2,2} & \cdots & x_{2,m} \\ \vdots & \vdots & & \vdots \\ x_{n,1} & x_{n,2} & & x_{n,m} \end{pmatrix}.$$

Note that multiplication of a vector $\mathbf{c} = (c_1, c_2, \ldots, c_m)^T$ by the matrix X results in a *linear combination* of the columns $\mathbf{x}_1, \mathbf{x}_2, \ldots, \mathbf{x}_m$ of X:

$$X\mathbf{c} = c_1\mathbf{x}_1 + c_2\mathbf{x}_2 + \cdots + c_m\mathbf{x}_m.$$

Therefore, $\mathbf{x}_1, \mathbf{x}_2, \ldots, \mathbf{x}_m$ are linearly independent if the equation $X\mathbf{c} = \mathbf{0}$ has only the trivial solution $\mathbf{c} = \mathbf{0}$, and they are linearly dependent if $X\mathbf{c} = \mathbf{0}$ has at least one nontrivial solution $\mathbf{c}$.

Consider now the case where $m = n$, in which X is an $n \times n$ (square) matrix, and let the symbol $|X|$ denote the determinant of X. (See Appendix I.B.) The following theorem from linear algebra states the simple connection between $|X|$ and the linear dependence/independence of the columns of X.

Theorem 1

Let X be an $n \times n$ matrix. The columns of X are linearly dependent if $|X| = 0$, and they are linearly independent if $|X| \neq 0$. ▲

Example 1

Consider the 3×3 matrix

$$X = \begin{pmatrix} 1 & 2 & 1 \\ 3 & 1 & 2 \\ 1 & 1 & 2 \end{pmatrix}.$$

By a Laplace expansion on the first row (see Appendix I.B.), the determinant of X is

$$\begin{aligned} |X| &= (1)\begin{vmatrix} 1 & 2 \\ 1 & 2 \end{vmatrix} - (2)\begin{vmatrix} 3 & 2 \\ 1 & 2 \end{vmatrix} + (1)\begin{vmatrix} 3 & 1 \\ 1 & 1 \end{vmatrix} \\ &= (1)((1)(2) - (2)(1)) - (2)((3)(2) - (2)(1)) + (1)((3)(1) - (1)(1)) \\ &= 0 - 8 + 2 = -6. \end{aligned}$$

Since $|X| \neq 0$, we conclude that the columns of X are linearly independent. ■

Example 2

Consider the 3×3 matrix

$$X = \begin{pmatrix} 1 & 2 & 3 \\ 3 & 1 & 4 \\ 1 & 1 & 2 \end{pmatrix}.$$

By a Laplace expansion on the first row, the determinant of X is

$$\begin{aligned} |X| &= (1)\begin{vmatrix} 1 & 4 \\ 1 & 2 \end{vmatrix} - (2)\begin{vmatrix} 3 & 4 \\ 1 & 2 \end{vmatrix} + (3)\begin{vmatrix} 3 & 1 \\ 1 & 1 \end{vmatrix} \\ &= (1)((1)(2) - (4)(1)) - (2)((3)(2) - (4)(1)) + (3)((3)(1) - (1)(1)) \\ &= -2 - 4 + 6 = 0, \end{aligned}$$

from which we conclude that the columns of X are linearly dependent. Note that the third column of X is equal to the sum of the first two. ■

Vector-Valued Functions and Linear Independence Let $\mathbf{y}_1, \mathbf{y}_2, \ldots, \mathbf{y}_m$ be vector-valued functions from an interval I into $\mathbb{R}^n$:

$$\mathbf{y}_j(t) = \begin{pmatrix} y_{1,j} & (t) \\ y_{2,j} & (t) \\ \vdots & \\ y_{n,j} & (t) \end{pmatrix}, \quad j = 1, 2, \ldots, m.$$

We say that these functions are **linearly dependent on** I if there are scalars c_1, $c_2, \ldots, c_m$, with at least one $c_i \neq 0$, such that

$$c_1\mathbf{y}_1(t) + c_2\mathbf{y}_2(t) + \cdots + c_m\mathbf{y}_m(t) = \mathbf{0} \quad \textit{for all} \quad t \text{ in } I.$$

When the only such scalars that exist are all zero, then $\mathbf{y}_1, \mathbf{y}_2, \ldots, \mathbf{y}_m$ are said to be **linearly independent on** I.

It is crucial to understand that the *values* $\mathbf{y}_1(t), \mathbf{y}_2(t), \ldots, \mathbf{y}_m(t)$ at certain t in I may form a linearly dependent set of numeric vectors, while the *functions* $\mathbf{y}_1$, $\mathbf{y}_2, \ldots, \mathbf{y}_m$ are linearly independent on I.

Example 3

Let $\mathbf{y}_1(t) = \begin{pmatrix} \sin t \\ \cos t \end{pmatrix}$, $\mathbf{y}_2(t) = \begin{pmatrix} \cos t \\ \sin t \end{pmatrix}$. Then

$$\mathbf{y}_1(\pi/4) - \mathbf{y}_2(\pi/4) = \mathbf{0},$$

which shows that $\mathbf{y}_1(\pi/4)$ and $\mathbf{y}_2(\pi/4)$ are linearly dependent vectors, but it is easy to see that $\mathbf{y}_1$ and $\mathbf{y}_2$ are linearly independent functions on $I = \mathbb{R}$, since there are no scalars c_1 and c_2 such that $c_1\mathbf{y}_1(t) + c_2\mathbf{y}_2(t) = \mathbf{0}$ for *all* t in $\mathbb{R}$, other than $c_1 = c_2 = 0$. ■

Now suppose that $m = n$, and let Y be the $n \times n$ matrix-valued function whose j^{th} column is $\mathbf{y}_j$; that is, the ij^{th} entry of Y is $y_{i,j}$. Then $\mathbf{y}_1, \mathbf{y}_2, \ldots, \mathbf{y}_n$ are linearly dependent on an interval I, if and only if there is a nonzero, constant vector $\mathbf{c} = (c_1, c_2, \ldots, c_n)^T$ such that

$$Y(t)\mathbf{c} = \mathbf{0} \quad \text{for all} \quad t \text{ in } I.$$

From this we can conclude that the following theorem is true.

Theorem 2

If $\mathbf{y}_1, \mathbf{y}_2, \ldots, \mathbf{y}_n$ are linearly dependent functions on an interval I, then $|Y(t)| = 0$ for all t in I, where $|Y(t)|$ is the determinant of $Y(t)$. ▲

We must be very careful here. *The converse of Theorem 2 is false*, as the following example shows.

Example 4

Let $\mathbf{y}_1(t) = \begin{pmatrix} t \\ t \end{pmatrix}$, $\mathbf{y}_2(t) = \begin{pmatrix} 1 \\ 1 \end{pmatrix}$. The determinant of the matrix

$$Y(t) = \begin{pmatrix} t & 1 \\ t & 1 \end{pmatrix}$$

is zero for all t. Let $\mathbf{c} = (c_1, c_2)^T$, where c_1, c_2 are constants. Then

$$Y(t)\mathbf{c} = \begin{pmatrix} c_1 t + c_2 \\ c_1 t + c_2 \end{pmatrix},$$

which clearly cannot equal $(0, 0)^T$ for all t unless $c_1 = c_2 = 0$. So the columns of $Y(t)$ are linearly independent functions on $\mathbb{R}$, in spite of the fact that $|Y(t)| = 0$ for all t. ■

While the converse of Theorem 2 is false in general, we are about to see that it *is* true, if $\mathbf{y}_1, \mathbf{y}_2, \ldots, \mathbf{y}_n$ are solutions of a linear system $\mathbf{y}' - A\mathbf{y} = \mathbf{0}$ on I.

Solutions of Homogeneous Linear Systems Let A be an $n \times n$ matrix with entries that are continuous functions of t (perhaps constants) on an interval I.

Definition Suppose that the vector-valued functions $\mathbf{y}_1, \mathbf{y}_2, \ldots, \mathbf{y}_n$ each satisfy the homogeneous differential equation

$$\mathbf{y}' - A\mathbf{y} = \mathbf{0}$$

on an interval I, and let $Y(t)$ be the matrix whose columns are $\mathbf{y}_1, \mathbf{y}_2, \ldots, \mathbf{y}_n$. Such a matrix is said to be a **solution matrix** for the equation $\mathbf{y}' - A\mathbf{y} = \mathbf{0}$ on I, or more simply, a *solution matrix generated by A*. ◆

Suppose that $Y(t)$ is a solution matrix generated by A and that $|Y(t_0)| = 0$, where t_0 is some number in I. By Theorem 1, since $|Y(t_0)| = 0$, there are numbers $c_1, c_2, \ldots, c_n$, with at least one $c_i \neq 0$, such that

$$c_1\mathbf{y}_1(t_0) + c_2\mathbf{y}_2(t_0) + \cdots + c_n\mathbf{y}_n(t_0) = \mathbf{0}.$$

Given this particular collection of numbers $c_1, c_2, \ldots, c_n$, let

$$\mathbf{y}_\mathbf{c}(t) = c_1\mathbf{y}_1(t) + c_2\mathbf{y}_2(t) + \cdots + c_n\mathbf{y}_n(t)$$

for all t in I. Because of linearity, $\mathbf{y}_c$ satisfies the differential equation. Therefore, since $\mathbf{y}_\mathbf{c}(t_0) = \mathbf{0}$, uniqueness of solutions implies that $\mathbf{y}_\mathbf{c}(t) = \mathbf{0}$ for *all* t in I. Thus $\mathbf{y}_1, \mathbf{y}_2, \ldots, \mathbf{y}_n$ are linearly dependent, and so $|Y(t)| = 0$ for all t in I. This argument is the basis of the proof of the following theorem.

Theorem 3

Let A be an $n \times n$ matrix with entries that are continuous functions of t on I, and let $Y(t)$ be a solution matrix generated by A. Then the following statements are true.

i) Either $|Y(t)| = 0$ for all t in I or else $|Y(t)| \neq 0$ for all t in I.

ii) The columns $\mathbf{y}_1, \mathbf{y}_2, \ldots, \mathbf{y}_n$ *of* $Y(t)$ *are linearly independent on* I *if and only if* $|Y(t_0)| \neq 0$ *for some* t_0 *in* I.

iii) The columns $\mathbf{y}_1, \mathbf{y}_2, \ldots, \mathbf{y}_n$ *of* $Y(t)$ *are linearly independent on* I *if and only if the numeric vectors* $\mathbf{y}_1(t_0), \mathbf{y}_2(t_0), \ldots, \mathbf{y}_n(t_0)$ *are linearly independent for some* t_0 *in* I. ▲

Example 5

The functions

$$\mathbf{y}_1(t) = \begin{pmatrix} e^t \\ 3e^t \end{pmatrix} \quad \text{and} \quad \mathbf{y}_2(t) = \begin{pmatrix} e^{-t} \\ 2e^{-t} \end{pmatrix}$$

satisfy the differential equation

$$\mathbf{y}' - \begin{pmatrix} -5 & 2 \\ -12 & 5 \end{pmatrix} \mathbf{y} = \mathbf{0},$$

and the entries of the matrix are continuous on any interval. Note that

$$|Y(t)| = \begin{vmatrix} e^t & e^{-t} \\ 3e^t & 2e^{-t} \end{vmatrix} = -1,$$

which is nonzero for all t, thus illustrating part (i) of Theorem 3. Part (ii) of Theorem 3 tells us that $\mathbf{y}_1$ and $\mathbf{y}_2$ are linearly independent on any interval. By part (iii) of Theorem 3, we could have arrived at the same conclusion from the linear independence of the vectors $\mathbf{y}_1(0) = (1, 3)^T$ and $\mathbf{y}_2(0) = (1, 2)^T$, for example. ■

Example 6

Consider the differential equation

$$\mathbf{y}' - \frac{1}{t^2 - 1} \begin{pmatrix} t & -1 \\ -1 & t \end{pmatrix} \mathbf{y} = \mathbf{0},$$

in which the entries of the matrix are continuous on any interval that does not contain $t = \pm 1$. It is easy to check that the functions

$$\mathbf{y}_1(t) = \begin{pmatrix} t \\ 1 \end{pmatrix} \quad \text{and} \quad \mathbf{y}_2(t) = \begin{pmatrix} 1 \\ t \end{pmatrix}$$

satisfy the differential equation on any interval that does not contain $t = \pm 1$. Now notice that

$$|Y(t)| = \begin{vmatrix} t & 1 \\ 1 & t \end{vmatrix} = t^2 - 1,$$

and so $|Y(t)| = 0$ at $t = \pm 1$ and is nonzero elsewhere. This is consistent with part (i) of the Theorem 3, because $\mathbf{y}_1$ and $\mathbf{y}_2$ are solutions of the differential equation only on intervals that do not contain $t = \pm 1$. ■

A simple consequence of linearity is that, if $\mathbf{y}_1, \mathbf{y}_2, \ldots, \mathbf{y}_n$ each satisfy the differential equation $\mathbf{y}' - A\mathbf{y} = \mathbf{0}$ on an interval I, then any linear combination of $\mathbf{y}_1, \mathbf{y}_2, \ldots \mathbf{y}_n$ is also a solution. It turns out, moreover, that *every* solution must be of that form, *provided* that $\mathbf{y}_1, \mathbf{y}_2, \ldots, \mathbf{y}_n$ are linearly independent on I. To see why, first suppose that $\mathbf{y}_1, \mathbf{y}_2, \ldots, \mathbf{y}_n$ are linearly independent solutions of $\mathbf{y}' - A\mathbf{y} = \mathbf{0}$ on I and that $\tilde{\mathbf{y}}$ is *any* solution whatsoever of $\mathbf{y}' - A\mathbf{y} = \mathbf{0}$ on the interval I. Now pick any number t_0 in I. We know that $|Y(t_0)| \neq 0$; so by Theorem 1 there are scalars $c_1, c_2, \ldots, c_m$, with at least one $c_i \neq 0$, such that

$$c_1\mathbf{y}_1(t_0) + c_2\mathbf{y}_2(t_0) + \cdots + c_m\mathbf{y}_n(t_0) = \tilde{\mathbf{y}}(t_0).$$

With this particular choice of $c_1, c_2, \ldots, c_m$, let

$$\mathbf{y_c} = c_1\mathbf{y}_1 + c_2\mathbf{y}_2 + \cdots + c_m\mathbf{y}_n.$$

Then $\mathbf{y_c}$ has the same value as $\tilde{\mathbf{y}}$ at t_0 and satisfies the same differential equation. Therefore, uniqueness of solutions implies that $\mathbf{y_c}(t) = \tilde{\mathbf{y}}(t)$ for all t in I, which tells us that $\tilde{\mathbf{y}}$ is a linear combination of $\mathbf{y}_1, \mathbf{y}_2, \ldots, \mathbf{y}_n$.

We summarize the result of this argument in the following theorem.

Theorem 4

Suppose that the $n \times n$ matrix A has entries that are continuous functions on an interval I and that $\mathbf{y}_1, \mathbf{y}_2, \ldots, \mathbf{y}_n$ are linearly independent solutions of $\mathbf{y}' - \mathrm{A}\mathbf{y} = \mathbf{0}$ on I. Then every solution of $\mathbf{y}' - \mathrm{A}\mathbf{y} = \mathbf{0}$ on I can be expressed as a linear combination of $\mathbf{y}_1, \mathbf{y}_2, \ldots, \mathbf{y}_n$; that is, every solution is described by the general solution

$$\mathbf{y} = c_1\mathbf{y}_1 + c_2\mathbf{y}_2 + \cdots + c_n\mathbf{y}_n.$$

Stated in terms of the solution matrix $Y(t) = (\mathbf{y}_1\ \mathbf{y}_2 \cdots \mathbf{y}_n)$, every solution of $\mathbf{y}' - \mathrm{A}\mathbf{y} = \mathbf{0}$ on I is of the form $\mathbf{y} = Y(t)\mathbf{c}$ for some constant vector $\mathbf{c} = (c_1, c_2, \ldots, c_n)^T$. ▲

PROBLEMS

1. Prove Theorem 1 for 2×2 matrices $X = \begin{pmatrix} p & h \\ k & q \end{pmatrix}$.

Problems 2 through 4: Compute the determinant in order to determine whether the columns of the matrix are linearly dependent or linearly independent.

2. $\begin{pmatrix} 1 & 2 & 0 \\ -1 & 3 & 1 \\ 0 & 2 & 2 \end{pmatrix}$ **3.** $\begin{pmatrix} 2 & 1 & 0 \\ -1 & 3 & 7 \\ 2 & -1 & -4 \end{pmatrix}$ **4.** $\begin{pmatrix} 0 & 1 & 1 & 1 \\ -2 & 1 & 1 & 0 \\ 1 & -1 & 0 & 2 \\ 1 & 0 & 2 & -2 \end{pmatrix}$

Problems 5 through 7: Compute each product as a linear combination of the columns of the matrix (rather than using dot products).

5. $\begin{pmatrix} 1 & 2 \\ -1 & 3 \end{pmatrix}\begin{pmatrix} 3 \\ 2 \end{pmatrix}$ **6.** $\begin{pmatrix} 2 & 1 & -1 \\ -1 & 3 & 1 \\ 2 & -1 & 3 \end{pmatrix}\begin{pmatrix} 3 \\ -1 \\ 2 \end{pmatrix}$ **7.** $\begin{pmatrix} 5 & 1 & 3 \\ 2 & -3 & 1 \\ 3 & -1 & 2 \end{pmatrix}\begin{pmatrix} 0 \\ 1 \\ 0 \end{pmatrix}$

8. Consider the functions $\mathbf{u}$ and $\mathbf{v}$ defined on $I = \mathbb{R}$ by

$$\mathbf{u}(t) = \begin{pmatrix} t \\ 3 \end{pmatrix} \quad \text{and} \quad \mathbf{v}(t) = \begin{pmatrix} 1 \\ t+4 \end{pmatrix}.$$

(a) Show that $\mathbf{u}$ and $\mathbf{v}$ are linearly independent functions on $\mathbb{R}$.

(b) Find all t in $\mathbb{R}$ at which the numeric vectors $\mathbf{u}(t)$ and $\mathbf{v}(t)$ are linearly dependent.

9. Consider the functions $\mathbf{u}$, $\mathbf{v}$, and $\mathbf{w}$ defined on $\mathbb{R}$ by

$$\mathbf{u}(t) = \begin{pmatrix} t \\ e^{-t} \\ 2e^{t} \end{pmatrix}, \quad \mathbf{v}(t) = \begin{pmatrix} 0 \\ e^{-t} \\ e^{t} \end{pmatrix}, \quad \mathbf{w}(t) = \begin{pmatrix} -t \\ e^{-t} \\ e^{t} \end{pmatrix}.$$

Use Theorem 2 to show that $\mathbf{u}$, $\mathbf{v}$, $\mathbf{w}$ are linearly independent on every interval I.

In each of Problems 10 through 14,

(a) verify the statement;

(b) explain the choice of the interval I;

(c) verify statement (i) in Theorem 3, and then use statement (ii) to conclude that the columns of $Y(t)$ are linearly independent on I;

(d) write the general solution on I in the form $\mathbf{y} = c_1\mathbf{y}_1 + c_2\mathbf{y}_2$.

10. $Y(t) = \begin{pmatrix} t & t^{-1} \\ t & t^{-2} \end{pmatrix}$ is a solution matrix on $I = (0, 1)$ for the equation

$$\mathbf{y}' - \frac{1}{t(t-1)}\begin{pmatrix} -t-1 & 2t \\ -3 & t+2 \end{pmatrix}\mathbf{y} = \mathbf{0}.$$

11. $Y(t) = \begin{pmatrix} \sin t & \cos t \\ \cos t & \sin t \end{pmatrix}$ is a solution matrix on $I = (-\frac{\pi}{4}, \frac{\pi}{4})$ for the equation

$$\mathbf{y}' - \sec 2t\begin{pmatrix} -\sin 2t & 1 \\ 1 & -\sin 2t \end{pmatrix}\mathbf{y} = \mathbf{0}.$$

12. $Y(t) = \tan t\begin{pmatrix} 1 & \sec t \\ \sec t & 1 \end{pmatrix}$ is a solution matrix on $I = (0, \frac{\pi}{2})$ for the equation

$$\mathbf{y}' - \sec t \csc t\begin{pmatrix} 2 & -\cos t \\ -\cos t & 2 \end{pmatrix}\mathbf{y} = \mathbf{0}.$$

13. $Y(t) = \begin{pmatrix} e^{-t} & 3e^{-2t} \\ 2e^{-t} & e^{-2t} \end{pmatrix}$ is a solution matrix on $I = \mathbb{R}$ for the equation

$$\mathbf{y}' - \frac{1}{5}\begin{pmatrix} -11 & 3 \\ -2 & -4 \end{pmatrix}\mathbf{y} = \mathbf{0}.$$

14. $Y(t) = \begin{pmatrix} -te^{-t} & e^{-t} \\ e^{-t} & te^{-t} \end{pmatrix}$ is a solution matrix on $I = \mathbb{R}$ for the equation

$$\mathbf{y}' - \frac{1}{1+t^2}\begin{pmatrix} -1+t-t^2 & -1 \\ 1 & -1+t-t^2 \end{pmatrix}\mathbf{y} = \mathbf{0}.$$

15. Use either the method of elimination or Laplace transforms to find a pair of linearly independent solutions on $\mathbb{R}$ of the equation

$$\mathbf{y}' - \begin{pmatrix} -3 & 1 \\ 2 & -2 \end{pmatrix}\mathbf{y} = \mathbf{0}.$$

Then form the corresponding solution matrix $Y(t)$ and verify that $|Y(t)| \neq 0$ on $\mathbb{R}$.

16. Suppose that the entries of the matrix A are continuous on I. Using Theorem 3, explain why $Y(t) = \begin{pmatrix} te^{-t} & t \\ e^{-t} & e^{-t} \end{pmatrix}$ cannot be a solution matrix for $\mathbf{y}' - A\mathbf{y} = \mathbf{0}$ on I if I contains 0.

17. Find the matrix A, given that $Y(t) = \begin{pmatrix} e^{t/2} & 2e^{-t} \\ 2e^{t/2} & -e^{-t} \end{pmatrix}$ is a solution matrix generated by A.

18. Suppose that u and v are solutions of the second-order equation $x'' + px' + qx = 0$ and that we convert this equation to a first-order system $\mathbf{y}' - A\mathbf{y} = \mathbf{0}$, where $\mathbf{y} = (x, x')^T$.

(a) What solution matrix $Y(t)$ generated by A corresponds to u and v?

(b) Show that $|Y(t)| = W(t)$, where $W(t)$ is the Wronskian of u and v. (See Section 5.4.)

8.4 Evolution Matrices and Variation of Constants

Throughout this section, A is assumed to be an $n \times n$ matrix whose entries are continuous functions of t (possibly constants) on an interval I. We are concerned mainly with the homogeneous problem:

$$\mathbf{y}' - A\mathbf{y} = \mathbf{0} \quad \text{on } I. \tag{1}$$

Definition Suppose that $Y(t)$ is a solution matrix generated by A. If the columns of $Y(t)$ are linearly independent on I, then $Y(t)$ is said to be a **fundamental matrix** generated by A. ◆

Note that, by Theorem 3 of Section 8.3, a solution matrix $Y(t)$ is a fundamental matrix generated by A, if $|Y(t_0)| \neq 0$ for some t_0 in I.

The following theorem states three important properties of any fundamental matrix. Part (i) follows from the fact that the columns of $Y(t)$ satisfy (1), and parts (ii) and (iii) follow from Theorem 4 in Section 8.3 and Theorem 1 in Section 8.1.

Theorem 1

Let $Y(t)$ be a fundamental matrix generated by A. Then

i) $Y'(t) - A(t)Y(t) = 0$ *for all t in I.*

ii) Every solution of (1) is of the form $Y(t)\mathbf{c}$ for some constant vector $\mathbf{c}$ in $\mathbb{R}^n$.

iii) For any t_0 in I and any $\mathbf{y}_0$ in $\mathbb{R}^n$, $\mathbf{y} = Y(t)Y(t_0)^{-1}\mathbf{y}_0$ is the unique solution of (1) satisfying $\mathbf{y}(t_0) = \mathbf{y}_0$. ▲

Next we will define, for each t_0 in I, a special fundamental matrix, which we will call the *evolution matrix* generated by A at t_0. This evolution matrix will be the *solution operator* for an initial-value problem with initial point t_0, governing how any solution "evolves" from its value at time $t = t_0$. For this purpose, we first let $\mathbf{b}_1, \mathbf{b}_2, \ldots, \mathbf{b}_n$ denote the **standard basis vectors** in $\mathbb{R}^n$:

$$\mathbf{b}_1 = \begin{pmatrix} 1 \\ 0 \\ \vdots \\ 0 \end{pmatrix}, \mathbf{b}_2 = \begin{pmatrix} 0 \\ 1 \\ \vdots \\ 0 \end{pmatrix}, \ldots, \mathbf{b}_n = \begin{pmatrix} 0 \\ 0 \\ \vdots \\ 1 \end{pmatrix}.$$

Note that $\mathbf{b}_1, \mathbf{b}_2, \ldots, \mathbf{b}_n$ are the columns of the $n \times n$ identity matrix $\mathcal{I}$.

Definition The **evolution matrix $\Phi_{t_0}(t)$ generated by A at t_0** in I is the fundamental matrix whose i^{th} column is the unique solution on I of the initial-value problem

$$\mathbf{y}' - A\mathbf{y} = \mathbf{0}, \quad \mathbf{y}(t_0) = \mathbf{b}_i;$$

that is, $\Phi_{t_0}(t)$ is the fundamental matrix generated by A such that $\Phi_{t_0}(t_0) = \mathcal{I}$. ◆

For each t_0 in I the evolution matrix $\Phi_{t_0}(t)$ generated by A is unique by Theorem 1 in Section 8.1. In addition to the properties stated in Theorem 1 in this section, $\Phi_{t_0}(t)$ also has the property stated in the following theorem.

Theorem 2

For any t_0 in I and any $\mathbf{y}_0$ in $\mathbb{R}^n$, the unique solution on I of the initial-value problem

$$\mathbf{y}' - A\mathbf{y} = \mathbf{0}, \quad \mathbf{y}(t_0) = \mathbf{y}_0,$$

is

$$\mathbf{y} = \Phi_{t_0}(t)\mathbf{y}_0. \quad ▲$$

The following theorem, whose proof is straightforward, is helpful for computing an evolution matrix.

Theorem 3

Suppose that $Y(t)$ is a fundamental matrix generated by A. Then the evolution matrix $\Phi_{t_0}(t)$ generated by A at t_0 in I is given by

$$\Phi_{t_0}(t) = Y(t)Y(t_0)^{-1}. \quad ▲$$

Following are two examples that illustrate Theorem 3. The first involves a nonautonomous 2×2 system, and the second involves an autonomous 3×3 system. In each of these examples, a fundamental matrix is given with no indication of how it was obtained. Do not worry about this. In fact, each system was computed from the solution—rather than the other way around—with the help of a computer algebra system. In the next section, we will begin to learn how to find fundamental matrices.

Example 1

Consider the differential equation

$$\mathbf{y}' - \frac{1}{(t+1)^2}\begin{pmatrix} t & 1 \\ -1 & 2+t \end{pmatrix}\mathbf{y} = \mathbf{0},$$

for the unknown $\mathbf{y} = (y_1, y_2)^T$, and note that the entries of the matrix are continuous on $I = (-1, \infty)$. It turns out that a simple fundamental matrix is

$$Y(t) = \begin{pmatrix} 2+t & t \\ 1 & 1+2t \end{pmatrix},$$

as is easily verified. Consequently, the evolution matrix at 0, for example, is

$$\begin{aligned}\Phi_0(t) = Y(t)Y(0)^{-1} &= \begin{pmatrix} 2+t & t \\ 1 & 1+2t \end{pmatrix}\begin{pmatrix} 2 & 0 \\ 1 & 1 \end{pmatrix}^{-1} \\ &= \begin{pmatrix} 2+t & t \\ 1 & 1+2t \end{pmatrix}\begin{pmatrix} 1/2 & 0 \\ -1/2 & 1 \end{pmatrix} = \begin{pmatrix} 1 & t \\ -t & 1+2t \end{pmatrix}.\end{aligned}$$

The evolution matrix at 1, on the other hand, is

$$\begin{aligned}\Phi_1(t) = Y(t)Y(1)^{-1} &= \begin{pmatrix} 2+t & t \\ 1 & 1+2t \end{pmatrix}\begin{pmatrix} 3 & 1 \\ 1 & 3 \end{pmatrix}^{-1} \\ &= \begin{pmatrix} 2+t & t \\ 1 & 1+2t \end{pmatrix}\begin{pmatrix} \frac{3}{8} & -\frac{1}{8} \\ -\frac{1}{8} & \frac{3}{8} \end{pmatrix} = \frac{1}{4}\begin{pmatrix} t+3 & t-1 \\ 1-t & 1+3t \end{pmatrix}.\end{aligned}$$

■

Example 2

Consider the differential equation

$$\mathbf{y}' - \frac{1}{2}\begin{pmatrix} -3 & -1 & 2 \\ 1 & -5 & 2 \\ 1 & 1 & -4 \end{pmatrix}\mathbf{y} = \mathbf{0},$$

for the unknown $\mathbf{y} = (y_1, y_2, y_3)^T$, and note that the entries of the matrix are continuous on $I = \mathbb{R}$. It turns out that a simple fundamental matrix is

$$Y(t) = \begin{pmatrix} e^{-t} & -e^{-2t} & e^{-3t} \\ e^{-t} & e^{-2t} & e^{-3t} \\ e^{-t} & e^{-2t} & -e^{-3t} \end{pmatrix},$$

as is easily verified. Consequently, the evolution matrix at 0, for example, is

$$\Phi_0(t) = Y(t)Y(0)^{-1} = \begin{pmatrix} e^{-t} & -e^{-2t} & e^{-3t} \\ e^{-t} & e^{-2t} & e^{-3t} \\ e^{-t} & e^{-2t} & -e^{-3t} \end{pmatrix} \begin{pmatrix} 1 & -1 & 1 \\ 1 & 1 & 1 \\ 1 & 1 & -1 \end{pmatrix}^{-1}$$

$$= \begin{pmatrix} e^{-t} & -e^{-2t} & e^{-3t} \\ e^{-t} & e^{-2t} & e^{-3t} \\ e^{-t} & e^{-2t} & -e^{-3t} \end{pmatrix} \frac{1}{2} \begin{pmatrix} 1 & 0 & 1 \\ -1 & 1 & 0 \\ 0 & 1 & -1 \end{pmatrix}$$

$$= \frac{1}{2} \begin{pmatrix} e^{-2t} + e^{-t} & e^{-3t} - e^{-2t} & -e^{-3t} + e^{-t} \\ -e^{-2t} + e^{-t} & e^{-3t} + e^{-2t} & -e^{-3t} + e^{-t} \\ -e^{-2t} + e^{-t} & -e^{-3t} + e^{-2t} & e^{-3t} + e^{-t} \end{pmatrix}.$$

■

Nonhomogeneous Systems Consider the nonhomogeneous system

$$\mathbf{y}' - A\mathbf{y} = \mathbf{f}, \tag{2}$$

where $\mathbf{f}$ is a continuous, vector-valued function on I, and let $Y(t)$ be any fundamental matrix generated on I by A—possibly some evolution matrix $\Phi_s(t)$.

Since every solution of the associated homogeneous system is of the form $\mathbf{y} = Y(t)\mathbf{c}$, where $\mathbf{c}$ is a constant vector, we will look at a particular solution $\tilde{\mathbf{y}}$ in the form $\tilde{\mathbf{y}} = Y(t)\mathbf{u}(t)$, where $\mathbf{u}(t)$ is a vector-valued function to be determined. Recall that such a technique is called *variation of constants*.

Substitution of $\tilde{\mathbf{y}} = Y(t)\mathbf{u}(t)$ into (2) produces

$$Y'\mathbf{u} + Y\mathbf{u}' - AY\mathbf{u} = \mathbf{f}$$

by the product rule for matrix-vector products. Since $Y' = AY$, this becomes

$$Y\mathbf{u}' = \mathbf{f}.$$

Since $|Y(t)| \neq 0$ for all t in I, $Y(t)$ is invertible for all t in I, and so we can express $\mathbf{u}'$ as $\mathbf{u}' = Y^{-1}\mathbf{f}$. Thus, using any t_0 in I, we can integrate to obtain

$$\mathbf{u}(t) = \int_{t_0}^{t} Y(s)^{-1}\mathbf{f}(s)ds.$$

So a particular solution of (2) is given by

$$\tilde{\mathbf{y}}(t) = Y(t) \int_{t_0}^{t} Y(s)^{-1}\mathbf{f}(s)ds. \tag{3}$$

Note that this particular solution satisfies $\tilde{\mathbf{y}}(t_0) = \mathbf{0}$.

Now consider the nonhomogeneous initial-value problem

$$\mathbf{y}' - A\mathbf{y} = \mathbf{f}, \quad \mathbf{y}(t_0) = \mathbf{y}_0, \tag{4}$$

under the same assumptions on A and $\mathbf{f}$ as stated previously. The solution satisfying $\mathbf{y}(t_0) = \mathbf{0}$ is given by (3). Thanks to linearity, we can superimpose $\tilde{\mathbf{y}}$ with the solution $Y(t)Y(t_0)^{-1}\mathbf{y}_0$ of the homogeneous problem to arrive at

$$\mathbf{y}(t) = Y(t)\left(Y(t_0)^{-1}\mathbf{y}_0 + \int_{t_0}^{t} Y^{-1}(s)\mathbf{f}(s)\,ds\right). \tag{5}$$

If we take the fundamental matrix $Y(t)$ to be the evolution matrix $\Phi_{t_0}(t)$ at t_0 in I, this becomes the slightly simpler formula

$$\mathbf{y}(t) = \Phi_{t_0}(t)\left(\mathbf{y}_0 + \int_{t_0}^{t} \Phi_{t_0}(s)^{-1}\mathbf{f}(s)\,ds\right). \tag{6}$$

We will refer to (5) and (6) as **variation of constants formulas.**

Example 3

The following system is a nonhomogeneous version of the equation from Example 1:

$$\mathbf{y}' - \frac{1}{(t+1)^2}\begin{pmatrix} t & 1 \\ -1 & 2+t \end{pmatrix}\mathbf{y} = \begin{pmatrix} 1 \\ t^2 \end{pmatrix}.$$

Let's compute the particular solution that satisfies $\mathbf{y}(0) = \mathbf{0}$. The evolution matrix at 0 turned out to be

$$\Phi_0(t) = \begin{pmatrix} 1 & t \\ -t & 1+2t \end{pmatrix}.$$

A straightforward calculation reveals that

$$\Phi_0(t)^{-1} = \frac{1}{(t+1)^2}\begin{pmatrix} 1+2t & -t \\ t & 1 \end{pmatrix}.$$

The quantity to be integrated in the variation of constants formula is then

$$\Phi_0(s)^{-1}\mathbf{f}(s) = \frac{1}{(s+1)^2}\begin{pmatrix} 1+2s & -s \\ s & 1 \end{pmatrix}\begin{pmatrix} 1 \\ s^2 \end{pmatrix} = \begin{pmatrix} 2 - s - \dfrac{1}{s+1} \\ 1 - \dfrac{1}{s+1} \end{pmatrix},$$

the integral of which, from 0 to t, is

$$\int_0^t \Phi_0(s)^{-1}\mathbf{f}(s)ds = \int_0^t \begin{pmatrix} 2 - s - \dfrac{1}{s+1} \\ 1 - \dfrac{1}{s+1} \end{pmatrix} ds = \begin{pmatrix} 2t - \dfrac{1}{2}t^2 - \ln(t+1) \\ t - \ln(t+1) \end{pmatrix}.$$

So the solution we seek is

$$\begin{aligned}\Phi_0(t)\int_0^t \Phi_0(s)^{-1}\mathbf{f}(s)ds &= \begin{pmatrix} 1 & t \\ -t & 1+2t \end{pmatrix}\begin{pmatrix} 2t - \dfrac{1}{2}t^2 - \ln(t+1) \\ t - \ln(t+1) \end{pmatrix} \\ &= \begin{pmatrix} \dfrac{1}{2}t^2 + 2t - (t+1)\ln(t+1) \\ t + \dfrac{1}{2}t^3 - (t+1)\ln(t+1) \end{pmatrix}.\end{aligned}$$

■

Example 4

Let's find the solution satisfying $\mathbf{y}(0) = \mathbf{0}$ of the following nonhomogeneous version of the equation from Example 2:

$$\mathbf{y}' - \frac{1}{2}\begin{pmatrix} -3 & -1 & 2 \\ 1 & -5 & 2 \\ 1 & 1 & -4 \end{pmatrix}\mathbf{y} = \begin{pmatrix} te^{-t} \\ 0 \\ 0 \end{pmatrix}.$$

In this case, it will be simpler to work with the fundamental matrix

$$Y(t) = \begin{pmatrix} e^{-t} & -e^{-2t} & e^{-3t} \\ e^{-t} & e^{-2t} & e^{-3t} \\ e^{-t} & e^{-2t} & -e^{-3t} \end{pmatrix},$$

rather than the more complicated evolution matrix $\Phi_0(t)$. A *CAS*-aided calculation reveals that

$$Y(t)^{-1} = \frac{1}{2}\begin{pmatrix} e^{t} & 0 & e^{t} \\ -e^{2t} & e^{2t} & 0 \\ 0 & e^{3t} & -e^{3t} \end{pmatrix}.$$

The quantity we need to integrate is then

$$Y(s)^{-1}\mathbf{f}(s) = \frac{1}{2}\begin{pmatrix} e^{s} & 0 & e^{s} \\ -e^{2s} & e^{2s} & 0 \\ 0 & e^{3s} & -e^{3s} \end{pmatrix}\begin{pmatrix} se^{-s} \\ 0 \\ 0 \end{pmatrix} = \frac{1}{2}\begin{pmatrix} s \\ -se^{-s} \\ 0 \end{pmatrix}.$$

The resulting integral is

$$\int_0^t Y(s)^{-1}\mathbf{f}(s)ds = \frac{1}{2}\int_0^t \begin{pmatrix} s \\ -se^{-s} \\ 0 \end{pmatrix} ds = \frac{1}{4}\begin{pmatrix} t^2 \\ 2(1-t)e^{t} - 2 \\ 0 \end{pmatrix}.$$

So the solution we seek is

$$\begin{aligned} Y(t)\int_0^t Y(s)^{-1}\mathbf{f}(s)ds &= \begin{pmatrix} e^{-t} & -e^{-2t} & e^{-3t} \\ e^{-t} & e^{-2t} & e^{-3t} \\ e^{-t} & e^{-2t} & -e^{-3t} \end{pmatrix}\frac{1}{4}\begin{pmatrix} t^2 \\ 2(1-t)e^{t} - 2 \\ 0 \end{pmatrix} \\ &= \frac{e^{-t}}{2}\begin{pmatrix} e^{-t} + \frac{1}{2}t^2 + t - 1 \\ -e^{-t} + \frac{1}{2}t^2 - t + 1 \\ -e^{-t} + \frac{1}{2}t^2 - t + 1 \end{pmatrix}. \end{aligned}$$

■

PROBLEMS

In Problems 1 through 4, use a solution matrix from one of Problems 10 through 14 in Section 8.3 to find the evolution matrix $\Phi_{t_0}(t)$ associated with the given equation at the indicated t_0.

1. $\mathbf{y}' - \frac{1}{t(t-1)}\begin{pmatrix} -t-1 & 2t \\ -3 & t+2 \end{pmatrix}\mathbf{y} = \mathbf{0}$ at $t_0 = 1/2$

2. $\mathbf{y}' - \sec 2t\begin{pmatrix} -\sin 2t & 1 \\ 1 & -\sin 2t \end{pmatrix}\mathbf{y} = \mathbf{0}$ at $t_0 = 0$

3. $\mathbf{y}' - \sec t \csc t\begin{pmatrix} 2 & -\cos t \\ -\cos t & 2 \end{pmatrix}\mathbf{y} = \mathbf{0}$ at $t_0 = \pi/4$

4. $\mathbf{y}' - \frac{1}{5}\begin{pmatrix} -11 & 3 \\ -2 & -4 \end{pmatrix}\mathbf{y} = \mathbf{0}$ at $t_0 = 0$

In Problems 5 through 8, use the variation of constants formula together with one of the evolution matrices from Problems 1 through 4 to find the solution of the initial-value problem.

5. $\mathbf{y}' - \frac{1}{t(t-1)}\begin{pmatrix} -t-1 & 2t \\ -3 & t+2 \end{pmatrix}\mathbf{y} = \begin{pmatrix} 8 \\ 8/t \end{pmatrix}$, $\mathbf{y}(1/2) = \mathbf{0}$

6. $\mathbf{y}' - \sec 2t\begin{pmatrix} -\sin 2t & 1 \\ 1 & -\sin 2t \end{pmatrix}\mathbf{y} = \begin{pmatrix} \sin t \\ \cos t \end{pmatrix}$, $\mathbf{y}(0) = \begin{pmatrix} 0 \\ 1 \end{pmatrix}$

7. $\mathbf{y}' - \sec t \csc t\begin{pmatrix} 2 & -\cos t \\ -\cos t & 2 \end{pmatrix}\mathbf{y} = 12\sin^4 t\begin{pmatrix} \sqrt{2} \\ -1 \end{pmatrix}$, $\mathbf{y}(\pi/4) = \mathbf{0}$

8. $\mathbf{y}' - \frac{1}{5}\begin{pmatrix} -11 & 3 \\ -2 & -4 \end{pmatrix}\mathbf{y} = \begin{pmatrix} 4e^{-s} \\ 3e^{-s} \end{pmatrix}$, $\mathbf{y}(0) = \begin{pmatrix} 0 \\ 5 \end{pmatrix}$

In each of Problems 9 through 11, the given $\tilde{\mathbf{y}}$ and $\hat{\mathbf{y}}$ are linearly independent solutions of $\mathbf{y}' - A\mathbf{y} = \mathbf{0}$ on an interval I containing $t = 0$. For the given function $\mathbf{f}$, find the solution on I of the initial-value problem $\mathbf{y}' - A\mathbf{y} = \mathbf{f}$, $\mathbf{y}(0) = \mathbf{0}$.

9. $\tilde{\mathbf{y}} = \begin{pmatrix} e^t \\ e^t \end{pmatrix}$, $\hat{\mathbf{y}} = \begin{pmatrix} e^{-2t} \\ -e^{-2t} \end{pmatrix}$, $\mathbf{f}(t) = \begin{pmatrix} 8 \\ 4 \end{pmatrix}$

10. $\tilde{\mathbf{y}} = \begin{pmatrix} t \\ t+1 \end{pmatrix}$, $\hat{\mathbf{y}} = \begin{pmatrix} t+1 \\ 0 \end{pmatrix}$, $\mathbf{f}(t) = \begin{pmatrix} 1 \\ (1+t)^2 \end{pmatrix}$

11. $\tilde{\mathbf{y}} = \begin{pmatrix} (t+1)^{3/2} \\ -(t+1)^{3/2} \end{pmatrix}$, $\hat{\mathbf{y}} = \begin{pmatrix} (t+1)^{5/2} \\ (t+1)^{5/2} \end{pmatrix}$, $\mathbf{f}(t) = \begin{pmatrix} (t+1)^{1/2} \\ -(t+1)^{1/2} \end{pmatrix}$

12. Let the entries of the matrix $Q = \begin{pmatrix} q_{11} & q_{12} \\ q_{21} & q_{22} \end{pmatrix}$ and the vector $\mathbf{x} = \begin{pmatrix} x_1 \\ x_2 \end{pmatrix}$ be differentiable functions of t. Prove the product rule: $(Q\mathbf{x})' = Q'\mathbf{x} + Q\mathbf{x}'$.

13. Let A be an $n \times n$ matrix with constant entries, and assume that the series

$$\Psi(t) = \mathcal{I} + tA + \frac{t^2}{2!}A^2 + \frac{t^3}{3!}A^3 + \cdots$$

converges (entrywise) for all t. ($\mathcal{I}$ is the $n \times n$ identity matrix.) Show that $\Psi' - A\Psi = \mathbf{0}$ for all t in $\mathbb{R}$ and that $\Psi(0) = \mathcal{I}$. Conclude therefore, by the uniqueness of solutions, that $\Psi(t) = \Phi_0(t)$, the evolution matrix for $\mathbf{y}' - A\mathbf{y} = \mathbf{0}$ at 0.

8.5 Autonomous Systems: Eigenvalues and Eigenvectors

In this section we will develop theory that characterizes the evolution matrix $\Phi_0(t)$ generated by a constant matrix A—that is, for an *autonomous* equation $\mathbf{y}' - A\mathbf{y} = \mathbf{0}$—in terms of certain properties of the matrix A. This theory provides one method for computing $\Phi_0(t)$, but more important, it provides an important means for understanding the qualitative behavior of solutions.

We will assume throughout this section that the equation $\mathbf{y}' - A\mathbf{y} = \mathbf{0}$ is autonomous. Thus we may assume that $I = \mathbb{R}$, and therefore we need make no further qualifying statements regarding I. Furthermore, if $\tilde{\mathbf{y}}(t)$ is a solution of $\mathbf{y}' - A\mathbf{y} = \mathbf{0}$, then any simple shift $\mathbf{y}(t) = \tilde{\mathbf{y}}(t - s)$ is also a solution, because if $\mathbf{y}(t) = \tilde{\mathbf{y}}(t - s)$, then

$$\mathbf{y}'(t) - A\mathbf{y}(t) = \tilde{\mathbf{y}}'(t - s) - A\tilde{\mathbf{y}}(t - s) = \mathbf{0}.$$

(This is not true when A depends on t. Why?) For this reason, we will only consider initial-value problems in which the initial point is $t_0 = 0$. An important related fact is that we need only consider the evolution matrix $\Phi_0(t)$, because for any s, we have $\Phi_s(t) = \Phi_0(t - s)$.

The following theorem states properties of $\Phi_0(t)$ that are interestingly analogous to properties of the exponential function e^{at}. Parts (i) and (iv) are true even when A depends on t, but parts (ii), (iii), and (v) are true only when A is constant.

Theorem 1

If the matrix A has constant entries, then the evolution matrix $\Phi_0(t)$ generated by A has the following properties:

i) $\Phi_0(0) = \mathcal{I}$;

ii) $\Phi_0(t + s) = \Phi_0(t)\Phi_0(s)$ *for all* t, s *in* $\mathbb{R}$;

iii) $\Phi_0(t)^{-1} = \Phi_0(-t)$ *for all* t *in* $\mathbb{R}$;

iv) $\Phi_0'(t) = A\Phi_0(t)$ *for all* t *in* $\mathbb{R}$;

v) $\Phi_0(t) = \mathcal{I} + tA + \frac{t^2}{2!}A^2 + \frac{t^3}{3!}A^3 + \cdots$ *for all* t *in* $\mathbb{R}$. ▲

Because $\Phi_0(t)$ has these properties that are analogous to properties of e^{at}, it is often called the **matrix exponential** and denoted by e^{At}. We will henceforth adopt that notation when A is constant:

$$e^{At} = \Phi_0(t).$$

Using this notation together with parts (ii) and (iii) of Theorem 1, we restate the variation of constants formula as follows.

Let $\mathbf{f}(t)$ be continuous on an interval I that contains $t_0 = 0$. The unique solution of the initial-value problem

$$\mathbf{y}' - A\mathbf{y} = \mathbf{f}(t), \quad \mathbf{y}(0) = \mathbf{y}_0$$

is

$$\mathbf{y} = e^{At}\mathbf{y}_0 + \int_0^t e^{A(t-s)}\mathbf{f}(s)ds. \tag{1}$$

Eigenvalues and Eigenvectors Let us begin by recalling the "scalar case." If a is a constant, then every solution of $y' - ay = 0$ is of the form $y = ce^{at}$, where c is a constant. So we will look for solutions of the system $\mathbf{y}' - A\mathbf{y} = 0$ in the form $\mathbf{y} = e^{\lambda t}\mathbf{p}$, where λ is a scalar constant and $\mathbf{p}$ is a nonzero constant vector. Substituting $\mathbf{y} = e^{\lambda t}\mathbf{p}$ into $\mathbf{y}' - A\mathbf{y} = \mathbf{0}$ results in

$$\lambda e^{\lambda t}\mathbf{p} - e^{\lambda t}A\mathbf{p} = \mathbf{0},$$

which, after we divide by $e^{\lambda t}$ and rearrange, becomes

$$A\mathbf{p} = \lambda\mathbf{p}.$$

This leads us to the following definition. (See also Appendix I.)

Definition Let A be an $n \times n$ with constant entries. A complex (possibly real) scalar λ is an **eigenvalue** of A if there is a nonzero vector $\mathbf{p}$ such that $A\mathbf{p} = \lambda\mathbf{p}$. Such a vector $\mathbf{p}$ is called an **eigenvector** of A corresponding to the eigenvalue λ. An eigenvalue λ and a corresponding eigenvector $\mathbf{p}$ are collectively called an **eigenpair** $(\lambda, \mathbf{p})$ of A. ◆

For any eigenvalue λ of A, there are infinitely many eigenvectors. In particular, given any eigenpair $(\lambda, \mathbf{p})$ and any scalar constant $c \neq 0$, $(\lambda, c\mathbf{p})$ is also an eigenpair of A. This is due to the fact that if $A\mathbf{p} = \lambda\mathbf{p}$, then $Ac\mathbf{p} = \lambda c\mathbf{p}$. Moreover, a given eigenvalue may have two (or more) linearly independent corresponding eigenvectors, resulting in eigenpairs $(\lambda, \mathbf{p}_1)$ and $(\lambda, \mathbf{p}_2)$ in which $\mathbf{p}_1$ and $\mathbf{p}_2$ are linearly independent.

Since λ is an eigenvalue, if and only if the equation $(A - \lambda\mathcal{I})\mathbf{x} = 0$ has a nonzero solution, it follows that $|A - \lambda\mathcal{I}| = 0$, if and only if λ is an eigenvalue of A. The determinant $|A - \lambda\mathcal{I}|$ is a polynomial in the variable λ and is called the **characteristic polynomial** of A. Its roots are precisely the eigenvalues of A. Once an eigenvalue λ of A is known, any nonzero solution of $(A - \lambda\mathcal{I})\mathbf{x} = \mathbf{0}$ is a corresponding eigenvector.

Obviously, the eigenpairs of A are potentially helpful for finding a solution matrix for $\mathbf{y}' - A\mathbf{y} = \mathbf{0}$. In fact, given a collection of n eigenpairs of A in which the eigenvectors are linearly independent, the corresponding solutions enable us to build a fundamental matrix $Y(t)$, from which we can then compute the matrix exponential as $e^{At} = Y(t)Y(0)^{-1}$.

Because linear independence of eigenvectors is a crucial issue, we state the following theorem (without proof) for reference.

Theorem 2

Let A be an $n \times n$ matrix with constant entries. If $(\lambda_i, \mathbf{p}_i)$ and $(\lambda_j, \mathbf{p}_j)$ are eigenpairs with $\lambda_i \neq \lambda_j$, then $\mathbf{p}_i$ and $\mathbf{p}_j$ are linearly independent. Moreover, any set of eigenvectors corresponding to distinct eigenvalues of A is linearly independent. In particular, if A has n distinct eigenvalues, then A has n linearly independent eigenvectors. ▲

Example 1

Consider the matrix $A = \begin{pmatrix} 1 & -2 \\ 1 & 4 \end{pmatrix}$. The characteristic polynomial of A is

$$|A - \lambda\mathcal{I}| = \begin{vmatrix} 1-\lambda & -2 \\ 1 & 4-\lambda \end{vmatrix} = (1-\lambda)(4-\lambda) + 2$$
$$= \lambda^2 - 5\lambda + 6 = (\lambda - 2)(\lambda - 3).$$

Therefore, the eigenvalues of A are $\lambda_1 = 2$ and $\lambda_2 = 3$. To find an eigenvector corresponding to $\lambda_1 = 2$, we first form the matrix

$$A - \lambda_1\mathcal{I} = \begin{pmatrix} -1 & -2 \\ 1 & 2 \end{pmatrix}$$

and then simply *observe** that $(2, -1)^T$ is a solution of $(A - \lambda_1\mathcal{I})\mathbf{x} = \mathbf{0}$. To find an eigenvector corresponding to $\lambda_2 = 3$, we form the matrix

$$A - \lambda_2\mathcal{I} = \begin{pmatrix} -2 & -2 \\ 1 & 1 \end{pmatrix}$$

and then observe that $(1, -1)^T$ is a solution of $(A - \lambda_2\mathcal{I})\mathbf{x} = \mathbf{0}$. Thus we have found the eigenpairs

$$\left(2, \begin{pmatrix} 2 \\ -1 \end{pmatrix}\right) \quad \text{and} \quad \left(3, \begin{pmatrix} 1 \\ -1 \end{pmatrix}\right).$$

We now have the following linearly independent solutions of $\mathbf{y}' - A\mathbf{y} = \mathbf{0}$:

$$e^{2t}\begin{pmatrix} 2 \\ -1 \end{pmatrix} \quad \text{and} \quad e^{3t}\begin{pmatrix} 1 \\ -1 \end{pmatrix},$$

from which it follows that

$$Y(t) = \begin{pmatrix} 2e^{2t} & e^{3t} \\ -e^{2t} & -e^{3t} \end{pmatrix}$$

is a fundamental matrix generated by A. The matrix exponential is therefore

$$e^{At} = Y(t)Y(0)^{-1} = \begin{pmatrix} 2e^{2t} & e^{3t} \\ -e^{2t} & -e^{3t} \end{pmatrix}\begin{pmatrix} 2 & 1 \\ -1 & -1 \end{pmatrix}^{-1}$$
$$= \begin{pmatrix} 2e^{2t} & e^{3t} \\ -e^{2t} & -e^{3t} \end{pmatrix}\frac{1}{3}\begin{pmatrix} 1 & -1 \\ 1 & 2 \end{pmatrix}$$
$$= \begin{pmatrix} 2e^{2t} - e^{3t} & 2e^{2t} - 2e^{3t} \\ -e^{2t} + e^{3t} & -e^{2t} + 2e^{3t} \end{pmatrix}.$$

■

* Note that we are merely looking for *some* x_1 and x_2 so that $x_1 + 2x_2 = 0$.

Complex Eigenvalues A matrix with real entries can have nonreal complex eigenvalues. An eigenvector corresponding to a nonreal complex eigenvalue must have at least one nonreal complex entry. Since the eigenvalues of A are the roots of a polynomial, nonreal complex eigenvalues must occur as conjugate pairs. It also turns out that the corresponding eigenvectors will be conjugate pairs as well; in particular, if$(\alpha + \beta i, \mathbf{r} + i\mathbf{q})$ is an eigenpair, then so is $(\alpha - \beta i, \mathbf{r} - i\mathbf{q})$.

The following theorem summarizes the relationship between complex eigenpairs of A and solutions of $\mathbf{y}' - A\mathbf{y} = \mathbf{0}$.

Theorem 3

Let A be an $n \times n$ matrix with constant entries, and suppose that $(\lambda, \mathbf{p})$ is an eigenpair of A. Then $e^{\lambda t}\mathbf{p}$ is a solution of $\mathbf{y}' - A\mathbf{y} = \mathbf{0}$. A complex conjugate pair of eigenvalues $\lambda_{\pm} = \alpha \pm \beta i$ with corresponding eigenvectors $\mathbf{p}_{\pm} = \mathbf{r} \pm i\mathbf{q}$ gives rise to a pair of linearly independent real solutions:

$$e^{\alpha t}(\cos \beta t\mathbf{r} - \sin \beta t\mathbf{q}) \quad \text{and} \quad e^{\alpha t}(\sin \beta t\mathbf{r} + \cos \beta t\mathbf{q}). \tag{2}$$

These are the real and imaginary parts, respectively, of the complex-valued solution $e^{\lambda_+}\mathbf{p}_+ = e^{(\alpha+\beta i)t}(\mathbf{r} + i\mathbf{q})$. ▲

REMARK The two solutions given in Theorem 3 are the columns of an $n \times 2$ matrix $e^{\alpha t}PR(t)$, where $P = (\mathbf{p}\,\mathbf{q})$ and $R(t)$ is a standard 2×2 "rotation" matrix:

$$R(t) = \begin{pmatrix} \cos \beta t & \sin \beta t \\ -\sin \beta t & \cos \beta t \end{pmatrix}.$$

Example 2

Consider the matrix $A = \begin{pmatrix} -2 & 6 \\ -3 & 4 \end{pmatrix}$. The characteristic polynomial of A is

$$|A - \lambda\mathcal{I}| = \begin{vmatrix} -2-\lambda & 6 \\ -3 & 4-\lambda \end{vmatrix} = (-2-\lambda)(4-\lambda) + 18 = \lambda^2 - 2\lambda + 10.$$

By the quadratic formula, the roots of $|A - \lambda\mathcal{I}|$ are the complex conjugate pair

$$\lambda_1 = 1 + 3i, \quad \lambda_2 = 1 - 3i.$$

To find an eigenvector corresponding to $\lambda_1 = 1 + 3i$, we first form the matrix

$$A - \lambda_1\mathcal{I} = \begin{pmatrix} -3-3i & 6 \\ -3 & 3-3i \end{pmatrix} = 3\begin{pmatrix} -1-i & 2 \\ -1 & 1-i \end{pmatrix}$$

and then observe (because of the entries in the second row of the matrix) that $(1-i, 1)^T$ is a solution of $(A - \lambda_1\mathcal{I})\mathbf{x} = 0$. Thus we have the eigenpair

$$\left(1+3i, \begin{pmatrix} 1-i \\ 1 \end{pmatrix}\right) = \left(1+3i, \begin{pmatrix} 1 \\ 1 \end{pmatrix} + i\begin{pmatrix} -1 \\ 0 \end{pmatrix}\right),$$

from which we use (2) to obtain the following pair of linearly independent solutions of $\mathbf{y}' - A\mathbf{y} = \mathbf{0}$:

$$e^t\left(\cos 3t\begin{pmatrix}1\\1\end{pmatrix} - \sin 3t\begin{pmatrix}-1\\0\end{pmatrix}\right) = e^t\begin{pmatrix}\cos 3t + \sin 3t\\ \cos 3t\end{pmatrix}$$

and

$$e^t\left(\sin 3t\begin{pmatrix}1\\1\end{pmatrix} + \cos 3t\begin{pmatrix}-1\\0\end{pmatrix}\right) = e^t\begin{pmatrix}\sin 3t - \cos 3t\\ \sin 3t\end{pmatrix}.$$

These produce the solution matrix

$$Y(t) = e^t\begin{pmatrix}\cos 3t + \sin 3t & \sin 3t - \cos 3t\\ \cos 3t & \sin 3t\end{pmatrix},$$

from which we compute the matrix exponential as follows:

$$\begin{aligned}
e^{At} &= e^t\begin{pmatrix}\cos 3t + \sin 3t & \sin 3t - \cos 3t\\ \cos 3t & \sin 3t\end{pmatrix}\begin{pmatrix}1 & -1\\ 1 & 0\end{pmatrix}^{-1}\\
&= e^t\begin{pmatrix}\cos 3t + \sin 3t & \sin 3t - \cos 3t\\ \cos 3t & \sin 3t\end{pmatrix}\begin{pmatrix}0 & 1\\ -1 & 1\end{pmatrix}\\
&= e^t\begin{pmatrix}\cos 3t - \sin 3t & 2\sin 3t\\ -\sin 3t & \cos 3t + \sin 3t\end{pmatrix}.
\end{aligned}$$

■

Deficient Eigenvalues and Generalized Eigenvectors It should be clear by now that if an $n \times n$ matrix A with constant entries has n linearly independent eigenvectors, then once the eigenvectors are computed, it is a relatively simple matter to form e^{At}. According to Theorem 3, this is the case any time A has n distinct eigenvalues. It is also possible, though not guaranteed, when A does not have n distinct eigenvalues. Our purpose now is to understand the solutions of $\mathbf{y}' - A\mathbf{y} = \mathbf{0}$ when A fails to have n linearly independent eigenvectors.

Definition An eigenvalue λ_i of A is said to have **multiplicity** m if λ_i is a root of multiplicity m of the characteristic polynomial $|A - \lambda\mathcal{I}|$—that is, $(\lambda - \lambda_i)^m$ is a factor of $|A - \lambda\mathcal{I}|$ while $(\lambda - \lambda_i)^{m+1}$ is not. ◆

According to the *fundamental theorem of algebra*, the sum of the multiplicities of all of the eigenvalues of A is n. If A fails to have n distinct eigenvalues, it is because one (or more) of its eigenvalues has multiplicity greater than 1. Often an eigenvalue with multiplicity $m > 1$ will have m corresponding linearly independent eigenvectors. If that is the case for all "repeated" eigenvalues of A, then A has n linearly independent eigenvectors. So our concern lies with the repeated eigenvalues of A for which the number of corresponding linearly independent eigenvectors is less than the eigenvalue's multiplicity.

Definition If λ_i is an eigenvalue of A with multiplicity m and A has only $m - k$ linearly independent eigenvectors corresponding to λ_i, then λ_i is said to be **deficient** with a deficiency of order k. ◆

Though its proof requires a far deeper foray into linear algebra than is appropriate here, we will state the following theorem, which describes the generalized eigenvectors of A and the associated solutions of $\mathbf{y}' - A\mathbf{y} = \mathbf{0}$. (See also Problem 31.)

Theorem 4

Suppose that λ is a deficient eigenvalue of A with multiplicity m and deficiency k. Then A has m linearly independent ***generalized eigenvectors*** *corresponding to λ. These include the $m - k$ eigenvectors of A corresponding to λ, plus k additional vectors $\mathbf{v}_1, \mathbf{v}_2, \ldots, \mathbf{v}_k$ that satisfy*

$$(A - \lambda\mathcal{I})^{j+1}\mathbf{v}_j = \mathbf{0}, \quad (A - \lambda\mathcal{I})^j\mathbf{v}_j \neq \mathbf{0}$$

for $j = 1, \ldots, k$. Moreover, each $\mathbf{v}_j$ produces a solution of $\mathbf{y}' - A\mathbf{y} = \mathbf{0}$ in the form

$$\mathbf{y} = e^{\lambda_i t}\left(\mathcal{I} + t(A - \lambda\mathcal{I}) + \frac{t^2}{2!}(A - \lambda\mathcal{I})^2 + \cdots + \frac{t^j}{j!}(A - \lambda\mathcal{I})^j\right)\mathbf{v}_j. \tag{3}$$

▲

Example 3

Consider the matrix

$$A = \begin{pmatrix} -2 & 1 & -1 \\ 1 & -2 & 1 \\ -1 & -5 & 1 \end{pmatrix}.$$

The characteristic polynomial of A is

$$|A - \lambda\mathcal{I}| = \begin{vmatrix} -2-\lambda & 1 & -1 \\ 1 & -2-\lambda & 1 \\ -1 & -5 & 1-\lambda \end{vmatrix} = -1 - 3\lambda - 3\lambda^2 - \lambda^3 = -(\lambda + 1)^3.$$

Thus A has only one eigenvalue, $\lambda = -1$, which has multiplicity $m = 3$. To determine corresponding eigenvectors, we form the matrix

$$A + \mathcal{I} = \begin{pmatrix} -1 & 1 & -1 \\ 1 & -1 & 1 \\ -1 & -5 & 2 \end{pmatrix},$$

which turns out to have the *row-echelon form*:

$$\begin{pmatrix} 1 & 0 & \frac{1}{2} \\ 0 & 1 & -\frac{1}{2} \\ 0 & 0 & 0 \end{pmatrix}.$$

Therefore, $\mathbf{p} = (-1, 1, 2)^T$ is an eigenvector corresponding to $\lambda = -1$. All other eigenvectors are simply scalar multiples of $\mathbf{p}$, so there can be no other eigenvector linearly independent from $\mathbf{p}$. This tells us that λ has a deficiency of order $k = 2$; so we need to find two generalized eigenvectors. We begin by computing

$$(A + \mathcal{I})^2 = \begin{pmatrix} 3 & 3 & 0 \\ -3 & -3 & 0 \\ -6 & -6 & 0 \end{pmatrix}.$$

By simple observation, a solution of $(A+\mathcal{I})^2\mathbf{x}=\mathbf{0}$ is the vector $\mathbf{v}_1=(1,-1,0)^T$, and it is easily verified that $(A+\mathcal{I})\mathbf{v}_1\neq\mathbf{0}$. Now after computing

$$\mathcal{I}+t(A+\mathcal{I})=\begin{pmatrix}1-t & t & -t\\ t & 1-t & t\\ -t & -5t & 1+2t\end{pmatrix},$$

we find the solution

$$e^{-t}\left(\mathcal{I}+t(A+\mathcal{I})\right)\mathbf{v}_1=e^{-t}\begin{pmatrix}1-2t\\ 2t-1\\ 4t\end{pmatrix}.$$

Computation now shows that $(A+\mathcal{I})^3=0$. Consequently, any vector $\mathbf{x}$ for which $(A+\mathcal{I})^2\mathbf{x}\neq\mathbf{0}$ may serve as $\mathbf{v}_2$. So we make the simple choice $\mathbf{v}_2=(1,0,0)^T$. We now compute the matrix

$$\mathcal{I}+t(A+\mathcal{I})+\frac{t^2}{2}(A+\mathcal{I})^2=\begin{pmatrix}1-t+\frac{3}{2}t^2 & t+\frac{3}{2}t^2 & -t\\ t-\frac{3}{2}t^2 & 1-t-\frac{3}{2}t^2 & t\\ -t-3t^2 & -5t-3t^2 & 1+2t\end{pmatrix}$$

followed by the solution

$$e^{-t}\left(\mathcal{I}+t(A+\mathcal{I})+\frac{t^2}{2}(A+\mathcal{I})^2\right)\mathbf{v}_2=e^{-t}\begin{pmatrix}1-t+\frac{3}{2}t^2\\ t-\frac{3}{2}t^2\\ -t-3t^2\end{pmatrix}.$$

We now have three linearly independent solutions of $\mathbf{y}'-A\mathbf{y}=\mathbf{0}$:

$$e^{-t}\begin{pmatrix}-1\\ 1\\ 2\end{pmatrix},\quad e^{-t}\begin{pmatrix}1-2t\\ 2t-1\\ 4t\end{pmatrix},\quad e^{-t}\begin{pmatrix}1-t+\frac{3}{2}t^2\\ t-\frac{3}{2}t^2\\ -t-3t^2\end{pmatrix}.$$

The resulting fundamental matrix generated by A is

$$Y(t)=e^{-t}\begin{pmatrix}-1 & 1-2t & 1-t+\frac{3}{2}t^2\\ 1 & 2t-1 & t-\frac{3}{2}t^2\\ 2 & 4t & -t-3t^2\end{pmatrix},$$

which produces the matrix exponential as follows:

$$\begin{aligned}e^{At}=Y(t)Y(0)^{-1}&=e^{-t}\begin{pmatrix}-1 & 1-2t & 1-t+\frac{3}{2}t^2\\ 1 & 2t-1 & t-\frac{3}{2}t^2\\ 2 & 4t & -t-3t^2\end{pmatrix}\begin{pmatrix}-1 & 1 & 1\\ 1 & -1 & 0\\ 2 & 0 & 0\end{pmatrix}^{-1}\\ &=e^{-t}\begin{pmatrix}-1 & 1-2t & 1-t+\frac{3}{2}t^2\\ 1 & 2t-1 & t-\frac{3}{2}t^2\\ 2 & 4t & -t-3t^2\end{pmatrix}\frac{1}{2}\begin{pmatrix}0 & 0 & 1\\ 0 & -2 & 1\\ 2 & 2 & 0\end{pmatrix}\\ &=e^{-t}\begin{pmatrix}1-t+\frac{3}{2}t^2 & t+\frac{3}{2}t^2 & -t\\ t-\frac{3}{2}t^2 & 1-t-\frac{3}{2}t^2 & t\\ -t-3t^2 & -5t-3t^2 & 1+2t\end{pmatrix}.\end{aligned}$$

■

PROBLEMS

1. List the properties of the exponential function e^{at} that are analogous to properties (i)–(v) of e^{At} in Theorem 1.
2. Assume that parts (i) and (ii) of Theorem 1 are true, and prove part (iii) by showing that $\Phi_0(t)\Phi_0(-t) = \Phi_0(-t)\Phi_0(t) = \mathcal{I}$. (*Hint*: Use $s = -t$ in part (ii).)
3. (abc) For each of the matrix exponentials found in Examples 1, 2, and 3, verify parts (i) through (iv) of Theorem 1.
4. Let $A = \begin{pmatrix} -3 & 9 \\ -1 & 3 \end{pmatrix}$.

   (a) Show that $A^2 = 0$. Then use part (v) of Theorem 1 to compute e^{At}.

   (b) Use the variation of constants formula (1) to find the solution of
$$\mathbf{y}' - A\mathbf{y} = \begin{pmatrix} t \\ 1 \end{pmatrix}, \quad \mathbf{y}(0) = \begin{pmatrix} 0 \\ 1 \end{pmatrix}.$$

5. Let $A = \begin{pmatrix} 0 & 1 & 0 \\ 2 & 0 & 1 \\ 0 & -2 & 0 \end{pmatrix}$.

   (a) Show that $A^3 = 0$. Then use part (v) of Theorem 1 to compute e^{At}.

   (b) Use the variation of constants formula (1) to find the solution of
$$\mathbf{y}' - A\mathbf{y} = (0, 1, 0)^T, \mathbf{y}(0) = \mathbf{0}.$$

6. Let A be the matrix in Example 1. Use the variation of constants formula (1) to find the solution of
$$\mathbf{y}' - A\mathbf{y} = \begin{pmatrix} 0 \\ e^{-t} \end{pmatrix}, \quad \mathbf{y}(0) = \begin{pmatrix} 1 \\ 0 \end{pmatrix}.$$

7. Find e^{At} given $A = \begin{pmatrix} 0 & 1 \\ -1 & 0 \end{pmatrix}$. What trigonometric identities follow from part (ii) of Theorem 1?
8. The **trace** of a square matrix A, written tr(A), is defined to be the sum of the diagonal entries of A; that is, $\text{tr}(A) = \sum_{i=1}^{n} a_{ii}$.

   (a) Show that the characteristic polynomial of a 2×2 matrix A is
$$|A - \lambda\mathcal{I}| = \lambda^2 - \text{tr}(A)\lambda + |A|.$$

   (b) Show that the characteristic polynomial of a 3×3 matrix A is
$$|A - \lambda\mathcal{I}| = -\lambda^3 + \text{tr}(A)\lambda^2 - c_1\lambda + |A|,$$
   where
$$c_1 = \begin{vmatrix} a_{11} & a_{12} \\ a_{21} & a_{22} \end{vmatrix} + \begin{vmatrix} a_{11} & a_{13} \\ a_{31} & a_{33} \end{vmatrix} + \begin{vmatrix} a_{22} & a_{23} \\ a_{32} & a_{33} \end{vmatrix}.$$

In Problems 9 through 15,

(a) find the eigenvalues of the given matrix A, using one of the results of Problem 8 to write down the characteristic polynomial;

(b) find a fundamental matrix generated by A;

(c) compute e^{At}.

9. $\begin{pmatrix} -1 & 3 \\ 1 & 1 \end{pmatrix}$ **10.** $\begin{pmatrix} -1 & -1 \\ 5 & 1 \end{pmatrix}$ **11.** $\begin{pmatrix} 1 & 3 \\ -2 & -4 \end{pmatrix}$ **12.** $\begin{pmatrix} -2 & 1 \\ 1 & -2 \end{pmatrix}$

13. $\begin{pmatrix} -3 & -1 & -1 \\ -2 & -2 & 2 \\ -1 & 1 & -3 \end{pmatrix}$ **14.** $\begin{pmatrix} 1 & 2 & -2 \\ -2 & -3 & 2 \\ -2 & -2 & 1 \end{pmatrix}$ **15.** $\begin{pmatrix} -4 & 2 & 2 \\ -3 & 1 & 2 \\ -2 & 1 & 0 \end{pmatrix}$

Compute e^{At} for each of the matrices A in Problems 16 through 24.

16. $\begin{pmatrix} -1 & 2 \\ 2 & -1 \end{pmatrix}$ **17.** $\begin{pmatrix} -1 & 2 \\ 2 & -4 \end{pmatrix}$ **18.** $\begin{pmatrix} 0 & 3 \\ -1 & -4 \end{pmatrix}$

19. $\begin{pmatrix} -4 & -9 \\ 4 & -4 \end{pmatrix}$ **20.** $\begin{pmatrix} 2 & -2 - \frac{\pi^2}{2} \\ 2 & -2 \end{pmatrix}$ **21.** $\begin{pmatrix} -10 & 33 \\ -1 & 4 \end{pmatrix}$

22. $\begin{pmatrix} -3 & -1 & -1 \\ -2 & -2 & 2 \\ -1 & 1 & -3 \end{pmatrix}$ **23.** $\begin{pmatrix} 1 & 2 & -2 \\ -2 & -3 & 2 \\ -2 & -2 & 1 \end{pmatrix}$ **24.** $\begin{pmatrix} -4 & 2 & 2 \\ -3 & 1 & 2 \\ -2 & 1 & 0 \end{pmatrix}$

In Problems 25 through 29, compute e^{At} for the given matrix A. In each case, A has at least one deficient eigenvalue.

25. $\begin{pmatrix} 8 & -9 \\ 16 & -16 \end{pmatrix}$ **26.** $\begin{pmatrix} 5 & -\frac{9}{16} \\ 36 & -4 \end{pmatrix}$ **27.** $\begin{pmatrix} -2 & -1 & 1 \\ -1 & 2 & -1 \\ -2 & 6 & -3 \end{pmatrix}$

28. $\begin{pmatrix} -2 & 2 & 2 \\ 2 & 1 & 1 \\ -4 & 1 & 1 \end{pmatrix}$ **29.** $\begin{pmatrix} 0 & 1 & -1 & 0 \\ -1 & 1 & 0 & 1 \\ -1 & 1 & 0 & 0 \\ 0 & -1 & 1 & -2 \end{pmatrix}$

30. This problem provides a partial proof of Theorem 8.5. Suppose that $(\lambda, \mathbf{p})$ is an eigenpair of A.

(a) Show that, if $\mathbf{v}$ satisfies $(A - \lambda\mathcal{I})^2\mathbf{v} = \mathbf{0}$, then $\mathbf{y} = e^{\lambda t}(\mathcal{I} + t(A - \lambda\mathcal{I}))\mathbf{v}$ is a solution of $\mathbf{y}' - A\mathbf{y} = \mathbf{0}$.

(b) Show that, if $\mathbf{v}$ satisfies $(A - \lambda\mathcal{I})^3\mathbf{v} = \mathbf{0}$, then

$$\mathbf{y} = e^{\lambda t}(\mathcal{I} + t(A - \lambda\mathcal{I}) + \frac{1}{2}t^2(A - \lambda\mathcal{I})^2)\mathbf{v}$$

is a solution of $\mathbf{y}' - A\mathbf{y} = \mathbf{0}$.

31. This problem describes a technique for generating the generalized eigenvectors described in Theorem 8.5. It is a technique that avoids the computation of the matrix powers $(A - \lambda\mathcal{I})^j$. Let $(\lambda, \mathbf{p})$ be an eigenpair of A in which λ is a deficient eigenvalue, and suppose that $\mathbf{v}_1$, $\mathbf{v}_2$, $\mathbf{v}_3, \ldots, \mathbf{v}_k$ are nonzero vectors satisfying

$$(A - \lambda\mathcal{I})\mathbf{v}_1 = \mathbf{p} \text{ and } (A - \lambda\mathcal{I})\mathbf{v}_j = \mathbf{v}_{j-1} \text{ for } j = 2, 3, \ldots, k.$$

(a) Show that, for $j = 1, \ldots, k$, $\mathbf{v}_j$ satisfies

$$(A - \lambda\mathcal{I})^j\mathbf{v}_j \neq \mathbf{0}, \quad (A - \lambda\mathcal{I})^{j+1}\mathbf{v}_j = \mathbf{0}.$$

(b) Show that the solution produced by $\mathbf{v}_j$ (see Theorem 8.5) can be written as

$$\mathbf{y} = e^{\lambda t}\left(\frac{t^j}{j!}\mathbf{p} + \frac{t^{j-1}}{(j-1)!}\mathbf{v}_1 + \cdots + t\mathbf{v}_{j-1} + \mathbf{v}_j\right).$$

Each of the matrices A in Problems 32 through 34 has only a single eigenvalue with multiplicity 3 and an order 2 deficiency. Use the method of Problem 31 to find a fundamental matrix generated by A.

32. $\begin{pmatrix} 5 & -1 & -3 \\ -2 & 0 & 2 \\ 1 & -3 & 1 \end{pmatrix}$ **33.** $\begin{pmatrix} -2 & 2 & 1 \\ 2 & -1 & -2 \\ -1 & 2 & 0 \end{pmatrix}$ **34.** $\begin{pmatrix} 1 & 6 & 3 \\ 2 & -2 & 2 \\ -3 & -6 & -5 \end{pmatrix}$

In Problems 35 through 38, you are given a matrix A along with its characteristic polynomial. Find a fundamental matrix generated by A.

35. $\begin{pmatrix} 0 & 0 & -2 & 3 \\ 2 & 1 & 1 & -2 \\ -1 & 0 & 0 & 2 \\ -1 & 0 & -1 & 3 \end{pmatrix}$; $(\lambda - 1)^4$ **36.** $\begin{pmatrix} 1 & -3 & 3 & -4 \\ -2 & 0 & -1 & 2 \\ 0 & -1 & 0 & -1 \\ 2 & -2 & 2 & -4 \end{pmatrix}$; $\lambda(\lambda + 1)^3$

37. $\begin{pmatrix} -1 & 1 & -1 & 1 \\ -1 & 1 & -2 & 1 \\ -1 & 0 & -1 & 0 \\ 0 & -1 & 1 & -1 \end{pmatrix}$; $\lambda^2(\lambda + 1)^2$ **38.** $\begin{pmatrix} 1 & -1 & -1 & -1 \\ 1 & 0 & 0 & -1 \\ 0 & -1 & 0 & 1 \\ 1 & 0 & -1 & -1 \end{pmatrix}$; $(\lambda^2 + 1)^2$

8.6 e^{At} and the Cayley-Hamilton Theorem

In this section we present a method for computing e^{At} that does not require the computation of eigenvectors—although we will still need to compute the eigenvalues of A. At the heart of this method lies the following centerpiece from linear algebra known as the *Cayley-Hamilton theorem.*

Theorem 1

Let A be a constant $n \times n$ matrix and let $\chi(\lambda) = |A - \lambda\mathcal{I}|$ be the characteristic polynomial of A. Then

$$\chi(A) = 0,$$

where the zero on the right side represents the $n \times n$ zero matrix ▲.

The characteristic polynomial χ always has degree n, so suppose that

$$\chi(\lambda) = \lambda^n + c_{n-1}\lambda^{n-1} + \cdots + c_1\lambda + c_0.$$

Theorem 1 says then that

$$A^n + c_{n-1}A^{n-1} + \cdots + c_1 A + c_0\mathcal{I} = 0,$$

which we may rewrite as

$$A^n = -c_{n-1}A^{n-1} - \cdots - c_1 A - c_0\mathcal{I}.$$

The upshot of this is that A^n (and therefore any higher power of A) can always be expressed as a linear combination of the lesser powers A^{n-1}, A^{n-2}, . . . , A, $\mathcal{I}$. (Note that $\mathcal{I} = A^0$.) As a result we have the following corollary.

COROLLARY 1 *Any linear combination of powers of A can be expressed as a linear combination of A^{n-1}, A^{n-2}, . . . , A, $\mathcal{I}$.*

Example 1

Let's compute $A^4 + A^3 + A^2 + A$ for the 2×2 matrix

$$A = \begin{pmatrix} 1 & 2 \\ 3 & -2 \end{pmatrix}.$$

The characteristic polynomial is

$$\chi(\lambda) = \begin{vmatrix} 1-\lambda & 2 \\ 3 & -2-\lambda \end{vmatrix} = \lambda^2 + \lambda - 8.$$

By Theorem 1,

$$A^2 + A - 8\mathcal{I} = 0,$$

and so $A^2 = -A + 8\mathcal{I}$, which is easily verified. Furthermore,

$$A^3 = -A^2 + 8A = -(-A + 8\mathcal{I}) + 8A = 9A - 8\mathcal{I},$$

and

$$A^4 = 9A^2 - 8A = 9(-A + 8\mathcal{I}) - 8A = -17A + 72\mathcal{I}.$$

Therefore,

$$A^4 + A^3 + A^2 + A = -8A + 72\mathcal{I} = \begin{pmatrix} 64 & -16 \\ -24 & 88 \end{pmatrix}.$$ ■

A limiting argument based on the Cayley-Hamilton theorem can be made in order to establish the following result.

Theorem 2

Let A be an $n \times n$ matrix and suppose that φ is a function on $\mathbb{R}$ whose Maclaurin series expansion

$$\varphi(x) = \varphi(0) + \varphi'(0)x + \frac{\varphi''(0)}{2!}x^2 + \frac{\varphi'''(0)}{3!}x^3 + \cdots$$

converges for all x in $\mathbb{R}$. Then for each t in $\mathbb{R}$ the matrix series

$$\varphi(0)\mathcal{I} + \varphi'(0)tA + \frac{\varphi''(0)}{2!}t^2A^2 + \frac{\varphi'''(0)}{3!}t^3A^3 + \cdots$$

converges to an $n \times n$ matrix, to which we assign the symbol $\varphi(tA)$. Moreover, there is a polynomial ψ of degree at most $n - 1$, whose coefficients depend on t, such that $\varphi(tA) = \psi(A)$ for all t.* ▲

For our purposes here, the most important implication of Theorem 2 is as follows:

COROLLARY 2 *There is a polynomial γ of degree at most $n - 1$, with coefficients depending upon t, such that $e^{At} = \gamma(A)$.*

So we now focus our efforts on finding the polynomial γ whose existence is asserted by Corollary 2. Let's begin by writing

$$\gamma(A) = c_0(t) + c_1(t)A + c_2(t)A^2 + \cdots + c_{n-1}(t)A^{n-1} = e^{At} \tag{1}$$

and assuming that A has eigenpairs $(\lambda_1, \mathbf{p}_1), (\lambda_2, \mathbf{p}_2), \ldots, (\lambda_n, \mathbf{p}_n)$, where the λ_i are distinct. For each i, we know that

$$e^{At}\mathbf{p}_i = e^{\lambda_i t}\mathbf{p}_i, \tag{2}$$

because each side of the equation is the solution of $\mathbf{y}' - A\mathbf{y} = \mathbf{0}$ with $\mathbf{y}(0) = \mathbf{p}_i$. Furthermore, for $k = 1, 2, 3, \ldots,$

$$A^k\mathbf{p}_i = \lambda_i^k\mathbf{p}_i, \tag{3}$$

* This means that the individual entries in the partial sums of the series each converge.

because $A^k\mathbf{p}_i = A^{k-1}\lambda_i\mathbf{p}_i = A^{k-2}\lambda_i^2\mathbf{p}_i = \cdots = \lambda_i^k\mathbf{p}_i$. We now equate $e^{At}\mathbf{p}_i$ with $\gamma(A)\mathbf{p}_i$, using (1), (2), and (3), to obtain n linear equations of the form

$$c_0(t) + c_1(t)\lambda_i + c_2(t)\lambda_i^2 + \cdots + c_{n-1}(t)\lambda_i^{n-1} = e^{\lambda_i t}, \quad i = 1, 2, \ldots, n \tag{4}$$

for the coefficients $c_0(t), c_1(t), \ldots, c_{n-1}(t)$. In matrix form, this system of equations becomes

$$\begin{pmatrix} 1 & \lambda_1 & \lambda_1^2 & \cdots & \lambda_1^{n-1} \\ 1 & \lambda_2 & \lambda_2^2 & \cdots & \lambda_2^{n-1} \\ 1 & \lambda_3 & \lambda_3^2 & \cdots & \lambda_3^{n-1} \\ \vdots & \vdots & \vdots & & \vdots \\ 1 & \lambda_n & \lambda_n^2 & \cdots & \lambda_n^{n-1} \end{pmatrix} \begin{pmatrix} c_0(t) \\ c_1(t) \\ c_2(t) \\ \vdots \\ c_{n-1}(t) \end{pmatrix} = \begin{pmatrix} e^{\lambda_1 t} \\ e^{\lambda_2 t} \\ e^{\lambda_3 t} \\ \vdots \\ e^{\lambda_n t} \end{pmatrix}.$$

The coefficient matrix, an example of a *Vandermonde matrix*, is guaranteed to be invertible, provided that the λ_i's are distinct. Therefore, the system has a unique solution $(c_0(t), c_1(t), \ldots, c_{n-1}(t))^T$ for each t.

Example 2

Suppose that A is a 2×2 matrix with characteristic polynomial $\chi(\lambda) = \lambda^2 + 3\lambda + 2$. The eigenvalues of A are $\lambda_1 = -1$ and $\lambda_2 = -2$; so the system of equations of interest is

$$\begin{pmatrix} 1 & -1 \\ 1 & -2 \end{pmatrix} \begin{pmatrix} c_0(t) \\ c_1(t) \end{pmatrix} = \begin{pmatrix} e^{-t} \\ e^{-2t} \end{pmatrix}.$$

Inversion of the matrix easily shows that

$$\begin{pmatrix} c_0(t) \\ c_1(t) \end{pmatrix} = \begin{pmatrix} 2 & -1 \\ 1 & -1 \end{pmatrix} \begin{pmatrix} e^{-t} \\ e^{-2t} \end{pmatrix} = \begin{pmatrix} 2e^{-t} - e^{-2t} \\ e^{-t} - e^{-2t} \end{pmatrix}.$$

Therefore, the polynomial γ is given by

$$\gamma(x) = 2e^{-t} - e^{-2t} - (e^{-t} - e^{-2t})x,$$

and the matrix exponential is

$$e^{At} = \gamma(A) = (2e^{-t} - e^{-2t})\mathcal{I} - (e^{-t} - e^{-2t})A,$$

which could be carried further, were we given the entries of A. (Note that all of this was done without any knowledge of A other than its eigenvalues.) With the help of Theorem 1, which tells us that $A^2 = -3A - 2\mathcal{I}$, this result can be easily verified by showing that $Y(t) = e^{At}$ satisfies $Y' - AY = 0$ and $Y(0) = \mathcal{I}$. ■

The argument preceding Example 2 gives a method for finding γ that amounts to solving an $n \times n$ system of equations. More important, however, it casts the problem as one of finding an *interpolating polynomial*. We summarize this result as follows:

Theorem 3

If A has distinct eigenvalues $\lambda_1, \lambda_2, \dots, \lambda_n$, then the coefficients of γ are uniquely determined by the equations

$$\gamma(\lambda_1) = e^{\lambda_1 t}, \quad \gamma(\lambda_2) = e^{\lambda_2 t}, \dots, \quad \gamma(\lambda_n) = e^{\lambda_n t}.$$

That is, for each fixed t, γ is the unique polynomial in x that agrees with e^{xt} at $x = \lambda_1, \dots, \lambda_n$. ▲

There are techniques for polynomial interpolation that are far more efficient in general than solving an $n \times n$ system of equations involving a Vandermonde matrix. A simple technique that will serve our present needs nicely is interpolation by *Lagrange polynomials.*

Lagrange Interpolation Consider for a moment the general problem of finding a polynomial ψ of degree $n-1$ (or less) whose graph contains the points (x_1, y_1), $(x_2, y_2), \dots, (x_n, y_n)$ in which $x_1, x_2, \dots, x_n$ are distinct. For each $i = 1, 2, \dots, n$ define a degree $n-1$ polynomial ℓ_i by

$$\ell_i(x) = \frac{(x - x_1)\cdots(x - x_{i-1})(x - x_{i+1})\cdots(x - x_n)}{(x_i - x_1)\cdots(x_i - x_{i-1})(x_i - x_{i+1})\cdots(x_i - x_n)}.$$

These are the **fundamental polynomials** associated with $x_1, x_2, \dots, x_n$. Notice that the numerator of $\ell_i(x)$ contains each of the factors $(x - x_j)$ except $(x - x_i)$ and that the denominator is just the numerator evaluated at x_i. Figure 1 shows the graph of ℓ_3, based on $x_1 = 1$, $x_2 = 2$, $x_3 = 3$, $x_4 = 4$, $x_5 = 5$:

$$\ell_3(x) = \frac{(x-1)(x-2)(x-4)(x-5)}{(3-1)(3-2)(3-4)(3-5)} = \frac{1}{4}(x-1)(x-2)(x-4)(x-5).$$

Lagrange observed that such polynomials are useful in the interpolation problem, because each ℓ_i has the property that

$$\ell_i(x_i) = 1, \quad \text{and} \quad \ell_i(x_j) = 0 \text{ if } j \neq i.$$

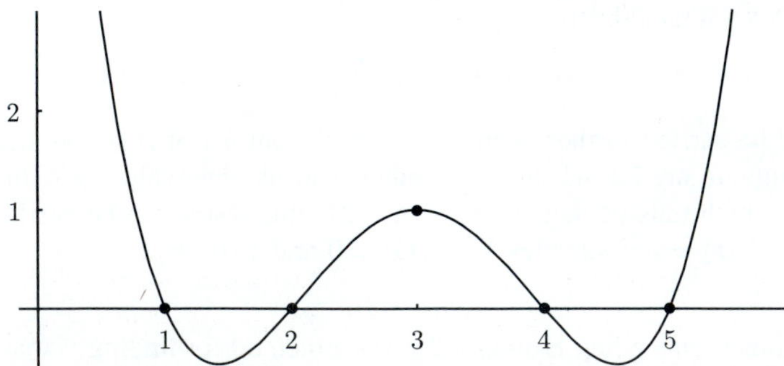

Figure 1

In fact, the polynomial we seek is precisely

$$\psi(x) = \mathbf{y}_1\ell_1(x) + y_2\ell_2(x) + \cdots + y_n\ell_n(x).$$

This is the so-called *Lagrange form* of the interpolating polynomial.

Example 3

Let's find the cubic polynomial ψ that interpolates the points $(0, 1)$, $(1, 0)$, $(3, 2)$, and $(4, -1)$. We first build fundamental polynomials:

$$\begin{aligned}
\ell_1(x) &= \frac{(x-1)(x-3)(x-4)}{-12}, \\
\ell_2(x) &= \frac{x(x-3)(x-4)}{6}, \\
\ell_3(x) &= \frac{x(x-1)(x-4)}{-6}, \\
\ell_4(x) &= \frac{x(x-1)(x-3)}{12},
\end{aligned}$$

and then compute

$$\begin{aligned}
\psi(x) &= 1\ell_1(x) + 0\ell_2(x) + 2\ell_3(x) - 1\ell_4(x) \\
&= -\frac{1}{12}(x-1)(x-3)(x-4) - \frac{1}{3}x(x-1)(x-4) - \frac{1}{12}x(x-1)(x-3) \\
&= -\frac{1}{6}(3x^3 - 16x^2 + 19x - 6).
\end{aligned}$$

■

In the next two examples, we'll use Lagrange interpolation to help us compute a matrix exponential.

Example 4

The 3×3 matrix

$$A = \frac{1}{4}\begin{pmatrix} -8 & -4 & 4 \\ 1 & -3 & 1 \\ 5 & 5 & -7 \end{pmatrix}$$

has the characteristic polynomial $\chi(\lambda) = -\frac{1}{2}(2\lambda^3 + 9\lambda^2 + 10\lambda + 3)$ and therefore eigenvalues $\lambda_1 = -1$, $\lambda_2 = -\frac{1}{2}$, and $\lambda_3 = -3$. To compute e^{At} we need a quadratic polynomial $\gamma(x)$ that satisfies

$$\gamma(-1) = e^{-t}, \quad \gamma(-1/2) = e^{-t/2}, \quad \gamma(-3) = e^{-3t}.$$

So we construct fundamental polynomials:

$$\ell_1(x) = \frac{(x+1/2)(x+3)}{-1} = -\frac{1}{2}(2x+1)(x+3),$$
$$\ell_2(x) = \frac{(x+1)(x+3)}{5/4} = \frac{4}{5}(x+1)(x+3),$$
$$\ell_3(x) = \frac{(x+1)(x+1/2)}{5} = \frac{1}{10}(x+1)(2x+1),$$

and then compute

$$\begin{aligned}\gamma(x) &= e^{-t}\ell_1(x) + e^{-t/2}\ell_2(x) + e^{-3t}\ell_3(x)\\ &= -\frac{1}{2}e^{-t}(2x+1)(x+3) + \frac{4}{5}e^{-t/2}(x+1)(x+3) + \frac{1}{10}e^{-3t}(x+1)(2x+1).\end{aligned}$$

Therefore,

$$\begin{aligned}e^{At} &= -\frac{e^{-t}}{2}(2A+\mathcal{I})(A+3\mathcal{I}) + \frac{4e^{-t/2}}{5}(A+\mathcal{I})(A+3\mathcal{I})\\ &\quad + \frac{e^{-3t}}{10}(A+\mathcal{I})(2A+\mathcal{I}),\end{aligned}$$

which after some (*CAS*-aided) computation becomes

$$\begin{aligned}e^{At} &= \frac{e^{-t}}{2}\begin{pmatrix} 1 & -1 & 1\\ -1 & 1 & -1\\ 0 & 0 & 0\end{pmatrix} + \frac{e^{-t/2}}{2}\begin{pmatrix} 0 & 0 & 0\\ 1 & 1 & 1\\ 1 & 1 & 1\end{pmatrix} + \frac{e^{-3t}}{2}\begin{pmatrix} 1 & 1 & -1\\ 0 & 0 & 0\\ -1 & -1 & 1\end{pmatrix}\\ &= \frac{1}{2}\begin{pmatrix} e^{-t}+e^{-3t} & -e^{-t}+e^{-3t} & e^{-t}-e^{-3t}\\ -e^{-t}+e^{-t/2} & e^{-t}+e^{-t/2} & -e^{-t}+e^{-t/2}\\ e^{-t/2}-e^{-3t} & e^{-t/2}-e^{-3t} & e^{-t/2}+e^{-3t}\end{pmatrix}.\end{aligned}$$

■

Example 5

The 2×2 matrix $A = \begin{pmatrix} 1 & -6\\ 3 & -5\end{pmatrix}$ has the characteristic polynomial $\chi(\lambda) = \lambda^2 + 4\lambda + 13$ and therefore nonreal complex eigenvalues $\lambda_1 = -2+3i$ and $\lambda_2 = -2-3i$. To compute e^{At} we need a linear polynomial $\gamma(x)$ that satisfies

$$\gamma(-2+3i) = e^{(-2+3i)t}, \quad \gamma(-2-3i) = e^{(-2-3i)t}.$$

So we construct the fundamental polynomials

$$\ell_1(x) = \frac{x-(-2-3i)}{6i} = -\frac{1}{6}i(x+2+3i),$$
$$\ell_2(x) = \frac{x-(-2+3i)}{-6i} = \frac{1}{6}i(x+2-3i),$$

and then compute

$$\begin{aligned}\gamma(x) &= e^{(-2+3i)t}\ell_1(x) + e^{(-2-3i)t}\ell_2(x)\\ &= -\frac{1}{6}e^{(-2+3i)t}i(x+2+3i) + \frac{1}{6}e^{(-2-3i)t}i(x+2-3i).\end{aligned}$$

We arrive at the following real form for $\gamma(x)$ by applying Euler's formula and simplifying:

$$\begin{aligned}\gamma(x) &= -\frac{ie^{-2t}}{6}((\cos 3t + i\sin 3t)(x+2+3i)\\ &\quad -(\cos 3t - i\sin 3t)(x+2-3i))\\ &= \frac{e^{-2t}}{3}(3\cos 3t + (x+2)\sin 3t)\end{aligned}$$

Therefore,

$$e^{At} = \frac{e^{-2t}}{3}(3\cos 3t\mathcal{I} + (A+2\mathcal{I})\sin 3t),$$

which after some computation becomes

$$e^{At} = e^{-2t}\begin{pmatrix}\cos 3t + \sin 3t & -2\sin 3t\\ \sin 3t & \cos 3t - \sin 3t\end{pmatrix}.$$

■

Repeated Eigenvalues When A has repeated eigenvalues (deficient or not), the preceding method fails, because there are not enough equations to determine the required polynomial. This can be remedied by including appropriate derivative information.

Theorem 4

Suppose that A has eigenvalues $\lambda_1, \lambda_2, \ldots, \lambda_k$, with multiplicities $m_1, m_2, \ldots, m_k$, respectively. Then the coefficients of γ are uniquely determined by n equations consisting, for $i = 1, 2, \ldots, k$, of

$$\gamma(\lambda_i) = e^{\lambda_i t}, \gamma'(\lambda_i) = te^{\lambda_i t}, \gamma''(\lambda_i) = t^2e^{\lambda_i t}, \ldots, \gamma^{(m_i-1)}(\lambda_i) = t^{m_i-1}e^{\lambda_i t}.$$

▲

Example 6

Consider the 3×3 matrix

$$A = \begin{pmatrix}-2 & -1 & 1\\ -1 & 2 & -1\\ -2 & 6 & -3\end{pmatrix},$$

whose characteristic polynomial turns out to be

$$\chi(\lambda) = (\lambda+1)^3.$$

So the only eigenvalue of A is $\lambda_1 = -1$, with multiplicity $m_1 = 3$. Therefore, the polynomial γ will be determined by

$$\gamma(-1) = e^{-t}, \quad \gamma'(-1) = te^{-t}, \quad \gamma''(-1) = t^2e^{-t}.$$

If we let c_0, c_1, c_2 be the (t-dependent) coefficients of γ, that is,

$$\gamma(x) = c_0 + c_1 x + c_2 x^2,$$

then those equations become

$$c_0 - c_1 + c_2 = e^{-t}, \quad c_1 - 2c_2 = te^{-t}, \quad 2c_2 = t^2 e^{-t},$$

which are easily solved by back-substitution to yield

$$c_0 = \frac{e^{-t}}{2}(t^2 + 2t + 2), \quad c_1 = e^{-t}(t^2 + t), \quad c_2 = \frac{e^{-t}}{2}t^2.$$

Therefore,

$$\gamma(x) = \frac{e^{-t}}{2}((t^2 + 2t + 2) + 2(t^2 + t)x + t^2 x^2),$$

and so

$$e^{At} = \frac{e^{-t}}{2}((t^2 + 2t + 2)\mathcal{I} + 2(t^2 + t)A + t^2 A^2).$$

Further computation reveals, finally, that

$$e^{At} = e^{-t}\begin{pmatrix} 1-t & -t+2t^2 & t-t^2 \\ -t & 1+3t+2t^2 & -t-t^2 \\ -2t & 6t+4t^2 & 1-2t-2t^2 \end{pmatrix}. \qquad \blacksquare$$

PROBLEMS

1. Let A be a 2×2 matrix with eigenvalues λ_1 and λ_2.

(a) Given that $\lambda_1 \neq \lambda_2$, derive the formula

$$e^{At} = \frac{1}{\lambda_1 - \lambda_2}(e^{\lambda_1 t}(A - \lambda_2 \mathcal{I}) - e^{\lambda_2 t}(A - \lambda_1 \mathcal{I})).$$

(b) Given that λ_1 and λ_2 are complex conjugates $\alpha \pm \beta i$, show that the formula in part (a) becomes

$$e^{At} = \frac{e^{\alpha t}}{\beta}(\beta \cos \beta t \mathcal{I} + \sin \beta t (A - \alpha \mathcal{I})).$$

(c) Given that $\lambda_1 = \lambda_2$, derive the formula $e^{At} = e^{\lambda_1 t}(tA + (1 - \lambda_1 t)\mathcal{I})$.

2. Let A be a 2×2 matrix with (possibly repeated) eigenvalues λ_1 and λ_2.

(a) Use the Cayley-Hamilton theorem to show that $A^2 = (\lambda_1 + \lambda_2)A - \lambda_1 \lambda_2 \mathcal{I}$.

(b) For each of the formulas in Problem 1, verify that $e^{A0} = \mathcal{I}$ and, with the help of part (a), verify that $\frac{d}{dt}e^{At} = Ae^{At}$.

For each matrix A in Problems 3 through 6, use the appropriate formula from Problem 1 to compute e^{At}.

3. $\begin{pmatrix} 1 & 3 \\ 3 & 1 \end{pmatrix}$

4. $\begin{pmatrix} -1 & 2 \\ -2 & -1 \end{pmatrix}$

5. $\begin{pmatrix} -4 & 1 \\ 1 & -4 \end{pmatrix}$

6. $\begin{pmatrix} 1 & -3 \\ 3 & -5 \end{pmatrix}$

7. Let $A = \begin{pmatrix} -1 & 2 \\ -2 & -1 \end{pmatrix}$. (See Problem 4.) Use the variation of constants formula (1) in Section 8.5 to find the solution of the initial-value problem

$$\mathbf{y}' - A\mathbf{y} = e^{-t}\begin{pmatrix} \sin 2t \\ \cos 2t \end{pmatrix}, \quad \mathbf{y}(0) = \begin{pmatrix} 1 \\ 0 \end{pmatrix}.$$

In Problems 8 through 11, find e^{At} in terms of A, given that A is a 3×3 matrix with the given eigenvalues.

8. $0, -1, -2$

9. $-1, -2 \pm i$

10. $-1, -2, -2$

11. $-2, -2, -2$

12. Suppose that A is a 3×3 matrix with eigenvalues $-1, -1, 2$. Use the Cayley-Hamilton theorem to express A^5 as a linear combination of A^2, A, and $\mathcal{I}$.

In Problems 13 through 16, assume that A satisfies the given equation. Compute e^{At} in terms of A by means of its series representation.

13. $A^2 = A$

14. $A^2 = -A$

15. $A^2 = -\mathcal{I}$

16. $A^3 = A$

8.7 Asymptotic Stability

We now turn our attention to questions concerning the *asymptotic behavior*—that is, the behavior as $t \to \infty$—of solutions of

$$\mathbf{y}' - A\mathbf{y} = \mathbf{f}, \tag{1}$$

where the entries of A and $\mathbf{f}$ are continuous for all t in $[0, \infty)$. We allow the case $n = 1$, in which $\mathbf{y} = y$, $A = a$, and $\mathbf{f} = f$ are scalar valued.

In order to state a definition of *asymptotic stability*, we first need to recall the notation $\|\mathbf{x} - \mathbf{a}\|$ for the distance from $\mathbf{x}$ to $\mathbf{a}$, which is defined by

$$\|\mathbf{x} - \mathbf{a}\| = \left(\sum_{i=1}^{n} (x_i - a_i)\right)^{1/2}.$$

Note that in $\mathbb{R}$ (i.e., when $n = 1$) $\|\mathbf{x} - \mathbf{a}\|$ reduces to $|x - a|$. In $\mathbb{R}^2$ and $\mathbb{R}^3$, $\|\mathbf{x} - \mathbf{a}\|$ is the usual distance formula.

Definition Let $\tilde{\mathbf{y}}$ be a particular solution of (1). If there is some number $\varepsilon > 0$ such that

$$\lim_{t \to \infty} \|\mathbf{y}(t) - \tilde{\mathbf{y}}(t)\| = 0$$

for all solutions $\mathbf{y}$ with $\|\mathbf{y}(0) - \tilde{\mathbf{y}}(0)\| < \varepsilon$, then $\tilde{\mathbf{y}}$ is said to be **asymptotically stable**. ◆

This definition essentially says that an asymptotically stable solution is one with the property that solutions sufficiently "close" to it at $t = 0$ become closer and closer as $t \to \infty$. *Because the system is linear*, it turns out that if a solution $\tilde{\mathbf{y}}$ is asymptotically stable, then

$$\lim_{t\to\infty} \|\mathbf{y}(t) - \tilde{\mathbf{y}}(t)\| = 0$$

for *all* solutions $\mathbf{y}$, not just solutions whose initial values are close to $\tilde{\mathbf{y}}(0)$. Moreover, if the system has one asymptotically stable solution, then *all* of its solutions are asymptotically stable. These facts are results of the following theorem, whose proof is the subject of Problem 15.

Theorem 1

If every solution of the homogeneous equation $\mathbf{y}' - A\mathbf{y} = \mathbf{0}$ approaches $\mathbf{0}$ as $t \to \infty$, then all solutions of equation (1) are asymptotically stable; otherwise no solution of (1) is asymptotically stable. ▲

Example 1

Consider the equation

$$y' - (\cos t - 1/4)y = -4\sin t.$$

Let $\tilde{y}$ be a given solution, and let y be any other solution. Their difference $u = y - \tilde{y}$ will satisfy

$$u' - (\cos t - 1/4)u = 0, \quad u(0) = u_0 = y(0) - \tilde{y}(0),$$

whose solutions is

$$u = u_0 e^{\sin t - t/4} = u_0 e^{\sin t} e^{-t/4}.$$

It follows that

$$\lim_{t\to\infty} |y(t) - \tilde{y}(t)| = \lim_{t\to\infty} |u(t)| = 0;$$

therefore, $\tilde{y}$ is asymptotically stable. Since $\tilde{y}$ represents an arbitrary solution, we conclude that *every* solution is asymptotically stable. This fact is illustrated by the solutions plotted in Figure 1, all of which apparently approach a certain nonconstant, periodic solution, which begs description as a *periodic attractor*. ■

The behavior that we observed in Example 1 is actually typical of linear, scalar, first-order equations with asymptotically stable solutions. Consider

$$y' - py = f, \tag{2}$$

where p and f are continuous functions on $[0, \infty)$. Given any two solutions y and $\tilde{y}$, their difference $u = y - \tilde{y}$ will satisfy the homogeneous equation

$$u' - pu = 0. \tag{3}$$

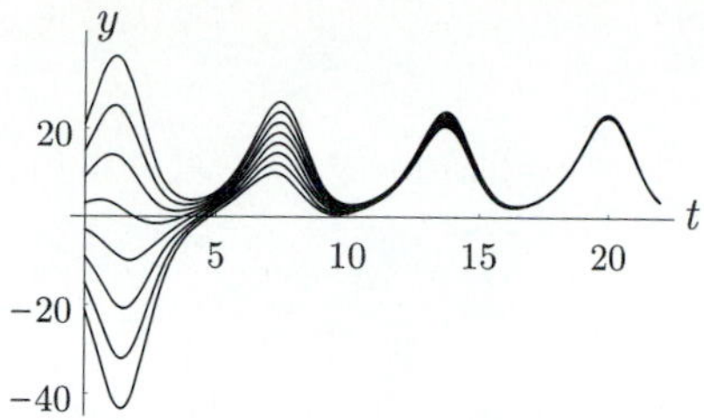

Figure 1

Therefore, every pair of solutions y and $\tilde{y}$ of (2) will have the property that

$$\lim_{t\to\infty} |y(t) - \tilde{y}(t)| = 0,$$

if and only if every solution of the homogeneous equation (3) approaches zero as $t \to \infty$. Since every solution of (2) is of the form

$$u = u(0)e^{P(t)}, \quad \text{where} \quad P(t) = \int_0^t p(s)ds,$$

it follows that every solution u of (3) approaches 0 as $t \to \infty$, if and only if $P(t) \to -\infty$ as $t \to \infty$. So we conclude that *every* solution of (2) is asymptotically stable, if

$$\lim_{t\to\infty} \int_0^t p(s)ds = -\infty;$$

otherwise *no* solution of (2) is asymptotically stable.

Example 2

Consider the system

$$\mathbf{y}' - A\mathbf{y} = \begin{pmatrix} \cos t \\ \sin 2t \end{pmatrix},$$

where

$$A = \frac{1}{1+t^2}\begin{pmatrix} t - t^2 - 1 & -1 \\ 1 & t - t^2 - 1 \end{pmatrix}.$$

It is simple to verify that the evolution matrix generated by A (at $t = 0$) is

$$\Phi_0(t) = \begin{pmatrix} e^{-t} & -te^{-t} \\ te^{-t} & e^{-t} \end{pmatrix}.$$

Since each entry in $\Phi_0(t)$ approaches 0 as $t \to \infty$, it follows that every solution of the homogeneous equation $\mathbf{y}' - A\mathbf{y} = \mathbf{0}$ approaches $\mathbf{0}$ as $t \to \infty$. (Why?) Therefore, all solutions of the nonhomogeneous system are asymptotically stable. Figure 2 shows phase-plane orbits of two solutions with initial points (0, 0.5) and (0.5, −0.5). Notice

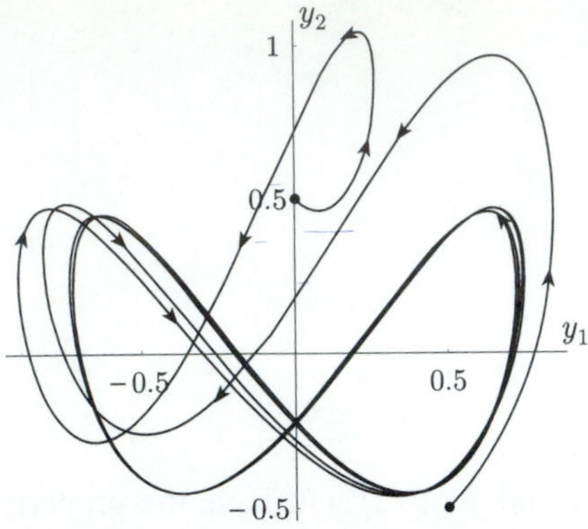

Figure 2

that, as $t \to \infty$, they both approach the same closed curve (which is not the orbit of any solution of the system). ■

We now turn our attention to equation (1) under the assumption that A is a *constant* $n \times n$ matrix and $\mathbf{f}$ is continuous on $[0, \infty)$. We have seen in this chapter that solutions of the homogeneous equation $\mathbf{y}' - A\mathbf{y} = \mathbf{0}$ can be constructed from the eigenvalues of A, which are the roots of the characteristic polynomial

$$\chi(\lambda) = |A - \lambda \mathcal{I}|.$$

Since the solution produced by each eigenvalue λ_i has $e^{\lambda_i t}$ as a factor that is dominant as $t \to \infty$, it is a relatively simple matter to prove the following:

COROLLARY 1 *Let A have constant entries. If the real part of every eigenvalue of A is negative, then all solutions of (1) are asymptotically stable; otherwise no solution of (1) is asymptotically stable.*

Example 3

Consider the system $\mathbf{y}' - A\mathbf{y} = \mathbf{f}$, where

$$A = \frac{1}{3}\begin{pmatrix} 8 & -18 \\ 9 & -10 \end{pmatrix}.$$

The eigenvalues of A are $-\frac{1}{3} \pm 3i$; hence each has negative real part. Therefore, solutions of $\mathbf{y}' - A\mathbf{y} = \mathbf{f}$ are asymptotically stable for any choice of $\mathbf{f}$. Figure 3 shows the graphs of y_1 and y_2 corresponding to the constant nonhomogeneous term $\mathbf{f}(t) = (2, 1)^T$ (and $\mathbf{y}(0) = (-1, 0)^T$). Note that y_1 and y_2 approach constants as $t \to \infty$. Figure 4 shows the phase-plane orbit of the same solution $\mathbf{y} = (y_1, y_2)^T$.

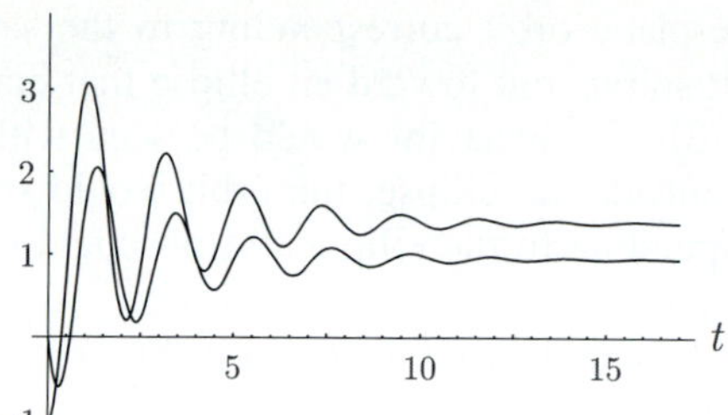

Figure 3

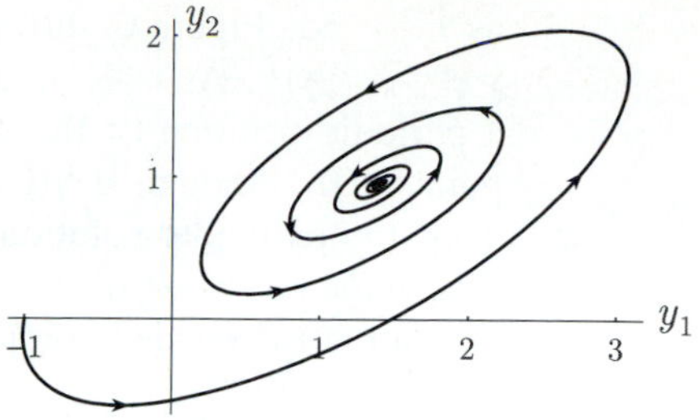

Figure 4

Here we see that the orbit spirals in toward an equilibrium point as $t \to \infty$. Similar behavior would be seen with any initial point $\mathbf{y}(0)$. ■

When the nonhomogeneous term $\mathbf{f}$ is constant, as was the case in Example 4, an isolated equilibrium point—if one exists—clearly must be the unique constant solution $\mathbf{y}^*$ satisfying

$$-A\mathbf{y}^* = \mathbf{f}.$$

(Note that if solutions are asymptotically stable, then A is invertible. Why?) In the case of a homogeneous system (i.e., $\mathbf{f} = \mathbf{0}$), this gives us $\mathbf{y}^* = \mathbf{0}$. When $\mathbf{f}$ is not constant, one does not expect any equilibrium points to exist.

Example 4

Consider the system

$$\mathbf{y}' - \frac{1}{3}\begin{pmatrix} 8 & -18 \\ 9 & -10 \end{pmatrix}\mathbf{y} = \begin{pmatrix} \cos 3t \\ \sin 3t \end{pmatrix}, \quad \mathbf{y}(0) = \mathbf{0}.$$

The matrix here is the same as in Example 4. Figure 5 shows the graphs of y_1 and y_2 corresponding to $\mathbf{y}(0) = (0, 0)^T$. Here, y_1 and y_2 approach periodic functions

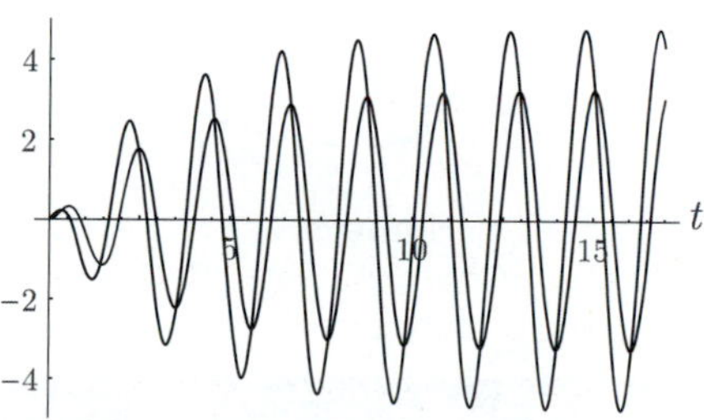

Figure 5

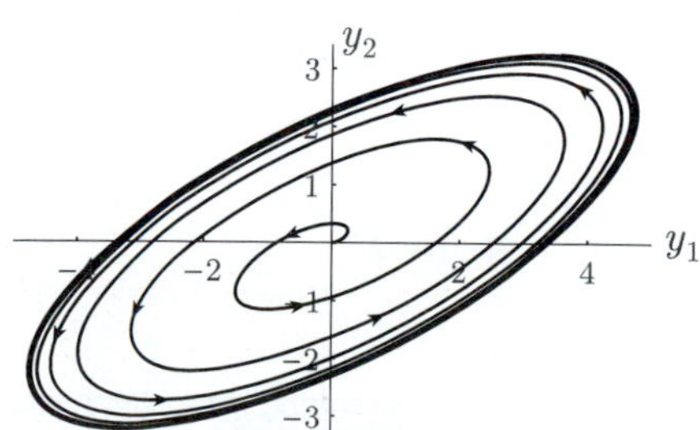

Figure 6

as $t \to \infty$. Figure 6 shows phase-plane orbit corresponding to the same solution $\mathbf{y} = (y_1, y_2)^T$. As $t \to \infty$ the orbit spirals out toward an ellipse that corresponds to a periodic solution of the system. Similar behavior would be seen with any initial point $\mathbf{y}(0)$, although if $\mathbf{y}(0)$ were outside the ellipse, the orbit would spiral inward. Thus the particular solution corresponding to the elliptical orbit could be viewed as a *periodic attractor* of all other solutions.

Some *CAS*-aided computation reveals that the solution in Figures 5 and 6 is

$$\mathbf{y} = \frac{3}{325}\begin{pmatrix} 496\left(1 - e^{-t/3}\right)\cos 3t + \left(153 - 172e^{-t/3}\right)\sin 3t \\ 162\left(1 - e^{-t/3}\right)\cos 3t + 2\left(158 - 167e^{-t/3}\right)\sin 3t \end{pmatrix}.$$

Since the eigenvalues of A are $-\frac{1}{3} \pm 3i$, we know that solutions of the homogeneous system $\mathbf{y}' - A\mathbf{y} = \mathbf{0}$ will have $e^{-t/3}$ as a factor. So we separate the solution into two terms as follows:

$$\mathbf{y} = -\frac{3e^{-t/3}}{325}\begin{pmatrix} 496\cos 3t + 172\sin 3t \\ 162\cos 3t + 334\sin 3t \end{pmatrix} + \frac{3}{325}\begin{pmatrix} 496\cos 3t + 153\sin 3t \\ 162\cos 3t + 316\sin 3t \end{pmatrix}.$$

The first term, which satisfies the homogeneous system $\mathbf{y}' - A\mathbf{y} = \mathbf{0}$, is sometimes called the *transient* part of the solution, and the second term is precisely the periodic attractor:

$$\mathbf{y}^* = \frac{3}{325}\begin{pmatrix} 496\cos 3t + 153\sin 3t \\ 162\cos 3t + 316\sin 3t \end{pmatrix}.$$

Figure 7 shows the graphs of y_1^* and y_2^* versus t, and Figure 8 shows the phase-plane orbit for $\mathbf{y}^*$. ■

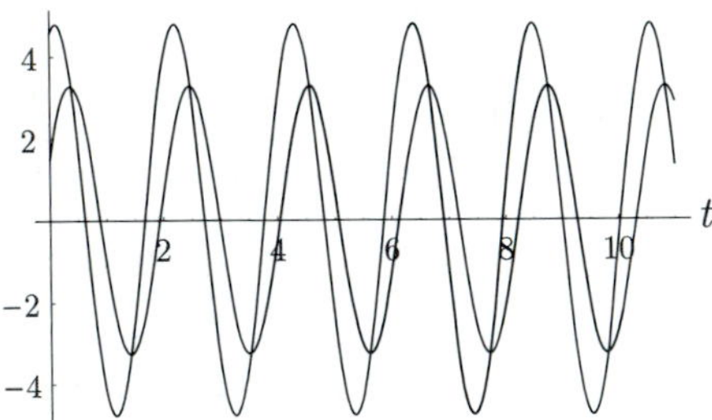

Figure 7

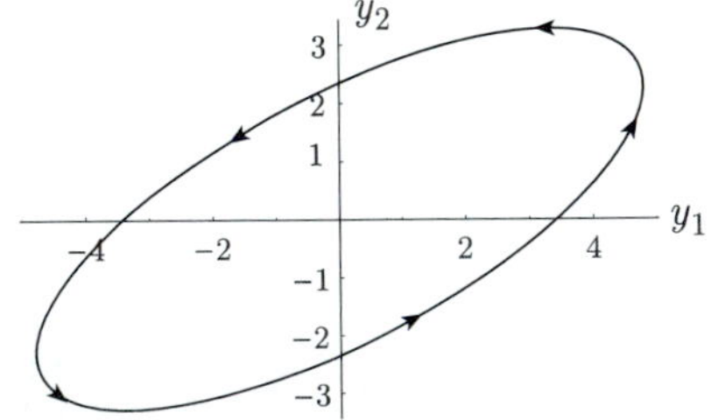

Figure 8

PROBLEMS

For each of the linear, first-order equations in Problems 1 through 6,

(a) find the general solution, and

(b) determine whether solutions are asymptotically stable. If so, describe their behavior as $t \to \infty$.

1. $y' + \dfrac{1}{t+1} y = 2$

2. $y' - \dfrac{1}{t+1} y = 2$

3. $y' + \dfrac{1}{\sqrt{t+1}} y = 2$

4. $y' - \dfrac{1}{\sqrt{t+1}} y = 2$

5. $y' - \dfrac{\sin t}{2 + \cos t} y = \cos t$

6. $(t+1)y' + \dfrac{y}{\ln(t+1)} = 1$

For each matrix A in Problems 7 through 9, determine all values of γ for which solutions of $\mathbf{y}' - A\mathbf{y} = \mathbf{f}$ are asymptotically stable.

7. $\begin{pmatrix} -1 & \gamma \\ 4 & -1 \end{pmatrix}$

8. $\begin{pmatrix} \gamma & b \\ c & \gamma \end{pmatrix}$

9. $\begin{pmatrix} -1 & 1 & \gamma \\ 1 & -1 & 1 \\ -\gamma & 1 & -1 \end{pmatrix}$

10. Show that solutions of $\mathbf{y}' - A\mathbf{y} = \mathbf{f}$, where $A = \begin{pmatrix} a & b \\ c & d \end{pmatrix}$, are asymptotically stable if and only if $a + d < 0$ and $ad - bc > 0$.

11. For the equation

$$\mathbf{y}' - \begin{pmatrix} -2 & 1 \\ 1 & -2 \end{pmatrix} \mathbf{y} = \begin{pmatrix} 10 \sin t \\ 0 \end{pmatrix},$$

(a) show that all solutions are asymptotically stable;

(b) find a periodic solution.

In Problems 12 and 13, show that every solution of the given equation approaches a certain periodic function w, and show that w is not a solution of the differential equation.

12. $y' + y = e^{-t} + \cos t$

13. $y' + \dfrac{t+1}{t} y = \cos t$

14. **(a)** Let a and b be continuous functions on $[0, \infty)$ (possibly constants). Show that if solutions of $y' - ay = f$ and $y' - by = f$ are asymptotically stable, then solutions of $y' - (a+b)y = f$ are also asymptotically stable.

(b) Find a pair of 2×2 matrices A and B for which solutions of $\mathbf{y}' - A\mathbf{y} = \mathbf{f}$ and $\mathbf{y}' - B\mathbf{y} = \mathbf{f}$ are asymptotically stable, yet solutions of $\mathbf{y}' - (A+B)\mathbf{y} = \mathbf{f}$ are not asymptotically stable.

15. Prove Theorem 1 as follows, noting first that for any initial point the existence of a unique solution on $[0, \infty)$ is guaranteed.

(a) Suppose that every solution of $\mathbf{y}' - A\mathbf{y} = \mathbf{0}$ approaches $\mathbf{0}$ as $t \to \infty$. Show that, for any two solutions $\mathbf{u}$ and $\mathbf{v}$ of $\mathbf{y}' - A\mathbf{y} = \mathbf{f}$, the difference $\mathbf{w} = \mathbf{u} - \mathbf{v}$ approaches $\mathbf{0}$ as $t \to \infty$. Conclude that every solution of $\mathbf{y}' - A\mathbf{y} = \mathbf{f}$ is asymptotically stable.

(b) Suppose that there is a solution of $\mathbf{y}' - A\mathbf{y} = \mathbf{0}$—call it $\mathbf{w}$—that does not approach $\mathbf{0}$ as $t \to \infty$. Let $\tilde{\mathbf{y}}$ be any solution of $\mathbf{y}' - A\mathbf{y} = \mathbf{f}$. Show that, for any number $c \neq 0$, $\mathbf{y} = \tilde{\mathbf{y}} + c\mathbf{w}$ is also a solution of $\mathbf{y}' - A\mathbf{y} = \mathbf{f}$ and that $\mathbf{y} - \tilde{\mathbf{y}}$ does not approach $\mathbf{0}$ as $t \to \infty$. Conclude that $\tilde{\mathbf{y}}$ is not asymptotically stable.

Problems 16 through 20 concern the stability of solutions of nonlinear, scalar equations. The definition of stability stated in this section applies to nonlinear problems as well as linear ones, provided we make an assumption about the existence of solutions for $t \geq 0$. Consider a single equation $y' = f(t, y)$, where f is continuous on all of $\mathbb{R}^2$. A solution $\tilde{y}$ is asymptotically stable if (1) it exists for all $t \geq 0$ and (2) there is some number $\varepsilon > 0$ such that every solution y with $|y(0) - \tilde{y}(0)| < \varepsilon$ exists for all $t \geq 0$ and

$$\lim_{t \to \infty} (y(t) - \tilde{y}(t)) = 0.$$

In other words, $\tilde{y}$ is asymptotically stable if for every solution y with a "nearby" initial value, the difference $y(t) - \tilde{y}(t)$ approaches zero as $t \to \infty$.

16. Show that each positive solution of the (nonlinear) logistic equation

$$y' = ky\left(1 - \frac{y}{M}\right), \quad \text{with } k, M > 0,$$

is asymptotically stable and that nonpositive solutions are not asymptotically stable.

For each equation in Problems 17 and 18, make a sketch of the solution curves (cf. Section 4.3), and describe the asymptotically stable solutions.

17. $y' = y(1-y)(2-y)$

18. $y' = -y(1-y)(2-y)$

In Problems 19 and 20, the given Bernoulli/logistic equation has a positive, asymptotically stable, periodic solution. Find it.

19. $y' - \left(1 + \dfrac{\cos t}{2+\sin t}\right) y = -y^2$

20. $y' - y = -(2+\cos t)y^2.$

CHAPTER

9

GEOMETRY OF AUTONOMOUS SYSTEMS IN THE PLANE

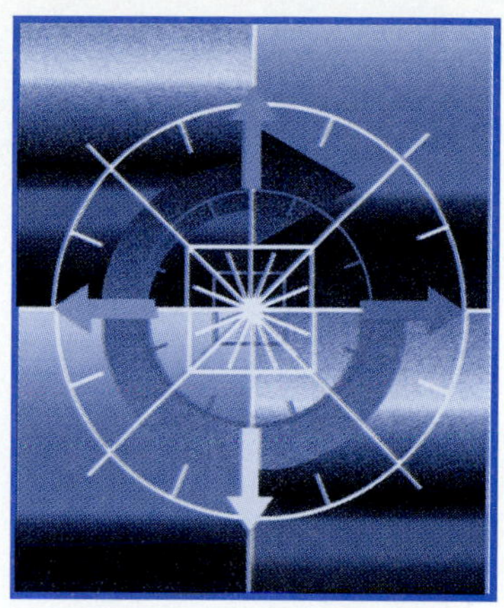

In this chapter we will study coupled pairs of autonomous, first-order differential equations (i.e., systems "in the plane"). Our primary goal is to develop the theory and methods that allow us to perform a type of geometric analysis on nonlinear systems—systems for which explicit solutions generally cannot be found. After a brief general discussion of the phase plane in Section 9.1, we will undertake a study of the geometry of autonomous *linear* systems in the plane, which will establish a necessary foundation for understanding nonlinear systems.

9.1 The Phase Plane

Recall from calculus that a pair of continuous functions x and y defined on an interval I gives rise to a "parametric curve" in the (x, y)-plane defined by

$$x = x(t), \quad y = y(t) \quad \text{for all } t \text{ in } I.$$

For instance, $x = \cos t$ and $y = \sin t$, for t in $[0, 2\pi)$, is the familiar, standard *parametrization* of the unit circle. Keep in mind that a parametric curve is more than simply a set of points forming a curve. The parametrization describes movement along the curve. In particular, it implies a *direction* of movement at each point along the curve.

A particular ellipse centered at $(2, 0)$ is parametrized by

$$x = \sin t, \quad y = 2 + \cos(t - 2/3) \quad \text{for all } t \text{ in } [0, 2\pi).$$

In Figure 1, dashed rectangles indicate the correspondence between pairs of points $(t, x(t))$ and $(t, y(t))$ on the graphs of x and y versus t and the point $(x(t), y(t))$ on the parametric curve. Arrows on the ellipse indicate the direction of the movement of $(x(t), y(t))$ as t increases.

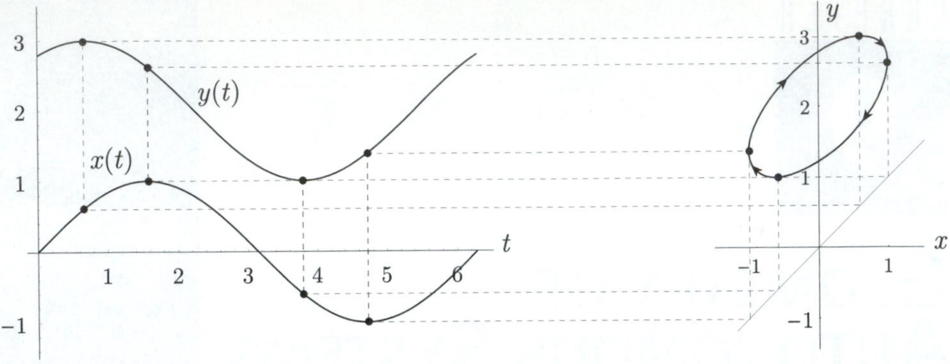

Figure 1

Parametric curves provide an interesting way to visualize the phase shift between a pair of sinusoidal functions. Consider the parametric equations

$$x = \sin(t - \phi) \quad \text{and} \quad y = \sin t,$$

where $0 \le \phi < 2\pi$. The number ϕ is the *phase angle* between $x(t)$ and $y(t)$. If $\phi = 0$, then x and y are "in phase." If $\phi = \pi$, then x and y are "completely out of phase." In general, we say that x and y are $180\phi/\pi$ *degrees out of phase*.

Figure 2 contains a sequence of pairs of plots. Each pair of plots shows the graphs of x and y versus t (for $0 \le t \le 2\pi$) together with the corresponding parametric curve for the indicated phase angle.* When x and y are 90° or 270° (i.e., "halfway") out of phase, the ellipse is actually a circle. When x and y are either in phase or completely out of phase, the ellipse collapses to a line segment. Also, the extent to which x and y are out of phase affects the *eccentricity* of the ellipse but not the angle of its tilt. It should be easy to see that the tilt angle would be affected by the difference in the amplitudes of the two curves. Equal amplitudes result in a tilt angle of either $\pm\pi/4$, as demonstrated in Figure 2.

The parametric curves in Figures 1 and 2 are *closed curves*, because in each case the functions $x(t)$ and $y(t)$ are periodic with a common period.

Phase Portraits Our primary interest is in parametric curves defined by solutions of a pair of autonomous differential equations:

$$x' = f(x, y), \quad y' = g(x, y). \tag{1}$$

We assume that the partial derivatives f_x, f_y, g_x, and g_y exist and are continuous throughout the (x, y)-plane. This assumption guarantees that for any initial point $(x(t_0), y(t_0))$, there is a unique solution of (1) on a maximal open interval I containing t_0. The interval I is maximal in the sense that if it has a finite left- or right-endpoint T, then $|x(t)| + |y(t)| \to \infty$ as t approaches T from within I.

In this context, we refer to the (x, y)-plane as the **phase plane**. A parametric curve defined by a maximal solution $(x(t), y(t))$ of (1) is called a **phase-plane orbit**

* These parametric curves are examples of *Lissajous curves*, which arise when the current in a simple electrical circuit is plotted versus an alternating source voltage.

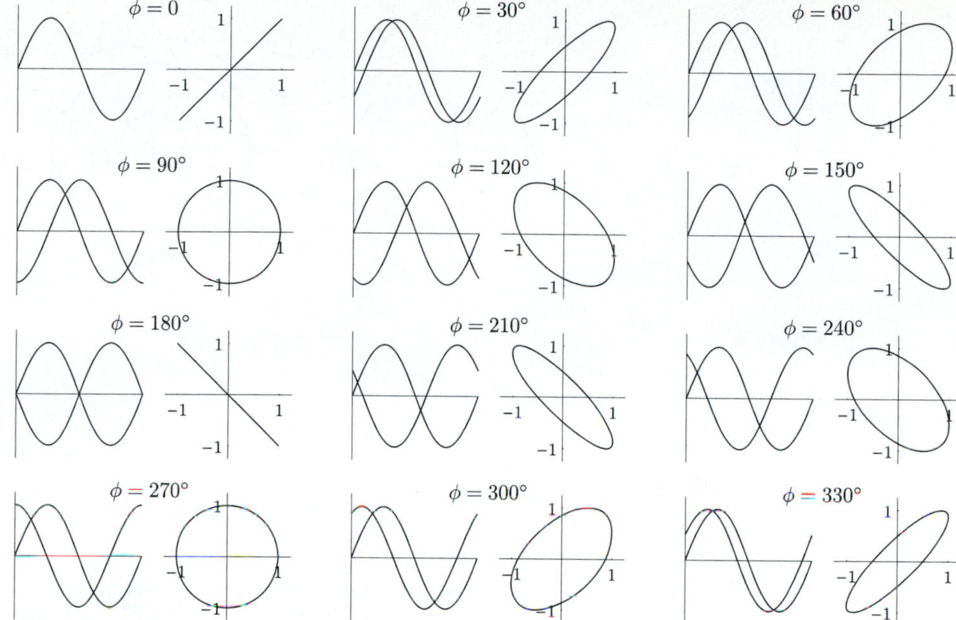

Figure 2

(or simply an orbit). The family of phase-plane orbits corresponding to solutions of (1) is called the **phase portrait** of (1).

Example 1

An instance of the equation of the simple pendulum is $x'' = -\sin x$, where x is the angle that the arm of the pendulum makes with the (downward) vertical axis. Upon conversion to a first-order system, with $y = x'$, this becomes

$$x' = y, \quad y' = -\sin x.$$

In Figure 3, the plot on the left shows the graphs of x and y versus t for a particular pair of initial values, and the plot on the right shows the phase portrait for the system. The particular orbit corresponding to the graphs on the left is indicated by a thick curve. ■

Example 2

A simple instance of the equation of motion for a damped spring-mass system is $x'' + x' + x = 0$. The corresponding first-order system, with $y = x'$, is

$$x' = y, \quad y' = -x - y.$$

The plot on the left in Figure 4 shows the graphs of x and y versus t for a particular pair of initial values, and the plot on the right shows the phase portrait

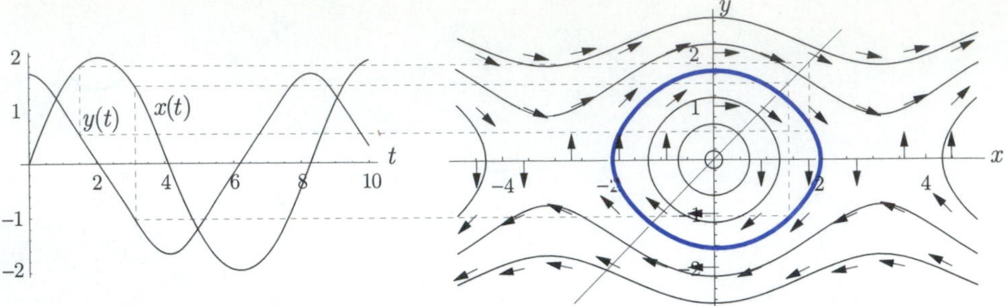

Figure 3

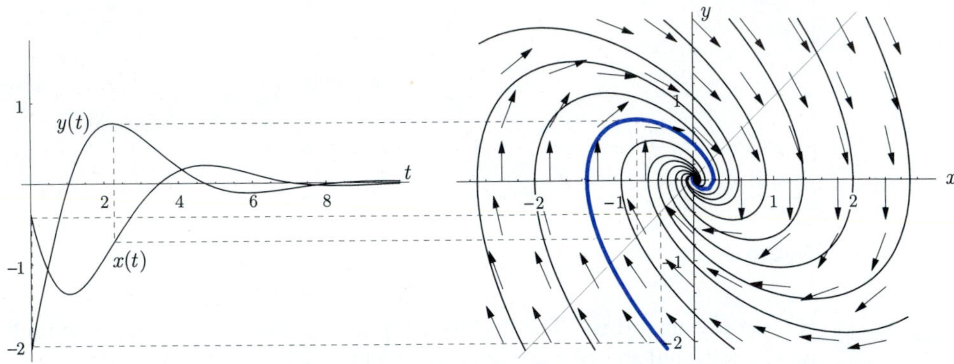

Figure 4

for the system. The thickened curve represents the particular orbit that corresponds to the graphs on the left. Notice that all of the phase-plane orbits spiral in toward the origin, corresponding to the fact that all (nontrivial) solutions oscillate while approaching 0 as $t \to \infty$. ■

Example 3

The equation $x'' + k(x^2 - 1)x' + x = 0$, where k is a constant, is known as *van der Pol's equation** and is related to the current in an RLC circuit with a nonlinear resistor. The corresponding first-order system with $k = 1$ is

$$x' = y, \quad y' = -x - (x^2 - 1)y,$$

where $y = x'$. Figure 5 is similar in its composition to Figures 3 and 4. Notice that all of the (nontrivial) phase-plane orbits are "attracted" to a particular closed curve that corresponds to a periodic solution. ■

---

* Balthasar van der Pol (1889–1959) was a Dutch engineer.

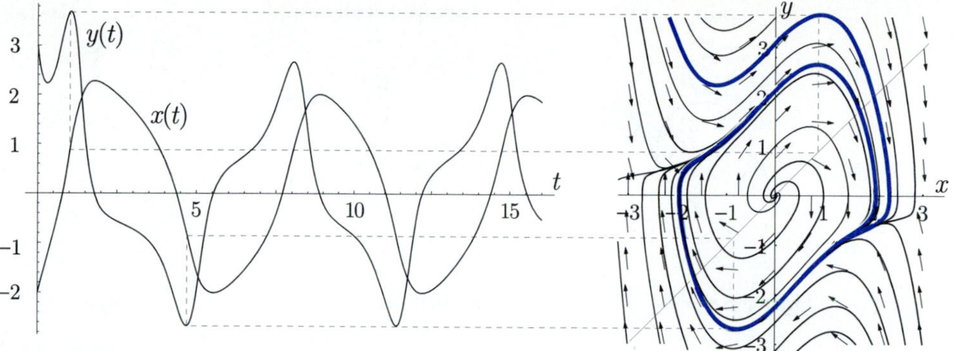

Figure 5

Equilibrium Points and Stability A constant solution of (1) gives rise to a phase-plane orbit consisting of a single point. Such a point is an **equilibrium point**,[†] and the corresponding constant solution of (1) is called an **equilibrium solution**. Note that (x^*, y^*) is an equilibrium point of (1) if and only if $f(x^*, y^*) = 0$ and $g(x^*, y^*) = 0$.

Much of what we will do in this chapter amounts to studying the local behavior of phase-plane orbits near equilibrium points. In particular, we will be interested in stability properties of equilibrium points, defined as follows.

Definition Let (x^*, y^*) be an equilibrium point of (1). Then (x^*, y^*) is

- **stable** if for any open disk Ω centered at (x^*, y^*) there is a smaller open disk $\widetilde{\Omega}$ centered at (x^*, y^*) with the property that $(x(t), y(t))$ remains in Ω for all $t \geq 0$ whenever $(x(0), y(0))$ is in $\widetilde{\Omega}$;
- **unstable** if it is not stable.

A stable equilibrium point is

- **asymptotically stable** if it is the center of an open disk Ω with the property that $(x(t), y(t)) \to (x^*, y^*)$ as $t \to \infty$ whenever $(x(0), y(0))$ is in Ω;
- **neutrally stable** if it is not asymptotically stable. ◆

These notions are illustrated in Figures 6a–c. In Figure 6a, every orbit whose initial point is inside the smaller disk $\widetilde{\Omega}$ remains inside Ω for all $t \geq 0$. In Figure 6b, all orbits that begin in (or enter) Ω approach (x^*, y^*) as $t \to \infty$. In Figure 6c, there are orbits that begin arbitrarily close (x^*, y^*) but eventually exit Ω.

Before we look at our next example, let's revisit Examples 1 through 3. The equilibrium points of the pendulum system in Example 1 lie along the x-axis at integer multiples of π. Figure 3 (as well as the physical interpretation) suggests that the equilibrium points at even multiples of π are neutrally stable, and the equilibrium points at odd multiples of π are unstable. In Example 2, the origin is the only equilibrium point, and it is asymptotically stable. In Example 3, the origin is the only equilibrium point, and it is unstable.

[†] Equilibrium points are often called *critical points* or *steady states*.

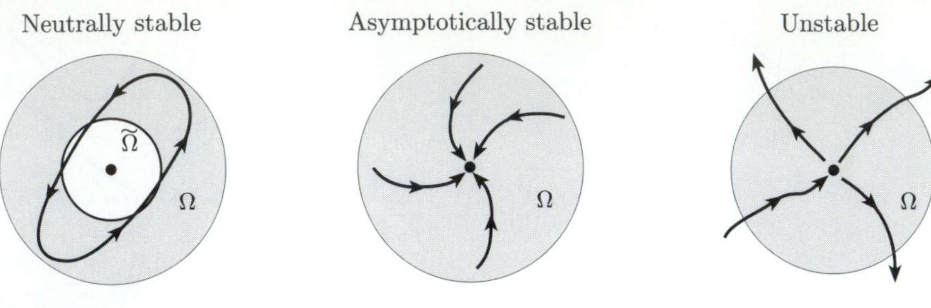

Figure 6a **Figure 6b** **Figure 6c**

Example 4

A variation on the usual *predator-prey* population model is

$$x' = x(1 - y), \quad y' = -y(1 - x + y).$$

Equilibrium points for this system are found by solving the equations

$$x(1 - y) = 0, \quad -y(1 - x + y) = 0.$$

There are three solutions: $(0, -1)$, $(0, 0)$, and $(2, 1)$. In the context of a population model, only $(0, 0)$ and $(2, 1)$ are of interest. The phase portrait in Figure 7 indicates that $(0, 0)$ is unstable and $(2, 1)$ is asymptotically stable. In fact, all orbits corresponding to positive solutions approach $(2, 1)$ as $t \to \infty$. ■

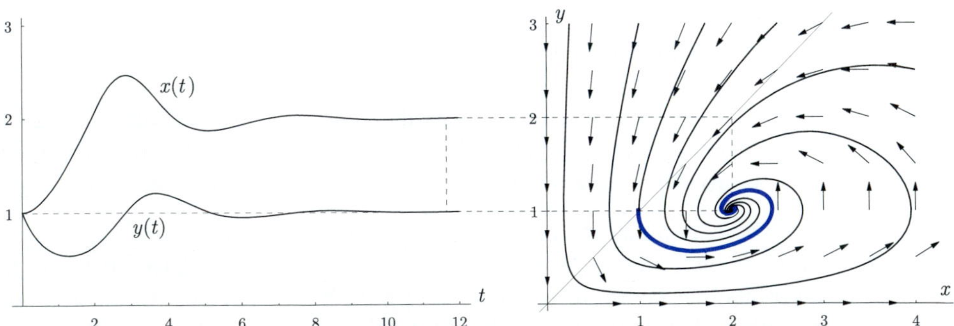

Figure 7

Isoclines and Nullclines Just as in our discussion of direction fields and solution curves for single equations in Chapter 2, the notion of an *isocline* is of interest in the phase plane. **Isoclines** in the phase plane for (1) are the family of curves described by

$$f(x, y) = 0 \quad \text{or} \quad g(x, y) = mf(x, y)$$

as the constant m varies over $\mathbb{R}$. Orbits satisfy $\frac{dx}{dy} = 0$ along an isocline given by $f(x, y) = 0$. Orbits satisfy $\frac{dy}{dx} = m$ along an isocline given by $g(x, y) = mf(x, y)$.

Example 5

The isoclines for the pendulum system in Example 1,

$$x' = y, \quad y' = -\sin x,$$

are the x-axis and the sinusoidal curves $-\sin x = my$. Figure 8 shows the phase portrait for this system together with isoclines corresponding to

$$m = \pm 1, \ \pm 1/2, \quad \text{and} \quad \pm 1/3.$$ ■

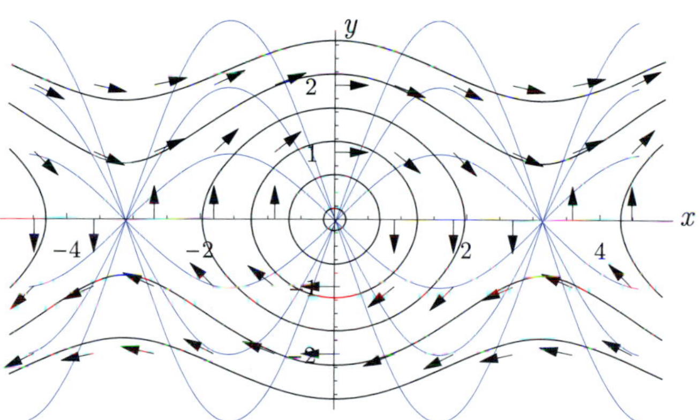

Figure 8

Nullclines are isoclines on which $x' = 0$ or $y' = 0$; that is, $f(x, y) = 0$ or $g(x, y) = 0$. A nullcline on which $f(x, y) = 0$ is called an *x-nullcline*, and a nullcline on which $g(x, y) = 0$ is called a *y-nullcline*. Orbits cross x-nullclines with vertical tangents and y-nullclines with horizontal tangents. Also, any point where an x-nullcline and a y-nullcline intersect is an equilibrium point. Note that in Examples 1 and 5, the x-axis is the x-nullcline, while each of the vertical lines $x = k\pi$, $k = 0, \pm 1, \pm 2, \ldots$, is a y-nullcline.

Example 6

For the van der Pol equation in Example 3, x- and y-nullclines are given by

$$y = 0 \quad \text{and} \quad -x - (x^2 - 1)y = 0,$$

respectively. So the x-nullcline is the x-axis, and the y-nullclines are the three parts of the graph of

$$y = x/(1 - x^2),$$

which has vertical asymptotes at $x = \pm 1$. These nullclines are shown in Figure 9. ■

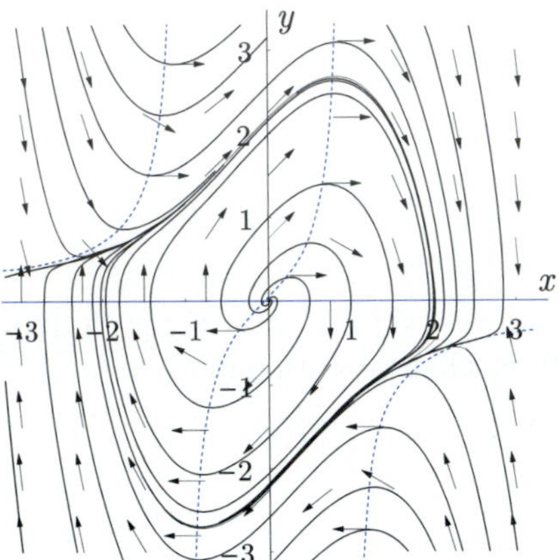

Figure 9

The Slope Equation Phase-plane orbits of (1) trace out curves that satisfy the "slope equation"

$$\frac{dy}{dx} = \frac{g(x, y)}{f(x, y)}.$$

This equation often provides useful information about the phase portrait of (1) and in some cases allows us to obtain a (usually implicit) general solution that describes the curves traced out by phase-plane orbits.

Example 7

Consider the system

$$x' = (y - 1)y, \quad y' = -xy.$$

The slope equation is

$$\frac{dy}{dx} = -\frac{xy}{(y - 1)y},$$

which is easily solved, resulting in

$$y^2 - 2y + x^2 = C.$$

Orbits therefore trace out circular arcs centered at the equilibrium point $(0, 1)$. However, this is far from the full story. In addition to $(0, 1)$, every point on the x-axis is an equilibrium point. Thus no orbit can surround the origin. Figure 10 shows four orbits that trace out three circles. Figure 11 shows plots of x and y versus t corresponding to each the four orbits in Figure 10. ■

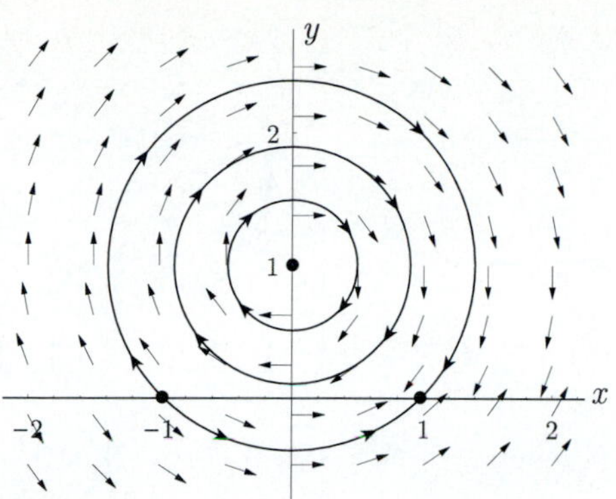

Figure 10

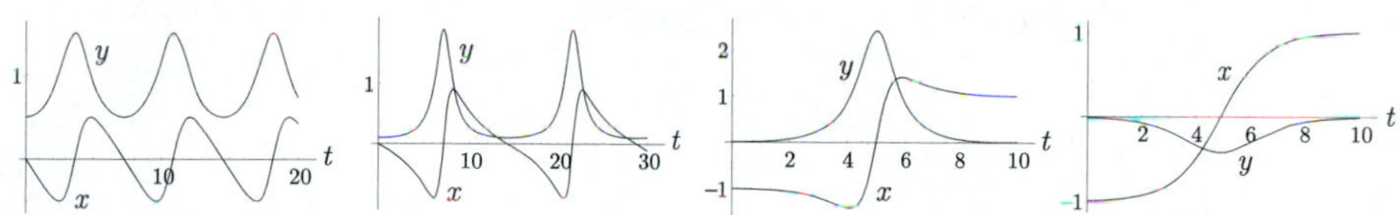

Figure 11

PROBLEMS

In each of the following pairs of plots, identify the particular phase-plane orbit on the right that corresponds to the solution-pair plotted versus t on the left. In the first two pairs, the graphs of x and y versus t are labeled; in the next two you must distinguish between them yourself.

1.

2.

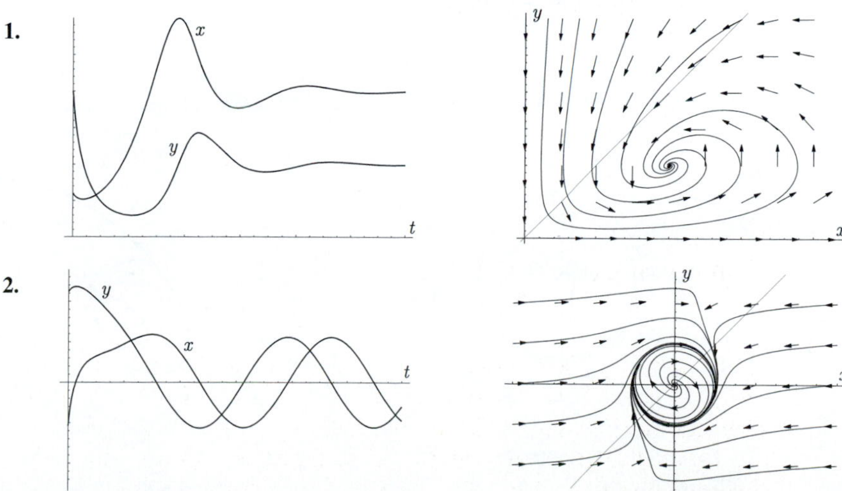

3.

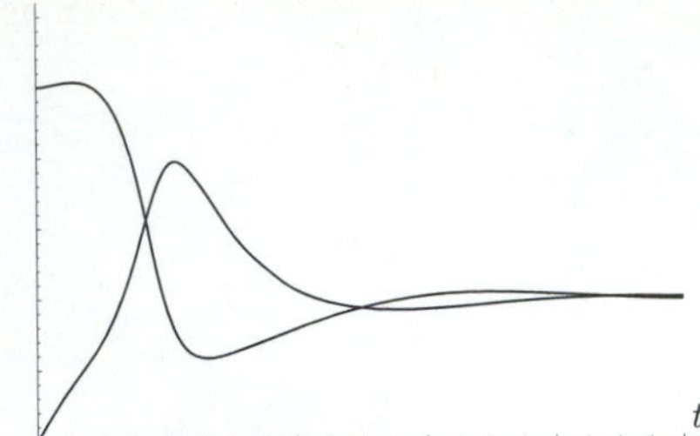

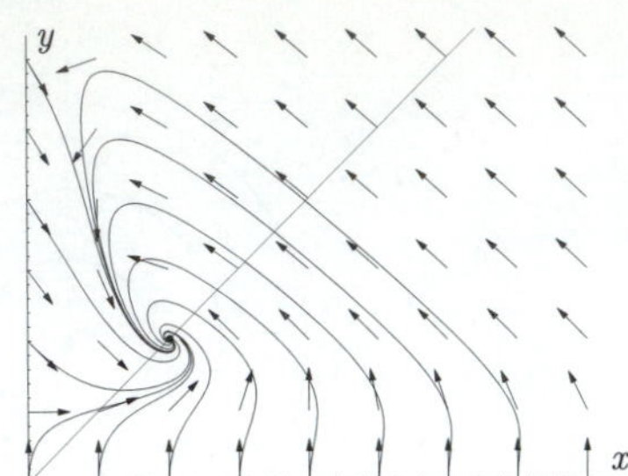

4.

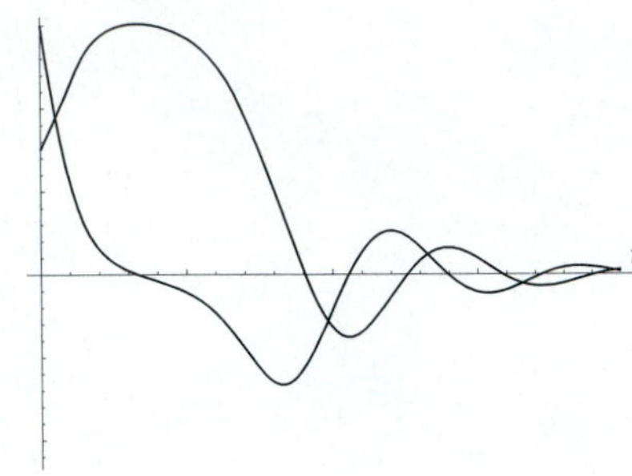

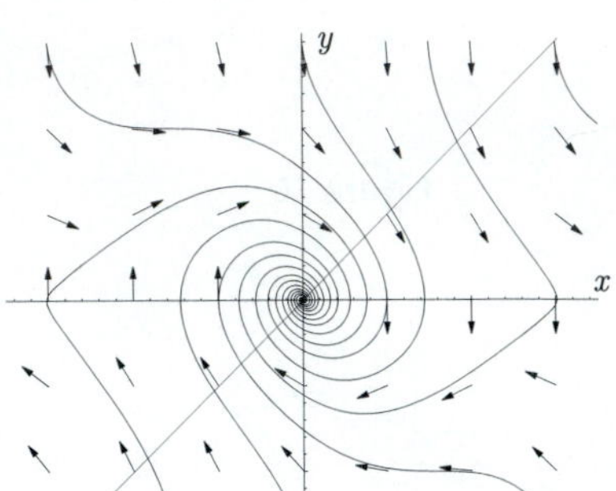

For each of the following parametric curves, sketch two possible graphs of x and y versus t. Give two reasons why there are numerous possibilities.

5.

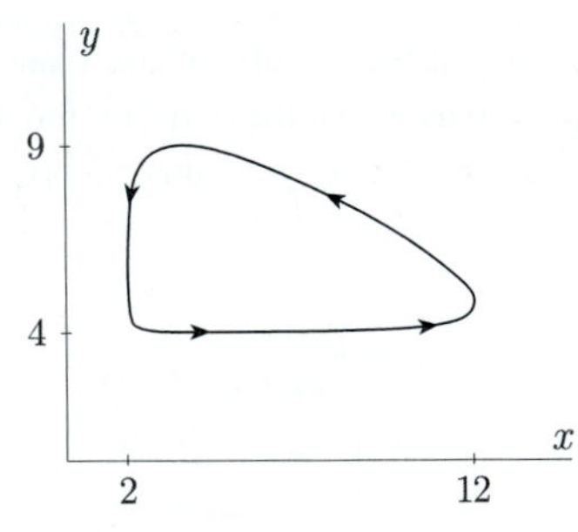

6.

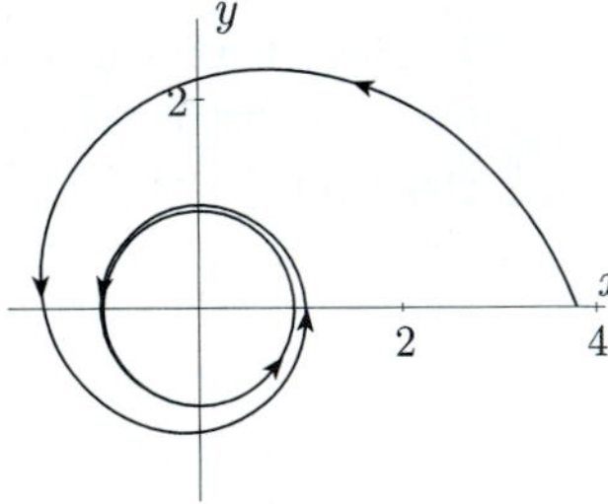

7.

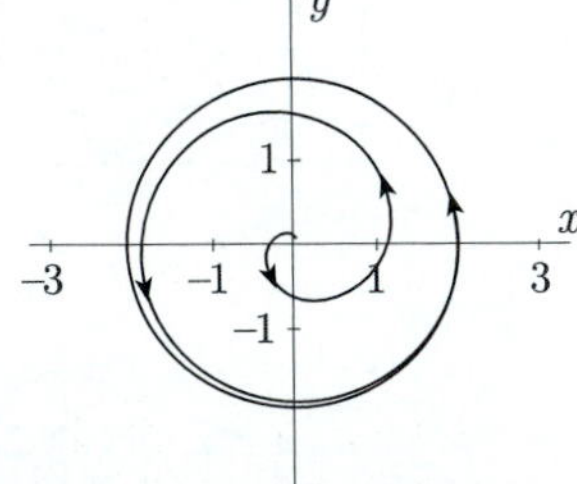

8.

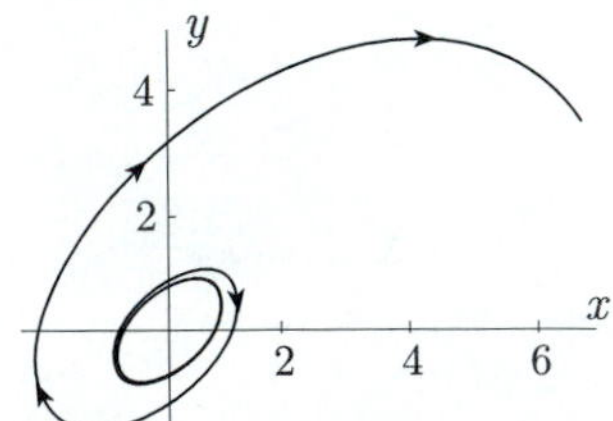

For each of the following graphs of x and y versus t, sketch the corresponding parametric curve for $t \geq 0$.

9.

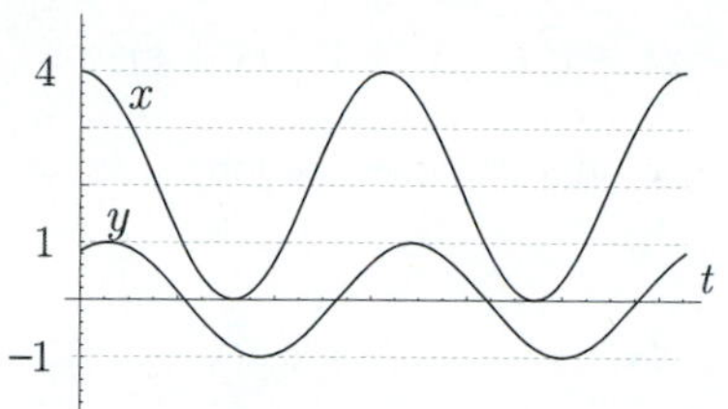

10.

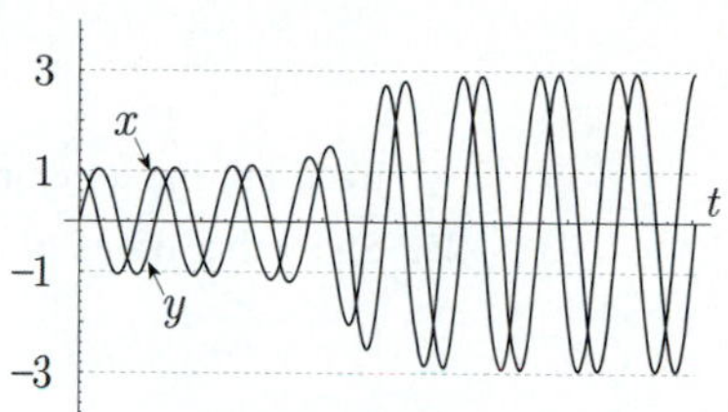

For each of the following graphs of x and y versus t, sketch the corresponding parametric curve for $-\infty < t < \infty$.

11.

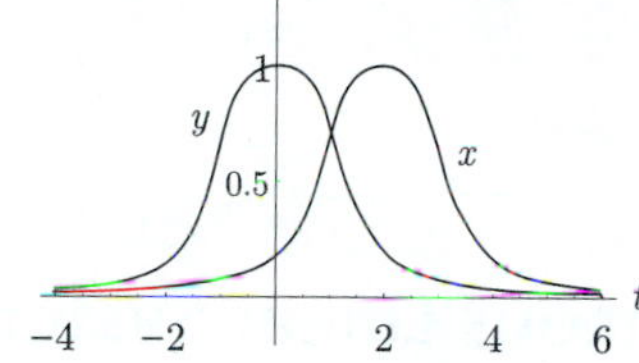

12.

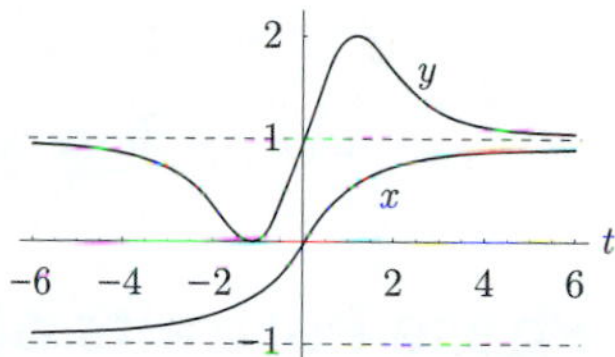

13. Refer to the phase portrait for the van der Pol equation in Example 3, and sketch the graphs of x and y versus t for the solution with initial values $x(0) = 0.25$ and $y(0) = 0.25$. (*Hint*: After a relatively brief time, the shapes of the curves become very close to those shown in Figure 5.)

14. Refer to the phase portrait in Example 4, and sketch the graphs of x and y versus t for the solution with initial values $x(0) = -0.1$ and $y(0) = 0$.

Each of the phase portraits below corresponds to a system for which $(0, 0)$, $(1, 0)$, $(2, 0)$, and $(3, 0)$ are the only equilibrium points. State whether each equilibrium point is unstable, asymptotically stable, or neutrally stable.

15.

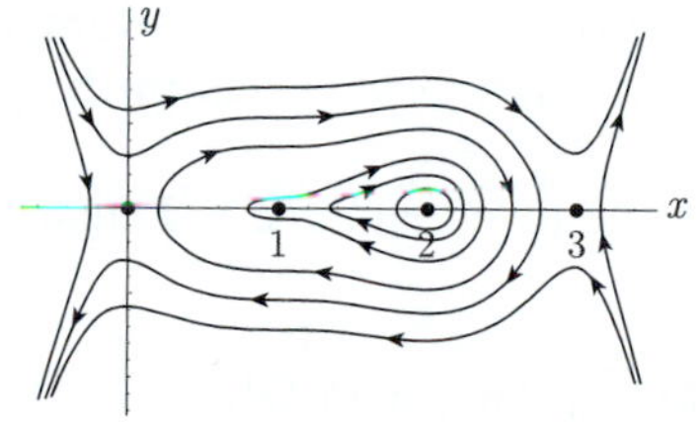

16.

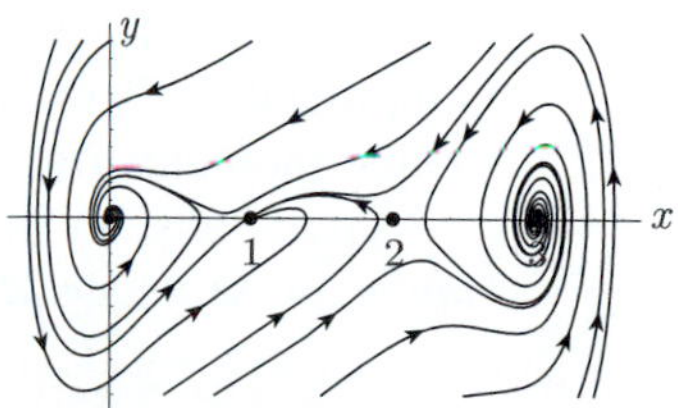

17. Describe the correspondence between the four plots in Figure 11 and the four orbits in Figure 10.

For each of the systems in Problems 18 through 22, solve the corresponding slope equation to find the curves traced out by orbits of the system. Then sketch the phase portrait.

18. $x' = y, \; y' = -x$

19. $x' = y, \; y' = x$

20. $x' = -xy, \; y' = xy$

21. $x' = y, \; y' = -xy$

22. $x' = xy, \; y' = -y^2 + 2x^2 y$

23. Find an equation for the curves traced out by orbits of the pendulum system in Example 1.

24. Show that orbits of the system

$$x' = 2y^2(2 - x^2 - y^2), \quad y' = 1 - (x + y)^2$$

trace out the unit circle. Is there a single orbit that does the job?

25. Show that orbits of the system

$$x' = x - 2y, \quad y' = 2x - y + x^2,$$

trace out curves described by $x^3 + 3x^2 - 3xy + 3y^2 = C$.

26. Let φ be a continuously differentiable function of two variables. Show that orbits of

$$x' = -\frac{\partial \varphi}{\partial y}, \quad y' = \frac{\partial \varphi}{\partial x}$$

trace out curves described by $\varphi(x, y) = C$.

9.2 Phase Portraits of Homogeneous Linear Systems

Let A be a 2×2 matrix and consider the system

$$\mathbf{w}' - A\mathbf{w} = \mathbf{0}, \tag{1}$$

where $\mathbf{w} = (x, y)^T$. Our goal here is to understand the geometry of the phase portrait of (1) in terms of properties of A. In particular, we want to understand the behavior of orbits near the origin, which is always an equilibrium point.

Linear Transformations in the Plane In order to simplify our study of these phase portraits, we will show that nearly all possibilities fall into one of only a handful of standard categories, and each of these standard categories can be understood to a large extent by understanding one or two very simple examples. The basis of this simplification is the notion of *linearly isomorphic* sets in the plane, which we will define momentarily.

Given any set of points Ω in the plane and a 2×2 matrix M, let $M\Omega$ be the *image of* Ω *under* M defined by

$$M\Omega = \{M\mathbf{w} \mid \mathbf{w} \text{ in } \Omega\}.$$

Example 1

Let $M = \begin{pmatrix} 1 & 2 \\ -1 & 0 \end{pmatrix}$. If Ω_1 is the circle described by $x^2 + y^2 = 4$, then $M\Omega_1$ is made up of points of the form

$$M\mathbf{w} = \begin{pmatrix} 1 & 2 \\ -1 & 0 \end{pmatrix}\begin{pmatrix} 2\cos s \\ 2\sin s \end{pmatrix} = \begin{pmatrix} 2\cos s + 4\sin s \\ -2\cos s \end{pmatrix}.$$

Therefore, points in $M\Omega_1$ satisfy the equation $x^2+5y^2+2xy = 16$, which describes an ellipse. If Ω_2 is the graph of $y = x$, then $M\Omega_2$ is made up of points of the form

$$M\mathbf{w} = \begin{pmatrix} 1 & 2 \\ -1 & 0 \end{pmatrix}\begin{pmatrix} s \\ s \end{pmatrix} = \begin{pmatrix} 3s \\ -s \end{pmatrix}.$$

Therefore, points in $M\Omega_2$ satisfy the equation $x = -3y$. Figure 1 shows Ω_1 and Ω_2; Figure 2 shows the images $M\Omega_1$ and $M\Omega_2$. ■

Definition Let Ω_1 and Ω_2 be a sets of points in the plane. If $\Omega_2 = M\Omega_1$, where M is an invertible matrix, then Ω_1 and Ω_2 are **linearly isomorphic**. ◆

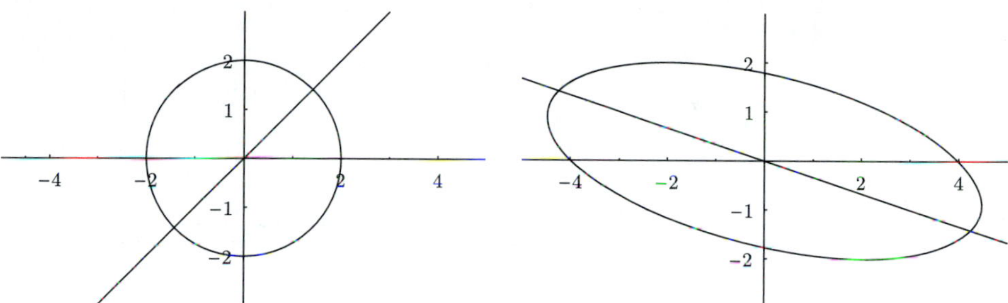

Figure 1 **Figure 2**

When two sets are linearly isomorphic, each set may be viewed as a description of the other with respect to a different set of basis vectors. In particular, the coordinates of a point $\mathbf{w}$ in Ω_1 with respect to the standard basis are precisely the coordinates of $M\mathbf{w} \in \Omega_2$ with respect to the basis consisting of the columns of M. Conversely, the coordinates of a point $\mathbf{w} \in \Omega_2$ with respect to the standard basis are precisely the coordinates of $M^{-1}\mathbf{w}$ in Ω_1 with respect to the basis consisting of the columns of M^{-1}.

Example 2

Let $M = \begin{pmatrix} 1 & 2 \\ -1 & 1 \end{pmatrix}$ and $\mathbf{w} = (3, 1)^T$. Then $M\mathbf{w} = (5, -2)^T$, whose coordinates with respect to the basis consisting of the columns of M are (3, 1), since

$$\begin{pmatrix} 5 \\ -2 \end{pmatrix} = 3 \cdot \begin{pmatrix} 1 \\ -1 \end{pmatrix} + 1 \cdot \begin{pmatrix} 2 \\ 1 \end{pmatrix}.$$

Furthermore, $M^{-1} = \begin{pmatrix} 1/3 & -2/3 \\ 1/3 & -1/3 \end{pmatrix}$, and the coordinates of $\mathbf{w}$ with respect to the basis consisting of the columns of M^{-1} are $(5, -2)$, since

$$\begin{pmatrix} 3 \\ 1 \end{pmatrix} = 5 \cdot \begin{pmatrix} 1/3 \\ 1/3 \end{pmatrix} + (-2) \cdot \begin{pmatrix} -2/3 \\ 1/3 \end{pmatrix}.$$

Now let Ω be the set consisting of the two hyperbolas $x^2 - y^2 = 1$ and $-x^2 + y^2 = 1$ along with their asymptotes, the lines $y = \pm x$. Then Ω has four hyperbolic "branches" that can be parametrized as $(\pm\cosh s,\ \sinh s)$ and $(\sinh s,\ \pm\cosh s)$, and their asymptotes can be parametrized as $(s, \pm s)$. So the six pieces of $M\Omega$ are made up of points of the forms

$$(\pm\cosh s + 2\sinh s,\ \mp\cosh s + \sinh s),$$

$$(\sinh s \pm 2\cosh s,\ -\sinh s \pm \cosh s),\ (3s, 0),\quad \text{and}\quad (-s, -2s).$$

Figure 3 shows Ω, and Figure 4 shows $M\Omega$. The dashed lines in Figure 4 are the images under M of the coordinate axes in Figure 3. The columns of M serve as direction vectors for these lines. ■

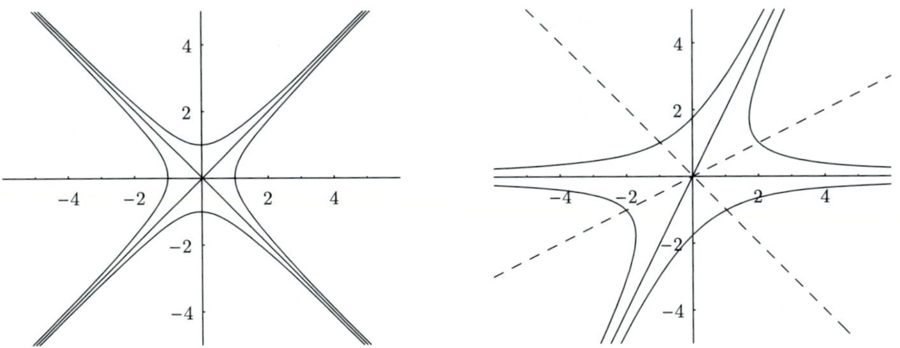

Figure 3 **Figure 4**

The important point to be made here is that when two sets are linearly isomorphic, their graphs are similar and can be thought of as "simple distortions" of each other. The following facts are not difficult to prove and say much about the similarity of linearly isomorphic sets.

i. If Ω is linearly isomorphic to a line through the origin, then Ω is a line through the origin.
ii. If Ω is linearly isomorphic to a circle or ellipse centered at the origin, then Ω is a circle or ellipse centered at the origin.
iii. If Ω is linearly isomorphic to a collection of nonintersecting curves, then Ω is a collection of nonintersecting curves.

Phase Portraits We now return our attention to determining the nature of the phase portrait of

$$\mathbf{w}' - A\mathbf{w} = \mathbf{0}. \tag{1}$$

Let $(\lambda_1, \mathbf{p}_1)$ and $(\lambda_2, \mathbf{p}_2)$ be the eigenpairs of A. Unless stated otherwise, we assume that neither eigenvalue is 0. As result, A is invertible, and the only equilibrium point of (1) is the origin.

Real Eigenvalues Suppose that λ_1 and λ_2 are real. It is worth noting right away that (1) has *straight-line orbits* given by

$$\mathbf{w}(t) = c_1 e^{\lambda_1 t}\mathbf{p}_1 \quad \text{and} \quad \mathbf{w}(t) = c_2 e^{\lambda_2 t}\mathbf{p}_2.$$

These orbits lie on lines through the origin with direction vectors $\mathbf{p}_1$ and $\mathbf{p}_2$, respectively. (If λ_1 and λ_2 are not real, then (1) has no straight-line orbits. Also, $\mathbf{p}_1$ and $\mathbf{p}_2$ may be parallel if $\lambda_1 = \lambda_2$.)

Suppose that $\mathbf{p}_1$ and $\mathbf{p}_2$ are linearly independent. Let $P = (\mathbf{p}_1\, \mathbf{p}_2)$, and let D be the diagonal matrix with diagonal entries λ_1, λ_2. Then it is easy to show that $AP = PD$ and consequently $A = PDP^{-1}$. Using this in $\mathbf{w}' - A\mathbf{w} = \mathbf{0}$, we obtain $\mathbf{w}' - PDP^{-1}\mathbf{w} = \mathbf{0}$. Left-multiplication by P^{-1} then produces

$$P^{-1}\mathbf{w}' - DP^{-1}\mathbf{w} = \mathbf{0}.$$

Now let $\mathbf{u} = P^{-1}\mathbf{w}$. Then $\mathbf{u}$ satisfies the *decoupled* system

$$\mathbf{u}' - D\mathbf{u} = \mathbf{0}.$$

Therefore, we conclude that the phase portraits of $\mathbf{w}' - A\mathbf{w} = \mathbf{0}$ and $\mathbf{u}' - D\mathbf{u} = \mathbf{0}$ are linearly isomorphic.

The system $\mathbf{u}' - D\mathbf{u} = \mathbf{0}$ is easily solved, yielding

$$\mathbf{u}(t) = \begin{pmatrix} e^{\lambda_1 t} & 0 \\ 0 & e^{\lambda_2 t} \end{pmatrix} \mathbf{u}(0) = \begin{pmatrix} u_1(0)e^{\lambda_1 t} \\ u_2(0)e^{\lambda_2 t} \end{pmatrix}.$$

Therefore, there are constants c_1 and c_2 such that

$$c_1|u_1|^{\lambda_2} = c_2|u_2|^{\lambda_1}.$$

This equation describes the curves traced out by the phase-plane orbits. The character of these curves depends mainly upon whether λ_1 and λ_2 have the same sign or opposite signs.

Distinct Real Eigenvalues, Same Sign This situation is illustrated nicely by $\lambda_1 = 1$ and $\lambda_2 = 3$. With those eigenvalues, there are constants c_1 and c_2 such that

$$c_1|u_1|^3 = c_2|u_2|.$$

Thus the phase-plane orbits of $\mathbf{u}' - D\mathbf{u} = \mathbf{0}$ trace out the coordinate axes and cubic curves as shown along with the direction field in Figure 5. Since λ_1 and λ_2 are positive, the direction of the "flow" along every orbit is away from the origin. If instead we had $\lambda_1 = -1$ and $\lambda_2 = -3$, the phase portrait would be identical, except that the flow along orbits would be *toward* the origin. This is shown in Figure 6.

The phase portraits in Figures 5 and 6 can serve as canonical portraits for all systems in which A has distinct real eigenvalues with the same sign. In such portraits, we call the origin a **biaxial node**.* In Figure 5 the origin is an unstable biaxial node, because orbits move away from the origin as t increases. In Figure 6 the origin is an

* The word *node* as used here is synonymous with *vertex*. Also, *biaxial* connotes the pair of axes comprising the two straight-line orbits.

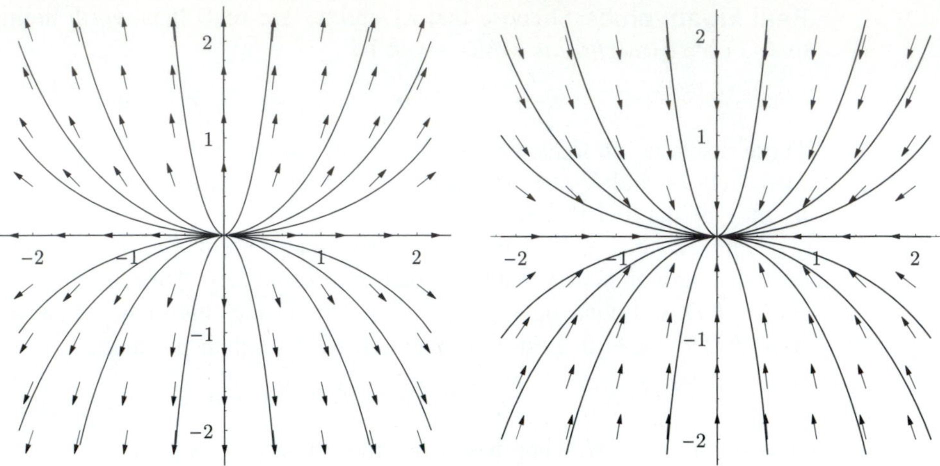

Figure 5 Figure 6

(asymptotically) stable biaxial node, since all orbits approach the origin as $t \to \infty$. Since every stable biaxial node is asymptotically stable, we will simply refer to biaxial nodes as stable or unstable.

Example 3

Figures 7 and 8 show phase portraits of $\mathbf{w}' - \mathbf{A}\mathbf{w} = \mathbf{0}$ for

$$A = \frac{1}{2}\begin{pmatrix} 4 & -4 \\ -1 & 4 \end{pmatrix} \quad \text{and} \quad A = \frac{1}{5}\begin{pmatrix} -11 & 2 \\ 12 & -9 \end{pmatrix},$$

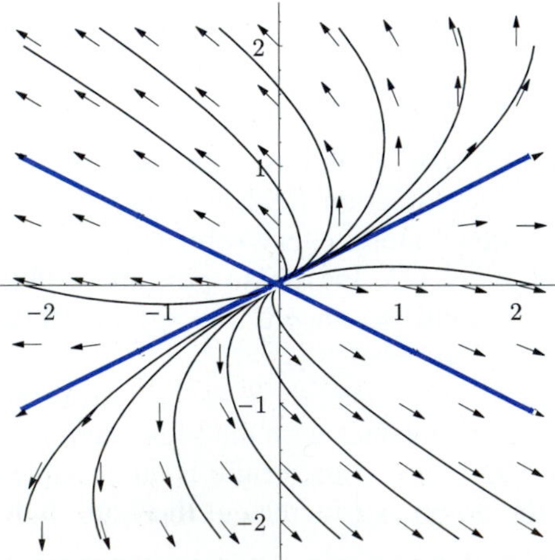

Figure 7

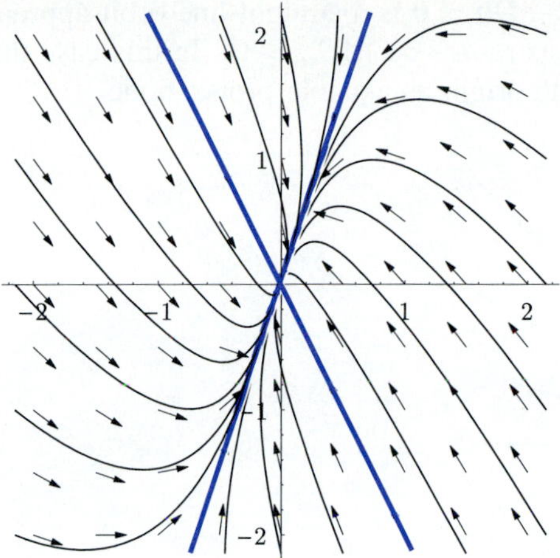

Figure 8

respectively, whose eigenpairs are, respectively,

$$\left(1, \begin{pmatrix} 2 \\ 1 \end{pmatrix}\right), \left(3, \begin{pmatrix} -2 \\ 1 \end{pmatrix}\right) \text{ and } \left(-1, \begin{pmatrix} 1 \\ 3 \end{pmatrix}\right), \left(-3, \begin{pmatrix} 1 \\ -2 \end{pmatrix}\right).$$

In Figure 7 the origin is an unstable biaxial node, and in Figure 8 the origin is a stable biaxial node. Notice the (blue) straight-line orbits that correspond to the eigenvectors of A in each case. ■

In a phase portrait in which the origin is a biaxial node, one of the eigenvectors has the property that almost every orbit is nearly parallel to it near the origin—the only exceptions being straight-line orbits that are parallel to the *other* eigenvector. (This may happen as $t \to \infty$ or as $t \to -\infty$, depending on whether the eigenvalues are positive or negative.) The eigenvector to which almost all orbits are nearly parallel near the origin is the one that corresponds to the eigenvalue that is closer to zero. (Why?) This is illustrated in Figures 5 and 6, in which almost all orbits are nearly parallel to the eigenvector $\mathbf{p}_1 = (1, 0)^T$ near the origin. Note also that in Figure 7 almost all orbits are nearly parallel near the origin to the eigenvector $\mathbf{p}_1 = (2, 1)^T$, which corresponds to the lesser of the two eigenvalues in absolute value, $\lambda_1 = 1$. In Figure 8, almost all orbits are nearly parallel near the origin to the eigenvector $\mathbf{p}_1 = (1, 3)^T$, which corresponds to the lesser of the two eigenvalues in absolute value, $\lambda_1 = -1$.

The eigenvector that corresponds to the greatest eigenvalue in absolute value has the property that every orbit is nearly (or exactly) parallel to it *far away* from the origin. This is also confirmed in each of Figures 5 through 8.

Repeated Real Eigenvalues If A has a single, repeated eigenvalue λ_1 with a pair of linearly independent eigenvectors, then $D = \lambda_1 \mathcal{I}$, and there are constants c_1 and c_2 such that

$$c_1|u_1| = c_2|u_2|.$$

So every (nontrivial) orbit of $\mathbf{u}' - D\mathbf{u} = \mathbf{0}$ is a straight-line orbit approaching $(0, 0)$ either as $t \to \infty$ (if $\lambda_1 < 0$) or as $t \to -\infty$ (if $\lambda_1 > 0$). In this case the origin is a **proper node**. Figure 9 shows the origin as a stable proper node.

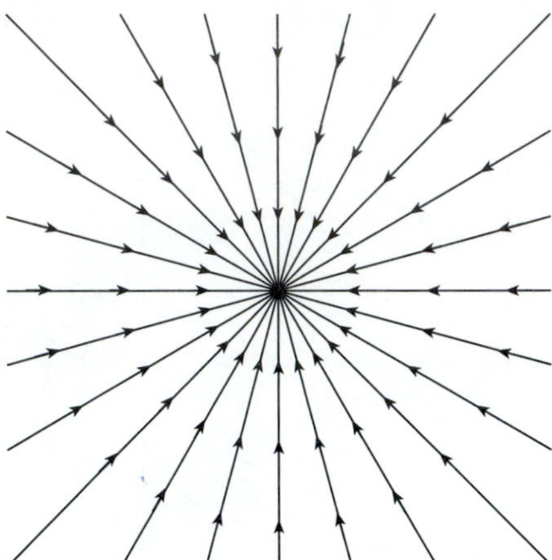

Figure 9

A proper node occurs only when A is a multiple of the identity matrix. In fact, $A = D$ in this case, since

$$A = PDP^{-1} = P\lambda_1 \mathcal{I} P^{-1} = \lambda_1 PP^{-1} = \lambda_1 \mathcal{I}.$$

So the phase portrait of $\mathbf{w}' - A\mathbf{w} = \mathbf{0}$ is precisely the same as that of $\mathbf{u}' - D\mathbf{u} = \mathbf{0}$.

The Coaxial Node Suppose that A has a single repeated eigenvalue λ_1 and does not have two linearly independent eigenvectors. Let $\tilde{A} = A + \begin{pmatrix} \varepsilon & 0 \\ 0 & 0 \end{pmatrix}$ with eigenvalues $\tilde{\lambda}_1$ and $\tilde{\lambda}_2$. In Problem 19, you will show that if ε has the same sign as $a_{11} - a_{22}$ (with no restriction on ε if $a_{11} - a_{22} = 0$) and if $|\varepsilon|$ is sufficiently small, then $\tilde{\lambda}_1$ and $\tilde{\lambda}_2$ will be real and have the same sign. Moreover, $\tilde{\lambda}_1 - \tilde{\lambda}_2 \to 0$ and the eigenvectors of $\tilde{A}$ become parallel as $|\varepsilon| \to 0$. Therefore, the corresponding straight-line orbits in the phase portrait of $\mathbf{w}' - \tilde{A}\mathbf{w} = \mathbf{0}$ become coincident in the limit as $|\varepsilon| \to 0$. The upshot is that the phase portrait for $\mathbf{w}' - A\mathbf{w} = \mathbf{0}$ looks essentially like a portrait in which the origin is a biaxial node whose axes are so nearly parallel as to be indistinguishable. In such a phase portrait we call the origin a **coaxial node**, since there is a single axis formed by straight-line orbits.

Example 4

Figures 10a and 10b show the phase portraits of $\mathbf{w}' - \tilde{A}\mathbf{w} = \mathbf{0}$ for

$$\tilde{A} = \begin{pmatrix} -3.1 & 1 \\ -1 & -1 \end{pmatrix} \quad \text{and} \quad \tilde{A} = \begin{pmatrix} -3.01 & 1 \\ -1 & -1 \end{pmatrix},$$

respectively. In each of these the origin is a stable biaxial node. Figure 10c shows the stable coaxial node in the phase portrait of

$$\mathbf{w}' - \begin{pmatrix} -3 & 1 \\ -1 & -1 \end{pmatrix} \mathbf{w} = \mathbf{0}.$$

In each of these plots, thick lines indicate the straight-line orbits. ■

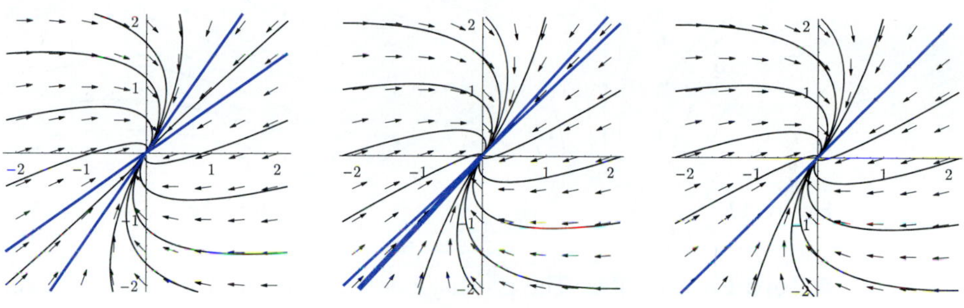

Figure 10a Figure 10b Figure 10c

Real Eigenvalues, Opposite Signs We will construct a canonical phase portrait for this case with $\lambda_1 = 1$ and $\lambda_2 = -1$. Given these eigenvalues, there are constants c_1 and c_2 such that

$$c_1|u_1|^{-1} = c_2|u_2|.$$

With the exception of those that lie on the coordinate axes, orbits lie on hyperbolas whose asymptotes are the coordinate axes. Here, the equilibrium point at the origin is called a **saddle point**. Such a portrait is shown in Figure 11. Figure 12 shows the phase portrait for

$$\mathbf{w}' - A\mathbf{w} = \mathbf{0}, \quad \text{where} \quad A = \frac{1}{3}\begin{pmatrix} 5 & 14 \\ 7 & -2 \end{pmatrix},$$

in which A has eigenpairs

$$\left(-3, \begin{pmatrix} -1 \\ 1 \end{pmatrix}\right) \quad \text{and} \quad \left(4, \begin{pmatrix} 2 \\ 1 \end{pmatrix}\right).$$

Again, notice the straight-line orbits corresponding to the eigenvectors of A.

Here, as in the previous case where the origin was a biaxial node, straight-line orbits form two distinct lines through the origin. In this case, however, orbits on each of the two lines behave differently. One of these lines is called the *stable axis*; orbits on it approach the origin as $t \to \infty$. The other is called the *unstable axis*; orbits on it move strictly away from the origin. The *only* orbits that approach the origin as $t \to \infty$ are orbits that lie on the stable axis. All other orbits are asymptotic to the unstable axis as $t \to \infty$.

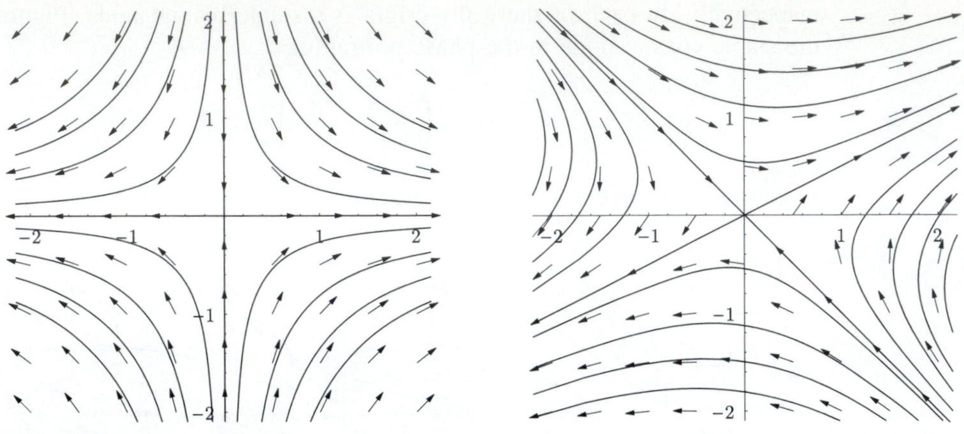

Figure 11 **Figure 12**

A Zero Eigenvalue The primary consequence of A *not* having a zero eigenvalue is that the origin is the only equilibrium point of $\mathbf{w}' - A\mathbf{w} = \mathbf{0}$. If 0 is an eigenvalue of A—and A itself is not the zero matrix—then $\mathbf{w}' - A\mathbf{w} = \mathbf{0}$ has an entire line of equilibrium points, since $A\mathbf{x} = \mathbf{0}$ has an entire line of solutions. (What if A *is* the zero matrix?) As we will see in the next example, the zero-eigenvalue situation can be thought of as a "critical case" in which any small change in A might cause the origin to become either a biaxial node or a saddle point.

Example 5

Consider the system

$$\mathbf{w}' - A\mathbf{w} = \mathbf{0}, \quad \text{where} \quad A = \begin{pmatrix} -.8 & .4 \\ .4 & -.2 \end{pmatrix}.$$

The eigenvalues of A are 0 and -1. The equilibrium points of this system are the solutions of $A\mathbf{x} = \mathbf{0}$, which are easily found to be all points on the line $y = 2x$. Also, since the first row of A is -2 times the second row, we have

$$(x + 2y)' = 0$$

for every solution $\mathbf{w} = (x, y)^T$. Therefore, every orbit lies on a line described by $x + 2y = c$, where c is a constant. The phase portrait of this system is shown in Figure 13b. If we change a_{22} from $-.2$ to $-.1$, the eigenvalues of A become approximately .0815 and $-.9815$, and so the origin is a saddle point. That phase portrait is shown in Figure 13a. With $a_{22} = -.3$, A has two negative eigenvalues, $-.0783$ and -1.022; so the origin is the stable biaxial node seen in Figure 13c. ■

The zero-eigenvalue case is but one of the "critical cases" in which a small change in A might result in more than one type of phase portrait. The case of repeated eigenvalues is also a critical case, because a small change in A might result in either distinct real or nonreal complex eigenvalues. (See Problem 19.) In the case

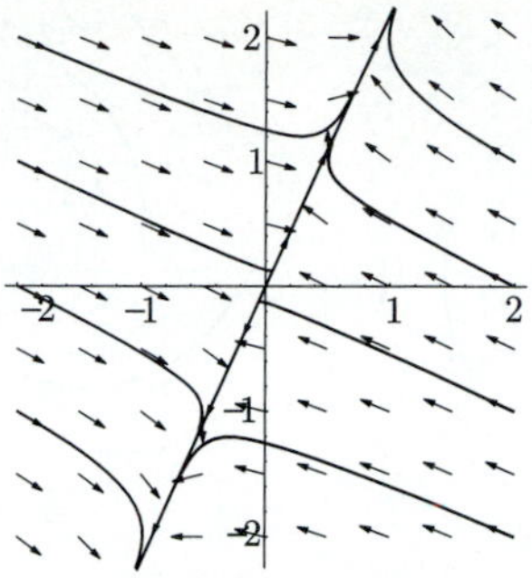

Figure 13a

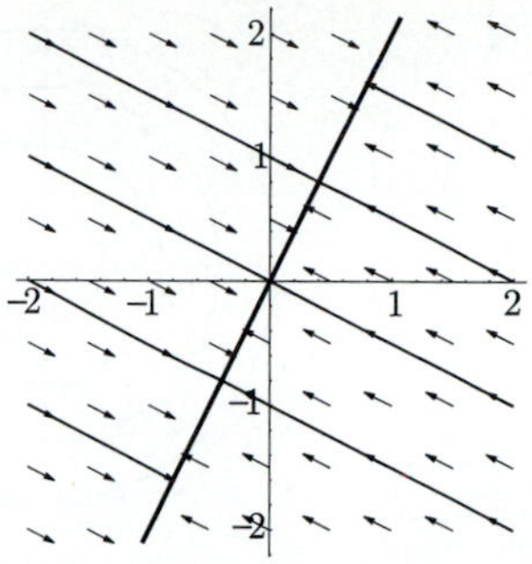

Figure 13b

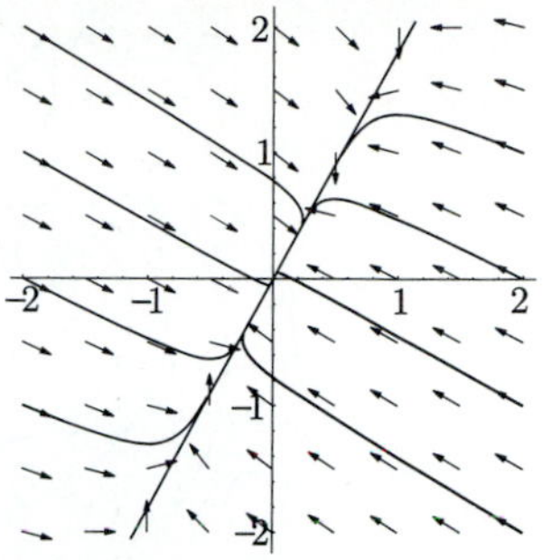

Figure 13c

of purely imaginary eigenvalues, a small change in A might result in nonreal complex eigenvalues with either negative or positive real parts.

Nonreal Complex Eigenvalues Suppose that the eigenvalues of A are $\alpha \pm \beta i$, where $\beta \neq 0$, with corresponding eigenvectors $\mathbf{r} \pm i\mathbf{q}$. Let $Q = (\mathbf{pq})$ and

$$\Gamma = \begin{pmatrix} \alpha & \beta \\ -\beta & \alpha \end{pmatrix}.$$

Then it is not difficult to verify that $AQ = Q\Gamma$ and, consequently, $A = Q\Gamma Q^{-1}$. Using this factorization in $\mathbf{w}' - A\mathbf{w} = \mathbf{0}$, we have $\mathbf{w}' - Q\Gamma Q^{-1}\mathbf{w} = \mathbf{0}$. Left multiplication by P^{-1} then produces

$$Q^{-1}\mathbf{w}' - \Gamma Q^{-1}\mathbf{w} = \mathbf{0}.$$

Now let $\mathbf{u} = Q^{-1}\mathbf{w}$. Then $\mathbf{u}$ satisfies the system

$$\mathbf{u}' - \Gamma\mathbf{u} = \mathbf{0}.$$

Therefore, we conclude that the phase portraits of $\mathbf{w}' - A\mathbf{w} = \mathbf{0}$ and $\mathbf{u}' - \Gamma\mathbf{u} = \mathbf{0}$ are linearly isomorphic.

The general solution of $\mathbf{u}' - \Gamma\mathbf{u} = \mathbf{0}$ is easily found to be

$$\mathbf{u}(t) = e^{\alpha t}\begin{pmatrix} c_1 \cos\beta t + c_2 \sin\beta t \\ c_2 \cos\beta t - c_1 \sin\beta t \end{pmatrix}.$$

From this, an easy calculation shows that

$$u_1^2 + u_2^2 = c_0 e^{2\alpha t},$$

where $c_0 = c_1^2 + c_2^2$. We can now conclude that

i. If $\alpha < 0$, then phase-plane orbits spiral in toward the origin.
ii. If $\alpha = 0$, then phase-plane orbits are circles centered at the origin.
iii. If $\alpha > 0$, then phase-plane orbits spiral out away from the origin.

These three possibilities (with $\beta > 0$) are illustrated, respectively, in Figures 14a–c.

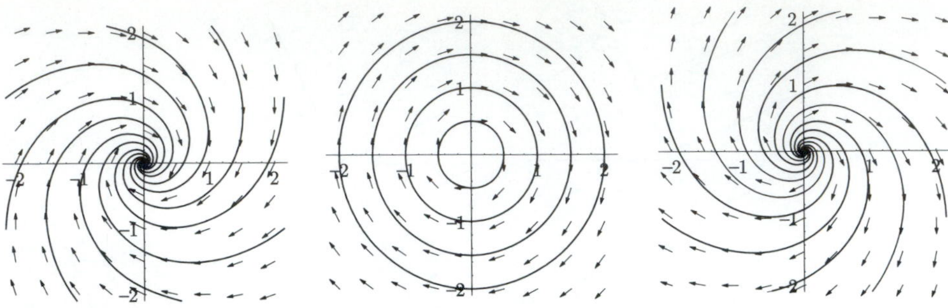

Figure 14a **Figure 14b** **Figure 14c**

For a system whose phase portrait is linearly isomorphic to one of Figures 14a–c, the origin is called a **stable spiral point**, a **center**, or an **unstable spiral point**, respectively. A stable spiral point is also asymptotically stable, and a center is a *neutrally stable* equilibrium point.

We should point out that although the flow in each of the portraits in Figures 14a–c is clockwise around the origin, a phase portrait that is linearly isomorphic to any of these may exhibit either clockwise or counterclockwise flow.

Example 6

The phase portraits shown in Figures 15 and 16, respectively, are for the systems

$$\mathbf{w}' - \frac{1}{6}\begin{pmatrix} -11 & 10 \\ -10 & 5 \end{pmatrix}\mathbf{w} = \mathbf{0} \quad \text{and} \quad \mathbf{w}' - \frac{1}{3}\begin{pmatrix} 4 & -5 \\ 5 & -4 \end{pmatrix}\mathbf{w} = \mathbf{0}.$$

The matrices in these systems have eigenvalues $-1/2 \pm i$ and $\pm i$, respectively. ■

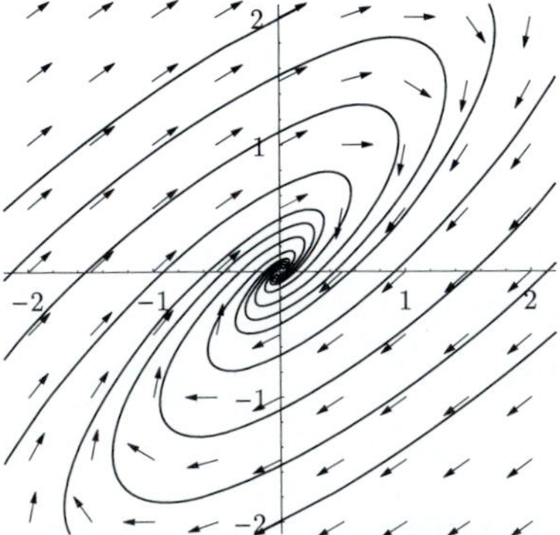

Figure 15

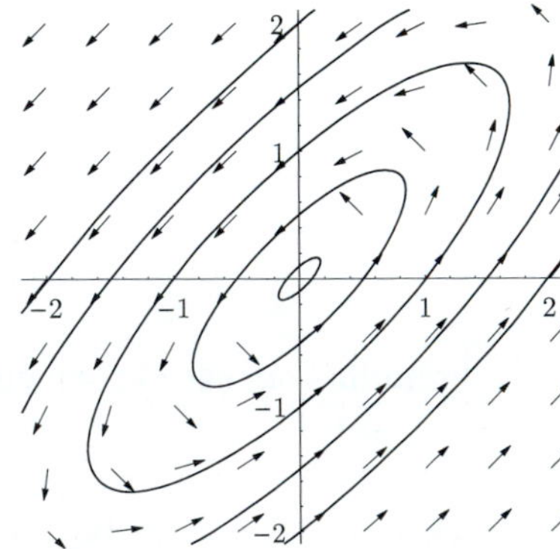

Figure 16

Summary: The Big Picture The array shown in Figure 17 depicts all of the ten possible configurations for the eigenvalues of a nonzero 2×2 matrix. The ten cases are arranged around a pentagon with one case at each edge and one case at each corner. Corresponding to each case, there is an "icon" consisting of perpendicular axes and one or two large dots that forms a schematic plot of the eigenvalues in the complex plane.* Along with each icon is a plot of a typical phase portrait. Because of the way the figures are arranged, any movement from case to case caused by continuous change in the entries of the matrix is always to a neighboring case in the array (as long as the zero matrix is avoided).

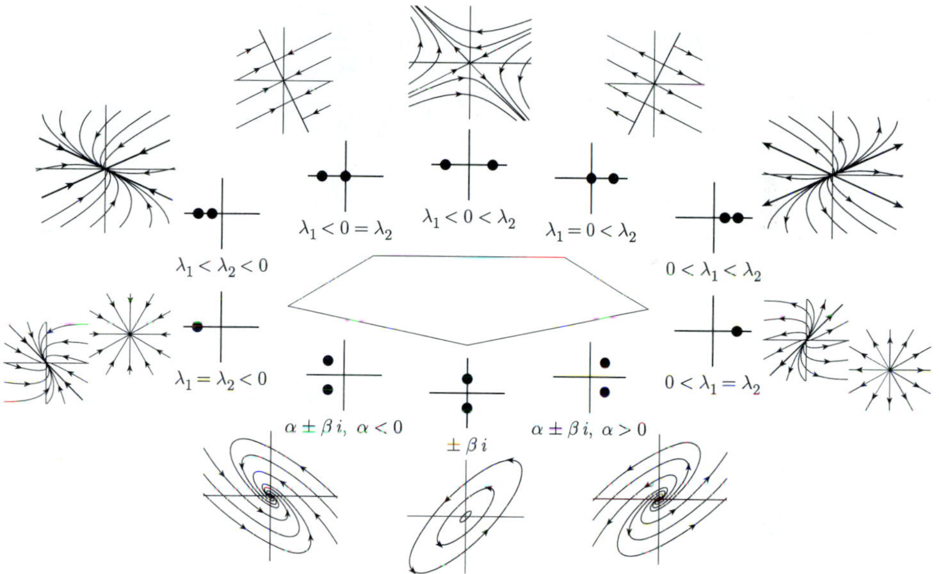

Figure 17

Another important point concerning Figure 17 is that each edge of the pentagon corresponds to a "noncritical" case in the sense that a sufficiently small change in the entries of the matrix will not result in a different case, while each corner of the pentagon corresponds to a "critical" case, meaning that any small change in the entries of the matrix can result in a different case. Also, the repeated-eigenvalue case is represented by two quite different phase portraits: the coaxial node and the proper node. (What determines which one occurs?)

PROBLEMS

For each phase portrait in Problems 1 through 6,

(a) classify the origin as stable or unstable, and as a biaxial node, coaxial node, saddle point, spiral point, or center;

(b) describe the eigenvalues of the corresponding matrix as real and distinct, real and repeated, or nonreal complex. If the eigenvalues are real, state whether they are positive,

---

* Real numbers are on the horizontal axis; imaginary numbers are on the vertical axis. Complex numbers with positive real part lie to the right of the vertical axis, while complex numbers with negative real part lie to the left of the vertical axis.

negative, or of opposite sign. If the eigenvalues are not real, state whether the real part is positive, negative, or zero.

1.

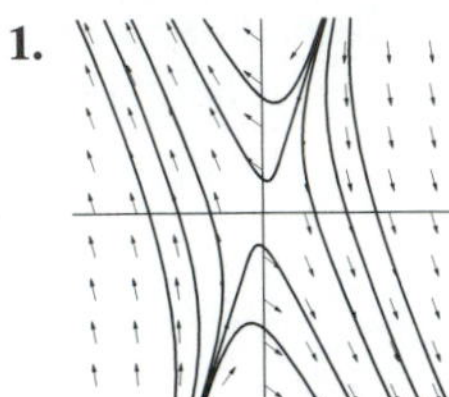

2.

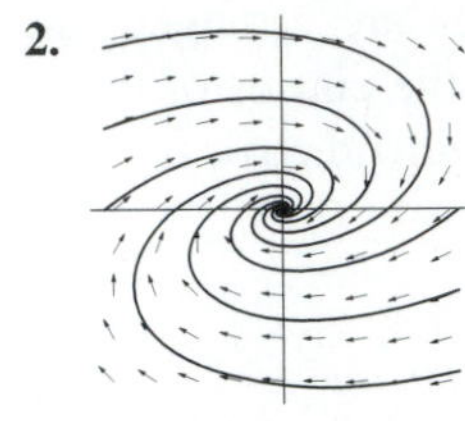

3.

4.

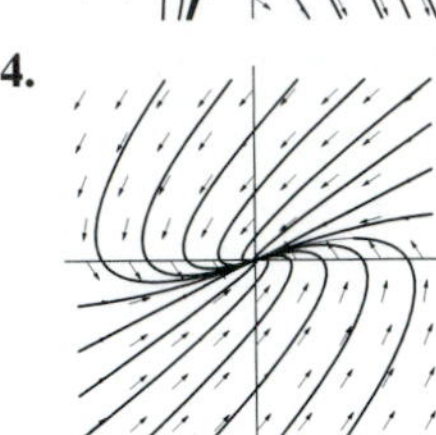

5.

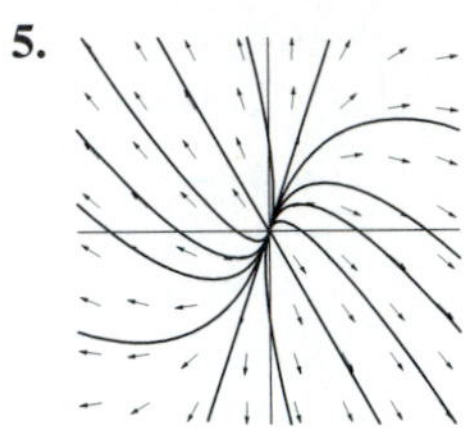

6.

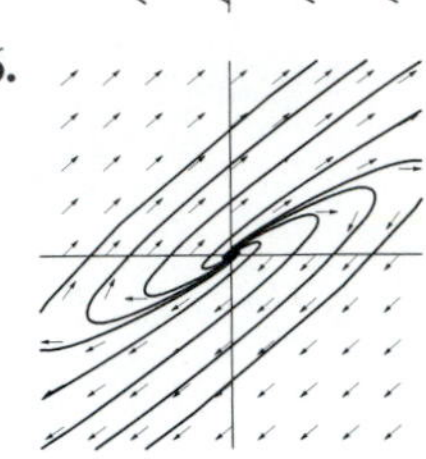

In Problems 7 through 10, sketch the phase portrait of $\mathbf{w}' - A\mathbf{w} = \mathbf{0}$, where A has the given eigenpairs. Include any straight-line orbits in the sketch.

7. $\left(-1, \begin{pmatrix} 1 \\ 2 \end{pmatrix}\right), \left(-2, \begin{pmatrix} 1 \\ -2 \end{pmatrix}\right)$

8. $\left(1, \begin{pmatrix} 1 \\ 2 \end{pmatrix}\right), \left(-2, \begin{pmatrix} 1 \\ -2 \end{pmatrix}\right)$

9. $\left(1, \begin{pmatrix} 1 \\ 1 \end{pmatrix}\right), \left(2, \begin{pmatrix} 2 \\ -1 \end{pmatrix}\right)$

10. $\left(-1, \begin{pmatrix} 1 \\ 1 \end{pmatrix}\right), \left(-1, \begin{pmatrix} 1 \\ 1 \end{pmatrix}\right)$

In Problems 11 through 14, suppose that the eigenvalues of A are nonreal complex conjugates with negative real parts. Sketch a typical orbit of $\mathbf{w}' - A\mathbf{w} = \mathbf{0}$ for the given sign pattern of the entries of A. How would the sketch be different if the real parts of the eigenvalues were positive?

11. $\begin{pmatrix} + & - \\ + & - \end{pmatrix}$

12. $\begin{pmatrix} + & + \\ - & - \end{pmatrix}$

13. $\begin{pmatrix} - & + \\ - & - \end{pmatrix}$

14. $\begin{pmatrix} - & - \\ + & - \end{pmatrix}$

15. Suppose that the x- and y-nullclines of the system $\mathbf{w}' - A\mathbf{w} = \mathbf{0}$ are $x - 2y = 0$ and $2x - y = 0$, respectively, and that $a_{12} < 0$. Sketch the phase portrait of the system, given that the eigenvalues of A are

(a) $\pm 3i$

(b) $-1 \pm 2i$

(c) $\frac{1}{4} \pm i$

(d) $-3/5, -1/3$

16. Consider the matrix $A = \begin{pmatrix} k & -k \\ 3 & -2 \end{pmatrix}$, where k is a parameter, noting that the isoclines of the system $\mathbf{w}' - A\mathbf{w} = \mathbf{0}$ are independent of k for $k \neq 0$. Describe the origin as an equilibrium point of $\mathbf{w}' - A\mathbf{w} = \mathbf{0}$ as k varies over $(-\infty, 0) \cup (0, \infty)$. What happens if $k = 0$?

For the matrices in Problems 17 and 18,

(a) graph (by computer or graphing calculator) the real part(s) of the eigenvalues as a function of k;

(b) use the graph to infer the values of k for which the eigenvalues of the matrix are real and distinct, real and repeated, imaginary, and nonreal complex, respectively;

(c) describe what type of equilibrium point $\mathbf{0}$ can be for $\mathbf{w}' - A\,\mathbf{w} = \mathbf{0}$ as k varies over $(-\infty, \infty)$.

17. $A = \begin{pmatrix} 0 & 1 \\ -1 & k \end{pmatrix}$

18. $A = \begin{pmatrix} -1 & 1 \\ -2 & k \end{pmatrix}$

19. Let A be a 2×2 matrix with a repeated eigenvalue, and let $\tilde{A} = A + \begin{pmatrix} \varepsilon & 0 \\ 0 & 0 \end{pmatrix}$. Show that if ε has the same sign as $a_{11} - a_{22}$ (with no restriction on ε if $a_{11} - a_{22} = 0$) and $|\varepsilon|$ is sufficiently small, then the eigenvalues of $\tilde{A}$ will be real, distinct, and of the same sign. Show further that if ε and $a_{11} - a_{22}$ have opposite signs and $|\varepsilon|$ is sufficiently small, then the eigenvalues of $\tilde{A}$ will be nonreal complex conjugates.

Nullclines and Straight-line Orbits One can usually make a reasonable sketch of a phase portrait after calculating eigenvalues and plotting nullclines and straight-line orbits. For a linear system $\mathbf{w}' - A\mathbf{w} = \mathbf{0}$, where $A = \begin{pmatrix} a & b \\ c & d \end{pmatrix}$, nullclines are simply lines with the equations

$$ax + by = 0 \quad \text{and} \quad cx + dy = 0.$$

The line $ax + by = 0$ is the x-nullcline (where $x' = 0$), and the the line $cx + dy = 0$ is the y-nullcline (where $y' = 0$). Straight-line orbits can be determined by calculating eigenvectors, but an interesting alternative approach is to determine the slope(s) m for which some solution satisfies $y = mx$. If $y = mx$, then

$$\frac{dy}{dx} = \frac{cx + dy}{ax + by} \quad \text{becomes} \quad m = \frac{c + dm}{a + bm},$$

which can be solved for m. If there are no straight-line solutions, then this equation will have no real solutions. For each of the following systems, use this technique to make a rough sketch of the phase portrait based on the eigenvalues, nullclines, and straight-line orbits (or lack thereof).

20. $\begin{cases} x' = y \\ y' = x + y \end{cases}$

21. $\begin{cases} x' = y \\ y' = -2x - 3y \end{cases}$

22. $\begin{cases} x' = -x + 3y \\ y' = x + y \end{cases}$

23. $\begin{cases} x' = -x + y \\ y' = -4x - y \end{cases}$

24. $\begin{cases} x' = -3x + 4y \\ y' = -x + y \end{cases}$

25. $\begin{cases} x' = x - y \\ y' = -x + y \end{cases}$

9.3 Phase Portraits of Nonlinear Systems

This section is a study the phase portraits of autonomous nonlinear systems

$$x' = f(x, y), \quad y' = g(x, y), \tag{1}$$

which we will have occasion to express more compactly as

$$\mathbf{w}' = F(\mathbf{w}),$$

where $\mathbf{w} = (x, y)^T$ and F is the *vector field* defined by $F(\mathbf{w}) = (f(x, y), g(x, y))^T$. We assume that f and g have continuous second-order partial derivatives. Examples 1, 3, and 4 in Section 9.1 dealt with systems of this type. The following example is conjured specifically to preview our further discussion here.

Example 1

Consider the system

$$x' = x(y - x), \quad y' = \frac{1}{2}y(x^2 - 7x - 2y + 14).$$

The equilibrium points for this system are found by solving the algebraic system

$$x(y - x) = 0, \quad y(x^2 - 7x - 2y + 14) = 0.$$

There are four solutions, namely

$$(0, 0), \ (0, 7), \ (2, 2), \quad \text{and} \quad (7, 7).$$

The first-quadrant portion of the phase portrait—along with a "magnification" of a small neighborhood of each equilibrium point—is shown in Figure 1.

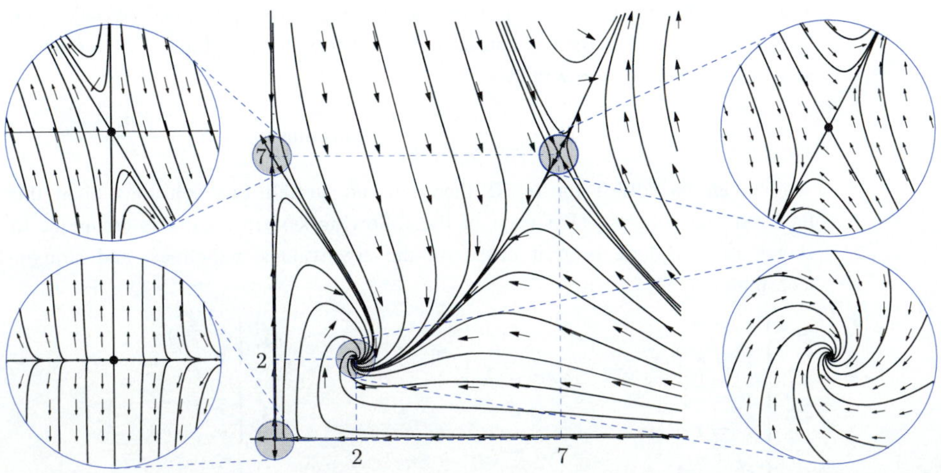

Figure 1

Comparing the magnified regions in Figure 1 to the phase portraits of linear systems discussed in the previous section, we observe the following. The points $(0, 7)$ and $(7, 7)$ exhibit the characteristics of a saddle point, and $(2, 2)$ appears to be a stable spiral point. The phase portrait near the origin does not match any of the types seen with linear systems. In fact, the right half of the magnified phase portrait near $(0, 0)$ shows behavior similar to that near a saddle point, while the left half shows behavior similar to that near an unstable biaxial node. ■

Remarkably, the phase portrait of (1) near an equilibrium point typically—though not always—shares characteristic qualitative properties with the phase-portrait of a linear system near the origin. Hence it often makes sense to describe an equilibrium point of (1) as being a stable biaxial node, a saddle point, an unstable spiral point, or any of the other specific categorizations of $\mathbf{0}$ as an equilibrium point of a linear system. For instance, in Example 1 we can say that $(0, 7)$ and $(7, 7)$ *are* saddle points, $(2, 2)$ *is* a stable spiral point, and so on. The role of straight-line orbits, which is always important at a node or saddle point of a linear system, is typically played in a nonlinear system by special orbits called *separatrices*,* which sometimes "connect" equilibrium points. Can you identify the separatrices in Figure 1?

Linearization at an Equilibrium Point Because f and g have continuous second-order partials, Taylor's theorem for functions of two variables implies the following expansions about (x_0, y_0):

$$f(x, y) = f(x_0, y_0) + f_x(x_0, y_0)(x - x_0) + f_y(x_0, y_0)(y - y_0) + R_1(x, y),$$
$$g(x, y) = g(x_0, y_0) + g_x(x_0, y_0)(x - x_0) + g_y(x_0, y_0)(y - y_0) + R_2(x, y),$$

where the subscripts on f and g denote partial derivatives and the "remainder" terms R_1 and R_2 have the property that, for $i = 1, 2$,

$$\lim_{\substack{x \to x_0 \\ y \to y_0}} \frac{R_i(x, y)}{\sqrt{(x - x_0)^2 + (y - y_0)^2}} = 0. \tag{2}$$

If (x_0, y_0) is an equilibrium point of (1), then $f(x_0, y_0) = g(x_0, y_0) = 0$, and so

$$f(x, y) = f_x(x_0, y_0)(x - x_0) + f_y(x_0, y_0)(y - y_0) + R_1(x, y),$$
$$g(x, y) = g_x(x_0, y_0)(x - x_0) + g_y(x_0, y_0)(y - y_0) + R_2(x, y).$$

Written in vector form, this becomes

$$\begin{pmatrix} f(x, y) \\ g(x, y) \end{pmatrix} = \begin{pmatrix} f_x(x_0, y_0) & f_y(x_0, y_0) \\ g_x(x_0, y_0) & g_y(x_0, y_0) \end{pmatrix} \begin{pmatrix} x - x_0 \\ y - y_0 \end{pmatrix} + \begin{pmatrix} R_1(x, y) \\ R_2(x, y) \end{pmatrix}.$$

With $F = (f, g)^T$ and $\mathbf{w} = (x, y)^T$, we then have

$$F(\mathbf{w}) = \mathcal{J}(\mathbf{w_0})(\mathbf{w} - \mathbf{w_0}) + R(\mathbf{w}),$$

where $\mathcal{J}(\mathbf{w_0})$ is the **Jacobian matrix** of F at $\mathbf{w_0}$, defined by

$$\mathcal{J}(\mathbf{w_0}) = \begin{pmatrix} f_x(x_0, y_0) & f_y(x_0, y_0) \\ g_x(x_0, y_0) & g_y(x_0, y_0) \end{pmatrix}.$$

* The singular is *separatrix*.

Since $(\mathbf{w} - \mathbf{w_0})' = \mathbf{w}'$, the system (1) can now be written as

$$(\mathbf{w} - \mathbf{w_0})' = \mathcal{J}(\mathbf{w_0})(\mathbf{w} - \mathbf{w_0}) + R(\mathbf{w}).$$

In order to shift the equilibrium point to the origin, we make the substitution $\mathbf{u} = \mathbf{w} - \mathbf{w_0}$ and obtain

$$\mathbf{u}' = \mathcal{J}(\mathbf{w_0})\mathbf{u} + R(\mathbf{u} + \mathbf{w_0}). \tag{3}$$

With these contortions done, we can now get to the point. When $\mathbf{u}$ is very close to $\mathbf{0}$ (i.e., $\mathbf{w}$ is very close to $\mathbf{w}_0$) solutions of (3) behave much the same as solutions of the *linear* equation

$$\mathbf{u}' = \mathcal{J}(\mathbf{w}_0)\mathbf{u},$$

provided that 0 is not an eigenvalue of $\mathcal{J}(\mathbf{w}_0)$. More specifically, the following theorem is true. We omit the proof, as it is highly technical and requires more mathematical subtlety than is appropriate here.

Theorem 1

Let $\mathbf{w}_0$ be an isolated[†] *equilibrium point of $\mathbf{w}' = F(\mathbf{w})$. Let λ_1 and λ_2 be the eigenvalues of $\mathcal{J}(\mathbf{w}_0)$, and suppose that neither λ_1 nor λ_2 is zero.*

- *If λ_1 and λ_2 are distinct and not purely imaginary, then $\mathbf{w}_0$ is the same type of equilibrium point as is $\mathbf{0}$ for the linear system $\mathbf{u}' = \mathcal{J}(\mathbf{w}_0)\mathbf{u}$.*
- *If λ_1 and λ_2 are purely imaginary, then $\mathbf{w}_0$ is either a center point, a stable spiral point, or an unstable spiral point.*
- *If $\lambda_1 = \lambda_2$, then $\mathbf{w}_0$ is either a node or a spiral point with stability determined by the sign of of the eigenvalues.* ▲

REMARK In the context of nonlinear systems, we will no longer attempt to make any distinction between biaxial, coaxial, and proper nodes.

Example 1 (*continued*)

We will examine the eigenvalues of the Jacobian matrix at each critical point. The Jacobian matrix at (x, y) is easily found to be

$$\mathcal{J}(x, y) = \begin{pmatrix} y - 2x & x \\ xy - \frac{7}{2}y & \frac{1}{2}(x^2 - 7x - 4y + 14) \end{pmatrix}.$$

At the equilibrium point $(0, 7)$, we have

$$\mathcal{J}(0, 7) = \begin{pmatrix} 7 & 0 \\ -\frac{49}{2} & -7 \end{pmatrix},$$

[†] An equilibrium point is **isolated** if it is the center of some disk that contains no other equilibrium points.

whose eigenvalues are ± 7. This confirms our graphical observation that $(0, 7)$ is a saddle point. At $(7, 7)$, we have

$$\mathcal{J}(7,7) = \begin{pmatrix} -7 & 7 \\ \frac{49}{2} & -7 \end{pmatrix},$$

for which the characteristic equation is $\lambda^2 + 14\lambda - 245/2 = 0$. Thus a quick computation shows the eigenvalues to be $7(-1 \pm \sqrt{7/2})$, confirming that $(7, 7)$ is a saddle point. At $(2, 2)$, we have

$$\mathcal{J}(2,2) = \begin{pmatrix} -2 & 2 \\ -3 & -2 \end{pmatrix}.$$

The characteristic equation is $\lambda^2 + 4\lambda + 10 = 0$, from which we find complex eigenvalues $-2 \pm i\sqrt{6}$ (with negative real parts). This confirms the observation that $(2, 2)$ is a stable spiral point. Finally, at $(0, 0)$ the Jacobian matrix is

$$\mathcal{J}(0,0) = \begin{pmatrix} 0 & 0 \\ 0 & 7 \end{pmatrix},$$

whose eigenvalues are 0 and 7. Since zero is an eigenvalue, Theorem 1 *does not apply*. This is fortunate, because the behavior of orbits near $(0, 0)$ is not consistent with any of the possible phase portraits for linear systems. However, the observed behavior is sensible in light of the critical nature of the zero-eigenvalue case for nontrivial linear systems. ■

Example 2

Consider the system

$$x' = \frac{1}{2}x(2y - (x-4)^2),$$

$$y' = \frac{1}{2}y(-x^2 + 8x - 2y - 8).$$

The equilibrium points for this system are found by solving the algebraic system

$$x(2y - (x-4)^2) = 0, \quad y(x^2 + 8x - 2y - 8) = 0,$$

which yields five solutions:

$$(0,0), \ (0,-4), \ (4,0), \ (2,2), \quad \text{and} \quad (6,2).$$

The Jacobian matrix at (x, y) is

$$\mathcal{J}(x,y) = \frac{1}{2}\begin{pmatrix} -16 + 16x - 3x^2 + 2y & 2x \\ 8y - 2xy & -8 + 8x - x^2 - 4y \end{pmatrix}$$

The following table indicates the results of the "linearization analysis" suggested by Theorem 1 at each equilibrium point.

(x_0, y_0)	$\mathcal{J}(x_0, y_0)$	λ_1, λ_2	Conclusion
$(0, 0)$	$\begin{pmatrix} -8 & 0 \\ 0 & -4 \end{pmatrix}$	$-8, -4$	Stable node
$(0, -4)$	$\begin{pmatrix} -12 & 0 \\ -16 & 4 \end{pmatrix}$	$-12, 4$	Saddle point
$(4, 0)$	$\begin{pmatrix} 0 & 4 \\ 0 & 4 \end{pmatrix}$	$0, 4$	?
$(2, 2)$	$\begin{pmatrix} 4 & 2 \\ 4 & -2 \end{pmatrix}$	$1 \pm \sqrt{17}$	Saddle point
$(6, 2)$	$\begin{pmatrix} -12 & 6 \\ -4 & -2 \end{pmatrix}$	$-8, -6$	Stable node

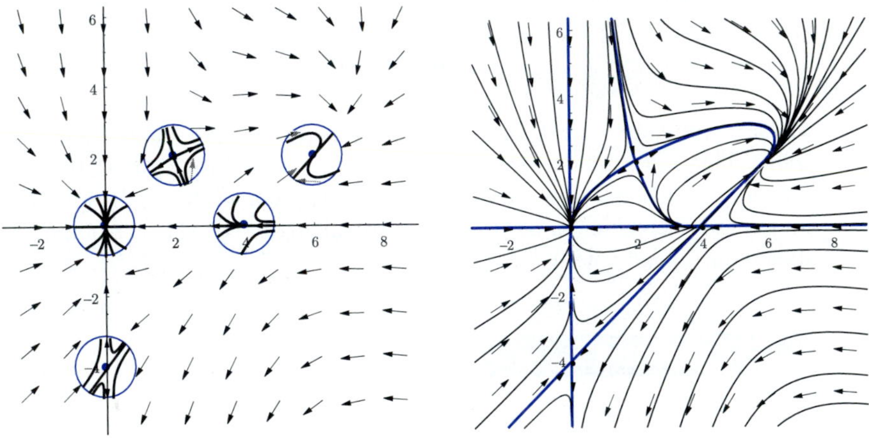

Figure 2a **Figure 2b**

The critical point at $(4, 0)$ cannot be categorized this way because $\mathcal{J}(4, 0)$ has a zero eigenvalue. However, since the nonzero eigenvalue is positive, $(4, 0)$ will be unstable, and it will likely share certain properties with an unstable node and/or a saddle point. The direction field and a rough sketch of expected orbit behavior near each equilibrium point are shown in Figure 2a, where both the direction field and the preceding linearization analysis are taken into account. Figure 2b is a rendering of the full phase portrait. Notice the interesting network of separatrices connecting the saddle points and nodes in Figure 2b. ■

Example 3

Consider the system

$$x' = x(y - 5)$$
$$y' = -y(y + (x - 1)(x - 7)).$$

The equilibrium points of this system are found by solving the algebraic system

$$x(y-5)=0,$$
$$y(y+(x-1)(x-7))=0,$$

which yields four solutions:

$$(0,0),\ (0,-7),\ (2,5),\ (6,5).$$

The Jacobian matrix of the system at (x, y) is

$$\mathcal{J}(x,y)=\begin{pmatrix} -5+y & x \\ -2(-4+x)y & -7+8x-x^2-2y \end{pmatrix}.$$

The following table indicates the results of the linearization analysis suggested by Theorem 1 at each equilibrium point. Since the numbers are not as tidy as in the previous example, we show some of the eigenvalues to only one decimal place.

(x_0, y_0)	$\mathcal{J}(x_0, y_0)$	λ_1, λ_2	Conclusion
$(0, 0)$	$\begin{pmatrix} -5 & 0 \\ 0 & -7 \end{pmatrix}$	$-5, -7$	Stable node
$(0, -7)$	$\begin{pmatrix} -12 & 0 \\ -56 & 7 \end{pmatrix}$	$-12, 7$	Saddle point
$(2, 5)$	$\begin{pmatrix} 0 & 2 \\ 20 & -5 \end{pmatrix}$	$-9.3, 4.3$	Saddle point
$(6, 5)$	$\begin{pmatrix} 0 & 6 \\ -20 & -5 \end{pmatrix}$	$2.5 \pm 10.7i$	Stable spiral point

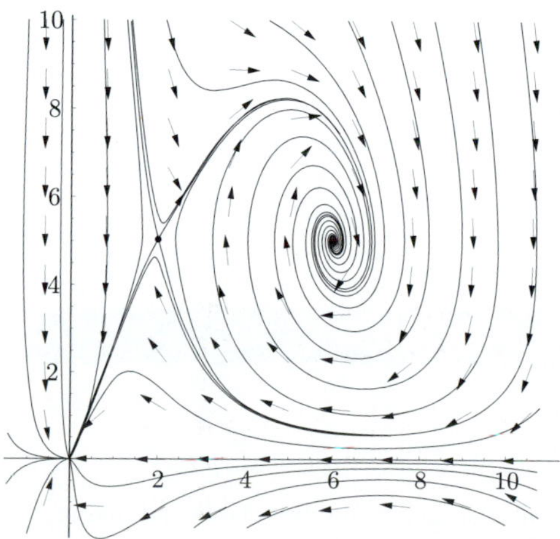

Figure 3

Our conclusion about the character of each of these equilibrium points supports the behavior indicated by Figure 3, which shows only the first-quadrant portion of the phase portrait. ■

Example 4

Consider the system

$$x' = -2x + 3y + 8x^2 - 5xy - 4y^2,$$
$$y' = -2x + 2y + 5x^2 + xy - 6y^2.$$

It is relatively easy to show that $(0, 0)$ and $(1, 1)$ are the only equilibrium points of this system and that the eigenvalues of $\mathcal{J}(x, y)$ at each of them are purely imaginary. So it is not obvious whether the two equilibrium points are centers, stable spiral points, or unstable spiral points. To discern the nature of these equilibrium points by precise analysis is difficult. However, a computer-generated phase portrait, as shown in Figure 4, reveals that $(0, 0)$ is an unstable spiral point and $(1, 1)$ is a stable spiral point. Note that the rotations about these spiral points are opposite in direction. (Could it possibly be otherwise?) ■

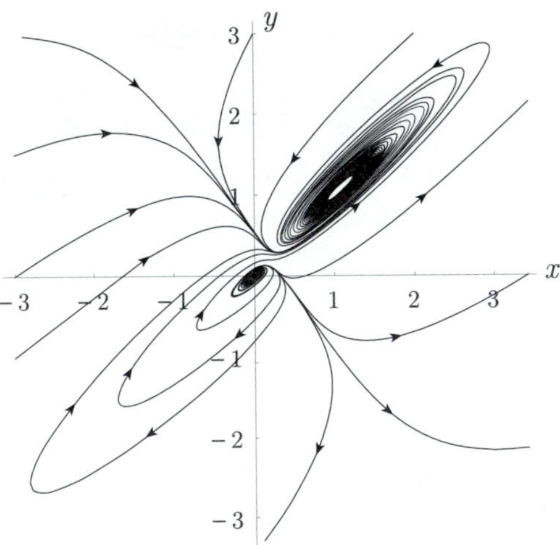

Figure 4

PROBLEMS

For each of the systems in Problems 1 through 9,

(a) find the equilibrium points;

(b) evaluate the Jacobian matrix at each equilibrium point and find its eigenvalues;

(c) state the nature of each equilibrium point for which it is possible to do so;

(d) sketch the nullclines and the phase portrait *by hand*;

(e) use suitable computer software or a graphing calculator to graph the nullclines and the phase portrait.

1. $\begin{cases} x' = x(y - x - 2) \\ y' = x^2 - y \end{cases}$

2. $\begin{cases} x' = x(1 - y) \\ y' = -y(1 - x) \end{cases}$

3. $\begin{cases} x' = x(1 - y) \\ y' = x - y \end{cases}$

4. $\begin{cases} x' = y \\ y' = x(4 - x) \end{cases}$

5. $\begin{cases} x' = 2 - x^2 - y^2 \\ y' = x - y \end{cases}$

6. $\begin{cases} x' = 2 - x^2 - y^2 \\ y' = y(x - y) \end{cases}$

7. $\begin{cases} x' = x^2 + y^2 - 2 \\ y' = x^2 - y^2 \end{cases}$

8. $\begin{cases} x' = -x\,y + 3 \\ y' = -x^2y + 2x + 1 \end{cases}$

9. $\begin{cases} x' = x^2y + y - 1 \\ y' = -x^2y + x - y \end{cases}$

10. The equilibrium points for the pendulum system

$$x' = y, \quad y' = -k \sin x \quad (k > 0)$$

are $(\pm n\pi, 0)$, $n = 0, 1, 2, \ldots$. Show that these are saddle points when n is odd. Give a physical interpretation of this fact.

11. A damped pendulum gives rise to the system

$$x' = y, \quad y' = -\rho(y) - k \sin x,$$

where $k > 0$ and ρ is a differentiable function for which $\rho(0) = 0$ and $\rho'(y) > 0$. Show that the equilibrium points are $(\pm n\pi, 0)$, $n = 0, 1, 2, \ldots$ (just as in the undamped case) and that these are saddle points when n is odd and either stable spiral points or stable nodes when n is even. Give a physical interpretation of these facts.

12. Show that the origin is the only equilibrium point of the van der Pol system

$$x' = y, \quad y' = -x - k(x^2 - 1)y,$$

and determine the nature of that equilibrium point as k varies over $(0, \infty)$.

13. Consider the system

$$x' = -xy^2 + 1, \quad y' = xy^2 - ky,$$

where k is a positive parameter. First find the equilibrium point in terms of k and then determine the nature of that equilibrium point as k varies over $(0, \infty)$. Create (by hand or computer) a typical phase portrait to illustrate each distinct case.

For the systems in 14 and 15, after plotting nullclines and determining the nature of equilibrium points by linearization, sketch (by hand) the first-quadrant portion of the phase portrait.

14. $\begin{cases} x' = x(1 - xy) \\ y' = y(x - y) \end{cases}$

15. $\begin{cases} x' = x(1 - x^2y) \\ y' = y\left(\dfrac{2}{1 + x^2} - y\right) \end{cases}$

Let $h(x) = x(x-1)^2(x-2)(x-3)$. The following phase portraits (which also appear in Problem 6 of Section 9.1) are for the following systems:

16. $\begin{cases} x' = y \\ y' = h(x) \end{cases}$

17. $\begin{cases} x' = -y \\ y' = h(x) - y \end{cases}$

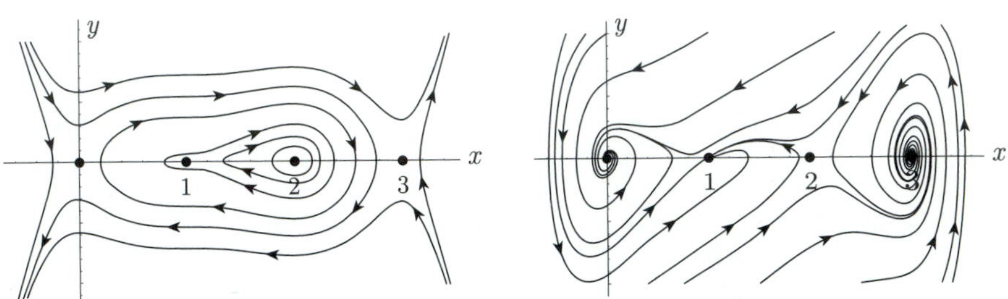

Perform a linearization analysis at each equilibrium point and describe how the analyses are consistent with the phase portrait shown.

18. The phase portrait below is for the system

$$x' = \frac{x(2-x-4y)(x+4y-3)}{1+(x+4y)^2},$$

$$y' = \frac{y}{2}(2-2x-y).$$

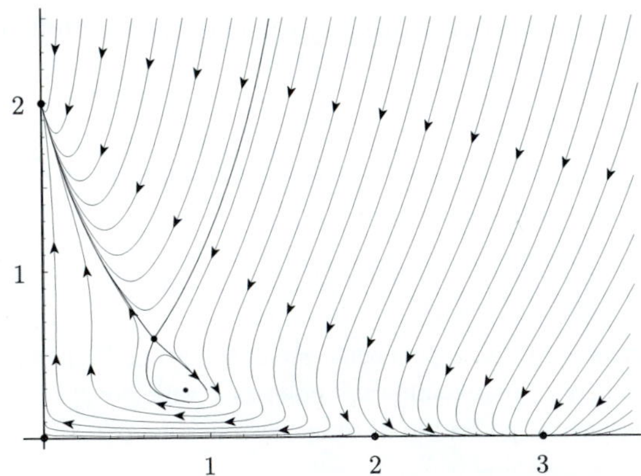

Perform a linearization analysis at each of the five equilibrium points and describe how the analyses are consistent with the phase portrait. Describe the "unusual" aspect of the phase portrait.

19. Consider the system

$$x' = 6y - 2y^2, \quad y' = 4x^2 + 2xy + 4y^2 - 10x - y.$$

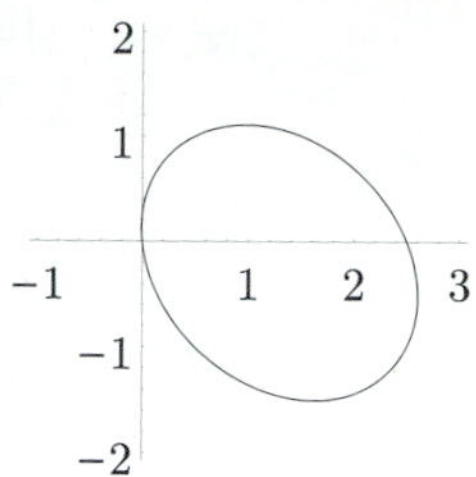

(a) Find the equilibrium points and determine their nature by linearization.

(b) The y-nullcline is the ellipse shown on the right. Using it, as well as the x-nullcline(s) and the results of part (a), sketch the phase portrait of the system.

(c) Show that the circle $(x - 1/2)^2 + y^2 = 1$ satisfies the slope equation

$$\frac{dy}{dx} = \frac{4x^2 + 2xy + 4y^2 - 10x - y}{6y - 2y^2}.$$

(*Suggestion*: Parametrize the circle by $x = 1/2 + \cos\theta$, $y = \sin\theta$, and look at $\frac{dy}{d\theta}/\frac{dx}{d\theta}$.) Finally, modify your phase-portrait sketch to take this into account.

20. Consider the system

$$x' = -2xy, \quad y' = x^2 - y^2.$$

(a) Show that $(0, 0)$ is the only equilibrium point and that $\mathcal{J}(0, 0)$ is the zero matrix.

(b) Make a rough sketch of the phase portrait based on the nullclines of the system.

(c) Show that circles of the form $x^2 + y^2 - 2ax = 0$ satisfy the slope equation

$$\frac{dy}{dx} = -\frac{x^2 - y^2}{2xy},$$

and use this fact to refine your sketch of the phase portrait. Do orbits near the origin bear resemblance to orbits near the origin in any linear system? Explain.

21. The curves traced out by orbits of a system

$$x' = f(x, y), \quad y' = g(x, y),$$

are solution curves of

$$\frac{dy}{dx} = \frac{g(x, y)}{f(x, y)}.$$

Suppose that an implicit solution of this equation is given by $\varphi(x, y) = C$.

(a) Describe the connection between the phase portrait of the system and the graph of φ (i.e., the surface $z = \varphi(x, y)$).

(b) Suppose that (x_0, y_0) is an isolated equilibrium point of the system. If (x_0, y_0) is an isolated critical point of φ at which φ attains either a local maximum or a local minimum, then what conclusions can be reached about the nearby orbits of the system and the nature of (x_0, y_0) as an equilibrium point? Explain.

22. Consider the system $x' = y$, $y' = 4x(1 - 2x^2)$.

(a) Show that the origin is a saddle point and that the other two equilibrium points might be centers or spiral points.

(b) Describe the curves traced out by orbits by finding, in the form $\varphi(x, y) = C$, an implicit general solution of the separable equation

$$\frac{dy}{dx} = \frac{4x(1 - 2x^2)}{y}.$$

(c) Show that $\varphi(x, y)$ attains a locally maximum value at two of the three equilibrium points. (Consult your calculus book as needed.) What conclusions can be reached about the nearby orbits and the nature of these equilibrium points?

(d) Show that the Lissajous curve parametrized by $x = \sin\theta$, $y = \sin 2\theta$, satisfies $\varphi(x, y) = C$ and thus is the separatrix traced out by orbits that approach the saddle point $(0, 0)$ as $t \to \pm\infty$. Include this curve in a sketch of the phase portrait.

23. Consider the pendulum system $x' = y$, $y' = -k \sin x$, where $k > 0$. The equilibrium points for this system are $(\pm n\pi, 0)$, $n = 0, 1, 2, \ldots$. Linearization analysis reveals that these are saddle points when n is odd (see Problem 10), but the method fails when n is even, because the Jacobian matrix has imaginary eigenvalues.

(a) Describe the curves traced out by orbits by finding, in the form $\varphi(x, y) = C$, an implicit general solution of the separable equation

$$\frac{dy}{dx} = \frac{-\sin x}{y}.$$

(b) Show that if n is even, then the value of $\varphi(x, y)$ at $(\pm n\pi, 0)$ is a local minimum. (Consult your calculus book as needed.) What conclusions can be reached about the nearby orbits and the nature of these equilibrium points?

9.4 Limit Cycles

Consider for a moment an autonomous, *scalar* equation

$$y' = f(y), \tag{1}$$

where f is a continuously differentiable function on $\mathbb{R}$. The solutions of such a problem are very limited in terms of possible behavior. This is mainly due to the fact that any nonconstant solution of (1) must be either strictly increasing or strictly decreasing. (Why?) As a result, for any maximal* solution of (1), one of the following statements must be true:

- $\lim_{t \to T} y(t) = \pm\infty$ for some finite time $T > 0$;
- $\lim_{t \to \infty} y(t) = \pm\infty$;
- $\lim_{t \to \infty} y(t) = b$, where $f(b) = 0$.

These alternatives are illustrated in Figures 1a–c.

* See Section 3.2.

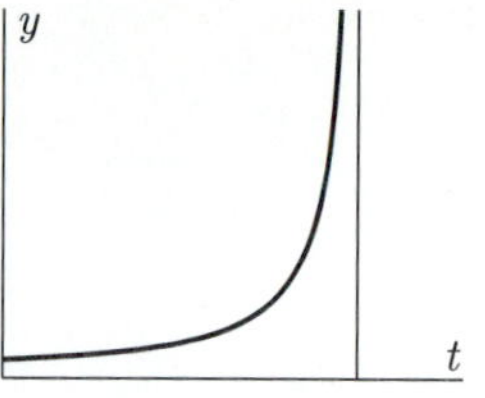

Figure 1a

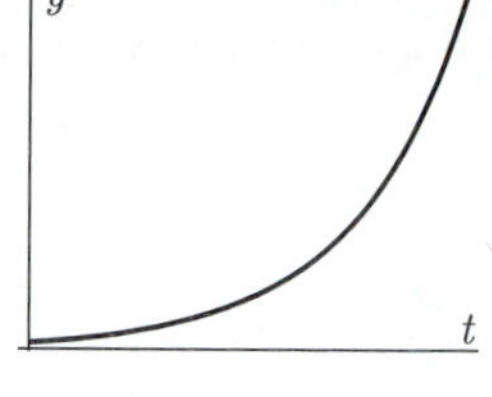

Figure 1b

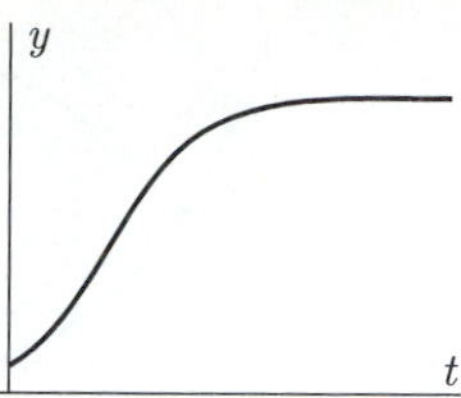

Figure 1c

As a result, a *bounded*[†] maximal solution of (1) must exist for all t and *can do only one thing as* $t \to \infty$: It must approach a constant equilibrium solution. So, in particular, it is impossible for (1) to have a (nonconstant) periodic solution.

Consider now an autonomous *system*

$$x' = f(x, y), \quad y' = g(x, y), \tag{2}$$

where f and g have continuous first partial derivatives. We have already seen that the extra dimension afforded by the plane allows such a system to have periodic solutions. Recall that linear systems in which the matrix has purely imaginary eigenvalues have *only* periodic solutions. The nonlinear van der Pol system of Example 3 in Section 9.1 has a unique *nonsinusoidal* periodic solution to which other solutions are attracted as $t \to \infty$. Many nonlinear systems of two equations have solutions that exhibit more complicated behavior. While such behavior is generally well understood in the planar case, a thorough study of all possible types of planar orbits is far more technical than is appropriate for this text. We will mainly limit our study here to periodic solutions.

Closed Orbits Since periodic solutions give rise to "closed orbits," let us proceed by first carefully defining that notion.

Definition A nontrivial[‡] orbit of (2) is said to be a **closed orbit** if either

i) there is some number $t_1 > 0$ for which $(x(t_1), y(t_1)) = (x(0), y(0))$, or
ii) $\lim_{t\to-\infty} (x(t), y(t))$ and $\lim_{t\to\infty} (x(t), y(t))$ exist and are equal. ◆

In case (i), where a closed curve is traced out completely as t varies over $[0, t_1)$, the closed orbit corresponds to a periodic solution. In case (ii), the solution approaches the same equilibrium point as $t \to \pm\infty$, forming what is known as a *homoclinic orbit*. Figure 2 illustrates a homoclinic orbit that approaches a saddle point as $t \to \pm\infty$.

Definition The closed orbit of a periodic solution is a **periodic orbit**. A periodic orbit is a **limit cycle** if it is *isolated* (i.e., if it lies in some open set that contains no other closed orbits). ◆

Note that the periodic orbits of a linear system with nonreal eigenvalues (which trace out concentric circles or ellipses) are not isolated and therefore not limit cycles.

[†] y is **bounded** if there is a constant M such that $|y(t)| \le M$ for all t in its domain.

[‡] An orbit is *trivial* if it consists of only an equilibrium point.

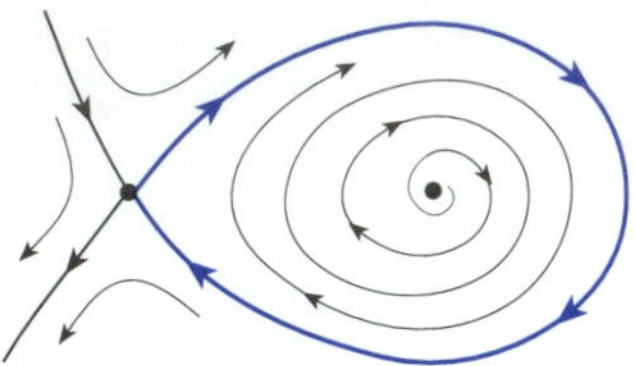

Figure 2

The following example presents two systems with simple limit cycles.

Example 1

The clockwise circular orbit $(\sin t, \cos t)$ is a limit cycle for

$$x' = y + x(1 - x^2 - y^2),$$
$$y' = -x + y(1 - x^2 - y^2),$$

and the counterclockwise circular orbit $(\cos t, \sin t)$ is a limit cycle for

$$x' = -y - x(1 - x^2 - y^2),$$
$$y' = x - y(1 - x^2 - y^2).$$

The phase portrait of the first of these is shown in Figure 3a. Note that the origin is the only equilibrium point, and it is an unstable spiral point. Figure 3b shows the phase portrait of the second system. There the origin is also the only equilibrium point, and it is a stable spiral point.

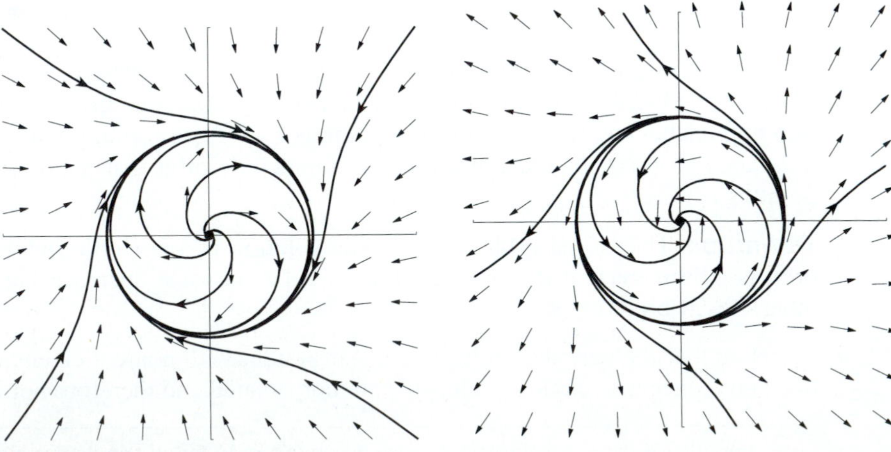

Figure 3a

Figure 3b

The phase portraits in Figures 3a and b suggest a notion of *orbital stability*. A limit cycle such as the one in Figure 3a is said to be **orbitally stable**, while a limit cycle such as the one in Figure 3b is said to be **orbitally unstable**.

The systems in Example 1 also provide illustration of the following theorem. Its proof is an application of Green's theorem and is the subject of Problem 9. ■

Theorem 1

Every periodic orbit must enclose at least one equilibrium point. ▲

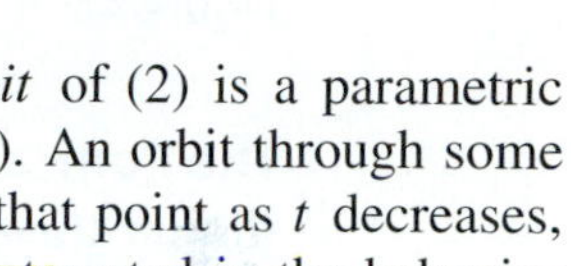

Semiorbits and Invariant Regions Recall that an *orbit* of (2) is a parametric curve $(x(t), y(t))$ associated with a maximal solution of (2). An orbit through some initial point $(x(t_0), y(t_0))$ can be followed backward from that point as t decreases, as well as forward as t increases. Since we are primarily interested in the behavior of solutions as $t \to \infty$ (i.e., for increasing t), we define the term *forward semiorbit* as follows.

Definition Given a maximal solution of (2) and any t_0 in its interval of existence I, the parametric curve $(x(t), y(t))$, for t in $I \cap [t_0, \infty)$, is said to be a **forward semiorbit** of (2). ◆

An important role in our main theorem concerning periodic orbits in the plane will be played by *forward-invariant regions*.

Definition A region $\mathcal{M}$ in the plane is said to be a **forward-invariant region** for (2) if it contains every forward semiorbit that begins in $\mathcal{M}$. ◆

Note that if $\mathcal{M}$ is a forward-invariant region, then every orbit for which $(x(t_0), y(t_0))$ is in $\mathcal{M}$—at *any* time t_0—remains in Ω for all $t \geq t_0$. So a forward-invariant region is simply a region from which no orbit of (2) can escape as t increases.

In Example 1, every open or closed disk centered at $(0, 0)$ with radius $r \geq 1$ is a forward-invariant region for the first system, whose phase portrait is shown in Figure 3a. For the second system in Example 1, whose phase portrait is shown in Figure 3b, every open or closed disk centered at $(0, 0)$ with radius $r \leq 1$ is a forward-invariant region.

The following theorem gives a sufficient condition for forward-invariance that is general enough to suit our present needs. For its statement we need to introduce another topological concept. Let Ω be a region in the plane. The boundary of Ω, denoted by $\partial\Omega$, consists of all points P in the plane for which every open disk centered at P intersects both Ω and its complement. The union of Ω and $\partial\Omega$ is the **closure** of Ω, which we denote by $\overline{\Omega}$. A region Ω is closed if and only if $\Omega = \overline{\Omega}$.

Theorem 2

Let $\mathcal{M}$ be an open region in the plane, with closure $\overline{\mathcal{M}}$. Suppose that $\partial\mathcal{M}$ consists of a finite number of smooth curves and that, at every point where $\partial\mathcal{M}$ is smooth, the vector field $(f(x, y), g(x, y))$ of (2) is either **0**, *tangent to $\partial\mathcal{M}$, or pointing into $\mathcal{M}$. Then $\overline{\mathcal{M}}$ is a forward-invariant region.* ▲

Figures 4a–c provide three illustrations of forward-invariant regions. Notice the boundary property stated in Theorem 2 in each case.

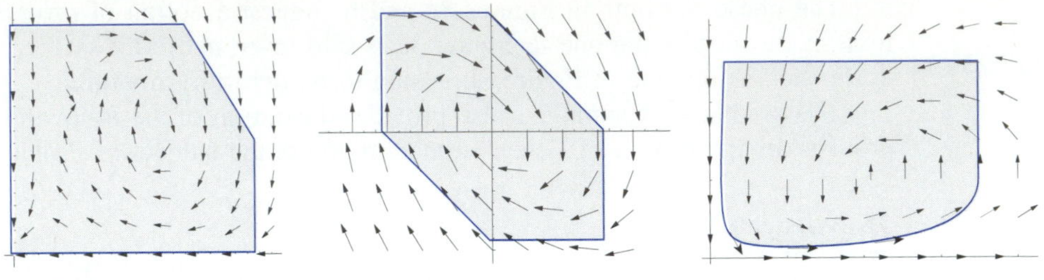

Figure 4a   Figure 4b   Figure 4c

The Poincaré-Bendixson Theorem The following version of the **Poincaré-Bendixson theorem*** provides one means of demonstrating the existence of a periodic solution to an autonomous system in the plane. The proof of this theorem is beyond the scope of this text and hence is omitted.

Theorem 3 *(Poincaré-Bendixson)*

Let $\mathcal{M}$ be a closed, forward-invariant region for (2). If no *forward semiorbit in $\mathcal{M}$ approaches an equilibrium point in $\mathcal{M}$ as $t \to \infty$, then* every *forward semiorbit in $\mathcal{M}$ either is part of a periodic orbit or approaches a periodic orbit as $t \to \infty$; in particular, there exists a periodic orbit in $\mathcal{M}$.* ▲

Example 2

Consider the system

$$x' = -y(4 - x^2 - y^2) + x(2 - 5x^2 - y^2),$$
$$y' = x(4 - x^2 - y^2) + y(3 - 2x^2 - y^2),$$

and the closed disk $\mathcal{M} = \{(x, y) \mid x^2 + y^2 \leq 4\}$, whose boundary is the circle $x^2 + y^2 = 4$. Whenever an orbit intersects that circle, we have

$$x' = x(2 - 5x^2 - y^2) = x(-2 - 4x^2)$$
$$y' = y(3 - 2x^2 - y^2) = y(-1 - x^2).$$

When these are not zero, their signs are opposite those of x and y, respectively. Thus, on the boundary of $\mathcal{M}$, the direction field of the system points into the interior of $\mathcal{M}$. Some *CAS*-aided computation shows that (0, 0) is the system's only equilibrium point and that it is an unstable spiral point; thus no orbit approaches it as $t \to \infty$. Therefore, the Poincaré-Bendixson theorem guarantees that the system has a periodic orbit in $\mathcal{M}$. That periodic orbit is indeed an orbitally stable limit cycle, as evidenced by Figure 5. ■

* The French mathematician Henri Poincaré (1854–1912) was arguably the preeminent mathematician of his time and is often regarded as the father of the modern theory of differential equations. Ivar Bendixson was a Swedish mathematician who in 1901 refined and generalized Poincaré's original theorem published in 1880.

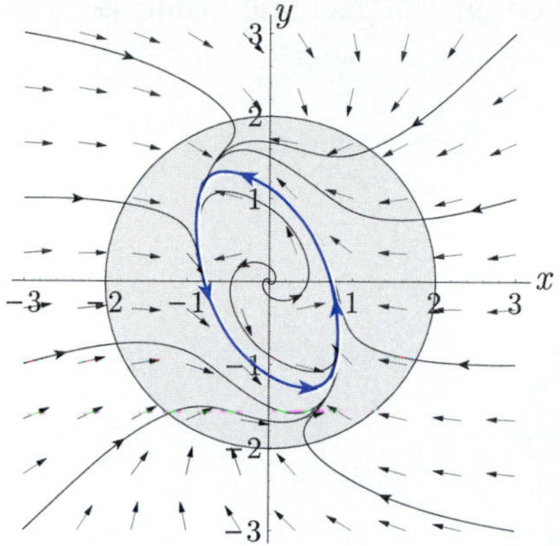

Figure 5

Example 3

Consider the van der Pol system

$$x' = y, \quad y' = -x - (x^2 - 1)y.$$

The Poincaré-Bendixson theorem implies that every orbit of this system is either periodic or approaches a period orbit as $t \to \infty$. The argument is as follows. Clearly the only equilibrium point for this system is $(0, 0)$. The Jacobian matrix at $(0, 0)$ is

$$\mathcal{J}(0,0) = \begin{pmatrix} 0 & 1 \\ -1 - 2xy & 1 - x^2 \end{pmatrix}\Bigg|_{\substack{x=0\\y=0}} = \begin{pmatrix} 0 & 1 \\ -1 & 1 \end{pmatrix},$$

whose eigenvalues are the roots of $\lambda^2 - \lambda + 1$, which are $(1 \pm i\sqrt{3})/2$. Since these have positive real parts, it follows that $(0, 0)$ is an *unstable* spiral point. So we can eliminate the possibility of any orbit approaching an equilibrium point as $t \to \infty$. To construct a forward-invariant region, we begin by making the following observations:

i) *The quantity $x^2 + y^2$, and hence the distance to the origin, is nonincreasing along any orbit whenever $|x| \geq 1$ and nondecreasing along any orbit whenever $-1 < x < 1$.* To see this, multiply the two equations by x and y, respectively, and add to obtain

$$xx' + yy' = -(x^2 - 1)y^2.$$

The assertion follows because $xx' + yy' = \frac{1}{2}\frac{d}{dt}(x^2 + y^2)$.

ii) *The slope $\frac{dy}{dx}$ satisfies $\frac{dy}{dx} \leq \frac{5}{4}$ when $|y| \geq 1$.* This follows from a simple analysis of the function

$$\frac{dy}{dx} = \frac{-x - (x^2 - 1)y}{y} = 1 - x^2 - \frac{x}{y}.$$

Observation (i) tells us that orbits generally move closer to the origin, except when passing through the region $-1 < x < 1$. Observation (ii) provides a limit to how much an orbit's distance from the origin can increase while the orbit is in that region. ■

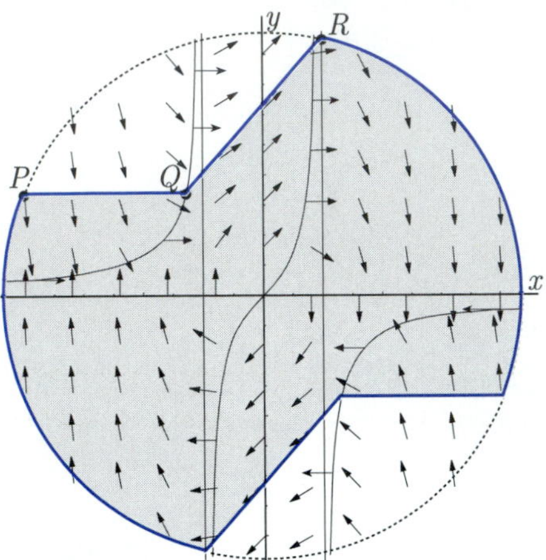

Figure 6

We now proceed with an informal construction of a forward-invariant region. The direction field and the y-nullcline(s) $y = x/(1 - x^2)$ are shown in Figure 6. We first choose a point Q, in the second quadrant, on the graph of $y = x/(1 - x^2)$ and above the line $y = 1$. Then we draw a segment with slope $5/4$ from Q to a point R on the line $x = 1$. We next draw a circle, centered at the origin, through R, followed by a horizontal segment from Q to a point P on the circle. Then we reflect the segments $\overline{PQ}$ and $\overline{QR}$ through the origin. The resulting region is the shaded region in Figure 6, on whose boundary the direction field either is tangent to the boundary or points into the region. This is true along the circular portions of the boundary because of observation (i). It is true along the slanted segments because of observation (ii) and along the horizontal segments because of the sign of y'. Since the point Q could have been chosen anywhere along the graph of $y = x/(1-x^2)$ and above the line $y = 1$ in the second quadrant, a closed forward-invariant region could be constructed to contain the initial point of any forward semiorbit. The Poincaré-Bendixson theorem therefore implies that every orbit is either periodic or approaches a periodic orbit as $t \to \infty$. In particular, it guarantees the existence of a periodic orbit inside the smallest such region we could construct, which corresponds to choosing Q on the line $y = 1$.

PROBLEMS

1. Consider the first system from Example 1,

$$x' = y + x(1 - x^2 - y^2),$$

$$y' = -x + y(1 - x^2 - y^2),$$

and let $p(t) = x(t)^2 + y(t)^2$.

(a) Show that $p'(t) = 2p(t)(1 - p(t))$.

(b) Use the result of part (a) to argue that every closed annulus

$$\mathcal{A}_{r,R} = \left\{(x, y) \mid r^2 \le x^2 + y^2 \le R^2\right\},$$

with $0 \le r \le 1 \le R$, is a forward-invariant region.

(c) Show that $(0, 0)$ is the only equilibrium point of the system, and use the result of part (a) to argue that no nontrivial orbits approach it as $t \to \infty$.

(d) What conclusion(s) can now be drawn with the help of the Poincaré-Bendixson theorem?

2. Show that every closed disk centered at $(0, 0)$ with radius $r \le 1$ is a forward-invariant region for the second system in Example 1. Do any of those disks contain a periodic orbit?

3. Let φ be a continuous function on $\mathbb{R}^2$ with $\varphi(0, 0) > 0$. Suppose that there is a number $R > 0$ such that $\varphi(x, y) < 0$ whenever

$$\sqrt{x(t)^2 + y(t)^2} \ge R.$$

With a constant $a \ne 0$, consider the system

$$x' = ay + x\varphi(x, y),$$

$$y' = -ax + y\varphi(x, y),$$

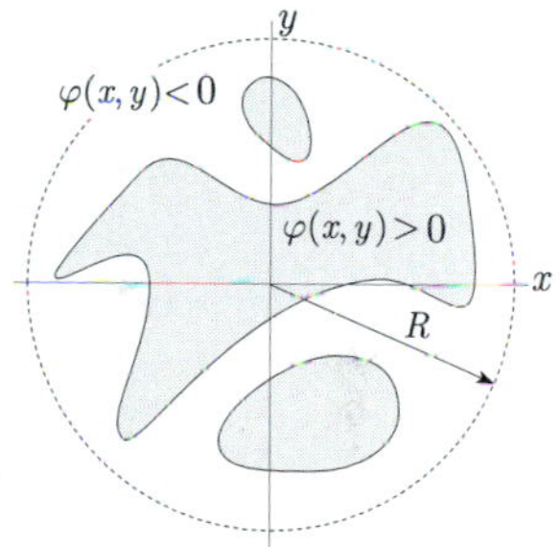

and let $p(t) = x(t)^2 + y(t)^2$.

(a) Show that $p'(t) = 2p(t)\varphi(x(t), y(t))$.

(b) Use the result of part (a) to argue that every closed disk centered at $(0, 0)$ with radius $r \ge R$ is a forward-invariant region.

(c) Show that $(0, 0)$ is the only equilibrium point of the system, and use the result of part (a) to argue that no nontrivial orbits approach it as $t \to \infty$.

(d) What conclusion(s) can now be drawn with the help of the Poincaré-Bendixson theorem?

4. Making appropriate assumptions on the coefficients a, b, and c, adapt the argument in Problem 3 to systems of the form

$$x' = ay + x\varphi(x, y), \quad y' = -bx + cy\varphi(x, y).$$

5. Let φ be a continuous function and $a, b, p, q, \alpha, \beta$ constants with $p, b, \alpha, \beta \ge 0$. Consider the system

$$x' = xy\varphi(x, y) + ax + py + \alpha,$$

$$y' = -xy\varphi(x, y) + bx + qy + \beta.$$

(a) Show that the closed first quadrant, $\{(x, y) | x \ge 0, y \ge 0\}$, is a forward-invariant region.

(b) Show that if $a + b < 0$, $p + q < 0$, and $\alpha + \beta \ge 0$, then the triangular region $\{(x, y) | x \ge 0, y \ge 0, x + y \le k\}$ is forward-invariant for any

$$k \ge \max\left\{-\frac{\alpha + \beta}{a + b}, -\frac{\alpha + \beta}{p + q}\right\}.$$

(c) Apply the results of (a) and (b) in an analysis of the first-quadrant portion of the phase portrait of

$$x' = xy - x + y/4, \quad y' = -xy - y + 1.$$

6. Carry out the following argument to show that the system

$$x' = x^2y - 2x + y/4, \quad y' = -x^2y + x + y$$

has a periodic orbit in the first quadrant.

(a) Show that $(0, 0)$ and $(3/2, 6/5)$ are the only equilibrium points with nonnegative entries and that they are both unstable. Moreover, no nontrivial orbit approaches $(3/2,\ 6/5)$ as $t \to \infty$.

(b) Show that $(x + 2y)' \leq 0$ when $y \geq 0$ and $x \geq 3/2$.

(c) Show that if $y > 0$, then

$$x^2y - 2x + \frac{1}{4}y \geq \frac{y^2 - 4}{4y} \quad \text{for all } x,$$

and that if $y > 2$, then

$$\frac{dy}{dx} \leq \frac{(-x^2y + x + y)4y}{y^2 - 4} \leq \frac{4y^2 + 1}{y^2 - 4} \quad \text{for all } x.$$

(d) Show that if $y > 5$, then $\frac{dy}{dx} < 5$ for all x.

(e) Let $\mathcal{U} = \{(x, y) \mid x \geq 0,\ y \geq 0,\ y \leq 5(x + 1),\ x + 2y \leq 53/2\}$. Make a sketch of $\mathcal{U}$ and use the results of parts (b) and (d) to argue (by Theorem 2) that $\mathcal{U}$ is a forward-invariant region for the system.

(f) The region $\mathcal{U}$ fulfills all but one the requirements of the Poincaré-Bendixson theorem. It contains the saddle point $(0, 0)$, which is approached by orbits on the x-axis as $t \to \infty$. To get around this difficulty, pick a point P on the line $x + 2y = 53/2$ and just above the x-axis. Follow the orbit beginning at P until it reaches the line $x = 3/2$ at a point Q. Then move horizontally to a point R on the y-axis. Let the resulting path be $y = \beta(x)$. Now, let $\widetilde{\mathcal{U}} = \{(x, y)$ in $j\mathcal{U} \mid y \geq \beta(x)\}$. Argue that $\widetilde{\mathcal{U}}$ fulfills all of the requirements necessary to conclude that $\widetilde{\mathcal{U}}$ contains a periodic orbit by the Poincaré-Bendixson theorem.

7. Show that the system

$$x' = x^2y - 4x + 1, \quad y' = -x^2y + 3x$$

has a periodic orbit in the first quadrant.

8. Let a and b be positive constants and consider the system

$$x' = x^2y - x + a, \quad y' = -x^2y + b.$$

(a) Construct a closed forward-invariant region in the first quadrant.

(b) For what values of a and b can you conclude that the system has a periodic orbit?

9. Prove Theorem 1 as follows, consulting your calculus book as needed.
Suppose that (2) has a periodic orbit $(x(t), y(t))$ with period T, so that $\mathbf{r}(t) = (x(t), y(t))$, $0 \leq t < T$, is a one-to-one parametrization of the resulting curve C. Let $\phi(t)$ denote the angle formed by the tangent vector $\mathbf{r}'(t)$ and the x-axis. To simplify notation, also let $u = f(x, y)$ and $v = g(x, y)$.

(a) Show that $\displaystyle\int_C d\phi = \int_0^T \phi'(t)\, dt = 2\pi$.

(b) Use the fact that $\tan\phi = v/u$ to show that $d\phi = \dfrac{udv - vdu}{u^2 + v^2}$.

(c) Assuming that $u^2 + v^2 \neq 0$ in the region $\mathcal{G}$ enclosed by C, use Green's theorem to convert the line integral

$$\int_C d\phi = \int_C \left(\frac{u}{u^2 + v^2} dv + \frac{-v}{u^2 + v^2} du \right)$$

to a double integral over the region $\mathcal{G}$.

(d) Show that the integrand in the double integral from part (c) is identically zero in $\mathcal{G}$, and conclude that $\int_C d\phi = 0$.

(e) Compare the results of parts (a) and (d). Identify the assumption that caused the contradiction and explain how this proves Theorem 1.

What conclusion can be drawn from Theorem 1 about the following systems?

10. $x' = 1 - 3xy^2, \quad y' = xy^2$

11. $x' = y, \; y' = x - 5xy$

12. $x' = xy - x, \; y' = y^2 - 2xy + 1$

13. $x' = (x - y)^2, \; y' = x^2 - y^2$

14. Suppose that every equilibrium point of (2) lies on the boundary of a forward-invariant region. What conclusion can then be drawn from Theorem 1? Give an example with at least two isolated equilibrium points and where the forward-invariant region is the first quadrant.

15. **(a)** Use Green's theorem to show that if (2) has a periodic orbit that encloses a region Ω, then

$$\int\int_\Omega (f_x + g_y)\, dx\, dy = 0,$$

and therefore either $f_x + g_y = 0$ throughout Ω or $f_x + g_y$ has both positive and negative values in Ω.

(b) Apply the result of (a) to show that the system

$$x' = y - x + x\cos y, \quad y' = x - y + y\sin x$$

has no periodic orbit.

(c) Apply the result of (a) to show that the second-order equation

$$x'' + \rho(x)x' + \mu(x) = 0,$$

where ρ and μ are continuously differentiable functions, has no periodic solution if $\rho(x) \geq 0$ for all x and $\rho(x) > 0$ for some x.

16. Let $p(x, y) = y^2 - x^2 + \frac{1}{2}x^4$ and consider the system

$$x' = y + (1 - x^2)p(x, y), \quad y' = x(1 - x^2) - yp(x, y).$$

(a) Show that $(0, 0)$, $(-1, 0)$, and $(1, 0)$ are equilibrium points. Show further that $(0, 0)$ is a saddle point, $(-1, 0)$ is a stable spiral point, and $(1, 0)$ is an unstable spiral point.

(b) Show that the flow about each of the spiral points $(\pm 1, 0)$ is clockwise. (*Hint*: Look at the sign of y' along the x-axis.)

(c) Show that the "figure-8" Lissajous curve parametrized by

$$x = \sqrt{2}\sin\theta, \; y = \sqrt{1/2}\,\sin 2\theta, \; 0 \leq \theta < 2\pi,$$

satisfies the slope equation

$$\frac{dy}{dx} = \frac{x(1+x^2) - yp(x,y)}{y + (1+x^2)p(x,y)},$$

and therefore conclude that each lobe of this curve is traced out by orbits that approach the saddle point $(0, 0)$ as $t \to \pm\infty$. These are *homoclinic orbits.*

(d) Include all of the above information in a rough sketch of the phase portrait for $-1.5 \le x \le 1.5$ and $-1 \le y \le 1$. (There are no other equilibrium points in that rectangle.) Then use a computer or graphing calculator to create a more accurate portrait.

17. Consider the system

$$\frac{dx}{dt} = -y + x^2 + y^2, \quad \frac{dy}{dt} = x - 2xy.$$

(a) Find the equilibrium points and make a sketch showing them along with the nullclines of the system.

(b) Describe the behaviors of all solutions for which $y(0) = 1/2$. Indicate this behavior in your sketch from part (a).

(c) Show that the ellipse $3x^2 + (y - \frac{1}{2})^2 = \frac{3}{4}$ satisfies the slope equation

$$\frac{dy}{dx} = \frac{x - 2xy}{-y + x^2 + y^2}$$

for $y \ne 1/2$. Add this ellipse to your sketch, indicating the direction of orbits along it. What can you now conclude about the nature of the equilibrium points that lie on this ellipse?

(d) Show that the quantity $\varphi(x, y) = \frac{1}{6}(6y^2 - 6x^2y - 2y^3 + 3x^2)$ is constant along orbits of the system and that this is consistent with parts (b) and (c).

(e) Show that the function $\varphi(x, y)$ attains a local maximum or minimum at each equilibrium point inside the ellipse of part (c). What does this imply about the orbits near these points? Include this information in your sketch.

(f) Add orbits to your sketch as needed in order to obtain a reasonably complete phase portrait.

18. Consider the system (2) and suppose that (x^*, y^*) is an isolated equilibrium point. Suppose further that there is a function $\varphi(x, y)$ that has a local minimum at (x^*, y^*) and the property that $\varphi_x f + \varphi_y g \le 0$ along orbits of (2) near (x^*, y^*). Show that the regions enclosed by level curves of φ near (x^*, y^*) are forward-invariant regions for (2). (*Hint*: The quantity $\varphi_x f + \varphi_y g$ has the same sign as the directional derivative of φ in the direction of the vector (f, g).) Argue further that (x^*, y^*) is stable, and that if $\varphi_x f + \varphi_y g < 0$ along orbits near (x^*, y^*), then (x^*, y^*) is asymptotically stable.

Apply the result(s) of Problem 18 at $(x^*, y^*) = (0, 0)$ with $\varphi(x, y) = x^2 + y^2$ to each of the systems:

19. $\begin{cases} x' = y \\ y' = -x \end{cases}$

20. $\begin{cases} x' = y - x \\ y' = -x - y \end{cases}$

21. $\begin{cases} x' = y^2 - x^3 \\ y' = -xy \end{cases}$

9.5 Beyond the Plane

In the preceding section, we saw that autonomous systems in the plane can exhibit asymptotic behavior that is not possible for autonomous scalar equations—specifically periodic solutions and convergence to limit cycles.

For autonomous systems of three equations, much more is in store. In particular, the bounded orbit of an autonomous system in three dimensions can exhibit "*chaotic*" behavior. In this section there will be no thorough description of local behavior near equilibrium points—as in Theorem 1 of Section 9.3—for systems of three or more equations. Our analysis will instead focus mainly upon stability of equilibrium points.

The Chaotic System of Lorenz In the 1950s, while trying to explain the difficulty in making accurate long-term weather forecasts, the meteorologist E. N. Lorenz set about finding a simple system of three differential equations with bounded *aperiodic* solutions that converge neither to an equilibrium point nor a limit cycle. He succeeded—and found a place in history—when he derived the following system from a complex model of thermal convection:

$$x' = -ax + ay,$$
$$y' = bx - y - xz,$$
$$z' = xy - cz,$$

where a, b, and c are positive constants. If $b > 1$, then the system has two nontrivial equilibrium points. The "usual" choice of parameters is $a = 10$, $b = 28$, and $c = 8/3$, for which the nontrivial equilibrium points are $(-6\sqrt{2}, -6\sqrt{2}, 27)$ and $(6\sqrt{2}, 6\sqrt{2}, 27)$. Figures 1a–c show the graphs of x, y, and z, respectively, versus t over the interval $[0, 40]$ from a numerically generated solution with initial point $(-8, 8, 27)$. Viewing the solution this way gives little insight into the dynamics of the system and even suggests that the behavior of solutions may be hopelessly chaotic. However, a plot of the solution's orbit in three-dimensional "phase space" reveals that the solution actually behaves in a very "orderly" way. (This is an extraordinary testament to the power of computer-aided visualization.) Figures 2a–c show three disjoint parts of the same numerically generated orbit as in Figures 1a–c, each corresponding to the time interval with which it is labeled. Notice that during each of these time intervals, the orbit spirals around each of the two equilibrium points a seemingly random number of times.

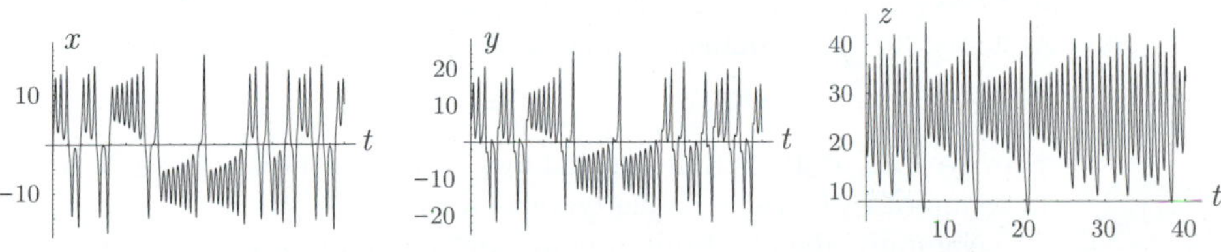

Figure 1a **Figure 1b** **Figure 1c**

Although the orbit seen in Figures 2a–c approaches neither an equilibrium point nor a periodic orbit as $t \to \infty$, it does approach *something* that is far more complex but nevertheless has a particular structure. Such peculiar sets of points, approached by orbits in three or higher dimensions, have become known by the trendy term "strange attractors."

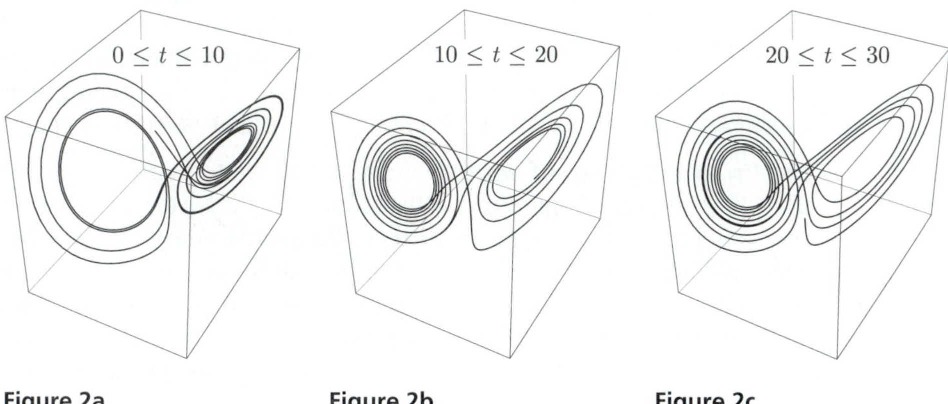

Figure 2a **Figure 2b** **Figure 2c**

The Lorenz system is extremely sensitive to changes in initial values. For example, with the numerical method that generated the plots in Figures 1a–c, the initial point $(-8, 8, 27)$ results in the point $(8.03, 2.61, 32.41)$ at time $t = 40$ (rounded to two decimal places). With the same numerical method, the initial point $(-8, 8.0000001, 27)$ results in the point $(-2.54, -0.64, 23.87)$ at time $t = 40$. This demonstrates that a very slight change in the initial point causes a very large change in the solution at $t = 40$. In fact, because of the round-off errors inherent in any numerical method, we cannot be sure that the orbit in Figures 1a–c bears any more than a general qualitative resemblance to the orbit of the true solution.

This type of extreme sensitivity to initial values is at the heart of what has become known as *chaos*. Lorenz coined the term *butterfly effect* to caricature this sensitivity in the context of weather prediction, suggesting that the state of the earth's atmosphere at any significantly later time in the future is so sensitive to changes its present state that the flutter of a butterfly's wings can make a difference. But the serious upshot of Lorenz's discovery is that no deterministic model (as opposed to a *stochastic* one) can ever accurately predict the weather beyond a very short period of time, because it is impossible to gather the data necessary to describe the present state of the atmosphere to a sufficient degree of accuracy.

Stability The remainder of this section will be specifically concerned with "3×3" systems; that is, systems of three equations in three unknowns:

$$x' = f(x, y, z), \quad y' = g(x, y, z), \quad z' = h(x, y, z). \tag{1}$$

However, all of the definitions and theorems—perhaps with some minor adjustments in terminology—will be equally valid for $n \times n$ systems for any $n \ge 1$.

Naturally, the equilibrium points of (1) are the solutions of the algebraic system

$$f(x, y, z) = 0, \quad g(x, y, z) = 0, \quad h(x, y, z) = 0.$$

Definitions of stability and asymptotic stability of equilibrium points in the plane were stated in Section 9.1. Those definitions require only minor modification to become the desired definitions for equilibrium points in three-dimensional space:

Definition An equilibrium point $E^* = (x^*, y^*, z^*)$ of (1) is said to be

- **stable** if for any ball Ω centered at E^* there is a smaller ball $\widetilde{\Omega}$ centered at E^* with the property that $(x(t), y(t), z(t))$ remains in Ω for all $t \geq \tau$ whenever $(x(\tau), y(\tau), z(\tau))$ is in $\widetilde{\Omega}$;
- **asymptotically stable** if there is a ball Ω centered at E^* with the property that $(x(t), y(t), z(t)) \to E^*$ as $t \to \infty$ whenever $(x(\tau), y(\tau), z(\tau))$ is in Ω for some τ;
- **neutrally stable** if it is stable and not asymptotically stable;
- **unstable** if it is not stable. ◆

In some cases, stability results can be obtained by linearization. The following discussion parallels that of the 2×2 case in Section 9.3. Letting $\mathbf{w} = (x, y, z)^T$ and $F(\mathbf{w}) = (f(x, y, z), g(x, y, z), h(x, y, z))^T$, the system (1) can be written as

$$\mathbf{w}' = F(\mathbf{w}),$$

and the linearization at E^*, shifted to $\mathbf{0}$, is

$$\mathbf{u}' = \mathcal{J}(E^*)\mathbf{u},$$

where $\mathcal{J}(E^*)$ is the **Jacobian matrix** of F at E^*, defined by

$$\mathcal{J}(E^*) = \begin{pmatrix} f_x & f_y & f_z \\ g_x & g_y & g_z \\ h_x & h_y & h_z \end{pmatrix}\Bigg|_{(x,y,z)=E^*}.$$

Theorem 1

Suppose that f, g, and h have continuous second-order partial derivatives and that E^* is an equilibrium point of (1).

i) If the real part of each eigenvalue of $\mathcal{J}(E^*)$ is negative, then E^* is asymptotically stable.

ii) If $\mathcal{J}(E^*)$ has an eigenvalue with positive real part, then E^* is unstable. ▲

Example 1

The Jacobian matrix for the Lorenz system

$$x' = -10x + 10y,$$

$$y' = 28x - y - xz,$$

$$z' = xy - 8z/3,$$

which produced Figures 1a–c and 2a–c, is

$$\mathcal{J}(x, y, z) = \begin{pmatrix} -10 & 10 & 0 \\ 28 - z & -1 & -x \\ y & x & -\frac{8}{3} \end{pmatrix}.$$

The following table summarizes the *CAS*-aided linearization analysis at each equilibrium point, revealing that all three of them are unstable, as expected.

E^*	$\mathcal{J}(E^*)$	$\lambda_1, \lambda_2, \lambda_3$	Conclusion
$(0, 0, 0)$	$\begin{pmatrix} -10 & 10 & 0 \\ 28 & -1 & 0 \\ 0 & 0 & -\frac{8}{3} \end{pmatrix}$	$-2.7, -22.8, 11.8$	Unstable
$(-6\sqrt{2}, -6\sqrt{2}, 27)$	$\begin{pmatrix} -10 & 10 & 0 \\ 1 & -1 & 6\sqrt{2} \\ -6\sqrt{2} & -6\sqrt{2} & -\frac{8}{3} \end{pmatrix}$	$-13.9, 0.1 \pm 10.2i$	Unstable
$(6\sqrt{2}, 6\sqrt{2}, 27)$	$\begin{pmatrix} -10 & 10 & 0 \\ 1 & -1 & -6\sqrt{2} \\ 6\sqrt{2} & 6\sqrt{2} & -\frac{8}{3} \end{pmatrix}$	$-13.9, 0.1 \pm 10.2i$	Unstable

■

Example 2

Consider the system

$$\begin{aligned} x' &= x(1 - x/4 - y), \\ y' &= y(-1 + x - 2z), \\ z' &= z(-1 + 2y), \end{aligned}$$

where x, y, and z represent nonnegative physical quantities. There are four equilibrium points of interest: $(0, 0, 0)$, $(4, 0, 0)$, $(3/4, 1, 0)$, and $(2, 1/2, 1/2)$. The Jacobian matrix at (x, y, z) is

$$\mathcal{J}(x, y, z) = \begin{pmatrix} 1 - \frac{x}{2} - y & -x & 0 \\ y & -1 + x - 2z & -2y \\ 0 & 2z & -1 + 2y \end{pmatrix}.$$

The following table summarizes the linearization analysis.

E^*	$\mathcal{J}(E^*)$	$\lambda_1, \lambda_2, \lambda_3$	Conclusion
$(0, 0, 0)$	$\begin{pmatrix} 1 & 0 & 0 \\ 0 & -1 & 0 \\ 0 & 0 & -1 \end{pmatrix}$	$-1, -1, 1$	Unstable
$(4, 0, 0)$	$\begin{pmatrix} -1 & -4 & 0 \\ 0 & 3 & 0 \\ 0 & 0 & -1 \end{pmatrix}$	$-1, -1, 3$	Unstable
$\left(\frac{3}{4}, 1, 0\right)$	$\frac{1}{4}\begin{pmatrix} 3 & -4 & 0 \\ 3 & 0 & -6 \\ 0 & 0 & 2 \end{pmatrix}$	$\frac{1}{2}, \frac{-1\pm i\sqrt{47}}{8}$	Unstable
$\left(2, \frac{1}{2}, \frac{1}{2}\right)$	$\frac{1}{2}\begin{pmatrix} -1 & -4 & 0 \\ 1 & 0 & -2 \\ 0 & 2 & 0 \end{pmatrix}$	$-0.26, -0.12 \pm 1.39i$	Asymptotically stable

Notice also that summing the equations in the system gives

$$(x + y + z)' = x(1 - x/4) - y - z,$$

from which we can observe that $x+y+z$ is decreasing whenever either $y+z > 1$ or $x > 4$, $y > 0$, $z > 0$. Therefore, nonnegative solutions are bounded. This, together with the knowledge that $(2, 1/2, 1/2)$ is asymptotically stable and the only stable equilibrium point, suggests that all solutions with positive initial values will approach $(2, 1/2, 1/2)$ as $t \to \infty$. ■

Conserved Quantities and Lyapunov Functions When the linearized system at an equilibrium point has imaginary eigenvalues or a zero eigenvalue, linearization analysis fails to determine stability/instability. In this case, a different type of analysis—one that deals directly with the nonlinear system rather than its linearization—is needed.

Consider the system (1), redisplayed here for convenience:

$$x' = f(x, y, z), \quad y' = g(x, y, z), \quad z' = h(x, y, z),$$

where the vector field $F = (f, g, h)^T$ is continuous on (for simplicity) all of $\mathbb{R}^3$. Suppose that ℓ is a differentiable function on $\mathbb{R}^3$ with the property that

$$\frac{d}{dt}\ell(x(t), y(t), z(t)) = 0$$

along any orbit of (1). We refer to such a function ℓ as a **conserved quantity** for solutions of (1). By the chain rule, $\frac{d}{dt}\ell(x, y, z) = \ell_x x' + \ell_y y' + \ell_z z'$, and so ℓ is a conserved quantity for solutions of (1) if and only if

$$\ell_x f + \ell_y g + \ell_z h = 0 \tag{2}$$

for all (x, y, z); that is, the gradient of ℓ, $\nabla\ell$, is always orthogonal to the vector field F. Since $\ell(x(t), y(t), z(t))$ is constant along any orbit of (1), it follows that each orbit of (1) lies on a particular level surface of ℓ.

Example 3

Consider the system

$$x' = xy - z - 2yz,$$
$$y' = -x^2 + z + 3xz,$$
$$z' = x - y - xy.$$

Notice that $xx' + yy' + zz' = 0$ along any orbit. Therefore, the quantity

$$\ell(x, y, z) = \frac{1}{2}(x^2 + y^2 + z^2)$$

is conserved. The level surfaces of ℓ are spheres centered at the origin; thus each orbit of this system lies on a particular sphere centered at the origin. This fact implies that $(0, 0, 0)$ is a neutrally stable equilibrium point. (Why?) Figure 3a shows the orbit (for $t \geq 0$) with initial point $(0, 0, 1)$, and Figure 3b shows the orbit (for $t \geq 0$) with initial point $(\sqrt{2}/2, \sqrt{2}/2, 0)$. Each of these orbits lies on the unit sphere $x^2 + y^2 + z^2 = 1$.

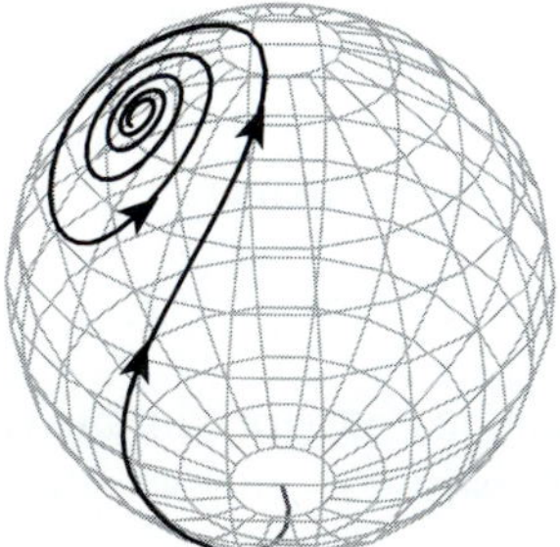

Figure 3a

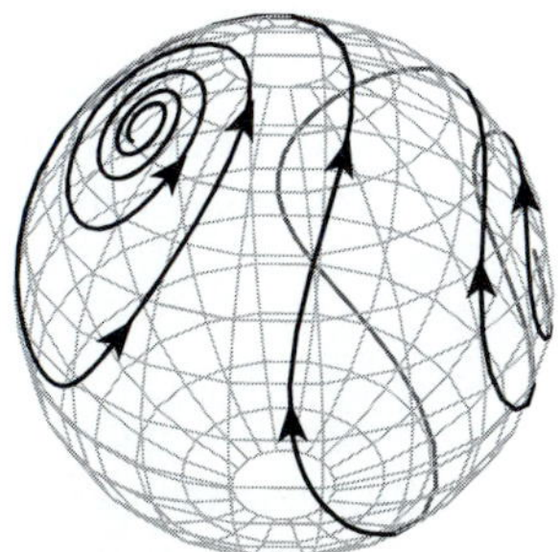

Figure 3b

It turns out that the equilibrium points of this system comprise the curve parametrized by $(s, s^2/(1 + 3s), s/(1 + s))$. Thus none of the equilibrium points is isolated, and there is an equilibrium point on every sphere centered at the origin, as illustrated by the orbits in Figures 3a and 3b. ■

Example 4

Consider the following modification of the system in Example 3:

$$\begin{aligned} x' &= xy - z - 2yz - x, \\ y' &= -x^2 + z + 3xz - y, \\ z' &= x - y - xy - z. \end{aligned}$$

Notice that $xx' + yy' + zz' = -(x^2 + y^2 + z^2)$ along any orbit. Thus if $\ell(x, y, z) = \frac{1}{2}(x^2 + y^2 + z^2)$, then

$$\frac{d}{dt}\ell(x(t), y(t), z(t)) < 0$$

along any nontrivial orbit, which indicates that the direction field points *into* spheres centered at the origin. We conclude that the origin is asymptotically stable and is approached by *every* orbit as $t \to \infty$. Figure 4 shows a few orbits with initial points on the unit sphere. ■

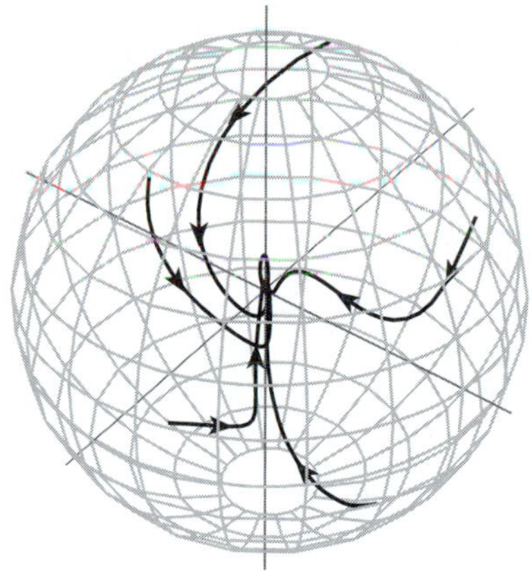

Figure 4

Each of the functions ℓ in Examples 3 and 4 is an example of a *Lyapunov function*.* The following definition gives the properties of a Lyapunov function.

Definition Let E^* be an equilibrium point of (1). A function $\ell(x, y, z)$ with continuous first-order partial derivatives is a **Lyapunov function** for (1) at E^* if the following are true:

i) The value of ℓ at E^* is a local minimum;
ii) $\ell_x x' + \ell_y y' + \ell_z z' \leq 0$ along all orbits inside of some ball $\mathcal{B}$ centered at E^*. ◆

---

* So named for the Russian mathematician A. M. Lyapunov (1857–1918), who was a student of Chebyshev.

Note that (i) implies that if ℓ is a Lyapunov function at E^*, then $\ell_x(E^*) = \ell_y(E^*) = \ell_z(E^*) = 0$; that is, $\nabla\ell(E^*) = \mathbf{0}$.

A Lyapunov function at E^* may be thought of geometrically as a function with level surfaces that enclose E^* and with the additional property that the system's direction field near E^* is always either tangent to a level surface or pointing into the region bounded by a level surface. This guarantees that no orbit can escape a region bounded by a level surface near E^*; that is, regions bounded by level surfaces near E^* are *forward-invariant regions*.

A Lyapunov function is often thought of as representing "energy" in the system. In models of mechanical systems in which energy is either conserved or dissipated, an expression of the total energy in the system will indeed be a Lyapunov function.

Our *Lyapunov stability theorem* is as follows:

Theorem 2

Let E^* be an equilibrium point of (1). If there exists a Lyapunov function ℓ at E^*, then E^* is stable. If, in addition, property (ii) holds with $\ell_x x' + \ell_y y' + \ell_z z' < 0$ for $(x, y, z) \neq E^*$, then E^* is asymptotically stable. ▲

Example 5

Consider the system

$$x' = -2z^3 - z\sin y,$$
$$y' = 2xz - \sin y,$$
$$z' = x - \sin^3 z,$$

for which the only equilibrium point is $(0, 0, 0)$. The Jacobian matrix at $(0, 0, 0)$ turns out to be

$$\mathcal{J}(0,0,0) = \begin{pmatrix} 0 & 0 & 0 \\ 0 & -1 & 0 \\ 1 & 0 & 0 \end{pmatrix},$$

whose eigenvalues are 0, 0, and -1. Consequently, linearization does not help us determine whether $(0, 0, 0)$ is stable. So consider the function

$$\ell(x, y, z) = x^2 - \cos y + z^4,$$

which is constructed* so that

$$\begin{aligned} \ell_x x' + \ell_y y' + \ell_z z' &= -\sin^2 y - 4z^3 \sin^3 z \\ &\leq 0 \text{ whenever } -\pi \leq z \leq \pi, \\ \nabla\ell(0,0,0) &= \mathbf{0}, \text{ and} \\ \ell(0,0,0) &= 0 \text{ is a local minimum.} \end{aligned}$$

* The truth is that this $\ell(x, y, z)$ came first, and the system was constructed around it.

So $\ell(x, y, z)$ is a Lyapunov function for the system at $(0, 0, 0)$, which implies that $(0, 0, 0)$ is stable. Since $\ell_x x' + \ell_y y' + \ell_z z' \leq 0$ near $(0, 0, 0)$, no orbit near $(0, 0, 0)$ can escape a level surface of ℓ. Moreover, since $\ell_x x' + \ell_y y' + \ell_z z' < 0$ at (x, y, z) near E^* except on the x-axis, we can conclude that $(0, 0, 0)$ is asymptotically stable. Figure 5 shows a "cut-away" view of three level surfaces of ℓ that enclose $(0, 0, 0)$ along with a typical orbit. On each of these surfaces, the direction field of the system points into the region bounded by the surface. ■

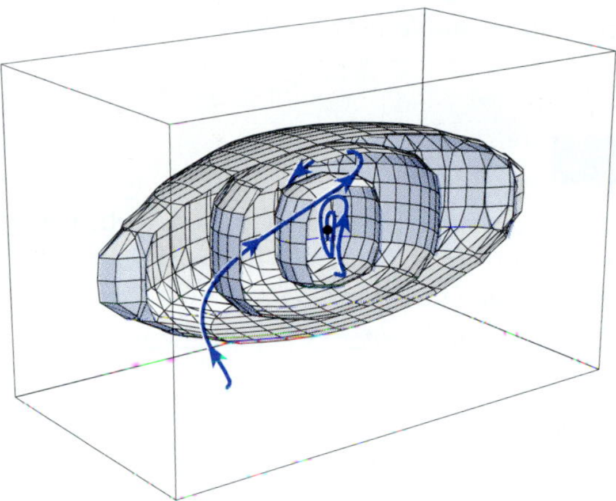

Figure 5

Conservation of Energy Suppose that the motion of a particle with mass $m = 1$ takes place within a three-dimensional, *conservative force field*; that is, the vector field describing the force on the particle is the negative gradient of some function $\varphi(x, y, z)$, which we interpret as *potential energy*. Thus, the second-order equations of motion are

$$x'' = -\varphi_x, \quad y'' = -\varphi_y, \quad z'' = -\varphi_z.$$

Converting these to a first-order system, we have six equations:

$$\begin{aligned} x' &= u, & u' &= -\varphi_x, \\ y' &= v, & v' &= -\varphi_y, \\ z' &= w, & w' &= -\varphi_z, \end{aligned}$$

where u, v, and w are the x-, y-, and z-components of the particle's velocity, respectively. The total (kinetic + potential) energy of the particle at any time is

$$\ell(x, u, y, v, z, w) = \frac{1}{2}(u^2 + v^2 + w^2) + \varphi(x, y, z).$$

Observe that along any orbit, the total energy is constant:

$$\begin{aligned} \frac{d\ell}{dt} &= \ell_x x' + \ell_u u' + \ell_y y' + \ell_v v' + \ell_z z' + \ell_w w' \\ &= \varphi_x u - u\varphi_x + \varphi_y v - v\varphi_y + \varphi_z w - w\varphi_w \\ &= 0. \end{aligned}$$

Now suppose that E^* is an equilibrium point. Then $E^* = (x^*, 0, y^*, 0, z^*, 0)$, where $\varphi_x = \varphi_y = \varphi_z = 0$ at (x^*, y^*, z^*). Notice also that

$$\nabla \ell(E^*) = (\varphi_x, u, \varphi_y, v, \varphi_z, w)^T|_{E^*} = \mathbf{0}.$$

Therefore, ℓ is a Lyapunov function at E^*—and thus E^* is stable—provided that ℓ has a local minimum at E^*. Because of its particular form, ℓ has a local minimum at $E^* = (x^*, 0, y^*, 0, z^*, 0)$ if and only if the potential φ has a local minimum at (x^*, y^*, z^*). (Why?) Thus, we have demonstrated that *in a conservative force field, the state of a stationary particle is stable if the potential energy at its position is a local minimum.*

Example 6

We have illustrated the ideas in the preceding discussion with a two-dimensional example. Suppose that the motion of a particle with mass $m = 1$ takes place in the plane under the influence of the conservative force field $-\nabla\varphi$, where $\varphi(x, y) = xe^{-(x^2/2+y^2)}$. After conversion to a first-order system, the equations of motion are

$$x' = u, \ \ u' = -(1 - x^2)e^{-(x^2/2+y^2)},$$
$$y' = v, \ \ v' = 2xye^{-(x^2/2+y^2)}.$$

There are two equilibrium points, $(\pm 1, 0, 0, 0)$, at which $(x^*, y^*) = (\pm 1, 0)$. It turns out that $\varphi(-1, 0)$ is a local minimum, and $\varphi(1, 0)$ is a local maximum. So $(-1, 0)$ is a stable "rest" position for the particle. Figure 6a shows the "potential surface" defined by the graph of φ, and Figure 6b shows the corresponding force field and a sample path of the particle near the stable rest point. (Note that the force field vectors—given by $-\nabla\varphi$—are *not* tangent to the path. The direction field for the system lives in four-dimensional space!) One can think of the motion of the particle either as that of a marble rolling along the (frictionless) three-dimensional surface in Figure 6a, or as a particle in the plane that is weakly attracted by a fixed particle at $(-1, 0)$ and repelled by another at $(1, 0)$, with no other forces present. ■

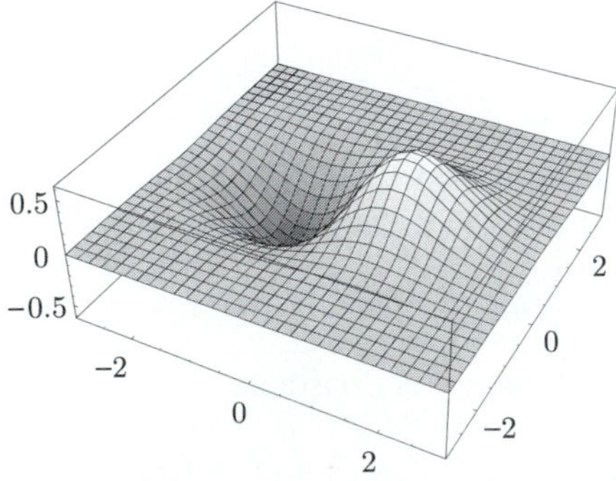

Figure 6a

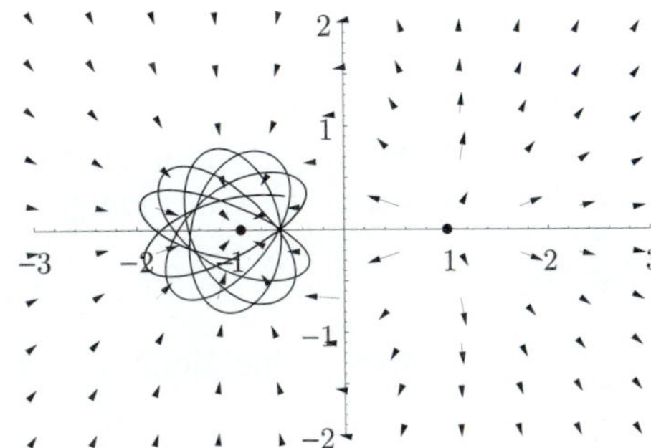

Figure 6b

REMARK If the force field $-\nabla\varphi$ is the *gravitational* force field exerted on a particle with mass m by a fixed particle with mass M located at $(0, 0, 0)$, then $\varphi(x, y, z) = -MmG/\sqrt{x^2 + y^2 + z^2}$, where G is the "universal gravitation constant." In this case, $\varphi \to -\infty$ as $(x, y, z) \to (0, 0, 0)$; yet we can still think of $(0, 0, 0)$ as a point where $-\varphi$ has a local minimum and to which the particle is attracted. In six-dimensional $xuyvzw$-space, the point $(0, 0, 0, 0, 0, 0)$ is stable in the sense that an orbit will remain as close as we like to that point for all $t > 0$, provided that the orbit is sufficiently close to it at time $t = 0$. This is not contradictory to the fact that in three-dimensional xyz-space, the particle can be extremely close to the mass at $(0, 0, 0)$ and still move far away, if its speed is sufficiently large.

PROBLEMS

1. Show that if $a > 0$, $0 < b \leq 1$, and $c > 0$, then the Lorenz system

$$x' = -ax + ay, \quad y' = bx - y - xz,$$
$$z' = xy - cz,$$

has only one equilibrium point and that it is asymptotically stable.

2. Show that all three equilibrium points of the Lorenz system with $a = 10$, $b = 28$, and $c = 3.2$ are unstable. (Use numerical approximations of the equilibrium points and the eigenvalues.)

3. Show that the nonzero equilibrium points of the Lorenz system with $a = 10$, $b = 28$, and $c = 3.24$ are asymptotically stable. (Use numerical approximations of the equilibrium points and the eigenvalues.)

For each of the following systems, determine whether $\ell(x, y, z) = x^2 + y^2 + z^2$ is

(a) conserved quantity

(b) a Lyapunov function at $(0, 0, 0)$

4. $x' = xyz - xz^2, \quad y' = xz^2 - x^2z,$
$z' = x^2z - xyz$

5. $x' = -x + yz, \quad y' = -yz^2 - 2xz,$
$z' = -x^2z + xy$

6. $x' = y - xz, \quad y' = -x - yz, \quad z' = xyz$

7. $x' = x - yz, \quad y' = -x - yz, \quad z' = x^2 + y^2$

8. Consider a system of the form

$$x' = h(x, y)f(y), \quad y' = h(x, y)g(x),$$

where h, f, and g are continuous.

(a) Show that

$$\ell(x, y) = \int f(y)\,dy - \int g(x)\,dx$$

is a conserved quantity.

(b) Use the result of part (a) to find a conserved quantity for

$$x' = (x^2 + y^3)y, \quad y' = -2(x^2 + y^3)x.$$

Show that the conserved quantity is a Lyapunov function at $(0, 0)$ and conclude that $(0, 0)$ is stable.

(c) Use the result of part (a) to find a conserved quantity for positive solutions of

$$x' = k_1x(a - y), \quad y' = -k_2y(b - x),$$

where k_1, k_2, $a, b > 0$. (*Hint*: Rewrite the system so that $h(x, y) = xy$.) Show that the conserved quantity is a Lyapunov function at (b, a) and conclude that (b, a) is stable.

9. Suppose that $w(x, y, z)$ is continuous and that $\varphi(x, y, z)$ has continuous first partial derivatives. Show that $\varphi(x, y, z)$ is a conserved quantity for the system

$$x' = w(x, y, z)(\varphi_y(x, y, z) - \varphi_z(x, y, z)),$$
$$y' = w(x, y, z)(\varphi_z(x, y, z) - \varphi_x(x, y, z)),$$
$$z' = w(x, y, z)(\varphi_x(x, y, z) - \varphi_y(x, y, z)).$$

CHAPTER

10

NONLINEAR SYSTEMS IN APPLICATIONS

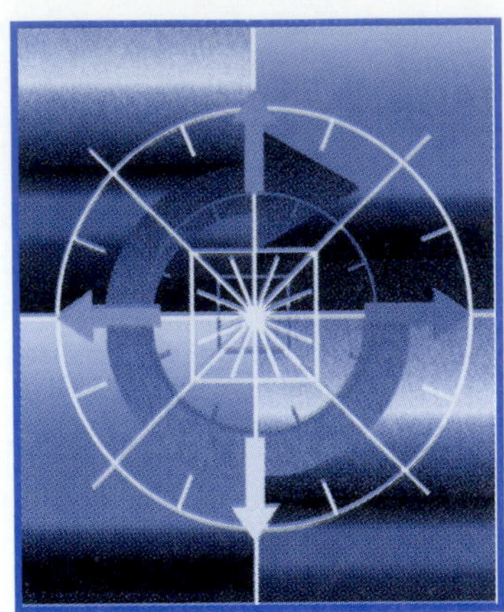

This chapter is a survey of a few nonlinear systems that arise in biology, chemistry, and mechanics. Our intent is to describe to some extent both the modeling process and the qualitative analysis of these systems.

10.1 Lotka-Volterra Systems in Ecology

One of the most interesting contexts in which nonlinear systems of differential equations arise is the modeling of interacting populations. Such *ecological* models first appeared in the independent work of Lotka and Volterra in the 1920s and comprise a cornerstone of *mathematical biology*, or *biomathematics*. In this section, we will investigate in detail two of the most fundamental types of interaction: competition and predation.

Competition Suppose that two species P and Q compete for resources that each requires in order to live, such as food or territory. Let $p(t)$ and $q(t)$ denote the respective sizes of the two populations at time t. We will assume the following.

i. In the absence of its competitor, each of the species is governed by a simple logistic equation with intrinsic growth rate k_i and environmental carrying capacity c_i for $i = 1,2$. (See Section 3.4.)
ii. The per capita growth rates, p'/p and q'/q, decrease proportionally with increasing competitor population.

These assumptions lead to

$$\frac{p'}{p} = k_1\left(1 - \frac{p}{c_1}\right) - \alpha_1 q \quad \text{and} \quad \frac{q'}{q} = k_2\left(1 - \frac{q}{c_2}\right) - \alpha_2 p, \tag{1}$$

in which k_1, k_2, c_1, c_2, α_1, and α_2 are positive constants. By substituting the *dimensionless* quantities

$$x(t) = \frac{1}{c_1}p(t), \quad y = \frac{1}{c_2}q(t), \quad a = \frac{\alpha_1}{k_1}c_2, \quad \text{and} \quad b = \frac{\alpha_2}{k_2}c_1,$$

we arrive at the tidier system

$$x' = k_1 x(1 - x - ay), \quad y' = k_2 y(1 - y - bx). \tag{2}$$

Note that x and y represent the population sizes as proportions of the respective environmental carrying capacities of P and Q. Also, we view the coefficients a and b as measures of the *competitive strengths* of Q and P, respectively. Assuming that $a \neq b$, we begin our analysis of (2) by determining the equilibrium points and the nullclines of the system. The nullclines are the lines given by

$$x = 0, \quad x + ay = 1,$$

along which the direction field is vertical, and

$$y = 0, \quad bx + y = 1,$$

along which the direction field is horizontal. There are three equilibrium points on the coordinate axes:

$$(0, 0), \quad (0, 1), \text{ and } (1, 0),$$

and another, (x^*, y^*), at the intersection of the lines $x + ay = 1$ and $bx + y = 1$, which is of interest only if x^* and y^* are both positive. An easy calculation gives

$$(x^*, y^*) = \left(\frac{1-a}{1-ab}, \frac{1-b}{1-ab}\right),$$

and so we observe that x^* and y^* are both positive if and only if

$$\textit{either} \quad a < 1 \text{ and } b < 1 \quad \textit{or} \quad a > 1 \text{ and } b > 1.$$

There are four main cases determined by the signs of $1 - a$ and $1 - b$. Figures 1a–d depict the equilibrium points and nullclines in each of these four cases.

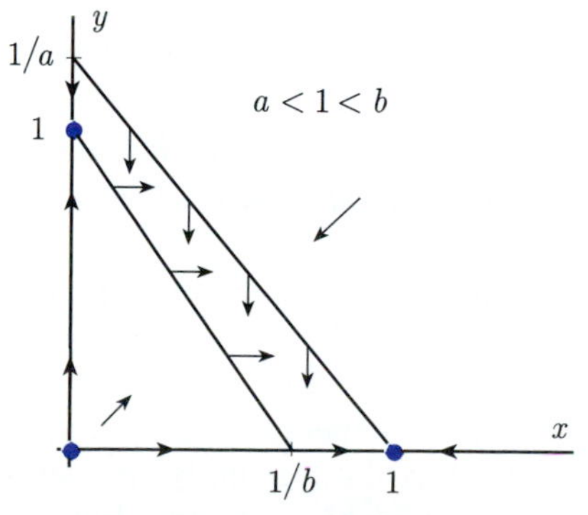

Figure 1a

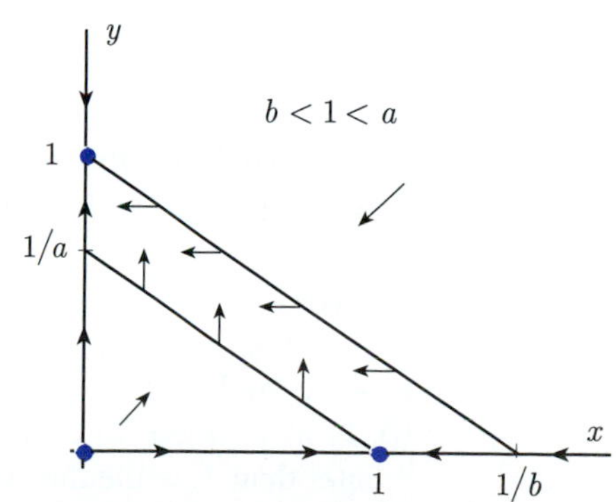

Figure 1b

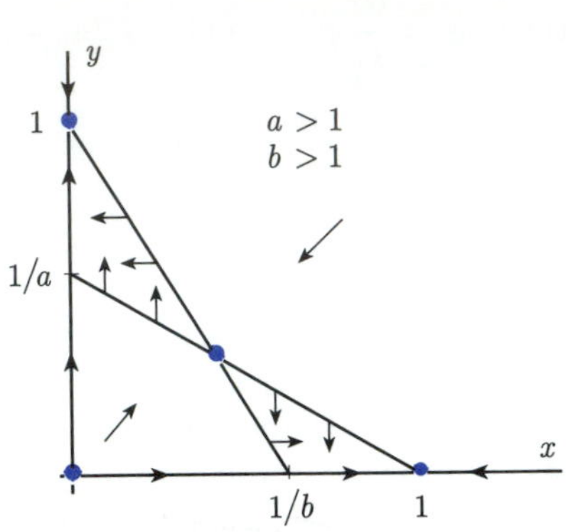

Figure 1c

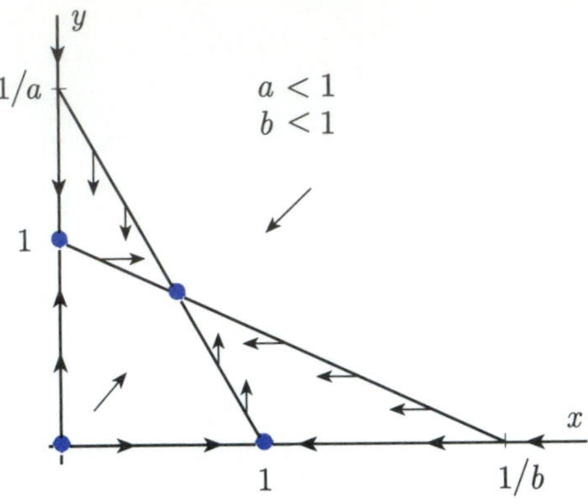

Figure 1d

The Jacobian matrix for the system (2) at (x, y) is

$$\mathcal{J}(x, y) = \begin{pmatrix} k_1(2x - 1 + ay) & -ak_1x \\ -bk_2y & k_2(2y - 1 + bx) \end{pmatrix},$$

and common to every case is the fact that the Jacobian matrix at the origin is

$$\mathcal{J}(0, 0) = \begin{pmatrix} k_1 & 0 \\ 0 & k_2 \end{pmatrix},$$

which implies that *the origin is unstable*, since k_1 and k_2 are each positive. If $k_1 \neq k_2$, the origin will be an unstable node; if $k_1 = k_2$, the origin will behave similarly, since orbits cannot spiral away from $(0, 0)$. (Why?) Also in every case, the Jacobian matrices at $(1, 0)$ and $(0, 1)$ are

$$\mathcal{J}(1, 0) = \begin{pmatrix} -k_1 & -k_1a \\ 0 & k_2(1 - b) \end{pmatrix},$$

whose eigenvalues are $-k_1$ and $k_2(1 - b)$, and

$$\mathcal{J}(0, 1) = \begin{pmatrix} k_1(1 - a) & 0 \\ -k_2b & -k_2 \end{pmatrix},$$

whose eigenvalues are $k_1(1 - a)$ and $-k_2$. So we can conclude that

- $(1, 0)$ *is a stable node when* $b > 1$ *and a saddle point when* $b < 1$;
- $(0, 1)$ *is a stable node when* $a > 1$ *and a saddle point when* $a < 1$.

We can now complete the phase portraits corresponding to Figures 1a and 1b. If $a < 1 < b$, then all orbits of (2) with $x(0) > 0$ and $y(0) \geq 0$ approach $(1, 0)$ as $t \to \infty$. If $b < 1 < a$, then all orbits of (2) with $x(0) \geq 0$ and $y(0) > 0$ approach $(0, 1)$ as $t \to \infty$. In each of these cases, the competitive strength of one species is greater than 1, while the other's is less than 1. Not surprisingly, the "stronger" species survives and the "weaker" species is driven toward extinction. Figure 2a illustrates the case where $a < 1 < b$, and Figure 2b illustrates the case where $b < 1 < a$.

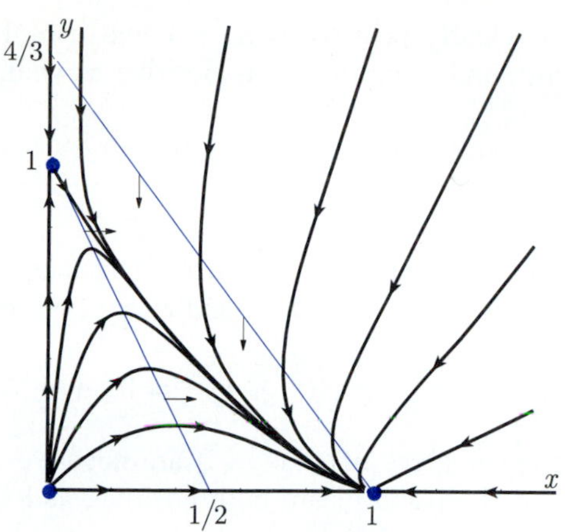

Figure 2a

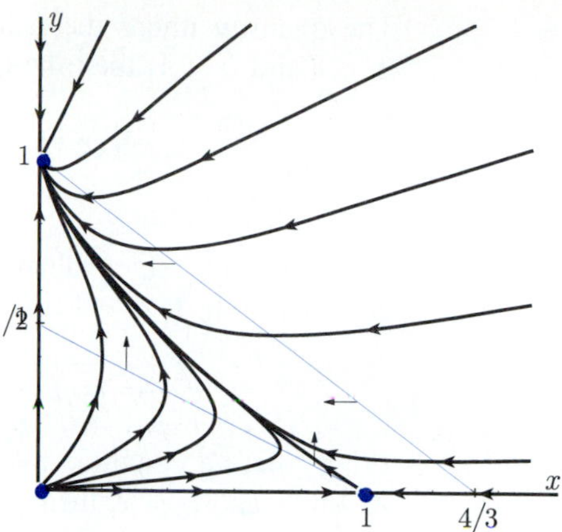

Figure 2b

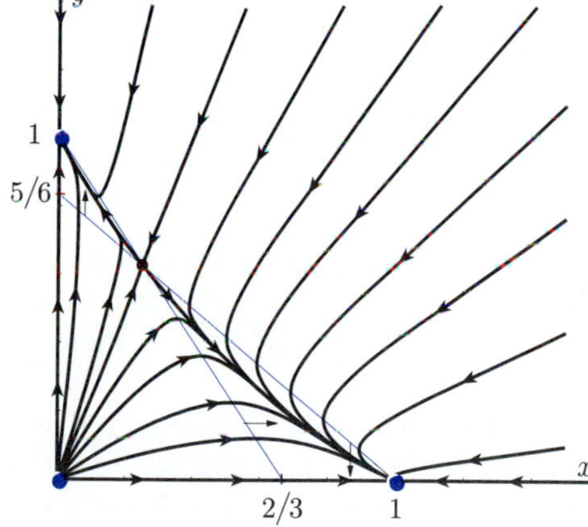

Figure 2c

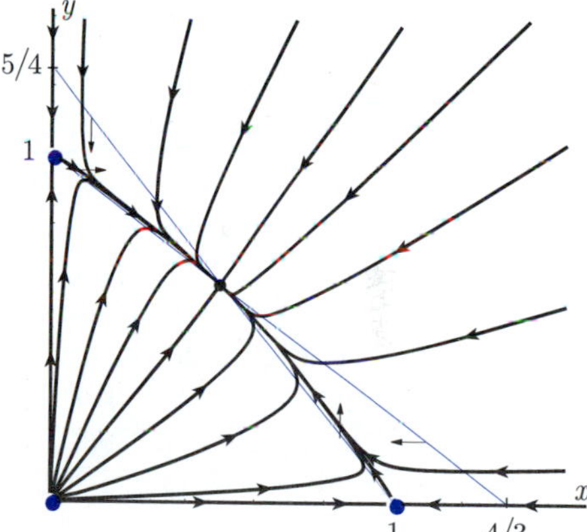

Figure 2d

We now look to the cases in which x^* and y^* are both positive. So we assume that one or the other of the inequalities in (2) holds. In these cases, (x^*, y^*) is an equilibrium that represents *steady, long-term coexistence* of the two species. The Jacobian matrix at (x^*, y^*) is

$$\mathcal{J}(x^*, y^*) = \mathcal{J}\left(\frac{1-a}{1-ab}, \frac{1-b}{1-ab}\right) = -\frac{1}{1-ab}\begin{pmatrix} k_1(1-a) & k_1a(1-a) \\ k_2b(1-b) & k_2(1-b) \end{pmatrix}.$$

Some *CAS*-aided computation shows that the eigenvalues of $\mathcal{J}(x^*, y^*)$ are

$$\frac{1}{2(1-ab)}\left(r_1 + r_2 \pm \sqrt{-4r_1r_2(1-ab) + (r_1+r_2)^2}\right).$$

where

$$r_1 = -k_1(1-a) \quad \text{and} \quad r_2 = -k_2(1-b).$$

The quantity under the radical is clearly positive if $a > 1$ and $b > 1$. (Why?) If $a < 1$ and $b < 1$, then the quantity under the radical is positive as well, because

$$\begin{aligned}-4r_1r_2(1-ab)+(r_1+r_2)^2 &\geq -4r_1r_2+(r_1+r_2)^2 \\ &= (r_1-r_2)^2.\end{aligned}$$

Therefore, the eigenvalues of $\mathcal{J}(x^*, y^*)$ are real in each of the cases now under consideration. If $a < 1$ and $b < 1$, then $1-ab > 0$, r_1 and r_2 are both negative, and

$$\sqrt{-4r_1r_2(1-ab)+(r_1+r_2)^2} < \sqrt{(r_1+r_2)^2} = |r_1+r_2|.$$

From this we conclude that the eigenvalues of $\mathcal{J}(x^*, y^*)$ are negative and distinct. If $a > 1$ and $b > 1$, then $1-ab < 0$, r_1 and r_2 are both positive, and

$$\sqrt{-4r_1r_2(1-ab)+(r_1+r_2)^2} > \sqrt{(r_1+r_2)^2} = |r_1+r_2|.$$

From this we conclude that the eigenvalues of $\mathcal{J}(x^*, y^*)$ have opposite signs. So finally we conclude that

- *if* $a > 1$ *and* $b > 1$, *then* (x^*, y^*) *is a saddle point*;
- *if* $a < 1$ *and* $b < 1$, *then* (x^*, y^*) *is a stable node.*

If we discount the extremely unlikely event that when $a > 1$ and $b > 1$ the initial point will lie on an orbit that converges to the saddle point, then we can say that the populations tend toward stable, long-term coexistence if and only if $a < 1$ and $b < 1$. Figure 2c illustrates the case where $a > 1$ and $b > 1$. Figure 2d illustrates the case where $a < 1$ and $b < 1$.

In conclusion, suppose that we describe a species as *weak* if its competitive strength is less than 1 and *strong* if its competitive strength is greater than 1. Then the main result of this analysis can be stated as follows: *Whether it is weak or strong, a species survives in the long run if its competitor is weak. If both species are strong, then the surviving species is determined by initial conditions.*

Predation Suppose that two species live within the same environment. One species, the *prey*, is the food source for the other species, the *predator*. The prey's food source is an abundant third organism. Let $p(t)$ and $q(t)$ denote the respective sizes of the prey and predator populations at time t. We will assume the following:

i. In the absence of predators, the prey population is governed by a simple logistic equation with intrinsic growth rate k_1 and environmental carrying capacity c_1.
ii. In the absence of prey, per capita growth rate of the predator population, q'/q, is a negative constant (i.e., the predator population declines exponentially).

iii. The per capita growth rate of the prey population, p'/p, decreases proportionally with increasing predator population.
iv. The per capita growth rate of the predator population, q'/q, increases proportionally with increasing prey population.

These assumptions lead to

$$\frac{p'}{p} = k_1\left(1 - \frac{p}{c_1}\right) - \alpha_1 q \quad \text{and} \quad \frac{q'}{q} = -k_2 + \alpha_2 p. \tag{3}$$

With the dimensionless quantities

$$x = \frac{\alpha_2 p}{k_2}, \quad y = \frac{\alpha_1 q}{k_1}, \quad \text{and} \quad \varepsilon = \frac{c_1 \alpha_2}{k_2},$$

we arrive at the tidier system

$$x' = k_1 x\left(1 - \frac{x}{\varepsilon} - y\right), \quad y' = -k_2 y(1 - x). \tag{4}$$

Here, x and y are just conveniently scaled population sizes. Note that if $p = c_1$ (the prey's stable equilibrium with no predators), then $q'/q = k_2(\varepsilon - 1)$. So we can think of ε as a measure of *predatory efficiency*—the efficiency with which consumption of prey results in predator reproduction. If $0 < \varepsilon < 1$, then $q'/q < 0$ when $p = c_1$, indicating that the predator is inefficient. If $\varepsilon > 1$, then $q'/q > 0$ when $p = c_1$, indicating that the predator is efficient.

We begin our analysis of (4) by determining the equilibrium points and the nature of the nullclines of the system. The nullclines are the lines given by

$$x = 0, \quad \frac{x}{\varepsilon} + y = 1,$$

along which the direction field is vertical, and

$$y = 0, \quad x = 1,$$

along which the direction field is horizontal. There are two equilibrium points on the coordinate axes:

$$(0, 0), \quad \text{and} \quad (\varepsilon, 0),$$

and another equilibrium point (x^*, y^*) at the intersection of the lines

$$\frac{x}{\varepsilon} + y = 1 \quad \text{and} \quad x = 1,$$

which is of interest only if x^* and y^* are both positive. An easy calculation gives

$$(x^*, y^*) = \left(1, \frac{\varepsilon - 1}{\varepsilon}\right),$$

and so we observe that x^* and y^* are both positive if and only if $\varepsilon > 1$. Figures 3a and b show the arrangement of equilibrium points and nullclines in each of the cases $0 < \varepsilon < 1$ and $\varepsilon > 1$, respectively.

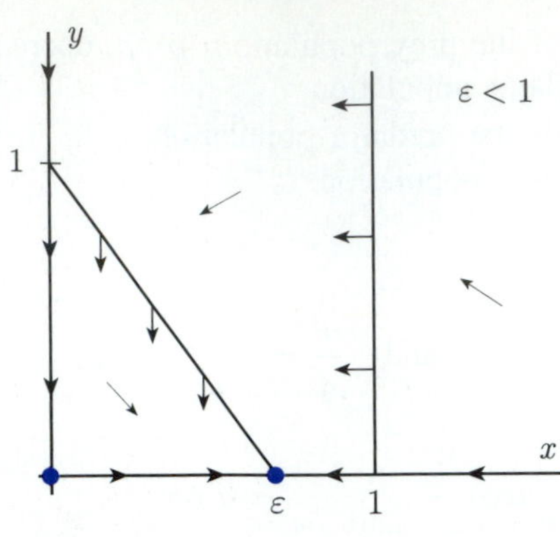

Figure 3a

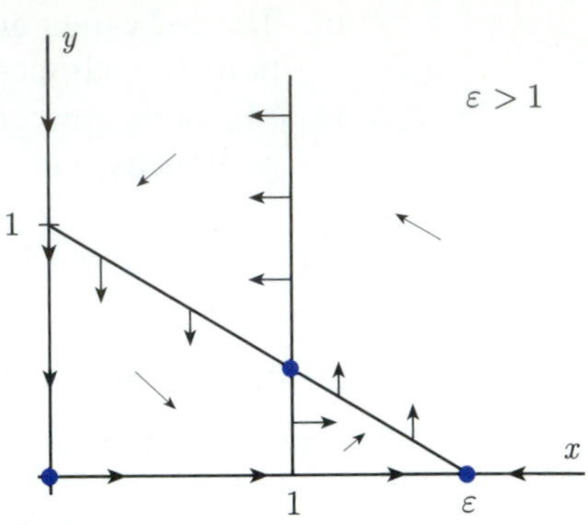

Figure 3b

In all cases the Jacobian matrix at (x, y) is

$$\mathcal{J}(x, y) = \begin{pmatrix} k_1\left(1 - y - \frac{2x}{\varepsilon}\right) & -k_1 x \\ k_2 y & -k_2(1-x) \end{pmatrix},$$

and so the Jacobian matrix at the origin is

$$\mathcal{J}(0, 0) = \begin{pmatrix} k_1 & 0 \\ 0 & -k_2 \end{pmatrix},$$

which implies that *the origin is a saddle point*. The Jacobian matrix at $(\varepsilon, 0)$ is

$$\mathcal{J}(\varepsilon, 0) = \begin{pmatrix} -k_1 & -k_1\varepsilon \\ 0 & -k_2(1-\varepsilon) \end{pmatrix},$$

whose eigenvalues are $-k_1$ and $-k_2(1-\varepsilon)$. So we can conclude that

- $(\varepsilon, 0)$ *is a stable node if* $0 < \varepsilon < 1$ *and a saddle point if* $\varepsilon > 1$.

The Jacobian matrix at $(x^*, y^*) = (1, (\varepsilon - 1)/\varepsilon)$ is

$$\mathcal{J}\left(1, \frac{\varepsilon - 1}{\varepsilon}\right) = \begin{pmatrix} -k_1/\varepsilon & -k_1 \\ k_2(\varepsilon - 1)/\varepsilon & 0 \end{pmatrix},$$

whose eigenvalues are the roots of $\varepsilon\lambda^2 + k_1\lambda + k_1k_2(\varepsilon - 1)$, which turn out to be

$$\frac{k_1}{2\varepsilon}\left(-1 \pm \sqrt{\frac{k_1 - 4k_2\varepsilon(\varepsilon - 1)}{k_1}}\right).$$

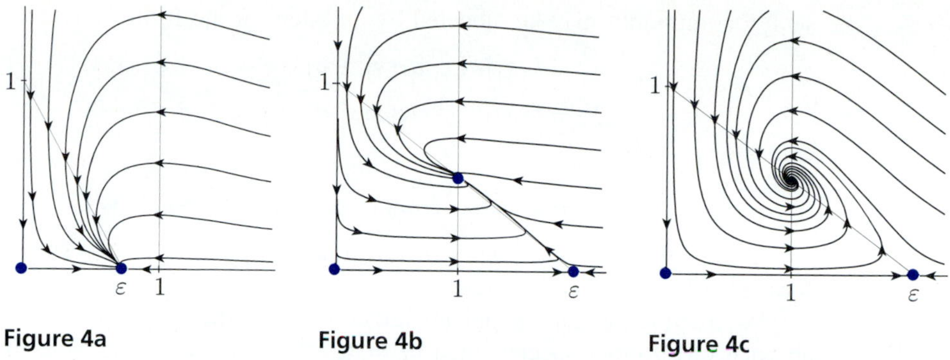

Figure 4a **Figure 4b** **Figure 4c**

From this we conclude that

- $\left(1, \dfrac{\varepsilon - 1}{\varepsilon}\right)$ *is a stable node if* $\varepsilon > 1$ *and* $4k_2\varepsilon(\varepsilon - 1) < k_1$;
- $\left(1, \dfrac{\varepsilon - 1}{\varepsilon}\right)$ *is a stable spiral point if* $\varepsilon > 1$ *and* $4k_2\varepsilon(\varepsilon - 1) > k_1$.

We can now complete the phase portraits corresponding to Figures 3a and 3b. Figure 4a is typical of the case where $0 < \varepsilon < 1$. Here, all orbits of (4) with $x(0) > 0$ and $y(0) > 0$ approach $(\varepsilon, 0)$ as $t \to \infty$, indicating that the predator tends toward extinction in the long run, while the prey survives. In the case where $\varepsilon > 1$, (x^*, y^*) may be either a stable node or a stable spiral point approached as $t \to \infty$ by all orbits of (4) with $x(0) > 0$ and $y(0) \geq 0$, indicating that the predator and prey coexist in the long run. These situations are illustrated in Figures 4b and 4c.

If ε is very large, then $(x^*, y^*) \approx (1, 1)$; the nullcline given by $x/\varepsilon + y = 1$ is nearly horizontal; and orbits with $x(0) > 0$ and $y(0) > 0$ are nearly closed—that is, there is a time T such that $(x(t), y(t)) \approx (x(0), y(0))$. In fact, in the limit as $\varepsilon \to \infty$ (a "perfectly efficient" predator), orbits with $x(0) > 0$ and $y(0) > 0$ are indeed closed, corresponding therefore to periodic solutions. This is the subject of Problem 7 at the end of this section.

We end with a simple ecological conclusion: *If the predator is* efficient—*as indicated by* $\varepsilon > 1$—*then the predator population survives in the long run. Otherwise the predator drives itself toward extinction by consuming too much and reproducing too little.*

Complex Communities The two-species competitive and predator-prey systems just described are models of very simple ecological communities. More complex communities may consist of any number of species $S_1, S_2, \ldots, S_n$, each of which is involved in a competitive, predator-prey, or *cooperative* interaction with at least one other species in the community. Letting y_i represent the size of the S_i population, such a community may be modeled by a system of the form

$$\begin{aligned}
y_1' &= k_1 y_1(b_1 + a_{11}y_1 + a_{12}y_2 + \cdots + a_{1n}y_n),\\
y_2' &= k_2 y_2(b_2 + a_{21}y_1 + a_{22}y_2 + \cdots + a_{2n}y_n),\\
&\vdots\\
y_n' &= k_n y_n(b_n + a_{n1}y_1 + a_{n2}y_2 + \cdots + a_{nn}y_n),
\end{aligned}$$

where each $b_i = \pm 1$, each $a_{ii} \le 0$, and if $a_{ij} = 0$, then $a_{ji} = 0$. This is the general form of a *Lotka-Volterra system.*

We assume that all coefficients are constant. If the ith species can survive in the absence of all other species, then $b_i = 1$; otherwise $b_i = -1$. Note that if $b_i = 1$, then $a_{ii} = -\frac{1}{c_i}$, where c_i is the environmental carrying capacity for S_i. For each distinct pair of species S_i and S_j, the signs of the coefficients a_{ij} and a_{ji} reflect the type of interaction that exists between S_i and S_j as follows.

$$\begin{aligned}
a_{ij} > 0,\ a_{ji} < 0 &: \quad \text{predator-prey}\\
a_{ij} < 0,\ a_{ji} < 0 &: \quad \text{competition}\\
a_{ij} > 0,\ a_{ji} > 0 &: \quad \text{cooperation}\\
a_{ij} = 0,\ a_{ji} = 0 &: \quad \text{independence}
\end{aligned}$$

Such a system may have many equilibrium points. An equilibrium point at which each $y_i > 0$ (all species survive) will be a solution of the linear algebraic system

$$A\mathbf{y} = -\mathbf{b},$$

where A is the $n \times n$ matrix with entries a_{ij}, $\mathbf{y} = (y_1, y_2, \ldots, y_n)^T$, and $\mathbf{b} = (b_1, b_2, \ldots, b_n)^T$. The Jacobian matrix $\mathcal{J}(\mathbf{y})$ associated with the system is often called the *community matrix* by ecologists.

Example 1

Consider a three-species community of pine trees, oak trees, and squirrels. Pine trees and oak trees compete for sunlight and nutrients, and squirrels feed upon acorns from the oak trees. An associated dimensionless system, with particular parameter values, might be

$$\begin{aligned}
x' &= x(1 - x - 2y),\\
y' &= y(1 - x - y - z),\\
z' &= z(-1 + 3y),
\end{aligned}$$

where x, y, and z represent dimensionless measures of population size for the pine trees, oak trees, and squirrels, respectively. This particular system has five equilibrium points of interest: (0, 0, 0), (0, 1/3, 2/3), (0, 1, 0), (1, 0, 0), and (1/3, 1/3, 1/3). The Jacobian (i.e., community) matrix at (x, y, z) is

$$\mathcal{J}(x, y, z) = \begin{pmatrix} 1 - 2x - 2y & -2x & 0 \\ -y & 1 - x - 2y - z & -y \\ 0 & 3z & -1 + 3y \end{pmatrix}.$$

The following table summarizes the linearization analysis.

E^*	$\mathcal{J}(E^*)$	$\lambda_1, \lambda_2, \lambda_3$	Conclusion
$(0, 0, 0)$	$\begin{pmatrix} 1 & 0 & 0 \\ 0 & 1 & 0 \\ 0 & 0 & -1 \end{pmatrix}$	$1, 1, -1$	Unstable
$\left(0, \frac{1}{3}, \frac{2}{3}\right)$	$\frac{1}{3}\begin{pmatrix} 1 & 0 & 0 \\ -1 & -1 & -1 \\ 0 & 6 & 0 \end{pmatrix}$	$\frac{1}{3}, \frac{1}{6}(-1 \pm i\sqrt{23})$	Unstable
$(0, 1, 0)$	$\begin{pmatrix} -1 & 0 & 0 \\ -1 & -1 & -1 \\ 0 & 0 & 2 \end{pmatrix}$	$-1, -1, 2$	Unstable
$\left(\frac{1}{3}, \frac{1}{3}, \frac{1}{3}\right)$	$\frac{1}{3}\begin{pmatrix} -1 & -2 & 0 \\ -1 & -1 & -1 \\ 0 & 3 & 0 \end{pmatrix}$	$-.60, -.03 \pm .43i$	Asymptotically stable
$(1, 0, 0)$	$\begin{pmatrix} -1 & -2 & 0 \\ 0 & 0 & 0 \\ 0 & 0 & -1 \end{pmatrix}$	$-1, -1, 0$	?

Note that $(1/3, 1/3, 1/3)$ is the only stable equilibrium point, and it is also asymptotically stable. The observation that $x + 3y + z$ is nonincreasing when $x > 1$ and $y > 1$ implies that nonnegative solutions are bounded. These facts indicate that with positive initial values the community will *typically* tend toward a steady state in which all three species thrive. ■

Example 2

Consider a three-species community of lions, zebra, and grass. Lions feed upon the zebra, which in turn feed upon the grass. The associated system, with particular parameter values, might be

$$\begin{aligned} x' &= x(1-y), \\ y' &= y(-1+2x-z), \\ z' &= z(-1+y), \end{aligned}$$

where x, y, and z represent dimensionless measures of population size for the lions, zebra, and grass, respectively. In addition to the equilibrium point at $(0, 0, 0)$, this system has a line of equilibrium points described by $(s, 1, 2s-1)$, of interest for $s \geq 1/2$. The Jacobian matrix at (x, y, z) is

$$\mathcal{J}(x, y, z) = \begin{pmatrix} 1-y & -x & 0 \\ 2y & 2x-z-1 & -y \\ 0 & z & y-1 \end{pmatrix}.$$

At each of the equilibrium points on the line $(s, 1, 2s - 1)$, $s \geq 1/2$, the Jacobian matrix becomes

$$\mathcal{J}(s, 1, 2s-1) = \begin{pmatrix} 0 & -s & 0 \\ 2 & 0 & -1 \\ 0 & -1+2s & 0 \end{pmatrix},$$

whose eigenvalues are 0 and $\pm i\sqrt{4s-1}$. Consequently, linearization does not help us determine whether these equilibrium points are stable. So consider the function

$$\ell(x, y, z) = 2(x - s) + y - 1 + z - (2s - 1) - \ln\left(\frac{x^2 yz}{s^2(2s-1)}\right),$$

which is cleverly constructed so that

$$\ell_x f + \ell_y g + \ell_z h = 0 \quad \text{for all} \quad x, y, z > 0$$

and

$$\ell(s, 1, 2s - 1) = 0 \quad \text{for all} \quad s > \frac{1}{2}.$$

Now observe also that

$$\nabla \ell(s, 1, 2s - 1) = \left(2 - \frac{2}{s}, 0, 1 - \frac{1}{2s-1}\right),$$

which implies that ℓ does not have a local minimum at $(s, 1, 2s - 1)$ if $s \neq 1$. When $s = 1$, the equilibrium point of interest is $(1, 1, 1)$, and ℓ does indeed have a local minimum there, as can be verified by examining the second-order partials of ℓ. So at $(1, 1, 1)$, the system has the Lyapunov function

$$\ell(x, y, z) = 2(x - 1) + y - 1 + z - 1 - \ln(x^2 yz),$$

which implies that $(1, 1, 1)$ is stable. Since $\ell_x f + \ell_y g + \ell_z h = 0$ for x, y, $z > 0$, each orbit remains on a particular level surface of ℓ for all t. Thus, we can conclude that $(1, 1, 1)$ is not asymptotically stable. Figure 5 shows two "cut-away" views of three level surfaces of ℓ that enclose $(1, 1, 1)$. ■

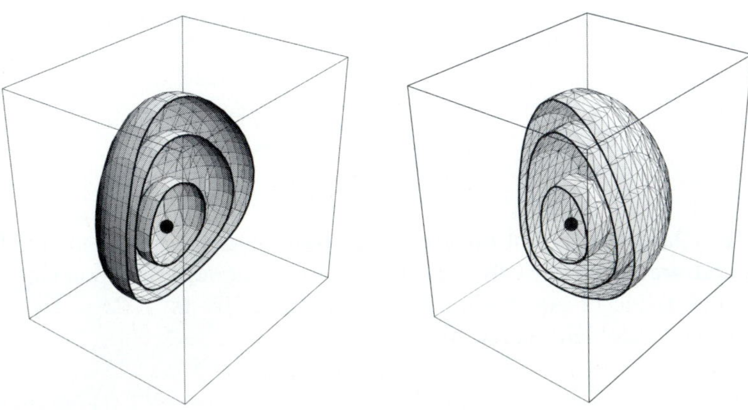

Figure 5

PROBLEMS

1. **(a)** What units do the coefficients k_1, k_2, c_1, c_2, α_1, and α_2 have in system (1)?

   (b) Check that the quantities x, y, a, and b in system (2) are indeed "dimensionless" (i.e., have no units).

   (c) Verify that system (2) follows from system (1) by the indicated substitution of dimensionless quantities.

2. **(a)** What units do the coefficients k_1, k_2, c_1, α_1, and α_2 have in system (3)?

   (b) Check that the quantities x, y, and ε in system (4) are indeed "dimensionless" (i.e., have no units).

   (c) Verify that system (4) follows from system (3) by the indicated substitution of dimensionless quantities.

Consider the competition model as in (1),

$$p' = k_1 p \left(1 - \frac{p}{c_1}\right) - \alpha_1 pq, \quad p(0) = p_0,$$

$$q' = k_2 q \left(1 - \frac{q}{c_2}\right) - \alpha_2 pq, \quad q(0) = q_0,$$

in which p and q are the population sizes of species P and Q, respectively. For each of the following sets of data/coefficients,

(a) form the corresponding dimensionless system for x and y as in (2), and make a rough sketch of the (x, y) phase portrait;

(b) determine which of P and Q survives in the long run and the steady long-term size of each surviving population.

3. $\begin{cases} k_1 = 5, & c_1 = 500, \alpha_1 = 100^{-1}, p_0 = 500 \\ k_2 = 20, & c_2 = 1000, \alpha_2 = 10^{-1}, q_0 = 1000 \end{cases}$

4. $\begin{cases} k_1 = 10, & c_1 = 400, \alpha_1 = 100^{-1}, p_0 = 300 \\ k_2 = 50, & c_2 = 2000, \alpha_2 = 20^{-1}, q_0 = 1000 \end{cases}$

5. $\begin{cases} k_1 = 10, & c_1 = 1000, \alpha_1 = 20^{-1}, p_0 = 800 \\ k_2 = 20, & c_2 = 100, \alpha_2 = 100^{-1}, q_0 = 50 \end{cases}$

6. $\begin{cases} k_1 = 10^{-1}, & c_1 = 100, \alpha_1 = 50000^{-1}, p_0 = 100 \\ k_2 = 30, & c_2 = 10000, \alpha_2 = 10^{-1}, q_0 = 9000 \end{cases}$

7. Figure 2c suggests that when $a > 1$ and $b > 1$ the competition system (2) has a straight-line separatrix along which orbits approach the saddle point as $t \to \infty$. Show that this is true if $k_1 = k_2$, but that the system has no straight-line orbits if $k_1 \neq k_2$.

Analyze the modified competition models in Problems 8 through 10.

8. The carrying capacity for each species is inversely proportional to the square root of the population size of the other species:

$$x' = x\left(1 - x\sqrt{y}\right), \quad y' = y\left(1 - y\sqrt{x}\right).$$

9. The competitive strength of the x-population decreases as x increases:

$$x' = x\left(1 - \frac{1}{2}x - y\right), \quad y' = y\left(1 - y - \frac{x}{1+x}\right).$$

10. In the absence of competition, the per capita rate of decline of large populations is bounded:

$$x' = k_1 x\left(\frac{1}{x} - 1 - ay\right), \quad y' = k_2 y\left(\frac{1}{y} - 1 - bx\right).$$

Consider the predator-prey model as in (3),

$$p' = k_1 p\left(1 - \frac{p}{c_1}\right) - \alpha_1 pq, \quad p(0) = p_0,$$
$$q' = -k_2 q + \alpha_2 pq, \quad q(0) = q_0,$$

in which p and q are the population sizes of prey species P and predator species Q, respectively. For each set of data/coefficients in Problems 11 through 13,

(a) form the corresponding dimensionless system for x and y as in (4), and make a rough sketch of the (x, y) phase portrait;

(b) determine the steady long-term size of each population.

11. $k_1 = 1$, $c_1 = 500$, $\alpha_1 = 100^{-1}$, $p_0 = 500$; $k_2 = 20$, $\alpha_2 = 10^{-1}$, $q_0 = 1000$

12. $k_1 = 10$, $c_1 = 400$, $\alpha_1 = 100^{-1}$, $p_0 = 300$; $k_2 = 50$, $\alpha_2 = 20^{-1}$, $q_0 = 1000$

13. $k_1 = 50$, $c_1 = 1000$, $\alpha_1 = 20^{-1}$, $p_0 = 800$; $k_2 = 5$, $\alpha_2 = 100^{-1}$, $q_0 = 50$

14. Consider the dimensionless predator-prey model with a perfectly efficient predator (i.e., $\varepsilon = \infty$):

$$x' = k_1 x(1 - y), \quad y' = -k_2 y(1 - x).$$

(a) Find an implicit solution, valid for $x, y > 0$, of the *separable* slope equation

$$\frac{dy}{dx} = -\frac{k_2 y(1 - x)}{k_1 x(1 - y)}.$$

(b) Write the implicit solution in the form $\varphi(x, y) = C$ and show that φ attains an extreme value at $(1, 1)$.

(c) Sketch the phase portrait of the system, making use of the result of part (b).

(d) Show that the same dimensionless system arises from

$$p' = k_1 p - \alpha_1 pq, \quad q' = -k_2 q + \alpha_2 pq.$$

What modification of the assumptions that lead to (3) produces this model? What characteristic of the prey results in the same type of system behavior as does perfect predatory efficiency?

Analyze the modified predator-prey models in Problems 15 through 19.

15. Predatory efficiency decreases with increasing numbers of predators:

$$x' = x\left(1 - \frac{1}{2}x - y\right), \quad y' = -\frac{1}{2}y(1 - x + y).$$

16. Predatory efficiency increases with y when y is small but eventually decreases as y increases:

$$x' = x\left(1 - x - \frac{1}{2}y\right), \quad y' = -\frac{1}{10}y(1 - x + (y - 2)y).$$

17. In the absence of predators, the per capita rate of decline of large prey populations is bounded:

$$x' = \frac{1}{2}x\left(-1 + \frac{2}{x} - y\right), \quad y' = -y(1 - x).$$

18. The carrying capacity for the predator population is proportional to the number of prey:

$$x' = \frac{1}{3}x(1-x-y), \quad y' = -\frac{1}{10}y\left(1-\frac{y}{2x}\right).$$

19. Predators migrate into the environment at a rate proportional to the number of prey:

$$x' = \frac{1}{3}x\left(1-\frac{1}{2}x-y\right), \quad y' = -y\left(1-\frac{1}{2}x\right)+\frac{1}{2}x.$$

20. Consider the modified predator-prey system

$$x' = k_1 x\left(\frac{x}{\varepsilon}\left(1-\frac{x}{\varepsilon}\right)-y\right), \quad y' = -k_2 y(1-x+2y^2).$$

(a) Give a geometric argument based on nullclines to show that, if $\varepsilon > 1$, there is a single equilibrium point (x^*, y^*) for which $x^*, y^* > 0$.

(b) Show that, if $\varepsilon > 1$, the rectangle in which $0 \le x \le \varepsilon$ and $0 \le y \le \sqrt{\frac{\varepsilon-1}{2}}$ is a forward-invariant region.

(c) Set $k_1 = k_2 = 1$, and determine the nature of (x^*, y^*) for $\varepsilon = 3$ and for $\varepsilon = 6$. Taking into account the result of (b) and the Poincaré-Bendixson theorem, sketch phase portraits (with nullclines) for each case.

21. Analyze the following model of two *cooperative* populations:

$$x' = k_1 x\left(1-\frac{x}{\varepsilon}+y\right), \quad y' = k_2 y\left(1-\frac{y}{\beta}+x\right).$$

22. Consider the following two-predator, one-prey system with competition between the predator species:

$$\begin{aligned} x' &= x\left(1-x-\frac{1}{2}y-z\right), \\ y' &= y(-1+2x-z), \\ z' &= z(-1+3x-2y). \end{aligned}$$

Find the equilibrium points at which $x, y, z \ge 0$, and determine whether each is stable or unstable.

23. Consider a three-species community modeled by

$$\begin{aligned} x' &= x(-1-y+2z), \\ y' &= y(-1+2x-2z), \\ z' &= z(-1-3x+2y). \end{aligned}$$

(a) Describe the relationships among the x-, y-, and z-species.

(b) Find the equilibrium points at which $x, y, z \ge 0$, and determine whether each is stable or unstable.

24. Consider a three-species community modeled by

$$\begin{aligned} x' &= x(1-x+y), \\ y' &= y(1-y+x-z), \\ z' &= z(-1+2y). \end{aligned}$$

(a) Describe the relationships among the x-, y-, and z-species.

(b) Find the equilibrium points at which $x, y, z \ge 0$, and determine whether each is stable or unstable.

10.2 Infectious Disease and Epidemics

This section describes three basic models for the transmission of an infectious disease in a population. It is assumed that each individual in the population falls into one of three classes: those susceptible to the disease ("susceptibles"), those infected with the disease ("infectives"), and those who have recovered from the disease. The sizes of these subpopulations are are denoted by S, I, and R, respectively. We assume that the spread of the disease is relatively rapid, allowing us to ignore natural birth and death rates in the population. We will, however, consider the possibility of infectives dieing from the disease. In all cases we make the following assumption.

ASSUMPTION *The rate at which susceptibles contract the disease is proportional to the rate at which susceptibles encounter infectives, which is proportional to the product of the sizes of the susceptible and infective populations and thus given by βSI, where β is a positive constant.*

SIR The *SIR* system models the spread of a disease in which individuals who recover (or die) from the disease become immune to reinfection. Thus an individual who contracts the disease (and survives) passes through three stages: $S \to I \to R$. In addition to the assumption stated previously, we assume that infectives recover from the disease at a rate of γI individuals per unit time and that the death rate among infectives is δI individuals per unit time, where γ and δ are positive constants. The resulting equations are

$$S' = -\beta SI, \quad I' = \beta SI - (\gamma + \delta)I, \quad R' = \gamma I. \tag{1}$$

The S- and I-equations are independent of R, so we may as well look only at those. In the phase portrait, the S- and I-axes are S-nullclines, and the I-nullclines are the line $S = (\gamma + \delta)/\beta$ and the S-axis. Every point on the S-axis is an equilibrium point, and there are no others. A typical phase portrait is indicated in Figure 1. The direction field makes it clear that equilibrium points $(S, 0)$ with $0 \le S \le (\gamma + \delta)/\beta$ are stable, while those with $S > (\gamma + \delta)/\beta$ are unstable. Moreover, $I \to 0$ as $t \to \infty$ for any (nonnegative) initial values of S and I. The epidemic always "runs its course," leaving fewer than $(\gamma + \delta)/\beta$ individuals unaffected.

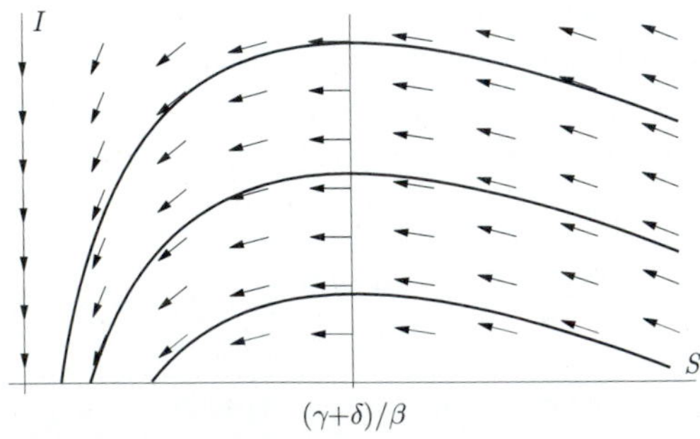

Figure 1

SIS The *SIS* system models the spread of a disease in which individuals who recover from the disease immediately become susceptible again. With β, γ, and δ representing the same proportionality constants as previously described for the *SIR* model, the *SIS* equations are

$$S' = -\beta SI + \gamma I, \quad I' = \beta SI - (\gamma + \delta)I. \tag{2}$$

The S-nullclines are the S-axis and the line $S = \gamma/\beta$, and the I-nullclines are the S-axis and the line $S = (\gamma + \delta)/\beta$. So there are two distinct cases of interest determined by whether the death rate δ is positive or zero—that is, whether the disease is fatal or not. If $\delta = 0$, then the I- and S-nullclines coincide to form two lines of equilibrium points, and nontrivial phase plane orbits are parallel lines, as shown in Figure 2a, due to the fact that $S + I$ is constant. Note that all equilibrium points $(\gamma/\beta, I)$ are stable. Equilibrium points on the S-axis are stable if $0 \leq S \leq \gamma/\beta$ and unstable otherwise. The phase portrait indicates further that if $S(0) > 0$, then the solution approaches a stable equilibrium at which there are positive numbers of both infectives and susceptibles, *unless* $S(0) + I(0) \leq \gamma/\beta$, in which case the number of infectives approaches zero.

The case in which $\delta > 0$ is illustrated by the phase portrait in Figure 2b. Here, the nullclines $S = \gamma/\beta$ and $S = (\gamma + \delta)/\beta$ are distinct, and so the only equilibrium points are those on the S-axis, which are stable if $S \leq (\gamma + \delta)/\beta$ and unstable otherwise. Since $I \to 0$ as $t \to \infty$ for any (nonnegative) initial values of S and I, The epidemic always runs its course, ultimately leaving fewer than $(\gamma + \delta)/\beta$ survivors.

SIRS The *SIRS* system models the spread of a disease in which individuals who recover from the disease become immune to reinfection, but thereafter return to the susceptible class at a rate of σR individuals per unit time, where σ is a positive constant. This leads to the equations

$$\begin{aligned} S' &= -\beta SI + \sigma R, \\ I' &= \beta SI - (\gamma + \delta)I, \\ R' &= \gamma I - \sigma R. \end{aligned} \tag{3}$$

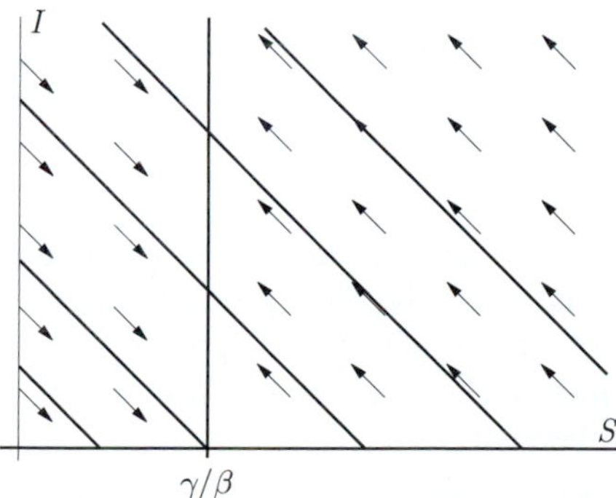

Figure 2a

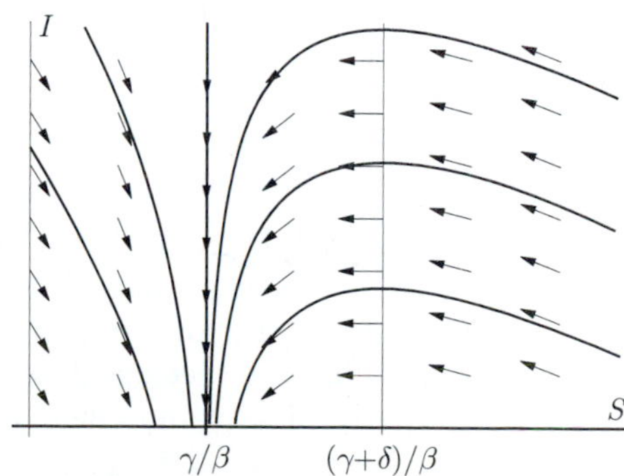

Figure 2b

If $\delta > 0$, it turns out that every point of the form $(S, 0, 0)$ is an equilibrium point, and the behavior of the system is difficult to analyze. To simplify the problem, we will assume that susceptibles enter the population at the same rate that infectives die, thus causing the total population $S + I + R$ to remain constant. Denoting that constant by N, we have $R = N - S - I$, and thus S and I satisfy

$$\begin{aligned} S' &= -\beta SI + \delta I + \sigma(N - S - I), \\ I' &= \beta SI - (\gamma + \delta)I. \end{aligned} \tag{4}$$

To simplify the analysis of this system, we let

$$x(t) = \ell\beta S(\ell t), \quad y(t) = \ell\beta I(\ell t), \quad a = (\delta - \sigma)\ell, \quad b = \sigma\ell, \quad c = N\beta\sigma\ell^2,$$

where $\ell = (\gamma + \delta)^{-1}$, and thus obtain the simpler system:

$$\begin{aligned} x' &= -xy + ay - bx + c, \\ y' &= xy - y. \end{aligned} \tag{5}$$

Note that the coefficients b and c are positive and that $a < 1$. This system has two equilibrium points, $(c/b, 0)$ and $\left(1, \frac{c-b}{1-a}\right)$. The second of these is of interest only if $c > b$. The results of some *CAS*-aided computation are indicated in the following table:

E^*	$\mathcal{J}(E^*)$	λ_1, λ_2
$(c/b, 0)$	$\begin{pmatrix} -b & \dfrac{ab-c}{b} \\ 0 & \dfrac{c-b}{b} \end{pmatrix}$	$-b, \dfrac{c-b}{b}$
$\left(1, \dfrac{c-b}{1-a}\right)$	$\begin{pmatrix} \dfrac{c-ab}{1-a} & a-1 \\ \dfrac{c-b}{1-a} & 0 \end{pmatrix}$	$\dfrac{ab-c}{2(1-a)} \pm \sqrt{\left(\dfrac{ab-c}{2(1-a)}\right)^2 - (c-b)}$

Thus we see that $(c/b, 0)$ is a stable node if $c < b$ and a saddle point if $c > b$. If $c > b$, then $ab - c < 0$ (since $a < 1$), and so the point $\left(1, \frac{c-b}{1-a}\right)$ is asymptotically stable and may be a node or a spiral, depending upon the sign of the quantity under the radical. Figures 3a and b show typical phase portraits in which $c < b$ and $c > b$, respectively. We conclude that the sign of $c - b$ determines whether or not the epidemic eventually runs its course. In terms of the original coefficients, the sign of $c - b$ is the same as the sign of $N\beta - (\gamma + \delta)$.

SIRS with Vaccination Consider again the *SIRS* model and assume that susceptibles are vaccinated—and thus removed from the susceptible class—at a rate of vS individuals per unit time. (Note that v represents the proportion of the susceptible class that are vaccinated per unit time; thus there may be a practical limit on how large v can be.) Our equations then become

$$\begin{aligned} S' &= -\beta SI + \delta I + \sigma(N - S - I) - vS, \\ I' &= \beta SI - (\gamma + \delta)I. \end{aligned} \tag{6}$$

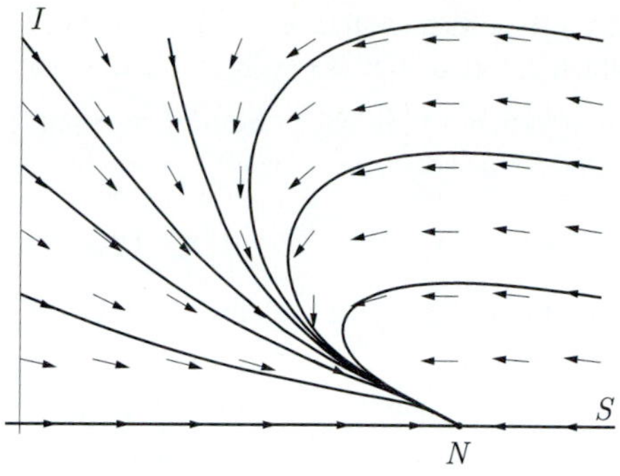

Figure 3a

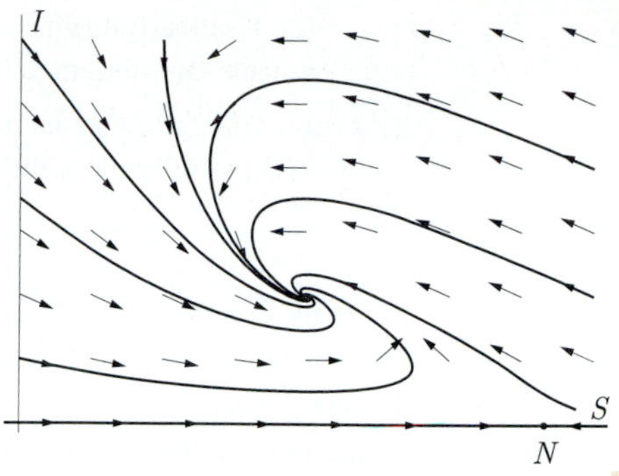

Figure 3b

We assume here that $N\beta - (\gamma + \delta) > 0$, so that without vaccination the epidemic is persistent; that is, the system in (3) has a stable equilibrium point at which $I > 0$. Our goal is to determine how large v must be in order for the vaccination effort to be successful. Intuitively, the vaccination effort will be successful if there is no stable equilibrium point at which $I > 0$.

As you will verify in Problem 3, if we let

$$x(t) = \ell\beta S(\ell t), \quad y(t) = \ell\beta I(\ell t), \quad a = (\delta - \sigma)\ell, \quad b = (\sigma + v)\ell, \quad c = N\beta\sigma\ell^2,$$

where $\ell = (\gamma + \delta)^{-1}$, we obtain the very same simplified system as (5):

$$\begin{aligned} x' &= -xy + ay - bx + c, \\ y' &= xy - y. \end{aligned}$$

Therefore, our previous analysis tells us that the vaccination effort will successfully eradicate the disease if $c < b$. In terms of the original coefficients, the inequality $c < b$ is equivalent to

$$v > \frac{\sigma}{\gamma + \delta}(N\beta - (\gamma + \delta)).$$

Note that the quantity on the right side of this inequality may be so large as make a successful vaccination effort impossible to implement.

PROBLEMS

1. The models presented in the text ignore natural birth and death rates and thus do not include *vital dynamics*. Consider the following *SIS* system with vital dynamics:

$$S' = (a - bS)(S + I) - \beta SI + \gamma I, \quad I' = -bI(S + I) + \beta SI - (\gamma + \delta)I.$$

(a) Verify and interpret the fact that

$$(S + I)' = (S + I)(a - b(S + I)) - \delta I.$$

(b) Describe the roles of the vital-dynamics terms, $(a - bS)(S + I)$ and $-bI(S + I)$, in the system.

(c) Use the following parameter values to show that, even if $\delta > 0$, the system can have a stable equilibrium point at which $I > 0$: $a = \beta = \gamma = \delta = 1$, $b = 1/10$.

2. Analyze the following modification of system (4), in which birth of susceptibles is accounted for with the kS-term in the S-equation:

$$S' = kS - \beta SI + \delta I + \sigma(N - S - I), \quad I' = \beta SI - (\gamma + \delta)I.$$

3. Verify that the system (5) is obtained from each of the systems (4) and (6) by means of the substitutions described in the text.

4. Analyze the following modification of the S- and I-equations of the *SIR* model, in which the rate of disease transmission decreases for large infective populations (due to increased awareness on the part of the susceptibles?) and the susceptible population grows exponentially in the absence of infectives:

$$S' = kS - \frac{\beta SI}{\eta + I^2}, \quad I' = \frac{\beta SI}{\eta + I^2} - (\gamma + \delta)I.$$

5. Consider the *SIR* model in (1).

(a) Solve the slope equation corresponding to the S- and I-equations to obtain a description of the curves traced out by phase-plane orbits in the (S, I) phase plane.

(b) Find the maximum number of infectives in terms of $I(0)$, $S(0)$, and $\rho = (\gamma + \delta)/\beta$.

6. Consider the full *SIRS* model in (3). Show that if the disease is nonfatal (i.e., $\delta = 0$), then $N = S + I + R$ is constant.

7. Consider the full *SIRS* model in (3), without the assumption of a constant total population:

$$\begin{aligned} S' &= -\beta SI + \sigma R, \\ I' &= \beta SI - (\gamma + \delta)I, \\ R' &= \gamma I - \sigma R. \end{aligned}$$

(a) Show that $(0, 0, 0)$ is the only equilibrium point.

(b) Use the fact that $(S+I+R)' = -\delta I$ to argue that if $S(0) \geq 0$, $I(0) > 0$, and $R(0) \geq 0$, then $(S(t), I(t), R(t)) \to (0, 0, 0)$ as $t \to \infty$. (In particular, $(0, 0, 0)$ is asymptotically stable with respect to nonnegative solutions for which $I(0) > 0$.)

8. Under the assumption that the population of susceptibles remains a constant $S = S_0$ due to births, deaths, and migration, the *SIRS* model in (3) gives rise to

$$I' = \beta S_0 I - (\gamma + \delta)I, \quad R' = \gamma I - \sigma R.$$

Analyze this linear model.

9. The following system is a simple *cell-level* model of a viral infection under treatment:

$$\begin{aligned} S' &= -\beta SV + \sigma - \delta S, \\ I' &= \beta SV - \tilde{\delta} I, \\ V' &= -\beta SV + \gamma I - \rho V. \end{aligned}$$

Here S is the number of uninfected cells, I is the number of infected cells, and V is the number of virions (i.e., the extracellular, infective form of the virus). In the V-equation the γI-term accounts for the production of virions by infected cells, and the $-\rho V$-term accounts for the removal of virions due to the treatment.

(a) Show that the change of variables

$$x(t) = \beta S(t/\gamma), \quad y(t) = \beta I(t/\gamma), \quad z(t) = \beta\sigma V(t/\gamma),$$

and substitution of

$$a = \delta/\gamma, \quad b = \tilde{\delta}/\gamma, \quad c = \rho/\gamma, \quad q = \beta\sigma/\gamma^2$$

produces the simplified system

$$x' = -xz + q - ax, \quad y' = xz - by, \quad z' = -xz + y - cz,$$

containing four parameters rather than six.

(b) Show that if $\rho = 0$ (i.e., there is no treatment), then the system has only one equilibrium point. Show that if $0 < b < 1$, then that equilibrium point is unstable.

(c) Determine conditions on a, b, c, and q guaranteeing that the system has an equilibrium point where x, y, and z are each positive.

(d) Choose simple values of a, b, c, and q for which the system has an equilibrium point with positive coordinates, and determine whether or not that equilibrium point is stable. What does this say about the success/failure of the treatment?

10.3 Other Biological Models

In this section we will study three disparate biological models. The first is a model known as the *chemostat*, in which a culture of bacteria feeds upon a steady inflow of nutrients. We then examine a crude model for the treatment of a liver tumor by chemotherapy, followed by a study of the Fitzhugh-Nagumo equations, which are related to the Nobel prize-winning work of Hodgkin and Huxley on the transmission of neural impulses.

The Chemostat A chemostat is a simple bacteria farm. It consists of a bacteria culture together with a reservoir for replenishing the nutrients upon which the bacteria feed. Fluid containing nutrients at concentration C_0 flows into the culture container at a steady rate ρ. Fluid from which bacteria are harvested flows out of the container at the same rate ρ, thus maintaining a constant volume of fluid in the culture container. The fluid in the culture contained is kept thoroughly mixed.* Figure 1 is a schematic illustration of a chemostat.

The goal in constructing a chemostat would be to produce a steady outflow of harvestable bacteria by maintaining a stable culture of bacteria in the container.

Let B denote the bacteria concentration (number per unit volume) in the container, C the nutrient concentration (mass per unit volume) in the container, and V the constant volume of fluid in the container. We make the following assumptions:

i. Nutrient consumption by the bacteria decreases the nutrient concentration at a rate given by $Bf(C)$, where f is some nonnegative, nondecreasing function such that $f(0) = 0$.

* Notice the similarity to the "mixing problems" of Section 2.2.3.

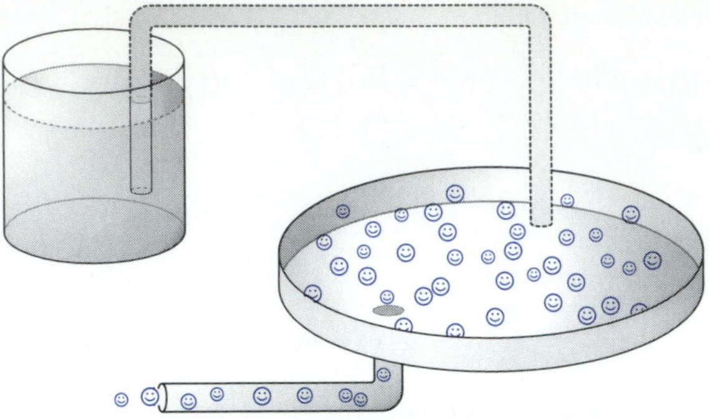

Figure 1

ii. The rate at which the bacteria concentration increases is proportional to the rate at which the bacteria consume nutrients and is thus given by $\beta B f(C)$, where β is a positive constant.

The resulting system is

$$\begin{aligned} B' &= \beta B f(C) - \frac{\rho}{V} B, \\ C' &= -B f(C) - \frac{\rho}{V} C + \frac{\rho}{V} C_0. \end{aligned} \tag{1}$$

The question now is, What type of function f would be appropriate for the model? A realistic consideration is that a bacteria culture of a given size can consume only so much nutrient mass, regardless of how much is available. One type of function that incorporates this notion is

$$f(C) = \frac{aC}{b + C}, \tag{2}$$

where a and b are positive constants.† A typical graph, indicating the significance of a and b, is shown in Figure 2.

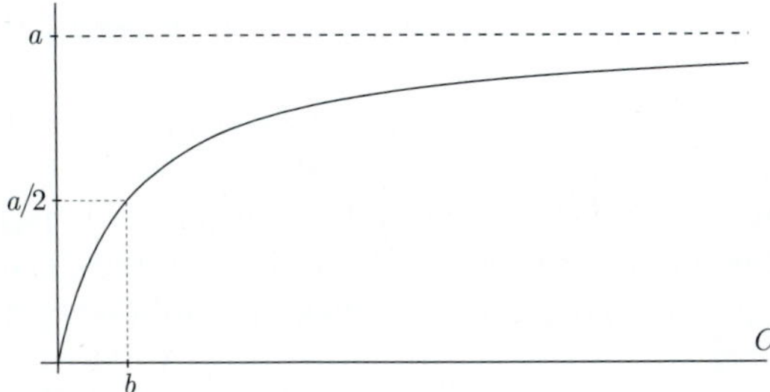

Figure 2

---

† This choice of f amounts to the assumption of *Michaelis-Menten kinetics*. For more information see Edelstein-Keshet [19].

It turns out that (1), with f given by (2), is equivalent to the following system, which contains only two dimensionless parameters (see Problem 1):

$$x' = k\frac{xy}{1+y} - x, \quad y' = -\frac{xy}{1+y} - y + q. \tag{3}$$

The quantities here are related to the original ones by

$$x(t) = \frac{a\beta V}{b\rho}B(\rho t/V), \quad y(t) = \frac{1}{b}C(\rho t/V), \quad k = \frac{aV}{\rho}, \quad \text{and} \quad q = \frac{C_0}{b}.$$

This system has two equilibrium points:

$$(0, q) \quad \text{and} \quad \left(kq - \frac{k}{k-1}, \frac{1}{k-1}\right).$$

Clearly the second of these is the interesting one. For the chemostat to have an equilibrium state with a positive number of bacteria, k and q must satisfy

$$k > 1 \quad \text{and} \quad q > \frac{1}{k-1}. \tag{4}$$

The eigenvalues of the linearized system at that point are

$$-1 \quad \text{and} \quad -\frac{(k-1)^2}{k}\left(q - \frac{1}{k-1}\right),$$

which are negative under the conditions in (4). Thus the equilibrium point of interest is always a stable node. We conclude that (4) characterizes the properly functioning chemostat. In terms of the original parameters, the conditions in (4)—after a little algebra—become sensible constraints on the flow rate ρ and the stock nutrient concentration C_0:

$$\rho < aV \quad \text{and} \quad C_0 > \frac{\rho b}{aV - \rho}.$$

A Chemotherapy Model We consider here a simple-minded model of the treatment of a liver tumor by continuous injection of a drug solution through the hepatic artery. For simplicity we view the liver as a simple well-mixed tank through which blood flows at a steady rate and the tumor as a ball composed of small cells. Let

c_0 = the drug concentration in the injected solution;
ρ = the volume flow rate of the injected solution;
V = the volume of blood in the liver;
R = the rate of blood flow in and out of the liver;
$T(t)$ = the number of tumor cells;
$A(t)$ = the amount of drug in the liver.

Our three main assumptions are as follows:

1. The rate at which tumor cells are killed by the drug is jointly proportional to the surface area of the tumor and a Michaelis-Menten–type function of the drug concentration.

2. The rate of "uptake" (i.e., absorption) of the drug by the tumor cells is jointly proportional to the surface area of the tumor and a Michaelis-Menten–type function of the drug concentration as in (2).
3. The natural growth rate of tumor cells is given by

$$f(T) = \gamma \ln(\mu/T)T,$$

reflecting the tendency of a tumor to grow more slowly as it progresses, due to oxygen and nutrient deprivation in the interior of the tumor.‡

We assume further that the volume of the tumor is proportional to T and, consequently, that the surface area is proportional to $T^{2/3}$. Using this, together with the three assumptions stated previously, we arrive at

$$\begin{aligned} T' &= \gamma \ln(\mu/T)T - k_1 \frac{aA}{bV + A} T^{2/3}, \\ A' &= c_0\rho - k_2 \frac{\alpha A}{\beta V + A} T^{2/3} - \frac{R}{V} A. \end{aligned} \tag{5}$$

The complexity of this model makes a thorough analysis difficult. So we will simply give an illustration with (not necessarily meaningful) parameter values as indicated in

$$\begin{aligned} T' &= 0.3 \ln\left(\frac{100}{T}\right) T - \frac{2A}{20 + A} T^{2/3}, \\ A' &= c_0\rho - \frac{A/2}{10 + A} T^{2/3} - A. \end{aligned} \tag{6}$$

The quantity $c_0\rho$ is the drug injection rate. Figure 3a shows the direction field and nullclines for $c_0\rho = 30$. The nullclines intersect at a stable equilibrium point where $T > 0$, indicating that the drug is injected at a rate insufficient to destroy the tumor. Figure 3b shows the direction field and nullclines for $c_0\rho = 75$, an injection rate that *is* high enough to destroy the tumor.

Neural Impulses The most prominent feature of a neuron is a long tubular *axon*, along which electrochemical impulses travel to a primary cell body, or *soma*. In 1952, A. L. Hodgkin and A. F. Huxley published a description of this neural mechanism that won them a Nobel prize. Hodgkin and Huxley used data collected from extensive experiments on the giant squid axon to construct a complex model consisting of four highly nonlinear differential equations.

When no stimulus is applied to an axon, it remains electrochemically "at rest." A small stimulus will cause a small perturbation from this rest state, but the axon will quickly return to rest. A sufficiently large stimulus will cause the axon to "fire," sending a series of electrochemical impulses along its length toward the cell body. This *excitability* was a key axon characteristic successfully reproduced by the equations of Hodgkin and Huxley.

---

‡ The assumption here is of *Gompertz* growth. See Edelstein-Keshet [19].

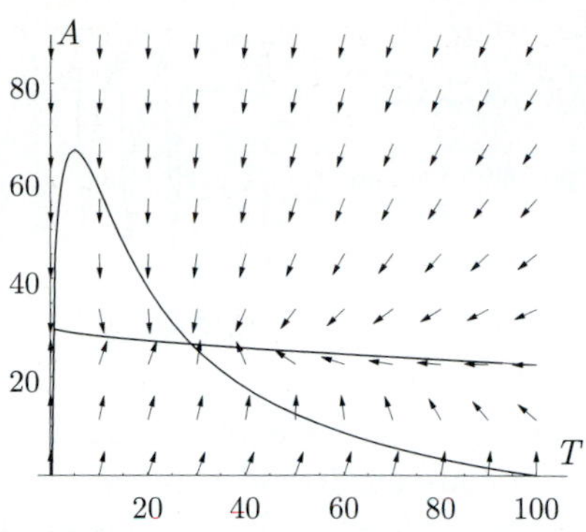

Figure 3a

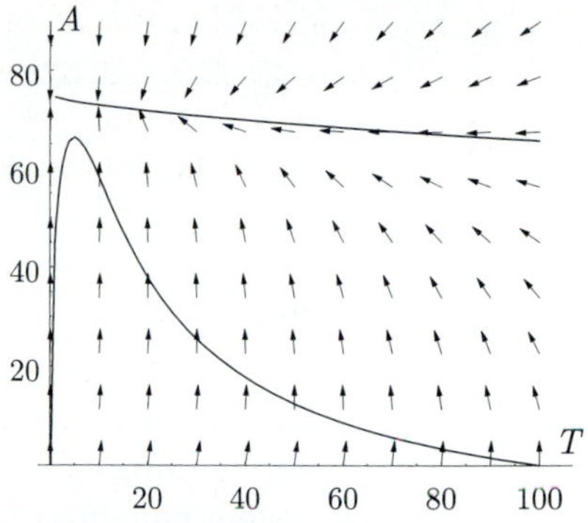

Figure 3b

For several years the *Hodgkin-Huxley equations* could only be studied numerically. Such work was limited due to the embryonic state of digital computing in the 1950s. Near the end of that decade, a breakthrough came when R. Fitzhugh successfully applied phase-plane methods to the Hodgkin-Huxley equations, arguing that a family of systems of two equations could explain some of the behavior of the Hodgkin-Huxley equations. Subsequent work by Fitzhugh, which was essentially duplicated independently by J. S. Nagumo et al., produced a relatively simple system of two equations whose solutions imitated those of the Hodgkin-Huxley equations. The *Fitzhugh-Nagumo equations* (with particular coefficients) are

$$\begin{aligned} v' &= 3w - v(v^2 - 3) + s(t), \\ w' &= \frac{1}{30}(7 - 10v - 8w). \end{aligned} \tag{7}$$

The v-component of the system plays the role of the membrane potential (voltage), while w may be viewed as "lumped" representation of the other three components in the Hodgkin-Huxley equations. The function $s(t)$ plays the role of an applied *stimulus* to the system.

In the case where $s(t) = 0$, the system has a single equilibrium point $(\tilde{v}_0, \tilde{w}_0) \approx (1.199, -0.624)$. To illustrate the interesting excitability that these equations can produce, let us suppose that the initial point is $(\tilde{v}_0, \tilde{w}_0)$ and that the stimulus is a step function of the form

$$s(t) = \begin{cases} 0, & \text{if } t < 20, \\ \sigma, & \text{if } t \geq 20 \end{cases},$$

where σ is a constant. The system will remain at rest until $t = 20$, at which time the stimulus comes into play. Figure 4a shows the graph of v versus t with $\sigma = -0.5$. Notice that one small "pulse" is generated by the stimulus, but the system quickly comes back to rest, even though the stimulus does not go away. Figure 4b is the graph of v versus t with $\sigma = -1$. The pulse is larger and is followed by some damped "ringing." In Figure 4c, in which $\sigma = -1.02$, we see a periodic train of pulses. Somewhere between $\sigma = -1$ and $\sigma = -1.02$ there is a "threshold" value at which periodic pulses begin to occur.

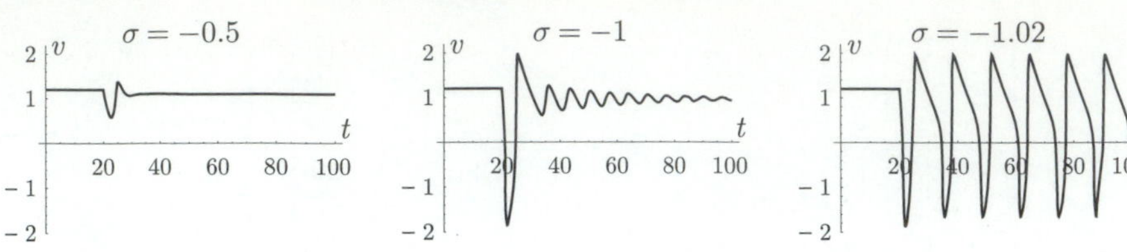

Figure 4a Figure 4b Figure 4c

The behavior illustrated by Figures 4a–c can be explained using phase-plane techniques. Suppose now that $s(t) = \sigma$, a constant, for all t. Figures 5a and b show the system nullclines for $\sigma = 0$ and $\sigma = -1$, respectively. We observe that

- on the cubic v-nullcline, w always has a local minimum at $v = 1$;
- decreasing σ shifts the cubic v-nullcline vertically upward;
- there will be exactly one equilibrium point for any σ;
- as the v-nullcline is shifted upward, the equilibrium point moves to the left.

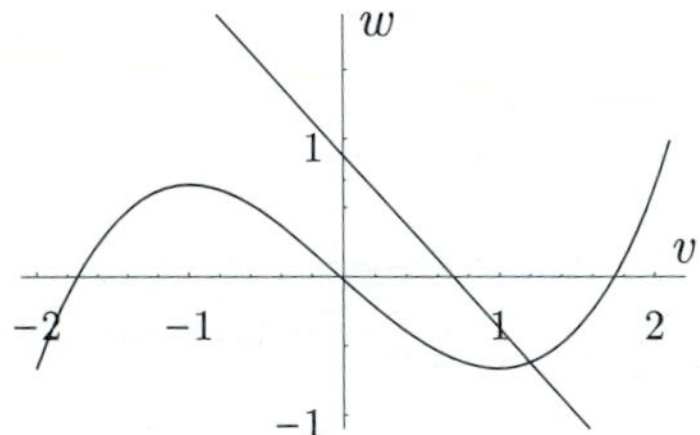

Figure 5a

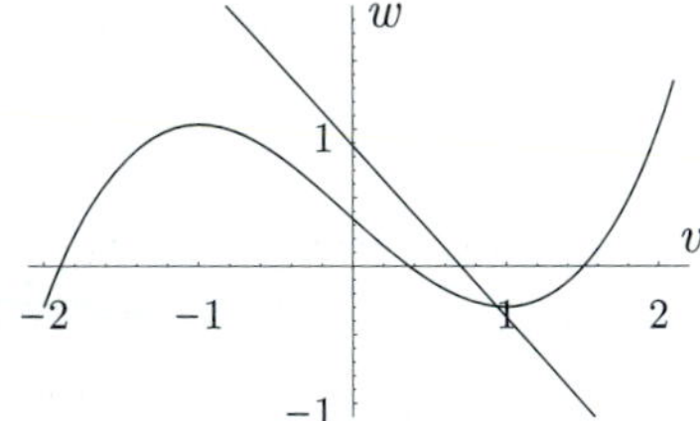

Figure 5b

Let (v_σ^*, w_σ^*) denote the equilibrium point corresponding to σ. The Jacobian matrix for the system at $(\tilde{v}_\sigma, \tilde{w}_\sigma)$ is

$$\mathcal{J}(\tilde{v}_\sigma, \tilde{w}_\sigma) = \begin{pmatrix} 3(1 - \tilde{v}_\sigma^2) & 3 \\ -\frac{1}{3} & -\frac{4}{15} \end{pmatrix},$$

whose eigenvalues turn out to be

$$\frac{1}{30}\left(41 - 45\tilde{v}_\sigma^2 \pm \sqrt{(41 - 45\tilde{v}_\sigma^2)^2 - 180(1 + 4\tilde{v}_\sigma^2)}\right).$$

Consequently, the equilibrium point is asymptotically stable when $\tilde{v}_\sigma^2 > 41/45 \approx 0.9545$ and unstable when $\tilde{v}_\sigma^2 < 41/45$. So, as we decrease σ, the value of $\tilde{v}_\sigma^2$ decreases until it crosses the critical value $\sqrt{41/45}$, causing the equilibrium point to become unstable. Figures 6a–6d show solutions corresponding to $\sigma = 0, -0.5, -1$, and -1.02, respectively, each with the same initial point (v_0^*, w_0^*), which is the equilibrium point in the $\sigma = 0$ case. Only in Figure 6d (where $\sigma = -1.02$) is $v_\sigma < \sqrt{41/45}$.

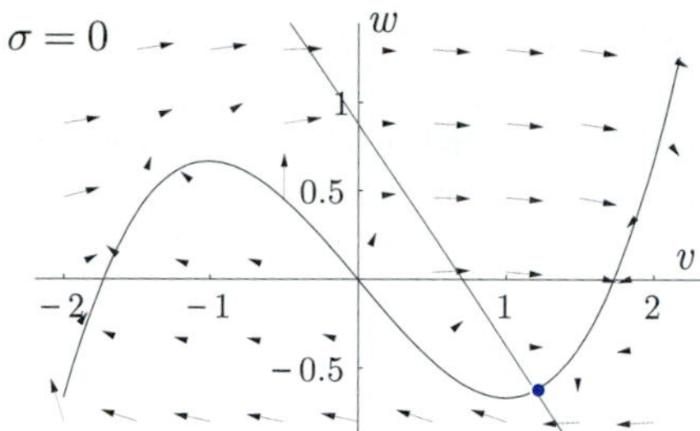

Figure 6a

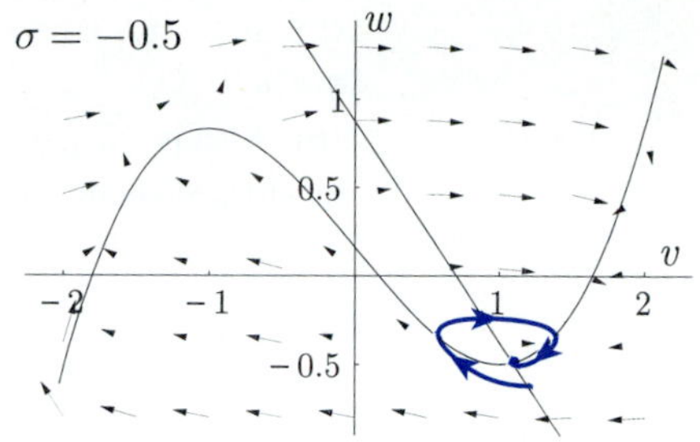

Figure 6b

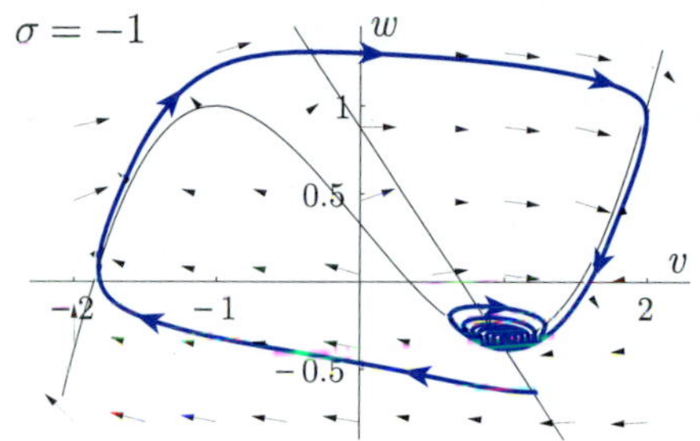

Figure 6c

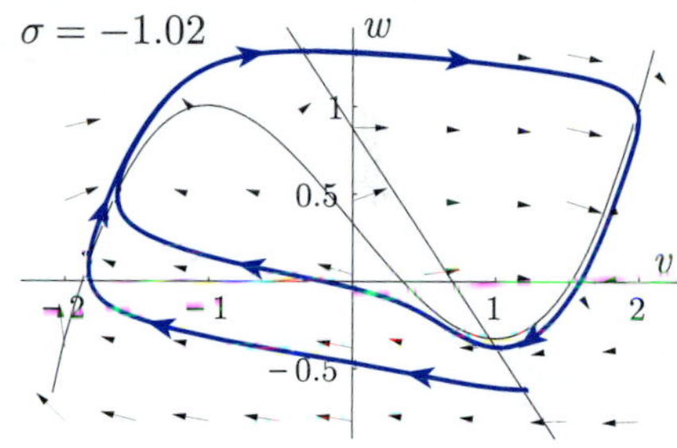

Figure 6d

PROBLEMS

1. Verify that system (3) arises from (1) by means of the indicated substitutions.

2. Consider the following variation on the chemostat, in which $f(C) = \alpha C$, where α is a positive constant:

$$B' = \beta BC - \frac{\rho}{V}B, \quad C' = -\alpha BC - \frac{\rho}{V}C + \frac{\rho}{V}C_0.$$

   (a) Show that appropriate substitutions lead to the simpler system

$$x' = kxy - x, \quad y' = -xy - y + q.$$

   (b) Under what conditions on k and q will the chemostat have a stable equilibrium point with a positive bacteria concentration?

3. Analyze the following system, which arises from a model of a chemostat containing two competing bacteria populations.

$$x' = \frac{\beta xz}{1+z}(1-xy) - x, \quad y' = \frac{\gamma yz}{1+z}(1-xy) - y,$$
$$z' = -\frac{z}{1+z}(x+y) - z + q.$$

4. A model for oxygen consumption in a bacterial culture gives rise to

$$x' = -\frac{xy}{1+x^2} - x + q, \quad y' = -\frac{xy}{1+x^2} + a,$$

where x and y represent nutrient and oxygen concentrations, respectively, and a and q are positive constants.

(a) Show that the system has a single equilibrium point (x^*, y^*) and that $x^*, y^* > 0$, if $q > a$.

(b) Assume that $q > a$ and construct a bounded forward-invariant region in the first quadrant, making use of

$$\begin{aligned} y' &< 0 \quad \text{if } xy > a(1+x^2); \\ (x+y)' &< 0 \quad \text{if } x > a+q; \\ (y-x)' &< 0 \quad \text{if } x < q-a. \end{aligned}$$

Conclude from the Poincaré-Bendixson theorem that if (x^*, y^*) is unstable, then the system has a periodic orbit.

(c) Set $q = 10$ and find particular values of $a < 10$ for which (x^*, y^*) is (i) a stable spiral point, and (ii) an unstable spiral point.

5. A simple model of hormone secretion leads to the system

$$x' = \frac{a}{b+y^2} - x, \quad y' = x - y.$$

Discuss the behavior of nonnegative solutions.

6. A simple cell-differentiation model leads to the system

$$x' = \frac{ay^2}{b+y^2} - x, \quad y' = x - y.$$

Discuss the behavior of nonnegative solutions.

7. One general form of the Fitzhugh-Nagumo equations is

$$v' = k\left(w + v - \frac{1}{3}v^3 + s(t)\right), \quad w' = \frac{1}{k}(a - v - bw),$$

where a, b, k are positive constants. In this problem, assume that $s(t) = 0$ for all t.

(a) Show that if $b < 1$, then the system has a single equilibrium point (v^*, w^*).

(b) Show that if $b < 1$ and $a < 1 - 2b/3$, then $0 < v^* < 1$. Conclude that (v^*, w^*) lies between the two extreme points on the cubic v-nullcline.

(c) Suppose that $b < 1$ and $a < 1 - 2b/3$. Show that (v^*, w^*) is asymptotically stable if $k^2(1 - v^{*2}) - b < 0$ and unstable if $k^2(1 - v^{*2}) - b > 0$.

8. Consider the Fitzhugh-Nagumo system in Problem 7, with $s(t) = \sigma$ (a constant). Show that if $v^4/3 + bw^2 \geq v^2 + \sigma v + aw$, then $(k^{-1}v^2 + kw^2)' \leq 0$. Conclude that if C is sufficiently large, then the ellipse $k^{-1}v^2 + kw^2 = C$ is a bounded, forward-invariant region containing all of the system's equilibrium points. What can you now conclude from the Poincaré-Bendixson theorem?

10.4 Chemical Systems

Mathematical models of chemical reactions provide some of the most interesting examples of nonlinear systems of differential equations. The primary principle used in chemical reaction modeling is the *law of mass action*. We will first discuss how this principle leads to differential equations and then explore a few examples, including two of the most well-known "chemical oscillators."

The Law of Mass Action Chemical reactions are composed of one or more elementary reaction steps. For example, the decomposition of ozone resulting from the absorption of ultraviolet light is the elementary reaction

$$O_3 \to O_2 + O,$$

while the formation of ozone from atmospheric oxygen,

$$3O_2 \to 2O_3,$$

is a complex reaction that consists of two elementary steps,

$$\begin{gathered} O_2 \to 2O \\ O_2 + O + N \to N + O_3. \end{gathered}$$

Typically the quantities of interest in the analysis of a chemical reaction are the concentrations (in moles per unit volume) of the chemical species involved. The common notation for the concentration of a species A is $[A]$. This becomes a bit cumbersome as mathematical notation, so we will usually follow the convention of using a lowercase letter to denote the concentration of the chemical species represented by the corresponding uppercase letter; for instance, $a = [A]$.

For a reaction step

$$n_1 R_1 + n_2 R_2 + \cdots + n_m R_m \to \text{ product(s)},$$

in which $R_1, R_2, \ldots, R_m$ are distinct reactant species, the quantity $\frac{1}{n_i} r_i'$, where $r_i = [R_i]$, is the same for all $i = 1, \ldots, m$. This quantity defines the *rate of the reaction*; that is,

$$\text{reaction rate } = \frac{1}{n_1} r_1' = \frac{1}{n_2} r_2' = \cdots = \frac{1}{n_m} r_m'.$$

Suppose that an *elementary* chemical reaction—or one elementary step of a complex reaction—is caused by a molecular event described by

$$Q_1 + Q_2 + \cdots + Q_m \to \text{ product(s)},$$

where $Q_1, Q_2, \ldots, Q_m$ represent single molecules of one or more, *not necessarily distinct*, reactant species. The **law of mass action** states that *the rate of the reaction is proportional to the product of the reactant concentrations,* $[Q_1][Q_2]\cdots[Q_m]$. Thus when the molecular event that causes the reaction involves n molecules of one of the reactants, the concentration of that reactant is raised to the nth power in the reaction rate. The reaction rate multiplied by the number of molecules of a species involved in each molecular event is equal to the rate of change in the concentration of that species.

In the following examples we assume that the law of mass action is in force (i.e., we assume *mass-action kinetics*) and that the concentrations of the reacting species are affected only by the reaction itself.

Example 1

If one molecule of A combines irreversibly with two molecules of B to form one molecule of C,

$$A + 2B \to C,$$

then the reaction rate is proportional to $[A][B]^2$. Letting $a = [A]$ and $b = [B]$, this leads to the system

$$a' = -kab^2, \quad b' = -2kab^2, \quad c' = kab^2. \qquad \blacksquare$$

Example 2

If one molecule of A combines *reversibly* with one molecule of B to form one molecule of C,

$$A + B \rightleftharpoons C,$$

then the forward reaction rate is proportional to $[A][B]$, and the reverse reaction rate is proportional to $[C]$. With $c = [C]$, we arrive at the system

$$a' = -k_1ab + k_2c, \quad b' = -k_1ab + k_2c, \quad c' = k_1ab - k_2c. \qquad \blacksquare$$

Example 3

The mechanism for the formation and decomposition of ozone mentioned previously can be written as

$$A \rightleftharpoons 2B$$
$$A + B \rightleftharpoons P,$$

since the nitrogen concentration is not affected by the reaction. In the first reaction, the rate of the forward step is proportional to $[A]$, while the rate of the reverse step is proportional to $[B]^2$. In the second reaction, the rate of the forward step is proportional to $[A][B]$, while the rate of the reverse step is proportional to $[P]$. So the resulting system of equations for $[A]$, $[B]$, and $[P]$ is

$$\begin{aligned} a' &= -k_1a + k_2b^2 - k_3ab + k_4p, \\ b' &= 2k_1a - 2k_2b^2 - k_3ab + k_4p, \\ p' &= k_3ab - k_4p. \end{aligned} \qquad \blacksquare$$

A Simple Enzymatic Reaction Consider the simple two-step, *enzymatic* reaction

$$A + C \rightleftharpoons B, \quad B \to P + C,$$

in which A is converted to a product P via an intermediate complex B that is formed by A and an enzyme *catalyst* C. The rate of the forward reaction in the first step is proportional to $[A][C]$. The rates of the reverse reaction in the first step and the reaction in the second step are each proportional to $[B]$. The resulting system of equations is

$$\begin{aligned} a' &= -k_1ac + k_2b, \\ b' &= k_1ac - (k_2 + k_3)b, \\ c' &= -k_1ac + (k_2 + k_3)b \\ p' &= k_3b, \end{aligned}$$

where k_1, k_2, $k_3 > 0$. The initial values of interest chemically are of the form

$$a(0) = a_0 > 0, \quad b(0) = 0, \quad c(0) = c_0 > 0, \quad p(0) = 0.$$

The equations for b and c imply that $b+c$ is constant. (Why?) Thus we can substitute $c = c_0 - b$ and $\gamma = k_2 + k_3$ into the equations for a and b to obtain

$$\begin{aligned} a' &= -k_1a(c_0 - b) + k_2b, \\ b' &= k_1a(c_0 - b) - \gamma b. \end{aligned} \tag{1}$$

Given any solution of this reduced system, the corresponding c- and p-components of the original system can easily be recovered from $c = c_0 - b$ and $p' = k_3b$. It turns out that, for the reduced system, $(0, 0)$ is a stable node to which all nonnegative solutions converge as $t \to \infty$. (See Problem 6.) From this we can conclude that $\lim_{t\to\infty} c = c_0$ and, since $a + b + p$ is constant, that $\lim_{t\to\infty} p = a_0$.

The Brusselator The *Brusselator*,* a hypothetical reaction mechanism proposed by I. Prigogene and R. Lefever in 1968, is described by

$$\begin{aligned} A &\to B, \\ B + D &\to C, \\ 2B + C &\to 3B, \\ B &\to P. \end{aligned}$$

It assumed that the concentrations $a = [A]$ and $d = [D]$ are constant. The resulting equations for $[B]$ and $[C]$ are

$$\begin{aligned} b' &= k_1a - k_2db - 2k_3b^2c + 3k_3b^2c - k_4b, \\ c' &= k_2db - k_3b^2c. \end{aligned}$$

With $x(t) = b(t/k_4)$, $y(t) = c(t/k_4)$, $\alpha = k_3k_1^2a^2/k_4^3$, and $\beta = k_2d/k_4$, these equations become

$$x' = x^2y - (\beta + 1)x + \alpha, \quad y' = -x^2y + \beta x, \tag{2}$$

where α and β are positive constants. This system has a single equilibrium point at $(\alpha, \beta/\alpha)$, which may be stable or unstable, depending upon the values of α and β. (See Problem 7.) It turns out that, because $x' > 0$ on the y-axis and $y' > 0$ on the positive x-axis, the direction field does not allow any orbit to leave the first quadrant; that is, any solution with nonnegative initial values must remain nonnegative for all $t \geq 0$. (This is usually a desirable property for a system that models some physical process involving naturally nonnegative quantities.) Solutions of this system also

* A hybrid of "Brussels" (Belgium) and "oscillator."

remain bounded for all $t \geq 0$. This is fairly easy to see, because adding the two equations results in

$$(x + y)' = \alpha - x, \tag{3}$$

which implies that the quantity $x + y$ is decreasing whenever $x > \alpha$, and because

$$\frac{dy}{dx} \leq \frac{\beta^2}{(1+\beta)^2} \quad \text{if } 0 \leq x \leq \alpha \text{ and } y \geq \frac{(\beta+1)^2}{2\alpha}. \tag{4}$$

(See Problem 8.) The shaded region in Figure 1 (drawn with $\alpha = 1$ and $\beta = 1.5$) is bounded by the coordinate axes, a line L_1 through $\big(0, (\beta+1)^2/(2\alpha)\big)$ with slope $\beta^2/(1+\beta)^2$, and a line L_2, which has slope -1 and intersects L_1 at $x = \alpha$. Notice that everywhere along the boundary of the shaded region, the direction field either is tangent to the boundary or else points into the shaded region. No orbit can leave the shaded region by crossing L_1 because of (4), and no orbit can leave the shaded region by crossing L_2 because of (3). Therefore, every orbit that begins in (or eventually enters) the shaded region is bounded. Since we may as well have constructed the shaded region with L_1 intersecting the y-axis at any point where $y \geq (\beta+1)^2/(2\alpha)$, the shaded region could be expanded to enclose the initial point of any orbit. Therefore, all orbits in the first quadrant are bounded. As result, if the equilibrium point $(\alpha, \beta/\alpha)$ is unstable, then the Poincaré-Bendixson theorem guarantees that every orbit must approach an orbitally stable limit cycle. Figure 2 illustrates this with $\alpha = 1$ and $\beta = 2.2$.

The Belousov-Zhabotinskii Reaction Around 1950, the Russian chemist B. P. Belousov discovered a certain chemical reaction in which a catalyst concentration exhibited sustained oscillations. Belousov's discovery was met with disbelief, since at the time such oscillations were generally believed to be impossible violations of

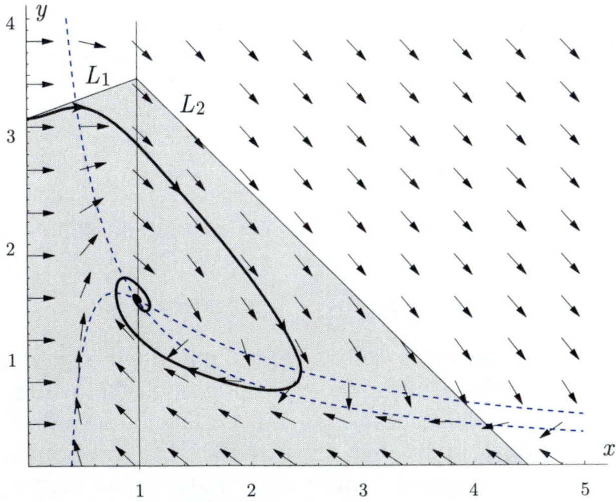

Figure 1

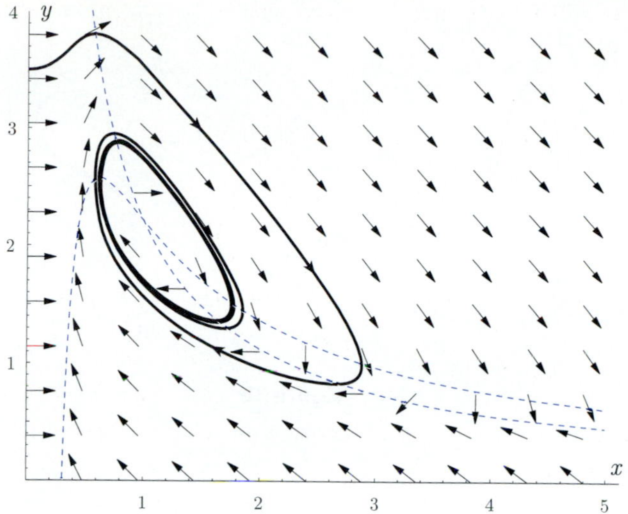

Figure 2

the second law of thermodynamics. His initial work was never published, but he eventually published an obscure article describing his work in 1959. The study of Belousov's reaction was later continued by A. M. Zhabotinskii in the 1960s, and scientists in the West finally learned of it in the 1970s. Since then, the *Belousov-Zhabotinskii* (BZ) reaction has become a well-known prototype for chemical and biochemical oscillators.

Belousov's reaction involved the oxidation of an acid by bromate ions, catalyzed by cerium ions. In the presence of certain dyes, oscillations in the cerium ion concentrations can be observed visually as color variations.

An important, mathematically tractable model of the reaction was developed by R. J. Field and R. M. Noyes at the University of Oregon in the early 1970s. The *Field-Noyes* model, also known as the *Oregonator*, is the following five-step sequence:

$$A + Y \to X + P$$
$$X + Y \to 2P$$
$$A + X \to 2X + 2Z$$
$$2X \to A + P$$
$$2Z \to Y$$

The species in the Field-Noyes model are analogous to species in the actual BZ reaction as follows:

$$A \sim \mathrm{BrO_3^-}, \quad X \sim \mathrm{HBrO_2}, \quad Y \sim \mathrm{Br^-}, \quad Z \sim \mathrm{Ce^{4+}}, \quad P \sim \mathrm{HOBr}.$$

With x, y, and z representing scaled versions of the concentrations $[X]$, $[Y]$, and $[Z]$, respectively, and $[A]$ assumed constant, one of the many possible nondimensionalized

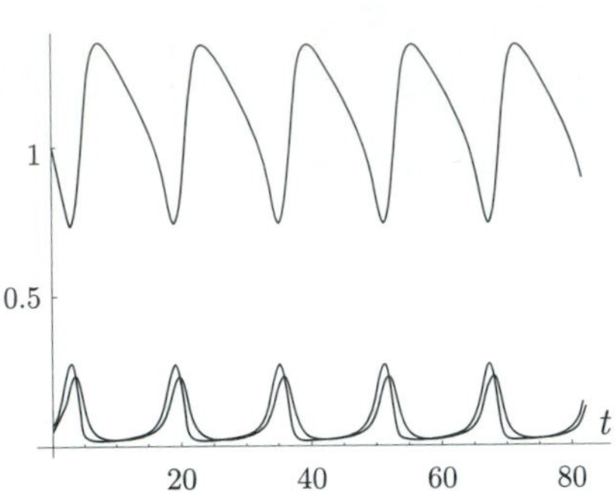

Figure 3a

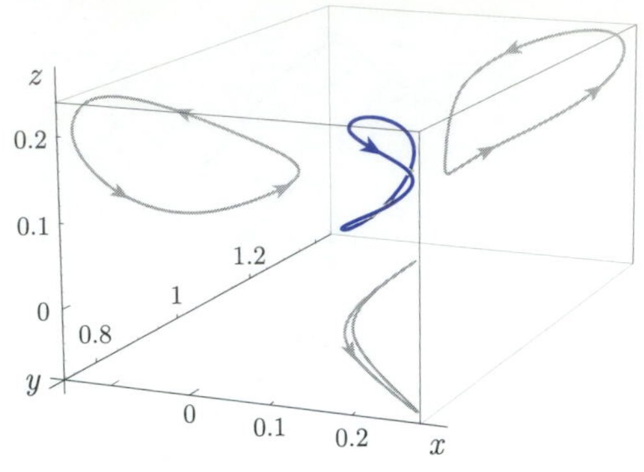

Figure 3b

systems that arise is

$$\begin{aligned} x' &= \epsilon^{-1}(qy - xy + x(1-x)), \\ y' &= \delta^{-1}(-qy - xy + z), \\ z' &= x - z. \end{aligned} \tag{5}$$

A thorough analysis of this system is more subtle than is appropriate here. The main result is that, with certain parameter values, the system possesses stable three-dimensional limit cycles. A stable limit cycle of the system with parameter values $\epsilon = \frac{1}{100}$, $\delta = \frac{1}{3}$, and $q = \frac{1}{200}$ is shown in Figures 3a and b. Figure 3a shows the graphs of x, y, and z versus t, while Figure 3b shows the three-dimensional closed orbit, as well as its projection on to each of the xy-, xz-, and yz-planes.

PROBLEMS

For each reaction described in Problems 1 through 5, write down a system of differential equations for the concentrations of the chemical species involved in the reaction.

1. $A + 2B \rightleftharpoons C$

2. $A + B \rightleftharpoons 2C$

3. $A + B \rightleftharpoons C,\ A + C \to P$

4. $A + B \to C,\ B + C \rightleftharpoons A$

5. $A + B + C \rightleftharpoons P$

6. Show that $(0, 0)$ is a stable node for system (1). Then explain why every orbit in the first quadrant approaches $(0, 0)$ as $t \to \infty$.

7. Determine conditions on α and β under which $(\alpha, \beta/\alpha)$ is a stable/unstable equilibrium point of the Brusselator system (2).

8. Show that the statement (4) is true for the Brusselator system (2).

9. The *Schnakenberg* reaction mechanism

$$2A + B \rightleftharpoons 3A, \quad C \to B, \quad A \to P,$$

with $[C]$ assumed constant, leads to the system

$$x' = x^2 y - x, \quad y' = \gamma - x^2 y.$$

(a) Find the system's equilibrium point and describe its stability/instability as γ varies over $(0, \infty)$.

(b) Sketch the nullclines in the first quadrant, and explain why a bounded forward-invariant region cannot be constructed.

Suppose now that the last step of the reaction is replaced with the reversible reaction step $A \rightleftharpoons P$, where $[P]$ is assumed constant, resulting in the modified system

$$x' = x^2y - x + \rho, \quad y' = \gamma - x^2y.$$

(a) Find the modified system's equilibrium point and describe its stability/instability as γ and ρ vary over $(0, \infty)$.

(b) For each case resulting from part (c), pick sample values of γ and ρ and sketch the nullclines in the first quadrant. Construct a bounded forward-invariant region in each case, and conclude that the modified system has a periodic orbit whenever its equilibrium point is unstable.

10. A reaction mechanism studied by Lotka is given by

$$C + A \to 2A, \quad A + B \to 2B, \quad B \to P.$$

Assuming that $[C] = c$ is constant, show that the related system of differential equations for the concentrations $a = [A]$ and $b = [B]$ is equivalent to the Lotka-Volterra predator-prey system with a perfectly efficient predator. Refer to Problem 7 in Section 10.1 and show that all nontrivial, nonnegative orbits are periodic.

11. Consider the Field-Noyes system (5) with the parameter values that produced Figures 3a and b. Find the equilibrium point with positive coordinates (numerically if you wish) and show that it is unstable.

12. Consider the Field-Noyes system (5) with the parameter values $\epsilon = 1/100$ and $\delta = 1/3$. Show that the equilibrium point with positive coordinates is unstable if $q = 0.0197$, but stable if $q = 0.0198$.

13. Show that if $q < 1$, then the region described by

$$q \le x \le 1, \quad \frac{q}{q+1} \le y \le \frac{1}{2q}, \quad q \le z \le 1.$$

is a forward-invariant region for the Field-Noyes system (5).

10.5 Mechanics

Many of the most interesting and historically significant applications of differential equations come from mechanical systems. The modeling process for mechanical systems involves application of physical laws such as Newton's laws of motion and conservation of energy. For complicated systems this can be quite subtle and complex. However, a profound principle attributed to Lagrange can greatly simplify the construction of a system's equations of motion. In this section we will first describe Lagrange's equations and illustrate their use with a couple of simple systems. Then we will model the double pendulum and derive equations for the "three-body" problem in orbital mechanics.

Lagrange's Equations Consider a mechanical system whose state at any time t is described in terms of state variables* q_i and $q_i' = \frac{dq_i}{dt}$ for $i = 1, \ldots, n$. The variables q_i are typically some collection of rectangular coordinates and/or various angles used to describe positions of masses in the system.

A force F acting on the system is *conservative* if $F = -\nabla\varphi$ for some function $\varphi(q_1, \ldots, q_n)$, which we call a *potential function*. The motivation for this terminology is that if F is conservative and accounts for all forces acting on the system, then the sum of the system's kinetic and potential energies is constant.

Let KE denote the system's kinetic energy and PE its potential energy. The **Lagrangian** L of the system is defined by

$$L(q_1, \ldots, q_n, q_1', \ldots, q_n') = KE - PE.$$

If all forces on the system are conservative, then *Lagrange's equations of motion* are described, for $i = 1, \ldots, n$, by the chain rule–like expression

$$\frac{d}{dt}\frac{\partial L}{\partial q_i'} = \frac{\partial L}{\partial q_i}. \tag{1}$$

We emphasize that $q_1, \ldots, q_n$ and $q_1', \ldots, q_n'$ are simply treated as $2n$ independent variables in the formulation of these equations.

The quantities on each side of (1) have units of force. (Why?) If there are nonconservative forces Φ_i (such as friction) in the coordinate directions q_i, then Lagrange's equations of motion become

$$\frac{d}{dt}\frac{\partial L}{\partial q_i'} = \frac{\partial L}{\partial q_i} + \Phi_i.$$

Example 1

Consider a frictionless spring-mass system consisting of a mass m attached to one end of massless, horizontal spring. Let x be the displacement of the mass from the equilibrium position $x = 0$. The Hooke's law force, $-kx$, exerted by the spring on the mass is conservative and corresponds to the potential function $\frac{1}{2}kx^2$. So the kinetic and the potential energies are

$$KE = \frac{1}{2}mx'^2 \quad \text{and} \quad PE = kx^2.$$

The Lagrangian is therefore

$$L(x, x') = \frac{1}{2}mx'^2 - kx^2.$$

Thus the equation of motion, $\frac{d}{dt}\frac{\partial L}{\partial x'} = \frac{\partial L}{\partial x}$, becomes

$$\frac{d}{dt}mx' = -kx,$$

which gives us the familiar equation $mx'' - kx = 0$. Now suppose that a friction force in the form of $-rx'$ is present in the system. Since that force is not conservative,

* Often called "generalized coordinates" by physicists.

we simply insert it into the equation of motion,

$$\frac{d}{dt}mx' = -kx - rx',$$

and obtain the familiar equation

$$mx'' + rx' + kx = 0.$$

■

Example 2

Consider the simple frictionless pendulum in Figure 1, in which a mass m is attached to a massless arm of length ℓ. State variables for the system are θ and θ'. Since the velocity of the mass is $\ell\theta'$, the kinetic energy is

$$KE = \frac{1}{2}m(\ell\theta')^2.$$

The potential energy, corresponding to the force $-mg$, is mgz, where z is the upward vertical displacement from the (arbitrarily chosen) position $\theta = 0$. Therefore, since $z = \ell - \ell\cos\theta$, we have

$$PE = mg\ell(1 - \cos\theta).$$

The Lagrangian is thus

$$L(\theta, \theta') = \frac{1}{2}m(\ell\theta')^2 - mg\ell(1 - \cos\theta),$$

and so the equation of motion is

$$\frac{d}{dt}(m\ell^2\theta') = -mg\ell\sin\theta,$$

which gives us the familiar $\theta'' + \frac{g}{\ell}\sin\theta = 0$.

■

The Double Pendulum Consider a frictionless, *double pendulum* as illustrated in Figure 2. State variables for the system are α, β, α', and β'. The contributions of m_1 to the system's kinetic and potential energies will be essentially the same as for

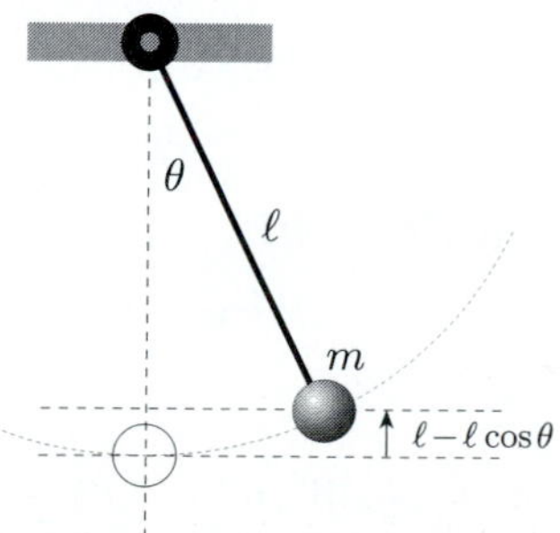

Figure 1

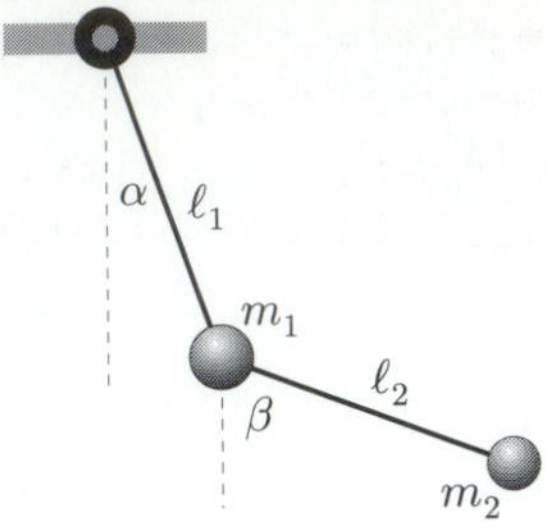

Figure 2

the mass in a simple pendulum:

$$\frac{1}{2}m_1\ell_1^2\alpha'^2 \quad \text{and} \quad m_1 g(\ell_1 - \ell_1\cos\alpha),$$

respectively. To determine the analogous contributions from m_2, first let $P(t)$ denote the position of m_2 in rectangular coordinates:

$$P(t) = (\ell_1\sin\alpha(t), \ell_1\cos\alpha(t)) + (\ell_2\sin\beta(t), \ell_2\cos\beta(t)).$$

Differentiation (and suppression of the t-dependence for convenience) then gives us the velocity:

$$P'(t) = (\ell_1\alpha'\cos\alpha + \ell_2\beta'\cos\beta, -\ell_1\alpha'\sin\alpha - \ell_2\beta'\sin\beta).$$

Straightforward computation of the square of the speed, plus some simplification, reveals that the contribution of m_2 to the system's kinetic energy is

$$\frac{1}{2}m_2\|P'(t)\|^2 = \frac{1}{2}m_2(\ell_1^2\alpha'^2 + \ell_2^2\beta'^2 + 2\ell_1\ell_2\alpha'\beta'\cos(\alpha-\beta)).$$

The contribution of m_2 to the system's potential energy is easier to observe:

$$m_2 g z = m_2 g(\ell_1 - \ell_1\cos\alpha + \ell_2 - \ell_2\cos\beta).$$

With these energies noted, we are ready to write down the Lagrangian of the system. Setting $\hat{m} = m_1 + m_2$ for notational convenience, we have

$$\begin{aligned} L(\alpha,\beta,\alpha',\beta') = \frac{1}{2}\hat{m}\ell_1^2\alpha'^2 + \frac{1}{2}m_2\ell_2^2\beta'^2 + m_2\ell_1\ell_2\alpha'\beta'\cos(\alpha-\beta) \\ - \hat{m}(\ell_1 - \ell_1\cos\alpha) - m_2 g(\ell_2 - \ell_2\cos\beta). \end{aligned}$$

Lagrange's equations of motion,

$$\frac{d}{dt}\frac{\partial L}{\partial\alpha'} = \frac{\partial L}{\partial\alpha} \quad \text{and} \quad \frac{d}{dt}\frac{\partial L}{\partial\beta'} = \frac{\partial L}{\partial\beta},$$

give us

$$\frac{d}{dt}\left(\hat{m}\ell_1^2\alpha' + m_2\ell_1\ell_2\beta'\cos(\alpha-\beta)\right) = -m_2\ell_1\ell_2\alpha'\beta'\sin(\alpha-\beta) - \hat{m}\ell_1\sin\alpha,$$

$$\frac{d}{dt}\left(m_2\ell_2^2\beta' + m_2\ell_1\ell_2\alpha'\cos(\alpha-\beta)\right) = m_2\ell_1\ell_2\alpha'\beta'\sin(\alpha-\beta) - m_2\ell_1\sin\beta.$$

After some expansion and simplification these become

$$\ell_1\alpha'' + \frac{m_2\ell_2}{\hat{m}}(\sin(\alpha-\beta)\beta'^2 + \cos(\alpha-\beta)\beta'') = -g\sin\alpha,$$
$$\ell_1\cos(\alpha-\beta)\alpha'' - \ell_1\sin(\alpha-\beta)\alpha'^2 + \ell_2\beta'' = -g\sin\beta,$$

which we solve for the accelerations α'' and β'' (with *CAS* assistance), obtaining

$$\alpha'' = -\frac{m_1 g\sin\alpha + m_2\sin(\alpha-\beta)(g\cos\beta + \ell_1\cos(\alpha-\beta)\alpha'^2 + \ell_2\beta'^2)}{\ell_1(m_1 + m_2\sin^2(\alpha-\beta))},$$
$$\beta'' = \frac{\sin(\alpha-\beta)(\hat{m}g\cos\alpha + \hat{m}\ell_1\alpha'^2 + m_2\ell_2\cos(\alpha-\beta)\beta'^2)}{\ell_1(m_1 + m_2\sin^2(\alpha-\beta))}.$$

So at last we have a pair of second-order equations that we can easily convert to a system of four first-order equations in the usual way and then solve numerically. Figure 3 shows four pieces of the path followed by m_2 from a numerically generated solution. The parameter values are $m_1 = m_2 = \frac{1}{16}$, $\ell_1 = 3$, $\ell_2 = 2$, and $g = 32$ and the initial values are $\alpha(0) = \beta(0) = 3$ and $\alpha'(0) = \beta'(0) = 0$. The circle of radius 3 indicates the path of m_1. Visualizing the actual motion of the system from these plots is nearly impossible; however, the same numerical solution can be used to produce a computer animation of the system's movement that looks quite realistic.

These bizarre plots suggest that the system behaves in a chaotic manner. Indeed, slight changes in initial conditions can cause large changes in subsequent behavior. In an actual physical model, it is essentially impossible to cause the same behavior twice, because initial conditions cannot be duplicated with sufficient precision.

The Three-Body Problem Consider a planar system consisting of three particles with masses m_0, m_1, and m_2; and let $\mathbf{r}_0(t)$, $\mathbf{r}_1(t)$, and $\mathbf{r}_2(t)$ be position vectors of the three particles, respectively, *relative to their center of mass*. Let $\mathbf{r}_0(t) = (x_0(t), y_0(t))$; and let

$$\mathbf{r}_1(t) = \mathbf{r}_0(t) + (x_1(t), y_1(t))$$

and

$$\mathbf{r}_2(t) = \mathbf{r}_0(t) + (x_2(t), y_2(t)),$$

so that $(x_1(t), y_1(t))$ and $(x_2(t), y_2(t))$ represent the positions of m_1 and m_2 *relative to* m_0. This set-up is depicted in Figure 4.

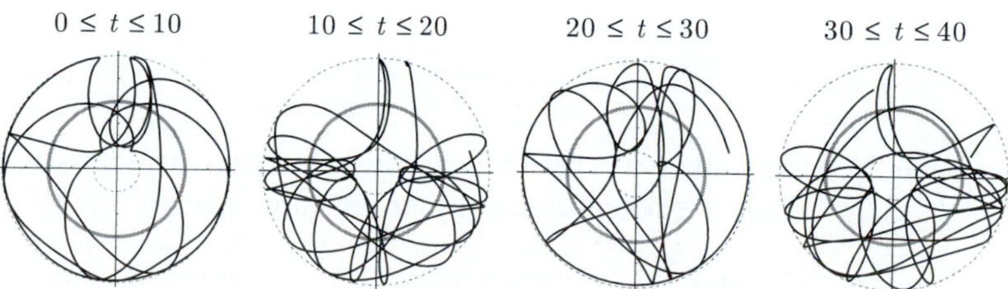

Figure 3

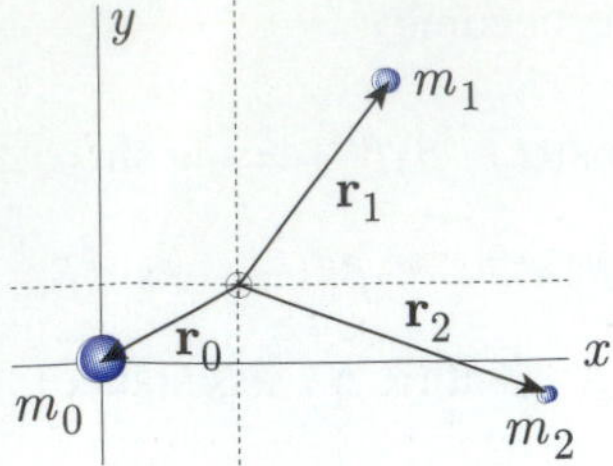

Figure 4

Since $m_0\mathbf{r}_0(t) + m_1\mathbf{r}_1(t) + m_2\mathbf{r}_2(t) = \mathbf{0}$, it follows that

$$\mathbf{r}_0(t) = -M^{-1}(m_1x_1(t) + m_2x_2(t), m_1y_1(t) + m_2y_2(t)),$$

where $M = m_0 + m_1 + m_2$. Therefore, the positions of m_1 and m_2 relative to the center of mass are, respectively,

$$\mathbf{r}_1(t) = M^{-1}\left((m_0 + m_2)x_1(t) - m_2x_2(t), (m_0 + m_2)y_1(t) - m_2y_2(t)\right)$$

and

$$\mathbf{r}_2(t) = M^{-1}\left((m_0 + m_1)x_2(t) - m_1x_1(t), (m_0 + m_1)y_2(t) - m_1y_1(t)\right).$$

In order to construct the Lagrangian for the system, we first observe that the system's kinetic energy is

$$KE = \frac{1}{2}m_1\mathbf{r}_0'(t)\cdot\mathbf{r}_0'(t) + \frac{1}{2}m_1\mathbf{r}_1'(t)\cdot\mathbf{r}_1'(t) + \frac{1}{2}m_2\mathbf{r}_2'(t)\cdot\mathbf{r}_2'(t),$$

which after some *CAS*-aided simplification becomes

$$KE = \frac{1}{2M}\Big(m_1(m_0 + m_2)x_1'^2 - 2m_1m_2x_1'x_2' + (m_0 + m_1)m_2x_2'^2 \\ + m_0m_1y_1'^2 + m_1m_2y_1'^2 - 2m_1m_2y_1'y_2' + m_0m_2y_2'^2 + m_1m_2y_2'^2\Big).$$

The potential energy follows from the *gravitational potential* between two particles with masses m and $\tilde{m}$ and separated by a distance r:

$$-\frac{m\tilde{m}G}{r}.$$

Here, G is Newton's universal gravitation constant, which is approximately 6.67×10^{-11} N·m/kg^2. Thus the potential energy of our three-particle system is

$$PE = -\frac{m_0m_1G}{\sqrt{x_1^2 + y_1^2}} - \frac{m_0m_2G}{\sqrt{x_2^2 + y_2^2}} - \frac{m_1m_2G}{\sqrt{(x_1 - x_2)^2 + (y_1 - y_2)^2}},$$

and so the Lagrangian is

$$L = \frac{1}{2M}\Big(m_1(m_0+m_2)x_1'^2 - 2m_1m_2x_1'x_2' + (m_0+m_1)m_2x_2'^2$$
$$+ m_0m_1y_1'^2 + m_1m_2y_1'^2 - 2m_1m_2y_1'y_2' + m_0m_2y_2'^2 + m_1m_2y_2'^2\Big)$$
$$+ \frac{m_0m_1G}{\sqrt{x_1^2+y_1^2}} + \frac{m_0m_2G}{\sqrt{x_2^2+y_2^2}} + \frac{m_1m_2G}{\sqrt{(x_1-x_2)^2+(y_1-y_2)^2}}.$$

The equations of motion therefore are

$$\frac{m_1}{M}((m_0+m_2)x_1'' - m_2x_2'') = -\frac{Gm_0m_1x_1}{(x_1^2+y_1^2)^{3/2}} - \frac{Gm_1m_2(x_1-x_2)}{\left((x_1-x_2)^2+(y_1-y_2)^2\right)^{3/2}},$$
$$\frac{m_2}{M}(m_1x_1'' - (m_0+m_1)x_2'') = -\frac{Gm_0m_2x_2}{(x_2^2+y_2^2)^{3/2}} + \frac{Gm_1m_2(x_1-x_2)}{((x_1-x_2)^2+(y_1-y_2)^2)^{3/2}},$$
$$\frac{m_1}{M}((m_0+m_2)y_1'' - m_2y_2'') = -\frac{Gm_0m_1y_1}{(x_1^2+y_1^2)^{3/2}} - \frac{Gm_1m_2(y_1-y_2)}{((x_1-x_2)^2+(y_1-y_2)^2)^{3/2}},$$
$$\frac{m_2}{M}(m_1y_1'' - (m_0+m_1)y_2'') = -\frac{Gm_0m_2y_2}{(x_2^2+y_2^2)^{3/2}} + \frac{Gm_1m_2(y_1-y_2)}{((x_1-x_2)^2+(y_1-y_2)^2)^{3/2}}.$$

Further algebraic maneuvers produce the system

$$m_1x_1'' = -\frac{m_1(m_0+m_1)Gx_1}{(x_1^2+y_1^2)^{3/2}} - \frac{m_1m_2Gx_2}{(x_2^2+y_2^2)^{3/2}} - \frac{m_1m_2G(x_1-x_2)}{\left((x_1-x_2)^2+(y_1-y_2)^2\right)^{3/2}},$$
$$m_1y_1'' = -\frac{m_1(m_0+m_1)Gy_1}{(x_1^2+y_1^2)^{3/2}} - \frac{m_1m_2Gy_2}{(x_2^2+y_2^2)^{3/2}} - \frac{m_1m_2G(y_1-y_2)}{\left((x_1-x_2)^2+(y_1-y_2)^2\right)^{3/2}},$$
$$m_2x_2'' = -\frac{m_2(m_0+m_2)Gx_2}{(x_2^2+y_2^2)^{3/2}} - \frac{m_2m_1Gx_1}{(x_1^2+y_1^2)^{3/2}} + \frac{m_2m_1G(x_1-x_2)}{\left((x_1-x_2)^2+(y_1-y_2)^2\right)^{3/2}},$$
$$m_2y_2'' = -\frac{m_2(m_0+m_2)Gy_2}{(x_2^2+y_2^2)^{3/2}} - \frac{m_2m_1Gy_1}{(x_1^2+y_1^2)^{3/2}} + \frac{m_2m_1G(y_1-y_2)}{\left((x_1-x_2)^2+(y_1-y_2)^2\right)^{3/2}}.$$

Figures 5a and 5b show numerically generated paths of m_1 and m_2 relative to m_0, where m_0 is fixed at the origin. In Figure 5a, the masses are such that $m_0G = 43.33$, $m_1G = 0.667$, and $m_2G = 6.67$. The initial positions are

$$(x_1(0), y_1(0)) = (10.75, 0) \quad \text{and} \quad (x_2(0), y_2(0)) = (10, 0)$$

with initial velocities

$$(x_1'(0), y_1'(0)) = (0, 4.5) \quad \text{and} \quad (x_2'(0), y_2'(0)) = (0, 1.5).$$

In Figure 5b the masses are such that $m_0G = 13.4$, $m_1G = 0.3335$, and $m_2G = 3.335$. The initial positions are

$$(x_1(0), y_1(0)) = (10, 0) \quad \text{and} \quad (x_2(0), y_2(0)) = (15, 5)$$

with initial velocities

$$(x_1'(0), y_1'(0)) = (0, 1) \quad \text{and} \quad (x_2'(0), y_2'(0)) = (0, 0).$$

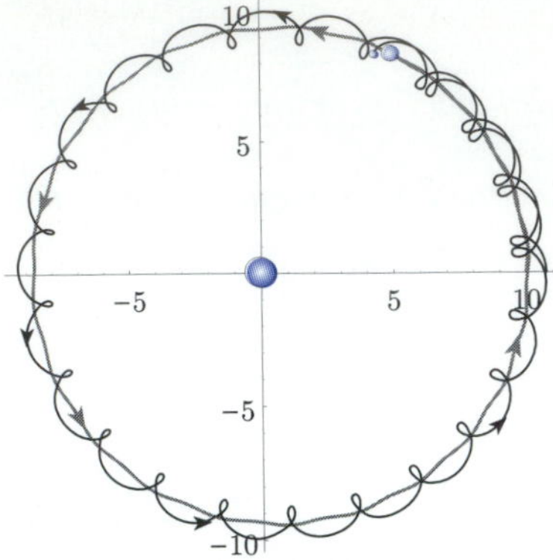

Figure 5a

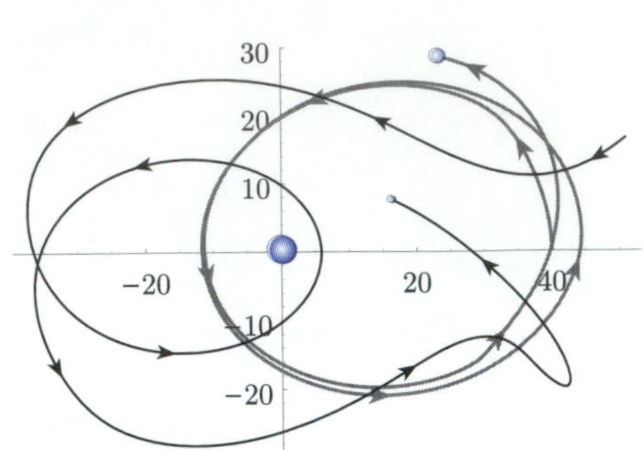

Figure 5b

PROBLEMS

1. The Lagrangian of a system consisting of three particles, each with mass m and connected around a circle by three springs with stiffness k, is

$$L = \frac{m}{2}(x_1'^2 + x_2'^2 + x_3'^2) - \frac{k}{2}((x_1 - x_2)^2 + (x_2 - x_3)^2 + (x_3 - x_1)^2).$$

Determine the first-order (linear) equations of motion, and find the eigenvalues associated with the system.

2. The figure below shows a pendulum consisting of a mass m_1 and an arm with length ℓ and negligible mass, attached to a pivot with mass m_2 that can move freely along a horizontal rail. Derive equations of motion for the two masses. Ignore friction, and assume that the motion takes place in a vertical plane.

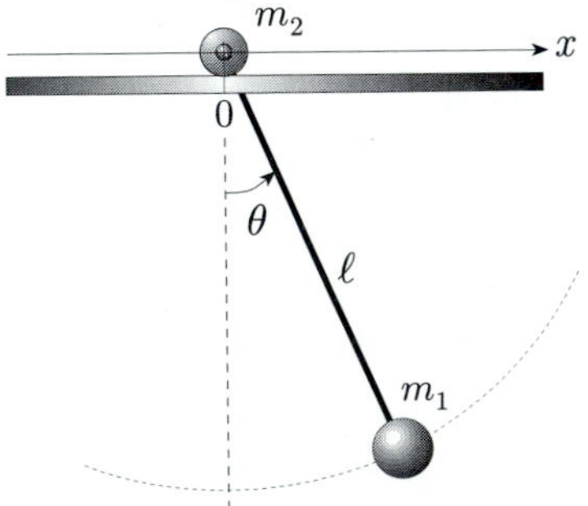

3. The figure below shows a spring-mass system consisting of a mass m and a spring with unstretched length ℓ, stiffness k, and negligible mass. The top of the spring is attached to a pivot, allowing the mass to swing in a pendulum-like manner. Ignoring friction and assuming that the motion takes place in a vertical plane, derive equations of motion for the mass.

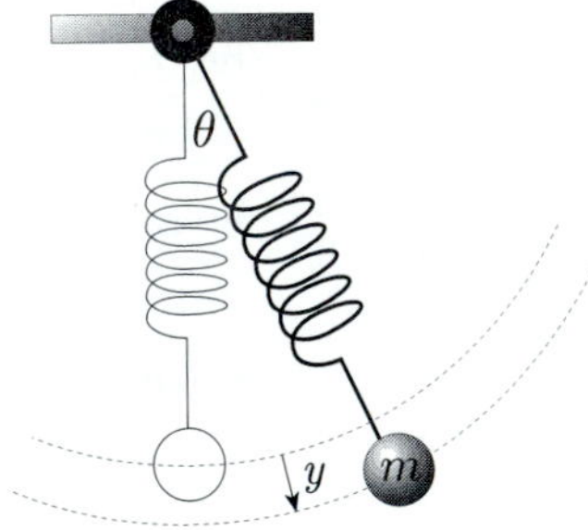

4. The figure on the next page a mass m hanging from two springs that are attached at the same height to supports separated by a distance 2ℓ. Each spring has the same unstretched length ℓ and negligible mass, but the two springs have different stiffnesses k_1 and k_2. The figure depicts the mass at its equilibrium position, which is h units below the horizontal line joining the attached ends of the springs and a units to the left of the vertical centerline between the supports. Let x and y denote the horizontal and vertical displacements of the mass from equilibrium, with downward as the positive direction for y. Relate the stiffnesses k_1 and k_2 to the other parameters in the figure,

and, ignoring friction, derive equations of motion for the mass. Assume that the motion takes place in a vertical plane.

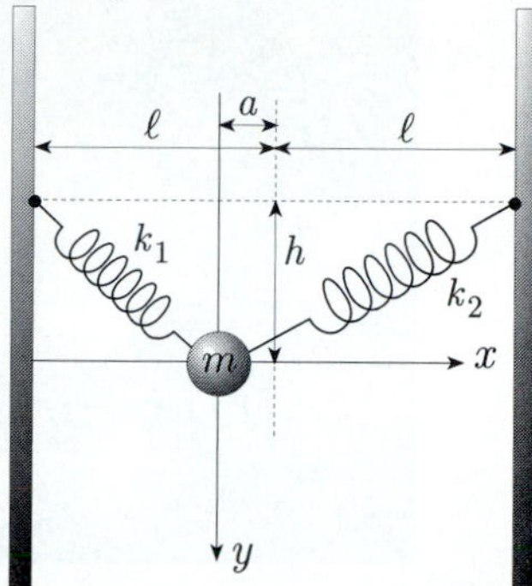

5. Consider a planar "two-body" problem in which masses m_1 and m_2 are attracted by gravitation. The first mass is constrained to move around the unit circle, and the second mass is constrained to move along the x-axis. Assuming no forces other than gravitational attraction, derive equations of motion for the two masses.

6. Consider a "two-body" problem for masses m_0 and m_1.

(a) Write the Lagrangian for the motion of m_1 relative to m_0 in polar coordinates $(r(t), \theta(t))$. (*Suggestion*: First show that the square of speed is given by $v^2 = r'^2 + r^2\theta'^2$.)

(b) Find the equations of motion.

(c) Show that $r^2\theta'$ is constant.

(d) Show that substitution of $r = u^{-1}$ leads to the equation $u'' + u = k$, where k is a constant. Use this to describe the possible orbits of m_1 relative to m_0.

7. Derive the equations of motion for a system of two identical pendulums whose pivots are separated by a distance equal to three times the length of each pendulum's arm and whose masses are connected by a spring (with negligible mass) whose unstretched length is twice the length of each pendulum's arm. Assume that the motion takes place in a vertical plane.

CHAPTER 11

DIFFUSION PROBLEMS AND FOURIER SERIES

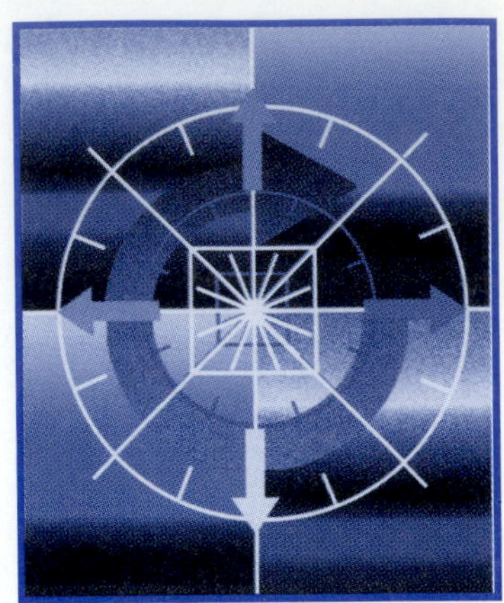

In previous chapters we have studied ordinary differential equations that govern the way in which a finite number of quantities evolve with respect to an independent variable t, usually thought of as time. We began by studying single equations governing the evolution of a single quantity and then studied coupled systems of n equations governing the simultaneous evolution of n quantities.

In this chapter and the next we undertake an introductory study of differential equations that govern the evolution in time of *functions* defined on some domain in space. Such equations involve derivatives with respect to one or more *spatial* variables (as well as time derivatives); thus they are *partial differential equations.*

Partial differential equations comprise a subject that is far too vast and far too complex for us to do more than scratch its surface here. Our purpose is only to introduce the most basic ideas. We will therefore restrict ourselves in this chapter to *one-dimensional* problems (i.e., problems involving only a single spatial variable x). In Chapter 12 we will look at a few problems involving two spatial variables.

11.1 The Basic Diffusion Problem

One of the most common processes in nature is the random movement of matter from regions of high concentration to regions of lower concentration. This "averaging-out" process is known as *diffusion.* Examples include a drop of ink diffusing in a glass of water and water vapor diffusing in air. Another quantity that diffuses is heat energy. For instance, when heat is applied to one end of a metal rod, it diffuses toward the other end of the rod. Diffusion is also a common part of population models that account for the tendency of species to disperse in order to avoid crowding.

Suppose that U is a quantity that diffuses in a one-dimensional, uniform medium—such as heat energy in a thin, laterally insulated iron rod or a drop of ink in a thin tube of water. Let $u(t, x)$ be the concentration of that quantity at time t and position x. Our fundamental postulate is that the spatial flow rate of U (per

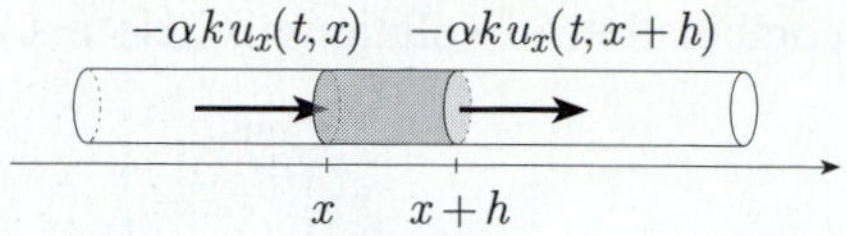

Figure 1

unit area) is proportional to its concentration gradient u_x*; that is,

$$\text{flow rate} = -ku_x. \tag{1}$$

This is known as *Fick's law*. In the context of heat conduction, it is *Fourier's law*. The coefficient k, generically called the *diffusivity*, is positive and in certain applications may depend upon x, t, or even the concentration u. However, for the sake of simplicity we will assume (for now) that k is constant. Note that the positivity of k implies that U flows in the direction of decreasing concentration.

Let us now derive a differential equation satisfied by u. Suppose that the rod has constant cross-sectional area α. Let $a < x < b$, and consider a small, interior section of the medium corresponding to the interval $[x, x+h]$. (See Figure 1.) Within this section at time t, the amount of U is

$$\alpha \int_x^{x+h} u(t, y)\, dy,$$

and the rate of "accumulation" of U is the time derivative of the amount:

$$\begin{matrix}\text{rate of} \\ \text{accumulation of } U\end{matrix} = \alpha \frac{d}{dt} \int_x^{x+h} u(t, y)\, dy = \alpha \int_x^{x+h} u_t(t, y)\, dy.$$

Furthermore, because of (1) we also have

$$\begin{aligned}\begin{matrix}\text{rate of} \\ \text{accumulation of } U\end{matrix} &= (\text{flow rate at } x) - (\text{flow rate at } x+h) \\ &= -\alpha k u_x(t, x) + \alpha k u_x(t, x+h).\end{aligned}$$

Subtracting the second of these two rate expressions from the first, and dividing by h, we arrive at

$$\frac{\alpha}{h} \int_x^{x+h} u_t(t, y)\, dy - \alpha k \frac{u_x(t, x+h) - u_x(t, x)}{h} = 0.$$

Dividing by α and taking the limit as $h \to 0$, we obtain the one-dimensional **diffusion equation** with constant diffusivity:

$$u_t - ku_{xx} = 0. \tag{2}$$

Note that this is a *partial differential equation*, since it involves partial derivatives with respect to each of the two independent variables. It is a *homogeneous, linear* partial differential equation, because the operator $\mathcal{H}$ associated with the left side of the equation, that is, defined by

$$\mathcal{H}u = u_t - ku_{xx},$$

* Subscripts signify partial derivatives: $u_x = \frac{\partial u}{\partial x}$, $u_{xx} = \frac{\partial^2 u}{\partial x^2}$, $u_t = \frac{\partial u}{\partial t}$, and so on.

is a linear operator and any linear combination of solutions is also a solution. (See Problem 10.)

Example 1

Each of the expressions

$$e^{-k\omega^2 t}\cos\omega x \quad \text{and} \quad e^{-k\omega^2 t}\sin\omega x,$$

where ω is any constant, is easily shown to satisfy (2) for x in any interval. Thus a family of solutions of (2) is given by

$$u(t, x) = e^{-k\omega^2 t}(A\cos\omega x + B\sin\omega x),$$

where A, B, and ω are constants. (See Problem 1.) It is interesting to note that any such solution, with $\omega \neq 0$, decays exponentially (in time) toward the zero solution. Also, the larger ω is (i.e., the higher the spatial frequency), the greater will be the rate of the exponential decay. With $k = 1$, a particular solution is

$$u(t, x) = e^{-t}\cos x,$$

which is plotted in Figures 2ab. Figure 2a shows the graph of u as a function of x (for $-2\pi \leq x \leq 2\pi$) at each of the discrete times $t = 0, 1, 2,$ and 3. Figure 2b shows the graph of $u(x, t)$ for $-2\pi \leq x \leq 2\pi$ and $0 \leq t \leq 3$. Notice the averaging out, or "smoothing," of the solution as t increases. ■

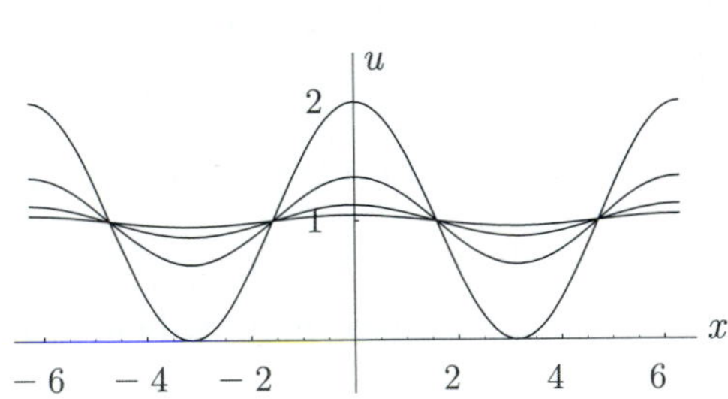

Figure 2a

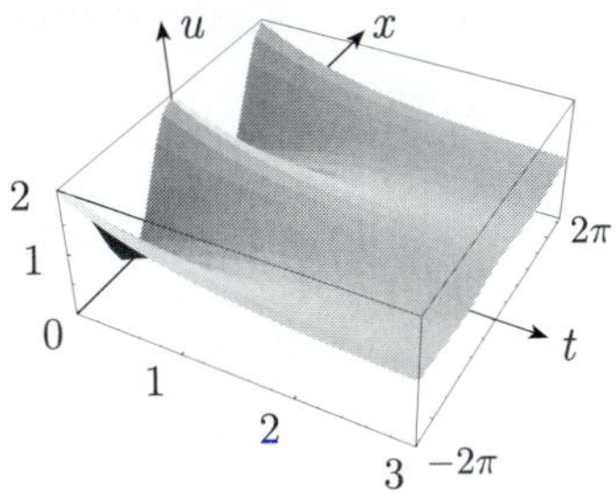

Figure 2b

Example 2

Routine calculation shows that

$$u(t, x) = \frac{1}{\sqrt{4\pi kt}}e^{-\frac{x^2}{4kt}}$$

is a solution of (2) for all $x \in \mathbb{R}$ and $t > 0$. (See Problem 2.) Figure 3a shows the graph of u (with $k = 1$) as a function of x, for $-1 \leq x \leq 1$, at each of the discrete times $t = 0.005, 0.055, \ldots, 0.255$. Figure 3b shows the graph of $u(x, t)$ for $-1 \leq x \leq 1$ and $0 \leq t \leq 0.25$. This solution is particularly interesting because, as $t \to 0^+$, $u(x, t)$ approaches the Dirac distribution (see Section 7.5), indicating that

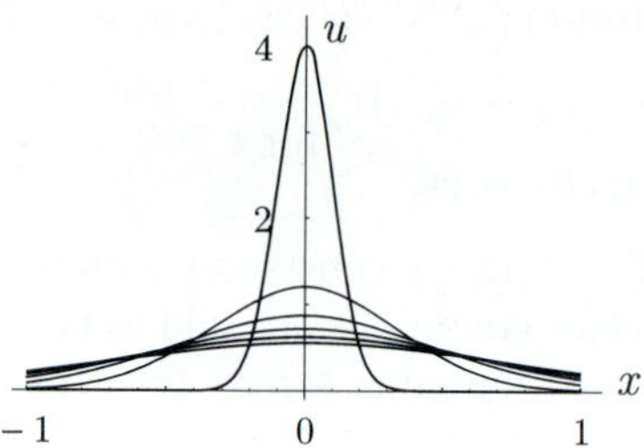

Figure 3a

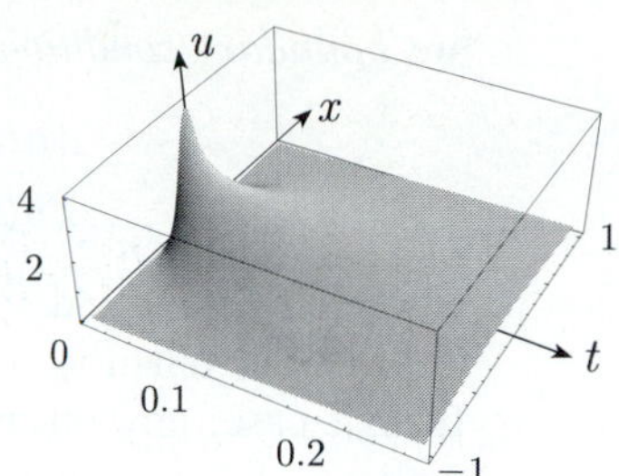

Figure 3b

$u(x,t)$ describes the diffusion of a unit mass concentrated at the point $x = 0$ at time $t = 0$. ■

x-Dependence and Sources It is often natural for the rate at which diffusion occurs to depend upon location. For instance, if a metal rod is made of a heterogeneous mixture of copper and tin, its ability to conduct heat will vary along its length. In such a case, the diffusivity coefficient k in postulate (1) becomes a positive function of x (and possibly t), and as a result the correct form of the diffusion equation is

$$u_t - (ku_x)_x = 0. \tag{3}$$

If the diffusing quantity U is generated or otherwise enters the medium at a rate per unit length given by $f(t,x)$, then (3) becomes

$$u_t - (ku_x)_x = f(t,x). \tag{4}$$

The term $f(t,x)$ in (4) generically represents a *source* of the quantity U. When $f = 0$, the equation is *homogeneous*; otherwise it is nonhomogeneous. When k and f are independent of t, the equation is *autonomous*. Each of these notions is thoroughly consistent with what we have seen previously in the context of ordinary differential equations.

Initial Data and Boundary Conditions Suppose that we are interested in solutions of the diffusion equation (4) for $t \geq 0$ and on some bounded (one-dimensional) spatial domain $a \leq x \leq b$. As one would expect, the diffusion equation itself will have infinitely many solutions. Our experience with ordinary differential equations suggests that it is appropriate to specify *initial data*, which in the present case will be of the form

$$u(0,x) = \phi(x), \quad \text{for } a \leq x \leq b, \tag{5}$$

where ϕ is a given function on $[a,b]$, providing an initial value for the solution at each x in $[a,b]$.

For each fixed t the diffusion equation is a second-order differential equation in x. Thus two independent conditions relative to the solution's x-dependence are appropriate. Conditions that are physically meaningful in a wide range of applications are *boundary conditions* of the general form

$$\begin{aligned} \alpha u(t,a) - (1-\alpha)u_x(t,a) &= \gamma_a, \\ \beta u(t,b) + (1-\beta)u_x(t,b) &= \gamma_b, \end{aligned} \qquad \text{for } t > 0 \tag{6}$$

where α, β, γ_a, and γ_b are functions of t (perhaps constants) with $0 \le \alpha \le 1$ and $0 \le \beta \le 1$. When $\gamma_a = \gamma_b = 0$, the boundary conditions are said to be *homogeneous*. In that case, any linear combination of functions that satisfy (6) will satisfy (6) as well.

Note that when $\alpha = \beta = 1$, (6) specifies endpoint values of the solution:

$$\begin{aligned} u(t,a) &= \gamma_a, \\ u(t,b) &= \gamma_b, \end{aligned} \qquad \text{for } t > 0,$$

as would be appropriate for a heated rod whose ends were held at fixed temperatures, for example. Such boundary conditions are known as **Dirichlet boundary conditions**.

When $\alpha = \beta = 0$, (6) specifies the *outward*† flow rate (cf. (1)), or outward *flux*, of the solution at each endpoint:

$$\begin{aligned} -u_x(t,a) &= \gamma_a, \\ u_x(t,b) &= \gamma_b, \end{aligned} \qquad \text{for } t > 0.$$

Such conditions are *flux boundary conditions* and are often called **Neumann boundary conditions**. The more specific case with $\gamma_a = \gamma_b = 0$ is appropriate for a heated rod with (perfectly) insulated ends or any other model in which the diffusing quantity is prohibited from flowing either into or out of the medium through its boundary.

For α and β with values in $(0, 1)$, the equations in (6) may be thought of as specifying the outward flux at each endpoint in terms of "exterior" concentrations *outside* of (a, b), which equate to the endpoint values $u(t, a)$ and $u(t, b)$ by continuity. An exterior concentration less than γ_a or γ_b results in a positive outward flux, and an exterior concentration greater than γ_a or γ_b results in a negative outward (or positive inward) flux. Boundary conditions of this type are called **Robin boundary conditions**.

It is not necessary that the boundary conditions at the two endpoints be of the same type. For instance, if $\alpha = \gamma_a = 1/2$ and $\beta = \gamma_b = 0$, then (6) becomes

$$\begin{aligned} -u_x(t,a) &= 1 - u(t,a), \\ u_x(t,b) &= 0, \end{aligned} \qquad \text{for } t > 0.$$

In summary, our *initial-boundary value problem* for the diffusion equation in one spatial dimension is

$$\left.\begin{aligned} u_t - (ku_x)_x = f \quad &\text{for } a < x < b, t > 0, \\ u(0,x) = \phi(x) \quad &\text{for } a \le x \le b, \\ \alpha u(t,a) - (1-\alpha)u_x(t,a) = \gamma_a \quad &\text{for } t > 0, \\ \beta u(t,b) + (1-\beta)u_x(t,b) = \gamma_b \quad &\text{for } t > 0. \end{aligned}\right\} \tag{7}$$

† This is the reason for the minus sign before $u_x(t, a)$. Without it, $u_x(t, a)$ represents the rate of flow *into* the interval.

In order to keep our present discussion relatively simple and to avoid some technical details that are more appropriately left for a course in partial differential equations, we make the following assumptions throughout the remainder of this section as well as the next section.

- The diffusivity k is a positive, continuously differentiable function of x for $a \le x \le b$ and is independent of t.
- The source term f is continuous for $t \ge 0$ and $a \le x \le b$.
- The function ϕ that specifies the initial data is continuous on $[a, b]$.
- The coefficients α, β, γ_a, and γ_b in the boundary conditions are constants.

We now state (without proof) the following existence and uniqueness theorem. We emphasize that the stated assumptions are merely *sufficient* for the conclusion of the theorem; they are not *necessary*.

Theorem 1

Under the preceding assumptions, (7) has a unique solution u, which is continuously differentiable in t and twice continuously differentiable in x for all $t > 0$ *and* $a \le x \le b$. ▲

The following are simple examples of diffusion problems and their solutions. (Bear in mind that we have yet to discuss any method for *finding* solutions.)

Example 3

Consider the problem

$$u_t - u_{xx} = 0 \quad \text{for } 0 < x < \pi, t > 0,$$
$$u(0, x) = 1 - \cos x \quad \text{for } 0 \le x \le \pi,$$
$$u_x(t, 0) = u_x(t, \pi) = 0 \quad \text{for } t > 0,$$

in which homogeneous Neumann boundary conditions require zero outward flux at each endpoint for all $t > 0$. Routine calculation verifies that the solution is

$$u(t, x) = 1 - e^{-t} \cos x.$$

Figure 4a shows the graph of u as a function of x at each of $t = 0, 1, 2,$ and 3, while Figure 4b shows the graph of u for $0 \le x \le \pi$ and $0 \le t \le 3$. ■

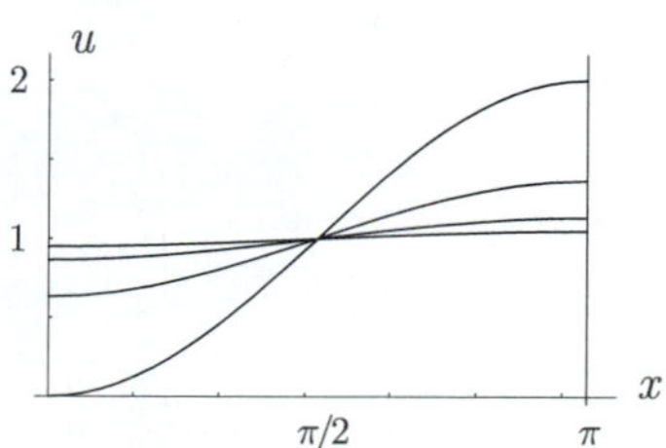

Figure 4a

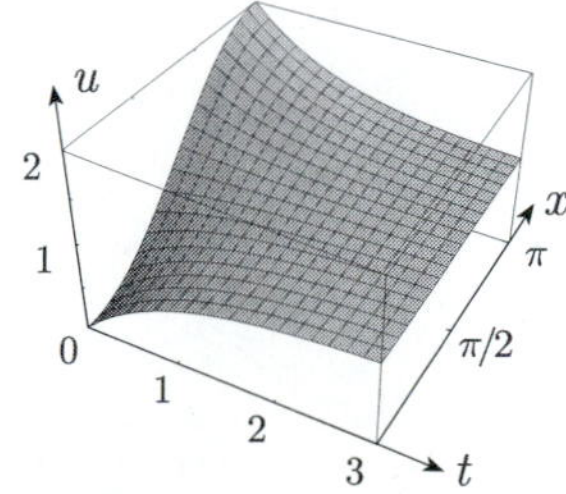

Figure 4b

Example 4

Consider the problem

$$\begin{aligned} u_t - u_{xx} &= -\frac{1}{8}\pi^2 \cos\frac{\pi x}{2} \quad \text{for } 0 < x < 2, t > 0, \\ u(0, x) &= \frac{1}{2}\left(1 - \cos\frac{\pi x}{2} + \sin\pi x\right) \quad \text{for } 0 \le x \le 2, \\ u(t, 0) &= 0, u(t, 2) = 1 \quad \text{for } t > 0, \end{aligned}$$

in which Dirichlet boundary conditions require that the solution's endpoint values be fixed for all $t > 0$. Routine calculation shows that the solution is

$$u(t, x) = \frac{1}{2}\left(1 - \cos\frac{\pi x}{2} + e^{-\pi^2 t}\sin\pi x\right),$$

which is illustrated in Figures 5a and b. ■

Example 5

Consider the problem

$$\begin{aligned} u_t - \frac{4}{81\pi^2}u_{xx} &= 8 \quad \text{for } 0 < x < 1, t > 0, \\ u(0, x) &= \pi^2(1 - x^2) + 5\cos\left(\frac{9\pi x}{2}\right) \quad \text{for } 0 \le x \le 1, \\ u_x(t, 0) &= 0, u(t, 1) = 0 \quad \text{for } t > 0, \end{aligned}$$

in which a homogeneous Neumann condition and a homogeneous Dirichlet condition require, respectively, that the solution's outward flux at $x = 0$ and value at $x = 1$ be zero for all $t > 0$. The solution,

$$u(t, x) = \pi^2(1 - x^2) + 5e^{-t}\cos\left(\frac{9\pi x}{2}\right),$$

is easily verified and is illustrated in Figures 6a and b. ■

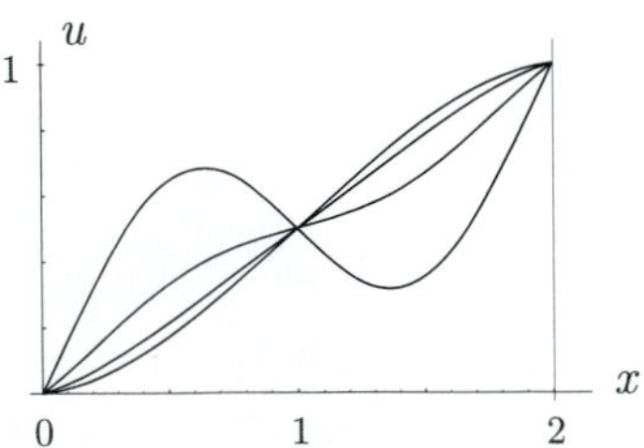

Figure 5a

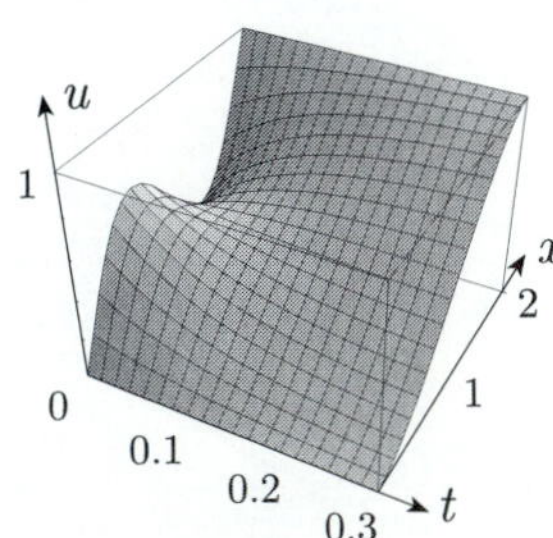

Figure 5b

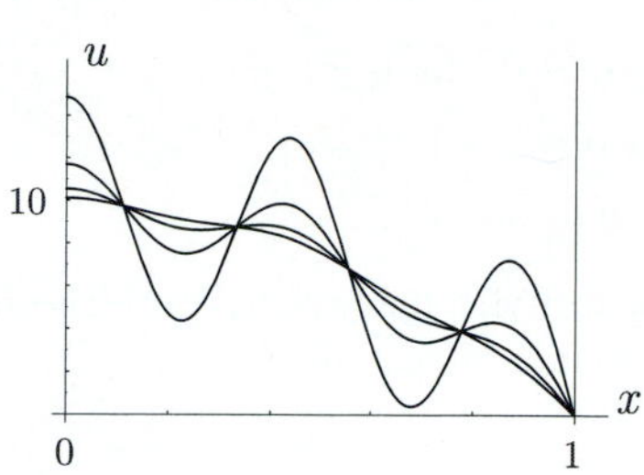

Figure 6a

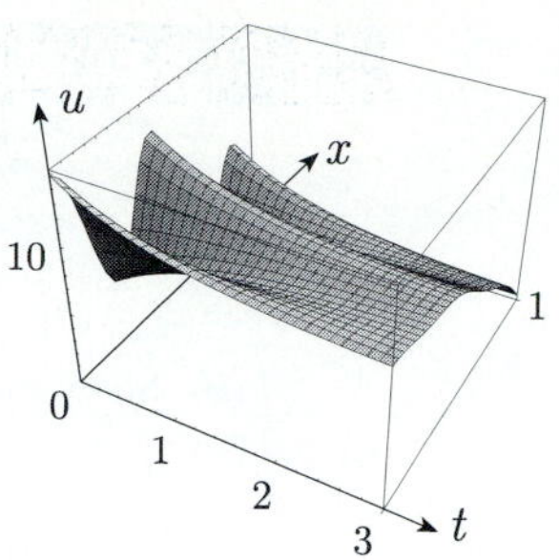

Figure 6b

PROBLEMS

1. Let k, ω, A, and B be constants. Verify that

$$u(t,x) = e^{-k\omega^2 t}(A\sin\omega x + B\cos\omega x)$$

satisfies $u_t - ku_{xx} = 0$ for $-\infty < x < \infty$ and $-\infty < t < \infty$.

2. Let k be a positive constant. Verify that

$$u(t,x) = \frac{1}{\sqrt{4\pi kt}} e^{-\frac{x^2}{4kt}}$$

satisfies $u_t - ku_{xx} = 0$ for $-\infty < x < \infty$ and $t > 0$.

3. Let k be a constant. Verify that, for any integer n, the function

$$u(t,x) = e^{-kn^2\pi^2 t}\cos(n\pi x)$$

satisfies the diffusion equation

$$u_t - ku_{xx} = 0 \quad \text{for } 0 < x < 1, t > 0,$$

and the boundary conditions

$$u_x(t,0) = u_x(t,1) = 0 \quad \text{for } t > 0.$$

4. Let k and ω be constants, and let u be defined by

$$u(t,x) = e^{-k\omega^2 t}(\omega\cos(\omega x) + \sin(\omega x)).$$

(a) Verify that u satisfies the diffusion equation

$$u_t - ku_{xx} = 0 \quad \text{for } 0 < x < 1, t > 0.$$

(b) Verify that, if ω satisfies $\omega\tan\omega = 1$, then u satisfies the boundary conditions

$$u(t,0) - u_x(t,0) = u_x(t,1) = 0 \quad \text{for } t > 0.$$

(c) Give a graphical argument that the equation $\omega\tan\omega = 1$ has infinitely many positive solutions. (*Suggestion*: Rewrite the equation as $\omega = \cot\omega$.)

5. (a) Verify that the function $u(t,x) = e^{-6t}(1-x^2)$ satisfies the diffusion problem

$$\begin{aligned} u_t - ((3-x^2)u_x)_x &= 0 \quad \text{for } 0 < x < 1, t > 0, \\ u(0,x) &= 1 - x^2 \quad \text{for } 0 \le x \le 1, \\ u_x(t,0) = 0, u(t,1) &= 0 \quad \text{for } t > 0. \end{aligned}$$

(b) Show that this solution is unique by verifying the hypotheses of Theorem 1.

6. Let

$$k(x) = \begin{cases} 1 & \text{if } x = 0; \\ \dfrac{(1+x^2)^2 \arctan x}{x} & \text{if } 0 < x \le 1. \end{cases}$$

(a) Verify that the function

$$u(t,x) = \frac{e^{-2t}}{1+x^2}$$

satisfies the diffusion problem

$$\begin{aligned} u_t - (ku_x)_x &= 0 \quad \text{for } 0 < x < 1, t > 0, \\ u(0,x) &= \frac{1}{1+x^2} \quad \text{for } 0 \le x \le 1, \\ u_x(t,0) = 0, u(t,1) + u_x(t,1) &= 0 \quad \text{for } t > 0. \end{aligned}$$

(b) Show that this solution is unique by verifying the hypotheses of Theorem 1.

7. Verify the solution in Example 3.
8. Verify the solution in Example 4.
9. Verify the solution in Example 5.
10. Let k be a differentiable function defined on $[a,b]$, and let $\mathcal{H}$ be the operator defined by

$$\mathcal{H}u = u_t - (ku_x)_x$$

for all functions u that are differentiable in t for $t > 0$ and twice differentiable in x for x in $[a,b]$. Show that $\mathcal{H}$ is a linear operator.

11. Let λ be a constant, and let X be a twice differentiable function of x (i.e., independent of t). Show that $u(t,x) = e^{\lambda t}X(x)$ satisfies the diffusion equation

$$u_t - (ku_x)_x = 0 \quad \text{for } a < x < b,\ t > 0,$$

if and only if X satisfies

$$(kX')' = \lambda X \quad \text{for } a < x < b.$$

12. Suppose that $\tilde{u}$ and $\tilde{v}$ each satisfy the following diffusion equation and boundary conditions:

$$\begin{aligned} u_t - (ku_x)_x &= f \quad \text{for } a < x < b, t > 0, \\ \alpha u(t,a) - (1-\alpha)u_x(t,a) &= \gamma_a \quad \text{for } t > 0, \\ \beta u(t,b) + (1-\beta)u_x(t,b) &= \gamma_b \quad \text{for } t > 0. \end{aligned}$$

Show that $\tilde{u} - \tilde{v}$ must satisfy the following *homogeneous* diffusion equation and boundary conditions:

$$\begin{aligned} u_t - (ku_x)_x &= 0 \quad \text{for } a < x < b, t > 0, \\ \alpha u(t,a) - (1-\alpha)u_x(t,a) &= 0 \quad \text{for } t > 0, \\ \beta u(t,b) + (1-\beta)u_x(t,b) &= 0 \quad \text{for } t > 0. \end{aligned}$$

A *steady-state* (i.e., *time-independent*) solution of (7) is a solution of the boundary-value problem

$$\begin{aligned} -(kz')' &= f \quad \text{for } a < x < b, \\ \alpha z(a) - (1-\alpha)z'(a) &= \gamma_a, \\ \beta z(b) + (1-\beta)z'(b) &= \gamma_b. \end{aligned}$$

Find the solution(s) of each of the boundary-value problems in 13 through 16.

13. $\begin{cases} -z'' = 6x \quad \text{for } 0 < x < 2, \\ z(0) - z'(0) = 1, \quad z'(2) = 0. \end{cases}$

14. $\begin{cases} -((x+1)z')' = 1 \quad \text{for } 0 < x < 1, \\ z(0) = 1, \quad z(1) = 0. \end{cases}$

15. $\begin{cases} -z'' = \sin x \quad \text{for } 0 < x < \pi, \\ z'(0) = z'(\pi) = 0. \end{cases}$

16. $\begin{cases} -(e^x z')' = e^{2x} \quad \text{for } 0 < x < 1, \\ z(0) - z'(0) = z(\ln 2) + z'(\ln 2) = 0. \end{cases}$

17. Suppose that z is a solution of the "steady-state" diffusion problem

$$\begin{aligned} -(kz')' &= f \quad \text{for } a < x < b, \\ \alpha z(a) - (1-\alpha)z'(a) &= \gamma_a, \\ \beta z(b) + (1-\beta)z'(b) &= \gamma_b, \end{aligned}$$

under the following assumptions:

(i) $k(x) > 0$ for $a < x < b$,
(ii) $f(x) > 0$ for $a < x < b$,
(iii) γ_a and γ_b are each nonnegative, and
(iv) α and β are each in $[0, 1]$.

Argue as follows to show that z is nonnegative on $[a, b]$.

(a) Rewrite the differential equation as $kz'' + k'z' = -f$, and argue that the minimum value of z for $a \le x \le b$ cannot be attained in (a, b) and therefore must be attained either at a or at b.

(b) For each of the three cases: $\alpha = 0$, $0 < \alpha < 1$, and $\alpha = 1$, argue that a *negative* minimum cannot be attained at a. (The case where $\alpha = \gamma_a = 0$ is the tricky one.)

(c) For each of the three cases: $\beta = 0$, $0 < \beta < 1$, and $\beta = 1$, argue that a *negative* minimum cannot be attained at b.

(d) Conclude that the minimum value of z for $a \le x \le b$ must be nonnegative.

11.2 Solutions by Separation of Variables

We now turn our attention to the following diffusion problem on the spatial interval $[0, \ell]$ and with constant diffusivity and fixed endpoint values:

$$\left.\begin{aligned} u_t - ku_{xx} &= f \quad \text{for } 0 < x < \ell, t > 0, \\ u(0, x) &= \phi(x) \quad \text{for } 0 \le x \le \ell, \\ u(t, 0) &= \gamma_a \quad \text{for } t > 0, \\ u(t, \ell) &= \gamma_b \quad \text{for } t > 0. \end{aligned}\right\} \qquad (1)$$

We assume that (1) is fully *autonomous*; that is, we assume that f (as well as k, γ_a, and γ_b as usual) does not depend upon t.

An important *boundary-value problem* related to (1) is satisfied by any *time-independent* solution z of (1):

$$\left.\begin{array}{rl} -kz_{xx} = f & \text{for } 0 < x < \ell, \\ z(0) = \gamma_0, & z(\ell) = \gamma_\ell. \end{array}\right\} \tag{2}$$

A solution of (2) is called a **steady-state solution** of (1). This problem is one of a class of boundary-value problems that we will study in more detail later in this chapter. It serves our present needs simply to observe that one can solve (2) for z by straightforward integration. (See Problem 13.) So suppose that the solution z of (2) is known, and set

$$\varphi = \phi - z,$$

where ϕ is the initial data function in (1). Because of the linearity of the differential equation, the solution of (1) may be obtained by adding z to the solution of the following *homogeneous* problem with zero source and zero endpoint values:

$$\left.\begin{array}{rl} w_t - kw_{xx} = 0 & \text{for } 0 < x < \ell, t > 0, \\ w(0, x) = \varphi(x) & \text{for } 0 \le x \le \ell, \\ w(t, 0) = w(t, \ell) = 0 & \text{for } t > 0. \end{array}\right\} \tag{3}$$

We now focus upon solving (3). A classical technique, dating back to the mid-eighteenth century, is commonly known as *separation of variables*. This technique begins with the supposition that certain simple solutions of the differential equation and boundary conditions can be expressed as the product of a function of t and a function of x—that is, in the form

$$w(t, x) = T(t)X(x).$$

Note that for such a product we have $w_t = T'(t)X(x)$ and $w_{xx} = T(t)X''(x)$. Thus, were such a product to satisfy the differential equation in (3), it would follow that

$$T'(t)X(x) - kT(t)X''(x) = 0,$$

and, consequently,

$$\frac{T'(t)}{kT(t)} = \frac{X''(x)}{X(x)}. \tag{4}$$

We now make the key observation that in order for (4) to be true it is necessary that *both sides of the equation be constant.* To see this, think of varying t while holding x fixed. It follows that the left side of (4) is constant; thus the right side is as well. So let us denote this common constant value by λ. Then (4) gives rise to a pair of ordinary differential equations:

$$T'(t) = \lambda kT(t) \tag{5}$$

and

$$X''(x) = \lambda X(x). \tag{6}$$

The upshot of the development so far is that if T and X satisfy (5) and (6), respectively, *for some constant* λ, then their product will satisfy $w_t - kw_{xx} = 0$. The next step is to enforce the boundary conditions in (3) upon $T(t)X(x)$. So we require that

$$T(t)X(0) = T(t)X(\ell) = 0 \quad \text{for all } t > 0,$$

which becomes

$$X(0) = X(\ell) = 0, \tag{7}$$

since we are not interested in the alternative that $T(t) = 0$ for all $t > 0$. Equations (6) and (7) together comprise a boundary-value problem for X. Our immediate goal now is to determine the value(s) of λ for which that problem has solutions other than the obvious trivial (i.e., identically zero) solution.

We begin by eliminating the case $\lambda = 0$. If $\lambda = 0$, then (6) becomes $X''(x) = 0$, which implies that $X(x) = ax + b$ for some constants a and b. From $X(0) = 0$ we conclude that $b = 0$, and then from $X(\ell) = 0$ we conclude that $a = 0$. Thus $X(x) = 0$ for all x.

In the case where $\lambda > 0$, every solution of $X'' = \lambda X$ is of the form

$$X(x) = Ae^{\sqrt{\lambda}x} + Be^{-\sqrt{\lambda}x},$$

where A and B are constants. The boundary conditions thus result in the system of equations

$$\begin{aligned} A + B &= 0, \\ Ae^{\sqrt{\lambda}\ell} + Be^{-\sqrt{\lambda}\ell} &= 0, \end{aligned}$$

of which $A = B = 0$ is the unique solution. This shows that a positive value of λ produces only the trivial solution of (6) and (7).

So we now consider (6) and (7) with λ negative. For convenience, we set $\lambda = -\sigma^2$, with $\sigma > 0$, and restate (6) and (7) as

$$\begin{aligned} X''(x) &= -\sigma^2 X(x) \\ X(0) &= X(\ell) = 0. \end{aligned} \tag{8}$$

We know that every solution of the differential equation can be written as

$$X(x) = A\cos\sigma x + B\sin\sigma x$$

for some choice of constants A and B. The boundary condition $X(0) = 0$ forces $A = 0$. The other boundary condition, $X(\ell) = 0$, then forces

$$B\sin\sigma\ell = 0;$$

that is, either $B = 0$ or $\sin\sigma\ell = 0$. With $B = 0$ (since A is also 0), we simply obtain the trivial solution $X = 0$. So it is the equation $\sin\sigma\ell = 0$ that reveals the interesting values of σ, which comprise the sequence

$$\sigma_n = \frac{n\pi}{\ell}, \quad n = 1, 2, 3, \ldots \tag{9}$$

and allow (8) to possess nontrivial solutions.

We can now state that, for any positive integer n and $\sigma_n = \frac{n\pi}{\ell}$, the function

$$w(t, x) = T(t) \sin \sigma_n x$$

satisfies both the differential equation and the boundary conditions in (3), *provided that* T satisfies (5) with $\lambda = -\sigma_n^2$, that is,

$$T'(t) = -\sigma_n^2 k T(t).$$

The solutions of this equation are all constant multiples of

$$T_n(t) = e^{-\sigma_n^2 kt}.$$

Thus we conclude that each of functions

$$w_n(t, x) = e^{-\sigma_n^2 kt} \sin \sigma_n x, \quad n = 1, 2, 3, \dots, \tag{10}$$

where $\sigma_n = \frac{n\pi}{\ell}$, satisfies the differential equation and the boundary conditions in (3). Moreover, so does any linear combination of them:

$$w(t, x) = \sum_{n=1}^{N} a_n e^{-\sigma_n^2 kt} \sin \sigma_n x, \tag{11}$$

as is easily verified. Indeed, if the initial data in (3) can be expressed as

$$\varphi(x) = \sum_{n=1}^{N} a_n \sin \sigma_n x,$$

then the solution of (3) is given precisely by (11), as is also easily verified. Thus we have solved (3) for a large class of initial functions φ—functions that are expressible as linear combinations of trigonometric functions that satisfy the boundary conditions.

Example 1

Consider the problem

$$\begin{aligned} w_t - k w_{xx} &= 0 && \text{for } 0 < x < 1, t > 0, \\ w(0, x) &= 2 \sin \pi x + 3 \sin 3\pi x + \sin 5\pi x && \text{for } 0 \le x \le 1, \\ w(t, 0) = w(t, 1) &= 0 && \text{for } t > 0. \end{aligned}$$

Since $\ell = 1$ here, we have $\lambda_n = n\pi$, and so we look for the solution in the form

$$w(t, x) = \sum_{n=1}^{N} a_n e^{-n^2 \pi^2 kt} \sin n\pi x.$$

This function already satisfies the differential equation and the boundary conditions; so we need only choose the coefficients $a_1, a_2, a_3, \dots, a_N$ so that the initial condition is met. At $t = 0$ we have

$$w(0, x) = \sum_{n=1}^{N} a_n \sin n\pi x,$$

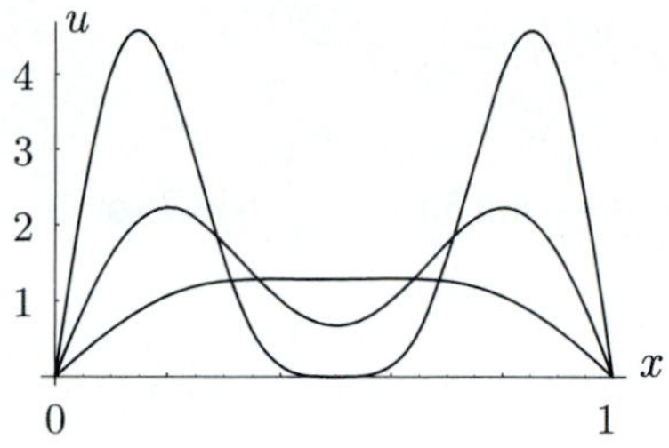

Figure 1a

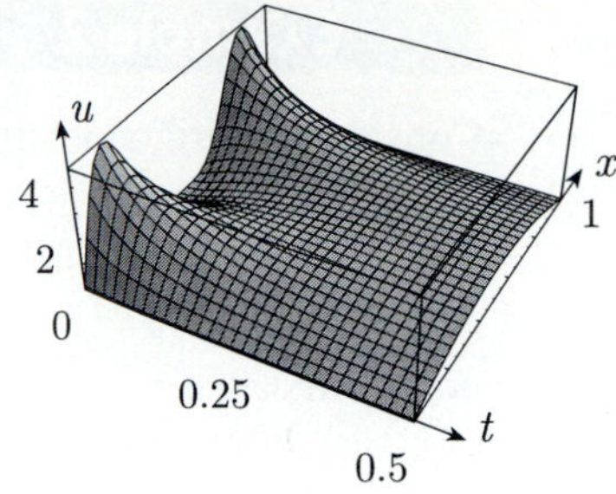

Figure 1b

which corresponds to the desired initial data if we let $N = 5$ and choose

$$a_1 = 2, a_3 = 3, a_5 = 1, \quad \text{and} \quad a_2 = a_4 = 0.$$

Therefore, the solution is

$$w(t, x) = 2e^{-\pi^2 kt} \sin \pi x + 3e^{-9\pi^2 kt} \sin 3\pi x + e^{-25\pi^2 kt} \sin 5\pi x.$$

Figures 1a and b indicate the behavior of the solution in the case where

$$k = \frac{1}{\pi^2},$$

with Figure 1a showing the solution at each of $t = 0$, 0.1, and 0.3. ■

The crux of our investigation so far is that the differential equation and boundary conditions in (3) are satisfied by any finite sum of the form

$$w(t, x) = \sum_{n=1}^{N} a_n e^{-\sigma_n^2 kt} \sin \sigma_n x,$$

where $\sigma_n = n\pi/\ell$, and thus we are able to solve (3) if the initial data are of the form

$$\varphi(x) = \sum_{n=1}^{N} a_n \sin \sigma_n x.$$

Let us now consider u given instead by an *infinite series* of the form

$$w(t, x) = \sum_{n=1}^{\infty} a_n e^{-\sigma_n^2 kt} \sin \sigma_n x,$$

while assuming that the series converges for all $t \geq 0$ and $0 \leq x \leq \ell$. Assuming also that term-by-term differentiation is valid, it is easy to see that such a function satisfies (3) with the initial data

$$\varphi(x) = \sum_{n=1}^{\infty} a_n \sin \sigma_n x.$$

The following example provides a simple illustration of this idea.

Example 2

Consider the trigonometric series

$$\varphi(x) = \frac{8}{\pi^2}\left(\sin \pi x - \frac{1}{9}\sin 3\pi x + \frac{1}{25}\sin 5\pi x - \frac{1}{49}\sin 7\pi x + \cdots\right),$$

noting first that

$$\varphi(x) = \sum_{n=1}^{\infty} a_n \sin(n\pi x),$$

where

$$a_n = \begin{cases} \dfrac{8(-1)^{\frac{n+1}{2}}}{\pi^2 n^2} & \text{if } n \text{ is odd,} \\ 0 & \text{if } n \text{ is even.} \end{cases}$$

At each x, the value of $\varphi(x)$ is, *by definition*, the limit of the sequence of partial sums:

$$\begin{aligned} &\frac{8}{\pi^2}\sin \pi x, \\ &\frac{8}{\pi^2}\left(\sin \pi x - \frac{1}{9}\sin 3\pi x\right), \\ &\frac{8}{\pi^2}\left(\sin \pi x - \frac{1}{9}\sin 3\pi x + \frac{1}{25}\sin 5\pi x\right), \\ &\frac{8}{\pi^2}\left(\sin \pi x - \frac{1}{9}\sin 3\pi x + \frac{1}{25}\sin 5\pi x - \frac{1}{49}\sin 7\pi x\right), \ldots, \end{aligned}$$

provided that the limit exists. For instance, at $x = 1/2$, the first five terms in this sequence are

$$0.810569, 0.900633, 0.933056, 0.949598, 0.959605,$$

and the 100th, 200th, ... , 500th terms are

$$0.997974, 0.998987, 0.999325, 0.999493, 0.999595,$$

suggesting that the value of $\varphi(1/2)$ may be 1. (This is indeed true.) Figure 2 shows the graph of each the first, second, third, tenth, and twentieth of the partial sums on the interval [0, 1].

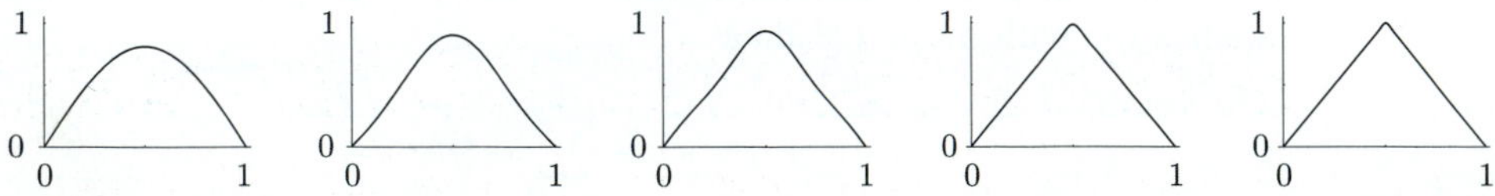

Figure 2

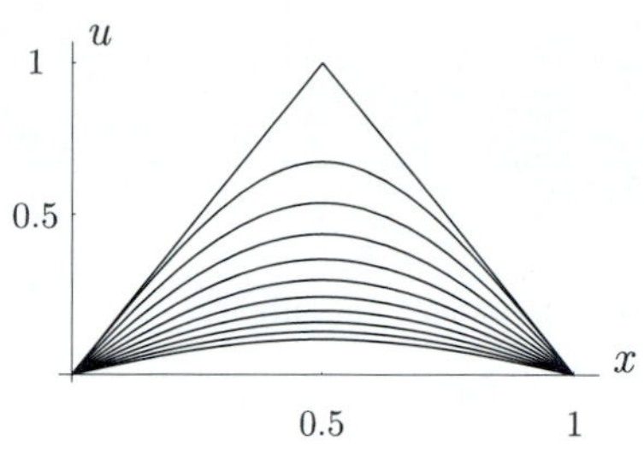

Figure 3a

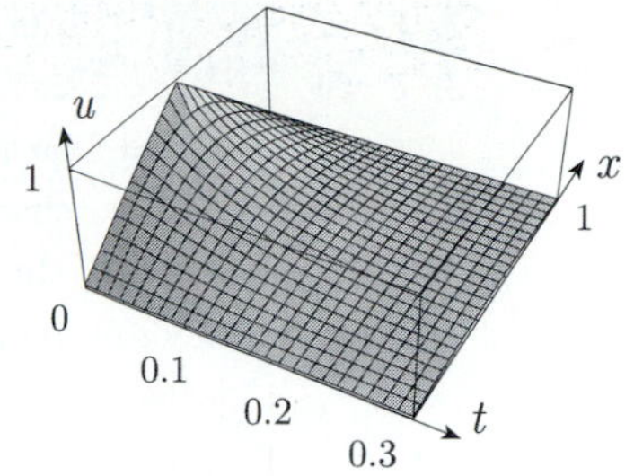

Figure 3b

The last of the graphs in Figure 2 is a reasonably good approximation of the graph of the series $\varphi(x)$, which has, in fact, the "closed form"

$$\varphi(x) = \begin{cases} 2x & \text{if } 0 \leq x \leq \frac{1}{2}, \\ 2(1-x) & \text{if } \frac{1}{2} < x \leq 1. \end{cases}$$

(We will learn in the next section how to find trigonometric series representations such as the one examined here.) The solution of (3) with initial data given by $\varphi(x)$ is

$$w(t, x) = \frac{8}{\pi^2}\left(e^{-\pi^2 kt} \sin \pi x - \frac{1}{9} e^{-9\pi^2 kt} \sin 3\pi x + \frac{1}{25} e^{-25\pi^2 kt} \sin 5\pi x + \cdots\right).$$

Figures 3a and b indicate the behavior of the solution in the case where $k = 1$, with Figure 3a showing the solution at each of $t = 0, 0.02, 0.04, \ldots, 0.20$. ■

The essential result of this section is that the solution of (3) is easily obtained (indeed may be written down by inspection) provided that we can express the initial data as a trigonometric series of the form

$$\varphi(x) = \sum_{n=1}^{\infty} a_n \sin \frac{n\pi}{\ell} x.$$

Similar results are true for problems with other types of boundary conditions, as you will be asked to show in the problems that follow. Thus we are faced with these questions:

What functions φ, defined on $[0, \ell]$, can be expressed as convergent trigonometric series of the appropriate form?

Given such a function φ, how can the coefficients of the series be calculated?

The next two sections will be devoted to answering these questions.

PROBLEMS

Find the solution of (3) in each of the instances described in Problems 1 through 6.

1. $\ell = 1$, $w(t, 0) = \sin 3\pi x$

2. $\ell = 2$, $w(t, 0) = \sin 3\pi x$

3. $\ell = 1$, $w(t, 0) = 2 \sin \pi x + \sin 2\pi x$

4. $\ell = 2$, $w(t, 0) = 2 \sin \pi x + \sin 2\pi x$

5. $\ell = \pi$, $w(t, 0) = 3 \sin x + \sin 2x - \frac{1}{2} \sin 3x$

6. $\ell = \frac{\pi}{2}$, $w(t, 0) = 5 \sin 2x - \sin 4x$

7. Consider the diffusion problem with zero-flux boundary conditions:

$$\begin{aligned} w_t - kw_{xx} &= 0 && \text{for } 0 < x < \ell, t > 0, \\ w(0, x) &= \varphi(x) && \text{for } 0 < x < \ell, \\ w_x(t, 0) = w_x(t, \ell) &= 0 && \text{for } t > 0. \end{aligned}$$

(a) Use the separation of variables technique to derive the following family of functions that satisfy the differential equation and boundary conditions:

$$w_n(t, x) = e^{-\sigma_n^2 kt} \cos \sigma_n x, \quad \sigma_n = \frac{n\pi}{\ell}, \quad n = 0, 1, 2, \dots .$$

(b) Let $\ell = 1$, and write down the solution $w(t, x)$ satisfying

$$w(0, x) = 2 + 5 \cos \pi x - \cos 3\pi x.$$

(c) Let $\ell = \pi$, and write down the solution $w(t, x)$ satisfying

$$w(0, x) = 1 + \cos 2x + \cos 3x.$$

(d) Write down the solution $w(t, x)$ satisfying

$$w(0, x) = \sum_{n=0}^{\infty} a_n \cos \sigma_n x, \quad \text{where } \sigma_n = n\pi/\ell.$$

8. Consider the diffusion problem with a zero-value condition at $x = 0$ and a zero-flux condition at $x = \ell$:

$$\begin{aligned} w_t - kw_{xx} &= 0 && \text{for } 0 < x < \ell, t > 0, \\ w(0, x) &= \varphi(x) && \text{for } 0 < x < \ell, \\ w(t, 0) = 0, \ w_x(t, \ell) &= 0 && \text{for } t > 0. \end{aligned}$$

(a) Use the separation of variables technique to derive the following family of functions that satisfy the differential equation and boundary conditions:

$$w_n(t, x) = e^{-\sigma_n^2 kt} \sin \sigma_n x, \quad \sigma_n = \frac{(2n-1)\pi}{2\ell} \quad n = 1, 2, 3, \dots .$$

(b) Let $\ell = 1$, and write down the solution $w(t, x)$ satisfying

$$w(0, x) = 5 \sin \frac{\pi x}{2} - \sin \frac{3\pi x}{2}.$$

(c) Let $\ell = \frac{\pi}{2}$, and write down the solution $w(t, x)$ satisfying

$$w(0, x) = \sin x + \frac{1}{2} \sin 3x - \frac{1}{10} \sin 7x.$$

(d) Write down the solution $w(t, x)$ satisfying

$$w(0, x) = \sum_{n=1}^{\infty} a_n \sin \sigma_n x, \quad \text{where } \sigma_n = \frac{(2n-1)\pi}{2\ell}.$$

9. Consider the diffusion problem with a zero-flux condition at $x = 0$ and a value-dependent flux condition at $x = \ell$:

$$\begin{aligned} w_t - k w_{xx} &= 0 \qquad \text{for } 0 < x < \ell, t > 0, \\ w(0, x) &= \varphi(x) \qquad \text{for } 0 < x < \ell, \\ w_x(t, 0) = 0,\ w(t, \ell) + w_x(t, \ell) &= 0 \qquad \text{for } t > 0. \end{aligned}$$

(a) Show that the differential equation and boundary conditions are satisfied by

$$w(t, x) = e^{-\sigma^2 kt} \cos \sigma x,$$

provided that σ is a solution of the equation

$$\sigma = \cot \sigma \ell.$$

(b) Give a graphical argument that the positive solutions of $\sigma = \cot \sigma \ell$ define an increasing, divergent sequence $\sigma_1, \sigma_2, \sigma_3, \ldots$. Conclude that there is an infinite family of functions, given by

$$w_n(t, x) = e^{-\sigma_n^2 kt} \cos \sigma_n x, \quad n = 1, 2, 3, \ldots,$$

that satisfy the differential equation and boundary conditions.

(c) Assuming that $\sigma_1, \sigma_2, \sigma_3, \ldots$ are known, write down the solution $w(t, x)$ satisfying

$$w(0, x) = \sum_{n=1}^{\infty} a_n \cos \sigma_n x.$$

10. Diffusion problems involving heat energy (measured as temperature) are often called heat conduction problems. In such problems, the diffusivity coefficient k is the *thermal diffusivity* and is related to material properties through $k = \frac{\kappa}{\rho c}$, where κ is thermal conductivity, ρ is density, and c is specific heat. For copper, the value of k is approximately 1.14.

Suppose that an 11-cm length of copper wire is perfectly insulated along its length, with its ends held at a fixed temperature of $0° C$. Suppose also that the initial temperature distribution along the wire is given by

$$w(0, x) = 50 \sin \frac{\pi x}{11} + 20 \sin \pi x, \quad 0 \le x \le 11.$$

(a) Write down the solution $w(t, x)$ of the relevant heat conduction problem.

(b) Graph the temperature distribution at each of the times $t = 0$, 1, and 10 seconds.

11. Consider the special case of (1):

$$\begin{aligned} u_t - \frac{1}{2} u_{xx} &= 3x \qquad \text{for } 0 < x < 1, t > 0, \\ u(0, x) &= x \qquad \text{for } 0 \le x \le 1, \\ u(t, 0) = 1,\ u(t, 1) &= 0 \qquad \text{for } t > 0 : . \end{aligned}$$

(a) Find the steady-state solution by solving the boundary-value problem

$$\begin{aligned} -\frac{1}{2} z_{xx} &= 3x \qquad \text{for } 0 < x < 1, \\ z(0) = 1, &\qquad z(1) = 0. \end{aligned}$$

(b) Write the homogeneous diffusion problem (as in (3)) satisfied by $w = u - z$.

12. Consider the special case of (1):

$$\begin{aligned} u_t - u_{xx} &= \sin x \quad && \text{for } 0 < x < \pi, t > 0, \\ u(0, x) &= 1 + \sin 2x \quad && \text{for } 0 \le x \le \pi, \\ u(t, 0) &= u(t, \pi) = 1 \quad && \text{for } t > 0, \end{aligned}$$

(a) Find the steady-state solution by solving the boundary-value problem

$$\begin{aligned} -z_{xx} &= \sin x \quad \text{for } 0 < x < \pi, \\ z(0) &= z(\pi) = 1. \end{aligned}$$

(b) Write *and solve* the homogeneous diffusion problem satisfied by $w = u - z$.

(c) Write the solution $u = w + z$.

13. Verify that the solution of (2) is given by

$$z(x) = \gamma_0 + \left(\gamma_\ell - \gamma_0 + \frac{1}{k}\int_0^\ell \int_0^\xi f(\eta)\, d\eta\, d\xi\right)\frac{x}{\ell} - \frac{1}{k}\int_0^x \int_0^\xi f(\eta)\, d\eta\, d\xi.$$

14. Consider the boundary-value problem satisfied by any steady-state solution of the diffusion problem in Problem 8:

$$\begin{aligned} -kz_{xx} &= f \quad \text{for } 0 < x < \ell, \\ z(0) &= z_x(\ell) = 0. \end{aligned}$$

Find the solution z in a form similar to that in Problem 13.

15. Consider the boundary-value problem satisfied by any steady-state solution of the diffusion problem in Problem 7:

$$\begin{aligned} -kz_{xx} &= f \quad \text{for } 0 < x < \ell, \\ z_x(0) &= z_x(\ell) = 0. \end{aligned}$$

(a) Show that, if $\int_0^\ell f(x)\, dx \ne 0$, then this boundary-value problem has no solution.

(b) Show that, if $\int_0^\ell f(x)\, dx = 0$, then this boundary-value problem has infinitely many solutions, and any two solutions differ by a constant.

16. **(a)** Suppose that the solution w of (3) is known and c is a constant. Show that v, defined by $v(t, x) = e^{-ct} w(t, x)$, satisfies the differential equation

$$v_t - kv_{xx} + cv = 0 \quad \text{for } 0 < x < \ell, t > 0,$$

and the same initial and boundary conditions as w.

(b) Find the solution of

$$\begin{aligned} v_t - v_{xx} + v &= 0 \quad && \text{for } 0 < x < \pi, t > 0, \\ v(0, x) &= 3 \sin x + \sin 3x \quad && \text{for } 0 \le x \le \pi, \\ v(t, 0) &= v(t, \pi) = 0 \quad && \text{for } t > 0. \end{aligned}$$

17. Let $\sigma_1, \sigma_2, \ldots$ be any sequence of real numbers. Use the comparison test (consult your calculus book) to show that if the series $\sum_{n=1}^\infty a_n$ converges absolutely, then each of the series

$$\sum_{n=1}^\infty a_n \sin \sigma_n x \quad \text{and} \quad \sum_{n=1}^\infty a_n \cos \sigma_n x$$

converges absolutely for all x.

18. Let $\sigma_n = n\pi/\ell$, $n = 1, 2, 3, \ldots$, and define the function

$$g(x) = \sum_{n=1}^{\infty} a_n \sin \sigma_n x.$$

Show that if the series defining g converges for all x, then g is periodic with period 2ℓ; that is,

$$g(x + 2\ell) = g(x) \quad \text{for all } x.$$

11.3 Fourier Series

We found in the preceding section that certain simple diffusion problems are easily solved, provided that the initial data are expressed as an appropriate series of trigonometric terms. So it is our goal in this section to explore the issue of expressing functions in that form.

First recall that a function f defined on $(-\infty, \infty)$ is **periodic** with period $p > 0$, if $f(x + p) = f(x)$ for all x. For brevity we will refer to a periodic function with period p as *p-periodic*.

Now let φ be a function defined on an interval $[-\ell, \ell]$. Corresponding to (the symbol) is to φ a unique 2ℓ-periodic function $\overline{\varphi}$ for which

$$\overline{\varphi}(x) = \varphi(x) \quad \text{for all } x \text{ in } (-\ell, \ell].$$

This function $\overline{\varphi}$ is called the **periodic extension** of φ. The idea is illustrated in Figures 1ab. A generic function on $[-\ell, \ell]$ is shown in Figure 1a, and its periodic extension is shown in Figure 1b. (Note that $\overline{\varphi}$ and φ will not agree at $x = -\ell$ unless $\varphi(-\ell) = \varphi(\ell)$.)

Let us now suppose that φ is a function defined on $[-\ell, \ell]$ with periodic extension $\overline{\varphi}$. Since each of the functions

$$1, \cos\frac{\pi x}{\ell}, \sin\frac{\pi x}{\ell}, \cos\frac{2\pi x}{\ell}, \sin\frac{2\pi x}{\ell}, \cos\frac{3\pi x}{\ell}, \sin\frac{3\pi x}{\ell}, \ldots \tag{1}$$

is periodic with common period $p = 2\ell$, we will investigate the possibility of expanding $\overline{\varphi}(x)$—and thus $\varphi(x)$ for x in $(-\ell, \ell]$—in an infinite series of the form*

$$\frac{a_0}{2} + \sum_{n=1}^{\infty}\left(a_n \cos\frac{n\pi x}{\ell} + b_n \sin\frac{n\pi x}{\ell}\right). \tag{2}$$

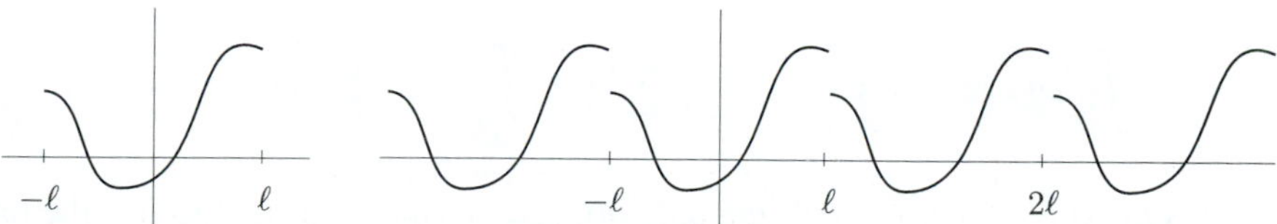

Figure 1a **Figure 1b**

* The reason for the factor of $\frac{1}{2}$ in the constant term is so that later a single formula can be given for all of $a_0, a_1, a_2, \ldots$.

Our first goal is to find formulae for the coefficients $a_0, a_1, a_2, \dots$ and $b_1, b_2, b_3, \dots$ in terms of computable properties of φ. For this purpose we will *assume* that the series in (2) does indeed converge to $\varphi(x)$ for all x in $(-\ell, \ell]$ and that φ has whatever properties are necessary to justify the steps in our calculations.

The key to this endeavor is found in a collection of integral properties possessed by the family of simple cosine and sine functions listed in (1).

Theorem 1

Let m and n be positive integers. Then

i) $$\int_{-\ell}^{\ell} \cos\frac{n\pi x}{\ell}\,dx = \int_{-\ell}^{\ell} \sin\frac{n\pi x}{\ell}\,dx = 0,$$

ii) $$\int_{-\ell}^{\ell} \cos\frac{m\pi x}{\ell}\sin\frac{n\pi x}{\ell}\,dx = 0,$$

iii) $$\int_{-\ell}^{\ell} \cos\frac{m\pi x}{\ell}\cos\frac{n\pi x}{\ell}\,dx = \begin{cases} 0 & \text{if } m \neq n, \\ \ell & \text{if } m = n, \end{cases}$$

iv) $$\int_{-\ell}^{\ell} \sin\frac{m\pi x}{\ell}\sin\frac{n\pi x}{\ell}\,dx = \begin{cases} 0 & \text{if } m \neq n, \\ \ell & \text{if } m = n. \end{cases}$$ ▲

Each of the formulae asserted in Theorem 1 can be derived by means of elementary integration techniques. This is the subject of Problem 13 at the end of this section. An alternative derivation of (ii)–(iv) with $m \neq n$ is the subject of Problem 14.

Our derivation of formulae for the coefficients $a_0, a_1, a_2, \dots$ and $b_1, b_2, b_3, \dots$ will require term-by-term integration over $[-\ell, \ell]$ of certain infinite series related to the one in (2). Such operations are not automatically justified; however they are justified for a large class of functions defined by series of the type in (2). So we proceed under the assumptions that

$$\varphi(x) = \frac{a_0}{2} + \sum_{n=1}^{\infty}\left(a_n\cos\frac{n\pi x}{\ell} + b_n\sin\frac{n\pi x}{\ell}\right) \quad \text{for all } x \text{ in } (-\ell, \ell], \tag{3}$$

and that the necessary term-by-term integrations over $[-\ell, \ell]$ are valid.

We begin by integrating φ over $[-\ell, \ell]$:

$$\int_{-\ell}^{\ell} \varphi(x)\,dx = \int_{-\ell}^{\ell} \frac{a_0}{2}\,dx + \sum_{n=1}^{\infty}\left(a_n\int_{-\ell}^{\ell}\cos\frac{n\pi x}{\ell}\,dx + b_n\int_{-\ell}^{\ell}\sin\frac{n\pi x}{\ell}\,dx\right).$$

By part (i) of Theorem 1, the only nonzero term on the right side is the one involving a_0. That term is $\int_{-\ell}^{\ell}\frac{a_0}{2}\,dx = a_0\ell$; thus it follows that

$$a_0 = \frac{1}{\ell}\int_{-\ell}^{\ell}\varphi(x)\,dx.$$

Next we let m be an arbitrary positive integer and multiply each side of (3) by $\cos\frac{m\pi x}{\ell}$ before integrating over $[-\ell, \ell]$:

$$\int_{-\ell}^{\ell} \varphi(x)\cos\frac{m\pi x}{\ell}\,dx = \int_{-\ell}^{\ell} \frac{a_0}{2}\cos\frac{m\pi x}{\ell}\,dx$$
$$+\sum_{n=1}^{\infty}\left(a_n\int_{-\ell}^{\ell}\cos\frac{n\pi x}{\ell}\cos\frac{m\pi x}{\ell}\,dx + b_n\int_{-\ell}^{\ell}\sin\frac{n\pi x}{\ell}\cos\frac{m\pi x}{\ell}\,dx\right).$$

By parts (i), (ii), and (iii) of Theorem 1, the only nonzero term on the right side is the one involving a_m. Thus

$$\int_{-\ell}^{\ell} \varphi(x)\cos\frac{m\pi x}{\ell}\,dx = a_m\int_{-\ell}^{\ell}\cos^2\frac{m\pi x}{\ell}\,dx$$
$$= \frac{a_m}{2}\int_{-\ell}^{\ell}\left(1+\cos\frac{2m\pi x}{\ell}\right)dx = a_m\ell,$$

and so we have

$$a_m = \frac{1}{\ell}\int_{-\ell}^{\ell}\varphi(x)\cos\frac{m\pi x}{\ell}\,dx.$$

Finally, we let m be an arbitrary positive integer and multiply each side of (3) by $\sin\frac{m\pi x}{\ell}$ before integrating over $[-\ell, \ell]$:

$$\int_{-\ell}^{\ell} \varphi(x)\sin\frac{m\pi x}{\ell}\,dx = \int_{-\ell}^{\ell} \frac{a_0}{2}\sin\frac{m\pi x}{\ell}\,dx$$
$$+\sum_{n=1}^{\infty}\left(a_n\int_{-\ell}^{\ell}\cos\frac{n\pi x}{\ell}\sin\frac{m\pi x}{\ell}\,dx + b_n\int_{-\ell}^{\ell}\sin\frac{n\pi x}{\ell}\sin\frac{m\pi x}{\ell}\,dx\right).$$

By parts (i), (ii), and (iv) of Theorem 1, the only nonzero term on the right side is the one involving b_m. Thus

$$\int_{-\ell}^{\ell} \varphi(x)\sin\frac{m\pi x}{\ell}\,dx = b_m\int_{-\ell}^{\ell}\sin^2\frac{m\pi x}{\ell}\,dx$$
$$= \frac{b_m}{2}\int_{-\ell}^{\ell}\left(1-\cos\frac{2m\pi x}{\ell}\right)dx = b_m\ell,$$

and so we have

$$b_m = \frac{1}{\ell}\int_{-\ell}^{\ell}\varphi(x)\sin\frac{m\pi x}{\ell}\,dx.$$

To summarize, we have found that if φ is given by (3) and if the term-by-term integrations in the preceding development are valid, then the coefficients in (3) are given by

$$a_n = \frac{1}{\ell}\int_{-\ell}^{\ell}\varphi(x)\cos\frac{n\pi x}{\ell}\,dx, \quad n = 0, 1, 2, \ldots \tag{4}$$

and

$$b_n = \frac{1}{\ell}\int_{-\ell}^{\ell} \varphi(x) \sin\frac{n\pi x}{\ell}\,dx, \quad n = 1, 2, 3, \ldots. \tag{5}$$

These formulae for a_n and b_n—obtained by formal calculations—motivate the following definition.

Definition Given a function φ defined on $[-\ell, \ell]$, the numbers $a_0, a_1, a_2, \ldots$ and $b_1, b_2, b_3, \ldots$ given by (4) and (5) are called the **Fourier coefficients** of φ (with respect to $[-\ell, \ell]$). The function $\underset{\sim}{\varphi}$ defined by

$$\underset{\sim}{\varphi}(x) = \frac{a_0}{2} + \sum_{n=1}^{\infty}\left(a_n \cos\frac{n\pi x}{\ell} + b_n \sin\frac{n\pi x}{\ell}\right) \tag{6}$$

is called the **Fourier series** associated with φ on $[-\ell, \ell]$. We also call $\underset{\sim}{\varphi}$ the *Fourier series representation* of φ *on the interval* $[-\ell, \ell]$. ◆

We point out that the definition of $\underset{\sim}{\varphi}$ is not dependent upon convergence properties and requires only that the coefficients $a_0, a_1, a_2, \ldots$ and $b_1, b_2, b_3, \ldots$ exist, which is the case if φ is integrable on $[-\ell, \ell]$. (Why?) It may be the case that, at certain x, the series defining $\underset{\sim}{\varphi}(x)$ either does not converge or converges to something other than $\varphi(x)$. The fundamental question regarding $\underset{\sim}{\varphi}$ is this: *For what values of* x, *if any, does* $\underset{\sim}{\varphi}(x) = \varphi(x)$? Before we state a theorem that answers this question, let's examine a couple of examples.

Example 1

Let φ be defined on $[-1, 1]$ by

$$\varphi(x) = \begin{cases} 0, & \text{if } -1 \le x \le 0; \\ 4x(1-x), & \text{if } 0 < x \le 1. \end{cases}$$

The graph of the periodic extension $\overline{\varphi}$ of φ is indicated in Figure 2. Note that $\overline{\varphi}$ is continuous. Since $\ell = 1$ and $\varphi(x) = 0$ for $-1 \le x \le 0$, the Fourier coefficients of φ are

$$a_n = \int_0^1 \cos(n\pi x)\,4x(1-x)\,dx, \quad n = 0, 1, 2, \ldots,$$

$$b_n = \int_0^1 \sin(n\pi x)\,4x(1-x)\,dx, \quad n = 1, 2, 3, \ldots.$$

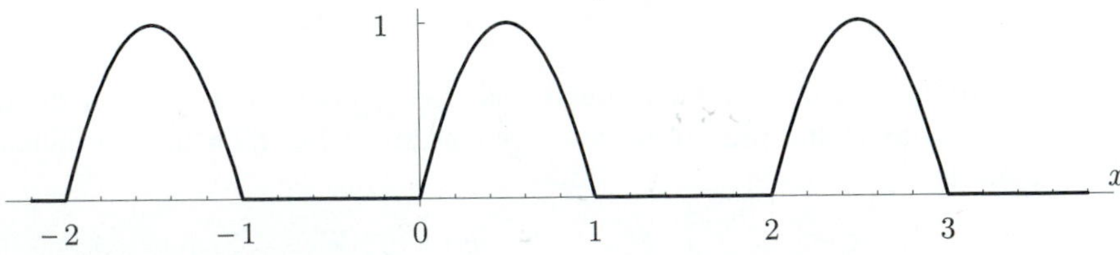

Figure 2

Further computation—either by hand using integration by parts or aided by a *CAS*—reveals that

$$a_0 = \frac{2}{3}, \quad a_n = -(1 + (-1)^n)\frac{4}{n^2\pi^2}, \quad n = 1, 2, 3, \ldots,$$
$$b_n = (1 - (-1)^n)\frac{8}{n^3\pi^3}, \quad n = 1, 2, 3, \ldots.$$

Thus the Fourier series associated with $\overline{\varphi}$—and the Fourier series representation of φ on $[-1, 1]$—is

$$\underset{\sim}{\varphi}(x) = \frac{1}{3} + \frac{8}{\pi^2}\left(\frac{2}{\pi}\sin \pi x - \frac{1}{2^2}\cos 2\pi x + \frac{2}{3^3\pi}\sin 3\pi x - \frac{1}{4^2}\cos 4\pi x + \cdots\right).$$

Figure 3 shows graphs of partial sums containing two, three, four, and five terms, each along with a dashed version of the graph of $\overline{\varphi}$. Graphs of subsequent partial sums show even more accurate approximations of $\overline{\varphi}$. So it is evident that the value of Fourier series associated with φ on $[-1, 1]$ indeed coincides with $\overline{\varphi}(x)$ for all x, and, in particular,

$$\underset{\sim}{\varphi}(x) = \varphi(x) \quad \text{for all } x \text{ in } [-1, 1].$$

■

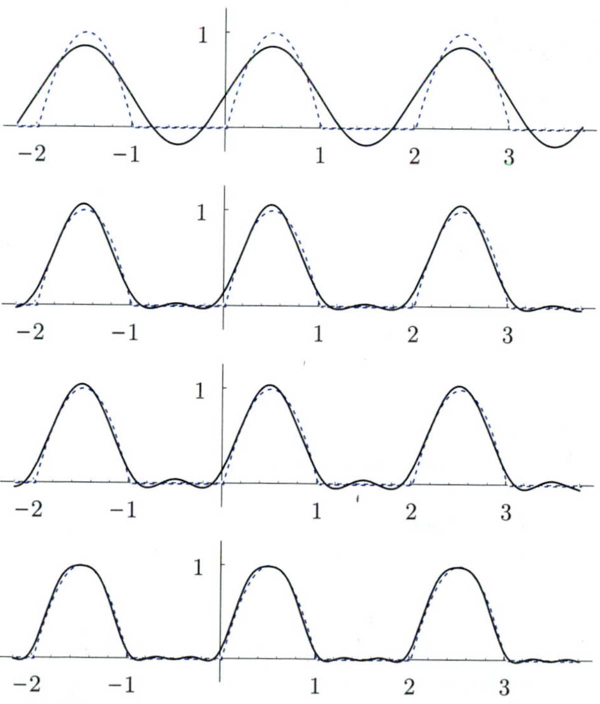

Figure 3

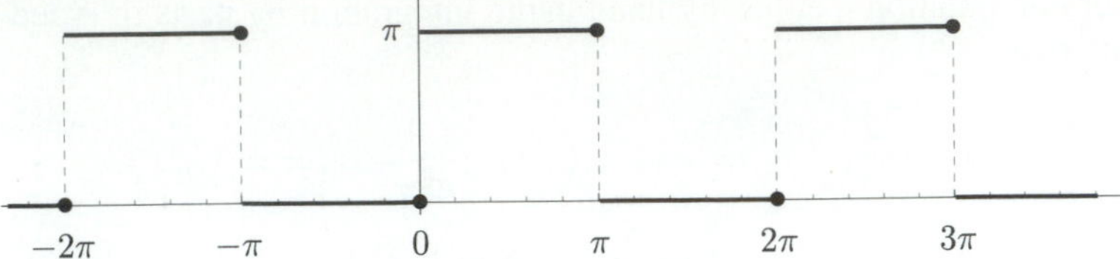

Figure 4

Example 2

Let φ be defined on $[-\pi, \pi]$ by

$$\varphi(x) = \begin{cases} 0, & \text{if } -\pi \leq x \leq 0; \\ \pi, & \text{if } 0 < x \leq \pi. \end{cases}$$

The graph of the periodic extension $\overline{\varphi}$ of φ is indicated in Figure 4. Note that $\overline{\varphi}$ is continuous except at each integer multiple of π.

The Fourier coefficients of φ are

$$a_n = \frac{1}{\pi}\int_{-\pi}^{\pi} \cos(nx)\,\varphi(x)\,dx = \frac{1}{\pi}\int_0^{\pi} \cos(nx)\,\pi\,dx, \quad n = 0, 1, 2, \ldots,$$

and

$$b_n = \frac{1}{\pi}\int_{-\pi}^{\pi} \sin(nx)\,\varphi(x)\,dx = \frac{1}{\pi}\int_0^{\pi} \sin(nx)\,\pi\,dx, \quad n = 1, 2, 3, \ldots,$$

from which we find that $a_0 = \pi$ and

$$a_n = 0, \quad b_n = \frac{1-(-1)^n}{n}, \quad n = 1, 2, 3, \ldots.$$

Thus the Fourier series associated with φ is

$$\underset{\sim}{\varphi}(x) = \frac{\pi}{2} + 2\left(\sin x + \frac{1}{3}\sin 3x + \frac{1}{5}\sin 5x + \cdots\right).$$

Figure 5a shows graphs of partial sums containing two, three, and four terms, and Figure 5b shows graphs of partial sums containing ten, twenty, and thirty terms. For comparison each of the graphs is superimposed over a dashed version of the graph of $\overline{\varphi}$.

These plots suggest that the partial sums defining $\underset{\sim}{\varphi}(x)$ converge to $\overline{\varphi}(x)$ at each x except those where $\overline{\varphi}$ is discontinuous. In fact, for any integer j, it is easy to see that

$$\underset{\sim}{\varphi}(j\pi) = \frac{1}{2\pi}, \quad \text{while } \overline{\varphi}(j\pi) = 0.$$ ■

A Convergence Theorem The preceding examples are illustrative of general convergence properties of Fourier series associated with *piecewise-continuous* periodic functions. Before stating these convergence properties as a theorem, we first need to give a precise definition of piecewise continuity.

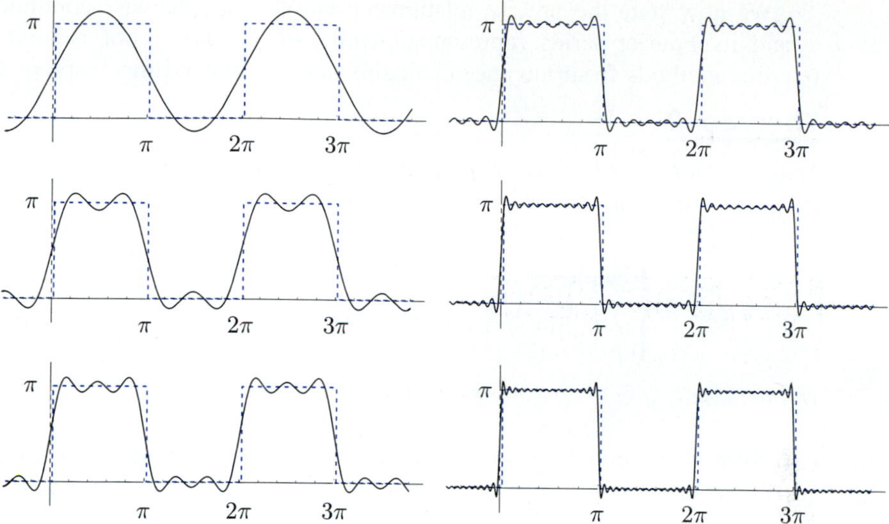

Figure 5a **Figure 5b**

Definition A function φ is **piecewise continuous** on a closed and bounded interval $[a, b]$ if there are numbers $x_0 = a < x_1 < x_2 < \cdots < x_{n-1} < x_n = b$ such that

i. φ is continuous on each open subinterval (x_{i-1}, x_i), $i = 1, 2, \ldots, n$,
ii. the left-sided limit $\lim_{x \to x_i^-} \varphi(x)$ exists for all $i = 1, 2, \ldots, n$, and
iii. the right-sided limit $\lim_{x \to x_i^+} \varphi(x)$ exists for all $i = 0, 1, \ldots, n-1$.

A function is piecewise continuous on $(-\infty, \infty)$ if it is piecewise continuous on every closed and bounded interval $[a, b]$. ◆

One may view piecewise continuous functions on $[a, b]$ as functions that either are continuous on $[a, b]$ or have finitely many discontinuities in $[a, b]$, each of which is either a removeable or "jump" discontinuity. Excluded are vertical asymptotes and oscillatory discontinuities such as that of $\sin(1/x)$ at $x = 0$.

Consequences of the preceding definition include the observations that

- a piecewise-continuous function on $[a, b]$ is also piecewise continuous on every closed subinterval of $[a, b]$, and
- a function that is piecewise continuous on both $[a, b]$ and $[b, c]$ is also piecewise continuous on $[a, c]$,

from which it follows that:

- the periodic extension $\overline{\varphi}$ of a piecewise-continuous function φ on $[-\ell, \ell]$ is piecewise continuous on $(-\infty, \infty)$.

We now state the precise relationship between a piecewise-continuous function φ and its Fourier series representation on $[-\ell, \ell]$. The proof is omitted, since it requires methods from advanced calculus that are beyond the scope of this text.

Theorem 2

Suppose that φ and φ' are each piecewise continuous on $[-\ell, \ell]$ and that $\overline{\varphi}$ is the periodic extension of φ. Then the Fourier series $\varphi_{\sim}$ defined by (6) converges for all x, and

$$\varphi_{\sim}(x) = \begin{cases} \overline{\varphi}(x), & \text{wherever } \overline{\varphi} \text{ is continuous;} \\ \dfrac{1}{2}\left(\lim_{\xi \to x^-} \overline{\varphi}(\xi) + \lim_{\xi \to x^+} \overline{\varphi}(\xi) \right), & \text{wherever } \overline{\varphi} \text{ is discontinuous.} \end{cases}$$

In particular, if $\overline{\varphi}$ is continuous, then $\varphi_{\sim} = \overline{\varphi}$. ▲

Note that according to Theorem 2 the Fourier series $\varphi_{\sim}$ associated with a piecewise-continuous function φ on $[-\ell, \ell]$ is a piecewise-continuous, 2ℓ-periodic function whose discontinuities in $[-\ell, \ell]$ coincide with those of φ and whose value at each of its discontinuities is the average of its two one-sided limits. (Note that this is also true at points where $\varphi_{\sim}$ is continuous.) This fact is illustrated by Figure 6, which shows the graph of the Fourier series computed in Example 2 and associated with the function φ whose 2π-periodic extension is plotted in Figure 4. Note that changing the value of φ at any or all of its discontinuities—or at any finite collection of points in $[-\ell, \ell]$—would have no effect on the resulting Fourier series.

REMARK We remark that, for any two distinct functions u, v from those listed in (1), Theorem 1 tells us that

$$\int_{-\ell}^{\ell} u(x)v(x)dx = 0.$$

Two functions with this property are said to be **orthogonal** on the interval $[-\ell, \ell]$. Thus Theorem 1 states that the functions listed in (1) comprise an *orthogonal family* of functions on $[-\ell, \ell]$. Indeed, orthogonality was the crucial property that enabled us to calculate the coefficients in (2) in a simple way.

The analogy with the notion of orthogonal vectors is worth noting. Recall that two vectors $\mathbf{u} = (u_1, u_2, \ldots, u_n)$ and $\mathbf{v} = (v_1, v_2, \ldots, v_n)$ in $\mathbb{R}^n$ are orthogonal if their *inner product* (or "dot product") is zero:

$$\langle \mathbf{u}, \mathbf{v} \rangle = \sum_{i=1}^{n} u_i v_i = 0.$$

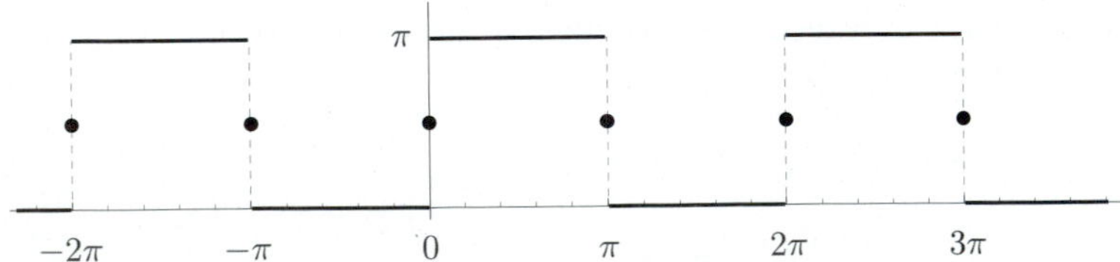

Figure 6

Furthermore, if $\mathbf{v}_1, \mathbf{v}_2, \ldots, \mathbf{v}_n$ are an orthogonal family of vectors in $\mathbb{R}^n$, then any vector $\mathbf{u}$ in $\mathbb{R}^n$ can be expressed as the sum of its *orthogonal projections* onto $\mathbf{v}_1, \mathbf{v}_2, \ldots, \mathbf{v}_n$:

$$\mathbf{u} = \sum_{i=1}^{n} \frac{\langle \mathbf{u}, \mathbf{v}_i \rangle}{\langle \mathbf{v}_i, \mathbf{v}_i \rangle} \mathbf{v}_i.$$

By defining the *inner product of two functions* u and v on $[-\ell, \ell]$ to be

$$\langle u, v \rangle = \int_{-\ell}^{\ell} u(x)v(x)dx,$$

we can analogously express the Fourier series of a function φ on $[-\ell, \ell]$ as

$$\underset{\sim}{\varphi} = \sum_{i=0}^{\infty} \frac{\langle \varphi, q_i \rangle}{\langle q_i, q_i \rangle} q_i,$$

where $q_1, q_2, q_3, \ldots$ are the orthogonal family of functions

$$1, \cos\frac{\pi x}{\ell}, \sin\frac{\pi x}{\ell}, \cos\frac{2\pi x}{\ell}, \sin\frac{2\pi x}{\ell}, \cos\frac{3\pi x}{\ell}, \sin\frac{3\pi x}{\ell}, \ldots.$$

PROBLEMS

For each function φ in Problems 1 through 8, (a) find $\underset{\sim}{\varphi}$, the Fourier series representation on $[-1, 1]$; (b) use a computer or graphing calculator to plot on $[-1, 1]$ the graph of the partial sum $\frac{a_0}{2} + \sum_{n=1}^{3}(a_n \cos\frac{n\pi x}{\ell} + b_n \sin\frac{n\pi x}{\ell})$; and (c) sketch the graph of $\underset{\sim}{\varphi}$ on $[-3, 3]$ (using Theorem 2).

1. $\varphi(x) = \begin{cases} 0, & \text{if } -1 \le x \le 1/2 \\ 1, & \text{if } 1/2 < x \le 1 \end{cases}$

2. $\varphi(x) = \begin{cases} 0, & \text{if } -1 \le x \le 0 \\ 1, & \text{if } 0 < x \le 1/2 \\ 0, & \text{if } 1/2 < x \le 1 \end{cases}$

3. $\varphi(x) = \begin{cases} 0, & \text{if } -1 \le x \le -1/2 \\ 1, & \text{if } -1/2 < x \le 1/2 \\ 0, & \text{if } 1/2 < x \le 1 \end{cases}$

4. $\varphi(x) = \begin{cases} 0, & \text{if } -1 \le x \le 0 \\ x, & \text{if } 0 < x \le 1 \end{cases}$

5. $\varphi(x) = x$

6. $\varphi(x) = |x|$

7. $\varphi(x) = \dfrac{x}{|x|}$

8. $\varphi(x) = e^{-|x|}$

For each function φ in Problems 9 through 12, let $\overline{\varphi}$ be the 2-periodic function that agrees with φ on $(-1, 1]$, and let $\underset{\sim}{\varphi}$ be the Fourier series representation of φ on $[-1, 1]$. Sketch the graphs of $\overline{\varphi}$ and $\underset{\sim}{\varphi}$ for $-3 \le x \le 3$, taking care to indicate the behavior at each discontinuity. (Do not compute the Fourier coefficients; use Theorem 2 to sketch the graph of $\underset{\sim}{\varphi}$.)

9. $\varphi(x) = \dfrac{x}{|x|} - x$

10. $\varphi(x) = \sin \pi x - \dfrac{x}{|x|}$

11. $\varphi(x) = x^3$

12. $\varphi(x) = \dfrac{x}{2}(x + |x|)$

13. Prove each of the properties listed in Theorem 1 directly by means of elementary integration techniques.

14. This problem concerns an alternative proof of the orthogonality properties stated in Theorem 1.

(a) Suppose that u and v are nonzero solutions on $[-\ell, \ell]$ of $u'' = \lambda u$ $v'' = \mu v$, respectively, where $\lambda \neq \mu$. Integrate by parts to show that

$$\int_{-\ell}^{\ell} (vu'' - uv'')dx = (u'v - uv')|_{-\ell}^{\ell}.$$

Then conclude that

$$\int_{-\ell}^{\ell} uv\,dx = \frac{1}{\lambda - \mu}(u'v - uv')|_{-\ell}^{\ell}.$$

(b) Let u and v be as in (a). Use the result of (a) to show that, if u and v are 2ℓ-periodic, then they are orthogonal on $[-\ell, \ell]$. (See Problems 15a and b.)

(c) Use the result of (b) to show that each of parts (ii), (iii), and (iv) of Theorem 1 is true when $m \neq n$. Finally, use the double-angle formula for sine to prove part (ii) in the case where $m = n$.

15. **(a)** Use the definition of the derivative to prove that, if a differentiable function f is p-periodic, then so is f'.

(b) Show that, if f and g are each p-periodic functions, then so is fg, and so is any linear combination $\alpha f + \beta g$.

(c) Suppose that f is p-periodic and g is q-periodic. Show that, if there are integers m and n such that $mp = nq$, then $f + g$ is mp-periodic.

(d) Find the smallest period of $f(x) = \sin 7x + \cos \frac{3x}{11}$.

(e) Prove that $f(x) = \cos x + \cos \pi x$ is not a periodic function. (*Hint*: Consider the equation $f(p) = f(0)$, which would be true if f were p-periodic.)

16. Use the Fourier series derived in Example 2 to show that

$$1 - \frac{1}{3} + \frac{1}{5} - \frac{1}{7} + \cdots = \frac{\pi}{4}.$$

17. Find the Fourier series representation of $|x|$ on $[-1, 1]$ and use it to show that

$$1 + \frac{1}{3^2} + \frac{1}{5^2} + \frac{1}{7^2} + \cdots = \frac{\pi^2}{8}.$$

18. The *Legendre polynomials*, P_0, P_1, $P_2, \ldots$, comprise an orthogonal family of functions on $[-1, 1]$. (See Problems 1 and 2 in Section 5.7.) The first four Legendre polynomials (scaled so that $P_n(1) = 1$) are

$$P_0(x) = 1, \quad P_1(x) = x, \quad P_2(x) = \frac{1}{2}(3x^2 - 1), \quad P_3(x) = \frac{1}{2}(5x^3 - 3x).$$

Let f be a given function defined on $[-1, 1]$, and suppose that we wish to approximate f with a polynomial p of degree N or less, in such a way that

$$\int_{-1}^{1} f(x)q(x)dx = \int_{-1}^{1} p(x)q(x)dx$$

for *all* polynomials q of degree N or less. Since any polynomial of degree N or less can be written as a linear combination of the first $N + 1$ Legendre polynomials, we will look for p the form

$$p(x) = \sum_{n=0}^{N} c_n P_n(x)$$

and choose the coefficients $c_0, c_1, c_2, \ldots$ so that

$$\int_{-1}^{1} f(x)P_m(x)dx = \int_{-1}^{1} p(x)P_m(x)dx, \quad m = 0, 1, 2, \ldots, N.$$

(a) Show that this plan produces coefficients

$$c_n = \frac{\displaystyle\int_{-1}^{1} f(x)P_n(x)dx}{\displaystyle\int_{-1}^{1} P_n(x)^2dx}, \quad n = 0, 1, 2, \ldots, N.$$

(b) Find a cubic polynomial approximation on $[-1, 1]$ for the function

$$f(x) = \begin{cases} 0, & \text{if } -1 \le x < 0; \\ 1, & \text{if } 0 \le x \le 1. \end{cases}$$

Sketch its graph along with the graph of f.

11.4 Fourier Sine and Cosine Series

Recall that in Section 11.2 we solved the diffusion problem

$$\left.\begin{array}{rl} w_t - kw_{xx} = 0 & \text{for } 0 < x < \ell, t > 0 \\ w(0, x) = \varphi(x) & \text{for } 0 \le x \le \ell \\ w(t, 0) = w(t, \ell) = 0 & \text{for } t > 0 \end{array}\right\} \tag{1}$$

in the case where φ is expressed as a Fourier series of the form

$$\underset{\sim}{\varphi}(x) = \sum_{n=1}^{\infty} b_n \sin \frac{n\pi x}{\ell}. \tag{2}$$

The solution of (1) then is

$$w(t, x) = \sum_{n=1}^{\infty} b_n e^{-n^2\pi^2 kt/\ell^2} \sin \frac{n\pi x}{\ell}.$$

For the analogous problem with homogeneous Neumann (i.e., zero-flux) boundary conditions,

$$\left.\begin{array}{rl} w_t - kw_{xx} = 0 & \text{for } 0 < x < \ell, t > 0 \\ w(0, x) = \varphi(x) & \text{for } 0 \le x \le \ell \\ w_x(t, 0) = w_x(t, \ell) = 0 & \text{for } t > 0, \end{array}\right\} \tag{3}$$

the desired Fourier series representation of φ is

$$\underset{\sim}{\varphi}(x) = \sum_{n=0}^{\infty} a_n \cos \frac{n\pi x}{\ell}, \tag{4}$$

and the consequent solution of (3) is

$$w(t, x) = \sum_{n=1}^{\infty} a_n e^{-n^2\pi^2 kt/\ell^2} \cos \frac{n\pi x}{\ell},$$

Figure 1a **Figure 1b**

since each of $\cos \frac{n\pi x}{\ell}$, $n = 0, 1, 2, \ldots$, satisfies the differential equation and the boundary conditions. (See Problem 7 in Section 11.2.)

Our goal in the present section is learn how to find Fourier series of the forms in (2) and (4) for a function φ defined on $[0, \ell]$. We refer to these as Fourier *sine* and *cosine* series, respectively. Since the functions φ we are considering are defined only on $[0, \ell]$, we are free to extend the definition of φ to the interval $[-\ell, \ell]$ by defining φ on $[-\ell, 0)$ however we wish. Indeed, the key to obtaining a Fourier sine or cosine series is the manner in which we extend the definition of φ to $[-\ell, \ell]$.

Odd Functions and Even Functions Recall that a function g defined on an interval of the form $[-\ell, \ell]$ is

- **odd** if $g(-x) = -g(x)$ for all x in $[-\ell, \ell]$;
- **even** if $g(-x) = g(x)$ for all x in $[-\ell, \ell]$.

For example, basic properties of sine and cosine tell us that $\sin \omega x$ is an odd function, and $\cos \omega x$ is an even function (each for any ω and on any interval $[-\ell, \ell]$). Recall also that the graph of an even function is symmetric about the y-axis, and the graph of an odd function is symmetric about the origin. These symmetries are illustrated in Figures 1a and b, where the graph of an odd function is shown in Figure 1a, and the graph of an even function is shown in Figure 1b.

Further important properties of odd functions and even functions are as follows. Verification of these properties is the subject of Problem 7.

Theorem 1

i) If f is odd and integrable on $[-\ell, \ell]$, then

$$\int_{-\ell}^{\ell} f(x)\,dx = 0.$$

ii) If f is even and integrable on $[-\ell, \ell]$, then

$$\int_{-\ell}^{\ell} f(x)\,dx = 2\int_{0}^{\ell} f(x)\,dx.$$

iii) If f and g are two functions that are either even or odd, then the product fg is also even or odd according to the following table.

f	g	fg
odd	*odd*	*even*
odd	*even*	*odd*
even	*even*	*even* ▲

Let us now consider the Fourier series representation of a function φ on $[-\ell, \ell]$, where φ is either even or odd. Recall that the Fourier series representation of any function φ on $[-\ell, \ell]$ is defined by

$$\underset{\sim}{\varphi}(x) = \frac{a_0}{2} + \sum_{n=1}^{\infty}\left(a_n \cos\frac{n\pi x}{\ell} + b_n \sin\frac{n\pi x}{\ell}\right),$$

where

$$a_n = \frac{1}{\ell}\int_{-\ell}^{\ell} \varphi(x)\cos\frac{n\pi x}{\ell}\,dx, \quad n = 0, 1, 2, \ldots$$

and

$$b_n = \frac{1}{\ell}\int_{-\ell}^{\ell} \varphi(x)\sin\frac{n\pi x}{\ell}\,dx, \quad n = 1, 2, 3, \ldots.$$

If φ is even, then $\varphi(x)\cos\frac{n\pi x}{\ell}$ is even and $\varphi(x)\sin\frac{n\pi x}{\ell}$ is odd for all n. Thus

$$a_n = \frac{2}{\ell}\int_{0}^{\ell} \varphi(x)\cos\frac{n\pi x}{\ell}\,dx, \quad n = 0, 1, 2, \ldots$$

and

$$b_n = 0, \quad n = 1, 2, 3, \ldots.$$

If φ is odd, then $\varphi(x)\cos\frac{n\pi x}{\ell}$ is odd and $\varphi(x)\sin\frac{n\pi x}{\ell}$ is even for all n. Thus

$$a_n = 0, \quad n = 1, 2, 3, \ldots$$

and

$$b_n = \frac{2}{\ell}\int_{0}^{\ell} \varphi(x)\sin\frac{n\pi x}{\ell}\,dx, \quad n = 1, 2, 3, \ldots.$$

The upshot of all this is that the Fourier series of an even function contains no nonzero sine terms, the Fourier series of an odd function contains no nonzero cosine terms, and in either of those cases the nonzero coefficients can be expressed as integrals over $[0, \ell]$. For reference we state the details of these conclusions as follows:

Theorem 2

If φ is an even function on $[-\ell, \ell]$, then the Fourier series representation of φ on $[-\ell, \ell]$ is given by

$$\underset{\sim}{\varphi}(x) = \frac{a_0}{2} + \sum_{n=1}^{\infty} a_n \cos\frac{n\pi x}{\ell},$$

where

$$a_n = \frac{2}{\ell}\int_0^{\ell} \varphi(x)\cos\frac{n\pi x}{\ell}\,dx, \quad n = 0, 1, 2, \ldots .$$

If φ is an odd function on $[-\ell, \ell]$, then the Fourier series representation of φ on $[-\ell, \ell]$ is given by

$$\underset{\sim}{\varphi}(x) = \sum_{n=1}^{\infty} b_n \sin\frac{n\pi x}{\ell},$$

where

$$b_n = \frac{2}{\ell}\int_0^{\ell} \varphi(x)\sin\frac{n\pi x}{\ell}\,dx, \quad n = 1, 2, 3, \ldots .$$ ▲

Functions on $[0, \ell]$ Suppose that φ is a given function defined on $[0, \ell]$. Then φ determines an *even extension* φ_e on $[-\ell, \ell]$ defined by

$$\varphi_e(x) = \begin{cases} \varphi(-x), & \text{if } -\ell \le x < 0; \\ \varphi(x), & \text{if } 0 \le x \le \ell. \end{cases}$$

We similarly define an *odd extension* φ_o of φ by

$$\varphi_o(x) = \begin{cases} -\varphi(-x), & \text{if } -\ell \le x < 0; \\ 0, & \text{if } x = 0; \\ \varphi(x), & \text{if } 0 < x \le \ell. \end{cases}$$

Example 1

Consider the function φ defined on [0, 1] by

$$\varphi(x) = x \quad \text{for } 0 \le x \le 1.$$

The graphs of the even extension φ_e and the odd extension φ_o on $[-1, 1]$ are shown in Figures 2a and b, respectively.

Given a function φ defined on $[0, \ell]$, the even periodic extension φ_e and the odd periodic extension φ_o provide us with two distinct Fourier series representations of φ on $[0, \ell]$, which we define as follows. ■

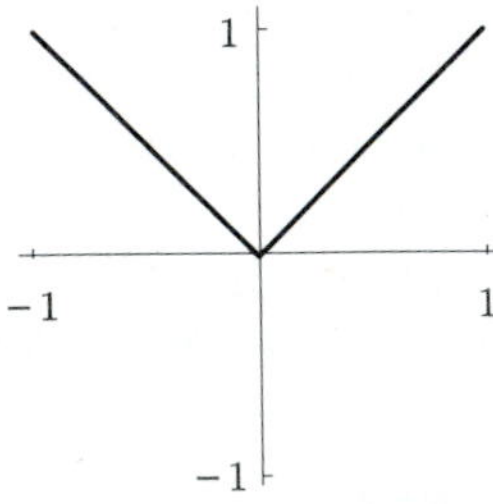

Figure 2a

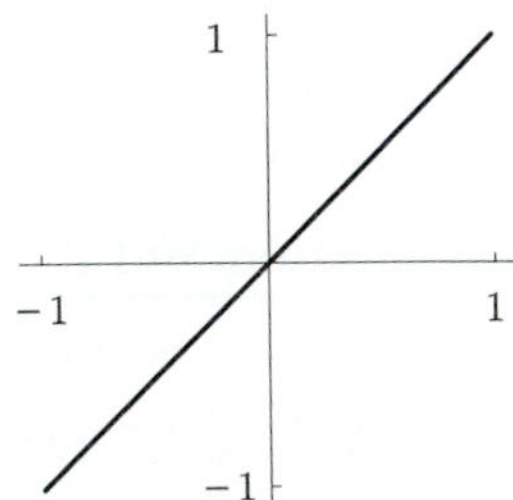

Figure 2b

Definition Let φ be defined on $[0, \ell]$, and let φ_e and φ_o be the even and odd extensions, respectively, of φ on $[-\ell, \ell]$. The **Fourier cosine series** representation of φ on $[0, \ell]$ is the Fourier series representation of φ_e on $[-\ell, \ell]$:

$$\varphi_{\cos}(x) = \varphi_{\underset{\sim}{e}}(x) = \frac{a_0}{2} + \sum_{n=1}^{\infty} a_n \cos\frac{n\pi x}{\ell},$$

where

$$a_n = \frac{2}{\ell}\int_0^\ell \varphi(x)\cos\frac{n\pi x}{\ell}\,dx, \quad n = 0, 1, 2, \ldots.$$

The **Fourier sine series** representation of φ on $[0, \ell]$ is the Fourier series representation of φ_o on $[-\ell, \ell]$:

$$\varphi_{\sin}(x) = \varphi_{\underset{\sim}{o}}(x) = \sum_{n=1}^{\infty} b_n \sin\frac{n\pi x}{\ell},$$

where

$$b_n = \frac{2}{\ell}\int_0^\ell \varphi(x)\sin\frac{n\pi x}{\ell}\,dx, \quad n = 1, 2, 3, \ldots. \qquad \blacklozenge$$

In the following example, we derive the Fourier cosine and sine series for a specific function on the interval [0, 1].

Example 2

Consider the function φ defined on [0, 1] by

$$\varphi(x) = x^2, \quad \text{for } 0 \le x \le 1.$$

The coefficients in the Fourier cosine series $\varphi_{\cos}$ are

$$a_n = 2\int_0^1 x^2\cos(n\pi x)dx, \quad n = 0, 1, 2, \ldots.$$

Integration—by parts or by *CAS*—reveals that

$$a_0 = \frac{2}{3} \quad \text{and} \quad a_n = \frac{4(-1)^n}{n^2\pi^2}, \quad n = 1, 2, 3, \ldots.$$

Therefore,

$$\varphi_{\cos}(x) = \frac{1}{2} + \frac{4}{\pi^2}\sum_{n=1}^{\infty}\frac{(-1)^n}{n^2}\cos(n\pi x).$$

The coefficients in the Fourier sine series $\varphi_{\sin}$ are

$$b_n = 2\int_0^1 x^2\sin(n\pi x)dx, \quad n = 1, 2, 3, \ldots.$$

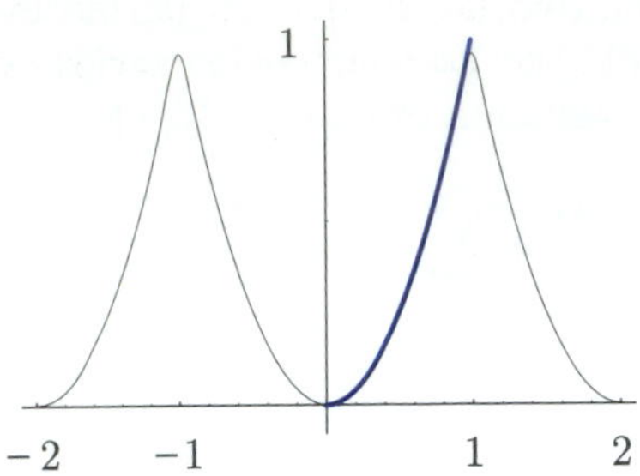

Figure 3a

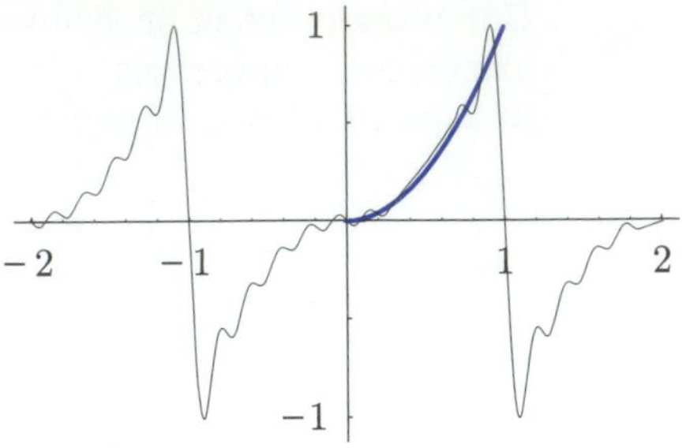

Figure 3b

Integration—again by parts or by *CAS*—reveals that

$$b_n = 4\frac{(-1)^n - 1}{n^3\pi^3} + \frac{(-1)^n}{n\pi}, \qquad n = 1, 2, 3, \ldots .$$

Therefore,

$$\varphi_{\sin}(x) = \sum_{n=1}^{\infty} \left(4\frac{(-1)^n - 1}{n^3\pi^3} + \frac{(-1)^n}{n\pi} \right) \sin(n\pi x).$$

Figure 3a shows the graph of the tenth cosine-series partial sum

$$\frac{1}{2} + \frac{4}{\pi^2} \sum_{n=1}^{10} \frac{(-1)^n}{n^2} \cos(n\pi x),$$

and Figure 3b shows the graph of the tenth sine-series partial sum

$$\sum_{n=1}^{10} \left(4\frac{(-1)^n - 1}{n^3\pi^3} + \frac{(-1)^n}{n\pi} \right) \sin(n\pi x).$$

It is apparent from Figures 3a and b that the tenth cosine-series partial sum gives a much better approximation of $\varphi(x)$ on $[0, 1]$ than does the tenth partial sum of the sine series. Note that this is to be expected, since the coefficients in the sine series are approximately proportional to $1/n$ for large n and thus decay more slowly than those in the cosine series, which are proportional to $1/n^2$. ■

Example 3

Consider the diffusion problem with zero-flux boundary conditions:

$$\begin{aligned} w_t - w_{xx} &= 0 && \text{for } 0 < x < 1, t > 0, \\ w(0, x) &= \sin(\pi x) && \text{for } 0 \le x \le 1, \\ w_x(t, 0) = w_x(t, 1) &= 0 && \text{for } t > 0. \end{aligned}$$

Because of the nature of the boundary conditions, we need to express the initial data as a Fourier cosine series on [0, 1]. The necessary coefficients for this purpose are

$$a_n = 2\int_0^1 \cos(n\pi x)\sin \pi x\, dx, \quad n = 0, 1, 2, \ldots .$$

CAS-aided computation reveals that

$$a_{2i} = \frac{-4}{(4i^2-1)\pi} \quad \text{and} \quad a_{2i+1} = 0, i = 0, 1, 2, \ldots .$$

Thus the Fourier cosine series for $\sin(\pi x)$ on [0, 1] is

$$\varphi_{\cos}(x) = \frac{2}{\pi} - \frac{4}{\pi}\sum_{n=1}^{\infty}\frac{\cos(2n\pi x)}{4n^2-1}.$$

Therefore, the solution of our diffusion problem can be expressed as

$$w(x,t) = \frac{2}{\pi} - \frac{4}{\pi}\sum_{n=1}^{\infty}\frac{e^{-4n^2\pi^2 t}\cos(2n\pi x)}{4n^2-1}.$$ ■

Figures 4a and b* illustrate this solution. Figure 4a shows the graphs of the solution (versus x) at each of $t = 0$, 0.01,0.02, and 0.05, while Figure 4b is a surface plot of $w(x,t)$ for $0 \le x \le 1$ and $0 \le t \le 0.75$. A couple of things are worth noting here. It is easy to see—both from the series representation of the solution and from Figure 4a—that the solution approaches the constant $2/\pi$ as $t \to \infty$. (This constant is the average value of the initial data. See Problem 16.) Also, we note that the initial data do not satisfy the boundary conditions, yet the solution does satisfy the (zero-flux) boundary conditions for all $t > 0$. This behavior is typical of diffusion problems: A solution will satisfy the boundary conditions for all $t > 0$, but it need not satisfy the boundary conditions at $t = 0$.

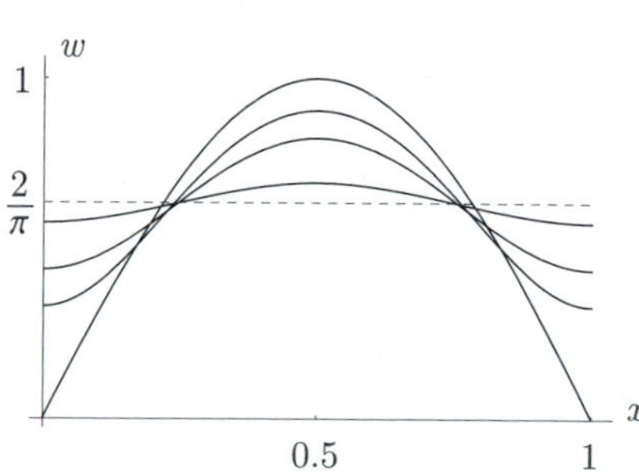

Figure 4a

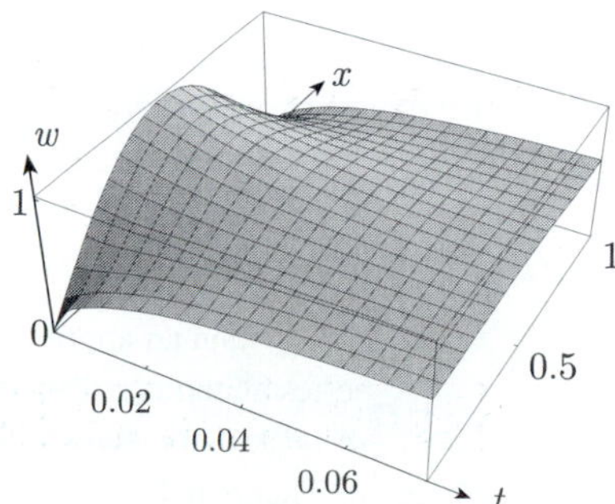

Figure 4b

* These plots were generated with the 21-term partial sum:
$\frac{2}{\pi} - \frac{4}{\pi}\left(\frac{1}{3}e^{-4\pi^2 t}\cos(2\pi x) + \frac{1}{15}e^{-16\pi^2 t}\cos(4\pi x) + \cdots + \frac{1}{399}e^{-400\pi^2 t}\cos(20\pi x)\right).$

PROBLEMS

For each function φ in Problems 1 through 4, (a) sketch the graph of the odd extension φ_o on $[-1, 1]$; (b) find $\varphi_{\sin}$, the Fourier sine series representation on $[0, 1]$; (c) with the help of a computer or graphing calculator, sketch the graph of the partial sum with three nonzero terms on $[-1, 1]$; and (d) sketch the graph of $\varphi_{\sin}$ on $[-1, 1]$.

1. $\varphi(x) = x$

2. $\varphi(x) = \begin{cases} 0, & \text{if } 0 \le x < 1/2 \\ 1, & \text{if } 1/2 \le x \le 1 \end{cases}$

3. $\varphi(x) = \begin{cases} 1, & \text{if } 0 \le x < 1/2 \\ 0, & \text{if } 1/2 \le x \le 1 \end{cases}$

4. $\varphi(x) = \begin{cases} -1, & \text{if } 0 \le x < 1/2 \\ 1, & \text{if } 1/2 \le x \le 1 \end{cases}$

For each function φ in Problems 5 through 8, (a) sketch the graph of the even extension φ_e on $[-1, 1]$; (b) find $\varphi_{\cos}$, the Fourier cosine series representation on $[0, 1]$; (c) with the help of a computer or graphing calculator, sketch the graph of the partial sum with three nonzero terms on $[-1, 1]$; and (d) sketch the graph of $\varphi_{\cos}$ on $[-1, 1]$.

5. φ from Problem 1

6. φ from Problem 2

7. φ from Problem 3

8. φ from Problem 4

9. Find the Fourier sine series representation of $\varphi(x) = \cos x$ on $[0, \pi]$.

10. Find the Fourier cosine series representation of $\varphi(x) = \sin x$ on $[0, \pi]$.

11. Find the Fourier sine series representation of $\varphi(x) = \sin x$ on $[0, \pi]$.

12. Find the Fourier sine series representation of $\varphi(x) = \sin x$ on $[0, \frac{\pi}{2}]$.

13. **(a)** Find the solution of the diffusion problem

$$\begin{aligned} w_t - w_{xx} &= 0 && \text{for } 0 < x < 1, t > 0, \\ w(0, x) &= 1 && \text{for } 0 \le x \le 1, \\ w(t, 0) = w(t, 1) &= 0 && \text{for } t > 0. \end{aligned}$$

(b) Obtain an approximation to $w(0.01, x)$ on $[0, 1]$ by discarding from its Fourier sine series all terms with coefficients less than 0.005 in absolute value. Graph this approximation to $w(0.01, x)$. Repeat for $w(0.1, x)$.

14. **(a)** Find the solution of the diffusion problem

$$\begin{aligned} w_t - w_{xx} &= 0 && \text{for } 0 < x < 1, t > 0, \\ w(0, x) &= \begin{cases} 0, & \text{if } 0 \le x < 1/2 \\ 1, & \text{if } 1/2 \le x \le 1 \end{cases}, \\ w_x(t, 0) = w_x(t, 1) &= 0 && \text{for } t > 0. \end{aligned}$$

(b) Obtain an approximation to $w(0.01, x)$ on $[0, 1]$ by discarding from its Fourier cosine series all terms with coefficients less than 0.005 in absolute value. Graph this approximation to $w(0.01, x)$. Repeat for $w(0.1, x)$.

15. Prove Theorem 1.

16. Consider the diffusion problem with homogeneous Neumann boundary conditions:

$$\begin{aligned} w_t - w_{xx} &= 0 && \text{for } 0 < x < \ell, t > 0, \\ w(0, x) &= \varphi(x) && \text{for } 0 \le x \le \ell, \\ w_x(t, 0) = w_x(t, \ell) &= 0 && \text{for } t > 0. \end{aligned}$$

(a) Let $\overline{w}(t)$ be the average value of w on $[0, \ell]$ at time t:

$$\overline{w}(t) = \frac{1}{\ell}\int_0^\ell w(t, x)dx \quad \text{for all } t \geq 0.$$

Integrate each side of the differential equation over $[0, \ell]$ and apply the boundary conditions to show that $\overline{w}(t)$ is constant; that is, show that

$$\overline{w}'(t) = 0 \quad \text{for all } t > 0.$$

(Assume that differentiation through the integral sign is valid, that is, $\frac{d}{dt}\int_0^\ell w\,dx = \int_0^\ell w_t\,dx$.) Thus conclude that

$$\overline{w}(t) = \frac{1}{\ell}\int_0^\ell \varphi(x)dx \quad \text{for all } t \geq 0.$$

(b) Use the Fourier cosine series representation of φ on $[0, \ell]$ to write the solution w. Then show that, as $t \to \infty$, w converges *pointwise* to the average value of φ; that is, show that

$$\lim_{t\to\infty} w(t, x) = \frac{1}{\ell}\int_0^\ell \varphi(x)dx \quad \text{for all } x \text{ in } [0, \ell].$$

17. Consider the diffusion problem with homogeneous Dirichlet boundary conditions:

$$\begin{aligned} w_t - w_{xx} &= 0 && \text{for } 0 < x < \ell, t > 0, \\ w(0, x) &= \varphi(x) && \text{for } 0 \leq x \leq \ell, \\ w(t, 0) = w(t, \ell) &= 0 && \text{for } t > 0. \end{aligned}$$

Use the Fourier sine series representation of φ on $[0, \ell]$ to write the solution w. Then show that, as $t \to \infty$, w converges *pointwise* to zero; that is, show that

$$\lim_{t\to\infty} w(t, x) = 0 \quad \text{for all } x \text{ in } [0, \ell].$$

18. Use the Fourier cosine series for x^2 from Example 2 to show that

$$\frac{\pi^2}{8} = 1 + \frac{1}{3^2} + \frac{1}{5^2} + \frac{1}{7^2} + \cdots.$$

19. Use the Fourier cosine series for $\sin \pi x$ from Example 3 to show that

$$\frac{\pi}{4} = \frac{1}{2} - \sum_{n=1}^{\infty}\frac{(-1)^n}{4n^2 - 1} = \frac{1}{2} + \frac{1}{3} - \frac{1}{15} + \frac{1}{35} - \frac{1}{63} + \cdots.$$

11.5 Sturm-Liouville Eigenvalue Problems

We begin this section by considering a general, linear, homogeneous, diffusion problem (cf. Section 11.1) of the form

$$\left.\begin{aligned} w_t - (kw_x)_x = 0 \quad & \text{for } a < x < b, t > 0, \\ w(0, x) = \varphi(x) \quad & \text{for } a \leq x \leq b, \\ \alpha w(t, a) - (1 - \alpha)w_x(t, a) = 0 \quad & \text{for } t > 0, \\ \beta w(t, b) + (1 - \beta)w_x(t, b) = 0 \quad & \text{for } t > 0, \end{aligned}\right\} \tag{1}$$

under the "usual" assumptions:

- k is positive and continuously differentiable on $[a, b]$;
- φ is piecewise continuous on $[a, b]$;
- α and β are constants in $[0, 1]$.

Our ultimate goal is to derive the solution of (1) based upon a *generalized* Fourier series representation for the initial data $\varphi(x)$.

It will serve the overall theme of this section for us to introduce the operator $\mathcal{S}$ defined by

$$\mathcal{S}\psi = (k\psi')'$$

for all twice-continuously differentiable functions ψ on $[a, b]$. (For brevity, we will henceforth refer to these functions as the C^2 *functions* on $[a, b]$.) Since for each $t > 0$ the solution w of (1) is such a function on $[a, b]$, it makes sense to write the differential equation in (1) as

$$w_t - \mathcal{S}w = 0,$$

with the understanding that in this context $\mathcal{S}w$ means $(kw_x)_x$; that is, derivatives are partial derivatives with respect to x.

Either separation of variables (as discussed in Section 11.2) or direct substitution reveals that the differential equation in (1) is satisfied by

$$w(t, x) = e^{\lambda t} X(x),$$

provided that λ is a constant and X satisfies

$$\mathcal{S}X = \lambda X.$$

If, in addition, X satisfies the boundary conditions in (1); that is, if

$$\alpha X(a) - (1 - \alpha)X'(a) = 0,$$
$$\beta X(b) + (1 - \beta)X'(b) = 0,$$

then $w = e^{\lambda t} X$ will satisfy both the differential equation and the boundary conditions in (1). Thus we seek nontrivial solutions of

$$\left.\begin{array}{l} \mathcal{S}X = \lambda X \quad \text{for } a < x < b, \\ \alpha X(a) - (1 - \alpha)X'(a) = 0, \\ \beta X(b) + (1 - \beta)X'(b) = 0. \end{array}\right\} \tag{2}$$

A problem such as (2) is called an *eigenvalue problem** for the operator $\mathcal{S}$. Any constant λ for which (2) has a nontrivial solution is called an **eigenvalue** of the operator $\mathcal{S}$ (with respect to the stated boundary conditions). A corresponding nontrivial solution X is called an **eigenfunction** of $\mathcal{S}$ (also with respect to the stated boundary conditions).

---

* Note the similarity to the matrix eigenvalue problem: $A\mathbf{x} = \lambda\mathbf{x}$, where A is an $n \times n$ matrix and $\mathbf{x}$ is a column vector of length n.

Example 1

Recall that in Section 11.2 we encountered the eigenvalue problem

$$kX'' = \lambda X \quad \text{for } 0 < x < \ell,$$
$$X(0) = X(\ell) = 0,$$

which was found to possess a nontrivial solution if and only if

$$\lambda = \lambda_n = -\frac{n^2\pi^2 k}{\ell^2} \quad \text{for some } n = 1, 2, 3, \ldots .$$

A nontrivial solution corresponding to each λ_n was determined to be

$$X_n(x) = \sin\left(\frac{n\pi x}{\ell}\right).$$

Thus, for the operator $\mathcal{S}$ defined by

$$\mathcal{S}\psi = k\psi'',$$

the described λ_n, $n = 1, 2, 3, \ldots$, are the eigenvalues of $\mathcal{S}$ with respect to the boundary conditions $X(0) = X(\ell) = 0$. Corresponding eigenfunctions are X_n, $n = 1, 2, 3, \ldots$, as described previously. ■

Example 2

Consider (1) with constant diffusivity k, $\alpha = 1/2$, and $\beta = 0$. Then (2) becomes

$$kX'' = \lambda X \quad \text{for } 0 < x < \ell,$$
$$X(0) - X'(0) = X'(\ell) = 0.$$

We first observe that nontrivial solutions exist only if $\lambda < 0$. (See Problem 12.) Setting $\omega = \sqrt{-\lambda/k}$ for economy of notation, we conclude that all nontrivial solutions will be of the form

$$X(x) = A\cos\omega x + B\sin\omega x$$

where A and B are constants. The boundary conditions then require that A and B satisfy

$$A - \omega B = 0 \quad \text{and} \quad -\omega A\sin\omega\ell + \omega B\cos\omega\ell = 0.$$

If we set $A = \omega B$, then the first of these equations is satisfied, and the second becomes

$$-\omega^2 B\sin\omega\ell + \omega B\cos\omega\ell = 0.$$

We dismiss the case of $B = 0$, since that leads to the trivial solution $X = 0$. Consequently, the desired values of ω satisfy the transcendental equation

$$\omega = \cot\omega\ell.$$

Figure 1 shows the graphs of ω and $\cot\omega\ell$ versus ω and reveals that there is a sequence of solutions $\omega_0, \omega_1, \omega_2, \ldots$, for which

$$n\pi < \omega_n < (n+1)\pi \quad \text{for all } n = 0, 1, 2, \ldots$$

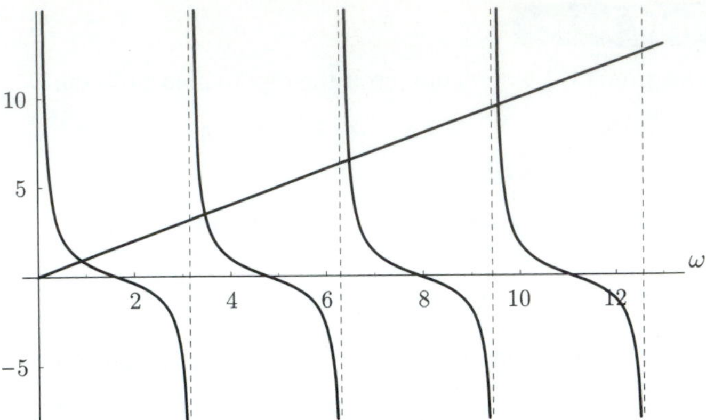

Figure 1

and

$$\frac{\omega_n}{n\pi} \to 1 \quad \text{as } n \to \infty.$$

Rounded to three decimal places, the first four of these are

$$\omega_0 = 0.860, \quad \omega_1 = 3.426, \quad \omega_2 = 6.437, \quad \omega_3 = 9.529.$$

The result is that, with respect to the stated boundary conditions, the operator $\mathcal{S}$ defined by $\mathcal{S}\psi = k\psi''$ has eigenvalues $\lambda_n = -k\omega_n^2$, $n = 0, 1, 2, 3, \ldots$, where the ω_n are as previously indicated. Moreover, corresponding eigenfunctions are given by

$$X_n(x) = B_n(\omega_n \cos \omega_n x + \sin \omega_n x),$$

where B_n is any nonzero constant. The constants B_n may be chosen to *scale* the eigenfunctions as desired. An important type of scaling, done to achieve

$$\int_0^\ell (X_n(x))^2 \, dx = 1, \quad n = 0, 1, 2, 3, \ldots,$$

uses

$$B_n = \left(\int_0^\ell (\omega_n \cos \omega_n x + \sin \omega_n x)^2 \, dx \right)^{-1/2}.$$

The resulting eigenfunctions are

$$X_n(x) = \frac{\omega_n \cos \omega_n x + \sin \omega_n x}{\sqrt{\displaystyle\int_0^\ell (\omega_n \cos \omega_n x + \sin \omega_n x)^2 \, dx}},$$

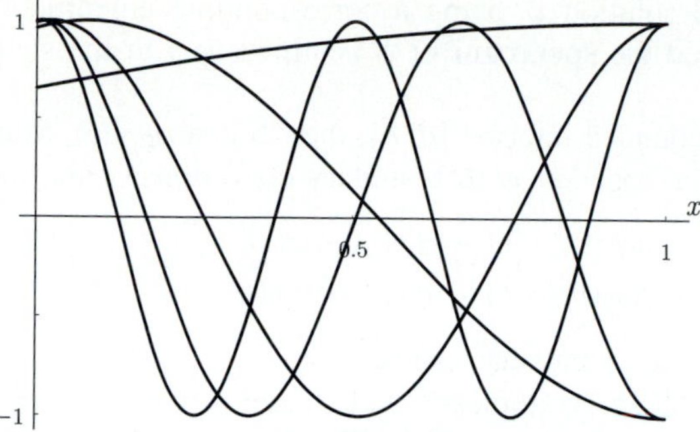

Figure 2

The first four of these are plotted in Figure 2. With coefficients rounded to four significant digits, they are

$$\begin{aligned}X_0(x) &= 0.8543(0.8603\cos(0.8603x) - \sin(0.8603x)),\\ X_1(x) &= 0.3816(3.426\cos(3.426x) - \sin(3.426x)),\\ X_2(x) &= 0.2146(6.437\cos(6.437x) - \sin(6.437x)),\\ X_3(x) &= 0.1468(9.529\cos(9.529x) - \sin(9.529x)).\end{aligned}$$ ■

Eigenvalue problems such as (2) comprise a subset of a class of problems known as *Sturm-Liouville*[†] *eigenvalue problems*. In order to give a general description of these problems, let us first define what we mean by "Sturm-Liouville operator." Let I be an interval, and suppose that the functions k, p, and q are such that

- k is positive and continuously differentiable on I;
- p is positive and continuous on I;
- q is continuous on I.

An operator $\mathcal{S}$, acting on $\mathcal{C}^2$ functions ψ on I, for which $\mathcal{S}\psi$ may be expressed as

$$\mathcal{S}\psi = \frac{1}{p}\big((k\psi')' + q\psi\big), \tag{3}$$

is a **Sturm-Liouville operator**. We note that Theorem 1 of Section 5.4 guarantees that, for *any* constant λ, the differential equation

$$\mathcal{S}\psi = \lambda\psi \tag{4a}$$

possesses a pair of linearly independent solutions on I, linear combinations of which provide the general solution on I.

Equation (4a) together with some suitable set of boundary conditions is a **Sturm-Liouville eigenvalue problem**. Any λ for which the problem possesses a nontrivial solution ψ is said to be an eigenvalue of $\mathcal{S}$ with respect to the stated boundary

[†] The work of the French mathematicians Charles Sturm and Joseph Liouville was published in the 1830s. (Note carefully the order of the vowels in "Liouville.")

conditions, the nontrivial solution ψ being a corresponding eigenfunction. The set of all eigenvalues is called the **spectrum** of $\mathcal{S}$ (with respect to the stated boundary conditions).

If I is a closed and bounded interval $[a, b]$, then $\mathcal{S}$ is a *regular* Sturm-Liouville operator, and equation (4a) together with boundary conditions of the form

$$\alpha\psi(a) - (1-\alpha)\psi'(a) = 0,$$
$$\beta\psi(b) + (1-\beta)\psi'(b) = 0, \tag{4b}$$

is a *regular* Sturm-Liouville eigenvalue problem.

Clearly the operator $\mathcal{S}$ in Examples 1 and 2, defined by $\mathcal{S}\psi = k\psi''$ for C^2 functions ψ on $[0, \ell]$, is a regular Sturm-Liouville operator (with $p = 1$ and $q = 0$), and the eigenvalue problems solved in those examples are regular Sturm-Liouville eigenvalue problems. The following example suggests that Sturm-Liouville operators are much wider in scope. (See also Problem 10.)

Example 3

Let the operator $\mathcal{S}$ be defined for C^2 functions ψ on $[0, 1]$ by

$$\mathcal{S}\psi = \psi'' + 4\psi' + \psi.$$

A simple computation shows that

$$\mathcal{S}\psi = e^{-4x}((e^{4x}\psi')' + e^{4x}\psi).$$

Thus $\mathcal{S}$ is a regular Sturm-Liouville operator as described by (3) with

$$p(x) = k(x) = q(x) = e^{4x},$$

and Problem (4a,b) is a regular Sturm-Liouville eigenvalue problem. Let us proceed now to solve this problem in the case where $\alpha = \beta = 1$:

$$\psi'' + 4\psi' + \psi = \lambda\psi \quad \text{for } 0 < x < 1,$$
$$\psi(0) = \psi(1) = 0.$$

First we write the differential equation as

$$\psi'' + 4\psi' + (1-\lambda)\psi = 0.$$

The associated characteristic equation is $z^2 + 4z + 1 - \lambda = 0$, whose solutions are are $-2 \pm \sqrt{3+\lambda}$. So, by the methods of Chapter 5, we find that ψ must be of the form

$$\psi(x) = e^{-2x}(Ae^{\eta x} + Be^{-\eta x}),$$

where $\eta = \sqrt{3+\lambda}$, and A and B are arbitrary (complex) constants. The boundary condition $\psi(0) = 0$ requires that $B = -A$; so

$$\psi(x) = Ae^{-2x}(e^{\eta x} - e^{-\eta x}).$$

The boundary condition $\psi(1) = 0$ now requires that

$$e^{\eta} = e^{-\eta},$$

whose only real solution, $\eta = 0$, leads to $\psi = 0$. Consequently, η must be imaginary. Thus we write $\eta = i\omega$, where $\omega = \sqrt{-3 - \lambda}$ is real, and note that

$$\begin{aligned}\psi(x) &= Ae^{-2x}(e^{i\omega x} - e^{-i\omega x}) \\ &= 2iAe^{-2x}\sin\omega x \\ &= Ce^{-2x}\sin\omega x,\end{aligned}$$

where C is an arbitrary constant that we may assume is real. The boundary condition $\psi(1) = 0$ now requires also that $\sin\omega = 0$; that is,

$$\omega = n\pi \quad \text{for some } n = 1, 2, 3, \ldots .$$

Therefore, since $\lambda = -3 - \omega^2$, the eigenvalues of $\mathcal{S}$ are

$$\lambda_n = -3 - n^2\pi^2, \quad n = 1, 2, 3, \ldots$$

with corresponding eigenfunctions

$$\psi_n(x) = C_n e^{-2x}\sin(n\pi x), \quad n = 1, 2, 3, \ldots .$$

The nonzero constants C_n may be chosen to scale the eigenfunctions as desired. Figure 3 is a plot of the first four of these eigenfunctions with the simple choice of $C_n = 1$ for all n. ■

Each of Examples 1 through 3 illustrates several fundamental properties of the eigenvalues and eigenfunctions of any regular Sturm-Liouville operator. The spectrum is a decreasing, infinite sequence of real numbers that diverges to $-\infty$. Moreover, for each eigenvalue, the corresponding set of eigenfunctions is "one dimensional;" that is, it consists of constant multiples of a single eigenfunction. These properties, as well as the crucial orthogonality property of the eigenfunctions, are stated for the record in the following theorem.

Theorem 1

Let $\mathcal{S}$ be a regular Sturm-Liouville operator defined for $\mathcal{C}^2$ functions on $[a, b]$. The following are true for the eigenvalue problem (4a,b).

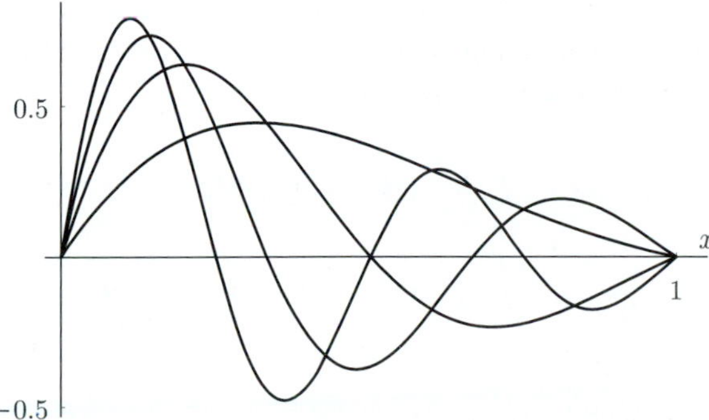

Figure 3

i) The eigenvalues are real and form an infinite, decreasing sequence

$$\cdots < \lambda_3 < \lambda_2 < \lambda_1 < \lambda_0$$

in which $\lambda_n \to -\infty$ *as* $n \to \infty$.

ii) Each eigenvalue is simple; that is, for each λ_i *the family of corresponding eigenfunctions consists of scalar multiples of a single eigenfunction* ψ_i.

iii) Eigenfunctions corresponding to distinct eigenvalues are orthogonal on $[a, b]$ *with respect to the "weight function"* p*; that is, if* λ_i *and* λ_j *are eigenvalues with corresponding eigenfunctions* ψ_i *and* ψ_j*, and if* $i \neq j$*, then*

$$\int_a^b \psi_i \psi_j p\, dx = 0. \quad \blacktriangle$$

The proofs of some parts of Theorem 1 are beyond the scope of this book—in particular, the existence of eigenvalues and the fact that they form an infinite sequence that diverges to $-\infty$. However, two assertions of the theorem, that there are no nonreal complex eigenvalues, and that eigenfunctions corresponding to distinct eigenvalues are orthogonal, are elementary consequences of the fact that $\mathcal{S}$ is a *symmetric* operator. The remainder of this section is devoted to the discussion of this concept and to the proof of the aforementioned parts of Theorem 1. We begin with the following theorem.

Theorem 2

A regular Sturm-Liouville operator $\mathcal{S}$ *is* ***symmetric***[‡] *(with respect to the weight function* p *and the boundary conditions in (4b)) in the sense that*

$$\int_a^b \gamma \mathcal{S}\psi p\, dx = \int_a^b \psi \mathcal{S}\gamma p\, dx$$

for all C^2 *functions* ψ *and* γ *on* $[a, b]$ *that satisfy the boundary conditions in (4b).* $\blacktriangle$

The proof of Theorem 2 is mainly an exercise in integration by parts and proceeds as follows. Suppose that ψ and γ are C^2 functions on $[a, b]$ and satisfy the boundary conditions in (4b). Then

$$\begin{aligned}\int_a^b \gamma \mathcal{S}\psi p\, dx &= \int_a^b \gamma \cdot \big((k\psi')' + q\psi\big) dx \\ &= \int_a^b \gamma \cdot (k\psi')' dx + \int_a^b \gamma q \psi\, dx.\end{aligned}$$

[‡] A real $n \times n$ matrix A is symmetric if $A^T = A$. An important consequence of symmetry in A (or an equivalent definition) is that $\mathbf{y}^T A\mathbf{x} = \mathbf{x}^T A\mathbf{y}$ for all real vectors of length n. Note the similarity between this and the conclusion of Theorem 2.

Next we use integration by parts on the first integral on the right side, obtaining

$$\int_a^b \gamma \mathcal{S}\psi p \, dx = \gamma k \psi' \Big|_{x=a}^{x=b} - \int_a^b \gamma' k \psi' \, dx + \int_a^b \gamma q \psi \, dx. \tag{5}$$

Similarly, it also follows that

$$\int_a^b \psi \mathcal{S}\gamma p \, dx = \psi k \gamma' \Big|_{x=a}^{x=b} - \int_a^b \psi' k \gamma' \, dx + \int_a^b \psi q \gamma \, dx. \tag{6}$$

Therefore, by subtracting (6) from (5), we find that

$$\int_a^b \gamma \mathcal{S}\psi p \, dx - \int_a^b \psi \mathcal{S}\gamma p \, dx = \left(\gamma k \psi' - \psi k \gamma'\right)\Big|_{x=a}^{x=b}. \tag{7}$$

Now we examine the right side of (7), first noting that

$$\begin{aligned}\left(\gamma k \psi' - \psi k \gamma'\right)\Big|_{x=a}^{x=b} &= k(b)\Big(\gamma(b)\psi'(b) - \psi(b)\gamma'(b)\Big) \\ &\quad - k(a)\Big(\gamma(a)\psi'(a) - \psi(a)\gamma'(a)\Big).\end{aligned} \tag{8}$$

If $\beta = 0$, then the boundary condition at $x = b$ in (4b) demands that $\psi'(b) = \gamma'(b) = 0$, and so $\gamma(b)\psi'(b) - \psi(b)\gamma'(b) = 0$. If $\beta \neq 0$, then the boundary condition at $x = b$ in (4b) demands that

$$\psi(b) = \frac{\beta - 1}{\beta}\psi'(b) \quad \text{and} \quad \gamma(b) = \frac{\beta - 1}{\beta}\gamma'(b).$$

These facts in turn imply that

$$\gamma(b)\psi'(b) - \psi(b)\gamma'(b) = \frac{\beta - 1}{\beta}\big(\gamma'(b)\psi'(b) - \psi'(b)\gamma'(b)\big) = 0.$$

Thus, for any β in [0, 1], we have that

$$\gamma(b)\psi'(b) - \psi(b)\gamma'(b) = 0.$$

Similarly, for any α in [0, 1], it also follows that

$$\gamma(a)\psi'(a) - \psi(a)\gamma'(a) = 0.$$

So we finally conclude from (7) and (8) that $\mathcal{S}$ is symmetric:

$$\int_a^b \gamma \mathcal{S}\psi p \, dx = \int_a^b \psi \mathcal{S}\gamma p \, dx.$$

In order to prove that $\mathcal{S}$ can have only real eigenvalues, we first need a complex version of Theorem 2, which we will state as Corollary 1. Its proof, which we omit, follows easily from Theorem 2 and the fact that $\mathcal{S}$ maps real functions to real functions. (A proof that closely parallels that of Theorem 3 is also straighforward.) For the statement of Corollary 1, we must recall that the *conjugate* of a complex number $z = x + iy$ is $\overline{z} = x - iy$; thus we define the conjugate of a complex-valued function $\zeta = u + iv$ to be

$$\overline{\zeta} = u - iv.$$

COROLLARY 1 *Let u, v, ρ, and μ be C^2 functions defined on $[a, b]$, each satisfying the boundary conditions in (4b). Also let $\psi = u + iv$ and $\gamma = \rho + i\mu$, and let $\mathcal{S}$ be a Sturm-Liouville operator as defined in (3). Then ψ and γ each satisfy the boundary conditions in (4b), and*

$$\int_a^b (\rho - i\mu)(\mathcal{S}u + i\mathcal{S}v)p\,dx = \int_a^b (u + iv)(\mathcal{S}\rho - i\mathcal{S}\mu)p\,dx.$$

Since $\mathcal{S}\psi = \mathcal{S}u + i\mathcal{S}v$ and $\mathcal{S}\gamma = \mathcal{S}\rho + i\mathcal{S}\mu$, it follows that

$$\int_a^b \overline{\gamma}\mathcal{S}\psi p\,dx = \int_a^b \psi\overline{\mathcal{S}\gamma}\,p\,dx.$$

To demonstrate that $\mathcal{S}$ has only real eigenvalues, we begin by supposing that $\lambda = \eta + i\omega$ is a complex eigenvalue[§] with corresponding (complex-valued) eigenfunction $\psi = u + iv$. Then, according to Corollary 1 with $Y = \psi$, we have

$$\int_a^b \overline{\psi}\mathcal{S}\psi p\,dx = \int_a^b \psi\overline{\mathcal{S}\psi}\,p\,dx.$$

But since $\mathcal{S}\psi = \lambda\psi$ and $\overline{\mathcal{S}\psi} = \overline{\lambda\psi}$, this becomes

$$\lambda\int_a^b \overline{\psi}\psi p\,dx = \overline{\lambda}\int_a^b \psi\overline{\psi}\,p\,dx.$$

Since $\psi\overline{\psi} = \overline{\psi}\psi = u^2 + v^2$, it follows that the two integrals are equal and neither is zero. Therefore, $\lambda = \overline{\lambda}$; that is, $\eta + i\omega = \eta - i\omega$. Consequently, $\omega = 0$, and so it follows that λ is a real number.

To show that eigenfunctions corresponding to distinct eigenvalues of $\mathcal{S}$ are orthogonal on $[a, b]$ with respect to the weight function p, we begin by supposing that λ_i and λ_j are distinct eigenvalues of $\mathcal{S}$ with corresponding eigenfunctions ψ_i and ψ_j, respectively.[¶] Then, by Theorem 2,

$$\int_a^b \psi_i\mathcal{S}\psi_j p\,dx = \int_a^b \psi_j\mathcal{S}\psi_i p\,dx.$$

Now since $\mathcal{S}\psi_i = \lambda_i\psi_i$ and $\mathcal{S}\psi_j = \lambda_j\psi_j$, we have

$$\lambda_j\int_a^b \psi_i\psi_j p\,dx = \lambda_i\int_a^b \psi_j\psi_i p\,dx.$$

Since $\lambda_i \neq \lambda_j$, the desired orthogonality property follows:

$$\int_a^b \psi_i\psi_j p\,dx = 0.$$

[§] Recall that the complex numbers include the real numbers.

[¶] Note that, since $\mathcal{S}$ can have only real eigenvalues, we may assume that all eigenfunctions are real-valued. (Why?)

PROBLEMS

1. Define the operator $\mathcal{S}$ by $\mathcal{S}\psi = \psi'' + \psi'$ for C^2 functions on $[0, 1]$. Show that $\mathcal{S}$ is a regular Sturm-Liouville operator.

For the operator $\mathcal{S}$ in Problem 1, find the eigenvalues and corresponding eigenfunctions of $\mathcal{S}$ with respect to the boundary conditions in Problems 2 through 4.

2. $\psi(0) = \psi(1) = 0$

3. $\psi'(0) = \psi'(1) = 0$

4. $\psi'(0) = \psi(1) = 0$

5. Let $\mathcal{S}$ defined by $\mathcal{S}\psi = k\psi''$ (with k a constant) for C^2 functions on $[0, \pi]$.

(a) Find the eigenvalues and corresponding eigenfunctions of $\mathcal{S}$ with respect to the boundary conditions

$$\psi(0) - \psi'(0) = \psi(\pi) + \psi'(\pi) = 0.$$

(b) Normalize the eigenfunctions (i.e., scaled them so that $\int_0^\pi \psi_n^2\, dx = 1$).

6. Let β be a constant, and define $\mathcal{S}$ by $\mathcal{S}\psi = \psi'' + \beta\psi'$ for C^2 functions on $[0, 1]$.

(a) Show that $\mathcal{S}$ is a regular Sturm-Liouville operator.

(b) Find, in terms of β, the eigenvalues and corresponding eigenfunctions of $\mathcal{S}$ with respect to the boundary conditions $\psi(0) = \psi(1) = 0$.

7. Let γ be a constant, and define $\mathcal{S}$ by $\mathcal{S}\psi = \psi'' + \gamma\psi$ for C^2 functions on $[0, 1]$.

(a) Show that $\mathcal{S}$ is a regular Sturm-Liouville operator.

(b) Find, in terms of γ, the eigenvalues and corresponding normalized eigenfunctions of $\mathcal{S}$ with respect to the boundary conditions $\psi(0) = \psi(1) = 0$.

8. Define the operator $\mathcal{S}$ by $\mathcal{S}\psi = \alpha\psi'' + \beta\psi' + \gamma\psi$ for C^2 functions ψ on $[a, b]$, where α, β, and γ are continuous functions on $[a, b]$. Show that

$$\mathcal{S}\psi = \alpha e^{-\int \beta/\alpha}\left((e^{\int \beta/\alpha}\psi')' + \gamma e^{\int \beta/\alpha}\psi\right),$$

and conclude, therefore, that $\mathcal{S}$ is a regular Sturm-Liouville operator, if α is positive on $[a, b]$.

In Problems 9 and 10, use the result of Problem 8 to write $\mathcal{S}\psi$ in the form of (3).

9. $\mathcal{S}\psi = (1 + x^2)\psi'' + 4x\psi'$

10. $\mathcal{S}\psi = \psi'' + x\psi'$

11. Let $\mathcal{S}$ be a regular Sturm-Liouville operator defined by

$$\mathcal{S}\psi = \frac{1}{p}((k\psi')' + q\psi)$$

for C^2 functions ψ on $[a, b]$, and consider the eigenvalue problem (4a,b). In this problem we will prove that, if $q(x) \le 0$ for all x in $[a, b]$, then

$$\int_a^b \psi\mathcal{S}\psi p\, dx \le 0$$

for all C^2 functions ψ on $[a, b]$ that satisfy (4b). Because of this property $\mathcal{S}$ is said to be **negative semidefinite.*** Furthermore, as a result, $\mathcal{S}$ has no positive eigenvalues.

---

* This is analogous to negative semidefiniteness for matrices. A square matrix A is negative semidefinite, if $\mathbf{x}^T A\mathbf{x} \le 0$ for all $\mathbf{x}$, which implies that A has no positive eigenvalues.

(a) Let ψ be a $\mathcal{C}^2$ function on $[a, b]$. Use integration by parts (as in the proof of Theorem 2) to obtain

$$\int_a^b \psi \mathcal{S}\psi p \, dx = k\psi\psi'\Big|_a^b - \int_a^b \left(k(\psi')^2 - q\psi^2\right) dx.$$

(b) Show that, if ψ satisfies the boundary conditions (4b), then $k\psi\psi'\big|_a^b \leq 0$ for any α and β in $[0, 1]$. Conclude that

$$\int_a^b \psi \mathcal{S}\psi p \, dx \leq -\int_a^b \left(k(\psi')^2 - q\psi^2\right) dx$$

and that, if $q(x) \leq 0$ for all x in $[a, b]$, then

$$\int_a^b \psi \mathcal{S}\psi p \, dx \leq 0.$$

(c) Now suppose that λ is an eigenvalue of $\mathcal{S}$ with corresponding eigenfunction ψ. Show that

$$\lambda \int_a^b \psi^2 p \, dx \leq -\int_a^b \left(k(\psi')^2 - q\psi^2\right) dx,$$

and conclude that, if $q(x) \leq 0$ for all x in $[a, b]$, then $\lambda \leq 0$.

12. Consider again the eigenvalue problem in Problem 11. Show that, if either of the following is true,

(a) $q(x) \leq 0$ for all x in $[a, b]$, and either α or β is not zero in (4a,b);

(b) $q(x) \leq 0$ for all x in $[a, b]$, $q(x) < 0$ for some x in $[a, b]$;

then, for all $\mathcal{C}^2$ functions ψ on $[a, b]$,

$$\int_a^b \psi \mathcal{S}\psi p \, dx < 0$$

and all eigenvalues of $\mathcal{S}$ are negative. In other words, $\mathcal{S}$ is **negative definite**.

13. Let $\mathcal{S}$ be defined for $\mathcal{C}^2$ functions ψ on $[0, e-1]$ by

$$\mathcal{S}\psi = ((1+x)^2\psi')'.$$

(a) Verify that, if $\lambda \leq -1/4$, then the general solution of $\mathcal{S}\psi = \lambda\psi$ is given by

$$\psi = \frac{1}{\sqrt{1+x}}\left(c_1 \cos\left(\sqrt{-\lambda - 1/4}\,\ln(1+x)\right) + c_2 \sin\left(\sqrt{-\lambda - 1/4}\,\ln(1+x)\right)\right).$$

(b) Verify that, if $\lambda > -1/4$, then the general solution of $\mathcal{S}\psi = \lambda\psi$ is given by

$$\psi = \frac{1}{\sqrt{1+x}}\left(c_1 \cosh\left(\sqrt{\lambda + 1/4}\,\ln(1+x)\right) + c_2 \sinh\left(\sqrt{\lambda + 1/4}\,\ln(1+x)\right)\right).$$

(c) Find the eigenvalues and corresponding eigenfunctions of $\mathcal{S}$ with respect to the boundary conditions $\psi(0) = \psi(e-1) = 0$.

14. Let the operator $\mathcal{S}$ be defined by

$$\mathcal{S}\psi = \psi'' - 2x\psi'$$

for $\mathcal{C}^2$ functions ψ on an interval $[a, b]$. This operator is associated with Hermite's equation. (See Problems 3 and 4 in Section 5.7.)

(a) Show that $\mathcal{S}$ is a regular Sturm-Liouville operator. (*Hint*: See Problem 8.)

(b) Consider the eigenvalue problem

$$\mathcal{S}\psi = \lambda\psi, \quad \psi(0) = \psi(1) = 0,$$

in which the differential equation is Hermite's equation

$$\psi'' - 2x\psi' + 2m\psi = 0, \quad \text{where } 2m = -\lambda.$$

Show that fundamental power-series solutions are

$$u_m = 1 - \frac{2mx^2}{2!} + \frac{2^2m(m-2)x^4}{4!} - \frac{2^3m(m-2)(m-4)x^6}{6!} + \cdots$$

$$\text{and } v_m = x - \frac{2(m-1)x^3}{3!} + \frac{2^2(m-1)(m-3)x^5}{5!} - \cdots,$$

and conclude that every solution of Hermite's equation is given by $c_1u_m + c_2v_m$ for some choice of c_1 and c_2.

(c) Show that $\lambda = -2m$ is an eigenvalue of $\mathcal{S}$, with corresponding eigenfunction $\psi = v_m$, if and only if m is a solution of

$$1 - \frac{2(m-1)}{3!} + \frac{2^2(m-1)(m-3)}{5!} - \frac{2^3(m-1)(m-3)(m-5)}{7!} + \cdots = 0. \qquad (*)$$

Conclude from Theorem 1 that this equation has an increasing, divergent, infinite sequence of solutions. Also, show that this equation has only positive solutions, and conclude that $\mathcal{S}$ has only negative eigenvalues.

(d) Truncate the series in (*) after five terms and use the resulting polynomial to approximate the least solution of (*). Compute the resulting (approximate) greatest eigenvalue λ_0, and graph the corresponding (approximate) eigenfunction.

15. Let the operator $\mathcal{S}$ be defined by $\mathcal{S}\psi = \psi'' + x\psi$ for C^2 functions ψ on $[0, 1]$.

(a) Show that $\mathcal{S}$ is a regular Sturm-Liouville operator.

(b) Treating λ as a parameter, find fundamental solutions u and v of $\mathcal{S}\psi = \lambda\psi$ in the form of power-series expansions (about $t = 0$).

(c) Let $\lambda_0, \lambda_1, \lambda_2, \ldots$ be the eigenvalues of $\mathcal{S}$ with respect to the boundary conditions $\psi(0) = \psi(1) = 0$. Find the corresponding eigenfunctions, and describe the eigenvalues as the zeros of a certain power series in λ. (cf. Problem 14c).

(d) Compute a numerical approximation to the greatest eigenvalue λ_0, and plot the graph of the corresponding (approximate) eigenfunction. (cf. Problem 14d).

16. Let $\mathcal{S}$ be a regular Sturm-Liouville operator defined by

$$\mathcal{S}\psi = \frac{1}{p}\left((k\psi')' + q\psi\right)$$

for C^2 functions ψ on $[a, b]$, and consider the eigenvalue problem

$$\mathcal{S}\psi = \lambda\psi, \quad \psi(a) = \psi(b) = 0.$$

In this problem we will show that the greatest eigenvalue λ_0 satisfies the inequality

$$\lambda_0 \le \max_{a \le x \le b} \frac{q(x)}{p(x)}. \qquad (**)$$

Begin by supposing that λ is a number for which

$$\lambda p(x) - q(x) > 0 \quad \text{for all } x \text{ in } [a, b]$$

and writing the equation $\mathcal{S}\psi = \lambda\psi$ in the form

$$k\psi'' + k'\psi' = (\lambda p - q)\psi.$$

(a) Show that, if a solution ψ attains its maximum value on $[a, b]$ at x^*, then $\psi(x^*) \leq 0$. Conclude that $\psi(x) \leq 0$ for all x in $[a, b]$.

(b) Show that, if a solution ψ attains its minimum value on $[a, b]$ at x^*, then $\psi(x^*) \geq 0$. Conclude that $\psi(x) \geq 0$ for all x in $[a, b]$.

(c) Conclude that $\psi = 0$ is the only solution and that therefore λ cannot be an eigenvalue.

(d) Conclude that, if λ *is* an eigenvalue, then $\lambda p(x) \leq q(x)$ for *some* x in $[a, b]$; consequently, $(**)$ is true.

17. Suppose that $\mathcal{S}$ is any linear operator with eigenvalues λ_n, $n = 0, 1, 2, \dots$, and corresponding eigenfunctions ψ_n, $n = 0, 1, 2, \dots$; that is,

$$\mathcal{S}\psi_n = \lambda_n \psi_n, \quad n = 0, 1, 2, \dots,$$

where each $\psi_n \neq 0$. Let c be a constant, and let $\mathcal{S}_c$ be defined by $\mathcal{S}_c\psi = \mathcal{S}\psi + c\psi$. Show that the eigenvalues of $\mathcal{S}_c$ are $\lambda_n + c$, $n = 0, 1, 2, \dots$, and that ψ_n, $n = 0, 1, 2, \dots$, are corresponding eigenfunctions.

18. Let $\mathcal{S}$ be a regular Sturm-Liouville operator defined by

$$\mathcal{S}\psi = \frac{1}{p}\left((k\psi')' + q\psi\right)$$

for C^2 functions ψ on $[a, b]$. Show that $\mathcal{S}$ is symmetric and negative semidefinite (see Problem 11) with respect to the weight function p and "periodic" boundary conditions

$$\psi(a) = \psi(b) \text{ and } k(a)\psi'(a) = k(b)\psi'(b).$$

19. Define $\mathcal{S}$ by $\mathcal{S}\psi = \psi''$ for C^2 functions ψ on $[0, \pi]$.

(a) Find the eigenvalues of $\mathcal{S}$ with respect to the boundary conditions

$$\psi(0) = \psi(\pi) \text{ and } \psi'(0) = \psi'(\pi).$$

(b) Show that each eigenvalue has two linearly independent eigenfunctions. (Thus part (ii) of Theorem 1 is not true for problems with periodic boundary conditions.)

11.6 Singular Sturm-Liouville Problems

Recall from the preceding section that an operator $\mathcal{S}$, acting on C^2 functions ψ on I, is a Sturm-Liouville operator if $\mathcal{S}\psi$ can be written in the form

$$\mathcal{S}\psi = \frac{1}{p}\left((k\psi')' + q\psi\right),$$

where

- k is positive and continuously differentiable on I;
- p is positive and continuous on I;
- q is continuous on I.

We have seen that, if I is a closed, bounded interval, then $\mathcal{S}$ is called a regular Sturm-Liouville operator. If, on the other hand, either

i. I is not bounded, or
ii. I is an open or half-open, bounded interval with endpoints a and b, and one of the above conditions on k, p, and q fails on $[a, b]$,

then $\mathcal{S}$ is called a **singular Sturm-Liouville operator**. A few simple examples are:

$$\begin{aligned} \mathcal{S}\psi &= (x\psi')', & I &= (0, 1]; \\ \mathcal{S}\psi &= ((1-x^2)\psi')', & I &= (-1, 1); \\ \mathcal{S}\psi &= x(x^{-1}\psi')', & I &= (0, 1]; \\ \mathcal{S}\psi &= e^{x^2}(e^{-x^2}\psi')', & I &= (-\infty, \infty). \end{aligned}$$

The important thing to notice about the first three of these operators is that if I were "closed up" by including the missing endpoint(s), then the coefficient k would no longer be positive and continuously differentiable on I.

A singular Sturm-Liouville eigenvalue problem consists of the differential equation

$$\mathcal{S}\psi = \lambda\psi \quad \text{on } I,$$

where $\mathcal{S}$ is a singular Sturm-Liouville operator, together with the "usual" boundary condition at any endpoint contained in I and some suitable side condition replacing the usual boundary condition at any "missing endpoint" of I (which may include ∞ or $-\infty$). In some cases, with appropriately chosen side conditions replacing the usual boundary conditions as necessary, the eigenvalues and eigenfunctions of a singular Sturm-Liouville eigenvalue problem will share crucial properties with the eigenvalues and eigenfunctions of regular Sturm-Liouville eigenvalue problems; namely, the spectrum is a sequence of real numbers (and therefore said to be *discrete*), and eigenfunctions comprise an orthogonal family. In other cases, however, the spectrum may fail to be discrete and, instead, forms an interval. When this happens we say the the operator has a *continuous spectrum*. (There are also operators with "mixed" spectra, which contain both intervals and discrete points.)

Since our primary aim here is to generalize the notion of Fourier series, our criteria for the suitability of side conditions will be that (1) the spectrum of $\mathcal{S}$ must be discrete, and (2) eigenfunctions corresponding to distinct eigenvalues must be orthogonal on I. Thus we require $\mathcal{S}$ to have the same symmetry property as in the regular case; that is,

$$\int_a^b \gamma \mathcal{S}\psi p \, dx = \int_a^b \psi \mathcal{S}\gamma p \, dx$$

for all C^2 functions ψ and γ on I that satisfy the boundary conditions. (Note that the integrals may be improper.) Examination of the proof of Theorem 2 in Section 11.6 reveals that what is crucial for this purpose is that side conditions be such that

$$\lim_{x\to a^+} k(x)\big(\gamma(x)\psi'(x) - \psi(x)\gamma'(x)\big) = \lim_{x\to b^-} k(x)\big(\gamma(x)\psi'(x) - \psi(x)\gamma'(x)\big) \quad (1)$$

for all C^2 functions ψ and γ that satisfy the boundary conditions.

A common and important type of singular problem occurs when I is bounded and k vanishes at one or both endpoints of I. If, for example, $I = (a, b)$ and $k(x) \to 0$ as $x \to a^+$ or $x \to b^-$, then (1) is true if the quantity $\gamma\psi' - \psi\gamma'$ is bounded on (a, b). Thus, symmetry of $\mathcal{S}$ is guaranteed by considering only bounded functions on (a, b) with bounded derivatives, and the eigenvalue problem for $\mathcal{S}$ becomes

$$\mathcal{S}\psi = \lambda\psi, \quad \psi \text{ and } \psi' \text{ bounded on } (a, b).$$

If $I = (a, b]$ and $k(x) \to 0$ as $x \to a^+$, then the appropriate condition at $x = b$ is the usual

$$\beta\psi(b) + (1-\beta)\psi'(b) = 0,$$

which we augment with the condition that ψ and ψ' be bounded on $(a, b]$. Similarly, if $I = [a, b)$ and $k(x) \to 0$ as $x \to b^-$, then the appropriate condition at $x = a$ is the usual

$$\alpha\psi(a) + (1-\alpha)\psi'(a) = 0,$$

which we also augment with the condition that ψ and ψ' be bounded on $[a, b)$.

Example 1

Let $\mathcal{S}$ be the singular Sturm-Liouville operator defined by

$$\mathcal{S}\psi = \sqrt{1-x^2}\left(\sqrt{1-x^2}\psi'\right)' = (1-x^2)\psi'' - x\psi'$$

for $\mathcal{C}^2$ functions ψ on $(-1, 1)$. This operator is associated with Chebyshev's equation. (See Section 5.6.) Since

$$k(x) = \sqrt{1-x^2} = 0 \text{ at } x = \pm 1,$$

the role of boundary conditions is played by the requirement that ψ and ψ' be bounded on $(-1, 1)$. Thus the eigenvalue problem associated with $\mathcal{S}$ is

$$\mathcal{S}\psi = \lambda\psi \text{ with } \psi \text{ and } \psi' \text{ bounded on } (-1, 1).$$

Let us now find the eigenvalues and corresponding eigenfunctions. The differential equation $\mathcal{S}\psi = \lambda\psi$ is equivalent to Chebyshev's equation

$$(1-x^2)\psi'' - x\psi' + m^2\psi = 0, \quad \text{where } m^2 = -\lambda. \tag{2}$$

When $m = 0$, any constant is a solution. So we note that $\lambda = 0$ is an eigenvalue with corresponding eigenfunction $\psi = 1$, and henceforth we may assume that $m \neq 0$. Section 5.7, Problem 7, tells us that, for any m, Chebyshev's equation has the general solution

$$\psi = c_1 \cos(m \cos^{-1} x) + c_2 \sin(m \cos^{-1} x)$$

on $[-1, 1]$, which is easily verified. All of these solutions are bounded; so, to determine the eigenvalues, we must rely on the boundedness requirement on ψ'. So we compute the derivative,

$$\psi'(x) = \frac{m(c_1 \sin(m \cos^{-1} x) - c_2 \cos(m \cos^{-1} x))}{\sqrt{1-x^2}},$$

and note that the denominator approaches 0 as $x \to \pm 1$ from within $(-1, 1)$. As $x \to 1^-$, the quantity in the numerator approaches

$$m(c_1 \sin(0) - c_2 \cos(0)) = -mc_2;$$

therefore, ψ' can be bounded near $x = 1$ only if $c_2 = 0$. So we set $c_2 = 0$ and $c_1 = 1$ and note then that

$$\psi'(x) = \frac{m \sin(m \cos^{-1} x)}{\sqrt{1 - x^2}}.$$

In order for ψ' to be bounded, it is necessary that

$$\sin(m \cos^{-1}(\pm 1)) = 0.$$

Observing that $\sin(m \cos^{-1}(1)) = \sin(m \cdot 0) = 0$ for any m, we then look to

$$\sin(m \cos^{-1}(-1)) = \sin(m\pi) = 0,$$

which requires that m be an integer (which we may as well assume is positive). Now, an application of l'Hôpital's rule reveals that, for any positive integer m,

$$\begin{aligned}
\lim_{x \to 1^-} \psi'(x) &= \lim_{x \to 1^-} \frac{m \sin(m \cos^{-1} x)}{\sqrt{1 - x^2}} \\
&= \lim_{x \to 1^-} \frac{m^2 \cos(m \cos^{-1} x)/\sqrt{1 - x^2}}{x/\sqrt{1 - x^2}} \\
&= \lim_{x \to 1^-} \frac{m^2 \cos(m \cos^{-1} x)}{x} \\
&= m^2 \cos(m \cdot 0) = m^2.
\end{aligned}$$

Similarly,

$$\begin{aligned}
\lim_{x \to -1^+} \psi'(x) &= \lim_{x \to -1^+} \frac{m^2 \cos(m \cos^{-1} x)}{x} \\
&= -m^2 \cos(m\pi) = -m^2(-1)^m.
\end{aligned}$$

So we have found that (2) has a nontrivial solution ψ on $(-1, 1)$, with ψ and ψ' bounded, precisely when m is a nonnegative integer. A nontrivial solution corresponding to any nonnegative integer m is $\psi_m = \cos(m \cos^{-1} x)$, which is, in fact, the nth Chebyshev polynomial. (See Section 5.7, Problems 8 and 9.) Therefore, we conclude that the eigenvalues of S are

$$\lambda_m = -m^2, \quad m = 0, 1, 2, 3, \ldots,$$

and corresponding eigenfunctions are the Chebyshev polynomials

$$T_m(x) = \cos(m \cos^{-1} x), \quad m = 0, 1, 2, 3, \ldots.$$

In particular, we note that the eigenvalues are real, and eigenfunctions corresponding to distinct eigenvalues are orthogonal on $(-1, 1)$. (See Section 5.7.) ■

PROBLEMS

1. Let $\mathcal{S}$ be defined (as in Example 1) by

$$\mathcal{S}\psi = \sqrt{1-x^2}\left(\sqrt{1-x^2}\,\psi'\right)'$$

for C^2 functions ψ on $[0, 1)$. Find the eigenvalues and corresponding eigenfunctions of $\mathcal{S}$ with respect to the conditions $\psi(0) = 0$ and ψ and ψ' bounded.

2. Let $\mathcal{S}$ be defined as in Problem 1. Find the eigenvalues and corresponding eigenfunctions of $\mathcal{S}$ with respect to the conditions $\psi'(0) = 0$ and ψ and ψ' bounded.

3. Let the operator $\mathcal{S}$ be defined by

$$\mathcal{S}\psi = (1-x^2)\psi'' - 2x\psi',$$

for C^2 functions ψ on $(-1, 1)$, and consider the eigenvalue problem

$$\mathcal{S}\psi = \lambda\psi \text{ with } \psi \text{ and } \psi' \text{ bounded on } (-1, 1).$$

The operator $\mathcal{S}$ is associated with Legendre's equation

$$(1-x^2)\psi'' - 2x\psi' + m(m+1)\psi = 0.$$

(See Section 5.7, Problems 1 and 2.)

(a) Show that $\mathcal{S}$ is a singular Sturm-Liouville operator. (*Hint*: Let $k = 1 - x^2$.)

(b) Show that, for each nonnegative integer m, $\lambda_m = -m(m+1)$ is an eigenvalue of $\mathcal{S}$ with corresponding eigenfunction P_m, where $P_0, P_1, P_2, \ldots$ are the Legendre polynomials.

(c) Give an informal argument that the derivative of every nonterminating power-series solution of $\mathcal{S}\psi = \lambda\psi$ approaches $\pm\infty$ as $x \to -1^+$ or $x \to 1^-$. Conclude that $\mathcal{S}$ has no eigenvalues other than those described in part (b).

4. Define $\mathcal{S}$ by $\mathcal{S}\psi = \psi''$ for C^2 functions ψ on $[0, \infty)$. Find the eigenvalues and corresponding eigenfunctions of $\mathcal{S}$ with respect to the conditions $\psi(0) = 0$ and ψ, ψ' bounded on $[0, \infty)$.

5. Let $\mathcal{S}$ be defined by $\mathcal{S}\psi = x^2(x^2\psi')'$ for C^2 functions on $(0, 1/\pi]$.

(a) Show that, if $y(t) = \psi(x)$, where $t = 1/x$, then $y''(t) = \mathcal{S}\psi(x)$. Conclude that ψ is a solution of

$$\mathcal{S}\psi = \lambda\psi \quad \text{on } (0, 1/\pi], \quad \psi(1/\pi) = 0, \tag{$*$}$$

if and only if $y(x) = \psi(1/x)$ is a solution of

$$y'' = \lambda y \quad \text{on } [\pi, \infty), \quad y(\pi) = 0. \tag{$**$}$$

(b) Show that $(**)$—and therefore $(*)$—has a nontrivial solution for any real λ. State a corresponding nontrivial solution ψ of $(*)$ for each λ.

(c) Show that if $\lambda \geq 0$, then the corresponding nontrivial solution of $(*)$ is unbounded on $(0, 1/\pi]$. Then show that if $\lambda < 0$, then the corresponding nontrivial solution of $(*)$ is bounded but has an unbounded derivative on $(0, 1/\pi]$.

11.7 Eigenfunction Expansions

The essence of Theorem 1 in Section 11.5 is that the eigenvalues and eigenfunctions associated with any regular Sturm-Liouville operator (with respect to suitable boundary conditions) share certain crucial properties with the familiar eigenvalues and eigenfunctions of the operator $\mathcal{S}$ defined by $\mathcal{S}X = X''$, which are (cf. Example 1 in Section 11.5)

$$\lambda_n = -\frac{n^2\pi^2}{\ell^2}, \quad X_n = \sin\left(\frac{n\pi x}{\ell}\right), \quad n = 1, 2, 3, \ldots .$$

Those properties—namely, that the spectrum is a sequence of real numbers and that eigenfunctions comprise a sequence of orthogonal functions on $[0, \ell]$—were the crucial tools in the development of (trigonometric) Fourier series representations of the form

$$\underset{\sim}{\varphi}(x) = \sum_{n=1}^{\infty} b_n \sin\left(\frac{n\pi x}{\ell}\right),$$

in which the coefficients are given by

$$b_n = \frac{\displaystyle\int_0^\ell \varphi(x)\sin\left(\frac{n\pi x}{\ell}\right)dx}{\displaystyle\int_0^\ell \sin^2\left(\frac{n\pi x}{\ell}\right)dx}, \quad n = 1, 2, 3, \ldots .$$

These observations motivate the following definition.

Definition Let $\psi_0, \psi_1, \psi_2, \ldots$ be the orthogonal sequence of eigenfunctions on $[a, b]$ of a regular Sturm-Liouville operator $\mathcal{S}$ with weight function p. Given a function φ defined on $[a, b]$, the **eigenfunction expansion** of φ relative to $\mathcal{S}$ is the series defined by

$$\Phi(x) = \sum_{n=0}^{\infty} c_n\psi_n(x), \quad \text{where } c_n = \frac{\displaystyle\int_a^b \varphi\psi_n p\,dx}{\displaystyle\int_a^b \psi_n^2 p\,dx}, \quad n = 0, 1, 2, \ldots . \tag{1}$$

◆

Note that, if the eigenfunctions $\psi_0, \psi_1, \psi_2, \ldots$ are scaled so that

$$\int_a^b \psi_n^2 p\,dx = 1$$

(cf. Example 2), then the formula for c_n in (1) becomes simply

$$c_n = \int_a^b \varphi\psi_n p\,dx, \quad n = 0, 1, 2, \ldots .$$

When scaled in this way, each ψ_n is said to be *normalized*, and the sequence ψ_0, $\psi_1, \psi_2, \ldots$ is called an *orthonormal* sequence.

The following theorem, a generalization of Theorem 1 in Section 11.3, describes convergence properties of eigenfunction expansions. Note that these are merely analogues of convergence properties of trigonometric Fourier series.

Theorem 1

Let $\mathcal{S}$ be a regular Sturm-Liouville operator defined for C^2 functions on $[a, b]$. Let $\lambda_0, \lambda_1, \lambda_2, \ldots$ be the eigenvalues of $\mathcal{S}$ with corresponding eigenfunctions $\psi_0, \psi_1, \psi_2, \ldots$. Suppose that φ and φ' are piecewise continuous functions on $[a, b]$, and let Φ be the eigenfunction expansion of φ relative to $\mathcal{S}$ given by (1). Then for each x in (a, b),

$$\Phi(x) = \frac{1}{2}\left(\lim_{\xi \to x^-} \varphi(\xi) + \lim_{\xi \to x^+} \varphi(\xi)\right).$$

In particular, $\Phi(x) = \varphi(x)$ at each x in (a, b) where φ is continuous. ▲

Example 1

Consider again the operator $\mathcal{S}$ in Example 3 of Section 11.5, defined for C^2 functions ψ on [0, 1] by

$$\mathcal{S}\psi = \psi'' + 4\psi' + \psi.$$

Recall that $\mathcal{S}$ is a regular Sturm-Liouville operator, since

$$\mathcal{S}\psi = e^{-4x}\left((e^{4x}\psi')' + e^{4x}\psi\right).$$

Recall also that, with respect to the boundary conditions $\psi(0) = \psi(1) = 0$, the eigenvalues of $\mathcal{S}$ are

$$\lambda_n = 3 - n^2\pi^2, \quad n = 1, 2, 3, \ldots,$$

with corresponding eigenfunctions given by

$$\psi_n(x) = A_n e^{-2x}\sin(n\pi x), \quad n = 1, 2, 3, \ldots.$$

Since the relevant weight function here is $p(x) = e^{4x}$, normalized eigenfunctions are obtained with

$$\begin{aligned} A_n &= \left(\int_0^1 (e^{-2x}\sin(n\pi x))^2 e^{4x}\,dx\right)^{-1/2} \\ &= \left(\int_0^1 \sin^2(n\pi x)\,dx\right)^{-1/2} = \sqrt{2}, \end{aligned}$$

resulting in the orthonormal sequence of eigenfunctions

$$\psi_n(x) = \sqrt{2}e^{-2x}\sin(n\pi x), \quad n = 1, 2, 3, \ldots.$$

Thus the coefficients in (1) for the eigenfunction expansion of φ take the form

$$\begin{aligned} c_n &= \int_0^1 \varphi(x)\sqrt{2}e^{-2x}\sin(n\pi x)e^{4x}\,dx \\ &= \sqrt{2}\int_0^1 \varphi(x)\sin(n\pi x)e^{2x}\,dx, \quad n = 1, 2, 3, \ldots. \end{aligned}$$

For the sake of illustration, let's consider the eigenfunction expansion of the function φ defined by

$$\varphi(x) = \begin{cases} 0, & \text{if } 0 \le x < \frac{1}{2}, \\ -\sin(2\pi x), & \text{if } \frac{1}{2} \le x \le 1, \end{cases}$$

for which the appropriate coefficients are

$$c_n = -\sqrt{2} \int_{\frac{1}{2}}^{1} \sin(2\pi x) \sin(n\pi x) e^{2x} \, dx, \quad n = 1, 2, 3, \dots .$$

With *CAS* assistance, we find that a "closed form" for c_n is

$$c_n = -2\sqrt{2} e\pi \frac{4n\pi \left((-1)^n e + \cos(\frac{n\pi}{2})\right) + \left((n^2-4)\pi^2 - 4\right) \sin(\frac{n\pi}{2})}{16 + 8(n^2+4)\pi^2 + (n^2-4)^2\pi^4}. \tag{2}$$

With these coefficients, and noting that $c_0 = 0$, we have the eigenfunction expansion

$$\Phi(x) = \sum_{n=1}^{\infty} c_n \sqrt{2} e^{-2x} \sin(n\pi x).$$

Since φ is continuous on $[0, 1]$ and φ' is piecewise continuous on $[0, 1]$, Theorem 1 tells us that

$$\Phi(x) = \varphi(x) \quad \text{for all } x \text{ in } (0, 1).$$

Inspection of the series itself reveals that $\Phi(x) = \varphi(x)$ at $x = 0$ and 1 as well. The convergence of the series is illustrated in Figure 1, where the four plots shown are graphs of partial sums

$$\sum_{n=1}^{N} c_n \sqrt{2} e^{-2x} \sin(n\pi x)$$

with $N = 3$, 6, 12, and 24, respectively. ∎

The Singular Case If $\mathcal{S}$ is a singular Sturm-Liouville operator defined for $\mathcal{C}^2$ functions on $(a, b]$, $[a, b)$, or (a, b), then Theorem 1 remains true, *provided* that the spectrum of $\mathcal{S}$ is a sequence and the corresponding sequence of eigenfunctions is an orthogonal sequence.

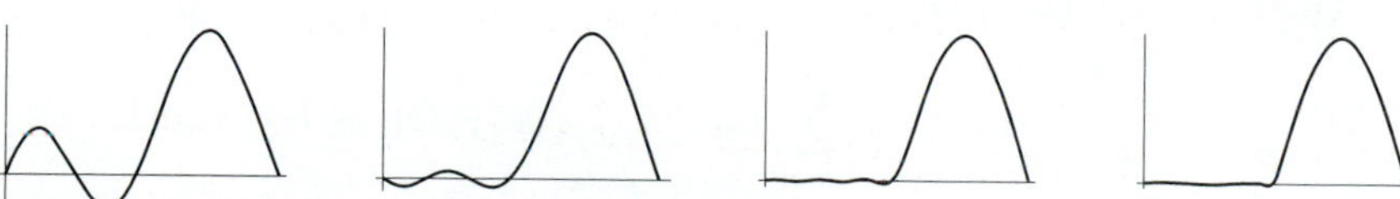

Figure 1

Example 2

Consider again the operator from Example 1 of Section 11.6:

$$\mathcal{S}\psi = \sqrt{1-x^2}\left(\sqrt{1-x^2}\psi'\right)'$$

for C^2 functions ψ on $(-1, 1)$. Recall that the eigenvalues of $\mathcal{S}$ are

$$\lambda_n = -n^2, \quad n = 0, 1, 2, 3, \dots,$$

and corresponding eigenfunctions are the Chebyshev polynomials T_n, $n = 0, 1, 2, 3, \dots$, which are

$$1, x, 2x^2 - 1, 4x^3 - 3x, 8x^4 - 8x^2 + 1, 16x^5 - 20x^3 + 5x, \dots.$$

CAS-aided computations reveal that

$$\int_{-1}^{1} T_0(x)^2 \frac{dx}{\sqrt{1-x^2}} = \pi$$

and

$$\int_{-1}^{1} T_n(x)^2 \frac{dx}{\sqrt{1-x^2}} = \frac{\pi}{2}, \quad n = 1, 2, 3, \dots;$$

thus the coefficients in (1) are

$$c_0 = \frac{1}{\pi}\int_{-1}^{1} \varphi(x) \frac{dx}{\sqrt{1-x^2}}$$

and

$$c_n = \frac{2}{\pi}\int_{-1}^{1} \varphi(x) T_n(x) \frac{dx}{\sqrt{1-x^2}}, \quad n = 1, 2, 3, \dots.$$

For the sake of illustration, let's find the eigenfunction expansion of $\varphi(x) = \sin \pi x$ on $[-1, 1]$. The coefficients in (1) are

$$c_0 = \frac{1}{\pi}\int_{-1}^{1} \sin \pi x \frac{dx}{\sqrt{1-x^2}}$$

and

$$c_n = \frac{2}{\pi}\int_{-1}^{1} \sin \pi x T_n(x) \frac{dx}{\sqrt{1-x^2}}, \quad n = 1, 2, 3, \dots.$$

When n is even, the integrand is an odd function; hence $c_n = 0$ when n is even. Therefore, by Theorem 1,

$$\sin \pi x = \sum_{n=0}^{\infty} c_{2n+1} T_{2n+1}(x) \quad \text{for } -1 < x < 1.$$

Numerical evidence shows that the series agrees with $\sin \pi x$ at $x = \pm 1$ as well. The first four nonzero coefficients, computed by *CAS* and rounded to six significant digits, are

$$c_1 = 0.569231, c_3 = -0.666917, c_5 = 0.104282, c_7 = -0.00684063, \ldots .$$

The degree-five partial sum (for example) of the resulting generalized Fourier series is

$$\begin{aligned} &0.569231T_1(x) - 0.666917T_3(x) + 0.104282T_5(x) \\ &\qquad = 3.09139x - 4.75331x^3 + 1.66852x^5. \end{aligned}$$

(The reader is invited to plot this polynomial on $[-1, 1]$ along with $\sin \pi x$ for comparison.) ■

Initial-Value Problems We complete our current discussion by returning to initial-value problems of the form

$$\left.\begin{aligned} &w_t - \mathcal{S}w = 0 \text{ for } x \text{ in } I \text{ and } t > 0, \\ &w(0, x) = \varphi(x) \text{ for } x \text{ in } I, \\ &w \text{ satisfies suitable side conditions on } I \text{ for } t > 0, \end{aligned}\right\} \tag{3}$$

where $\mathcal{S}$ is a Sturm-Liouville operator defined for $\mathcal{C}^2$ functions on the interval I. We assume that the side conditions guarantee that the spectrum of $\mathcal{S}$ is a divergent sequence of real numbers $\lambda_0, \lambda_1, \lambda_2, \ldots$, and corresponding eigenfunctions $\psi_0, \psi_1, \psi_2, \ldots$, form an orthogonal sequence. We also assume that φ and φ' are piecewise continuous on the closed interval consisting of all points in I together with any finite endpoints of I. (For instance, if $I = (0, 1)$, we assume that φ and φ' are piecewise continuous on $[0, 1]$.) Then φ has the eigenfunction expansion

$$\Phi(x) = \sum_{n=0}^{\infty} c_n \psi_n(x),$$

where the coefficients are as given in (1), and the solution of the initial-value problem (3) for $t > 0$ is given by

$$w(t, x) = \sum_{n=0}^{\infty} c_n e^{\lambda_n t} \psi_n(x).$$

Example 3

Consider the problem

$$\begin{aligned} w_t - (w_{xx} + 4w_x + w) &= 0 \quad \text{for } -1 \le x \le 1 \text{ and } t > 0, \\ w(0, x) &= \begin{cases} 0, & \text{if } 0 \le x < \frac{1}{2}, \\ -\sin(2\pi x), & \text{if } \frac{1}{2} \le x \le 1, \end{cases} \\ w(t, 0) &= w(t, 1) = 0 \quad \text{for } t > 0, \end{aligned}$$

which involves the regular Sturm-Liouville operator $\mathcal{S}$ from Example 1 in this section and Example 3 in Section 11.5. In Example 1 it was shown that the initial data have the eigenfunction expansion

$$\Phi(x) = \sum_{n=1}^{\infty} c_n \sqrt{2} e^{-2x} \sin(n\pi x)$$

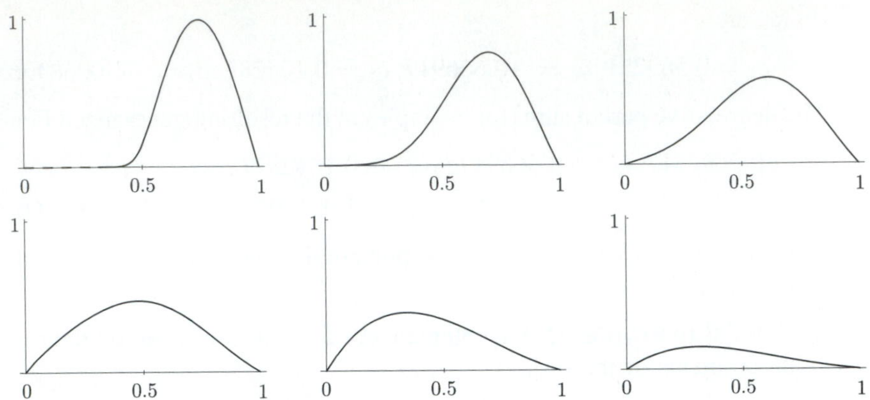

Figure 2

where c_n is given by formula (2). (See Example 1.) Since the eigenvalues of $\mathcal{S}$ are

$$\lambda_n = 3 - n^2\pi^2, \quad n = 1, 2, 3, \ldots ,$$

it follows that the solution is

$$w(t, x) = \sum_{n=1}^{\infty} c_n \sqrt{2} e^{(3-n^2\pi^2)t} e^{-2x} \sin(n\pi x),$$

Snapshots of the solution at $t = 0.001, 0.1, 0.025, 0.05, 0.1$, and 0.25 are shown in Figure 2. (Can you explain why the "wave" moves to the left?) ■

Example 4

Consider the problem

$$\begin{aligned} &w_t - ((1 - x^2)w_{xx} - xw_x) = 0 \quad \text{for } -1 < x < 1 \text{ and } t > 0, \\ &w(0, x) = -16x + 185x^3 - 394x^5 + 225x^7 \quad \text{for } -1 < x < 1, \\ &w(t, x),\ w_x(t, x) \text{ bounded on } (-1, 1) \text{ for each } t > 0, \end{aligned}$$

which involves the singular Sturm-Liouville operator $\mathcal{S}$ from Example 2. Recall that the eigenvalues of $\mathcal{S}$ are $\lambda_n = -n^2, n = 0, 1, 2, \ldots$ and corresponding eigenfunctions are the Chebyshev polynomials $T_n(x)$, $n = 0, 1, 2, \ldots$. The initial data have the terminating eigenfunction expansion

$$w(0, x) = -\frac{1}{64}\left(29T_1(x) + 195T_3(x) + T_4(x) - 225T_7(x)\right).$$

Therefore, the solution for $t \geq 0$ is

$$w(t, x) = -\frac{1}{64}\left(29e^{-t}T_1(x) + 195e^{-9t}T_3(x) + e^{-16t}T_4(x) - 225e^{-49t}T_7(x)\right).$$

Figure 3 shows snapshots of the solution at $t = 0, 0.02, 0.03, 0.5, 0.2$, and 0.5. ■

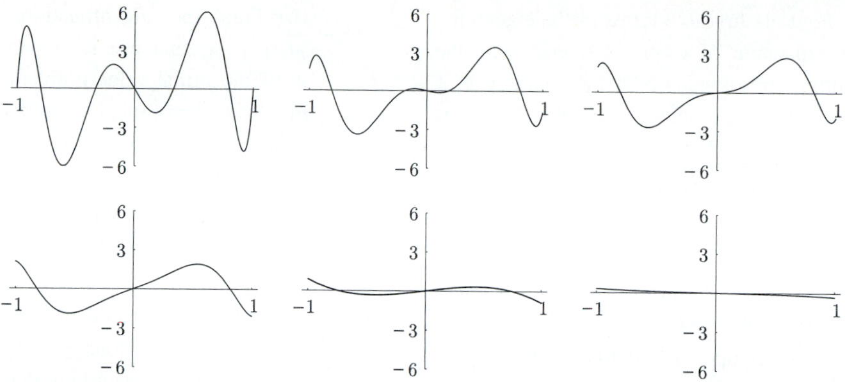

Figure 3

PROBLEMS

1. Let $\mathcal{S}$ be the operator defined for C^2 functions ψ on $[0, \pi]$ by

$$\mathcal{S}\psi = e^{2x}\left((e^{-2x}\psi')' + e^{-2x}\psi\right) = \psi'' - 2\psi' + \psi.$$

(a) Find the eigenvalues and corresponding *normalized* eigenfunctions of $\mathcal{S}$ with respect to the boundary conditions $\psi(0) = \psi(\pi) = 0$.

(b) Find the eigenfunction expansion of $\varphi(x) = e^x$.

2. Rework Problem 1 with the boundary conditions $\psi'(0) = \psi'(\pi) = 0$.

3. Rework Problem 1 with the boundary conditions $\psi'(0) = \psi(\pi) = 0$.

4. Let $\mathcal{S}$ be the operator defined for C^2 functions ψ on $[0, \pi]$ by

$$\mathcal{S}\psi = e^{-4x}(e^{4x}\psi')' = \psi'' + 4\psi'.$$

(a) Find the eigenvalues and corresponding *normalized* eigenfunctions of $\mathcal{S}$ with respect to the boundary conditions $\psi(0) = \psi(\pi) = 0$.

(b) Find the eigenfunction expansion of $\varphi(x) = e^{-2x}$.

5. Rework Problem 4 with the boundary conditions $\psi'(0) = \psi'(\pi) = 0$.

6. Rework Problem 4 with the boundary conditions $\psi'(0) = \psi(\pi) = 0$.

7. Let $\mathcal{S}$ be the operator defined for C^2 functions ψ on $[0, 1]$ by

$$\mathcal{S}\psi = e^{-2x}(e^{2x}\psi')' = \psi'' + 2\psi'.$$

(a) Find the eigenvalues and corresponding *normalized* eigenfunctions of $\mathcal{S}$ with respect to the boundary conditions $\psi(0) = \psi(1) = 0$.

(b) Find (numerically) the first four terms in the eigenfunction expansion of $\varphi(x) = 4x(1 - x)$. Plot the resulting approximation of φ along with φ on $[0, 1]$.

In each of Problems 8 through 10 find the eigenfunction expansion of the polynomial relative to the operator $\mathcal{S}$ in Example 2. (In other words, express the polynomial as a sum of Chebyshev polynomials. *Hint*: This requires no integration!)

8. $\varphi(x) = x^3$

9. $\varphi(x) = x^3 - x^2 + 1$

10. $\varphi(x) = x^5 + x$

11. Let $\mathcal{S}$ be the operator defined by $\mathcal{S}\psi = \psi''$ for C^2 functions ψ on $[0, \pi]$, and consider the initial-value problem

$$\begin{aligned} w_t - \mathcal{S}w &= 0 \quad \text{for } 0 < x < \pi, t > 0, \\ w(0, x) &= 1 \quad \text{for } 0 \le x \le \pi, \\ w(t, 0) &= w(t, \pi) + w_x(t, \pi) = 0 \quad \text{for } t > 0. \end{aligned}$$

(a) Find the first five eigenvalues $\lambda_0, \lambda_1, \lambda_2, \lambda_3, \lambda_4$ (rounded to three significant digits) and corresponding eigenfunctions $\psi_0, \psi_1, \psi_2, \psi_3, \psi_4$ of $\mathcal{S}$ with respect to the boundary conditions $\psi(0) = \psi(\pi) + \psi'(\pi) = 0$. Normalize each ψ_n so that $\int_0^\pi \psi_n^2 dx = 1$.

(b) Find the first five terms of the eigenfunction expansion of $\varphi(x) = 1$, and write the resulting approximation of the solution w of the initial-value problem. Then plot the graphs of $w(0.1, x)$, $w(1, x)$, $w(2, x)$, and $w(3, x)$.

12. Rework Problem 11 with the boundary conditions

$$w_x(t, 0) = w(t, \pi) + w_x(t, \pi) = 0$$

(and $\psi'(0) = \psi(\pi) + \psi'(\pi) = 0$). How and why does w behave differently?

13. Let $\mathcal{S}$ be the operator defined by $\mathcal{S}\psi = e^{2x}\left(e^{-2x}\psi'\right)'$ for C^2 functions ψ on $[0, \pi]$, and consider the initial-value problem

$$\begin{aligned} w_t - \mathcal{S}w = 0 \quad & \text{for } 0 < x < \pi, t > 0, \\ w(0, x) = 1 \quad & \text{for } 0 \le x \le \pi, \\ w(t, 0) = w_x(t, \pi) = 0 \quad & \text{for } t > 0. \end{aligned}$$

(a) Find the first five eigenvalues λ_0, λ_1, λ_2, λ_3, λ_4 (rounded to three significant digits) and the corresponding eigenfunctions ψ_0, ψ_1, ψ_2, ψ_3, ψ_4 of $\mathcal{S}$ with respect to the boundary conditions $\psi(0) = \psi(\pi) + \psi'(\pi) = 0$. Normalize each ψ_n so that $\int_0^\pi \psi_n^2 e^{-2x}\,dx = 1$.

(b) Find the first five terms of the eigenfunction expansion of $\varphi(x) = 1$, and write the resulting approximation of the solution w of the initial-value problem. Then plot the graphs of $w(0.2, x)$, $w(1, x)$, $w(2, x)$, and $w(3, x)$.

14. Rework Problem 13 with the boundary conditions

$$w_x(t, 0) = w(t, \pi) = 0$$

(and $\psi'(0) = \psi(\pi) = 0$). How and why does w behave differently?

15. Let $\mathcal{S}$ be the operator defined by $\mathcal{S}\psi = \psi'' + \psi$ for C^2 functions ψ on $[0, \pi]$, and consider the initial-value problem

$$\begin{aligned} w_t - \mathcal{S}w = 0 \quad & \text{for } 0 < x < \pi, t > 0, \\ w(0, x) = x \quad & \text{for } 0 \le x \le \pi, \\ w(t, 0) = w(t, \pi) = 0 \quad & \text{for } t > 0. \end{aligned}$$

(a) Find the eigenvalues $\lambda_0, \lambda_1, \lambda_2, \ldots$ and their corresponding eigenfunctions $\psi_0, \psi_1, \psi_2, \ldots$ of $\mathcal{S}$ with respect to the boundary conditions $\psi(0) = \psi(\pi) = 0$. Normalize each ψ_n so that $\int_0^\pi \psi_n^2\,dx = 1$.

(b) Find the eigenfunction expansion of $\varphi(x) = x$, and use it to write the solution w of the initial-value problem. Then use the first five terms to plot graphs of $w(0.1, x)$, $w(0.25, x)$, $w(0.5, x)$, and $w(1, x)$. What is $\lim_{t\to\infty} w(t, x)$?

16. Rework Problem 15 with the operator $\mathcal{S}$ defined by $\mathcal{S}\psi = \psi'' + 2\psi$. Explain the difference in the behavior of w.

17. Let $\mathcal{S}$ be the operator defined by $\mathcal{S}\psi = ((1 - x^2)\psi')'$ for C^2 functions ψ on $(-1, 1)$. With respect to the side conditions that ψ and ψ' be bounded on $(-1, 1)$, the eigenvalues of $\mathcal{S}$ are

$$\lambda_n = -n(n+1), \quad n = 0, 1, 2, 3, \ldots,$$

and corresponding eigenfunctions $\psi_0, \psi_1, \psi_2, \ldots$ are the Legendre polynomials

$$1, x, \ \frac{1}{2}\left(3x^2 - 1\right), \frac{1}{2}\left(5x^3 - 3x\right), \ \frac{1}{8}\left(3 - 30x^2 + 35x^4\right), \ldots.$$

(a) Find the eigenfunction expansion of $\varphi(x) = x^4 - x^3 + x$ relative to $\mathcal{S}$.

(b) Write down the solution w of the initial-value problem

$$\begin{aligned} w_t - \mathcal{S}w = 0 \quad & \text{for } -1 < x < 1, \\ & t > 0, \\ w(0, x) = x^4 - x^3 + x \quad & \text{for } -1 < x < 1, \\ w(t, x), w_x(t, x) \quad & \text{bounded on } (-1, 1) \\ & \text{for } t > 0. \end{aligned}$$

(c) Plot the graphs of $w(0, x)$, $w(0.1, x)$, $w(0.5, x)$, and $w(1, x)$.

18. Let $\mathcal{S}$ be the operator from Example 2, and consider the initial-value problem

$$\begin{aligned} w_t - \mathcal{S}w = 0 \quad & \text{for } -1 < x < 1, t > 0, \\ w(0, x) = \sin \pi x \quad & \text{for } -1 < x < 1, \\ w(t, x), w_x(t, x) \quad & \text{bounded on } (-1, 1) \\ & \text{for } t > 0. \end{aligned}$$

Find, and plot on $[-1, 1]$, the degree-five partial sum of the eigenfunction expansion of $w(x, t)$ at each of $t = 0$, 0.1, and 0.5.

19. Let $\ell = e^2 - 1$, and let $\mathcal{S}$ be defined for C^2 functions ψ on $[0, \ell]$ by

$$\mathcal{S}\psi = ((1+x)^2\psi')'.$$

(a) Verify that the general solution of $\mathcal{S}\psi = \lambda\psi$ on $[0, \ell]$ is given by

$$\psi = \frac{1}{\sqrt{1+x}}\Big(c_1 \cos\big(\sqrt{-\lambda - 1/4}\ln(1+x)\big) + c_2 \sin\big(\sqrt{-\lambda - 1/4}\ln(1+x)\big)\Big),$$

if $\lambda \leq -1/4$; otherwise (if $\lambda > -1/4$) it is given by

$$\psi = \frac{1}{\sqrt{1+x}}\Big(c_1 \cosh\big(\sqrt{\lambda + 1/4}\ln(1+x)\big) + c_2 \sinh\big(\sqrt{\lambda + 1/4}\ln(1+x)\big)\Big).$$

(b) Find the eigenvalues and corresponding eigenfunctions of $\mathcal{S}$ with respect to the boundary conditions

$$\psi(0) = 0, \quad \psi(\ell) = 0.$$

Normalize each ψ_n so that $\int_0^\ell \psi_n^2 dx = 1$.

(c) Write down the solution w of the initial-value problem

$$\begin{aligned}
&w_t - \mathcal{S}w = 0 \\
&\quad \text{for } 0 \leq x \leq \ell, t > 0, \\
&w(0, x) = \psi_0(x) - \tfrac{1}{3}\psi_2(x) + \tfrac{1}{2}\psi_4(x) \\
&\quad \text{for } 0 \leq x \leq \ell, \\
&w(t, 0) = w(t, \ell) = 0 \\
&\quad \text{for } t > 0.
\end{aligned}$$

(d) Plot the graph of $w(t, x)$ on $[0, \ell]$ for each of $t = 0, 0.02, 0.06, 0.2$, and 0.5.

20. Rework Problem 19 with the boundary conditions $\psi'(0) = \psi'(\ell) = 0$ (and $w_x(t, 0) = w_x(t, \ell) = 0$). How and why does w behave differently?

CHAPTER

12

FURTHER TOPICS IN PDEs

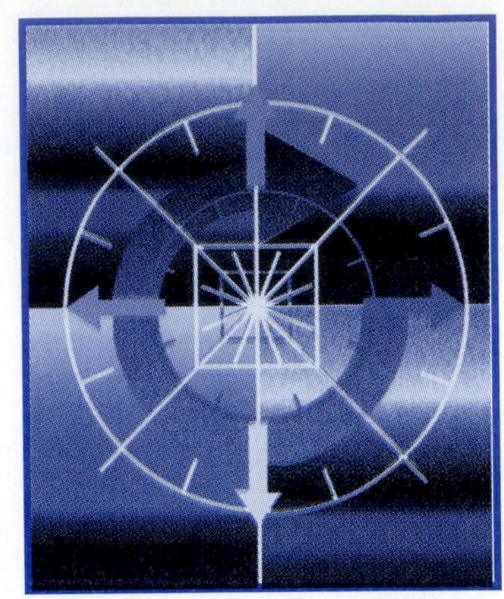

This chapter consists of a handful of topics related to some of the basic partial differential equations of physics. We will first study the one-dimensional wave equation, after which we will study a few problems involving two spatial dimensions.

12.1 The Wave Equation

We begin with a derivation of the wave equation, one of the fundamental partial differential equations of physics. The wave equation arises in many settings, including the propagation of sound and electromagnetic waves. Our derivation and subsequent discussion will be in the (mechanical) context of a vibrating string.

Suppose that a uniform, flexible string with mass density δ (per unit length) is pulled taut horizontally with tension τ between fixed ends at $x = 0$ and $x = \ell$. When displaced vertically, the string will vibrate in a vertical (x, u)-plane. Let $u(t, x)$ denote the vertical displacement of the string from its horizontal rest configuration at time $t \geq 0$ at a horizontal distance x from the left end of the string. Note that $u(t, x) = 0$ for $0 \leq x \leq \ell$ when the string is at rest.

At each point along the string there are equal and opposite tangential tension forces. We assume that the horizontal component of the tension is a constant τ for all x and t, so that only vertical motion can occur. If the string is not at its horizontal rest state, then the magnitude of the tension's vertical component will vary along the string. With

$$\theta(t, x) = \tan^{-1} u_x(t, x)$$

denoting the angle formed by the tangent line and the horizontal (as indicated in Figure 1), the vertical component of the tension is

$$\tau \tan(\theta(t, x)) = \tau u_x(t, x).$$

Now we apply Newton's second law to a small piece of the string between x_1 and x_2. (See Figure 2.) Its mass is $\delta(x_2 - x_1)$, its vertical acceleration is $u_{tt}(t, \overline{x})$ for

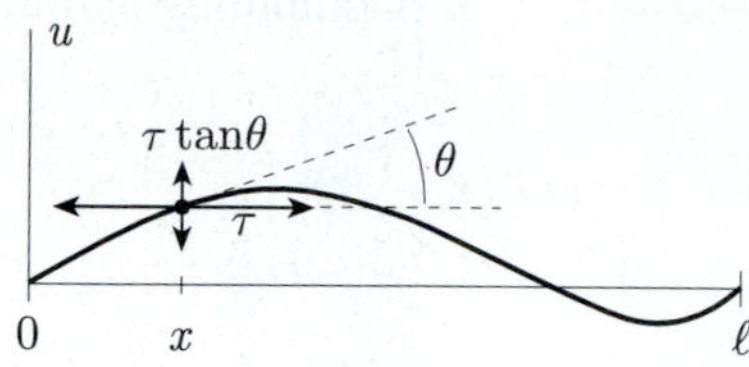

Figure 1

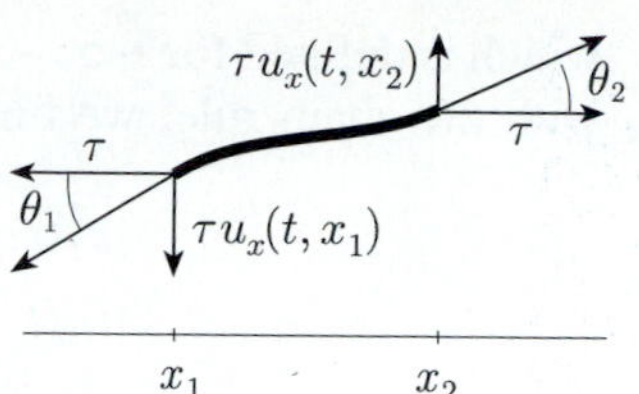

Figure 2

some $\overline{x}$ between x_1 and x_2, and the vertical component of the total force acting on it is the difference in the vertical components of the tensions at the ends: $\tau u_x(t, x_2) - \tau u_x(t, x_1)$. Therefore,

$$\delta(x_2 - x_1)u_{tt}(t, x_1) = \tau u_x(t, x_2) - \tau u_x(t, x_1),$$

which we rewrite as

$$u_{tt}(t, \overline{x}) = c^2 \frac{u_x(t, x_1) - u_x(t, x_2)}{x_2 - x_1},$$

where $c^2 = \tau/\delta$. Letting $x_2 \to x_1$, we arrive finally at the *one-dimensional wave equation*:

$$u_{tt} - c^2 u_{xx} = 0, \quad \text{where } c^2 = \tau/\delta. \tag{1}$$

As the wave equation is second order in t as well as in x, initial-value problems will involve two sets of initial data specifying both the initial "position" $u(0, x)$ and the initial velocity $u_t(0, x)$ on the x-interval of interest. This is illustrated in the following two problems, to which we will confine our attention.

- An infinite string that is initially perturbed from rest and gently released:

$$\left.\begin{aligned} u_{tt} - c^2 u_{xx} = 0, &\quad t > 0, \quad -\infty < x < \infty, \\ u(0, x) = \varphi(x), &\quad -\infty < x < \infty, \\ u_t(0, x) = 0, &\quad -\infty < x < \infty, \end{aligned}\right\} \tag{2}$$

  where φ is a continuous function on $(-\infty, \infty)$.
- A string of length ℓ with fixed ends that is initially perturbed from rest and gently released:

$$\left.\begin{aligned} u_{tt} - c^2 u_{xx} = 0, &\quad t > 0, \ 0 < x < \ell, \\ u(t, 0) = u(t, \ell) = 0, &\quad t \geq 0, \\ u(0, x) = \varphi(x), &\quad 0 \leq x \leq \ell, \\ u_t(0, x) = 0, &\quad 0 < x < \ell, \end{aligned}\right\} \tag{3}$$

  where φ is a continuous function on $[0, \ell]$ with $\varphi(0) = \varphi(\ell) = 0$.

Analogous problems with nontrivial initial velocities $u_t(0, x)$ are the subject of Problems 13 and 14 at the end of this section.

d'Alembert's Solution on $(-\infty, \infty)$ Suppose that F and G are twice differentiable functions on $(-\infty, \infty)$, and consider the function

$$u(t, x) = F(x + ct) + G(x - ct),$$

which is defined for $-\infty < t < \infty$ and $-\infty < x < \infty$. Computing partial derivatives with the chain rule, we find that

$$u_t(t,x) = cF'(x+ct) - cG'(x-ct),$$
$$u_{tt}(t,x) = c^2F''(x+ct) + c^2G''(x-ct),$$

and

$$u_x(t,x) = F'(x+ct) + G'(x-ct),$$
$$u_{xx}(t,x) = F''(x+ct) + G''(x-ct).$$

Thus it is easy to see that u satisfies $u_{tt} - c^2u_{xx} = 0$. This is *d'Alembert's solution** of (2). Notice that, if $G = 0$, the solution is a continuously shifted version of F, whose graph maintains its shape as it travels to the left along the x-axis with speed c. If $F = 0$, the solution is a continuously shifted version of G, whose graph maintains its shape as it travels to the right with the same speed c. In general, the solution is a superposition of these two types of functions.

We now look for the solution of the initial-value problem (2) in the form of

$$u(t,x) = F(x+ct) + G(x-ct).$$

The functions F and G will be determined by the initial conditions

$$u(0,x) = F(x) + G(x) = \varphi(x),$$
$$u_t(0,x) = cF'(x) - cG'(x) = 0.$$

A few moments of thought reveals that, if φ is twice differentiable, then the desired functions are

$$F = G = \frac{1}{2}\varphi,$$

and so the solution of (2) is

$$u(t,x) = \frac{1}{2}(\varphi(x+ct) + \varphi(x-ct)). \tag{4}$$

Recall that in our statement of problem (2) the function φ was not assumed to be twice differentiable, yet that assumption was needed to verify d'Alembert's solution. Though we will not go into the details, it is possible to define a kind of generalized, or "weak," notion of solution that will allow d'Alembert's solution to be valid under very mild assumptions on φ. For our purposes here, we will simply *define* (4) to be the solution of (2) for any continuous φ. (One justification of this is the fact that any such φ can be approximated arbitrarily well by a twice-differentiable function.)

---

* Jean le Rond d'Alembert was an illustrious French scholar whose work on the wave equation was published in 1747. His solution of the wave equation was discovered independently, and only slightly later, by Euler.

Example 1

Consider the problem

$$\begin{aligned} u_{tt} - u_{xx} &= 0, && t > 0, \quad -\infty < x < \infty, \\ u(0, x) &= e^{-x^2}, && -\infty < x < \infty, \\ u_t(0, x) &= 0, && -\infty < x < \infty. \end{aligned}$$

Using (4), we write the solution as

$$\begin{aligned} y &= \frac{1}{2}(e^{-(x+t)^2} + e^{-(x-t)^2}) = \frac{1}{2}e^{-(x^2+t^2)}(e^{-2xt} + e^{2xt}) \\ &= e^{-(x^2+t^2)} \cosh(2xt). \end{aligned}$$

The solution is plotted in Figures 3a and b. Figure 3a shows snapshots of the solution at times $t = 0, 0.5, 1, 2, 3$, and 4 (curves darken as t increases), while the surface in Figure 3b is the graph of y as a function of both x and t. ■

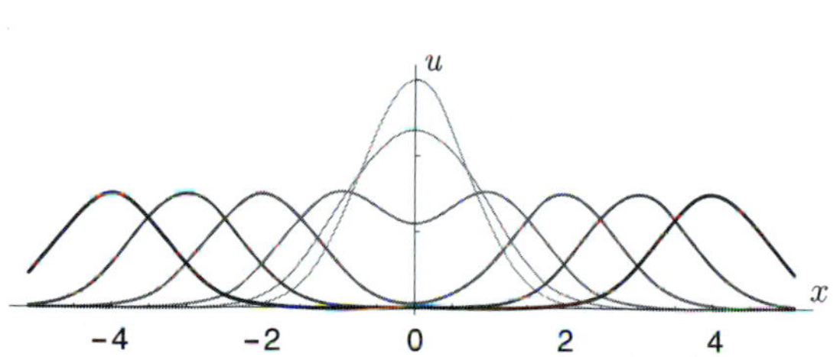

Figure 3a

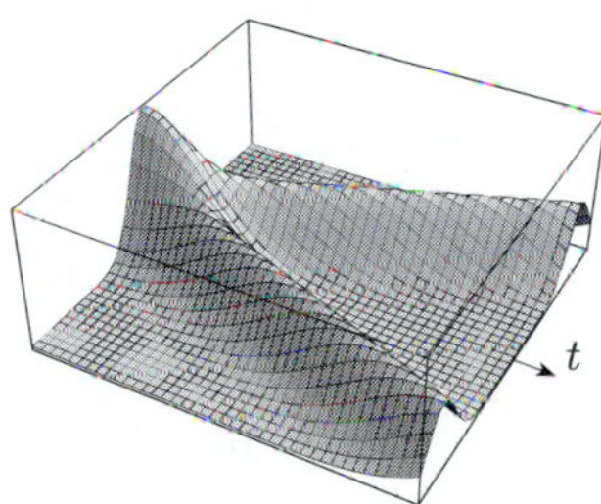

Figure 3b

Example 2

Consider the problem

$$\begin{aligned} u_{tt} - u_{xx} &= 0, && t > 0, \quad -\infty < x < \infty, \\ u(0, x) &= \sin x, && -\infty < x < \infty, \\ u_t(0, x) &= 0, && -\infty < x < \infty. \end{aligned}$$

Using (4), we write the solution as

$$u = \frac{1}{2}(\sin(x + t) + \sin(x - t)),$$

which can be rewritten as

$$u = \cos t \sin x$$

with the help of the sum and difference formulas for the sine function. Thus the wave motion is disguised as a stationary vibration, as illustrated in Figures 4a and b.

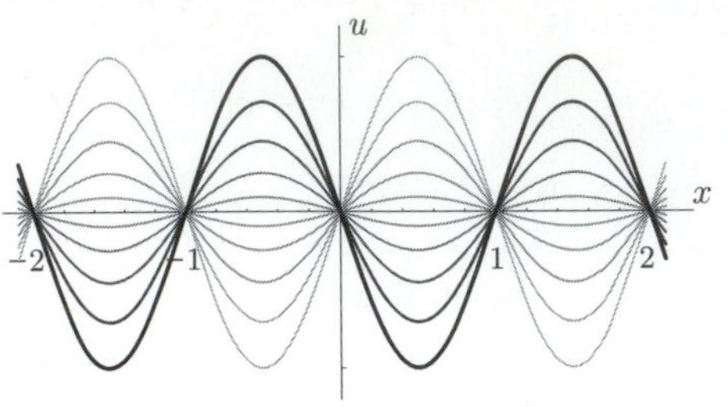

Figure 4a

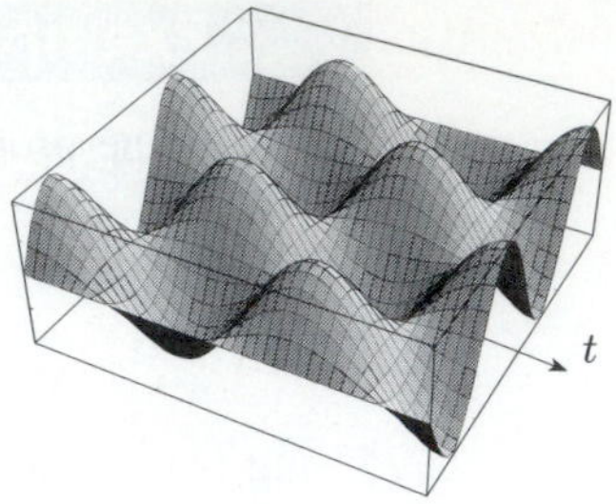

Figure 4b

Figure 4a shows snapshots of the solution at several times between $t = 0$ and $t = 1$. (Curves darken as t increases.) The surface in Figure 4b is the graph of y as a function of both x and t. ■

d'Alembert's Solution on $[0, \ell]$ Before addressing problem (3), we make a few important observations concerning d'Alembert's solution of (2),

$$u(t, x) = \frac{1}{2}(\varphi(x + ct) + \varphi(x - ct)).$$

i. If φ is an odd function, then u is an odd function of x for each fixed t.
ii. If φ is an even function, then u is an even function of x for each fixed t.
iii. If φ is p-periodic, then u is p-periodic in x for all t and $\frac{p}{c}$-periodic in t for all x.

These observations are easy to prove; you will be asked to do so in Problem 19. We point out, though, that (ii) is illustrated by Example 1, and (i) and (iii) are illustrated by Example 2.

Let us now consider problem (3), which we restate here for convenience:

$$\left.\begin{aligned} u_{tt} - c^2 u_{xx} &= 0, \quad t > 0, \quad 0 < x < \ell, \\ u(t, 0) = u(t, \ell) &= 0, \quad t \ge 0, \\ u(0, x) &= \varphi(x), \quad 0 \le x \le \ell, \\ u_t(0, x) &= 0, \quad 0 < x < \ell. \end{aligned}\right\} \tag{3}$$

We assume that φ is a continuous function on $[0, \ell]$ with $\varphi(0) = \varphi(\ell) = 0$. In the present context we will assume further that φ is twice differentiable on $[0, \ell]$. The idea is to extend φ to $(-\infty, \infty)$ in such a way that the resulting solution of (2) on $(-\infty, \infty)$ satisfies $u(t, 0) = u(t, \ell) = 0$ for all t. Then the solution of (3) is simply the restriction to $[0, \ell]$ of the solution of (2). The keys to this are observations (i) and (iii). Let φ_o be the odd, 2ℓ-periodic extension of φ; that is,

$$\varphi_o(x) = \begin{cases} \varphi(x), & \text{if } 0 \le x \le \ell, \\ -\varphi(-x), & \text{if } -\ell \le x < 0, \\ \varphi_o(x + 2\ell), & \text{for all } x, \end{cases}$$

noting that φ_o is continuous on $(-\infty, \infty)$. Then the function

$$u(t,x) = \frac{1}{2}(\varphi_o(x+ct) + \varphi_o(x-ct)) \tag{5}$$

satisfies (2) with initial data $\varphi_o(x)$. Moreover, u is an odd, 2ℓ-periodic function of x for each fixed t, from which it follows that $u(t,-\ell) = u(t,0) = u(t,\ell) = 0$ for all t. Therefore, the restriction of u to the interval $0 \le x \le \ell$ is precisely the solution of (3). In addition, it follows from observation (iii) that the solution is $\frac{2\ell}{c}$-periodic in t. In other words, the solution of (3) is a temporal vibration with period $\frac{2\ell}{c}$.

A brief remark on the practical construction of φ_o: Given φ defined on $[0,\ell]$, we may as well simply extend φ to $[-\ell,\ell]$ by defining

$$\varphi(x) = -\varphi(-x) \quad \text{for } -\ell \le x < 0.$$

That done, we next need a way to "shift" any x into the interval $[-\ell,\ell]$. This can be accomplished by means of the function

$$\sigma(x) = x - 2\ell\left\lfloor \frac{x+\ell}{2\ell} \right\rfloor, \tag{6}$$

in which $\lfloor\ \rfloor$ denotes the "floor" function, also known as the "greatest integer" function. (For instance, $\lfloor 2.3 \rfloor = 2$ and $\lfloor -\pi \rfloor = -4$.) The crucial property of σ is that, if n is an integer and $-\ell \le \xi < \ell$, then

$$\sigma(2n\ell + \xi) = \xi.$$

(See Problem 20.) Thus the odd, 2ℓ-periodic extension of φ is simply

$$\varphi_o(x) = \varphi(\sigma(x)).$$

Example 3

Consider the problem

$$\begin{aligned} u_{tt} - 3u_{xx} &= 0, & t > 0,\ 0 < x < 1, \\ u(t,0) = u(t,1) &= 0, & t \ge 0, \\ u(0,x) &= x^5(1-x), & 0 \le x \le 1, \\ u_t(0,x) &= 0, & 0 < x < 1. \end{aligned}$$

First we extend the initial data to $[-1,1]$ by defining

$$\varphi(x) = \begin{cases} x^5(1-x), & \text{if } 0 \le x \le 1; \\ x^5(1+x), & \text{if } -1 \le x < 0. \end{cases}$$

The odd, 2-periodic extension of the initial data is then given by

$$\varphi_o(x) = \varphi\left(x - 2\left\lfloor \frac{x+1}{2} \right\rfloor\right),$$

and the solution is

$$u(t,x) = \frac{1}{2}\left(\varphi_o(x+\sqrt{3}t) + \varphi_o(x-\sqrt{3}t)\right).$$

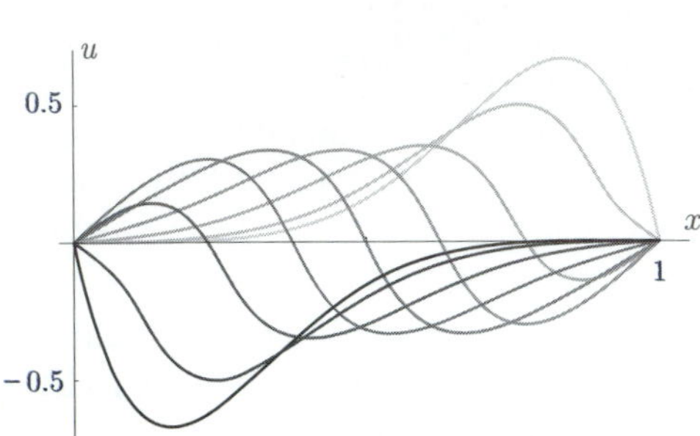

Figure 5a

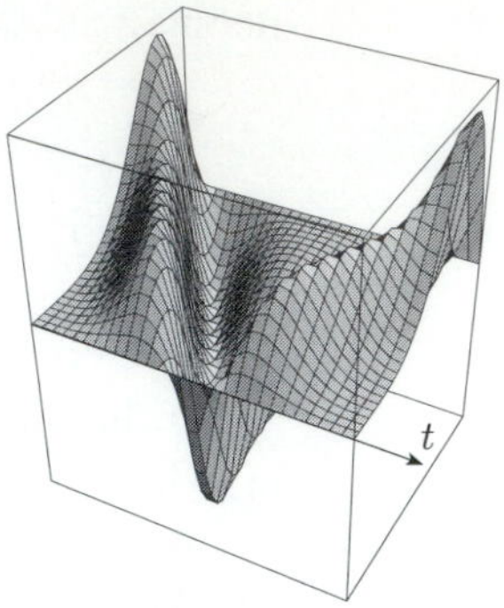

Figure 5b

The solution, which is periodic in t with period $2/\sqrt{3}$, is plotted in Figures 5a and b. Figure 5a shows snapshots of the solution at several times between $t = 0$ and $t = 1/\sqrt{3}$. The surface in Figure 5b is the graph of y as a function of both x and t, for $0 \le t \le 2/\sqrt{3}$. ■

Fourier Series Consider the problem

$$\begin{aligned} &u_{tt} - c^2 u_{xx} = 0, && t > 0, \quad 0 < x < \ell, \\ &u(t,0) = u(t,\ell) = 0, && t \ge 0, \\ &u(0,x) = \sin \frac{n\pi x}{\ell}, && 0 \le x \le \ell, \\ &u_t(0,x) = 0, && 0 < x < \ell, \end{aligned}$$

where n is some positive integer. Here the odd, 2ℓ-periodic extension of the initial data is given very easily by $\varphi_o(x) = \sin(n\pi x/\ell)$. Thus d'Alembert's solution is

$$u(t,x) = \frac{1}{2}\left(\sin\left(\frac{n\pi}{\ell}(x+ct)\right) + \sin\left(\frac{n\pi}{\ell}(x-ct)\right)\right).$$

Using the sum and difference formulas for sine, we rewrite this as

$$u(t,x) = \cos\left(\frac{n\pi}{\ell}ct\right)\sin\left(\frac{n\pi}{\ell}x\right). \tag{7}$$

Thus the solution represents a simple vibration with period $2\ell/(nc)$ (time units per cycle). Thus its *frequency* is $nc/(2\ell)$ (cycles per unit time).

Now let's suppose that the initial data are given by a Fourier sine series:

$$\begin{aligned} &u_{tt} - c^2 u_{xx} = 0, && t > 0, \quad 0 < x < \ell, \\ &u(t,0) = u(t,\ell) = 0, && t \ge 0, \\ &u(0,x) = \sum_{n=1}^{\infty} b_n \sin \frac{n\pi x}{\ell}, && 0 \le x \le \ell, \\ &u_t(0,x) = 0, && 0 < x < \ell. \end{aligned}$$

Because of linearity, we can use (7) to assemble the solution by superposition. Each term $b_n \sin(n\pi x/\ell)$ in the series for $u(0, x)$ contributes a term

$$b_n \cos\left(\frac{n\pi}{\ell}ct\right) \sin\left(\frac{n\pi}{\ell}x\right)$$

to the solution. Thus the solution† is

$$u(t, x) = \sum_{n=1}^{\infty} b_n \cos\left(\frac{n\pi}{\ell}ct\right) \sin\left(\frac{n\pi}{\ell}x\right). \tag{8}$$

For each nonzero b_n, the solution has a component that vibrates with frequency $nc/(2\ell)$. With $n = 1$, this gives the *fundamental frequency*, which is $c/(2\ell)$. All the other component frequencies, or *harmonics*, are integer multiples of the fundamental frequency. (This is why a single vibrating piano or guitar string makes a "harmonious," rather than dissonant, sound.) Note that the possible harmonics are determined solely by c and ℓ, which are physical characteristics of the string itself; the initial configuration, or shape, of the string determines only the "blend" of those harmonics. Recall that in our derivation of the wave equation c was defined by $c^2 = \tau/\delta$, where τ and δ are the tension and mass density of the string, respectively. Thus the fundamental frequency is

$$\frac{k}{2\ell} = \frac{1}{2\ell}\sqrt{\frac{\tau}{\delta}}.$$

So we observe that increased tension, decreased length, and decreased mass density each result in a greater fundamental frequency (i.e., higher *pitch*).

Example 4

Consider the problem

$$\begin{aligned}
&u_{tt} - \pi^2 u_{xx} = 0, && t > 0,\ 0 < x < \pi, \\
&u(t, 0) = u(t, \pi) = 0, && t \geq 0, \\
&u(0, x) = \sin x - \frac{1}{7}\sin 7x, && 0 \leq x \leq \pi, \\
&u_t(0, x) = 0, && 0 < x < \pi.
\end{aligned}$$

By inspection we write the solution

$$u(t, x) = \cos \pi t \sin x - \frac{1}{7}\cos(7\pi t)\sin(7x).$$

The fundamental frequency is $1/2$ (since the period of $\cos \pi t$ is 2). The only harmonic present in this case has a frequency of $7/2$. Figure 6 shows the graph of the solution at fixed times between $t = 0$ and $t = 1$. Notice how the harmonic is superimposed over the fundamental frequency. ■

---

† One can also arrive at this by the method of separation of variables.

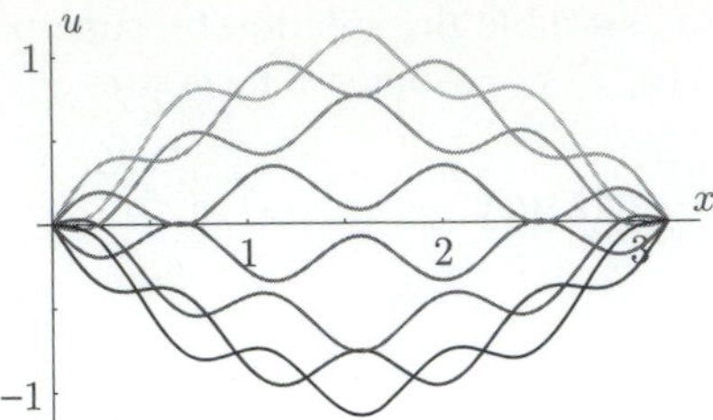

Figure 6

Example 5

Consider the problem

$$\begin{aligned}
&u_{tt} - c^2 u_{xx} = 0, && t > 0,\ 0 < x < 2,\\
&u(t,0) = u(t,2) = 0, && t \geq 0,\\
&u(0,x) = 1 - |x-1|, && 0 \leq x \leq 2,\\
&u_t(0,x) = 0, && 0 < x < 2.
\end{aligned}$$

Using the methods of Section 11.5, it is a fairly routine task to determine that the coefficients in the Fourier sine-series expansion

$$1 - |x-1| = \sum_{n=1}^{\infty} b_n \sin\left(\frac{n\pi x}{2}\right), \quad 0 \leq x \leq 2,$$

are, for even and odd indices respectively, given by

$$b_{2k} = 0 \quad \text{and} \quad b_{2k+1} = \frac{8(-1)^k}{(2k+1)^2\pi^2}.$$

Writing only the nonzero terms, we therefore have

$$1 - |x-1| = \sum_{k=0}^{\infty} \frac{8(-1)^k}{(2k+1)^2\pi^2} \sin\left(\frac{(2k+1)\pi x}{2}\right) \quad \text{for } 0 \leq x \leq 2.$$

Consequently, the solution is given as a Fourier series by

$$u(t,x) = \sum_{n=0}^{\infty} \frac{8(-1)^n}{(2n+1)^2\pi^2} \cos\left(\frac{(2n+1)\pi ct}{2}\right) \sin\left(\frac{(2n+1)\pi x}{2}\right).$$

Thus the harmonic frequencies present are the odd multiples of the fundamental frequency $c/4$, that is, $3c/4$, $5c/4$, ... The solution with $c = 1$ is plotted in Figures 7a and b. Figure 7a shows snapshots of the solution at several times between $t = 0$ and $t = 2$. The surface in Figure 7b is the graph of y as a function of both x and t, for $0 \leq t \leq 4$. Notice that this "solution" is not differentiable and hence is actually a type of generalized solution. This behavior is typical of solutions of the wave equation when the initial data are not differentiable. (Recall that solutions of the diffusion equation $u_t - ku_{xx}$ behave quite differently, in that nondifferentiable initial data are immediately "smoothed out," becoming differentiable for all $t > 0$.) ■

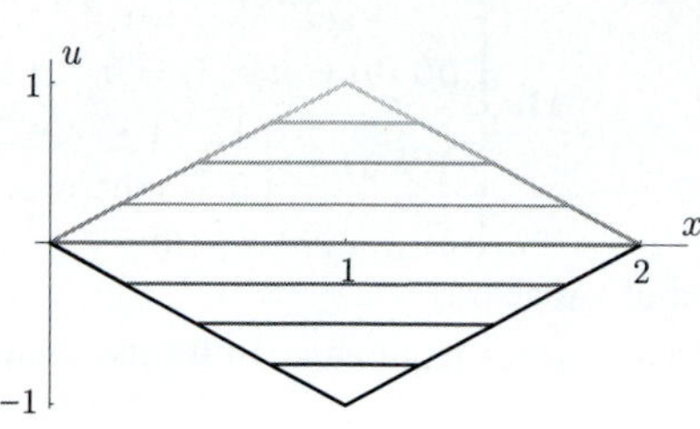

Figure 7a

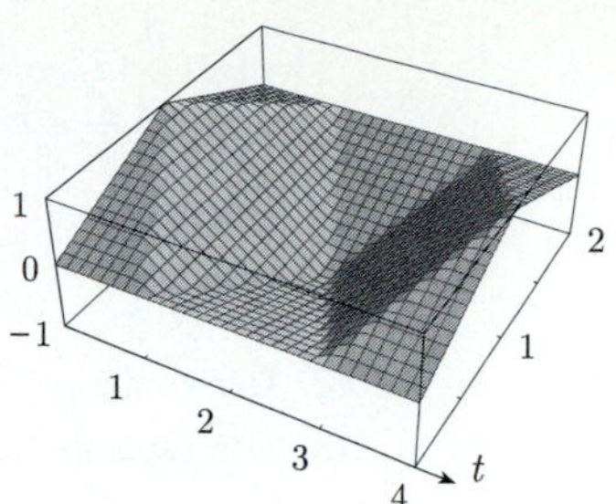

Figure 7b

PROBLEMS

For the given c and φ in each of Problems 1 through 5, sketch the graph of d'Alembert's solution of

$$u_{tt} - c^2 u_{xx} = 0, \quad t > 0, \quad -\infty < x < \infty,$$
$$u(0, x) = \varphi(x), \quad -\infty < x < \infty,$$
$$u_t(0, x) = 0, \quad -\infty < x < \infty,$$

at each of $t = 0, 1, 2, 3$.

1. $c = 1,\ \varphi(x) = \begin{cases} 2\cos\left(\frac{\pi x}{2}\right), & \text{if } -1 \le x \le 1 \\ 0, & \text{otherwise} \end{cases}$

2. $c = 1,\ \varphi(x) = \begin{cases} 2\sin\left(\frac{\pi x}{2}\right), & \text{if } -2 \le x \le 2 \\ 0, & \text{otherwise} \end{cases}$

3. $c = \frac{1}{2},\ \varphi(x) = \begin{cases} 2|\sin \pi x|, & \text{if } -1 \le x \le 1 \\ 0, & \text{otherwise .} \end{cases}$

4. $c = 1,\ \varphi(x) = \begin{cases} 2\sin \pi x, & \text{if } -2 \le x \le -1 \text{ or } 1 \le x \le 2 \\ 0, & \text{otherwise} \end{cases}$

5. $c = \frac{1}{3},\ \varphi(x) = \begin{cases} 1 - |x|, & \text{if } -1 \le x \le 1 \\ 0, & \text{otherwise} \end{cases}$

In Problems 6 through 11, find the Fourier series representation of the solution.

6. $\begin{cases} u_{tt} - u_{xx} = 0, & t > 0, \quad 0 < x < \pi, \\ u(t, 0) = u(t, \pi) = 0, & t > 0, \\ u(0, x) = \sin x - \dfrac{1}{8}\sin 4x, & 0 \le x \le \pi, \\ u_t(0, x) = 0, & 0 \le x \le \pi. \end{cases}$

7. $\begin{cases} u_{tt} - 4u_{xx} = 0, & t > 0, \quad 0 < x < \pi, \\ u(t, 0) = u(t, \pi) = 0, & t > 0, \\ u(0, x) = \displaystyle\sum_{n=1}^{\infty} \frac{2^n}{n!}\sin nx, & 0 \le x \le \pi, \\ u_t(0, x) = 0, & 0 \le x \le \pi \end{cases}$

8. $\begin{cases} u_{tt} - u_{xx} = 0, & t > 0, \quad 0 < x < 4, \\ u(t, 0) = u(t, 4) = 0, & t > 0, \\ u(0, x) = \begin{cases} 1 - |x - 2|, & 1 \le x \le 3 \\ 0, & \text{otherwise,} \end{cases} \\ u_t(0, x) = 0, & 0 < x < 4 \end{cases}$

9. $\begin{cases} u_{tt} - 9u_{xx} = 0, & t > 0, \quad 0 < x < 3\pi, \\ u(t, 0) = u(t, 3\pi) = 0, & t > 0, \\ u(0, x) = \begin{cases} -\sin x, & \pi \le x \le 2\pi \\ 0, & \text{otherwise,} \end{cases} \\ u_t(0, x) = 0, & 0 \le x \le 3\pi. \end{cases}$

10. $\begin{cases} u_{tt} - 4u_{xx} = 0, & t > 0, \quad 0 < x < 2\pi, \\ u(t,0) = u(t,2\pi) = 0, & t > 0, \\ u(0,x) = \begin{cases} \sin x, & 0 \le x \le \pi \\ 0, & \text{otherwise,} \end{cases} \\ u_t(0,x) = 0, & 0 \le x \le 2\pi. \end{cases}$

11. $\begin{cases} u_{tt} - \dfrac{1}{\pi^2} u_{xx} = 0, & t > 0, \quad 0 < x < 3, \\ u(t,0) = u(t,3) = 0, & t > 0, \\ u(0,x) = \begin{cases} 1, & 1 \le x \le 2 \\ 0, & \text{otherwise,} \end{cases} \\ u_t(0,x) = 0, & 0 \le x \le 3. \end{cases}$

12. Derive the solution in (8) by separation of variables.

13. Use separation of variables to find a Fourier series representation for the solution of

$$\begin{aligned} u_{tt} - c^2 u_{xx} &= 0, \quad && t > 0, \quad 0 < x < \ell, \\ u(t,0) = u(t,\ell) &= 0, && t > 0, \\ u(0,x) = 0, \quad u_t(0,x) &= v(x), && 0 \le x \le \ell, \end{aligned}$$

in terms of the coefficients of the Fourier sine series of v on $[0, \ell]$.

14. Consider an initially straight, infinite string with given initial velocity, whose subsequent motion satisfies

$$\begin{aligned} u_{tt} - c^2 u_{xx} &= 0, \quad && t > 0, \quad -\infty < x < \infty, \\ u(0,x) = 0, \quad u_t(0,x) &= v(x), && -\infty < x < \infty. \end{aligned}$$

(a) Show that a solution is given by

$$u(t,x) = \frac{1}{2c} \int_{x-ct}^{x+ct} v(s)ds.$$

(b) Use superposition and the result of part (a) to write d'Alembert's solution of

$$\begin{aligned} u_{tt} - c^2 u_{xx} &= 0, \quad && t > 0, \quad -\infty < x < \infty, \\ u(0,x) &= \varphi(x), && -\infty < x < \infty, \\ u_t(0,x) &= v(x), && -\infty < x < \infty. \end{aligned}$$

Use the results of Problem 14 to solve Problems 15 and 16.

15. $\begin{cases} u_{tt} - u_{xx} = 0, \quad t > 0, \quad -\infty < x < \infty, \\ u(0,x) = \sin x, \quad u_t(0,x) = \cos x, \quad -\infty < x < \infty \end{cases}$

16. $\begin{cases} u_{tt} - u_{xx} = 0, \quad t > 0, & -\infty < x < \infty, \\ u(0,x) = 0, u_t(0,x) = 2xe^{-x^2}, & -\infty < x < \infty \end{cases}$

17. Let $y = x + ct$ and $z = x - ct$, where $c \ne 0$.

(a) Use the chain rule ($f_\eta = f_\xi \xi_\eta$) to show that

$$u_{xx} = u_{yy} + u_{yz} + u_{zy} + u_{zz} \quad \text{and} \quad u_{tt} = c^2 u_{yy} - c^2 u_{yz} - c^2 u_{zy} + c^2 u_{zz}.$$

Conclude that

$$u_{tt} - c^2 u_{xx} = -4c^2 u_{yz}$$

and therefore that $u_{tt} - c^2 u_{xx} = 0$ is equivalent to $u_{yz} = 0$.

(b) By successive antidifferentiations, show that every solution of $u_{yz} = 0$, and hence every solution of $u_{tt} - c^2 u_{xx} = 0$, is of the form

$$u = F(y) + G(z).$$

18. Consider the *first-order wave equation*

$$u_t + cu_x = 0. \qquad (*)$$

(a) Show that, regardless of the sign of c, every solution of $(*)$ that is twice differentiable in both t and x also satisfies the second-order wave equation

$$u_{tt} - c^2 u_{xx} = 0.$$

(b) Show that the problem

$$u_t + cu_x = 0, \quad t > 0, \quad -\infty < x < \infty,$$
$$u(0, x) = \varphi(x), \quad -\infty < x < \infty,$$

is satisfied by

$$u(t, x) = \varphi(x - ct).$$

(c) Explain how parts (a) and (b) give rise to d'Alembert's solution of problem (2).

19. Prove the following assertions about

$$u(t, x) = \frac{1}{2}(\varphi(x + ct) + \varphi(x - ct)).$$

(a) If φ is an odd function, then u is an odd function of x for each fixed t.

(b) If φ is an even function, then u is an even function of x for each fixed t.

(c) If φ is p-periodic, then u is p-periodic in x for all t and $\frac{p}{c}$-periodic in t for all x.

20. Let $\ell > 0$, and consider the "shift function"

$$\sigma(x) = x - 2\ell \left\lfloor \frac{x + \ell}{2\ell} \right\rfloor, \quad -\infty < x < \infty.$$

(a) Show that any real number x can be written in the form $x = 2n\ell + \xi$, where n is an integer and $-\ell \le \xi < \ell$.

(b) Show that, if $x = 2n\ell + \xi$, where n is an integer and $-\ell \le \xi < \ell$, then $\sigma(x) = \xi$.

21. The problem

$$u_{tt} + \beta u_t - c^2 u_{xx} = 0, \quad t > 0, \quad 0 < x < \ell,$$
$$u(t, 0) = u(t, \ell) = 0, \quad t > 0,$$
$$u(0, x) = \varphi(x), \quad 0 \le x \le \ell,$$
$$u_t(0, x) = 0, \quad 0 \le x \le \ell,$$

where $\beta > 0$, models the *damped* motion of a string. Assume that the initial data have the Fourier series expansion

$$\varphi(x) = \sum_{n=1}^{\infty} b_n \sin \frac{n\pi x}{\ell}.$$

(a) Solve this problem by separation of variables under the assumption that $0 < \beta < 2\pi c/\ell$. Describe how the behavior of this solution differs from that with $\beta = 0$ (no damping).

(b) Solve again, assuming that $2\pi c/\ell < \beta < 2\sqrt{2}\,\pi c/\ell$. How does this solution behave?

22. Suppose that $u(t, x)$ satisfies

$$\begin{aligned} u_{tt} - c^2 u_{xx} &= 0, \quad t > 0, \quad 0 < x < \ell, \\ u(t, 0) = u(t, \ell) &= 0, \quad t > 0, \\ u(0, x) = \varphi(x), \quad u_t(0, x) &= v(x), \quad 0 \le x \le \ell, \end{aligned}$$

where $c > 0$. Also define dimensionless variables $\tau = ct/\ell$ and $\xi = x/\ell$, and let $w(\tau, \xi) = u(\ell\tau/c, \ell\xi)$. Show that $w(\tau, \xi)$ satisfies

$$\begin{aligned} w_{\tau\tau} - w_{\xi\xi} &= 0, \quad \tau > 0, 0 < \xi < 1, \\ w(\tau, 0) = w(\tau, 1) &= 0, \quad \tau > 0, \\ w(0, \xi) = \varphi(\ell\xi), \quad w_\tau(0, \xi) &= c\ell v(\ell\xi), \quad 0 \le \xi \le 1. \end{aligned}$$

12.2 The 2-D Laplace Equation

Suppose that Ω is an open, connected region in the plane whose boundary is a piecewise-smooth curve, which we denote by $\partial\Omega$. (See Figure 1.) For instance, Ω might be the interior of a polygon or a disk. In general, we will refer to such a region as a *domain*. Also let $\overline{\Omega}$ denote the *closure* of Ω, which consists of all points in Ω together with all points on $\partial\Omega$.

Now imagine that a thin plate occupies $\overline{\Omega}$ and that the faces of the plate are perfectly insulated, so that no heat may flow in the direction perpendicular to the plane. Suppose further that the thermal diffusivity of the plate is given by $k(x, y)$ for each (x, y) in $\overline{\Omega}$. Now let $u(t, x, y)$ denote the temperature at location (x, y) in $\overline{\Omega}$ at time t. A straightforward extension of the derivation of the one-dimensional diffusion equation in Section 11.1 reveals that u satisfies the *two-dimensional diffusion equation*

$$u_t - \big((ku_x)_x + (ku_y)_y\big) = 0 \quad \text{for } t > 0 \text{ and } (x, y) \text{ in } \Omega. \tag{1}$$

We assume from now on that k is constant. Thus (1) becomes

$$u_t - k(u_{xx} + u_{yy}) = 0 \quad \text{for } t > 0 \text{ and } (x, y) \text{ in } \Omega. \tag{2}$$

Time-dependent problems associated with (2) will be considered in the next section. In this section we concentrate on time-independent, or *steady-state*, problems

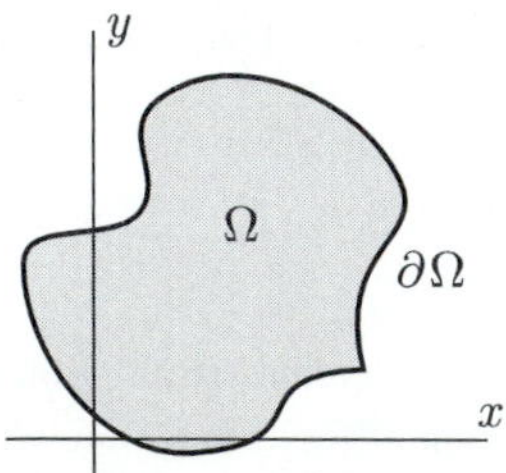

Figure 1

associated with (2). When u is independent of t, equation (2) becomes the two-dimensional *Laplace equation*:

$$u_{xx} + u_{yy} = 0 \quad \text{for } (x, y) \text{ in } \Omega. \tag{3}$$

In addition to steady-state diffusion problems, this partial differential equation and its three-dimensional analogue,

$$u_{xx} + u_{yy} + u_{zz} = 0,$$

arise in electrostatics and fluid mechanics, as well as other applications in physics. The Laplace equation is sometimes called the *potential equation*, and solutions of it are often called *harmonic functions*.

The Dirichlet Problem It should not be surprising that, in order for a problem involving the Laplace equation to have a unique solution, there must be some form of boundary condition imposed at each point on $\partial\Omega$. If the value of the solution is specified at each point on $\partial\Omega$, then we have what is known as a *Dirichlet problem*. With a function g describing the given boundary values, a Dirichlet problem has the form

$$\begin{aligned} u_{xx} + u_{yy} &= 0 \quad \text{in } \Omega, \\ u(x, y) &= g(x, y) \quad \text{on } \partial\Omega. \end{aligned} \tag{4}$$

In a heat conduction application, u is the steady-state temperature distribution in Ω, if the temperature along the boundary is fixed and given by g.

Except in a few special cases in which Ω has a simple and accommodating shape, we have no hope finding the exact solution of (4) in any elementary form. There are numerical procedures that may be used to compute approximate solutions; however, we will not attempt to describe any of them in this text. We will instead restrict our attention to two simple geometries that allow the use of Fourier series methods.

A Rectangular Domain Suppose that $\overline{\Omega}$ is the rectangle in which

$$0 \le x \le \ell \quad \text{and} \quad 0 \le y \le h,$$

and let $g_1, \ldots, g_4$ to denote the boundary data on the four pieces of $\partial\Omega$ as indicated in Figure 2. Then problem (4) can be expressed as

$$\begin{aligned} & & u_{xx} + u_{yy} &= 0 \quad \text{in } \Omega, \\ u(x, 0) &= g_1(x), & u(x, h) &= g_2(x) \quad \text{for } 0 \le x \le \ell, \\ u(0, y) &= g_3(y), & u(\ell, y) &= g_4(y) \quad \text{for } 0 < y < h. \end{aligned} \tag{5}$$

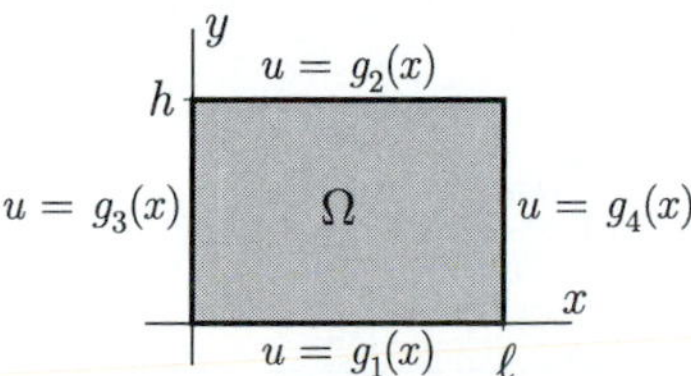

Figure 2

It will be convenient to construct the solution of (5) by superimposing the solutions of one or more simpler problems in which all but one of $g_1, \ldots, g_4$ are zero. Let's begin by considering (5) with $g_1 \neq 0$ and $g_2 = g_3 = g_4 = 0$. Using the separation-of-variables method, we start out looking for solutions of the Laplace equation in the form

$$u(x, y) = X(x)Y(y).$$

Such a product satisfies the Laplace equation if

$$X''Y + XY'' = 0.$$

This implies that there is a constant λ such that

$$\frac{X''}{X} = -\frac{Y''}{Y} = \lambda;$$

that is,

$$X'' - \lambda X = 0 \quad \text{and} \quad Y'' + \lambda Y = 0. \tag{6}$$

The boundary conditions $u(0, y) = 0$ and $u(\ell, y) = 0$ for $0 \le y \le h$ demand that X satisfy

$$X'' - \lambda X = 0, \quad X(0) = 0, \quad X(\ell) = 0.$$

From this we obtain a sequence of eigenvalues and eigenfunctions:

$$\lambda_n = -\frac{n^2\pi^2}{\ell^2} \quad \text{and} \quad X_n(x) = \sin\left(\frac{n\pi x}{\ell}\right), \quad n = 1, 2, 3, \ldots.$$

Now, for each n, we require a solution Y_n of

$$Y_n'' - \frac{n^2\pi^2}{\ell^2} Y_n = 0,$$

which yields*

$$Y_n(y) = a_n \cosh\left(\frac{n\pi y}{\ell}\right) + b_n \sinh\left(\frac{n\pi y}{\ell}\right),$$

where a_n and b_n are constants that will be determined by the boundary conditions along the other two edges. So far we have found that the Laplace equation and the boundary conditions $u(0, y) = 0$ and $u(\ell, y) = 0$ for $0 \le y \le h$ are satisfied by each of

$$X_n(x)Y_n(y) = \sin\left(\frac{n\pi x}{\ell}\right)\left(a_n \cosh\left(\frac{n\pi y}{\ell}\right) + b_n \sinh\left(\frac{n\pi y}{\ell}\right)\right),$$

for $n = 1, 2, 3, \ldots$, and therefore also by

$$u(x, y) = \sum_{n=1}^{\infty} \sin\left(\frac{n\pi x}{\ell}\right)\left(a_n \cosh\left(\frac{n\pi y}{\ell}\right) + b_n \sinh\left(\frac{n\pi y}{\ell}\right)\right). \tag{7}$$

---

* Recall that $\sinh x = (e^x - e^{-x})/2$ and $\cosh x = (e^x + e^{-x})/2$. The other hyperbolic functions, tanh, coth, sech, and csch, are defined by analogy with the trigonometric functions.

The boundary condition $u(x, 0) = g_1(x)$ will be satisfied if

$$\sum_{n=0}^{\infty} a_n \sin\left(\frac{n\pi x}{\ell}\right) = g_1(x).$$

Thus the a_n are the coefficients in the Fourier sine series of g_1, which by Theorem 2 in Section 11.4 are

$$a_n = \frac{2}{\ell}\int_0^{\ell} g_1(x)\sin\left(\frac{n\pi x}{\ell}\right)dx, \quad n = 1, 2, 3, \ldots. \tag{8}$$

From the boundary condition $u(x, h) = 0$ we obtain

$$\sum_{n=0}^{\infty}\left(a_n \cosh\left(\frac{n\pi h}{\ell}\right) + b_n \sinh\left(\frac{n\pi h}{\ell}\right)\right)\sin\left(\frac{n\pi x}{\ell}\right) = 0,$$

from which we conclude that, for each n,

$$a_n \cosh\left(\frac{n\pi h}{\ell}\right) + b_n \sinh\left(\frac{n\pi h}{\ell}\right) = 0.$$

Solving for b_n and using (8), we find

$$b_n = -a_n \coth\left(\frac{n\pi h}{\ell}\right), \quad n = 1, 2, 3, \ldots. \tag{9}$$

By combining (7) and (9), we arrive at

$$u(x, y) = \sum_{n=1}^{\infty} a_n \sin\left(\frac{n\pi x}{\ell}\right)\left(\cosh\left(\frac{n\pi y}{\ell}\right) - \coth\left(\frac{n\pi h}{\ell}\right)\sinh\left(\frac{n\pi y}{\ell}\right)\right),$$

where the a_n are given by (8). Some simplification gives us

$$u(x, y) = \sum_{n=1}^{\infty} a_n \sin\left(\frac{n\pi x}{\ell}\right)\operatorname{csch}\left(\frac{n\pi h}{\ell}\right)\sinh\left(\frac{n\pi(h-y)}{\ell}\right) \tag{10}$$

as the solution of (5) with $g_1 \neq 0$ and $g_2 = g_3 = g_4 = 0$.

Let us now closely examine the solution u in (10). It is easy to verify that each term in the series satisfies the Laplace equation; thus so does the full series. Also, it is easy to see that u is zero whenever $x = 0$, $x = \ell$, or $y = h$, and that $u(x, 0)$ has the form of a Fourier sine series whose coefficients are chosen so that u will satisfy the boundary condition along the edge of Ω where $y = 0$.

Now we make a useful observation. If we replace $h - y$ with y in (10), then the Laplace equation is still satisfied, and u is still zero whenever $x = 0$ or $x = \ell$, but now u is zero whenever $y = 0$. Also, $u(x, h)$ has the form of a Fourier sine series whose coefficients can be chosen so that u will satisfy the boundary condition along the edge of Ω where $y = h$. Therefore, the solution of (5) with $g_2 \neq 0$ and $g_1 = g_3 = g_4 = 0$ is

$$u(x, y) = \sum_{n=1}^{\infty} a_n \sin\left(\frac{n\pi x}{\ell}\right)\operatorname{csch}\left(\frac{n\pi h}{\ell}\right)\sinh\left(\frac{n\pi y}{\ell}\right), \tag{11}$$

where the a_n are the coefficients in the Fourier sine series of g_2.

The solution of (5) with $g_3 \neq 0$ and $g_1 = g_2 = g_4 = 0$ can be obtained from (10) by exchanging x and y and exchanging ℓ and h. This results in

$$u(x, y) = \sum_{n=1}^{\infty} a_n \sin\left(\frac{n\pi y}{h}\right) \operatorname{csch}\left(\frac{n\pi \ell}{h}\right) \sinh\left(\frac{n\pi(\ell - x)}{h}\right), \tag{12}$$

where the a_n are the coefficients in the Fourier sine series of g_3. Finally, the solution of (5) with $g_4 \neq 0$ and $g_1 = g_2 = g_3 = 0$ can be obtained by making the same exchanges in (11). This results in

$$u(x, y) = \sum_{n=1}^{\infty} a_n \sin\left(\frac{n\pi y}{h}\right) \operatorname{csch}\left(\frac{n\pi \ell}{h}\right) \sinh\left(\frac{n\pi x}{h}\right), \tag{13}$$

where the a_n are the coefficients in the Fourier sine series of g_4.

In summary, the solution of (5) can be constructed by first computing the coefficients in the Fourier sine series of each of $g_1, \dots, g_4$ and then superimposing (10)–(13).

Example 1

A particularly simple instance of (5) is

$$\begin{aligned} u_{xx} + u_{yy} &= 0 \quad \text{for} \quad 0 < x < \pi, 0 < y < \pi, \\ u(x, 0) = \sin x, \quad u(x, \pi) &= 0 \quad \text{for} \quad 0 \le x \le \pi, \\ u(0, y) = \sin y, \quad u(\pi, y) &= 0 \quad \text{for} \quad 0 \le y \le \pi. \end{aligned}$$

Here we have contributions only from the edge where $x = 0$ and the edge where $y = 0$. Each of the corresponding boundary functions are already Fourier sine series in which $a_1 = 1$ and $a_n = 0$ for $n \ge 2$. Thus we use (10) and (12) (and the fact that $\ell = h = \pi$) to obtain

$$u(x, y) = \sin x \operatorname{csch} \pi \sinh(\pi - y) + \sin y \operatorname{csch} \pi \sinh(\pi - x),$$

that is,

$$u(x, y) = \operatorname{csch} \pi \left(\sin x \sinh(\pi - y) + \sin y \sinh(\pi - x)\right).$$

Figure 3a shows the graph of this solution. Figure 3b is a contour plot of the solution. Recall that the solution is constant along each contour. In a heat conduction problem, for instance, these contours are called *isotherms*. ■

Example 2

Consider the Dirichlet problem

$$\begin{aligned} u_{xx} + u_{yy} &= 0 \quad \text{for} \quad 0 < x < 1, 0 < y < 2, \\ u(x, 0) = 0, \quad u(x, 2) &= 0 \quad \text{for} \quad 0 \le x \le 1, \\ u(0, y) = g_3(y), \quad u(1, y) &= 0 \quad \text{for} \quad 0 \le y \le 2, \end{aligned}$$

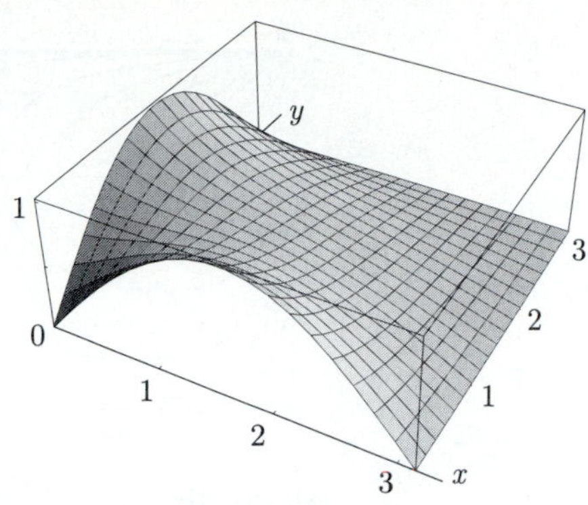

Figure 3a

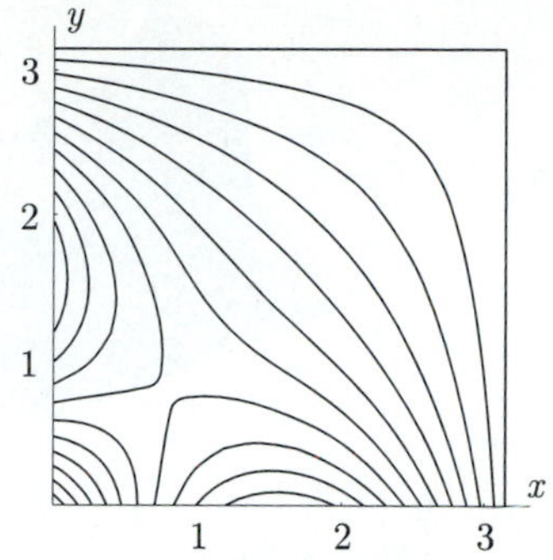

Figure 3b

where the function g_3 is defined by

$$g_3(y) = \begin{cases} 0, & \text{if } 0 \le y \le 1; \\ 1, & \text{if } 1 < y \le 2. \end{cases}$$

We begin by computing the coefficients in the Fourier sine series of g_3:

$$a_n = \frac{2}{2}\int_0^2 g_3(y)\sin\left(\frac{n\pi y}{2}\right)dy = \int_1^2 \sin\left(\frac{n\pi y}{2}\right)dy = \frac{2c_n}{n\pi},$$

where, for convenience, we have set

$$c_n = \cos\left(\frac{n\pi}{2}\right) - \cos(n\pi), \quad n = 1, 2, 3, \dots .$$

(Note that the sequence $c_1, c_2, c_3, \dots$ cycles thus: $1, -2, 1, 0, 1, -2, 1, 0, \dots$.) The Fourier sine series of g_3 is now

$$g_3(y) = \frac{2}{\pi}\sum_{n=1}^{\infty}\frac{c_n}{n}\sin\left(\frac{n\pi y}{2}\right),$$

and so, by (12), the solution of the Dirichlet problem is

$$u(x, y) = \frac{2}{\pi}\sum_{n=1}^{\infty}\frac{c_n}{n}\sin\left(\frac{n\pi y}{2}\right)\operatorname{csch}\left(\frac{n\pi}{2}\right)\sinh\left(\frac{n\pi(1-x)}{2}\right).$$

The solution is plotted in Figure 4a, and a contour plot of the solution is shown in Figure 4b. ■

A Circular Domain We will next examine the Dirichlet problem (4) in the case where $\overline{\Omega}$ is a disk with radius ρ centered at the origin:

$$\begin{aligned} u_{xx} + u_{yy} &= 0 \quad \text{for } x^2 + y^2 < \rho^2, \\ u(x, y) &= g(x, y) \quad \text{for } x^2 + y^2 = \rho^2. \end{aligned} \tag{14}$$

As one might expect, it will be convenient to work in polar coordinates. So recall that polar coordinates r, θ are defined by

$$x = r\cos\theta \quad \text{and} \quad y = r\sin\theta.$$

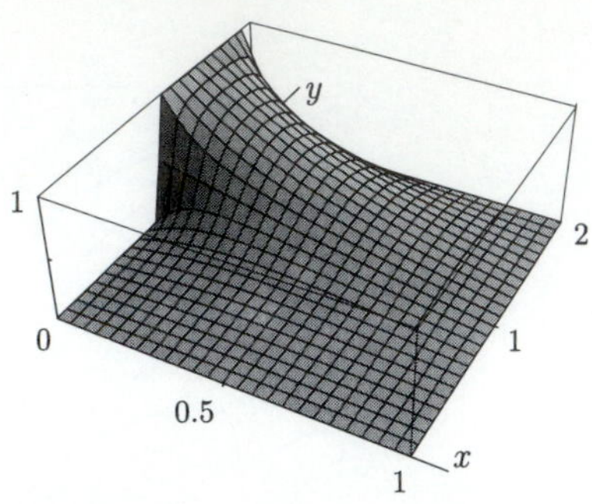

Figure 4a

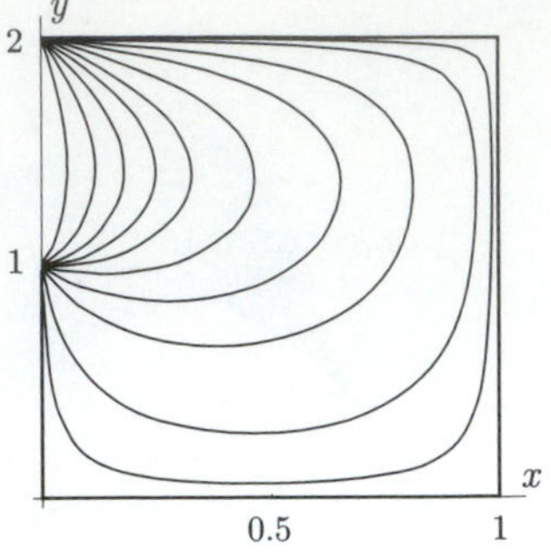

Figure 4b

Now suppose that u satisfies (14), and let w be defined by

$$w(r,\theta) = u(r\cos\theta, r\sin\theta) \quad \text{for } 0 \le r \le \rho \text{ and } -\infty < \theta < \infty.$$

Using the chain rule, we compute the first-order partial derivatives of w,

$$w_r = u_x \cos\theta + u_y \sin\theta, \quad w_\theta = -u_x r \sin\theta + u_y r \cos\theta,$$

followed by the second-order partial derivatives of w,

$$\begin{aligned} w_{rr} &= u_{xx}\cos^2\theta + u_{yy}\sin^2\theta, \\ w_{\theta\theta} &= u_{xx} r^2 \sin^2\theta - u_x r\cos\theta + u_{yy} r^2 \cos^2\theta - u_y r \sin\theta. \end{aligned}$$

Next we notice that

$$\begin{aligned} w_{rr} + \frac{1}{r^2} w_{\theta\theta} &= u_{xx} + u_{yy} - \frac{1}{r}(u_x \cos\theta + u_y \sin\theta) \\ &= 0 - \frac{1}{r} u_r. \end{aligned}$$

Thus we arrive at *the Laplace equation in polar coordinates*:

$$w_{rr} + \frac{1}{r} w_r + \frac{1}{r^2} w_{\theta\theta} = 0, \tag{15a}$$

which is satisfied by w for $0 < r < \rho$ and all θ. The boundary condition in (14) becomes

$$w(\rho, \theta) = \gamma(\theta), \quad -\infty < \theta < \infty, \tag{15b}$$

where $\gamma(\theta) = g(\rho\cos\theta, \rho\sin\theta)$. In addition, w is 2π-periodic,

$$w(r,\theta) = w(r, \theta + 2\pi) \quad \text{for } 0 \le r \le \rho \text{ and } -\infty < \theta < \infty, \tag{15c}$$

and satisfies the boundedness condition,

$$w \text{ and } w_r \text{ are bounded for } 0 \le r \le \rho \text{ and } -\infty < \theta < \infty. \tag{15d}$$

Our task now is to find w such that (15a–d) are satisfied.

Using separation of variables once again, we begin by looking for solutions of (15a) in the form

$$w(r,\theta) = R(r)\Theta(\theta).$$

Substitution of $R(r)\Theta(\theta)$ into (15a) yields

$$R''\Theta + \frac{1}{r}R'\Theta + \frac{1}{r^2}R\Theta'' = 0.$$

Next we multiply by r^2 and divide by $R\Theta$, getting

$$\frac{r^2R'' + rR'}{R} + \frac{\Theta''}{\Theta} = 0.$$

Therefore, there must be a constant λ such that

$$\frac{r^2R'' + rR'}{R} = -\frac{\Theta''}{\Theta} = \lambda.$$

This gives birth to two ordinary differential equations:

$$r^2R'' + rR' - \lambda R = 0 \quad \text{and} \quad \Theta'' + \lambda\Theta = 0.$$

The second of these must yield a 2π-periodic solution, because of (12c). Thus we rule out the hyperbolic solutions that arise with $\lambda < 0$ and conclude that nontrivial solutions may exist only if $\lambda \geq 0$. Indeed, there is a sequence of nonnegative eigenvalues and corresponding 2π-periodic eigenfunctions:

$$\lambda_n = n^2, \quad \Theta_n(\theta) = a_n \cos n\theta + b_n \sin n\theta, \quad n = 0, 1, 2, \ldots .$$

(Note that a constant eigenfunction $\Theta_0(\theta) = a_0$ corresponds to the eigenvalue $\lambda_0 = 0$.) Having found the eigenvalues, the next job is to solve

$$r^2R_n'' + rR_n' - n^2R_n = 0, \quad n = 0, 1, 2, \ldots . \tag{16}$$

With $n = 0$ this becomes

$$r^2R_0'' + rR_0' = r(rR_0')' = 0,$$

from which we get linearly independent, particular solutions 1 and $\ln r$. Because of (15d), we discard $\ln r$ and thus set $R_0(r) = 1$. Now let n be a positive integer. Substitution of $R_n(r) = r^p$ $(p \neq 0)$ into (16) yields

$$r^2p(p-1)r^{p-2} + rpr^{p-1} - n^2r^p = 0,$$

which is true for all $r > 0$, if p satisfies

$$p(p-1) + p - n^2 = 0.$$

This gives $p = \pm n$, from which we get linearly independent, particular solutions r^n and r^{-n}. We discard r^{-n} because of (15d), thus arriving at

$$R_n(r) = r^n, \quad n = 0, 1, 2, \ldots .$$

The result so far is that (15a), (15c), and (15d) are satisfied by each of

$$R_n(r)\Theta_n(\theta) = r^n(a_n \cos n\theta + b_n \sin n\theta), \quad n = 0, 1, 2, \ldots ,$$

and therefore also satisfied by

$$w(r, \theta) = a_0 + \sum_{n=1}^{\infty} r^n(a_n \cos n\theta + b_n \sin n\theta). \tag{17}$$

Finally, the boundary condition (15b) becomes

$$w(\rho, \theta) = a_0 + \sum_{n=1}^{\infty} \rho^n (a_n \cos n\theta + b_n \sin n\theta) = \gamma(\theta),$$

from which we conclude that the coefficients a_n and b_n can be obtained by dividing the nth Fourier coefficients of γ by ρ^n. Thus, given γ as a Fourier series

$$\gamma(\theta) = \alpha_0 + \sum_{n=1}^{\infty} (\alpha_n \cos n\theta + \beta_n \sin n\theta),$$

the boundary condition (15b) is satisfied if the coefficients a_n and b_n in (17) are chosen to be

$$a_n = \frac{\alpha_n}{\rho^n} \quad \text{and} \quad b_n = \frac{\beta_n}{\rho^n}.$$

Example 3

Let $\overline{\Omega}$ be the unit disk (i.e., radius 1, center (0, 0)). Also let Γ_1 be the portion of $\partial\Omega$ on which $x \geq \sqrt{2}/2$, and let Γ_2 be the rest of $\partial\Omega$. Note that Γ_1 is also the portion of $\partial\Omega$ corresponding to angles $-\pi/4 \leq \theta \leq \pi/4$. Now consider the Dirichlet problem

$$\begin{aligned} u_{xx} + u_{yy} &= 0 \quad \text{in } \Omega, \\ u(x, y) &= 1 \quad \text{on } \Gamma_1, \\ u(x, y) &= 0 \quad \text{on } \Gamma_2. \end{aligned}$$

Translated to polar coordinates, the problem becomes

$$\begin{aligned} w_{rr} + \frac{1}{r} w_r + \frac{1}{r^2} w_{\theta\theta} &= 0 \quad \text{for } 0 < r < 1, \quad -\infty < \theta < \infty, \\ w(1, \theta) &= \gamma(\theta) \quad \text{for } -\infty < \theta < \infty, \end{aligned}$$

where $\gamma(\theta)$ is the 2π-periodic extension of the function defined on $[-\pi, \pi]$ by

$$\gamma_0(\theta) = \begin{cases} 0, & \text{if } -\pi \leq \theta < -\dfrac{\pi}{4}; \\ 1, & \text{if } -\dfrac{\pi}{4} \leq \theta \leq \dfrac{\pi}{4}; \\ 0, & \text{if } \dfrac{\pi}{4} < \theta \leq \pi. \end{cases}$$

Since γ_0 is an even function, the coefficient of each sine term in its Fourier series will be zero. The coefficients for the cosine terms are

$$a_0 = \frac{1}{\pi} \int_{-\pi}^{\pi} \gamma_0(\theta)\, d\theta = \frac{1}{\pi} \int_{-\pi/4}^{\pi/4} d\theta = \frac{1}{2}$$

and, for $n \geq 1$,

$$a_n = \frac{1}{\pi} \int_{-\pi}^{\pi} \gamma_0(\theta) \cos(n\theta)\, d\theta = \frac{1}{\pi} \int_{-\pi/4}^{\pi/4} \cos(n\theta)\, d\theta = \frac{2}{n\pi} \sin\left(\frac{n\pi}{4}\right).$$

Thus the Fourier series representation of γ_0 is

$$\begin{aligned}\gamma(\theta) &= \frac{1}{4} + \frac{2}{\pi}\sum_{n=1}^{\infty} \frac{1}{n}\sin\left(\frac{n\pi}{4}\right)\cos(n\theta) \\ &= \frac{1}{4} + \frac{1}{\pi}\Big(\sqrt{2}\cos\theta + \cos 2\theta + \frac{\sqrt{2}}{3}\cos 3\theta \\ &\quad - \frac{\sqrt{2}}{5}\cos 5\theta - \frac{2}{6}\cos 6\theta - \frac{\sqrt{2}}{7}\cos 7\theta + \cdots\Big).\end{aligned}$$

Since the radius of $\overline{\Omega}$ is 1, the coefficients in (17) are the same as the Fourier coefficients of γ. Therefore, we have

$$\begin{aligned}w(r,\theta) &= \frac{1}{4} + \frac{2}{\pi}\sum_{n=1}^{\infty} \frac{r^n}{n}\sin\left(\frac{n\pi}{4}\right)\cos(n\theta) \\ &= \frac{1}{4} + \frac{1}{\pi}(\sqrt{2}r\cos\theta + r^2\cos 2\theta + \frac{\sqrt{2}}{3}r^3\cos 3\theta \\ &\quad - \frac{\sqrt{2}}{5}r^5\cos 5\theta - \frac{2}{6}r^6\cos 6\theta - \frac{\sqrt{2}}{7}r^7\cos 7\theta + \cdots).\end{aligned}$$

A graph of the solution is shown in Figure 5a, and a contour plot is shown in Figure 5b. ■

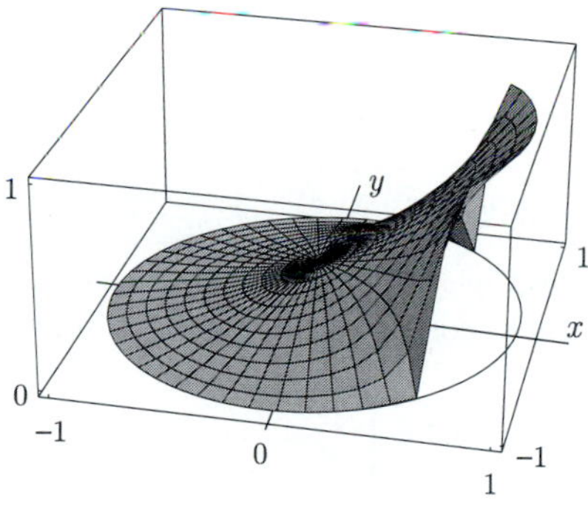

Figure 5a

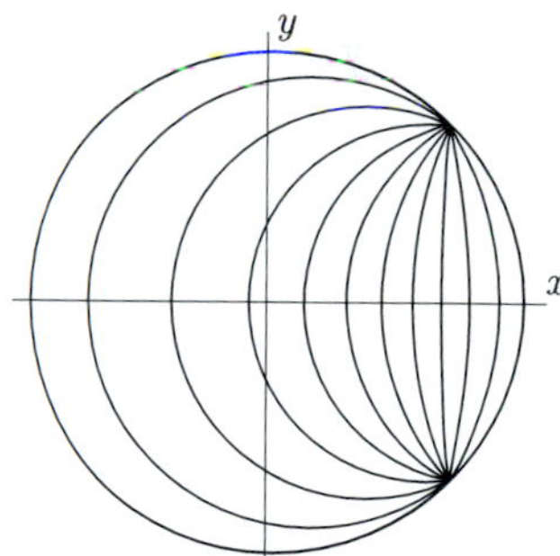

Figure 5b

PROBLEMS

For the given functions in each of Problems 1 through 4, solve the Dirichlet problem

$$\begin{aligned} u_{xx} + u_{yy} = 0, \quad 0 < x < 1, \quad 0 < y < 1, \\ u(x,0) = g_1(x), u(x,1) = g_2(x), \quad 0 \le x \le 1, \\ u(0,y) = g_3(y), u(1,y) = g_4(y), \quad 0 < y < 1. \end{aligned}$$

1. $g_1(x) = 0, \quad g_2(x) = 0, \quad g_3(y) = \sin 2\pi y, \quad g_4(y) = -\sin 2\pi y$

2. $g_1(x) = 1, \quad g_2(x) = 0, \quad g_3(y) = g_4(y) = 0$

3. $g_1(x) = 0, \quad g_2(x) = 1, \quad g_3(y) = 0, \quad g_4(y) = y$

4. $g_1(x) = x, \quad g_2(x) = 1 - x, \quad g_3(y) = g_4(y) = 0$

5. Consider the following problem involving a nonhomogeneous Laplace equation (a.k.a. *Poisson's equation*) and homogeneous Dirichlet boundary conditions:

$$\begin{aligned} u_{xx} + u_{yy} &= f(x, y), \quad 0 < x < \ell, \quad 0 < y < h, \\ u(x, 0) = 0, \quad u(x, h) &= 0, \quad 0 \le x \le \ell, \\ u(0, y) = 0, \quad u(\ell, y) &= 0, \quad 0 < y < h. \end{aligned} \tag{$*$}$$

Assume that f is given by a double Fourier sine series

$$f(x, y) = \sum_{m=1}^{\infty} \sum_{n=1}^{\infty} c_{mn} \sin\left(\frac{m\pi x}{\ell}\right) \sin\left(\frac{n\pi y}{h}\right).$$

(a) Show that, for any coefficients a_{mn}, the double Fourier sine series

$$u(x, y) = \sum_{m=1}^{\infty} \sum_{n=1}^{\infty} a_{mn} \sin\left(\frac{m\pi x}{\ell}\right) \sin\left(\frac{n\pi y}{h}\right)$$

satisfies the boundary conditions in $(*)$.

(b) Solve $(*)$ by relating the coefficients a_{mn} in the expansion of u to the coefficients c_{mn} in the expansion of f.

(c) Show that the coefficients in the expansion of f are

$$c_{mn} = \frac{4}{\ell h} \int_0^h \int_0^\ell f(x, y) \sin\left(\frac{m\pi x}{\ell}\right) \sin\left(\frac{n\pi y}{h}\right) dx dy, \quad m, n = 1, 2, 3, \ldots.$$

6. **(a)** Consider the problem

$$\begin{aligned} u_{xx} + u_{yy} &= f \quad \text{in } \Omega, \\ u &= g \quad \text{on } \partial\Omega, \end{aligned}$$

which involves both a nonhomogeneous Laplace equation and a nonhomogeneous boundary condition. Show that the solution u can be constructed by adding the solutions v and w of

$$\left\{ \begin{aligned} v_{xx} + v_{yy} &= f \quad \text{in } \Omega, \\ v &= 0 \quad \text{on } \partial\Omega; \end{aligned} \right. \quad \text{and} \quad \left\{ \begin{aligned} w_{xx} + w_{yy} &= 0 \quad \text{in } \Omega, \\ w &= g \quad \text{on } \partial\Omega. \end{aligned} \right.$$

(b) Solve

$$\begin{aligned} u_{xx} + u_{yy} &= \sin x \sin 2y, \quad 0 < x < \pi, \quad 0 < y < \pi, \\ u(x, 0) = \sin 3x, \quad u(x, \pi) &= 0, \quad 0 \le x \le \pi, \\ u(0, y) = 0, \quad u(\pi, y) &= \sin y, \quad 0 < y < \pi. \end{aligned}$$

In Problems 7 through 10, find the solution of the Laplace equation on the unit disk that satisfies the given Dirichlet boundary condition on the unit circle.

7. $w(1, \theta) = 1 + \sin\theta$

8. $w(1, \theta) = \theta/\pi, \ -\pi < \theta \le \pi$

9. $w(1, \theta) = \begin{cases} 0, & \text{if } -\pi < \theta < 0 \\ 1, & \text{if } 0 \le \theta \le \pi \end{cases}$

10. $w(1, \theta) = \begin{cases} 0, & \text{if } -\pi < \theta < 0 \\ \theta/\pi, & \text{if } 0 \le \theta \le \pi \end{cases}$

11. Show that, if w is a bounded, 2π-periodic solution of (15a), and if $w_{\theta\theta} = 0$, then w is constant. Conclude that the only "radially symmetric" (i.e., θ-independent) solutions of (15a–d) are constant.

For the given boundary conditions in Problems 12 through 15, use separation of variables to find the solution of the Laplace equation $u_{xx} + u_{yy} = 0$ on the rectangle $0 < x < \ell, 0 < y < h$. Each set of boundary conditions involves a Neumann-type (or "zero flux") boundary condition on one or more edges of the rectangle.

12. $\begin{cases} u(x,0) = g_1(x), \quad u_y(x,h) = 0, & 0 \le x \le \ell, \\ u(0,y) = 0, \quad u(\ell,y) = 0, & 0 < y < h \end{cases}$

13. $\begin{cases} u(x,0) = g_1(x), \quad u_y(x,h) = 0, & 0 \le x \le \ell, \\ u_x(0,y) = 0, \quad u(\ell,y) = 0, & 0 < y < h \end{cases}$

14. $\begin{cases} u(x,0) = g_1(x), \quad u(x,h) = 0, & 0 \le x \le \ell, \\ u_x(0,y) = 0, \quad u(\ell,y) = 0, & 0 < y < h \end{cases}$

15. $\begin{cases} u(x,0) = g_1(x), \quad u_y(x,h) = 0, & 0 \le x \le \ell, \\ u_x(0,y) = 0, \quad u_x(\ell,y) = 0, & 0 < y < h \end{cases}$

16. Consider the following problem involving a nonhomogeneous Laplace equation and homogeneous Neumann boundary conditions:

$$\begin{aligned} u_{xx} + u_{yy} &= f(x,y), \quad 0 < x < \ell, \quad 0 < y < h, \\ u_y(x,0) = 0, \quad u_y(x,h) &= 0, \quad 0 \le x \le \ell, \\ u_x(0,y) = 0, \quad u_x(\ell,y) &= 0, \quad 0 < y < h. \end{aligned} \tag{$**$}$$

(a) Show that, *if* $(**)$ has a solution u, then

$$\int_0^h \int_0^\ell f(x,y)\, dx\, dy = 0.$$

Conclude therefore that, if $\int_0^h \int_0^\ell f(x,y)\, dxdy \neq 0$, then $(**)$ has no solution.

(b) Show that, if u is a solution of $(**)$, then $u + C$ is a solution for any constant C.

(c) Assume that f is given by the double Fourier cosine series

$$f(x,y) = \sum_{m=0}^{\infty} \sum_{n=0}^{\infty} c_{mn} \cos\left(\frac{m\pi x}{\ell}\right) \cos\left(\frac{n\pi y}{h}\right),$$

and observe that $c_{00} = \int_0^h \int_0^\ell f(x,y) dx\, dy$. Then show that $(**)$ has a solution of the form

$$u(x,y) = \sum_{m=0}^{\infty} \sum_{n=0}^{\infty} a_{mn} \cos\left(\frac{m\pi x}{\ell}\right) \cos\left(\frac{n\pi y}{h}\right),$$

if and only if $c_{00} = 0$. (Note that the "only if" part of this is the same result as in part (a). Observe also that a_{00} is a free parameter in the solution, which implies the result of part (b).)

17. Let α be a nonzero constant, and let

$$f(x,y) = \sum_{m=0}^{\infty} \sum_{n=0}^{\infty} c_{mn} \cos\left(\frac{m\pi x}{\ell}\right) \cos\left(\frac{n\pi y}{h}\right).$$

Show that the problem

$$\begin{aligned} u_{xx} + u_{yy} + \alpha u &= f(x, y), & 0 < x < \ell, 0 < y < h, \\ u_y(x, 0) = 0, u_y(x, h) &= 0, & 0 \le x \le \ell, \\ u_x(0, y) = 0, u_x(\ell, y) &= 0, & 0 < y < h, \end{aligned}$$

has a unique solution of the form

$$u(x, y) = \sum_{m=0}^{\infty} \sum_{n=0}^{\infty} a_{mn} \cos\left(\frac{m\pi x}{\ell}\right) \cos\left(\frac{n\pi y}{h}\right).$$

18. Let F be a twice differentiable function of one variable, and let z be the complex-valued function of two variables defined by

$$z(x, y) = F(x + iy).$$

(a) Show formally that z satisfies the Laplace equation.

(b) Let u and v be the real and imaginary parts of z; that is, let

$$z(x, y) = u(x, y) + iv(x, y),$$

where u and v are real-valued. Show that u and v individually satisfy the Laplace equation.

Find, and verify, the solutions u and v of the Laplace equation that arise from each of the following functions F.

(c) $F(z) = z^2$

(d) $F(z) = \frac{1}{z}$

(e) $F(z) = \frac{1}{z^2}$

(f) $F(z) = e^{kz}$

(g) $F(z) = e^{-z^2}$

19. **(a)** Use the MacLaurin series of sin, sinh, cos, and cosh to justify the definitions

$$\cos iy = \cosh y \quad \text{and} \quad \sin iy = \sinh y.$$

(b) Use addition formulas and the definitions in part (a) to justify the definitions

$$\begin{aligned} \sin(x + iy) &= \sin x \cosh y + i \cos x \sinh y, \\ \cos(x + iy) &= \cos x \cosh y - i \sin x \sinh y. \end{aligned}$$

(c) Show that, consistent with Problem 18, the real and imaginary parts of each of $\sin(x + iy)$ and $\cos(x + iy)$ satisfy the Laplace equation.

A theorem known as the *maximum principle* states that *a nonconstant solution of the Laplace equation on a domain Ω cannot attain a maximum (or a minimum) value in Ω.* In particular, a solution of the Dirichlet problem (4) must attain its maximum value (and minimum value) on $\partial\Omega$. Problems 20 through 22 explore this result.

20. Prove the one-dimensional version of the maximum principle: *A nonconstant solution of $u''(x) = 0$, for x in (a, b), attains neither a maximum nor a minimum value in (a, b).*

21. Let u be a solution of (4). Argue as follows to prove that u must attain its maximum value on $\partial\Omega$. Begin by supposing that u attains a value at some point (x^*, y^*) in Ω that is greater than its maximum value on $\partial\Omega$; that is,

$$u(x^*, y^*) > \max_{\partial\Omega} u(x, y).$$

Now let ϵ be a positive number, and define

$$v(x, y) = u(x, y) + \epsilon\left((x - x^*)^2 + (y - y_0)^2\right), \qquad (x, y) \text{ in } \overline{\Omega}.$$

(a) Show that v satisfies

$$v_{xx} + v_{yy} = 6\epsilon > 0 \quad \text{in } \Omega. \tag{†}$$

(b) Argue that, if ϵ is small enough, then

$$v(x^*, y^*) > \max_{\partial\Omega} v(x, y).$$

(c) Conclude from part (b) that (for sufficiently small ϵ) v attains its maximum value (and minimum value) at some point $(x^\sharp, y^\sharp)$ in Ω. But this is impossible because of (†) and the second derivative test for functions of two variables. Explain. Then conclude finally that the point (x^*, y^*) cannot exist; therefore, u must attain its maximum value on $\partial\Omega$.

(d) Conclude that u must also attain its minimum value on $\partial\Omega$ from the fact that $-u$ must attain its maximum value on $\partial\Omega$.

22. **(a)** Use the result of Problem 21 to prove that the Dirichlet problem

$$u_{xx} + u_{yy} = 0 \text{ in } \Omega, \qquad u = 0 \text{ on } \partial\Omega,$$

has only the trivial solution $u = 0$.

(b) Use the result of part (a) to prove that, if a solution of

$$u_{xx} + u_{yy} = f(x, y) \text{ in } \Omega, \qquad u = g(x, y) \text{ on } \partial\Omega,$$

exists, then it is unique.

12.3 The 2-D Diffusion Equation

In this section we will study two-dimensional diffusion problems of the form

$$u_t - k(u_{xx} + u_{yy}) = 0 \qquad \text{for } t > 0, \quad (x, y) \text{ in } \Omega, \tag{1a}$$

$$\alpha u + (1 - \alpha)\frac{\partial u}{\partial \mathbf{n}} = 0 \qquad \text{for } t > 0, \quad (x, y) \text{ on } \partial\Omega, \tag{1b}$$

$$u(0, x, y) = \varphi(x, y) \qquad \text{for } (x, y) \text{ in } \overline{\Omega}. \tag{1c}$$

Here the diffusivity k is assumed constant, and we assume that Ω is a *domain* as described in Section 12.2. In the boundary condition (1b) the symbol $\frac{\partial u}{\partial \mathbf{n}}$ denotes the (outward) *normal derivative* of u along $\partial\Omega$, which in two dimensions is defined by

$$\frac{\partial u}{\partial \mathbf{n}} = n_1 u_x + n_2 u_y,$$

where $\mathbf{n} = (n_1, n_2)$ is the outward unit normal vector at each point along $\partial\Omega$ (wherever such a vector exists). The normal derivative is the *outward flux* of u on $\partial\Omega$.

Also in (1b), α may be any piecewise-continuous function defined on $\partial\Omega$ with values in $[0, 1]$. However, we will restrict our discussions to two simple cases: $\alpha = 1$ and $\alpha = 0$. When $\alpha = 1$, the boundary condition is

$$u = 0 \quad \text{for } t > 0 \text{ and } (x, y) \text{ on } \partial\Omega$$

and is said to be of *homogeneous Dirichlet* type. When $\alpha = 0$, the boundary condition is

$$\frac{\partial u}{\partial \mathbf{n}} = 0 \quad \text{for } t > 0 \text{ and } (x, y) \text{ on } \partial\Omega$$

and is said to be of *homogeneous Neumann* type.

Initial data are specified in (1c), where we assume that φ is a piecewise-continuous function on $\overline{\Omega}$.

A Rectangular Domain Suppose that $\overline{\Omega}$ is the rectangle in which $0 \le x \le \ell$ and $0 \le y \le h$. Then problem (1abc) with a homogeneous Dirichlet boundary condition becomes

$$u_t - k(u_{xx} + u_{yy}) = 0 \quad \text{for } t > 0, \quad 0 < x < \ell, \quad 0 < y < h, \tag{2a}$$

$$\begin{aligned} u(t, x, 0) = u(t, x, h) &= 0 \quad \text{for } t > 0, \quad 0 \le x \le \ell, \\ u(t, 0, y) = u(t, \ell, y) &= 0 \quad \text{for } t > 0, \quad 0 \le y \le h, \end{aligned} \tag{2b}$$

$$u(0, x, y) = \varphi(x, y) \quad \text{for } 0 \le x \le \ell, \quad 0 \le y \le h. \tag{2c}$$

For reasons that will soon be clear, we assume that the initial data $\varphi(x, y)$ can be represented as a *double Fourier sine series*,

$$\varphi(x, y) = \sum_{n=1}^{\infty} \sum_{m=1}^{\infty} b_{mn} \sin\left(\frac{m\pi x}{\ell}\right) \sin\left(\frac{n\pi y}{h}\right). \tag{3a}$$

It can be shown (see Problem 5) that the coefficients in this representation are given by

$$b_{mn} = \frac{4}{\ell h} \int_0^h \int_0^\ell \varphi(x, y) \sin\left(\frac{m\pi x}{\ell}\right) \sin\left(\frac{n\pi y}{h}\right) dx\, dy. \tag{3b}$$

Using the idea of separation of variables, we begin solving (2a–c) by looking for solutions of (2a) and (2b) in the form

$$u(t, x, y) = T(t)X(x)Y(y).$$

Substitution of this into (2a) gives

$$T'XY = k(TX''Y + TXY''),$$

which we rearrange to get

$$\frac{T'}{kT} = \frac{X''}{X} + \frac{Y''}{Y}.$$

Since the left side depends only upon t and the right side is independent of t, we conclude that both sides are equal to some constant λ. This gives rise to an ordinary differential equation for T,

$$T' = k\lambda T, \tag{4}$$

along with

$$\lambda - \frac{X''}{X} = \frac{Y''}{Y}.$$

Each side of this equation must be constant as well, so we conclude that there is a constant μ such that

$$X'' - (\lambda - \mu)X = 0 \quad \text{and} \quad Y'' - \mu Y = 0.$$

Now we apply the boundary conditions in (2b) to arrive at a pair of eigenvalue problems for X and Y. The first of these is

$$X'' - \sigma X = 0, \quad X(0) = X(\ell) = 0,$$

where for convenience we have set $\sigma = \lambda - \mu$. From this we obtain the sequence of eigenvalues and eigenfunctions

$$\sigma_n = -\frac{n^2\pi^2}{\ell^2} \quad \text{and} \quad X_n(x) = \sin\left(\frac{n\pi x}{\ell}\right), \quad n = 1, 2, 3, \dots .$$

The eigenvalue problem for Y is

$$Y'' - \mu Y = 0, \quad Y(0) = Y(h) = 0,$$

which produces the eigenvalues and eigenfunctions

$$\mu_n = -\frac{n^2\pi^2}{h^2} \quad \text{and} \quad Y_n(x) = \sin\left(\frac{n\pi y}{h}\right), \quad n = 1, 2, 3, \dots .$$

Now recall that we had set $\sigma = \lambda - \mu$, that is, $\lambda = \sigma + \mu$. Since we have found separate sequences of eigenvalues $\sigma_1, \sigma_2, \sigma_3, \dots$ and $\mu_1, \mu_2, \mu_3, \dots$, we obtain a doubly indexed sequence of values for λ:

$$\lambda_{mn} = \sigma_m + \mu_n = -\left(\frac{m^2}{\ell^2} + \frac{n^2}{h^2}\right)\pi^2, \quad m, n = 1, 2, 3, \dots .$$

Each of these gives rise to a solution of (4) given by

$$T_{mn}(t) = e^{-\left(\frac{m^2}{\ell^2} + \frac{n^2}{h^2}\right)\pi^2 t}.$$

Thus we have found that (2ab) are satisfied by each of the functions

$$T_{mn}X_mY_n = e^{-\left(\frac{m^2}{\ell^2} + \frac{n^2}{h^2}\right)\pi^2 t} \sin\left(\frac{m\pi x}{\ell}\right) \sin\left(\frac{n\pi y}{h}\right), \quad m, n = 1, 2, 3, \dots .$$

Any linear combination of these also satisfies (2ab); therefore,

$$u(t, x, y) = \sum_{n=1}^{\infty}\sum_{m=1}^{\infty} b_{mn} e^{-\left(\frac{m^2}{\ell^2} + \frac{n^2}{h^2}\right)\pi^2 t} \sin\left(\frac{m\pi x}{\ell}\right) \sin\left(\frac{n\pi y}{h}\right) \tag{5}$$

is a solution of (2ab) for any choice of coefficients b_{mn}. Since

$$u(0, x, y) = \sum_{n=1}^{\infty}\sum_{m=1}^{\infty} b_{mn} \sin\left(\frac{m\pi x}{\ell}\right) \sin\left(\frac{n\pi y}{h}\right),$$

we need only match the coefficients b_{mn} to the coefficients in (3b) in order for the initial condition (2c) to be satisfied.

Example 1

Consider (5) with $k = 1$ and with $\overline{\Omega}$ described by

$$0 \le x \le \pi \quad \text{and} \quad 0 \le y \le \pi.$$

Also suppose that the initial data are given by

$$\varphi(x, y) = \begin{cases} 1, & \text{if } \frac{\pi}{4} \le x \le \frac{\pi}{2} \text{ and } \frac{\pi}{3} \le y \le \frac{2\pi}{3}; \\ 0, & \text{otherwise.} \end{cases}$$

The Fourier coefficients given by (3b) are thus

$$\begin{aligned} b_{mn} &= \frac{4}{\pi^2} \int_0^{\pi} \int_0^{\pi} \varphi(x, y) \sin(mx) \sin(ny)\, dx\, dy \\ &= \frac{4}{\pi^2} \int_{\pi/3}^{2\pi/3} \int_{\pi/4}^{\pi/2} \sin(mx) \sin(ny)\, dx\, dy \\ &= \frac{4}{mn\pi^2} \left(\cos\left(\frac{m\pi}{4}\right) - \cos\left(\frac{m\pi}{2}\right)\right) \left(\cos\left(\frac{n\pi}{3}\right) - \cos\left(\frac{2n\pi}{3}\right)\right) \\ &= \frac{4\alpha_m \beta_n}{\pi^2 mn}, \end{aligned}$$

where we define

$$\alpha_j = \cos\left(\frac{j\pi}{4}\right) - \cos\left(\frac{j\pi}{2}\right) \quad \text{and} \quad \beta_j = \cos\left(\frac{j\pi}{3}\right) - \cos\left(\frac{2j\pi}{3}\right).$$

Note that the sequence $\alpha_1, \alpha_2, \alpha_3, \ldots$ cycles repeatedly through eight numbers,

$$\frac{\sqrt{2}}{2}, 1, -\frac{\sqrt{2}}{2}, -2, -\frac{\sqrt{2}}{2}, 1, \frac{\sqrt{2}}{2}, 0,$$

and the sequence $\beta_1, \beta_2, \beta_3, \ldots$ cycles repeatedly through six numbers,

$$1, 0, -2, 0, 1, 0.$$

We now use these coefficients (and (5)) to write the solution as

$$u(t, x, y) = \frac{4}{\pi^2} \sum_{n=1}^{\infty} \sum_{m=1}^{\infty} \frac{\alpha_m \beta_n}{mn} e^{-(m^2+n^2)t} \sin(mx) \sin(ny).$$

Because of the factored form of the coefficients, the solution can actually be written as a product of series. Bringing factors that do not depend on m out of the inner sum gives us

$$u(t, x, y) = \frac{4}{\pi^2} \sum_{n=1}^{\infty} \left(\frac{\beta_n}{n} e^{-n^2 t} \sin(ny) \sum_{m=1}^{\infty} \frac{\alpha_m}{m} e^{-m^2 t} \sin(mx) \right).$$

We then bring the entire inner sum out of the outer sum, obtaining

$$u(t, x, y) = \frac{4}{\pi^2}\left(\sum_{n=1}^{\infty} \frac{\beta_n}{n} e^{-n^2 t} \sin(ny)\right)\left(\sum_{m=1}^{\infty} \frac{\alpha_m}{m} e^{-m^2 t} \sin(mx)\right).$$

Figure 1 shows snapshots of the solution at several (unequally spaced) times between $t = 0$ and $t = 0.15$. ■

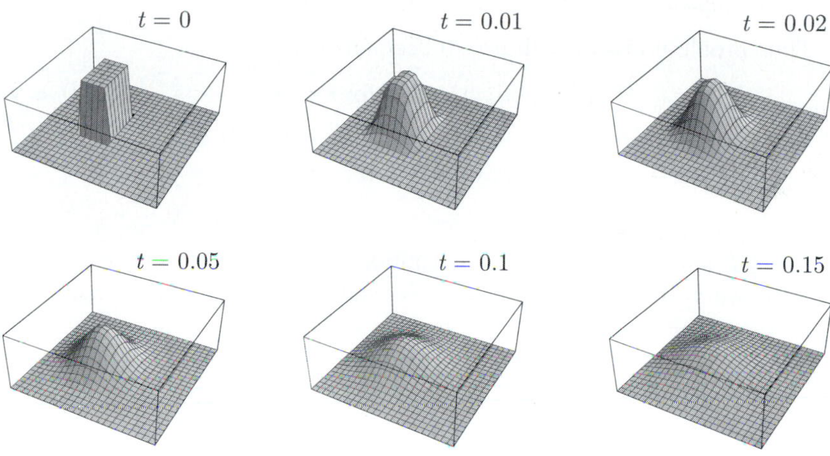

Figure 1

Neumann Boundary Condition Recall that the homogeneous Neumann (or zero-flux) boundary condition is

$$\frac{\partial u}{\partial \mathbf{n}} = 0 \quad \text{for } t > 0,\ (x, y) \text{ on } \partial\Omega,$$

where the normal derivative $\frac{\partial u}{\partial \mathbf{n}}$ is given (in two dimensions) by

$$\frac{\partial u}{\partial \mathbf{n}} = \mathbf{n} \cdot \nabla u = n_1 u_x + n_2 u_y.$$

In the present case, the outward unit normal vector $\mathbf{n}$ on $\partial\Omega$ always points in a vertical or horizontal direction, as illustrated in Figure 2. We indicate the particular form of the normal derivative along each edge of the rectangle in the following table.

Edge	$\mathbf{n}$	$\frac{\partial u}{\partial \mathbf{n}}$
$x = 0$	$(-1, 0)$	$-u_x$
$x = \ell$	$(1, 0)$	u_x
$y = 0$	$(0, -1)$	$-u_y$
$y = h$	$(0, 1)$	u_y

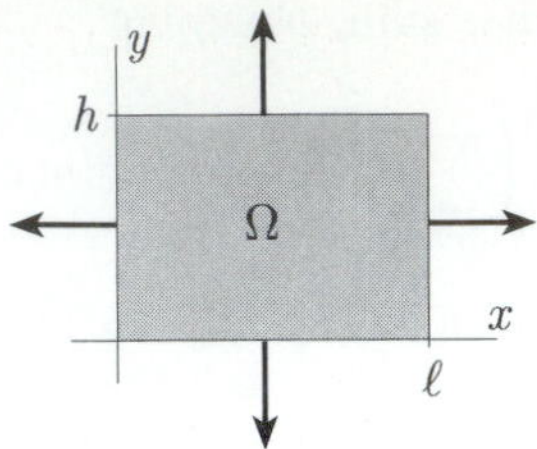

Figure 2

Thus problem (1a–c) with $\alpha = 0$ becomes

$$u_t - k(u_{xx} + u_{yy}) = 0 \quad \text{for } t > 0, \quad 0 < x < \ell, \quad 0 < y < h, \tag{6a}$$

$$\begin{aligned} u_y(t, x, 0) = u_y(t, x, h) &= 0 \quad \text{for } t > 0, \quad 0 \le x \le \ell, \\ u_x(t, 0, y) = u_x(t, \ell, y) &= 0 \quad \text{for } t > 0, \quad 0 \le y \le h, \end{aligned} \tag{6b}$$

$$u(0, x, y) = \varphi(x, y) \quad \text{for } 0 \le x \le \ell, \quad 0 \le y \le h. \tag{6c}$$

The solution of (6a–c) by separation of variables closely parallels that of (2abc). Thus we will only sketch the process and highlight the differences. Beginning with

$$u(t, x, y) = T(t)X(x)Y(y),$$

we find that T must satisfy

$$T' = k\lambda T$$

and that X and Y must satisfy, respectively, the eigenvalue problems

$$X'' - \sigma X = 0, \quad X'(0) = X'(\ell) = 0$$

and

$$Y'' - \mu Y = 0, \quad Y'(0) = Y'(h) = 0,$$

where $\sigma + \mu = \lambda$. (Note that the homogeneous Neumann boundary condition (6b) shows up as homogeneous Neumann boundary conditions in each of these eigenvalue problems.) The resulting eigenvalues and eigenfunctions are

$$\sigma_n = -\frac{n^2\pi^2}{\ell^2}, \quad X_n(x) = \cos\left(\frac{n\pi x}{\ell}\right), \quad n = 0, 1, 2, \ldots,$$

and

$$\mu_n = -\frac{n^2\pi^2}{h^2}, \quad Y_n(x) = \cos\left(\frac{n\pi y}{h}\right), \quad n = 0, 1, 2, \ldots.$$

(Note that the eigenfunctions are now cosines rather than sines and that zero is now an eigenvalue with corresponding eigenfunction 1.) These eigenvalues give rise to

$$\lambda_{mn} = \sigma_m + \mu_n = -\left(\frac{m^2}{\ell^2} + \frac{n^2}{h^2}\right)\pi^2, \quad m, n = 1, 2, 3, \ldots,$$

from which follows

$$T_{mn}(t) = e^{-\left(\frac{m^2}{\ell^2} + \frac{n^2}{h^2}\right)\pi^2 t}, \quad m, n = 0, 1, 2, \ldots.$$

We then conclude that, for any constants a_{mn}, a formal solution of (6a) and (6b) is given by

$$u(t, x, y) = \sum_{n=0}^{\infty}\sum_{m=0}^{\infty} a_{mn} T_{mn}(t) X_m(x) Y_n(y). \tag{7}$$

The initial condition (6c) will be satisfied if we choose the coefficients a_{mn} to match the coefficients in the double Fourier cosine series expansion of φ, which is

$$\varphi(x, y) = \sum_{n=0}^{\infty}\sum_{m=0}^{\infty} a_{mn} \cos\left(\frac{m\pi x}{\ell}\right) \cos\left(\frac{n\pi y}{h}\right),$$

where

$$\begin{aligned}
a_{0,0} &= \frac{1}{\ell h}\int_0^h \int_0^\ell \varphi(x, y)\,dx\,dy, \\
a_{m0} &= \frac{2}{\ell h}\int_0^h \int_0^\ell \varphi(x, y)\cos\left(\frac{m\pi x}{\ell}\right)dx\,dy, \quad m \geq 1, \\
a_{0n} &= \frac{2}{\ell h}\int_0^h \int_0^\ell \varphi(x, y)\cos\left(\frac{n\pi y}{h}\right)dx\,dy, \quad n \geq 1, \\
a_{mn} &= \frac{4}{\ell h}\int_0^h \int_0^\ell \varphi(x, y)\cos\left(\frac{m\pi x}{\ell}\right)\cos\left(\frac{n\pi y}{h}\right)dx\,dy, \quad m, n \geq 1.
\end{aligned} \tag{8}$$

A Circular Domain Let's now consider the diffusion problem (1a–c) where $\overline{\Omega}$ is a disk of radius ρ centered at the origin. For the case of a homogeneous Dirichlet boundary condition, conversion to polar coordinates (cf. Section 12.2) produces the problem

$$w_t - k\left(w_{rr} + \frac{1}{r}w_r + \frac{1}{r^2}w_{\theta\theta}\right) = 0 \quad \text{for } t > 0,\ 0 < r < \rho,\ -\infty < \theta < \infty, \tag{9a}$$

$$w(t, \rho, \theta) = 0 \quad \text{for } t > 0,\quad -\infty < \theta < \infty, \tag{9b}$$

$$w(0, r, \theta) = \psi(r, \theta) \quad \text{for } 0 \leq r \leq \rho,\quad -\infty < \theta < \infty. \tag{9c}$$

where $w(t, r, \theta) = u(t, r\cos\theta, r\sin\theta)$ and $\psi(r, \theta) = \varphi(r\cos\theta, r\sin\theta)$. The problem is naturally augmented by periodicity and boundedness conditions:

$$w \text{ is } 2\pi\text{-periodic in } \theta, \text{ and } w, w_r \text{ are bounded.} \tag{9d}$$

We begin by looking for solutions of (9a), (9b), and (9d) in the form

$$w(t, r, \theta) = T(t)R(r)\Theta(\theta).$$

Substitution of this into the diffusion equation (9a) yields

$$T'R\Theta - k\left(TR''\Theta + \frac{1}{r}TR'\Theta + \frac{1}{r^2}TR\Theta''\right) = 0.$$

After dividing by $kTR\Theta$, this becomes

$$\frac{T'}{kT} = \frac{R'' + R'/r}{R} + \frac{\Theta''}{r^2\Theta}.$$

Since the left side of this equation depends only on t and the right side is independent of t, we conclude that both sides must equal the same constant λ. Thus T must satisfy

$$T' = k\lambda T, \tag{10}$$

and R and Θ must satisfy

$$\frac{R'' + R'/r}{R} + \frac{\Theta''}{r^2\Theta} = \lambda,$$

which after some rearrangement becomes

$$-\frac{\Theta''}{\Theta} = \frac{r^2R'' + rR'}{R} - \lambda r^2.$$

Since the left side of this equation depends only on θ and the right side depends only on r, we conclude that both sides of this equation must equal the same constant μ. After rearranging a bit, we thus have two more ordinary differential equations,

$$\Theta'' = -\mu\Theta \quad \text{and} \quad r^2R'' + rR' - \lambda r^2R = \mu R.$$

The first of these and the 2π-periodicity requirement combine to produce the eigenvalue problem

$$-\Theta'' = \mu\Theta, \quad \Theta(\theta + 2\pi) = \Theta(\theta), \quad -\infty < \theta < \infty,$$

which is solved by

$$\mu_m = m^2 \quad \text{and} \quad \Theta_n(\theta) = a_m \cos m\theta + b_n \sin m\theta, \quad m = 0, 1, 2, \ldots . \tag{11}$$

The differential equation for R now gives rise to the sequence of equations

$$r^2R'' + rR' - \lambda r^2R = m^2R, \quad m = 0, 1, 2, \ldots .$$

Recalling the zero boundary condition and the boundedness requirement, we now have *for each* m the eigenvalue problem

$$R'' + \frac{1}{r}R' - \frac{m^2}{r^2}R = \lambda R; \quad R(\rho) = 0; \quad R, R' \text{ bounded.} \tag{12}$$

For each m this is a singular Sturm-Liouville eigenvalue problem (see Sections 11.5 and 11.6) involving the operator $\mathcal{S}_m$ defined by

$$\mathcal{S}_m R = R'' + \frac{1}{r}R' - \frac{m^2}{r^2}R = \frac{1}{r}\left((rR')' - \frac{m^2}{r}R\right).$$

It can be shown that $\mathcal{S}$ has no positive eigenvalues corresponding to problem (12); that is, (12) has only the trivial solution if $\lambda > 0$. So let $-\sigma^2 = \lambda$. Then the differential equation in (12) can be written as

$$r^2R'' + rR' + (\sigma^2r^2 - m^2)R = 0, \tag{13}$$

which is the standard form for *Bessel's equation*. It is well-known (see Churchill and Brown [14]) that for each nonnegative integer m, every bounded solution of (13) on $(0, \rho]$ is a constant multiple of the particular solution

$$R_m(r) = J_m(\sigma r), \tag{14}$$

where J_m is the *nth-order Bessel function of the first kind.*

There are numerous ways of representing J_m. These include the power series

$$J_m(x) = \sum_{i=0}^{\infty} \frac{(-1)^i}{i!(i+m)!} \left(\frac{x}{2}\right)^{2i+m}$$

and the integral form

$$J_m(x) = \frac{1}{\pi} \int_0^{\pi} \cos(mz - x \sin z)\, dz.$$

Moreover, computer algebra systems typically include built-in Bessel functions. Figure 3 shows the graphs of $J_0(x)$, $J_1(x)$, and $J_2(x)$ for $x \geq 0$, illustrating their oscillatory nature.

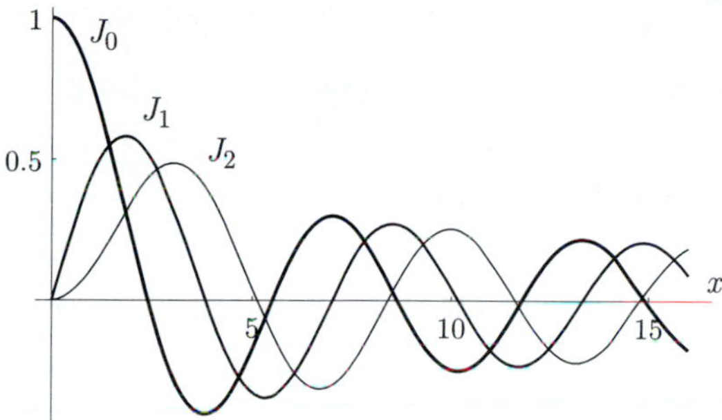

Figure 3

So let us assume that J_m is a known function and that the function R given by (14) satisfies the differential equation and the boundedness condition in (12). We now need to determine σ so that the boundary condition $R(\rho) = 0$ is satisfied. Fortunately, each J_m has an infinite sequence of simple, positive zeros, which we will denote by $x_{m1}, x_{m2}, x_{m3}, \ldots$ and assume are ordered so that $x_{m,j+1} > x_{mj}$.

Since $R_m(\rho) = J_m(\sigma\rho)$, it now follows that $R_m(\rho) = 0$, if σ is any of

$$\sigma_{mn} = \frac{x_{mn}}{\rho}, \qquad n = 1, 2, 3, \ldots .$$

Therefore, for each m, (11) is solved by the sequence of eigenvalues and eigenfunctions given by

$$\lambda_{mn} = -\sigma_{mn}^2, \quad R_{mn}(r) = J_m(\sigma_{mn} r), \quad n = 1, 2, 3, \ldots . \tag{15}$$

Now we revisit (10) and (11) and write

$$T_{mn}(t) = e^{-k\sigma_{mn}^2 t} \quad \text{and} \quad \Theta_{mn}(\theta) = a_{mn} \cos m\theta + b_{mn} \sin m\theta$$

for each $m \geq 0$ and $n \geq 1$. We conclude that each product $T_{mn} R_{mn} \Theta_{mn}$ satisfies the diffusion equation (9a) and the boundary condition (9b), and thus so does the formal series expansion

$$w(t, r, \theta) = \sum_{m=0}^{\infty} \sum_{n=1}^{\infty} e^{-k\sigma_{mn}^2 t} J_m(\sigma_{mn} r) \left(a_{mn} \cos m\theta + b_{mn} \sin m\theta\right). \tag{16}$$

In order that the initial condition (9c) be satisfied, the coefficients a_{mn} and b_{mn} must be such that

$$\psi(r,\theta) = \sum_{m=0}^{\infty}\sum_{n=1}^{\infty} J_m(\sigma_{mn} r)\,(a_{mn}\cos m\theta + b_{mn}\sin m\theta)\,. \tag{17}$$

With some rearrangement this becomes

$$\psi(r,\theta) = \sum_{n=1}^{\infty} a_{0n} J_0(\sigma_{0n} r) + \sum_{m=1}^{\infty}\left(\left(\sum_{n=1}^{\infty} a_{mn} J_m(\sigma_{mn} r)\right)\cos m\theta \right.$$
$$\left. + \left(\sum_{n=1}^{\infty} b_{mn} J_m(\sigma_{mn} r)\right)\sin m\theta\right).$$

For each fixed r, this is a Fourier series expansion on $[-\pi,\pi]$ whose coefficients are

$$\sum_{n=1}^{\infty} a_{0n} J_0(\sigma_{0n} r) = \frac{1}{2\pi}\int_{-\pi}^{\pi} \psi(r,\theta)\,d\theta,$$
$$\sum_{n=1}^{\infty} a_{mn} J_m(\sigma_{mn} r) = \frac{1}{\pi}\int_{-\pi}^{\pi} \psi(r,\theta)\cos m\theta\,d\theta, \quad m = 1,2,3,\ldots,$$
$$\sum_{n=1}^{\infty} b_{mn} J_m(\sigma_{mn} r) = \frac{1}{\pi}\int_{-\pi}^{\pi} \psi(r,\theta)\sin m\theta\,d\theta, \quad m = 1,2,3,\ldots.$$

The left side of each of these equations is an eigenfunction expansion of the right side, where, for each m, the eigenfunctions are $J_m(\sigma_{mn} r)$, $n = 1,2,3,\ldots$. These eigenfunctions are indeed orthogonal on $(0,\rho]$ with respect to the weight function $p(r) = r$. (See Problem 18.) So the results of Section 11.7 tell us that the coefficients in these expansions are given by

$$a_{0n} = \frac{\displaystyle\int_0^{\rho}\int_{-\pi}^{\pi} J_0(\sigma_{0n} r)\psi(r,\theta)\;r\,d\theta\,dr}{\displaystyle 2\pi\int_0^{\rho}(J_0(\sigma_{0n} r))^2\;r\,dr}, \quad n = 1,2,3,\ldots,$$

$$a_{mn} = \frac{\displaystyle\int_0^{\rho}\int_{-\pi}^{\pi} J_m(\sigma_{mn} r)\psi(r,\theta)\cos m\theta\;r\,d\theta\,dr}{\displaystyle \pi\int_0^{\rho}(J_m(\sigma_{mn} r))^2\;r\,dr}, \quad m,n = 1,2,3,\ldots, \tag{18}$$

$$b_{mn} = \frac{\displaystyle\int_0^{\rho}\int_{-\pi}^{\pi} J_m(\sigma_{mn} r)\psi(r,\theta)\sin m\theta\;r\,d\theta\,dr}{\displaystyle \pi\int_0^{\rho}(J_m(\sigma_{mn} r))^2\;r\,dr}, \quad m,n = 1,2,3,\ldots.$$

If ψ, ψ_x, and ψ_y are each piecewise continuous on $\overline{\Omega}$, then the resulting expansion (17) has the usual convergence properties. (See Theorem 1 in Section 11.7.)

It certainly should be of no surprise that one would not expect to compute the coefficients in (18) in any sort of exact, elementary form—either by hand or with a *CAS*. In fact, even the required zeros of the Bessel functions must be calculated numerically. However, numerical computation of the coefficients is relatively simple to accomplish with any software that has built-in Bessel functions and routines for computing their zeros.

We remark that some effort can be saved in the computation of these coefficients by using in each denominator the formula

$$\int_0^{\rho} (J_m(\sigma_{mn}r))^2 r\,dr = \frac{1}{2}\rho^2 (J_{m+1}(x_{mn}))^2, \tag{19}$$

which you are asked to derive in Problem 19.

Neumann Boundary Condition On the boundary of a disk centered at the origin, the outward unit normal vector always points in the radial direction. As a result, the normal derivative on the boundary is simply the partial derivative with respect to r. Thus, by replacing the boundary condition in (9b) with a homogeneous Neumann condition, we obtain the problem

$$w_t - k(w_{rr} + \frac{1}{r}w_r + \frac{1}{r^2}w_{\theta\theta}) = 0 \quad \text{for } t > 0, \quad 0 < r < \rho, -\infty < \theta < \infty, \tag{20a}$$

$$w_r(t, \rho, \theta) = 0 \quad \text{for } t > 0, \quad -\infty < \theta < \infty, \tag{20b}$$

$$w(0, r, \theta) = \psi(r, \theta) \quad \text{for } 0 \leq r \leq \rho, \quad -\infty < \theta < \infty. \tag{20c}$$

The solution of this problem is accomplished in basically the same way as with (9a–c). The essential difference is that the eigenvalue problem for R in (12) becomes instead

$$R'' + \frac{1}{r}R' - \frac{m^2}{r^2}R = \lambda R; \quad R'(\rho) = 0; \quad R, R' \text{ bounded.} \tag{21}$$

As a result, zero is an eigenvalue with corresponding eigenfunction $R(r) = 1$, and the numbers σ_{mn} in (15) are computed from the positive zeros of J_m' rather than those of J_m.

PROBLEMS

In Problems 1 through 4, with $h = \ell = \pi$ and the given initial data, find the solution of

(a) problem (2a–c); (b) problem (6a–c).

1. $\varphi(x, y) = 5 \sin x \sin y - \sin 2x \sin 3y$
2. $\varphi(x, y) = 1$
3. $\varphi(x, y) = \begin{cases} 1, & \text{if } 0 \leq x \leq \frac{\pi}{2} \text{ and } 0 \leq y \leq \frac{\pi}{2} \\ 0, & \text{otherwise} \end{cases}$
4. $\varphi(x, y) = xy$
5. Show that the coefficients in the double Fourier sine series

$$\varphi(x, y) = \sum_{n=1}^{\infty}\sum_{m=1}^{\infty} b_{mn} \sin\left(\frac{m\pi x}{\ell}\right) \sin\left(\frac{n\pi y}{h}\right)$$

are given by

$$b_{mn} = \frac{4}{\ell h}\int_0^h \int_0^{\ell} \varphi(x, y) \sin\left(\frac{m\pi x}{\ell}\right) \times \sin\left(\frac{n\pi y}{h}\right) dx dy.$$

(*Suggestion:* First write the series as

$$\varphi(x, y) = \sum_{n=1}^{\infty}\left(\sum_{m=1}^{\infty} b_{mn} \sin\left(\frac{m\pi x}{\ell}\right)\right) \sin\left(\frac{n\pi y}{h}\right)$$

and observe that, for each n and each fixed x, the inner sum is a coefficient in the Fourier sine series expansion of $\varphi(x, y)$ in the variable y.)

6. Show that the coefficients in the double Fourier cosine series

$$\varphi(x, y) = \sum_{n=0}^{\infty}\sum_{m=0}^{\infty} a_{mn} \cos\left(\frac{m\pi x}{\ell}\right) \cos\left(\frac{n\pi y}{h}\right)$$

are as given in (8).

7. Show that the coefficients in the double Fourier cosine-sine series

$$\varphi(x, y) = \sum_{n=1}^{\infty}\sum_{m=0}^{\infty} c_{mn} \cos\left(\frac{m\pi x}{\ell}\right) \sin\left(\frac{n\pi y}{h}\right)$$

are given by

$$c_{0n} = \frac{2}{\ell h}\int_0^h \int_0^\ell \varphi(x, y) \sin\left(\frac{n\pi y}{h}\right) dxdy, \quad n = 1, 2, 3, \ldots,$$

$$c_{mn} = \frac{4}{\ell h}\int_0^h \int_0^\ell \varphi(x, y) \cos\left(\frac{m\pi x}{\ell}\right) \times \sin\left(\frac{n\pi y}{h}\right) dxdy, \quad m, n = 1, 2, 3, \ldots.$$

Each of Problems 8 through 10 involves a mixture of Dirichlet and Neumann boundary conditions. Also, $h = \ell = \pi$ for convenience. Solve the problem by separation of variables.

8. $$\begin{cases} u_t - k(u_{xx} + u_{yy}) = 0 \\ \quad \text{for } t > 0,\ 0 < x < \pi,\ 0 < y < \pi, \\ u(t, x, 0) = u(t, x, \pi) = 0 \\ \quad \text{for } t > 0,\ 0 \le x \le \pi, \\ u_x(t, 0, y) = u_x(t, \pi, y) = 0 \\ \quad \text{for } t > 0,\ 0 < y < \pi, \\ u(0, x, y) = \varphi(x, y) \\ \quad \text{for } 0 \le x \le \pi,\ 0 \le y \le \pi \end{cases}$$

9. $$\begin{cases} u_t - k(u_{xx} + u_{yy}) = 0 \\ \quad \text{for } t > 0,\ 0 < x < \pi,\ 0 < y < \pi, \\ u(t, x, 0) = u(t, x, \pi) = 0 \\ \quad \text{for } t > 0,\ 0 \le x \le \pi, \\ u_x(t, 0, y) = u(t, \pi, y) = 0 \\ \quad \text{for } t > 0,\ 0 < y < \pi, \\ u(0, x, y) = \varphi(x, y) \\ \quad \text{for } 0 \le x \le \pi,\ 0 \le y \le \pi \end{cases}$$

10. $$\begin{cases} u_t - k(u_{xx} + u_{yy}) = 0 \\ \quad \text{for } t > 0,\ 0 < x < \pi,\ 0 < y < \pi, \\ u_y(t, x, 0) = u(t, x, \pi) = 0 \\ \quad \text{for } t > 0,\ 0 \le x \le \pi, \\ u_x(t, 0, y) = u(t, \pi, y) = 0 \\ \quad \text{for } t > 0,\ 0 < y < \pi, \\ u(0, x, y) = \varphi(x, y) \\ \quad \text{for } 0 \le x \le \pi,\ 0 \le y \le \pi \end{cases}$$

11. Consider the following diffusion problem involving both a nonhomogeneous diffusion equation and a nonhomogeneous boundary condition. Assume that the functions f, g, and φ are functions of x and y only.

$$\begin{aligned} u_t - k(u_{xx} + u_{yy}) &= f, \quad t > 0, (x, y) \text{ in } \Omega, \\ \alpha u + (1 - \alpha)\frac{\partial u}{\partial \mathbf{n}} &= g, \quad t > 0, (x, y) \text{ on } \partial\Omega, \\ u(0, x, y) &= \varphi, \quad (x, y) \text{ in } \overline{\Omega} \end{aligned} \tag{$*$}$$

Show that the solution of $(*)$ can be constructed by adding the solutions v, w, and z of the following problems. Note that v and w are independent of t.

$$\begin{aligned} v_{xx} + v_{yy} &= -\frac{1}{k}f, \quad (x, y) \text{ in } \Omega, \\ \alpha v + (1 - \alpha)\frac{\partial v}{\partial \mathbf{n}} &= 0, \quad (x, y) \text{ on } \partial\Omega; \end{aligned} \tag{i}$$

$$\begin{aligned} w_{xx} + w_{yy} &= 0, \quad (x, y) \text{ in } \Omega, \\ \alpha w + (1 - \alpha)\frac{\partial w}{\partial \mathbf{n}} &= g, \quad (x, y) \text{ on } \partial\Omega; \end{aligned} \tag{ii}$$

$$\begin{aligned} z_t - k(z_{xx} + z_{yy}) &= 0, \quad t > 0, (x, y) \text{ in } \Omega, \\ \alpha z + (1 - \alpha)\frac{\partial z}{\partial \mathbf{n}} &= 0, \quad t > 0, (x, y) \text{ on } \partial\Omega, \\ z(0, x, y) &= \varphi - v - w, \quad (x, y) \text{ in } \overline{\Omega} \end{aligned} \tag{iii}$$

12. Let m, n be positive integers. Solve

$$\begin{aligned} &u_t - k(u_{xx} + u_{yy}) = \sin mx \sin ny, \\ &\quad t > 0, \quad 0 < x < \pi, \quad 0 < y < \pi, \\ &u(t, x, 0) = u(t, x, \pi) = 0, \quad t > 0,\ 0 \le x \le \pi, \\ &u(t, 0, y) = u(t, \pi, y) = 0, \quad t > 0,\ 0 < y < \pi, \\ &u(0, x, y) = 0, \quad 0 \le x \le \pi, \quad 0 \le y \le \pi \quad (**) \end{aligned}$$

by adding the solutions of

$$v_{xx} + v_{yy} = -\frac{1}{k}\sin m\,x \sin n\,y,$$
$$0 < x < \pi, \quad 0 < y < \pi,$$
$$v(x,0) = v(x,\pi) = 0, \quad 0 \le x \le \pi,$$
$$v(0,y) = v(\pi,y) = 0, \quad 0 < y < \pi$$

and

$$z_t - k(z_{xx} + z_{yy}) = 0, \quad t > 0, \quad 0 < x < \pi, \quad 0 < y < \pi,$$
$$z(t,x,0) = z(t,x,\pi) = 0, \quad t > 0, \quad 0 \le x \le \pi,$$
$$z(t,0,y) = z(t,\pi,y) = 0, \quad t > 0, \quad 0 < y < \pi,$$
$$z(0,x,y) = -v, \quad 0 \le x \le \pi, \quad 0 \le y \le \pi.$$

13. Solve $(**)$ directly by substituting $u(t,x,y) = \alpha_{mn}(t)\sin mx \sin ny$ and solving the initial-value problem that arises for $\alpha_{mn}(t)$.

14. Solve

$$u_t - (u_{xx} + u_{yy}) = \sin x \sin 2y,$$
$$t > 0, \quad 0 < x < \pi, \quad 0 < y < \pi,$$
$$u(t,x,0) = \sin x, \quad u(t,x,\pi) = 0,$$
$$t > 0, 0 \le x \le \pi,$$
$$u(t,0,y) = u(t,\pi,y) = 0, \ t > 0, \ 0 < y < \pi,$$
$$u(0,x,y) = \sin 3x \sin y, \ 0 \le x \le \pi, \ 0 \le y \le \pi.$$

15. Suppose that $R(r)$ satisfies Bessel's equation

$$r^2 R'' + r R' + (\sigma^2 r^2 - m^2)R = 0.$$

Let $x = \sigma r$ and $y(x) = R(r) = R(x/\sigma)$. Show that y satisfies

$$x^2 y'' + xy' + (x^2 - m^2)y = 0.$$

16. Let m be a nonnegative integer. Use the method oulined below to find a *Frobenius series* solution of Bessel's equation

$$x^2 y'' + xy' + (x^2 - m^2)y = 0.$$

(a) Show that substitution of

$$y(x) = x^p \sum_{i=0}^{\infty} a_i x^i = \sum_{i=0}^{\infty} a_i x^{i+p}$$

produces

$$\sum_{i=0}^{\infty}((i+p)^2 - m^2)a_i x^i + \sum_{i=0}^{\infty} a_i x^{i+2} = 0,$$

which in turn yields

$$(p^2 - m^2)a_0 + ((1+p)^2 - m^2)a_1 x + \sum_{i=2}^{\infty}(((i+p)^2 - m^2)a_i + a_{i-2})x^i = 0.$$

(b) Show that the choice of $p = m$ leads to coefficients given by

$$a_1 = 0 \text{ and } a_i = -\frac{a_{i-2}}{i(i+2m)}, \quad i = 2, 3, \ldots.$$

(c) Show that, if we choose $a_0 = 2^{-m}/m!$, then the resulting coefficients are

$$a_{2j+1} = 0 \quad \text{and} \quad a_{2j} = \frac{(-1)^j}{j!(j+m)!2^{2j+m}}, \quad j = 0, 1, 2, \ldots.$$

Name the resulting solution J_m, and conclude that

$$J_m(x) = \sum_{j=0}^{\infty} \frac{(-1)^j}{j!(j+m)!}\left(\frac{x}{2}\right)^{2j+m}.$$

17. In this problem, we derive a recurrence formula that allows $J_m(x)$, $m = 2, 3, 4\ldots$, to be expressed in terms of $J_0(x)$ and $J_1(x)$.

(a) From the series form of $J_m(x)$, write

$$x^{-m} J_m(x) = 2^{-m} \sum_{j=0}^{\infty} \frac{(-1)^j}{j!(j+m)!}\left(\frac{x}{2}\right)^{2j},$$

and then show that

$$\frac{d}{dx}(x^{-m} J_m(x)) = -x^{-m} \sum_{j=0}^{\infty} \frac{(-1)^j}{j!(j+m+1)!}\left(\frac{x}{2}\right)^{2j+m+1} = -x^{-m} J_{m+1}(x), \quad m = 0, 1, 2, \ldots.$$

(b) Similarly show that

$$\frac{d}{dx}(x^m J_m(x)) = x^m J_{m-1}(x), \quad m = 1, 2, 3, \ldots.$$

(c) From parts (a) and (b), respectively, conclude that

$$x J_m'(x) = m J_m(x) - x J_{m+1}(x),\quad m = 1, 2, 3, \ldots,$$

and

$$x J_m'(x) = -m J_m(x) + x J_{m-1}(x),\quad m = 1, 2, 3, \ldots.$$

(d) Combine the results of part (c) to show that

$$x J_{m+1}(x) = 2m J_m(x) - x J_{m-1}(x),\quad m = 1, 2, 3, \ldots.$$

(e) Write each of $J_2(x)$ and $J_3(x)$ in terms of $J_0(x)$ and $J_1(x)$.

18. Define the operator $\mathcal{S}$ by

$$\mathcal{S}X = \frac{1}{r}\left((rX')' - \frac{m^2}{r}X\right)$$

for C^2 functions $X = X(r)$ on the interval $(0, \rho]$.

(a) Let R and Q be twice-differentiable functions on $(0, \rho]$ such that $R(\rho) = Q(\rho) = 0$ and R, R', Q, and Q' are each bounded on $(0, \rho]$. Show that

$$\int_0^\rho (R\mathcal{S}Q - Q\mathcal{S}R)\, r\, dr = 0.$$

(b) Consider the Sturm-Liouville eigenvalue problem

$$\mathcal{S}X = \lambda X;\quad X(\rho) = 0;\quad X, X' \text{ bounded.}$$

Suppose that R and Q are eigenfunctions corresponding to eigenvalues λ_1 and λ_2, respectively. Show that, if $\lambda_1 \neq \lambda_2$, then

$$\int_0^\rho QR\, r\, dr = 0.$$

19. Let $\rho > 0$, and let m a nonnegative integer. Let σ be any positive number such that $J_m(\sigma\rho) = 0$, and let $R(r) = J_m(\sigma r)$, which satisfies

$$r^2 R'' + rR' + (\sigma^2 r^2 - m^2)R = 0.$$

(a) Show that

$$\frac{d}{dr}(rR')^2 + (\sigma^2 r^2 - m^2)\frac{d}{dr}(R^2) = 0$$

(b) Integrate from $r = 0$ to $r = \rho$ (using "parts" on the second term) to arrive at

$$\rho^2 R'(\rho)^2 + (\sigma^2\rho^2 - m^2)R(\rho)^2 + m^2 R(0)^2 - 2\sigma^2 \int_0^\rho R(r)^2 r\, dr = 0,$$

and show that this reduces to

$$2\int_0^\rho J_m(\sigma r)^2 r\, dr = \rho^2 J_m'(\sigma\rho)^2.$$

(c) Use the first of the relations in Problem 17c to conclude that

$$\int_0^\rho J_m(\alpha r)^2 r\, dr = \frac{\rho^2}{2} J_{m+1}(\alpha r)^2.$$

20. Use separation of variables to solve the problem

$$w_t - k\left(w_{rr} + \frac{1}{r}w_r\right) = 0,\quad t > 0, 0 < r < \rho,$$
$$w(t, \rho) = 0,\quad t > 0,$$
$$w(0, r) = \gamma(r),\quad 0 \le r \le \rho,$$

which arises when radial symmetry (θ-independence) is assumed in problem (9a–d).

21. Use separation of variables to solve the problem

$$w_{tt} - c^2\left(w_{rr} + \frac{1}{r}w_r\right) = 0,\quad t > 0, 0 < r < 1,$$
$$w(t, 1) = 0,\quad t > 0,$$
$$w(0, r) = \gamma(r),\quad 0 < r < 1,$$
$$w_t(0, r) = 0,\quad 0 < r < 1,$$

for the radially symmetric vibrations of a circular membrane. (Assume the usual boundedness.)

22. Use separation of variables to solve the problem

$$u_{tt} - c^2(u_{xx} + u_{yy}) = 0, \quad t > 0, 0 < x < \pi,$$
$$0 < y < \pi,$$
$$u(t, x, 0) = u(t, x, \pi) = 0, \quad t > 0, 0 \le x \le \pi,$$
$$u(t, 0, y) = u(t, \pi, y) = 0, \quad t > 0, 0 < y < \pi,$$
$$u(0, x, y) = \varphi(x, y),$$
$$0 \le x \le \pi, 0 \le y \le \pi,$$
$$u_t(0, x, y) = 0,$$
$$0 \le x \le \pi, 0 \le y \le \pi,$$

for the vibrations of a square membrane.

APPENDIX

I

LINEAR ALGEBRA

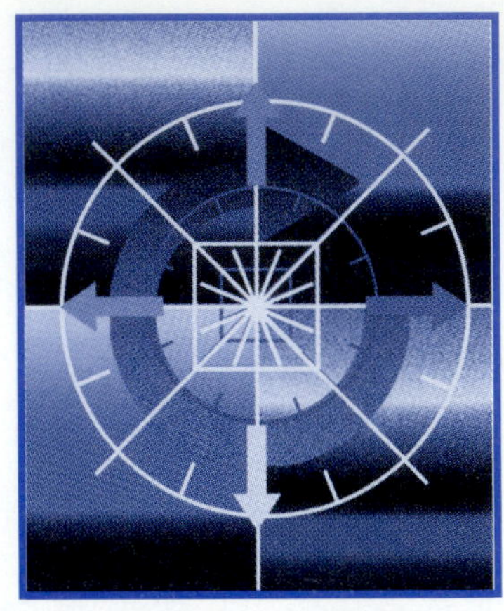

A. Linear Algebraic Equations

In this section we will study systems of m linear algebraic equations in n variables $x_1, x_2, \ldots, x_n$. Such a system may be written as

$$\begin{aligned} a_{1,1}x_1 + a_{1,2}x_2 + \cdots + a_{1,n}x_n &= b_1, \\ a_{2,1}x_1 + a_{2,2}x_2 + \cdots + a_{2,n}x_n &= b_2, \\ &\vdots \\ a_{m,1}x_1 + a_{m,2}x_2 + \cdots + a_{m,n}x_n &= b_m, \end{aligned} \tag{1}$$

where the coefficients $a_{i,j}$ and b_i are given. When each $b_i = 0$, we say that the system is *homogeneous*; otherwise it is *nonhomogeneous*. The following are three instances of the type of problem covered by (1):

$$\begin{aligned} x_1 + 3x_2 &= 0 \\ x_1 - x_2 &= 3 \end{aligned} \qquad \begin{aligned} x_1 + \phantom{x_2 - {}} 3x_3 &= 6 \\ x_1 + x_2 - 2x_3 &= 3 \\ 2x_1 - 3x_2 + x_3 &= 1 \end{aligned} \qquad \begin{aligned} x_1 + x_2 - x_3 &= 0 \\ 2x_1 - x_2 + x_3 &= 0 \end{aligned}$$

Note that the third of these is a homogeneous system, while the first two are nonhomogeneous.

If (1) has at least one solution, it is said to be a *consistent* system; if it has no solution, it is *inconsistent*. It is worth noting that any homogeneous system is consistent, since it will have the trivial solution $x_1 = x_2 = \cdots = x_n = 0$.

With any linear system (1), we associate an $m \times (n+1)$ *augmented matrix*

$$\begin{pmatrix} a_{1,1} & a_{1,2} & \cdots & a_{1,n} & b_1 \\ a_{2,1} & a_{2,2} & \cdots & a_{2,n} & b_2 \\ \vdots & \vdots & & \vdots & \vdots \\ a_{m,1} & a_{m,2} & \cdots & a_{m,n} & b_m. \end{pmatrix}.$$

(It is "$m \times (n+1)$" because it has m rows and $n+1$ columns.) The augmented matrix serves as a proxy for the system of equations, in the sense that manipulations of the equations for the purpose of solving the system may be accomplished by analogous manipulations of the rows of the augmented matrix.

Linear systems can be solved by a process known as **Gaussian elimination**. Each step in Gaussian elimination involves one of three kinds of "elementary row operations." These operations are listed—in terms of the equations of the system and in terms the rows of the augmented matrix—as follows.

Operations on Equations	**Row Operations**
Multiply an equation by a nonzero constant.	Multiply a row by a nonzero constant.
Add a multiple of one equation to another equation.	Add a multiple of one row to another row.
Interchange two equations.	Interchange two rows.

The goal of Gaussian elimination is to produce an equivalent system (i.e., one with the same solution(s)) that is easy to solve. Before we discuss this goal more precisely, let's illustrate the elimination process with the following two examples.

Example 1

Consider the system of equations on the left, whose augmented matrix is shown on the right:

$$\begin{aligned} x_1 + \phantom{x_2 -{}} 3x_3 &= 6 \\ x_1 + x_2 - 2x_3 &= 3 \\ 2x_1 - 3x_2 + x_3 &= 1 \end{aligned} \qquad \begin{pmatrix} 1 & 0 & 3 & 6 \\ 1 & 1 & -2 & 3 \\ 2 & -3 & 1 & 1 \end{pmatrix}.$$

Throughout the following steps, we will continue to display both the system of equations and the corresponding augmented matrix—primarily for the sake of illustrating the fact that the entire process need only be done on the augmented matrix. Our first goal is to eliminate x_1 from the second and third equations. To do this, we subtract the first equation from the second, and subtract 2 times the first equation from the third:

$$\begin{aligned} x_1 + \phantom{3x_2 -{}} 3x_3 &= 6 \\ x_2 - 5x_3 &= -3 \\ -\ 3x_2 - 5x_3 &= -11 \end{aligned} \qquad \begin{pmatrix} 1 & 0 & 3 & 6 \\ 0 & 1 & -5 & -3 \\ 0 & -3 & -5 & -11 \end{pmatrix}.$$

Our next goal is to eliminate x_2 from the third equation. So we add 3 times the second equation to the third:

$$\begin{aligned} x_1 + \phantom{x_2 -{}} 3x_3 &= 6 \\ x_2 - 5x_3 &= -3 \\ -\ 20x_3 &= -20 \end{aligned} \qquad \begin{pmatrix} 1 & 0 & 3 & 6 \\ 0 & 1 & -5 & -3 \\ 0 & 0 & -20 & -20 \end{pmatrix}.$$

Next we simplify the third equation by dividing it by -20:

$$\begin{aligned} x_1 + \quad\quad 3x_3 &= 6 \\ x_2 - 5x_3 &= -3 \\ x_3 &= 1 \end{aligned} \qquad \begin{pmatrix} 1 & 0 & 3 & 6 \\ 0 & 1 & -5 & -3 \\ 0 & 0 & 1 & 1 \end{pmatrix}.$$

Now* we want to eliminate x_3 from the first and second equations. So we subtract 3 times the third equation from the first, and add 5 times the third equation from the second:

$$\begin{aligned} x_1 \quad\quad &= 3 \\ x_2 \quad &= 2 \\ x_3 &= 1 \end{aligned} \qquad \begin{pmatrix} 1 & 0 & 0 & 3 \\ 0 & 1 & 0 & 2 \\ 0 & 0 & 1 & 1 \end{pmatrix}.$$

Thus, by eliminating all but the ith variable from the ith equation ($i = 1, 2, 3$), we have solved the system. Notice the final form of the augmented matrix. The "diagonal" entries are ones, the last column contains the solution, and all other entries are zeros. ■

Example 2

In this example we consider a system of only two equations in three unknowns:

$$\begin{aligned} x_1 + x_2 + x_3 &= -1 \\ x_1 - x_2 + 3x_3 &= 1 \end{aligned} \qquad \begin{pmatrix} 1 & 1 & 1 & -1 \\ 1 & -1 & 3 & 1 \end{pmatrix}.$$

This time we will only operate on the augmented matrix—without carrying along the equivalent equations. Let R_1 and R_2 denote the first and second rows of the augmented matrix, respectively. Our first step in reducing the system of equations would be to eliminate x_1 from the second equation. So we begin the reduction of the augmented matrix by replacing R_2 with $R_2 - R_1$ in order to "zero out" the first entry in R_2. We then multiply (the new) R_2 by $-1/2$. These steps are indicated thus:

$$\begin{pmatrix} 1 & 1 & 1 & -1 \\ 1 & -1 & 3 & 1 \end{pmatrix} \quad \begin{matrix} R_1 \\ R_2 - R_1 \end{matrix} \begin{pmatrix} 1 & 1 & 1 & -1 \\ 0 & -2 & 2 & 2 \end{pmatrix}$$
$$\begin{matrix} R_1 \\ -\frac{1}{2}R_2 \end{matrix} \begin{pmatrix} 1 & 1 & 1 & -1 \\ 0 & 1 & -1 & -1 \end{pmatrix}$$

Next we eliminate x_2 from the first equation by zeroing out the second entry of R_1 as follows:

$$\begin{pmatrix} 1 & 1 & 1 & -1 \\ 0 & 1 & -1 & -1 \end{pmatrix} \quad \begin{matrix} R_1 - R_2 \\ R_2 \end{matrix} \begin{pmatrix} 1 & 0 & 2 & 0 \\ 0 & 1 & -1 & -1 \end{pmatrix}.$$

---

* At this point we could proceed directly to the solution by a process known as *back-substitution*, in which we obtain x_3 from the third equation, then x_2 from the second, and then x_1 from the first; however, for the sake of illustration, we will instead continue the elimination process.

So we see that the original system is equivalent to the reduced system

$$\begin{aligned} x_1 \quad + 2x_3 &= 0, \\ x_2 - x_3 &= -1. \end{aligned}$$

To describe the (infinitely many) solutions of the system in a simple way, we let x_3 be a "free variable" which determines x_1 and x_2 as follows:

$$\begin{aligned} x_1 &= -2x_3, \\ x_2 &= x_3 - 1, \\ x_3 &= \text{ any real number.} \end{aligned}$$

So, for instance, each of the following are solutions of the system (in vector form):

$$\begin{pmatrix} 0 \\ -1 \\ 0 \end{pmatrix} \quad \begin{pmatrix} -2 \\ 0 \\ 1 \end{pmatrix} \quad \begin{pmatrix} 2 \\ -2 \\ -1 \end{pmatrix} \quad \begin{pmatrix} -10 \\ -6 \\ 5 \end{pmatrix}.$$

We point out that any of x_1, x_2, x_3 could have been used as the free variable, and the same set of solutions would result. ■

For the purpose of discussing Gaussian elimination in more detail, it is convenient to define the notion of a *pivot*. An entry in a matrix is said to be a **pivot**, if it is the first nonzero entry of its row. For instance, in

$$\begin{pmatrix} 0 & 2 & 4 & 5 \\ 1 & -1 & 0 & 0 \\ 0 & 3 & 1 & 1 \end{pmatrix},$$

2 is the pivot in the first row, 1 is the pivot in the second row, and 3 is the pivot in the third row. Note that every nonzero row of a matrix will contain exactly one pivot, and a row of zeros would contain no pivot.

The goal of Gaussian elimination is to obtain an augmented matrix that is in *echelon form*. A matrix is in **echelon form**, if

- the pivot in any nonzero row is to the left of the pivot in each nonzero row below it;
- all zero rows are below every nonzero row.

The following matrices are in echelon form:

$$\begin{pmatrix} 1 & 1 & 2 & 0 \\ 0 & 2 & 0 & 3 \\ 0 & 0 & -1 & 5 \end{pmatrix} \quad \begin{pmatrix} 1 & 2 & 1 & 3 \\ 0 & 0 & 3 & 2 \end{pmatrix} \quad \begin{pmatrix} 0 & 2 & 2 & 0 & 1 \\ 0 & 0 & 1 & -1 & -1 \\ 0 & 0 & 0 & 0 & 0 \end{pmatrix}.$$

The goal of the **Gauss-Jordan** process is to achieve an augmented matrix that is in *reduced* echelon form. A matrix is in **reduced echelon form**, if it is in echelon form, *and*

- each pivot is a 1;
- each pivot is the only nonzero entry in its column.

The following matrices are in reduced echelon form:

$$\begin{pmatrix} 1 & 0 & 0 & 0 \\ 0 & 1 & 0 & 3 \\ 0 & 0 & 1 & 5 \end{pmatrix} \qquad \begin{pmatrix} 1 & 2 & 0 & 3 \\ 0 & 0 & 1 & 2 \end{pmatrix} \qquad \begin{pmatrix} 0 & 1 & 0 & 0 & 1 \\ 0 & 0 & 1 & -1 & -1 \\ 0 & 0 & 0 & 0 & 0 \end{pmatrix}.$$

It turns out that a given matrix may have more than one echelon form; however, its *reduced* echelon form is unique.

Once the reduced echelon form of the augmented matrix is computed, it is a simple matter to describe the solution(s) of the original system. We illustrate with an example.

Example 3

Each of the matrices

$$\begin{pmatrix} 1 & 0 & 0 & -2 \\ 0 & 1 & 0 & 3 \\ 0 & 0 & 1 & 5 \end{pmatrix} \qquad \begin{pmatrix} 1 & 2 & 0 & 3 \\ 0 & 0 & 1 & 2 \\ 0 & 0 & 0 & 1 \end{pmatrix} \qquad \begin{pmatrix} 1 & 2 & 0 & 0 \\ 0 & 0 & 1 & -1 \\ 0 & 0 & 0 & 0 \end{pmatrix}$$

is the reduced echelon form of a system of three linear equations in three unknowns. In the first case, the original system is equivalent to

$$x_1 = -2, \quad x_2 = 3, \quad x_3 = 5,$$

and thus has a unique solution. In the second case, the equivalent system represented by the reduced matrix is

$$x_1 = 3 - 2x_2, \quad x_3 = 2, \quad 0x_3 = 1,$$

which clearly has no solution. Thus the system is inconsistent. In the third case, the equivalent system represented by the reduced matrix is

$$x_1 = -x_2, \quad x_3 = -1.$$

Here, x_2 is a "free variable." Thus there are infinitely many solutions, which can be described by

$$x_1 = -x_2, \quad x_2 = \text{any real number}, \quad x_3 = -1.$$

■

Example 3 illustrates essentially all the structural possibilities for the set of solutions of a linear system. The following theorem summarizes these possibilities and characterizes each in terms of the structure of the reduced echelon form of the augmented matrix.

Theorem 1

The linear system (1) has either a unique solution, no solution, or else infinitely many solutions. In particular, the following statements are true, where U denotes the reduced echelon form of the system's $m \times (n+1)$ augmented matrix.

i) The system is consistent, if and only if the last column of U does not contain a pivot; that is, U has no row of the form

$$R_j = \begin{pmatrix} 0 & 0 & \cdots & 0 & \tilde{b}_j \end{pmatrix} \quad \text{with } b_j \neq 0.$$

ii) The system has infinitely many solutions, if and only if it is consistent and at least one of the first n columns of U contains no pivot. If the jth *column of U contains no pivot, then* x_j *may be designated as a "free variable."*

iii) If $m < n$*, then the system is either inconsistent or else it has infinitely many solutions. If* $m = n$*, then the system has a unique solution, if and only if each "diagonal" entry* $u_{i,i}$ *of U is 1. If* $m > n$*, then the system has a unique solution, if and only if each diagonal entry* $u_{i,i}$ *in the first n rows of U is 1, and each row below the* nth *row contains only zeros. Whenever there is a unique solution, it is contained in the first n entries of the last column of U.* ▲

Example 4

Consider the following four augmented matrices in reduced echelon form:

$$\begin{pmatrix} 1 & 0 & 0 & 4 \\ 0 & 1 & 0 & -3 \\ 0 & 0 & 1 & 2 \end{pmatrix} \quad \begin{pmatrix} 1 & 2 & 0 & 3 \\ 0 & 0 & 1 & 2 \end{pmatrix} \quad \begin{pmatrix} 1 & 0 & 2 & 0 \\ 0 & 1 & 1 & 2 \\ 0 & 0 & 0 & 3 \\ 0 & 0 & 0 & 0 \end{pmatrix} \quad \begin{pmatrix} 1 & 0 & 2 \\ 0 & 1 & 3 \\ 0 & 0 & 0 \end{pmatrix}.$$

The first represents a system of $m = 3$ equations in $n = 3$ unknowns. Since $m = n$ and each diagonal entry of the matrix is a 1, the final column contains the solution:

$$x_1 = 4, \quad x_2 = -3, \quad x_3 = 2.$$

The second augmented matrix represents a system of $m = 2$ equations in $n = 3$ unknowns. The system is consistent, since the final column contains no pivot. Also, the second column contains no pivot; therefore the system has infinitely many solutions that can be described in terms of the free variable x_2:

$$x_1 = 3 - 2x_2, \quad x_2 = \text{ any real number }, \quad x_3 = 2.$$

The third augmented matrix represents a system of $m = 4$ equations in $n = 4$ unknowns. The system is inconsistent, since the final column contains a pivot (in the third row). Finally, the fourth augmented matrix represents a system of $m = 3$ equations in $n = 2$ unknowns. The system is consistent, since the final column contains no pivot. Also, each diagonal entry in the first 2 rows is 1; therefore there is a unique solution:

$$x_1 = 2, \quad x_2 = 3.$$ ■

The Gauss-Jordan Algorithm We now describe, as an algorithm in two phases, the Gauss-Jordan process for generating the reduced echelon form of a matrix. Phase 1 (sometimes called the "forward" phase) produces an echelon form. Phase 2 then produces the reduced echelon form.

1. Let M_i denote the submatrix consisting of the ith through the mth rows of the current augmented matrix. Repeat the following for $i = 1, 2, \ldots, m$.

a) Begin with the leftmost nonzero column of M_i. Exchange rows if necessary so that the top row of M_i contains a pivot.
b) Use row operations to zero out all entries below the pivot in the top row of M_i.

2. Determine whether there is a pivot in the last column. If so, stop, because the system is inconsistent. Otherwise, beginning with the rightmost pivot, work upward and to the left, using row operations to zero out all entries above each pivot and make each pivot 1.

The first phase of the algorithm is known as **Gaussian elimination**. The two phases together comprise the Gauss-Jordan process.

Example 5

Consider the system on the left, whose augmented matrix is shown on the right:

$$\begin{aligned} x_1 + x_2 + 3x_3 + x_4 &= 1 \\ x_1 + x_2 - 2x_3 &= 0 \\ -3x_2 + x_3 + x_4 &= 1 \end{aligned} \qquad \begin{pmatrix} 1 & 1 & 3 & 1 & 1 \\ 1 & 1 & -2 & 0 & 0 \\ 0 & -3 & 1 & 1 & 1 \end{pmatrix}.$$

We begin the Gauss-Jordan process by observing that the first entry in the first row is already a pivot and then using that pivot to zero out all entries below it:

$$\begin{matrix} R_1 \\ R_2 - R_1 \\ R_3 \end{matrix} \begin{pmatrix} 1 & 1 & 3 & 1 & 1 \\ 0 & 0 & -5 & -1 & -1 \\ 0 & -3 & 1 & 1 & 1 \end{pmatrix}.$$

Next we ignore the first row and concentrate on the second and third rows. (This is M_2.) We bring a pivot to the second row by exchanging it with the third row:

$$\begin{matrix} R_1 \\ R_3 \\ R_2 \end{matrix} \begin{pmatrix} 1 & 1 & 3 & 1 & 1 \\ 0 & -3 & 1 & 1 & 1 \\ 0 & 0 & -5 & -1 & -1 \end{pmatrix}.$$

Since there is no nonzero entry below that pivot, we are done with Step 1. Observing that the system is consistent, we now proceed with Step 2. Using the pivot in the third row, we zero out all other entries above it:

$$\begin{matrix} R_1 + \frac{3}{5}R_3 \\ R_2 + \frac{1}{5}R_3 \\ R_3 \end{matrix} \begin{pmatrix} 1 & 1 & 0 & \frac{2}{5} & \frac{2}{5} \\ 0 & -3 & 0 & \frac{4}{5} & \frac{4}{5} \\ 0 & 0 & -5 & -1 & -1 \end{pmatrix}.$$

Next we use the pivot in the second row to zero out all other entries above it:

$$\begin{matrix} R_1 + \frac{1}{3}R_2 \\ R_2 \\ R_3 \end{matrix} \begin{pmatrix} 1 & 0 & 0 & \frac{2}{3} & \frac{2}{3} \\ 0 & -3 & 0 & \frac{4}{5} & \frac{4}{5} \\ 0 & 0 & -5 & -1 & -1 \end{pmatrix}.$$

Now we finish up by making each pivot be a 1. (We held this for last simply for the sake of clarity.)

$$\begin{matrix} R_1 \\ -\frac{1}{3}R_2 \\ -\frac{1}{5}R_3 \end{matrix} \begin{pmatrix} 1 & 0 & 0 & \frac{2}{3} & \frac{2}{3} \\ 0 & 1 & 0 & -\frac{4}{15} & -\frac{4}{15} \\ 0 & 0 & 1 & \frac{1}{5} & \frac{1}{5} \end{pmatrix}.$$

Thus we have the reduced echelon form of the augmented matrix, from which we "read" the solution of the system:

$x_1 = \frac{2}{3}(1-x_4)$, $x_2 = \frac{4}{15}(x_4-1)$, $x_3 = \frac{1}{5}(1-x_4)$, $x_4 =$ any real number. ■

Homogeneous Systems For any homogeneous system of equations, the corresponding augmented matrix has only zeros in its last column. Moreover, this property persists throughout the Gauss-Jordan process. Therefore it is not necessary to carry that extra column; that is, the Gauss-Jordan process need be carried out only on the $m \times n$ *coefficient matrix*

$$A = \begin{pmatrix} a_{1,1} & a_{1,2} & \cdots & a_{1,n} \\ a_{2,1} & a_{2,2} & \cdots & a_{2,n} \\ \vdots & \vdots & & \vdots \\ a_{m,1} & a_{m,2} & \cdots & a_{m,n} \end{pmatrix}.$$

Example 6

Consider the homogeneous system on the left, whose coefficient matrix is on the right:

$$\begin{aligned} x_1 - x_2 + x_3 - x_4 &= 0 \\ x_1 + x_2 - x_3 &= 0 \\ 3x_1 + x_2 - x_3 - x_4 &= 0 \end{aligned} \qquad \begin{pmatrix} 1 & -1 & 1 & -1 \\ 1 & 1 & -1 & 0 \\ 3 & 1 & -1 & -1 \end{pmatrix}.$$

The Gauss-Jordan process proceeds as follows:

$$\begin{matrix} R_1 \\ R_2 - R_1 \\ R_3 - 3R_1 \end{matrix} \begin{pmatrix} 1 & -1 & 1 & -1 \\ 0 & 2 & -2 & 1 \\ 0 & 4 & -4 & 2 \end{pmatrix} \qquad \begin{matrix} R_1 \\ R_2 \\ R_3 - 2R_2 \end{matrix} \begin{pmatrix} 1 & -1 & 1 & -1 \\ 0 & 2 & -2 & 1 \\ 0 & 0 & 0 & 0 \end{pmatrix}$$

$$\begin{matrix} R_1 + \frac{1}{2}R_2 \\ R_2 \\ R_3 \end{matrix} \begin{pmatrix} 1 & 0 & 0 & -\frac{1}{2} \\ 0 & 2 & -2 & 1 \\ 0 & 0 & 0 & 0 \end{pmatrix} \qquad \begin{matrix} R_1 \\ \frac{1}{2}R_2 \\ R_3 \end{matrix} \begin{pmatrix} 1 & 0 & 0 & -\frac{1}{2} \\ 0 & 1 & -1 & \frac{1}{2} \\ 0 & 0 & 0 & 0 \end{pmatrix}.$$

Therefore, the original system is equivalent to

$$\begin{aligned} x_1 - \tfrac{1}{2}\, x_4 &= 0, \\ x_2 - x_3 + \tfrac{1}{2}\, x_4 &= 0. \end{aligned}$$

So we use free variables x_3 and x_4 to describe the set of solutions as follows:

$$x_1 = \tfrac{1}{2}x_4, \quad x_2 = x_3 - \tfrac{1}{2}x_4,$$
$$x_3 = \text{ any real number}, \quad x_4 = \text{ any real number}. \quad ■$$

Since every homogeneous system has the trivial solution ($x_1 = x_2 = \cdots = x_n = 0$), a primary concern is whether a given system has other solutions. The following theorem settles this issue in terms of any echelon form of the system's coefficient matrix.

Theorem 2

Suppose that the linear system (1) is homogeneous. Then the following statements are true, where A^* denotes any echelon form of the system's $m \times n$ coefficient matrix.

i) The trivial solution is unique, if and only if every column of A^* contains a pivot. Otherwise the system has infinitely many solutions. If the jth column of A^* contains no pivot, then x_j may be designated as a "free variable."

ii) The trivial solution is unique, if and only if A^* has exactly n nonzero rows. Otherwise the system has infinitely many solutions. ▲

Example 7

Each of the following is an echelon form of the coefficient matrix of a homogeneous linear system. Alongside each is the result of Theorem 2.

$$\begin{pmatrix} 1 & 2 \\ 0 & 3 \\ 0 & 0 \end{pmatrix} \quad \text{The trivial solution is unique.}$$

$$\begin{pmatrix} 1 & 2 & 0 & 4 \\ 0 & 0 & 1 & -3 \\ 0 & 0 & 0 & 0 \end{pmatrix} \quad \text{Infinitely many solutions; } x_2 \text{ and } x_4 \text{ are free variables.}$$

$$\begin{pmatrix} 1 & 0 & 2 \\ 0 & 1 & 1 \\ 0 & 0 & 0 \\ 0 & 0 & 0 \end{pmatrix} \quad \text{Infinitely many solutions; } x_3 \text{ is a free variable.}$$

$$\begin{pmatrix} 1 & 2 & 1 \\ 0 & 2 & 1 \\ 0 & 0 & 3 \end{pmatrix} \quad \text{The trivial solution is unique.}$$

■

Gaussian Elimination with Back-Substitution For "number-crunching" purposes, that is, for the numerical solution of $n \times n$ linear systems that arise in applications, a method known as *Gaussian elimination with back-substitution* is generally preferred over the Gauss-Jordan process. The difference between these methods lies only in the second (backward) phase, after an echelon form of the augmented matrix has been computed by Gaussian elimination. We state the algorithm as follows. The first phase of the algorithm is known as **Gaussian elimination**; the second phase is **back-substitution**.

1. Let M_i denote the submatrix consisting of the ith through the nth rows of the current augmented matrix. Repeat the following for $i = 1, 2, \dots, n$.
   a) Begin with the leftmost nonzero column of M_i. Exchange rows if necessary so that the top row of M_i contains a pivot.
   b) Use row operations to zero out all entries below the pivot in the top row of M_i.
2. Work from the last row of the reduced matrix up to the first, solving the ith equation of the corresponding reduced system for x_i, using the already computed values of $x_{i+1}, x_{i+2}, \dots, x_n$.

A more explicit description of back-substitution is relatively simple. Suppose that the matrix resulting from the first (forward) phase of the algorithm has entries $\tilde{a}_{i,j}$ and, in the last column, $\tilde{b}_i$. Then the details of back-substitution are:

First compute

$$x_n = \frac{\tilde{b}_n}{\tilde{a}_{n,n}}.$$

Then, for $i = n-1, n-2, \dots, 2, 1$, compute

$$x_i = \frac{1}{\tilde{a}_{i,i}}\left(\tilde{b}_i - \sum_{j=i+1}^{n} \tilde{a}_{i,j}x_j\right).$$

Example 8

Consider the system on the left, whose augmented matrix is shown on the right:

$$\begin{aligned} x_1 - x_2 + 3x_3 &= 4 \\ x_1 + 2x_2 - x_3 &= 5 \\ x_1 + x_2 + x_3 &= 6 \end{aligned} \qquad \begin{pmatrix} 1 & -1 & 3 & 4 \\ 1 & 2 & -1 & 5 \\ 1 & 1 & 1 & 6 \end{pmatrix}.$$

Gaussian elimination proceeds as follows:

$$\begin{matrix} R_1 \\ R_2 - R_1 \\ R_3 - R_1 \end{matrix} \begin{pmatrix} 1 & -1 & 3 & 4 \\ 0 & 3 & -4 & 1 \\ 0 & 2 & -2 & 2 \end{pmatrix} \qquad \begin{matrix} R_1 \\ R_2 \\ R_3 - \frac{2}{3}R_2 \end{matrix} \begin{pmatrix} 1 & -1 & 3 & 4 \\ 0 & 3 & -4 & 1 \\ 0 & 0 & \frac{2}{3} & \frac{4}{3} \end{pmatrix}$$

So the original system is equivalent to

$$\begin{aligned} x_1 - x_2 + 3x_3 &= 4, \\ 3x_2 - 4x_3 &= 1, \\ \frac{2}{3}x_3 &= \frac{4}{3}. \end{aligned}$$

From we last equation we find $x_3 = 2$; thus the second equation becomes

$$3x_2 - (4)(2) = 1,$$

which easily gives us $x_2 = (1+8)/3 = 3$. Now the first equation becomes

$$x_1 - 3 + (3)(2) = 4,$$

which yields $x_1 = 4 + 3 - 6 = 1$. Thus we have the solution:

$$x_1 = 1, \quad x_2 = 3, \quad x_3 = 2.$$ ■

PROBLEMS

In each of Problems 1 through 12, use the Gauss-Jordan process to solve the linear system.

1. $\begin{cases} x_1 - x_2 + x_3 = 1 \\ x_1 + x_2 - x_3 = 3 \\ -x_1 + x_2 + x_3 = 1 \end{cases}$

2. $\begin{cases} -x_1 + x_2 + x_3 = 1 \\ x_1 \qquad + x_3 = 4 \end{cases}$

3. $\begin{cases} -x_1 + x_2 + x_3 = 1 \\ x_1 \qquad + x_3 = 4 \\ x_1 + x_2 - x_3 = 1 \end{cases}$

4. $\begin{cases} x_1 + x_2 - x_3 = 3 \\ x_1 + 2x_2 + x_3 = 1 \end{cases}$

5. $\begin{cases} x_1 + x_2 - x_3 = 3 \\ x_1 + 2x_2 + x_3 = 1 \\ x_1 + x_2 + x_3 = -1 \end{cases}$

6. $\begin{cases} x_1 + 2x_2 + 3x_3 = 0 \\ 2x_1 + 3x_2 + x_3 = 0 \end{cases}$

7. $\begin{cases} x_1 + 2x_2 + 3x_3 = 0 \\ 3x_1 + x_2 + 2x_3 = 18 \\ 2x_1 + 3x_2 + x_3 = 0 \end{cases}$

8. $\begin{cases} x_1 - x_2 + x_3 = 3 \\ x_1 + 2x_2 + x_3 = -1 \\ x_1 + x_2 + x_3 = 1 \end{cases}$

9. $\begin{cases} x_1 - x_2 + x_3 = -1 \\ x_1 + 2x_2 + x_3 = 5 \\ x_1 + x_2 + x_3 = 3 \end{cases}$

10. $\begin{cases} -x_1 + x_2 \qquad = 3 \\ x_1 - x_2 + x_3 = 2 \\ x_2 - x_3 = 1 \end{cases}$

11. $\begin{cases} x_1 + x_2 + x_3 = 2 \\ x_1 + x_2 + 2x_3 = 1 \\ x_2 - x_3 = 3 \end{cases}$

12. $\begin{cases} -3x_1 + 2x_2 + 3x_3 = 2 \\ 5x_1 - x_2 + 4x_3 = 0 \\ 7x_1 + 2x_2 + 19x_3 = 0 \end{cases}$

In each of Problems 13 through 18, you are given an augmented matrix in reduced echelon form. Find all solutions of the corresponding linear system.

13. $\begin{pmatrix} 1 & 2 & 0 & 2 \\ 0 & 0 & 1 & 3 \\ 0 & 0 & 0 & 0 \end{pmatrix}$

14. $\begin{pmatrix} 1 & 0 & 2 & 0 & 2 \\ 0 & 1 & 3 & 0 & 0 \\ 0 & 0 & 0 & 1 & 1 \end{pmatrix}$

15. $\begin{pmatrix} 1 & -1 & 0 & 1 & 2 \\ 0 & 0 & 1 & -1 & 0 \\ 0 & 0 & 0 & 0 & 1 \end{pmatrix}$

16. $\begin{pmatrix} 1 & 0 & 2 & 0 \\ 0 & 1 & 0 & 0 \\ 0 & 0 & 0 & 1 \end{pmatrix}$

17. $\begin{pmatrix} 1 & 0 & 2 & 0 & 0 \\ 0 & 1 & 0 & 1 & 1 \\ 0 & 0 & 0 & 0 & 0 \end{pmatrix}$

18. $\begin{pmatrix} 1 & -1 & 0 & 1 & 2 \\ 0 & 0 & 1 & -1 & 0 \\ 0 & 0 & 0 & 0 & 0 \end{pmatrix}$

In Problems 19 through 24, each matrix is an echelon form of the augmented matrix of a system of four equations in four unknowns. Asterisks represent nonzero entries. Indicate whether the system has a unique solution, infinitely many solutions, or no solution.

19. $\begin{pmatrix} * & 0 & * & 0 & * \\ 0 & * & * & 0 & 0 \\ 0 & 0 & 0 & * & * \\ 0 & 0 & 0 & 0 & 0 \end{pmatrix}$ **20.** $\begin{pmatrix} * & 0 & * & * & * \\ 0 & * & * & 0 & * \\ 0 & 0 & * & * & * \\ 0 & 0 & 0 & * & * \end{pmatrix}$ **21.** $\begin{pmatrix} * & 0 & * & 0 & 0 \\ 0 & * & * & * & * \\ 0 & 0 & 0 & * & * \\ 0 & 0 & 0 & 0 & * \end{pmatrix}$

22. $\begin{pmatrix} * & * & 0 & 0 & * \\ 0 & * & * & * & * \\ 0 & 0 & 0 & 0 & * \\ 0 & 0 & 0 & 0 & 0 \end{pmatrix}$ **23.** $\begin{pmatrix} * & 0 & * & 0 & 0 \\ 0 & * & 0 & * & * \\ 0 & 0 & * & 0 & * \\ 0 & 0 & 0 & * & * \end{pmatrix}$ **24.** $\begin{pmatrix} * & 0 & * & 0 & * \\ 0 & 0 & * & * & 0 \\ 0 & 0 & 0 & 0 & 0 \\ 0 & 0 & 0 & 0 & 0 \end{pmatrix}$

In Problems 25 through 28, solve the homogeneous system whose coefficient matrix is given.

25. $\begin{pmatrix} 1 & 1 & 1 \\ 0 & 2 & 3 \\ 1 & 2 & 1 \end{pmatrix}$ **26.** $\begin{pmatrix} 1 & 2 & 1 \\ 2 & 0 & -1 \\ 1 & 6 & 4 \end{pmatrix}$ **27.** $\begin{pmatrix} 2 & -3 & -1 \\ 5 & -2 & 1 \\ 1 & 4 & 3 \end{pmatrix}$ **28.** $\begin{pmatrix} 1 & 2 & 3 & 1 \\ 1 & 3 & 1 & 2 \\ 1 & 5 & -3 & 4 \\ 1 & 0 & 7 & -1 \end{pmatrix}$.

In each of Problems 29 through 31, you are given an echelon form of the augmented matrix of a linear system. Use back-substitution to finish solving the system.

29. $\begin{pmatrix} 1 & 3 & 0 & 5 \\ 0 & 2 & 1 & 1 \\ 0 & 0 & 2 & 6 \end{pmatrix}$ **30.** $\begin{pmatrix} 1 & 1 & 1 & 1 & 0 \\ 0 & 1 & 1 & 1 & 1 \\ 0 & 0 & 1 & 1 & -1 \\ 0 & 0 & 0 & 1 & 1 \end{pmatrix}$ **31.** $\begin{pmatrix} 3 & 7 & 1 & 2 & 1 & 0 \\ 0 & 1 & 1 & 1 & 1 & 4 \\ 0 & 0 & 1 & 3 & 2 & 5 \\ 0 & 0 & 0 & 2 & 1 & 1 \\ 0 & 0 & 0 & 0 & -1 & 3 \end{pmatrix}$.

B. Matrices

We begin with a brief discussion of vectors. No distinction will be made between vectors of length n and points in $\mathbb{R}^n$; each is identified with an $n \times 1$ array:

$$\mathbf{x} = \begin{pmatrix} x_1 \\ x_2 \\ \vdots \\ x_n \end{pmatrix}.$$

However, usually the term "vector" will be used. It is traditional to use the term "scalar" when referring to a real or complex number. A vector $\mathbf{x}$ may be multiplied by a scalar c as follows:

$$c\,\mathbf{x} = c \begin{pmatrix} x_1 \\ x_2 \\ \vdots \\ x_n \end{pmatrix} = \begin{pmatrix} cx_1 \\ cx_2 \\ \vdots \\ cx_n \end{pmatrix}.$$

Furthermore, addition of two vectors in $\mathbb{R}^n$ is defined by

$$\mathbf{x}+\mathbf{y} = \begin{pmatrix} x_1 \\ x_2 \\ \vdots \\ x_n \end{pmatrix} + \begin{pmatrix} y_1 \\ y_2 \\ \vdots \\ y_n \end{pmatrix} = \begin{pmatrix} x_1+y_1 \\ x_2+y_2 \\ \vdots \\ x_n+y_n \end{pmatrix}.$$

Given vectors $\mathbf{x}_1, \mathbf{x}_2, \ldots, \mathbf{x}_k$ in $\mathbb{R}^n$ and scalars $c_1, c_2, \ldots, c_k$, the vector

$$c_1\mathbf{x}_1 + c_2\mathbf{x}_2 + \cdots + c_k\mathbf{x}_k$$

is said to be a **linear combination** of $\mathbf{x}_1, \mathbf{x}_2, \ldots, \mathbf{x}_k$.

The **standard basis vectors** in $\mathbb{R}^n$ are

$$\mathbf{e}_1 = \begin{pmatrix} 1 \\ 0 \\ \vdots \\ 0 \end{pmatrix}, \quad \mathbf{e}_2 = \begin{pmatrix} 0 \\ 1 \\ \vdots \\ 0 \end{pmatrix}, \ldots, \mathbf{e}_n = \begin{pmatrix} 0 \\ 0 \\ \vdots \\ 1 \end{pmatrix}.$$

Every vector (or point) $\mathbf{x}$ in $\mathbb{R}^n$ can be expressed uniquely as a linear combination of $\mathbf{e}_1, \ldots, \mathbf{e}_n$:

$$\begin{pmatrix} x_1 \\ x_2 \\ \vdots \\ x_n \end{pmatrix} = x_1\begin{pmatrix} 1 \\ 0 \\ \vdots \\ 0 \end{pmatrix} + x_2\begin{pmatrix} 0 \\ 1 \\ \vdots \\ 0 \end{pmatrix} + \cdots + x_n\begin{pmatrix} 0 \\ 0 \\ \vdots \\ 1 \end{pmatrix}$$
$$= x_1\mathbf{e}_1 + x_2\mathbf{e}_2 + \cdots + x_n\mathbf{e}_n.$$

Given a vector $\mathbf{a}$ in $\mathbb{R}^n$, we define the **transpose** of $\mathbf{a}$ to be the "row vector"

$$\mathbf{a}^T = \begin{pmatrix} a_1 & a_2 & \cdots & a_n \end{pmatrix}.$$

The "product" of $\mathbf{a}^T$ with $\mathbf{x}$ is defined as

$$\mathbf{a}^T\mathbf{x} = \begin{pmatrix} a_1 & a_2 & \cdots & a_n \end{pmatrix} \begin{pmatrix} x_1 \\ x_2 \\ \vdots \\ x_n \end{pmatrix} = a_1x_1 + a_2x_2 + \cdots + a_nx_n.$$

Such a product defines a linear transformation from $\mathbb{R}^n$ to $\mathbb{R}$:

$$\mathbf{x} \longmapsto \mathbf{a}^T\mathbf{x} = a_1x_1 + a_2x_2 + \cdots + a_nx_n,$$

which is *linear* by virtue of these easily verified facts:

$$\mathbf{a}^T(c\,\mathbf{x}) = c\mathbf{a}^T\mathbf{x} \quad \text{and} \quad \mathbf{a}^T(\mathbf{x}+\mathbf{y}) = \mathbf{a}^T\mathbf{x} + \mathbf{a}^T\mathbf{y}.$$

Indeed, *every* linear transformation from $\mathbb{R}^n$ to $\mathbb{R}$ can be expressed as $\mathbf{x} \longmapsto \mathbf{a}^T\mathbf{x}$ for some $\mathbf{a}$.

Example 1

Consider the linear transformation from $\mathbb{R}^5$ into $\mathbb{R}$ defined by

$$\begin{pmatrix} x_1 \\ \vdots \\ x_5 \end{pmatrix} \longmapsto x_1 + 2x_2 - x_3 - 3x_4 + x_5.$$

This is equivalent to $\mathbf{x} \longmapsto \mathbf{a}^T\mathbf{x}$, where $\mathbf{a}^T = \begin{pmatrix} 1 & 2 & -1 & -3 & 1 \end{pmatrix}$. ■

An $m \times n$ **matrix** A is a two-dimensional array of numbers

$$A = \begin{pmatrix} a_{11} & a_{12} & \cdots & a_{1n} \\ a_{21} & a_{22} & \cdots & a_{2n} \\ \vdots & \vdots & & \vdots \\ a_{m1} & a_{m2} & \cdots & a_{mn} \end{pmatrix}.$$

It is designated as "$m \times n$" because it has m rows and n columns. Notice that the first subscript of each entry signifies its row, and the second subscript of each entry signifies its column. In other words, the entry of the matrix that lives in row i and column j (i.e., the "ijth" entry) is a_{ij}.

Note that an $m \times 1$ matrix is a (column) vector, and a $1 \times n$ matrix is a row vector.

It will be useful to establish notation for the individual rows and columns of a matrix A. The symbol $\mathbf{a}_{i\cdot\cdot}$ will be used here to represent the ith row of A:

$$\mathbf{a}_{i\cdot\cdot} = \begin{pmatrix} a_{i1} & a_{i2} & \cdots & a_{in} \end{pmatrix},$$

and the symbol $\mathbf{a}_{:j}$ will represent the jth column of A:

$$\mathbf{a}_{:j} = \begin{pmatrix} a_{1j} \\ a_{2j} \\ \vdots \\ a_{mj} \end{pmatrix}.$$

Thus A can be represented by either its rows or its columns:

$$A = \begin{pmatrix} \mathbf{a}_{1\cdot\cdot} \\ \mathbf{a}_{2\cdot\cdot} \\ \vdots \\ \mathbf{a}_{m\cdot\cdot} \end{pmatrix} = \begin{pmatrix} \mathbf{a}_{:1} & \mathbf{a}_{:2} & \cdots & \mathbf{a}_{:n} \end{pmatrix}.$$

Matrix-Vector Products The product $A\mathbf{x}$ of an $m \times n$ matrix A and a column vector $\mathbf{x}$ in $\mathbb{R}^n$ is *defined* as

$$\begin{pmatrix} a_{11} & a_{12} & \cdots & a_{1n} \\ a_{21} & a_{22} & \cdots & a_{2n} \\ \vdots & \vdots & \vdots & \\ a_{m1} & a_{m2} & \cdots & a_{mn} \end{pmatrix} \begin{pmatrix} x_1 \\ x_2 \\ \vdots \\ x_n \end{pmatrix} = \begin{pmatrix} a_{11}x_1 + a_{12}x_2 + \cdots + a_{1n}x_n \\ a_{21}x_1 + a_{22}x_2 + \cdots + a_{2n}x_n \\ \vdots \\ a_{m1}x_1 + a_{m2}x_2 + \cdots + a_{mn}x_n \end{pmatrix}.$$

Note that $A\mathbf{x}$ is a vector in $\mathbb{R}^m$. Observe also that the ith entry in the product $A\mathbf{x}$ is the product of the ith row of A with $\mathbf{x}$:

$$(A\mathbf{x})_i = \sum_{j=1}^{n} a_{i,j}x_j = \mathbf{a}_{i..}\mathbf{x}.$$

Therefore, we may write

$$A\mathbf{x} = \begin{pmatrix} \mathbf{a}_{1..} \\ \mathbf{a}_{2..} \\ \vdots \\ \mathbf{a}_{m..} \end{pmatrix} \mathbf{x} = \begin{pmatrix} \mathbf{a}_{1..}\mathbf{x} \\ \mathbf{a}_{2..}\mathbf{x} \\ \vdots \\ \mathbf{a}_{m..}\mathbf{x} \end{pmatrix}. \tag{1}$$

The product $A\mathbf{x}$ is also a linear combination of the columns of A:

$$\begin{aligned} A\mathbf{x} &= x_1 \begin{pmatrix} a_{11} \\ a_{21} \\ \vdots \\ a_{m1} \end{pmatrix} + x_2 \begin{pmatrix} a_{12} \\ a_{22} \\ \vdots \\ a_{m2} \end{pmatrix} + \cdots + x_m \begin{pmatrix} a_{1n} \\ a_{2n} \\ \vdots \\ a_{mn} \end{pmatrix} \\ &= \sum_{j=1}^{n} x_j \mathbf{a}_{:j}. \end{aligned} \tag{2}$$

While (1) is the most convenient way to perform particular computations "by hand," it turns out that (2) is usually the most beneficial way to view the matrix-vector product.

Example 2

Consider the 2×3 matrix

$$A = \begin{pmatrix} 1 & 2 & 3 \\ -1 & 1 & 2 \end{pmatrix}.$$

For any vector $\mathbf{x}$ in $\mathbb{R}^3$,

$$A\mathbf{x} = \begin{pmatrix} 1 & 2 & 3 \\ -1 & 1 & 2 \end{pmatrix} \begin{pmatrix} x_1 \\ x_2 \\ x_3 \end{pmatrix} = \begin{pmatrix} x_1 + 2x_2 + 3x_3 \\ -x_1 + x_2 + 2x_3 \end{pmatrix}.$$

Notice that the two entries in $A\mathbf{x}$ are, respectively,

$$\begin{pmatrix} 1 & 2 & 3 \end{pmatrix} \begin{pmatrix} x_1 \\ x_2 \\ x_3 \end{pmatrix} \quad \text{and} \quad \begin{pmatrix} -1 & 1 & 2 \end{pmatrix} \begin{pmatrix} x_1 \\ x_2 \\ x_3 \end{pmatrix}.$$

On the other hand, $A\mathbf{x}$ can be viewed as a linear combination of the columns of A:

$$A\mathbf{x} = \begin{pmatrix} 1 & 2 & 3 \\ -1 & 1 & 2 \end{pmatrix} \begin{pmatrix} x_1 \\ x_2 \\ x_3 \end{pmatrix} = x_1 \begin{pmatrix} 1 \\ -1 \end{pmatrix} + x_2 \begin{pmatrix} 2 \\ 1 \end{pmatrix} + x_3 \begin{pmatrix} 3 \\ 2 \end{pmatrix}.$$

■

The matrix product $A\mathbf{x}$ defines a linear transformation from $\mathbb{R}^n$ into $\mathbb{R}^m$ by virtue of the following (easily verified) properties:

$$A(c\mathbf{x}) = cA\mathbf{x} \text{ for all } \mathbf{x} \text{ in } \mathbb{R}^n \text{ and all scalars } c;$$
$$A(\mathbf{x}+\mathbf{y}) = A\mathbf{x} + A\mathbf{y} \text{ for all } \mathbf{x} \text{ and } \mathbf{y} \text{ in } \mathbb{R}^n.$$

In fact, *every* linear transformation from $\mathbb{R}^n$ into $\mathbb{R}^m$ can be represented as $A\mathbf{x}$ for some $m \times n$ matrix A.

Linear Independence of Vectors A collection of vectors $\mathbf{x}_1, \mathbf{x}_2, \ldots, \mathbf{x}_k$ in $\mathbb{R}^n$ is said to be **linearly dependent** if there are scalars $c_1, c_2, \ldots, c_k$, not all zero, such that

$$c_1\mathbf{x}_1 + c_2\mathbf{x}_2 + \cdots + c_k\mathbf{x}_k = \mathbf{0}, \tag{3}$$

where $\mathbf{0}$ denotes the zero vector (or the origin). If the only such scalars are $c_1 = c_2 = \cdots = c_k = 0$, then the collection of vectors is said to be **linearly independent**. For instance, the standard basis vectors $\mathbf{e}_1, \ldots, \mathbf{e}_n$ clearly comprise a linearly independent set.

It is important to observe that (3) actually constitutes a system of linear equations for the scalars $c_1, c_2, \ldots, c_k$. The columns of the coefficient matrix of this system are precisely the vectors $\mathbf{x}_1, \mathbf{x}_2, \ldots, \mathbf{x}_k$, as illustrated by the following example.

Example 3

Consider the following three vectors in $\mathbb{R}^3$:

$$\mathbf{x}_1 = \begin{pmatrix} 1 \\ 1 \\ 1 \end{pmatrix}, \quad \mathbf{x}_2 = \begin{pmatrix} 2 \\ -1 \\ 3 \end{pmatrix}, \quad \mathbf{x}_3 = \begin{pmatrix} 1 \\ -5 \\ 3 \end{pmatrix}.$$

Equation (3) becomes

$$c_1 \begin{pmatrix} 1 \\ 1 \\ 1 \end{pmatrix} + c_2 \begin{pmatrix} 2 \\ -1 \\ 3 \end{pmatrix} + c_3 \begin{pmatrix} 1 \\ -5 \\ 3 \end{pmatrix} = \begin{pmatrix} 0 \\ 0 \\ 0 \end{pmatrix};$$

that is,

$$\begin{pmatrix} 1 & 2 & 1 \\ 1 & -1 & -5 \\ 1 & 3 & 3 \end{pmatrix} \begin{pmatrix} c_1 \\ c_2 \\ c_3 \end{pmatrix} = \begin{pmatrix} 0 \\ 0 \\ 0 \end{pmatrix}.$$

To solve this system, we row-reduce the coefficient matrix as follows:

$$\begin{pmatrix} 1 & 2 & 1 \\ 1 & -1 & -5 \\ 1 & 3 & 3 \end{pmatrix} \begin{matrix} R_1 \\ R_2 - R_1 \\ R_3 - R_1 \end{matrix} \begin{pmatrix} 1 & 2 & 1 \\ 0 & -3 & -6 \\ 0 & 1 & 2 \end{pmatrix} \begin{matrix} R_1 \\ R_3 \\ R_2 + 3R_3 \end{matrix} \begin{pmatrix} 1 & 2 & 1 \\ 0 & 1 & 2 \\ 0 & 0 & 0 \end{pmatrix}.$$

Since we have produced an echelon form of the matrix that contains a row of zeros, we know that the system has (infinitely many) nonzero solutions, one of which is

$$c_1 = 3, \quad c_2 = -2, \quad c_3 = 1.$$

Therefore, $\mathbf{x}_1$, $\mathbf{x}_2$, and $\mathbf{x}_3$ comprise a linearly *dependent* set. ■

Example 3 shows that the question of linear dependence/independence for a collection of vectors $\mathbf{x}_1, \mathbf{x}_2, \ldots, \mathbf{x}_k$ in $\mathbb{R}^n$ can be settled by computing an echelon form of the $n \times k$ matrix

$$\begin{pmatrix} \mathbf{x}_1 & \mathbf{x}_2 & \cdots & \mathbf{x}_k \end{pmatrix}.$$

We state this result, which follows from Theorem 2 in Appendix I.A, as

Theorem 1

A collection of vectors $\mathbf{x}_1, \mathbf{x}_2, \ldots, \mathbf{x}_k$ in $\mathbb{R}^n$ is linearly independent, if and only if some (and therefore every) echelon form of

$$\begin{pmatrix} \mathbf{x}_1 & \mathbf{x}_2 & \cdots & \mathbf{x}_k \end{pmatrix}$$

has exactly k nonzero rows. In particular, if $k > n$, then $\mathbf{x}_1, \mathbf{x}_2, \ldots, \mathbf{x}_k$ form a linearly dependent set. ▲

Matrix Algebra The product of a scalar c and a matrix A is defined by

$$cA = \begin{pmatrix} ca_{11} & ca_{12} & \cdots & ca_{1n} \\ ca_{21} & ca_{22} & \cdots & ca_{2n} \\ \vdots & \vdots & & \vdots \\ ca_{m1} & ca_{m2} & \cdots & ca_{mn} \end{pmatrix};$$

that is, the ijth entry of cA is ca_{ij}. For instance,

$$3\begin{pmatrix} 1 & 3 \\ -2 & 1 \end{pmatrix} = \begin{pmatrix} 3 & 9 \\ -6 & 3 \end{pmatrix} \quad \text{and} \quad \begin{pmatrix} \frac{1}{2} & \frac{1}{6} \\ \frac{1}{3} & 1 \end{pmatrix} = \frac{1}{6}\begin{pmatrix} 3 & 1 \\ 2 & -6 \end{pmatrix}.$$

The sum of two matrices A and B *with the same dimensions* $(m \times n)$ is similarly defined *entrywise*; that is, the ijth entry of $A + B$ is $a_{ij} + b_{ij}$. For instance,

$$\begin{pmatrix} 1 & 2 \\ 0 & 1 \\ 1 & -1 \end{pmatrix} + \begin{pmatrix} 1 & 1 \\ 2 & -1 \\ 3 & -1 \end{pmatrix} = \begin{pmatrix} 1 & 3 \\ 2 & 0 \\ 4 & -2 \end{pmatrix}.$$

Matrices with different dimensions cannot be added.

Subtraction of two matrices A and B with the same dimensions is naturally defined as the sum of A and $-B = (-1)B$. The net result is that the ijth entry of $A - B$ is $a_{ij} - b_{ij}$. Also, having defined scalar multiplication and addition, it is possible to form linear combinations of two matrices with the same dimensions. In particular, the ijth entry of $c_1A + c_2B$ is $c_1a_{ij} + c_2b_{ij}$.

Example 4

Let A and B be the following 2×3 matrices:

$$A = \begin{pmatrix} 1 & -3 & 0 \\ 5 & 2 & 1 \end{pmatrix} \quad \text{and} \quad B = \begin{pmatrix} 2 & -1 & 1 \\ -1 & 2 & 0 \end{pmatrix}.$$

The linear combination $2A - 3B$ is computed as follows:

$$\begin{aligned} 2A - 3B &= 2\begin{pmatrix} 1 & -3 & 0 \\ 5 & 2 & 1 \end{pmatrix} - 3\begin{pmatrix} 2 & -1 & 1 \\ -1 & 2 & 0 \end{pmatrix} \\ &= \begin{pmatrix} 2 & -6 & 0 \\ 10 & 4 & 2 \end{pmatrix} + \begin{pmatrix} -6 & 3 & -3 \\ 3 & -6 & 0 \end{pmatrix} \\ &= \begin{pmatrix} -4 & -3 & -3 \\ 13 & -2 & 2 \end{pmatrix}. \end{aligned}$$

■

Matrix Products Suppose that A is an $n \times m$ matrix and B is an $m \times k$ matrix. Then for any $\mathbf{x}$ in $\mathbb{R}^k$, $B\mathbf{x}$ is in $\mathbb{R}^m$, and $A(B\mathbf{x})$ is in $\mathbb{R}^n$. In other words, the *composition* of the linear transformations represented by A and B is well defined as a transformation from $\mathbb{R}^k$ into $\mathbb{R}^n$. Moreover, it is linear. We define the matrix "product" AB to the the matrix associated with this composition; that is, AB is defined to be the $n \times k$ matrix for which

$$(AB)\mathbf{x} = A(B\mathbf{x}) \text{ for all } \mathbf{x} \text{ in } \mathbb{R}^k.$$

Example 5

Let A be the 3×2 matrix

$$A = \begin{pmatrix} 1 & 2 \\ 0 & -1 \\ 1 & 3 \end{pmatrix},$$

and let B be the 2×4 matrix

$$B = \begin{pmatrix} 1 & 2 & 0 & 1 \\ 0 & -1 & 3 & 2 \end{pmatrix}.$$

Then AB is the 3×4 matrix for which

$$\begin{aligned} (AB)\mathbf{x} &= \begin{pmatrix} 1 & 2 \\ 0 & -1 \\ 1 & 3 \end{pmatrix}\begin{pmatrix} 1 & 2 & 0 & 1 \\ 0 & -1 & 3 & 2 \end{pmatrix}\begin{pmatrix} x_1 \\ x_2 \\ x_3 \\ x_4 \end{pmatrix} \\ &= \begin{pmatrix} 1 & 2 \\ 0 & -1 \\ 1 & 3 \end{pmatrix}\begin{pmatrix} x_1 + 2x_2 + 0x_3 + x_4 \\ 0x_1 - x_2 + 3x_3 + 2x_4 \end{pmatrix} \\ &= \begin{pmatrix} x_1 + 0x_2 + 6x_3 + 5x_4 \\ 0x_1 + x_2 - 3x_3 - 2x_4 \\ x_1 - x_2 + 9x_3 + 7x_4 \end{pmatrix}. \end{aligned}$$

Therefore,

$$AB = \begin{pmatrix} 1 & 0 & 6 & 5 \\ 0 & 1 & -3 & -2 \\ 1 & -1 & 9 & 7 \end{pmatrix}.$$

Close inspection of the computations in Example 5 reveals that the ijth entry in AB is the product of the ith row of A with the jth column of B:

$$\mathbf{a}_{i..}\mathbf{b}_{:j} = \sum_{\ell=1}^{n} a_{i\ell}b_{\ell j}.$$

This is indeed true in general. For any $n \times m$ matrix A and any $m \times k$ matrix B, the matrix product AB is the $n \times k$ matrix

$$AB = \begin{pmatrix} \mathbf{a}_{1..} \\ \mathbf{a}_{2..} \\ \vdots \\ \mathbf{a}_{m..} \end{pmatrix} \begin{pmatrix} \mathbf{b}_{:1} & \mathbf{b}_{:2} & \cdots & \mathbf{b}_{:m} \end{pmatrix}$$

$$= \begin{pmatrix} \mathbf{a}_{1..}\mathbf{b}_{:1} & \mathbf{a}_{1..}\mathbf{b}_{:2} & \cdots & \mathbf{a}_{1..}\mathbf{b}_{:k} \\ \mathbf{a}_{2..}\mathbf{b}_{:1} & \mathbf{a}_{2..}\mathbf{b}_{:2} & \cdots & \mathbf{a}_{2..}\mathbf{b}_{:k} \\ \vdots & \vdots & & \vdots \\ \mathbf{a}_{n..}\mathbf{b}_{:1} & \mathbf{a}_{n..}\mathbf{b}_{:2} & \cdots & \mathbf{a}_{n..}\mathbf{b}_{:k} \end{pmatrix}.$$

The jth column of AB is the matrix-vector product of A with the jth column of B:

$$A\mathbf{b}_{:j} = \sum_{k=1}^{n} \mathbf{a}_{:k}b_{kj}.$$ ■

It must be emphasized that the product AB is defined only when the number of columns in A is the same as the number of rows in B. For instance, the product AB makes sense if A is 5×3 and B is 3×2, but not if A is 3×2 and B is 5×3. In particular, it should not be surprising for AB to be defined while BA is not.

Example 6

Let A be the 2×3 matrix

$$A = \begin{pmatrix} 1 & 2 & 1 \\ 2 & -1 & 3 \end{pmatrix}.$$

and let B be the 3×3 matrix

$$B = \begin{pmatrix} 1 & 2 & 0 \\ 0 & -1 & 3 \\ 1 & 0 & 1 \end{pmatrix},$$

Then AB is the 2×3 matrix

$$AB = \begin{pmatrix} 1 & 2 & 1 \\ 2 & -1 & 3 \end{pmatrix} \begin{pmatrix} 1 & 2 & 0 \\ 0 & -1 & 3 \\ 1 & 0 & 1 \end{pmatrix} = \begin{pmatrix} 2 & 0 & 7 \\ 5 & 5 & 0 \end{pmatrix}.$$

The product BA is undefined, since the number of columns in B is different from the number of rows in A. ■

Another important point that must be emphasized is that, even when both products AB and BA are defined, they are generally not the same. In other words, *matrix multiplication is not commutative*. For instance, consider the 2×2 matrices

$$A = \begin{pmatrix} 1 & 1 \\ 2 & 2 \end{pmatrix} \quad \text{and} \quad B = \begin{pmatrix} 0 & 1 \\ 1 & 0 \end{pmatrix}.$$

The products AB and BA are, respectively,

$$\begin{pmatrix} 1 & 1 \\ 2 & 2 \end{pmatrix}\begin{pmatrix} 0 & 1 \\ 1 & 0 \end{pmatrix} = \begin{pmatrix} 1 & 1 \\ 2 & 2 \end{pmatrix} \quad \text{and} \quad \begin{pmatrix} 0 & 1 \\ 1 & 0 \end{pmatrix}\begin{pmatrix} 1 & 1 \\ 2 & 2 \end{pmatrix} = \begin{pmatrix} 2 & 2 \\ 1 & 1 \end{pmatrix}.$$

It is also possible that the product of two nonzero matrices may be a zero matrix. For instance, consider the matrices

$$A = \begin{pmatrix} 1 & 1 \\ 2 & 2 \end{pmatrix} \quad \text{and} \quad B = \begin{pmatrix} -1 & 1 \\ 1 & -1 \end{pmatrix}.$$

The products AB and BA are, respectively,

$$\begin{pmatrix} 1 & 1 \\ 2 & 2 \end{pmatrix}\begin{pmatrix} -1 & 1 \\ 1 & -1 \end{pmatrix} = \begin{pmatrix} 0 & 0 \\ 0 & 0 \end{pmatrix} \quad \text{and} \quad \begin{pmatrix} -1 & 1 \\ 1 & -1 \end{pmatrix}\begin{pmatrix} 1 & 1 \\ 2 & 2 \end{pmatrix} = \begin{pmatrix} 1 & 1 \\ -1 & -1 \end{pmatrix}.$$

Thus $AB = 0$, and again $AB \neq BA$.

With the two notable exceptions just noted, algebraic manipulations involving matrices are much the same as algebraic manipulations involving numbers. The following associative and distributive properties hold whenever the dimensions of A, B, and C are such that the expressions involved make sense:

$$\begin{aligned} A(BC) &= (AB)C \\ (cA)B &= A(cB) = c(AB) \\ A(B + C) &= AB + AC \\ (A + B)C &= AC + BC \end{aligned}$$

Note in particular the need for two distributive laws—a "left" distributive law and a "right" distributive law—since multiplication is generally not commutative.

Square Matrices, Operators, and Inverses A matrix is "square" if it has the same number of rows and columns. Thus, a square matrix A is $n \times n$ for some n. For instance, these are 2×2, 3×3, and 4×4 square matrices, respectively:

$$\begin{pmatrix} 1 & 2 \\ 2 & -1 \end{pmatrix} \quad \begin{pmatrix} 1 & 2 & 1 \\ 2 & -1 & 3 \\ 1 & 1 & 2 \end{pmatrix} \quad \begin{pmatrix} -1 & 1 & 1 & 0 \\ 1 & 2 & -1 & 3 \\ 1 & 0 & 1 & 2 \\ 1 & 3 & -1 & 0 \end{pmatrix}.$$

The linear transformation represented by an $n \times n$ matrix A "maps" each vector $\mathbf{x}$ in $\mathbb{R}^n$ onto some other vector $A\mathbf{x}$ which is also in $\mathbb{R}^n$. We call such a linear transformation a **linear operator** on $\mathbb{R}^n$.

The $n \times n$ **identity matrix** is

$$\mathcal{I} = \begin{pmatrix} 1 & 0 & \cdots & 0 \\ 0 & 1 & \cdots & 0 \\ \vdots & \vdots & \ddots & \vdots \\ 0 & 0 & \cdots & 1 \end{pmatrix};$$

that is, each *diagonal* entry of $\mathcal{I}$ is a 1, and all other entries are zeros. It should be easy to see that $\mathcal{I}$ has the property that

$$\mathcal{I}\mathbf{x} = \mathbf{x} \text{ for all } \mathbf{x} \text{ in } \mathbb{R}^n$$

as well as the property that

$$\mathcal{I}A = A\mathcal{I} = A \text{ for all } n \times n \text{ matrices } A.$$

An $n \times n$ matrix A is said to be **invertible** if there is an $n \times n$ matrix M with the property that

$$A(M\mathbf{x}) = M(A\mathbf{x}) = \mathbf{x} \text{ for all } \mathbf{x} \text{ in } \mathbb{R}^n,$$

which is equivalent to

$$AM = MA = \mathcal{I}.$$

The matrix M, if it exists, is the **inverse** of A, and is denoted by A^{-1}. Invertible square matrices are sometimes called *nonsingular*, while non-invertible square matrices are said to be *singular*.

Example 7

We will verify that the matrix

$$A = \begin{pmatrix} 3 & 2 & 1 \\ 2 & 2 & 1 \\ 1 & 1 & 1 \end{pmatrix}$$

is invertible with

$$A^{-1} = \begin{pmatrix} 1 & -1 & 0 \\ -1 & 2 & -1 \\ 0 & -1 & 2 \end{pmatrix}.$$

This requires only that we compute the following products, obtaining the identity matrix in each case:

$$\begin{pmatrix} 3 & 2 & 1 \\ 2 & 2 & 1 \\ 1 & 1 & 1 \end{pmatrix} \begin{pmatrix} 1 & -1 & 0 \\ -1 & 2 & -1 \\ 0 & -1 & 2 \end{pmatrix} = \begin{pmatrix} 1 & 0 & 0 \\ 0 & 1 & 0 \\ 0 & 0 & 1 \end{pmatrix},$$

and

$$\begin{pmatrix} 1 & -1 & 0 \\ -1 & 2 & -1 \\ 0 & -1 & 2 \end{pmatrix} \begin{pmatrix} 3 & 2 & 1 \\ 2 & 2 & 1 \\ 1 & 1 & 1 \end{pmatrix} = \begin{pmatrix} 1 & 0 & 0 \\ 0 & 1 & 0 \\ 0 & 0 & 1 \end{pmatrix}.$$

■

Example 8

Consider the 2×2 matrix

$$A = \begin{pmatrix} 1 & 1 \\ 1 & 2 \end{pmatrix}.$$

Let's look for the inverse of A in the form of

$$M = \begin{pmatrix} p & q \\ r & s \end{pmatrix}$$

by first requiring that $AM = \mathcal{I}$:

$$AM = \begin{pmatrix} 1 & 1 \\ 1 & 2 \end{pmatrix} \begin{pmatrix} p & q \\ r & s \end{pmatrix} = \begin{pmatrix} 1 & 0 \\ 0 & 1 \end{pmatrix},$$

which is equivalent to two linear systems:

$$\begin{pmatrix} 1 & 1 \\ 1 & 2 \end{pmatrix} \begin{pmatrix} p \\ r \end{pmatrix} = \begin{pmatrix} 1 \\ 0 \end{pmatrix} \quad \text{and} \quad \begin{pmatrix} 1 & 1 \\ 1 & 2 \end{pmatrix} \begin{pmatrix} q \\ s \end{pmatrix} = \begin{pmatrix} 0 \\ 1 \end{pmatrix}.$$

Gauss-Jordan elimination on the first system's augmented matrix proceeds thus:

$$\begin{pmatrix} 1 & 1 & 1 \\ 1 & 2 & 0 \end{pmatrix} \quad \begin{matrix} R_1 \\ R_2 - R_1 \end{matrix} \begin{pmatrix} 1 & 1 & 1 \\ 0 & 1 & -1 \end{pmatrix} \quad \begin{matrix} R_1 - R_2 \\ R_2 \end{matrix} \begin{pmatrix} 1 & 0 & 2 \\ 0 & 1 & -1 \end{pmatrix}.$$

So we find that $p = 2$ and $r = -1$. Similarly, we solve the second system:

$$\begin{pmatrix} 1 & 1 & 0 \\ 1 & 2 & 1 \end{pmatrix} \quad \begin{matrix} R_1 \\ R_2 - R_1 \end{matrix} \begin{pmatrix} 1 & 1 & 0 \\ 0 & 1 & 1 \end{pmatrix} \quad \begin{matrix} R_1 - R_2 \\ R_2 \end{matrix} \begin{pmatrix} 1 & 0 & -1 \\ 0 & 1 & 1 \end{pmatrix},$$

finding that $q = -1$ and $s = 1$. Thus, if the inverse of A exists, it must be

$$M = \begin{pmatrix} 2 & -1 \\ -1 & 1 \end{pmatrix}.$$

Now we need only check that $MA = \mathcal{I}$:

$$\begin{pmatrix} 2 & -1 \\ -1 & 1 \end{pmatrix} \begin{pmatrix} 1 & 1 \\ 1 & 2 \end{pmatrix} = \begin{pmatrix} 1 & 0 \\ 0 & 1 \end{pmatrix}.$$

Therefore, A is invertible with $A^{-1} = M$. ■

Notice that the two Gauss-Jordan processes that were used in Example 8 to solve for the two columns of A^{-1} involved *precisely the same row-operations*. This should not be surprising, since the augmented matrices differ only in their last columns. This observation suggests a significant simplification of the procedure by which we computed A^{-1}. The columns of A^{-1} can be found *simultaneously* by performing the Gauss-Jordan elimination process on the $n \times 2n$ augmented matrix $(A\ \mathcal{I})$. The plan is to use row operations to "move" the identity matrix from the last n columns to the first n columns of the augmented matrix, as indicated schematically by:

$$(A\ \mathcal{I}) \to \cdots \to (\mathcal{I}\ A^{-1}).$$

If that can be done, then the last n columns will contain A^{-1}. In the case of the matrix in Example 6, we find A^{-1} this way as follows:

$$\begin{pmatrix} 1 & 1 & 1 & 0 \\ 1 & 2 & 0 & 1 \end{pmatrix} \quad \begin{matrix} R_1 \\ R_2 - R_1 \end{matrix} \begin{pmatrix} 1 & 1 & 1 & 0 \\ 0 & 1 & -1 & 1 \end{pmatrix}$$
$$\begin{matrix} R_1 - R_2 \\ R_2 \end{matrix} \begin{pmatrix} 1 & 0 & 2 & -1 \\ 0 & 1 & -1 & 1 \end{pmatrix}.$$

An important fact from linear algebra (with a surprisingly complicated proof) is that, if $AM = \mathcal{I}$, then it automatically follows that $MA = \mathcal{I}$, and vice-versa. This gives us the following theorem:

Theorem 2

Let A be an $n \times n$ matrix. Either of the following conditions is sufficient for A to be invertible with $A^{-1} = M$:

i) $AM = \mathcal{I}$;
ii) $MA = \mathcal{I}$. ▲

As a result of Theorem 1, we are guaranteed that, if the Gauss-Jordan process described above succeeds, then it produces the inverse—there is no need to check that $MA = \mathcal{I}$. Furthermore, if it does not succeed, then the inverse does not exist.

Example 9

Consider the matrix

$$A = \begin{pmatrix} 1 & -1 & 1 \\ 2 & -1 & 2 \\ 1 & 0 & 2 \end{pmatrix}.$$

First we form $(A\ \mathcal{I})$:

$$(A\ \mathcal{I}) = \begin{pmatrix} 1 & -1 & 1 & 1 & 0 & 0 \\ 2 & -1 & 2 & 0 & 1 & 0 \\ 1 & 0 & 2 & 0 & 0 & 1 \end{pmatrix}.$$

Then the row operations indicated below produce $(\mathcal{I}\ A^{-1})$:

$$\begin{matrix} R_1 \\ R_2 - 2R_1 \\ R_3 - R_1 \end{matrix} \begin{pmatrix} 1 & -1 & 1 & 1 & 0 & 0 \\ 0 & 1 & 0 & -2 & 1 & 0 \\ 0 & 1 & 1 & -1 & 0 & 1 \end{pmatrix}$$

$$\begin{matrix} R_1 + R_2 \\ R_2 \\ R_3 - R_2 \end{matrix} \begin{pmatrix} 1 & 0 & 1 & -1 & 1 & 0 \\ 0 & 1 & 0 & -2 & 1 & 0 \\ 0 & 0 & 1 & 1 & -1 & 1 \end{pmatrix}$$

$$\begin{matrix} R_1 - R_3 \\ R_2 \\ R_3 \end{matrix} \begin{pmatrix} 1 & 0 & 0 & -2 & 2 & -1 \\ 0 & 1 & 0 & -2 & 1 & 0 \\ 0 & 0 & 1 & 1 & -1 & 1 \end{pmatrix} = (\mathcal{I}\ A^{-1}).$$

Therefore, A is invertible, and

$$A^{-1} = \begin{pmatrix} -2 & 2 & -1 \\ -2 & 1 & 0 \\ 1 & -1 & 1 \end{pmatrix}.$$

■

Example 10

Let's look for the inverse of

$$A = \begin{pmatrix} 1 & 1 \\ 2 & 2 \end{pmatrix}.$$

Forming $(A\ \mathcal{I})$ and reducing the first column produces

$$\begin{pmatrix} 1 & 1 & 1 & 0 \\ 2 & 2 & 0 & 1 \end{pmatrix} \begin{matrix} R_1 \\ R_2 - 2R_1 \end{matrix} \begin{pmatrix} 1 & 1 & 1 & 0 \\ 0 & 0 & -2 & 1 \end{pmatrix}.$$

At this point we see that it is impossible to obtain $(\mathcal{I}\ A^{-1})$; therefore A is not invertible. ■

We conclude our discussion of the matrix inverse with a theorem regarding the existence of the inverse.

Theorem 3

Let A be a square matrix. The following statements are equivalent; that is, they are either all true or all false.

i) A is invertible.

ii) The reduced echelon form of A is $\mathcal{I}$.

iii) Some, and therefore every, echelon form of A has no zeros on its diagonal.

iv) The equation $A\mathbf{x} = \mathbf{0}$ has only the trivial solution $\mathbf{x} = \mathbf{0}$.

v) The columns of A are linearly independent.

vi) The equation $A\mathbf{x} = \mathbf{b}$ has a unique solution $\mathbf{x}$ for each vector $\mathbf{b}$ in $\mathbb{R}^n$. ▲

That (i) and (ii) are equivalent follows from the observation that the process described above for computing A^{-1} succeeds precisely when the reduced echelon form of A is the identity matrix. The equivalence of (ii), (iii), and (iv) is easy to see by considering the Gauss-Jordan process for solving $A\mathbf{x} = \mathbf{0}$. It similarly follows that, if (ii) is true, then so is (vi). The equivalence of (iv) and (v) follows easily from the definition of linear independence. Finally, by taking $\mathbf{b} = \mathbf{0}$, it is easy to see that, if (vi) is true, then so is (iv). This chain of implications is indicated schematically by:

$$\begin{matrix} \text{(i)} \longleftrightarrow \text{(ii)} \longleftrightarrow & \text{(iii)} & \longleftrightarrow \text{(iv)} \longleftrightarrow \text{(v)} \\ \searrow & & \nearrow \\ & \text{(vi)} & \end{matrix}$$

Thus the truth of any one of (i)–(vi) implies that all the others are true as well.

Determinants Suppose that we wish to find a formula for the inverse of a 2×2 matrix

$$A = \begin{pmatrix} a & b \\ c & d \end{pmatrix},$$

assuming that the inverse exists. We proceed by forming $(A\,\mathcal{I})$ and row-reducing formally as follows:

$$\begin{pmatrix} a & b & 1 & 0 \\ c & d & 0 & 1 \end{pmatrix} \quad \begin{matrix} \frac{1}{a}R_1 \\ R_2 - \frac{c}{a}R_1 \end{matrix} \quad \begin{pmatrix} 1 & \frac{b}{a} & \frac{1}{a} & 0 \\ 0 & \frac{ad - bc}{a} & -\frac{c}{a} & 1 \end{pmatrix}$$

$$\begin{matrix} R_1 \\ \frac{a}{ad - bc}R_2 \end{matrix} \quad \begin{pmatrix} 1 & \frac{b}{a} & \frac{1}{a} & 0 \\ 0 & 1 & \frac{-c}{ad - bc} & \frac{a}{ad - bc} \end{pmatrix}$$

$$\begin{matrix} R_1 - \frac{b}{a}R_2 \\ R_2 \end{matrix} \quad \begin{pmatrix} 1 & 0 & \frac{d}{ad - bc} & \frac{-b}{ad - bc} \\ 0 & 1 & \frac{-c}{ad - bc} & \frac{a}{ad - bc} \end{pmatrix}.$$

This shows that A is invertible, if and only if $ad - bc \neq 0$. Furthermore, if $ad - bc \neq 0$, then the inverse of A is*

$$A^{-1} = \frac{1}{ad - bc}\begin{pmatrix} d & -b \\ -c & a \end{pmatrix}.$$

For instance,

$$\text{if } A = \begin{pmatrix} 2 & 1 \\ 1 & 3 \end{pmatrix}, \quad \text{then } |A| = (2)(3) - A^{-1} = \frac{1}{5}\begin{pmatrix} 3 & -1 \\ -1 & 2 \end{pmatrix},$$

as is easy to check.

The quantity $ad - bc$ is called the *determinant* of A; that is, we define the determinant of a 2×2 matrix A to be

$$|A| = \begin{vmatrix} a & b \\ c & d \end{vmatrix} = ad - bc.$$

For instance,

$$\begin{vmatrix} 1 & 2 \\ 2 & 3 \end{vmatrix} = (1)(3) - (2)(2) = -1, \quad \text{and} \quad \begin{vmatrix} 3 & 1 \\ 6 & 2 \end{vmatrix} = (3)(2) - (1)(6) = 0.$$

Thus, the first of these matrices is invertible, and the second is not.

The determinant of an $n \times n$ matrix, for $n \geq 3$, will be defined inductively. That is, we will describe the computation of the determinant of an $n \times n$ matrix in terms of determinants of certain $(n-1) \times (n-1)$ matrices. Then, in principle, the computation

---

* Even though we tacitly assumed that $a \neq 0$ in the first row-reduction step indicated above, it is easy to check that this result is valid even when $a = 0$.

of the determinant of any square matrix can be reduced to the computation of the determinants of a number of 2×2 matrices. So suppose that A is an $n \times n$ matrix

$$A = \begin{pmatrix} a_{11} & a_{12} & \cdots & a_{1n} \\ a_{21} & a_{22} & \cdots & a_{2n} \\ \vdots & \vdots & & \vdots \\ a_{n1} & a_{n2} & \cdots & a_{nn} \end{pmatrix}.$$

For each $i = 1, \ldots, n$, let M_{1j} be the determinant of the $(n-1) \times (n-1)$ matrix obtained by deleting the first row and the jth column from A. Each M_{1j} is called the *minor* of the entry a_{1j} in A. The determinant of A is then defined by the *Laplace expansion*

$$\begin{aligned} |A| &= \sum_{j=1}^{n} (-1)^{j+1} a_{1j} M_{1j} \\ &= a_{11}M_{11} - a_{12}M_{12} + \cdots + (-1)^{n+1} a_{1n} M_{1n} \end{aligned}$$

Example 11

Consider the 3×3 matrix

$$A = \begin{pmatrix} 1 & 2 & 1 \\ 3 & 2 & 3 \\ 2 & 0 & 1 \end{pmatrix}.$$

The minors described above are

$$M_{11} = \begin{vmatrix} 2 & 3 \\ 0 & 1 \end{vmatrix}, \quad M_{12} = \begin{vmatrix} 3 & 3 \\ 2 & 1 \end{vmatrix}, \quad \text{and} \quad M_{13} = \begin{vmatrix} 3 & 2 \\ 2 & 0 \end{vmatrix}.$$

So the determinant of A is

$$\begin{aligned} |A| &= (1)\begin{vmatrix} 2 & 3 \\ 0 & 1 \end{vmatrix} - (2)\begin{vmatrix} 3 & 3 \\ 2 & 1 \end{vmatrix} + (1)\begin{vmatrix} 3 & 2 \\ 2 & 0 \end{vmatrix} \\ &= (1)(2 \cdot 1 - 3 \cdot 0) - (2)(3 \cdot 1 - 3 \cdot 2) + (1)(3 \cdot 0 - 2 \cdot 2) \\ &= 1 + 6 - 4 = 3. \end{aligned}$$

■

Although our definition of $|A|$ is in terms of the minors of the entries in the first row of A, it turns out that $|A|$ can be computed using the minors of the entries in any row or column of A. With M_{ij} denoting the determinant of the $(n-1) \times (n-1)$ matrix obtained by deleting the ith row and the jth column from A, we can compute $|A|$ by means of a Laplace expansion on the ith row:

$$|A| = \sum_{j=1}^{n} (-1)^{i+j} a_{ij} M_{ij}$$

or on the jth column:

$$|A| = \sum_{i=1}^{n} (-1)^{i+j} a_{ij} M_{ij}.$$

Example 12

Consider the 4×4 matrix

$$A = \begin{pmatrix} 1 & 2 & 1 & 0 \\ 3 & 1 & 3 & 1 \\ 2 & 0 & 1 & 0 \\ 1 & 0 & 1 & 1 \end{pmatrix}.$$

We can take advantage of the fact that the second column contains two zeros by expanding on that column:

$$\begin{aligned} |A| &= (-1)^{1+2}2M_{12} + (-1)^{2+2}1M_{22} + (-1)^{3+2}0M_{32} + (-1)^{4+2}0M_{42} \\ &= -2M_{12} + 1M_{22} \\ &= -2\begin{vmatrix} 3 & 3 & 1 \\ 2 & 1 & 0 \\ 1 & 1 & 1 \end{vmatrix} + \begin{vmatrix} 1 & 1 & 0 \\ 2 & 1 & 0 \\ 1 & 1 & 1 \end{vmatrix} \end{aligned}$$

Next we'll continue to take advantage of zeros by expanding each of these 3×3 determinants on its third column:

$$\begin{aligned} |A| &= -2\left((-1)^{1+3}1\begin{vmatrix} 2 & 1 \\ 1 & 1 \end{vmatrix} + (-1)^{2+3}0\begin{vmatrix} 3 & 3 \\ 1 & 1 \end{vmatrix} + (-1)^{3+3}1\begin{vmatrix} 3 & 3 \\ 2 & 1 \end{vmatrix}\right) \\ &\quad + \left((-1)^{1+3}0\begin{vmatrix} 2 & 1 \\ 1 & 1 \end{vmatrix} + (-1)^{2+3}0\begin{vmatrix} 1 & 1 \\ 1 & 1 \end{vmatrix} + (-1)^{3+3}1\begin{vmatrix} 1 & 1 \\ 2 & 1 \end{vmatrix}\right) \\ &= -2((2-1) - 0 + (3-6)) + (0 - 0 + (1-2)) \\ &= 3. \end{aligned}$$

■

Properties of the Determinant Now that we have defined the determinant of a square matrix, we next present a series of crucial properties.

- A. *If two rows or columns of A are interchanged, then $|A|$ is multiplied by -1.*

Proof The proof of this assertion requires a mathematical induction argument. First, we observe that it is true for 2×2 matrices. Then we show that truth for $(n-1) \times (n-1)$ matrices implies truth for $n \times n$ matrices. An expansion of the determinant on any row (or column) *not* involved in the row (or column) exchange will involve n minors, each of which is the determinant of an $(n-1) \times (n-1)$ matrix upon which the same row (or column) exchange has been done. Therefore, if the assertion is true for $(n-1) \times (n-1)$ matrices, then it is true for $n \times n$ matrices. This, together with truth for $n = 2$, proves that the assertion is true for square matrices of any dimension. ●

- B. *If A has a zero row or a zero column, then $|A| = 0$.*
- C. *If a row or column of A is multiplied by c, then $|A|$ is multiplied by c.*
- D. *If a row vector $\mathbf{r}^T = (\, r_1 \quad r_2 \quad \cdots \quad r_n \,)$ is added to the ith row of A, then the determinant becomes the sum of the determinants of A and $\widetilde{A}$, where $\widetilde{A}$ is*

the matrix obtained by replacing the ith *row of* A *with* $\mathbf{r}^T$. *For instance,*

$$\begin{vmatrix} a_{11} & a_{12} & a_{13} \\ a_{21}+r_1 & a_{22}+r_2 & a_{23}+r_3 \\ a_{31} & a_{32} & a_{33} \end{vmatrix} = \begin{vmatrix} a_{11} & a_{12} & a_{13} \\ a_{21} & a_{22} & a_{23} \\ a_{31} & a_{32} & a_{33} \end{vmatrix} + \begin{vmatrix} a_{11} & a_{12} & a_{13} \\ r_1 & r_2 & r_3 \\ a_{31} & a_{32} & a_{33} \end{vmatrix}.$$

Proof Each of Properties B–D is easily proved by expanding the determinant on the row or column in question. ●

- E. *If one row or column is a multiple of another row or column, then* $|A| = 0$.

Proof First observe that we may as well assume that neither of the two rows or columns in question contain only zeros, for otherwise the result follows from Property 2. Now suppose that the two rows or columns in question are interchanged. Then by Property 1, the determinant is multiplied by -1. On the other hand, the interchange is equivalent to multiplying one of the rows or columns in question by some c and the other by $\frac{1}{c}$. So by Property 3, the determinant remains unchanged. Therefore, it must be the case that $|A| = 0$. ●

- F. *If a multiple of one row of* A *is added to another row, then* $|A|$ *is unchanged.*

Proof Let A^* be the matrix that results from adding a multiple of row i of A to row k, and let $\widetilde{A}$ be the matrix that results from replacing row k with the same multiple of row i. By Property 4, $|A^*| = |A| + |\widetilde{A}|$. But $|\widetilde{A}| = 0$ by Property 5. Therefore, $|A^*| = |A|$. ●

- G. *If any row of* A *is a linear combination of the other rows of* A, *then* $|A| = 0$.

Proof Suppose that row i of A is a linear combination of the other rows. Then there is a sequence of row operations, each consisting of adding a multiple of another row to row i, that will produce zeros in each entry of row i. By Property 6, each of these row operations leaves the determinant unchanged, and the determinant of the end result is 0 by Property 2. Therefore, the determinant of A was already zero. ●

- H. *The determinant of any lower-triangular or upper-triangular square matrix is the product of its diagonal entries. In particular, the determinant of any square matrix in echelon form is the product of its diagonal entries.*

Proof First note that the assertion is easily verified for 2×2 matrices. Now suppose that it is true for $(n-1) \times (n-1)$ matrices, and consider the case of an $n \times n$ lower-triangular matrix A. The expansion of $|A|$ on the first row contains a single nonzero term $a_{11}M_{11}$. The minor M_{11} is the determinant of an $(n-1)\times(n-1)$ lower-triangular matrix whose diagonal entries are $a_{22}, a_{33}, \ldots, a_{nn}$. So by our assumption, $M_{11} = a_{22}, a_{33}, \ldots, a_{nn}$, and therefore $|A| = a_{11}a_{22}a_{33}\ldots a_{nn}$. If A is upper-triangular, then expansion on the first column produces the same result. Thus, truth of the assertion for $(n-1) \times (n-1)$ matrices implies truth for $n \times n$ matrices. This, together with truth for 2×2 matrices, proves the result for square matrices of any dimension. ●

Properties 1, 3, and 6 tell us how the determinant of A is affected by the row operations of the Gaussian elimination process. Suppose that an echelon form A^* of A is produced without scaling rows, that is, using only row exchanges and steps in which a multiple of one row is added to another. Then $|A| = (-1)^k|A^*|$, where k is the number of row exchanges used in the production of A^*. This, together with Property 8, suggests an alternative method for computing the determinant.

Example 13

Consider the matrix

$$A = \begin{pmatrix} 0 & 1 & 3 & 1 \\ 1 & 0 & 1 & 3 \\ 2 & 3 & 1 & 0 \\ 3 & 0 & 1 & 1 \end{pmatrix}.$$

Our goal is to use row operations to transform A into an echelon form without scaling rows. The reduction of the first two columns is as follows:

$$\begin{matrix} R_2 \\ R_1 \\ R_3 - 2R_2 \\ R_4 - 3R_2 \end{matrix} \begin{pmatrix} 1 & 0 & 1 & 3 \\ 0 & 1 & 3 & 1 \\ 0 & 3 & -1 & -6 \\ 0 & 0 & -2 & -8 \end{pmatrix} \quad \begin{matrix} R_1 \\ R_2 \\ R_3 - 3R_2 \\ R_4 \end{matrix} \begin{pmatrix} 1 & 0 & 1 & 3 \\ 0 & 1 & 3 & 1 \\ 0 & 0 & -10 & -9 \\ 0 & 0 & -2 & -8 \end{pmatrix}.$$

Note that the row operations involved one row exchange, which multiplies the determinant by -1 (by Property 1). The other row operations leave the determinant unchanged (by Property 6). One more step, which also leaves the determinant unchanged, produces an echelon form:

$$\begin{matrix} R_1 \\ R_2 \\ R_3 \\ R_4 - \frac{1}{5}R_3 \end{matrix} \begin{pmatrix} 1 & 0 & 1 & 3 \\ 0 & 1 & 3 & 1 \\ 0 & 0 & -10 & -9 \\ 0 & 0 & 0 & -\frac{31}{5} \end{pmatrix}.$$

The determinant of this echelon form is the product of its diagonal entries, which is $(1)(1)(-10)(-\frac{31}{5}) = 62$. Since one row exchange was performed, and none of the other row operations changed the determinant, the determinant of A is -62. ■

This method of computing the determinant by row-reduction is not only very practical, it also gives rise to an important connection between the determinant of A and the invertibility of A. Indeed, we know that A is invertible if and only if it has an echelon form with no zeros on its diagonal—that is, an echelon form with a nonzero determinant! But the determinant of A is nonzero, if and only if an echelon form of A has a nonzero determinant. So, just as was so easily shown in the 2×2 case, an $n \times n$ matrix A is invertible, if and only if its determinant is not zero.

We add this important result to the equivalent statements of Theorem 3.

Theorem 3$^+$

Let A be a square matrix. The following statements are equivalent; that is, they are either all true or all false.

i) *A is invertible.*

ii) *The reduced echelon form of A is $\mathcal{I}$.*

iii) *Some (and therefore every) echelon form of A has no zeros on its diagonal.*

iv) *The equation $A\mathbf{x} = \mathbf{0}$ has only the trivial solution $\mathbf{x} = \mathbf{0}$.*

v) *The columns of A are linearly independent.*

vi) *The equation $A\mathbf{x} = \mathbf{b}$ has a unique solution $\mathbf{x}$ for each vector $\mathbf{b}$ in $\mathbb{R}^n$.*

vii) $|A| \neq 0$. ▲

PROBLEMS

In Problems 1 through 3, compute the indicated matrix-vector product in two ways:

(a) by multiplying the vector by each row of the matrix;

(b) as a linear combination of the columns of the matrix.

1. $\begin{pmatrix} 1 & -1 & 2 \\ 2 & 3 & -1 \end{pmatrix}\begin{pmatrix} 2 \\ 3 \\ 1 \end{pmatrix}$ **2.** $\begin{pmatrix} 5 & 1 & 2 \\ 1 & 0 & 1 \\ 3 & 2 & 1 \end{pmatrix}\begin{pmatrix} 2 \\ 1 \\ -3 \end{pmatrix}$ **3.** $\begin{pmatrix} 0 & 1 & 3 \\ 1 & 0 & 1 \\ 2 & 3 & 1 \\ 3 & 0 & 1 \end{pmatrix}\begin{pmatrix} 1 \\ 2 \\ 3 \end{pmatrix}$

In Problems 4 through 6, compute the indicated matrix product.

4. $\begin{pmatrix} 1 & -1 \\ 2 & 3 \\ -2 & 1 \end{pmatrix}\begin{pmatrix} 2 & 1 \\ 1 & 3 \end{pmatrix}$ **5.** $\begin{pmatrix} 5 & 1 & 2 \\ 1 & 0 & 1 \\ 3 & 2 & 1 \end{pmatrix}\begin{pmatrix} 1 & -1 & 2 \\ 0 & 2 & 1 \\ 1 & 1 & 1 \end{pmatrix}$ **6.** $\begin{pmatrix} 1 & 2 \\ 3 & -2 \end{pmatrix}\begin{pmatrix} 0 & 1 & 3 \\ 1 & 0 & 1 \end{pmatrix}$

In Problems 7 through 11, compute the indicated matrix, given that

$$A = \begin{pmatrix} 1 & 2 \\ 2 & 1 \end{pmatrix}, \quad B = \begin{pmatrix} 3 & 2 \\ 3 & -1 \end{pmatrix}, \quad \text{and} \quad C = \begin{pmatrix} 2 & 0 \\ 1 & 1 \end{pmatrix}.$$

7. $A(B + C)$

8. $AB + C$

9. $(A - 2B)C$

10. $3A - BC$

11. $AB - BA$

In Problems 12 through 15, verify that the given matrices are inverses of each other.

12. $\begin{pmatrix} 1 & 2 \\ 2 & 3 \end{pmatrix}, \begin{pmatrix} -3 & 2 \\ 2 & -1 \end{pmatrix}$

13. $\begin{pmatrix} -1 & 1 & 0 \\ 1 & -1 & 1 \\ 0 & 1 & -1 \end{pmatrix}, \begin{pmatrix} 0 & 1 & 1 \\ 1 & 1 & 1 \\ 1 & 1 & 0 \end{pmatrix}$

14. $\begin{pmatrix} \sqrt{2} & 1 \\ 1 & \sqrt{2} \end{pmatrix}, \begin{pmatrix} \sqrt{2} & -1 \\ -1 & \sqrt{2} \end{pmatrix}$

15. $\begin{pmatrix} 1 & 1 & 0 & -1 \\ 1 & 1 & 0 & 0 \\ 0 & 0 & 1 & 1 \\ -1 & 0 & 1 & 1 \end{pmatrix}, \begin{pmatrix} 0 & 0 & 1 & -1 \\ 0 & 1 & -1 & 1 \\ 1 & -1 & 1 & 0 \\ -1 & 1 & 0 & 0 \end{pmatrix}$

In Problems 16 through 23, find the inverse of the given matrix, if it exists.

16. $\begin{pmatrix} 1 & 1 \\ 1 & 0 \end{pmatrix}$ **17.** $\begin{pmatrix} 1 & 1 & 0 \\ 1 & 0 & 1 \\ 0 & 1 & 0 \end{pmatrix}$ **18.** $\begin{pmatrix} 1 & 1 & 1 \\ 1 & 1 & 0 \\ 1 & 0 & 1 \end{pmatrix}$

19. $\begin{pmatrix} 2 & -1 & 0 \\ -1 & 2 & -1 \\ 0 & -1 & 2 \end{pmatrix}$ **20.** $\begin{pmatrix} 1 & 1 \\ 1 & 2 \end{pmatrix}$ **21.** $\begin{pmatrix} 1 & 1 & 0 \\ 1 & 0 & 1 \\ 2 & 1 & 1 \end{pmatrix}$

22. $\begin{pmatrix} -2 & 1 & 1 \\ 1 & 1 & 0 \\ 1 & 0 & 0 \end{pmatrix}$ **23.** $\begin{pmatrix} 1 & 3 & 0 \\ 0 & 1 & 3 \\ 0 & 0 & 1 \end{pmatrix}$

In Problems 24 through 29, determine whether the given collection of vectors is linearly dependent or independent in two ways:

(a) by computing an echelon form of a matrix;
(b) by computing the determinant of a matrix.

24. $\begin{pmatrix} 1 \\ 2 \end{pmatrix}, \begin{pmatrix} 3 \\ 4 \end{pmatrix}$ **25.** $\begin{pmatrix} 1 \\ 2 \\ 1 \end{pmatrix}, \begin{pmatrix} 3 \\ 3 \\ 0 \end{pmatrix}, \begin{pmatrix} 1 \\ 4 \\ 3 \end{pmatrix}$ **26.** $\begin{pmatrix} -1 \\ 1 \\ 2 \end{pmatrix}, \begin{pmatrix} 1 \\ -1 \\ -3 \end{pmatrix}, \begin{pmatrix} -2 \\ 2 \\ 1 \end{pmatrix}$

27. $\begin{pmatrix} -2 \\ -3 \end{pmatrix}, \begin{pmatrix} 4 \\ 6 \end{pmatrix}$ **28.** $\begin{pmatrix} 0 \\ 1 \\ 2 \end{pmatrix}, \begin{pmatrix} 2 \\ 0 \\ 1 \end{pmatrix}, \begin{pmatrix} 1 \\ 2 \\ 0 \end{pmatrix}$ **29.** $\begin{pmatrix} 2 \\ -1 \\ 3 \end{pmatrix}, \begin{pmatrix} -1 \\ 2 \\ 4 \end{pmatrix}, \begin{pmatrix} 5 \\ 3 \\ -2 \end{pmatrix}$

In Problems 30 through 33, compute the determinant of the given matrix.

30. $\begin{pmatrix} 1 & 3 \\ 2 & 2 \end{pmatrix}$ **31.** $\begin{pmatrix} 2 & 1 & 0 \\ 1 & 2 & 3 \\ 2 & 3 & 1 \end{pmatrix}$ **32.** $\begin{pmatrix} -2 & 1 & 1 \\ 1 & 1 & 0 \\ 1 & 0 & 1 \end{pmatrix}$ **33.** $\begin{pmatrix} 2 & -1 & 0 \\ -1 & 2 & -1 \\ 0 & -1 & 2 \end{pmatrix}$

In Problems 34 through 37, find all values of x for which the given matrix is not invertible.

34. $\begin{pmatrix} x & 1 & 0 \\ 1 & 2 & 1 \\ 2 & 1 & 1 \end{pmatrix}$ **35.** $\begin{pmatrix} x & 1 & 0 \\ 1 & 4 & 1 \\ 0 & 1 & x \end{pmatrix}$ **36.** $\begin{pmatrix} 2 & 1 & 1 \\ 1 & x & 1 \\ 1 & 1 & 2 \end{pmatrix}$ **37.** $\begin{pmatrix} x & 0 & 2 & 3 \\ 0 & x & 0 & 2 \\ 2 & 0 & x & 0 \\ 3 & 2 & 0 & x \end{pmatrix}$

38. Show that if A and B are invertible, then AB is invertible with

$$(AB)^{-1} = B^{-1}A^{-1}.$$

39. Concoct simple examples using 2×2 matrices to show that

(a) if A and B are invertible, then $A + B$ may not be invertible;
(b) even if A, B, and $A + B$ are invertible, then it may happen that

$$(A + B)^{-1} \neq A^{-1} + B^{-1}.$$

40. Let A be a 2×2 matrix with entries a_{ij}. Verify that

$$A^2 = (a_{11} + a_{22})A - |A|\,\mathcal{I},$$

where A^2 denotes the product AA.

C. Eigenvalues and Eigenvectors

Let A be an $n \times n$ matrix, and let λ be a (real or complex) scalar. If the equation

$$A\mathbf{x} = \lambda\mathbf{x} \tag{1}$$

has a nontrivial solution $\mathbf{x}$ (with real or complex entries), then λ is said to be an **eigenvalue** of A with corresponding **eigenvector** $\mathbf{x}$. Also, $(\lambda, \mathbf{x})$ is said to be an **eigenpair**. Noting that (1) is equivalent to

$$(A - \lambda\mathcal{I})\mathbf{x} = \mathbf{0},$$

it follows that λ is an eigenvalue of A, if and only if the determinant of $A - \lambda\mathcal{I}$ is zero, that is,

$$|A - \lambda\mathcal{I}| = 0. \tag{2}$$

This is the **characteristic equation** of A. It turns out that $|A - \lambda\mathcal{I}|$ is actually an nth-degree polynomial in λ, which we refer to as the **characteristic polynomial** of A. Consequently, A will have n complex eigenvalues (including those that are real), if each is duplicated according to its multiplicity as a zero of $|A - \lambda\mathcal{I}|$.

Example 1

Consider the 2×2 matrix

$$A = \begin{pmatrix} -2 & 9 \\ 1 & -2 \end{pmatrix}.$$

The characteristic polynomial of A is

$$\begin{vmatrix} -2-\lambda & 9 \\ 1 & -2-\lambda \end{vmatrix} = (2+\lambda)^2 - 9,$$

whose zeros are easily found to be $\lambda = 1, -5$. These are the eigenvalues of A. To find an eigenvector corresponding to $\lambda = 1$, we substitute $\lambda = 1$ into the equation $(A - \lambda\mathcal{I})\mathbf{x} = 0$, which gives us

$$(A - \mathcal{I})\mathbf{x} = \begin{pmatrix} -3 & 9 \\ 1 & -3 \end{pmatrix}\begin{pmatrix} x_1 \\ x_2 \end{pmatrix} = \begin{pmatrix} 0 \\ 0 \end{pmatrix}.$$

Since each row of the matrix here is just a multiple of the other, we simply seek x_1 and x_2 for which

$$x_1 - 3x_2 = 0.$$

A "particular" solution is easily seen to be $x_1 = 3$ and $x_2 = 1$; that is, $\mathbf{x} = (3, 1)^T$. (Note that *every* solution $\mathbf{x}$ is some scalar multiple of this particular solution.) Thus A has the eigenpair

$$\left(1, \begin{pmatrix} 3 \\ 1 \end{pmatrix}\right).$$

Now, with $\lambda = -5$ the equation $(A - \lambda\mathcal{I})\mathbf{x} = \mathbf{0}$ becomes

$$(A - \mathcal{I})\mathbf{x} = \begin{pmatrix} 3 & 9 \\ 1 & 3 \end{pmatrix}\begin{pmatrix} x_1 \\ x_2 \end{pmatrix} = \begin{pmatrix} 0 \\ 0 \end{pmatrix}.$$

Again the rows of the matrix are multiples of each other; so we simply seek x_1 and x_2 for which

$$x_1 + 3x_2 = 0.$$

A particular solution is easily seen to be $x_1 = 3$ and $x_2 = -1$; that is, $\mathbf{x} = (3, -1)^T$. Thus A has a second eigenpair

$$\left(-5, \begin{pmatrix} 3 \\ -1 \end{pmatrix}\right).$$

■

Example 2

Consider the 3×3 matrix

$$A = \begin{pmatrix} 0 & 1 & 1 \\ 1 & 0 & -1 \\ -1 & 1 & 2 \end{pmatrix}.$$

The characteristic polynomial of A is

$$\begin{aligned} \begin{vmatrix} -\lambda & 1 & 1 \\ 1 & -\lambda & -1 \\ -1 & 1 & 2-\lambda \end{vmatrix} &= -\lambda \begin{vmatrix} -\lambda & -1 \\ 1 & 2-\lambda \end{vmatrix} - \begin{vmatrix} 1 & -1 \\ -1 & 2-\lambda \end{vmatrix} + \begin{vmatrix} 1 & -\lambda \\ -1 & 1 \end{vmatrix} \\ &= -\lambda(-\lambda(2-\lambda)+1) - (2-\lambda-1) + (1-\lambda) \\ &= -\lambda^3 + 2\lambda^2 - \lambda \\ &= -\lambda(\lambda-1)^2. \end{aligned}$$

Therefore, the eigenvalues of A (duplicated according to multiplicity) are

$$\lambda = 0, 1, 1.$$

To find an eigenvector corresponding to $\lambda = 0$, we examine the equation

$$(A - 0\mathcal{I})\mathbf{x} = \begin{pmatrix} 0 & 1 & 1 \\ 1 & 0 & -1 \\ -1 & 1 & 2 \end{pmatrix}\begin{pmatrix} x_1 \\ x_2 \\ x_3 \end{pmatrix} = \begin{pmatrix} 0 \\ 0 \\ 0 \end{pmatrix},$$

which is guaranteed to have a nontrivial solution. Gaussian elimination produces the equivalent equation

$$\begin{pmatrix} 1 & 0 & -1 \\ 0 & 1 & 1 \\ 0 & 0 & 0 \end{pmatrix} \begin{pmatrix} x_1 \\ x_2 \\ x_3 \end{pmatrix} = \begin{pmatrix} 0 \\ 0 \\ 0 \end{pmatrix}.$$

Using x_3 as a free variable, we set $x_3 = 1$ and obtain the nontrivial solution $\mathbf{x} = (1, -1, 1)^T$. The resulting eigenpair is

$$\left(0, \begin{pmatrix} 1 \\ -1 \\ 1 \end{pmatrix}\right).$$

To find an eigenvector corresponding to $\lambda = 1$, we examine the equation

$$(A - \mathcal{I})\mathbf{x} = \begin{pmatrix} -1 & 1 & 1 \\ 1 & -1 & -1 \\ -1 & 1 & 1 \end{pmatrix} \begin{pmatrix} x_1 \\ x_2 \\ x_3 \end{pmatrix} = \begin{pmatrix} 0 \\ 0 \\ 0 \end{pmatrix},$$

which is guaranteed to have a nontrivial solution. Gaussian elimination produces the equivalent equation

$$\begin{pmatrix} 1 & -1 & -1 \\ 0 & 0 & 0 \\ 0 & 0 & 0 \end{pmatrix} \begin{pmatrix} x_1 \\ x_2 \\ x_3 \end{pmatrix} = \begin{pmatrix} 0 \\ 0 \\ 0 \end{pmatrix}.$$

Since there are two rows of zeros, the system has two free variables, and so there will be two linearly independent solutions. Setting $x_2 = 1$ and $x_3 = 0$ leads to the nontrivial solution $\mathbf{x} = (1, 1, 0)^T$. The resulting eigenpair is

$$\left(1, \begin{pmatrix} 1 \\ 1 \\ 0 \end{pmatrix}\right).$$

By setting $x_2 = 0$ and $x_3 = 1$, we obtain the nontrivial solution $\mathbf{x} = (1, 0, 1)^T$. The resulting eigenpair is

$$\left(1, \begin{pmatrix} 1 \\ 0 \\ 1 \end{pmatrix}\right).$$

Note that, even though A has only two distinct eigenvalues, it does have a "full set" of three linearly independent eigenvectors. This is somewhat typical; however, it does not always happen, as the next example will show. ■

Example 3

Consider the 3×3 matrix

$$A = \begin{pmatrix} 2 & 0 & -2 \\ -1 & 1 & 2 \\ 2 & 1 & -1 \end{pmatrix}.$$

The characteristic polynomial of A is

$$\begin{vmatrix} 2-\lambda & 0 & -2 \\ -1 & 1-\lambda & 2 \\ 2 & 1 & -1-\lambda \end{vmatrix} = (2-\lambda)\begin{vmatrix} 1-\lambda & 2 \\ 1 & -1-\lambda \end{vmatrix} - 0 + (-2)\begin{vmatrix} -1 & 1-\lambda \\ 2 & 1 \end{vmatrix}$$

$$\begin{aligned} &= (2-\lambda)((1-\lambda)(-1-\lambda)-2) - 2(-1-2(1-\lambda)) \\ &= \cdots = -\lambda^3 + 2\lambda^2 - \lambda \\ &= -\lambda(\lambda-1)^2. \end{aligned}$$

Therefore the eigenvalues of A (duplicated according to multiplicity) are the same as those of the matrix in Example 2:

$$\lambda = 0, 1, 1.$$

To find an eigenvector corresponding to $\lambda = 0$, we examine the equation

$$(A - 0\mathcal{I})\mathbf{x} = \begin{pmatrix} 2 & 0 & -2 \\ -1 & 1 & 2 \\ 2 & 1 & -1 \end{pmatrix}\begin{pmatrix} x_1 \\ x_2 \\ x_3 \end{pmatrix} = \begin{pmatrix} 0 \\ 0 \\ 0 \end{pmatrix}.$$

Gaussian elimination produces the equivalent equation

$$\begin{pmatrix} 1 & 0 & -1 \\ 0 & 1 & 1 \\ 0 & 0 & 0 \end{pmatrix}\begin{pmatrix} x_1 \\ x_2 \\ x_3 \end{pmatrix} = \begin{pmatrix} 0 \\ 0 \\ 0 \end{pmatrix}.$$

By setting $x_3 = 1$, we obtain the nontrivial solution $\mathbf{x} = (1, -1, 1)^T$. The resulting eigenpair is

$$\left(0, \begin{pmatrix} 1 \\ -1 \\ 1 \end{pmatrix}\right).$$

To find an eigenvector corresponding to $\lambda = 1$, we examine the equation

$$(A - \mathcal{I})\mathbf{x} = \begin{pmatrix} 1 & 0 & -2 \\ -1 & 0 & 2 \\ 2 & 1 & -2 \end{pmatrix}\begin{pmatrix} x_1 \\ x_2 \\ x_3 \end{pmatrix} = \begin{pmatrix} 0 \\ 0 \\ 0 \end{pmatrix}.$$

Gaussian elimination produces the equivalent equation

$$\begin{pmatrix} 1 & 0 & -2 \\ 0 & 1 & 2 \\ 0 & 0 & 0 \end{pmatrix}\begin{pmatrix} x_1 \\ x_2 \\ x_3 \end{pmatrix} = \begin{pmatrix} 0 \\ 0 \\ 0 \end{pmatrix}.$$

Since there is only one row of zeros in the reduced matrix, there is only one free variable; therefore there cannot be two linearly independent solutions. By setting $x_3 = 1$, we obtain the nontrivial solution $\mathbf{x} = (2, -2, 1)^T$. The resulting eigenpair is

$$\left(1, \begin{pmatrix} 2 \\ -2 \\ 1 \end{pmatrix}\right).$$

So we see that, even though it is 3×3, A does not have three linearly independent eigenvectors. Since the eigenvalue $\lambda = 1$ has multiplicity 2 but does not have two linearly independent eigenvectors, we say that it is a *deficient* eigenvalue. ■

Our next example illustrates complex eigenvalues and eigenvectors.

Example 4

Consider the 3×3 matrix

$$\begin{pmatrix} 2 & 1 & -3 \\ -1 & -1 & 2 \\ 3 & 2 & -3 \end{pmatrix}.$$

The characteristic polynomial of A is

$$\begin{vmatrix} 2-\lambda & 1 & -3 \\ -1 & -1-\lambda & 2 \\ 3 & 2 & -3-\lambda \end{vmatrix} = \cdots = -\lambda^3 - 2\lambda^2 - \lambda - 2$$

$$= -(\lambda + 2)(\lambda^2 + 1).$$

Therefore the eigenvalues of A are

$$\lambda = -2, i, -i.$$

To find an eigenvector corresponding to $\lambda = -2$, we examine the equation

$$(A + 2\mathcal{I})\mathbf{x} = \begin{pmatrix} 4 & 1 & -3 \\ -1 & 1 & 2 \\ 3 & 2 & -1 \end{pmatrix} \begin{pmatrix} x_1 \\ x_2 \\ x_3 \end{pmatrix} = \begin{pmatrix} 0 \\ 0 \\ 0 \end{pmatrix}.$$

Gaussian elimination produces the equivalent equation

$$\begin{pmatrix} 1 & 0 & -1 \\ 0 & 1 & 1 \\ 0 & 0 & 0 \end{pmatrix} \begin{pmatrix} x_1 \\ x_2 \\ x_3 \end{pmatrix} = \begin{pmatrix} 0 \\ 0 \\ 0 \end{pmatrix}.$$

By setting $x_3 = 1$, we obtain the nontrivial solution $\mathbf{x} = (1, -1, 1)^T$. The resulting eigenpair is

$$\left(-2, \begin{pmatrix} 1 \\ -1 \\ 1 \end{pmatrix}\right).$$

To find an eigenvector corresponding to $\lambda = i$, we examine the equation

$$(A - i\mathcal{I})\mathbf{x} = \begin{pmatrix} 2-i & 1 & -3 \\ -1 & -1-i & 2 \\ 3 & 2 & -3-i \end{pmatrix} \begin{pmatrix} x_1 \\ x_2 \\ x_3 \end{pmatrix} = \begin{pmatrix} 0 \\ 0 \\ 0 \end{pmatrix}.$$

Gaussian elimination produces the equivalent equation

$$\begin{pmatrix} 1 & 0 & -1-i \\ 0 & 1 & i \\ 0 & 0 & 0 \end{pmatrix} \begin{pmatrix} x_1 \\ x_2 \\ x_3 \end{pmatrix} = \begin{pmatrix} 0 \\ 0 \\ 0 \end{pmatrix}.$$

By setting $x_3 = 1$, we obtain the nontrivial solution $\mathbf{x} = (1+i, -i, 1)^T$. The resulting eigenpair is

$$\left(i, \begin{pmatrix} 1+i \\ -i \\ 1 \end{pmatrix}\right).$$

To find an eigenvector corresponding to $\lambda = -i$, we examine the equation

$$(A + i\mathcal{I})\mathbf{x} = \begin{pmatrix} 2+i & 1 & -3 \\ -1 & -1+i & 2 \\ 3 & 2 & -3+i \end{pmatrix} \begin{pmatrix} x_1 \\ x_2 \\ x_3 \end{pmatrix} = \begin{pmatrix} 0 \\ 0 \\ 0 \end{pmatrix}.$$

Gaussian elimination produces the equivalent equation

$$\begin{pmatrix} 1 & 0 & -1+i \\ 0 & 1 & -i \\ 0 & 0 & 0 \end{pmatrix} \begin{pmatrix} x_1 \\ x_2 \\ x_3 \end{pmatrix} = \begin{pmatrix} 0 \\ 0 \\ 0 \end{pmatrix}.$$

Setting $x_3 = 1$ leads to the nontrivial solution $\mathbf{x} = (1 - i, i, 1)^T$. The resulting eigenpair is

$$\left(-i, \begin{pmatrix} 1-i \\ i \\ 1 \end{pmatrix}\right).$$

■

Facts about Eigenvalues

Complex Conjugates Note that in Example 4 the eigenvectors corresponding to the complex conjugate eigenvalues $\pm i$ are themselves complex conjugates. This is true in general; that is:

- A. *If $\lambda = \alpha + \beta i$ is an eigenvalue of A with eigenvector $\mathbf{p} + i\mathbf{q}$, then $\overline{\lambda} = \alpha - \beta i$ is an eigenvalue of A with eigenvector $\mathbf{p} - i\mathbf{q}$.*

Proof That nonreal complex eigenvalues come in conjugate pairs is a simple consequence of the fact that they are roots of a polynomial. If $\lambda = \alpha + \beta i$ is an eigenvalue of A with eigenvector $\mathbf{p} + i\mathbf{q}$, then

$$A(\mathbf{p} + i\mathbf{q}) = (\alpha + \beta i)(\mathbf{p} + i\mathbf{q}) = \alpha\mathbf{p} - \beta\mathbf{q} + i(\beta\mathbf{p} + \alpha\mathbf{q}).$$

Thus, by equating real and imaginary parts, we have

$$A\mathbf{p} = \alpha\mathbf{p} - \beta\mathbf{q} \text{ and } A\mathbf{q} = \beta\mathbf{p} + \alpha\mathbf{q}.$$

Therefore,

$$A(\mathbf{p} - i\mathbf{q}) = \alpha\mathbf{p} - \beta\mathbf{q} - i(\beta\mathbf{p} + \alpha\mathbf{q}) = (\alpha - \beta i)(\mathbf{p} - i\mathbf{q}),$$

and so $\mathbf{p} - i\mathbf{q}$ is an eigenvector corresponding to $\overline{\lambda} = \alpha - \beta i$. ●

Products and Sums We will prove the following facts concerning the product and the sum of the eigenvalues of A.

- B. *The product of the eigenvalues of A equals the determinant of A:*
$$\lambda_1\lambda_2\cdots\lambda_n = |A|.$$
- C. *The sum of the eigenvalues of A is equal to the sum of its diagonal entries:*
$$\sum_{i=1}^{n}\lambda_i = \sum_{i=1}^{n} a_{ii}.$$

The sum of the diagonal entries of A is called the **trace** of A and is denoted by $\text{tr}(A)$. Thus the sum of the eigenvalues of A is equal to $\text{tr}(A)$.

Proof of B Consider the characteristic polynomial of A:

$$|A - \lambda\mathcal{I}| = \begin{vmatrix} a_{11}-\lambda & a_{22} & \cdots & a_{1n} \\ a_{21} & a_{22}-\lambda & \cdots & a_{2n} \\ \vdots & \vdots & \ddots & \vdots \\ a_{n1} & a_{n2} & \cdots & a_{nn}-\lambda \end{vmatrix},$$

and suppose that $\lambda_1, \lambda_2, \ldots, \lambda_n$ are the eigenvalues of A. We will first show that the complete factorization of the characteristic polynomial is

$$|A - \lambda\mathcal{I}| = (\lambda_1 - \lambda)(\lambda_2 - \lambda)\cdots(\lambda_n - \lambda). \tag{3}$$

Since $\lambda_1, \lambda_2, \ldots, \lambda_n$ are the roots of $|A - \lambda\mathcal{I}|$, the only thing we need to do here is show that the leading coefficient $|A - \lambda\mathcal{I}|$ is $(-1)^n$. This is easily seen for 2×2 matrices. Now suppose that it is true for $(n-1)\times(n-1)$ matrices, where $n \geq 2$, and let A be $n \times n$. Expanding $|A - \lambda\mathcal{I}|$ on its first row gives

$$|A - \lambda\mathcal{I}| = (a_{11} - \lambda)M_{11} + \text{terms of degree less than } n.$$

The minor M_{11} is the characteristic polynomial of the $(n-1)\times(n-1)$ matrix obtained by deleting the first row and column from A. Thus, its leading coefficient is $(-1)^{n-1}$, by our supposition. Therefore, the leading coefficient in $|A - \lambda\mathcal{I}|$ is $-(-1)^{n-1} = (-1)^n$. This proves (3). Now, by substituting $\lambda = 0$ into (3), we obtain Property B. ●

Proof of C It also follows from (3) that the coefficient of $(-\lambda)^{n-1}$ in the characteristic polynomial is the sum of the eigenvalues of A; that is,

$$|A - \lambda\mathcal{I}| = (-\lambda)^n + \left(\sum_{i=1}^{n}\lambda_i\right)(-\lambda)^{n-1} + \text{lower degree terms.} \tag{4}$$

We will now argue, again by induction, that the coefficient of λ^{n-1} in $|A - \lambda\mathcal{I}|$ is also the sum of the diagonal entries of A. First, it is easy to prove the assertion for 2×2 matrices. Now suppose that $n > 2$, and assume that the assertion is true for $(n-1)\times(n-1)$ matrices. This means that for any $(n-1)\times(n-1)$ matrix the coefficient of $(-\lambda)^{n-2}$ in its characteristic polynomial is the sum of the diagonal entries in the matrix. Now suppose that A is $n \times n$. Expanding the determinant on the first row, we observe that

$$|A - \lambda \mathcal{I}| = (a_{11} - \lambda)M_{11} + \sum_{j=2}^{n}(-1)^{j+1}a_{1j}M_{1j}.$$

Each of the minors $M_{12}, \dots, M_{1n}$ are polynomials of degree $n-2$, since each is the determinant of a matrix in which λ appears only in $n-2$ entries, which are of the form $a_{ii} - \lambda$. The first minor

$$M_{11} = \begin{vmatrix} a_{22} - \lambda & \cdots & a_{2n} \\ \vdots & \ddots & \vdots \\ a_{n2} & \cdots & a_{nn} - \lambda \end{vmatrix}$$

is the characteristic polynomial of an $(n-1) \times (n-1)$ matrix $\widetilde{A}$ obtained by deleting the first row and column from A. So, by our assumption, the coefficient of $(-\lambda)^{n-2}$ in M_{11} is the sum of the diagonal entries in $\widetilde{A}$. Moreover, the leading coefficient of M_{11} is $(-1)^{n-1}$. Thus we have

$$|A - \lambda \mathcal{I}| = (a_{11} - \lambda)\left((-\lambda)^{n-1} + \left(\sum_{i=1}^{n} a_{ii}\right)(-\lambda)^{n-2} + \cdots\right) + \cdots,$$

where all terms not shown have degree $n-2$ or less. From this we conclude that

$$|A - \lambda \mathcal{I}| = (-\lambda)^{n} + \left(\sum_{i=1}^{n} a_{ii}\right)(-\lambda)^{n-1} + \text{lower degree terms.} \tag{5}$$

Thus, by induction, we have shown that (5) is true for *all* n. Comparing this with (4) yields Property C. ●

Example 5

Consider the 2×2 matrix

$$A = \begin{pmatrix} 1 & -1 \\ 2 & 3 \end{pmatrix},$$

whose determinant and trace are

$$|A| = (1)(3) - (-1)(2) = 5 \text{ and } \operatorname{tr}(A) = 1 + 3 = 4.$$

The characteristic polynomial of A is

$$\begin{vmatrix} 1 - \lambda & -1 \\ 2 & 3 - \lambda \end{vmatrix} = (1 - \lambda)(3 - \lambda) + 2 = \lambda^2 - 4\lambda + 5.$$

Thus the eigenvalues of A are

$$\frac{4 \pm \sqrt{(-4)^2 - (4)(5)}}{2} = 2 \pm i.$$

So the product of the eigenvalues is $(2+i)(2-i) = 4 - (-1) = 5$, which equals $|A|$, and the sum is $(2+i) + (2-i) = 4$, which equals $\operatorname{tr}(A)$. ■

Similarity and Diagonalization A matrix B is said to be **similar** to a matrix A, if there is an invertible matrix S such that

$$B = S^{-1}AS.$$

Note that, if $B = S^{-1}AS$, then $A = (S^{-1})^{-1}BS^{-1}$. Therefore, if B is similar to A, then A is also similar to B, and so it makes sense to say simply that A and B are similar. A basic fact about similar matrices is the following.

- D. *Similar matrices have the same eigenvalues.*

Proof Suppose that A and B are similar matrices with $B = S^{-1}AS$. Suppose also that λ is an eigenvalue of A with corresponding eigenvector x. Then $A\mathbf{x} = \lambda\mathbf{x}$ implies that

$$S^{-1}A\mathbf{x} = S^{-1}\lambda\mathbf{x} = \lambda S^{-1}\mathbf{x}.$$

But

$$S^{-1}A\mathbf{x} = S^{-1}ASS^{-1}\mathbf{x} = BS^{-1}\mathbf{x};$$

therefore,

$$BS^{-1}\mathbf{x} = \lambda S^{-1}\mathbf{x}.$$

Since $S^{-1}\mathbf{x} \neq \mathbf{0}$ (why?), this means that λ is also an eigenvalue of B (with corresponding eigenvector $S^{-1}x$), which proves Property D. ●

A matrix A is said to be **diagonalizable**, if it is similar to a diagonal matrix; that is, if there is a diagonal matrix D and an invertible matrix S such that

$$S^{-1}AS = D.$$

- E. *A is diagonalizable, if and only if A has n linearly independent eigenvectors.*

Proof Let's examine the equation $S^{-1}AS = D$ more closely. We first multiply both sides by S, obtaining

$$AS = SD.$$

Now let $\mathbf{s}_{:1}, \mathbf{s}_{:2}, \ldots, \mathbf{s}_{:n}$ denote the columns of S, noting that no $\mathbf{s}_{:j}$ is $\mathbf{0}$, since S is invertible. Also let $d_1, d_2, \ldots, d_n$ be the diagonal entries of D. The jth column of the product AS is the matrix-vector product $A\mathbf{s}_{:j}$. Likewise, the jth column of the product SD is the matrix-vector product of A with the jth column of D, of which d_j is the only (possibly) nonzero entry. Thus the jth column of SD is simply $d_j\mathbf{s}_{:j}$. Consequently, the equation $AS = SD$ becomes

$$\begin{pmatrix} A\mathbf{s}_{:1} & A\mathbf{s}_{:2} & \cdots & A\mathbf{s}_{:n} \end{pmatrix} = \begin{pmatrix} d_1\mathbf{s}_{:1} & d_2\mathbf{s}_{:2} & \cdots & d_n\mathbf{s}_{:n} \end{pmatrix}.$$

Therefore, $d_1, d_2, \ldots, d_n$ are eigenvalues of A, and the columns of S are corresponding eigenvectors. From this we can conclude that, if A has n linearly independent eigenvectors and those eigenvectors comprise the columns of S, then S is invertible and $S^{-1}AS$ is a diagonal matrix. Conversely, if $S^{-1}AS$ is a diagonal matrix, then the columns of S are eigenvectors of A, which are linearly independent by virtue of the invertibility of S. ●

From the observation that $S^{-1}AS = D$ is equivalent to $A = SDS^{-1}$, we conclude (with a slight change in notation) that:

- F. *If A has n linearly independent eigenvectors $\mathbf{x}_1, \mathbf{x}_2, \ldots, \mathbf{x}_n$, corresponding to eigenvalues $\lambda_1, \lambda_2, \ldots, \lambda_n$, respectively, then*

$$A = X\Lambda X^{-1},$$

where

$$X = \begin{pmatrix} \mathbf{x}_1 & \mathbf{x}_2 & \cdots & \mathbf{x}_n \end{pmatrix} \quad \textit{and} \quad \Lambda = \begin{pmatrix} \lambda_1 & 0 & \cdots & 0 \\ 0 & \lambda_2 & \ddots & \\ \vdots & \ddots & \ddots & 0 \\ 0 & & 0 & \lambda_n \end{pmatrix}.$$

Example 6

Suppose that we wish to find a 2×2 matrix A with eigenpairs

$$\left(1+i, \begin{pmatrix} 2 \\ i \end{pmatrix}\right) \text{ and } \left(1-i, \begin{pmatrix} 2 \\ -i \end{pmatrix}\right).$$

Note that the eigenvalues are conjugate pairs, as are the eigenvectors. To determine A, we simply need to compute the product

$$A = \begin{pmatrix} 2 & 2 \\ i & -i \end{pmatrix} \begin{pmatrix} 1+i & 0 \\ 0 & 1-i \end{pmatrix} \begin{pmatrix} 2 & 2 \\ i & -i \end{pmatrix}^{-1}.$$

After computing the inverse required for the third factor, we proceed thus:

$$\begin{aligned} A &= \begin{pmatrix} 2 & 2 \\ i & -i \end{pmatrix} \begin{pmatrix} 1+i & 0 \\ 0 & 1-i \end{pmatrix} \frac{1}{-4i} \begin{pmatrix} -i & -2 \\ -i & 2 \end{pmatrix} \\ &= \frac{1}{-4i} \begin{pmatrix} 2 & 2 \\ i & -i \end{pmatrix} \begin{pmatrix} 1-i & -2(1+i) \\ -1-i & 2(1-i) \end{pmatrix} \\ &= \frac{1}{-4i} \begin{pmatrix} -4i & -8i \\ 2i & -4i \end{pmatrix} \\ &= \frac{1}{2} \begin{pmatrix} 2 & 4 \\ -1 & 2 \end{pmatrix}. \end{aligned}$$

■

Our next result provides a sufficient (though not necessary) condition for the diagonalizability of A.

- G. *Eigenvectors corresponding to distinct eigenvalues are linearly independent. Therefore, if A has n distinct eigenvalues, then A is diagonalizable.*

Proof Let $\lambda_1, \lambda_2, \ldots, \lambda_k$ be any k distinct eigenvalues of A with corresponding eigenvectors $\mathbf{x}_1, \mathbf{x}_2, \ldots, \mathbf{x}_k$. If we assume that $\{\mathbf{x}_1, \mathbf{x}_2, \ldots, \mathbf{x}_k\}$ is linearly *dependent*, then we can choose the least index ℓ, with $1 \leq \ell \leq k-1$, for which $\{x_1, x_2, \ldots, x_\ell\}$ is linearly independent and $\{\mathbf{x}_1, \mathbf{x}_2, \ldots, \mathbf{x}_{\ell+1}\}$ is linearly dependent.* Then $\mathbf{x}_{\ell+1}$ is a linear combination of the vectors $\mathbf{x}_1, \mathbf{x}_2, \ldots, \mathbf{x}_\ell$:

$$c_1\mathbf{x}_1 + c_2\mathbf{x}_2 + \cdots + c_\ell\mathbf{x}_\ell = \mathbf{x}_{\ell+1}, \tag{6}$$

where $c_1, c_2, \ldots, c_k$ are not all zero, since $\mathbf{x}_{\ell+1} \neq \mathbf{0}$. Multiplying each side by A, we have

$$c_1 A\mathbf{x}_1 + c_2 A\mathbf{x}_2 + \cdots + c_\ell A\mathbf{x}_\ell = A\mathbf{x}_{\ell+1},$$

and consequently

$$c_1\lambda_1\mathbf{x}_1 + c_2\lambda_2\mathbf{x}_2 + \cdots + c_\ell\lambda_\ell\mathbf{x}_\ell = \lambda_{\ell+1}\mathbf{x}_{\ell+1}. \tag{7}$$

Now, multiplying each side of (6) by $\lambda_{\ell+1}$ and subtracting the result from (7), we find that

$$c_1(\lambda_{\ell+1} - \lambda_1)\mathbf{x}_1 + c_2(\lambda_{\ell+1} - \lambda_2)\mathbf{x}_2 + \cdots + c_\ell(\lambda_{\ell+1} - \lambda_\ell)\mathbf{x}_\ell = 0.$$

Since $\lambda_{\ell+1}$ differs from each of $\lambda_1, \lambda_2, \ldots, \lambda_\ell$, this contradicts the linear independence of $\{\mathbf{x}_1, \mathbf{x}_2, \ldots, \mathbf{x}_\ell\}$. Therefore it must be the case that $\{\mathbf{x}_1, \mathbf{x}_2, \ldots, \mathbf{x}_k\}$ is linearly independent. ●

The Transpose and Matrix Symmetry The **transpose** of a (not necessarily square) $n \times m$ matrix A is the $m \times n$ matrix A^T whose ijth entry is a_{ji}. In effect, each row of A becomes a column of A^T, and each column of A becomes a row of A^T. For instance,

$$\text{if } A = \begin{pmatrix} 1 & 2 & 3 \\ 4 & 5 & 6 \end{pmatrix}, \text{ then } A^T = \begin{pmatrix} 1 & 4 \\ 2 & 5 \\ 3 & 6 \end{pmatrix}.$$

When A is square, A^T is also square. For instance,

$$\text{if } A = \begin{pmatrix} 1 & 2 & 3 \\ 4 & 5 & 6 \\ 7 & 8 & 9 \end{pmatrix}, \text{ then } A^T = \begin{pmatrix} 1 & 4 & 7 \\ 2 & 5 & 8 \\ 3 & 6 & 9 \end{pmatrix}.$$

The following facts about the transpose are nearly trivial:

$$(A^T)^T = A, \quad (A+B)^T = A^T + B^T, \quad (cA)^T = cA^T. \tag{8}$$

A less obvious fact is that the transpose of a product is the product of the transposes in reverse order:

$$(AB)^T = B^T A^T. \tag{9}$$

To see this, first observe that, since the ijth entry of AB is $\mathbf{a}_{i..}\mathbf{b}_{:j}$, it follows that the ijth entry of $(AB)^T$ is $\mathbf{a}_{j..}\mathbf{b}_{:i}$. Next observe that the ijth entry of $B^T A^T$ is the product of the ith row of B^T and the jth column of A^T, which is the same as the

* Note that ℓ may be 1. A set of one nonzero vector is automatically linearly independent.

product of the jth row of A with the ith column of B. Thus $\mathbf{a}_{j..}\mathbf{b}_{:i}$ is the ijth entry of both $(AB)^T$ and $B^T A^T$.

If A is invertible, then it turns out that the transpose is also invertible, and the inverse of the transpose is the transpose of the inverse:

$$(A^T)^{-1} = (A^{-1})^T. \tag{10}$$

To prove this, we need to show that $A^T(A^{-1})^T = \mathcal{I}$. To do this, we will examine the transpose of $A^T(A^{-1})^T$. Using (9), we find that

$$(A^T(A^{-1})^T)^T = ((A^{-1})^T)^T(A^T)^T = A^{-1}A = \mathcal{I}.$$

Since $\mathcal{I}^T = \mathcal{I}$, we conclude that $A^T(A^{-1})^T = \mathcal{I}$.

Suppose that we compute the determinant of A using only row expansions. Corresponding column expansions will produce precisely the value for the determinant of A^T. Thus,

$$|A^T| = |A|.$$

It follows from this and (8) that A and A^T have the same characteristic polynomial:

$$\left|A^T - \lambda\mathcal{I}\right| = \left|(A - \lambda\mathcal{I})^T\right| = |A - \lambda\mathcal{I}|.$$

Therefore, A and A^T have precisely the same eigenvalues.

If $A^T = A$, then A is said to be **symmetric**. For instance, these matrices are symmetric:

$$\begin{pmatrix} 1 & 2 \\ 2 & -1 \end{pmatrix} \quad \begin{pmatrix} 1 & 2 & 3 \\ 2 & 4 & 5 \\ 3 & 5 & 6 \end{pmatrix} \quad \begin{pmatrix} 1 & -1 & 0 \\ -1 & 1 & -1 \\ 0 & -1 & 1 \end{pmatrix}.$$

Symmetric matrices arise in numerous applications and have the following highly significant property.

- H. *If A is symmetric, then the eigenvalues of A are real.*

Proof The proof is based upon the identity

$$\mathbf{y}^T A\mathbf{x} = \mathbf{x}^T A^T \mathbf{y},$$

which holds for any matrix A and vectors $\mathbf{x}$ and $\mathbf{y}$. The left and right sides of this identity are, respectively,

$$\sum_{i=1}^{n} y_i \sum_{j=1}^{n} a_{ij}x_j \quad \text{and} \quad \sum_{j=1}^{n} x_j \sum_{i=1}^{n} a_{ij}y_i,$$

each of which is equal to

$$\sum_{i=1}^{n}\sum_{j=1}^{n} a_{ij}x_j y_i.$$

Now suppose that A is symmetric. Then the identity above becomes

$$\mathbf{y}^T A\mathbf{x} = \mathbf{x}^T A\mathbf{y}.$$

In particular, if $\mathbf{x}$ is a complex vector, with conjugate $\overline{\mathbf{x}}$, then

$$\overline{\mathbf{x}}^T A\mathbf{x} = \mathbf{x}^T A\overline{\mathbf{x}},$$

and, moreover, the quantity $\overline{x}^T Ax$ is a real number, because

$$\begin{aligned}(\mathbf{p} - i\mathbf{q})^T A(\mathbf{p} + i\mathbf{q}) &= \mathbf{p}^T A\mathbf{p} + i\mathbf{p}^T A\mathbf{q} - i\mathbf{q}^T A\mathbf{p} + \mathbf{q}^T A\mathbf{q} \\ &= \mathbf{p}^T A\mathbf{p} + \mathbf{q}^T A\mathbf{q}\end{aligned}$$

for any real vectors $\mathbf{p}$ and $\mathbf{q}$. Now suppose that λ is an eigenvalue of A with corresponding eigenvector $\mathbf{x}$. Then

$$\overline{\mathbf{x}}^T Ax = \overline{\mathbf{x}}^T \lambda x = \lambda \overline{\mathbf{x}}^T \mathbf{x}.$$

Since $\overline{\mathbf{x}}^T A\mathbf{x}$ and $\overline{\mathbf{x}}^T \mathbf{x}$ are both real, it follows that λ is real. ●

PROBLEMS

In Problems 1 and 2, check that $A\mathbf{x} = \lambda\mathbf{x}$ for each of the given eigenpairs.

1. $A = \begin{pmatrix} 1 & -2 & 2 \\ -1 & 1 & 1 \\ -1 & -2 & 4 \end{pmatrix}$; $(\lambda, \mathbf{x}) = \left(1, \begin{pmatrix} 1 \\ 1 \\ 1 \end{pmatrix}\right)$, $\left(2, \begin{pmatrix} 0 \\ 1 \\ 1 \end{pmatrix}\right)$, $\left(3, \begin{pmatrix} 1 \\ 0 \\ 1 \end{pmatrix}\right)$

2. $A = \begin{pmatrix} 2 & -4 & 2 \\ 1 & 2 & -2 \\ 1 & -1 & 1 \end{pmatrix}$; $(\lambda, \mathbf{x}) = \left(2, \begin{pmatrix} 1 \\ 0 \\ 1 \end{pmatrix}\right)$, $\left(1+i, \begin{pmatrix} 1+i \\ 1 \\ 1 \end{pmatrix}\right)$, $\left(1-i, \begin{pmatrix} 1-i \\ 1 \\ 1 \end{pmatrix}\right)$

In Problems 3–14 , find the eigenvalues and corresponding eigenvectors of the matrix.

3. $\begin{pmatrix} 1 & 2 \\ 2 & 1 \end{pmatrix}$ **4.** $\begin{pmatrix} 1 & 2 \\ 2 & -2 \end{pmatrix}$

5. $\begin{pmatrix} -6 & 2 \\ -6 & 1 \end{pmatrix}$ **6.** $\begin{pmatrix} 1 & -1 \\ 2 & -1 \end{pmatrix}$

7. $\begin{pmatrix} 3 & -5 \\ 1 & -2 \end{pmatrix}$ **8.** $\begin{pmatrix} -3 & -1 & 6 \\ 6 & 4 & -12 \\ 1 & 1 & -2 \end{pmatrix}$

9. $\begin{pmatrix} -3 & -1 & 8 \\ 10 & 8 & -20 \\ 2 & 2 & -3 \end{pmatrix}$ **10.** $\begin{pmatrix} -3 & 4 & -2 \\ -1 & 2 & -1 \\ 2 & -2 & 1 \end{pmatrix}$

11. $\begin{pmatrix} 1 & 1 & -1 \\ 0 & 0 & 1 \\ 2 & 1 & 0 \end{pmatrix}$ **12.** $\begin{pmatrix} 1 & 0 & 1 \\ 0 & 3 & 1 \\ 1 & 0 & 1 \end{pmatrix}$

13. $\begin{pmatrix} 1 & 0 & 1 \\ 0 & 2 & 1 \\ 1 & 0 & 1 \end{pmatrix}$ **14.** $\begin{pmatrix} 1 & 0 & -1 \\ 0 & 1 & 1 \\ 1 & 0 & 1 \end{pmatrix}$

In Problems 15–18, A is a 3×3 matrix. Find the eigenvalues λ_2 and λ_3 from the given information.

15. $\lambda_1 = 3, |A| = -12, \text{tr}(A) = 3$

16. $\lambda_1 = 2 - i, \ \text{tr}(A) = 0$

17. $\lambda_1 = -3 - 2i, |A| = -26$

18. $\lambda_1 = -2, \ |A| = 60, \ \text{tr}(A) = 9$

In Problems 19 through 23, determine whether or not the given matrices are similar.

19. $\begin{pmatrix} 4 & -6 \\ 3 & -5 \end{pmatrix}, \begin{pmatrix} 1 & 1 \\ 0 & -2 \end{pmatrix}$

20. $\begin{pmatrix} 3 & 4 \\ -2 & -3 \end{pmatrix}, \begin{pmatrix} 2 & -3 \\ 1 & -2 \end{pmatrix}$

21. $\begin{pmatrix} -4 & 10 \\ -3 & 7 \end{pmatrix}$, $\begin{pmatrix} 4 & 1 \\ 2 & 5 \end{pmatrix}$

22. $\begin{pmatrix} 1 & 1 & -1 \\ -1 & -1 & 1 \\ 1 & 0 & 0 \end{pmatrix}$, $\begin{pmatrix} -1 & -1 & -1 \\ 1 & 1 & 1 \\ 1 & 0 & 0 \end{pmatrix}$

23. $\begin{pmatrix} 1 & 1 & -1 \\ -1 & -1 & 1 \\ 1 & 0 & 0 \end{pmatrix}$, $\begin{pmatrix} -1 & -1 & -1 \\ 1 & 1 & 1 \\ -1 & 0 & 0 \end{pmatrix}$

In Problems 24 and 25, find the matrix with the given eigenpairs.

24. $\left(1 - i, \begin{pmatrix} 2 \\ i \end{pmatrix}\right)$, $\left(1 + i, \begin{pmatrix} 2 \\ -i \end{pmatrix}\right)$

25. $\left(2, \begin{pmatrix} 1 \\ -1 \\ 0 \end{pmatrix}\right)$, $\left(-1, \begin{pmatrix} 0 \\ 2 \\ 1 \end{pmatrix}\right)$, $\left(1, \begin{pmatrix} 1 \\ 0 \\ 1 \end{pmatrix}\right)$

26. Give a direct proof that a symmetric, 2×2 matrix $A = \begin{pmatrix} a & b \\ b & c \end{pmatrix}$ has real eigenvalues.

27. Find an example of a 2×2 matrix with repeated eigenvalues $\lambda_1 = \lambda_2 = 1$ and without two linearly independent eigenvectors.

28. Prove that a square matrix A is invertible, if and only if 0 is not an eigenvalue of A.

29. Show that, if A has eigenvalues $\lambda_1, \lambda_2, \dots, \lambda_n$, and if A is invertible, then the eigenvalues of A^{-1} are $1/\lambda_1, 1/\lambda_2, \dots, 1/\lambda_n$.

30. Show that, for any square matrix A, each of the matrices $A^T A$ and AA^T is symmetric.

31. Find an example of an invertible 2×2 matrix A with the property that $A^{-1} = A^T$. Then find a 3×3 example. What can be said about the vectors that comprise the columns of such a matrix?

CONTINUITY AND DIFFERENTIABILITY ON INTERVALS

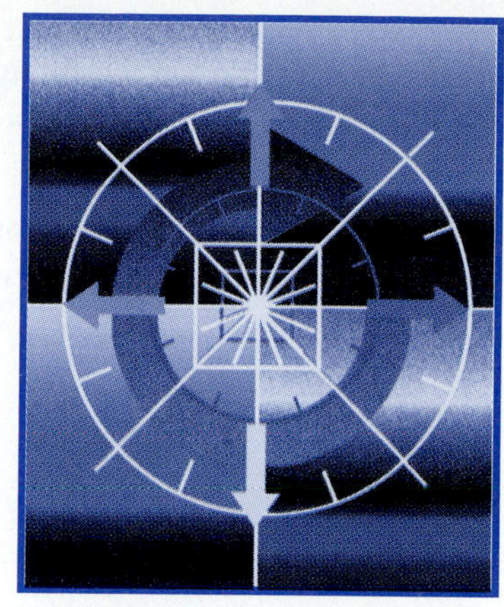

There are many statements in this book that contain the phrase "on an interval I." For instance, we might say that a function f is continuous on an interval I, or that a differential equation has a solution on an interval I containing 0. *Unless stated more specifically, the interval I may be open, closed, or half-open, as well as bounded or unbounded.* The following table summarizes the possible forms that I may have.

	Bounded	Unbounded
Open:	(a, b)	$[a, \infty)$ $(-\infty, b]$ $(-\infty, \infty)$
Half-open:	$[a, b)$ $(a, b]$	
Closed:	$[a, b]$	$[a, \infty)$ $(-\infty, b]$ $(-\infty, \infty)$

The **interior** of an interval is its largest open subinterval, which can be obtained by deleting whatever endpoints the interval has. So the interior of $[a, b]$ is (a, b), the interior of (a, b) is (a, b), and so on.

Our goal here is to state definitions of continuity and differentiability on an interval that are independent of any properties of f outside the interval and apply regardless of what type of interval is involved.

Continuity Let f be a function whose domain contains some interval I. Recall that f is continuous at a number τ in the interior of I, if and only if $\lim_{t\to\tau} f(t) = f(\tau)$.* Also recall that, in the case where I is an *open* interval, f is continuous on I, if and only if f is continuous at each number in I; that is, if and only if f is defined at each number in I and

$$\lim_{t\to\tau} f(t) = f(\tau) \text{ for each } \tau \text{ in } I.$$

For an interval I that is not open, the notion of continuity on I needs to be more carefully defined in order to reflect the desired behavior of f at each endpoint

---

* Implicit in this statement is that $\lim_{t\to\tau} f(t)$ must exist; otherwise f is not continuous at τ.

contained in I. In particular, if I contains a left endpoint a, then we will demand only that f be "right-continuous" at a in the sense that

$$\lim_{t\to a^+} f(t) = f(a).$$

Similarly, if I contains a right endpoint b, then we will demand only that f be "left-continuous" at b in the sense that

$$\lim_{t\to b^-} f(t) = f(b).$$

With these issues in mind, let us define the **limit relative to** I as

$$\lim_{\substack{t\to\tau\\ t\in I}} f(t) = \begin{cases} \lim\limits_{t\to\tau} f(t), & \text{if } \tau \text{ is in the interior of } I;\\ \lim\limits_{t\to\tau^+} f(t), & \text{if } \tau \text{ is a left endpoint of } I;\\ \lim\limits_{t\to\tau^-} f(t), & \text{if } \tau \text{ is a right endpoint of } I. \end{cases}$$

We can now make a simple statement that defines continuity on any type of interval. *A function f is continuous on an interval I, if and only if f is defined at each number in I and*

$$\lim_{\substack{t\to\tau\\ t\in I}} f(t) = f(\tau) \text{ for each } \tau \text{ in } I.$$

If I is the domain of f, and if f is continuous on I, then we say simply that f is continuous.

The Derivative Just as in the case of continuity, a satisfactory notion of differentiability on an interval I cannot be obtained by simply requiring differentiability at each point in I—except in the case where I is open. For a function f defined on an interval I, we define the **derivative of** f **relative to** I by

$$f_I'(t) = \lim_{\substack{s\to t\\ s\in I}} \frac{f(s)-f(t)}{s-t}$$

for all t in I at which the defining limit exists. This brief definition essentially encapsulates the following:

i) If t is in the interior of I, then $f_I'(t) = f'(t)$.

ii) If t is a left endpoint of I contained in I, then

$$f_I'(t) = \lim_{s\to t^+} \frac{f(s)-f(t)}{s-t} = \lim_{h\to 0^+} \frac{f(t+h)-f(t)}{h}.$$

iii) If t is a right endpoint of I contained in I, then

$$f_I'(t) = \lim_{s\to t^-} \frac{f(s)-f(t)}{s-t} = \lim_{h\to 0^-} \frac{f(t+h)-f(t)}{h}.$$

Note that if I is an open interval, then the derivative of f relative to I is just the usual derivative f'.

If t is in I and $f_I'(t)$ exists, we say that f is differentiable relative to I at t. If $f_I'(t)$ exists for all t in I, we simply say that f is differentiable on I. It follows that

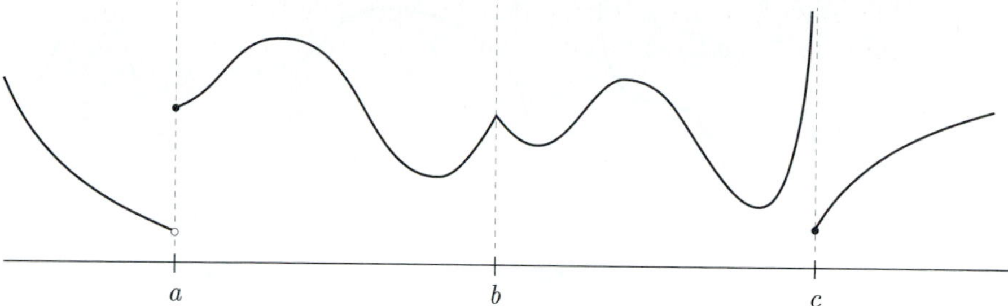

Figure 1

if f is differentiable on I, then f is continuous on I as well as differentiable (and continuous) on any subinterval of I.

Example 1

Consider the function graphed in Figure 1, which is defined on $(-\infty, \infty)$. This function is continuous and differentiable on $(-\infty, a)$ but is neither continuous nor differentiable on $(-\infty, a]$. It is continuous and differentiable on $[a, b]$, and is continuous but not differentiable on $[a, c)$. It is continuous and differentiable on $[b, c)$ as well as continuous and differentiable on $[c, \infty)$.

An important observation to make here is that the function in Figure 1 is differentiable on $[a, b]$ and on $[b, c)$ in spite of the fact that f is differentiable at neither a nor b. ■

Example 2

Consider the absolute value function, $f(t) = |t|$, recalling that f is not differentiable at $t = 0$ and

$$f'(t) = \frac{t}{|t|} \text{ for all } t \neq 0.$$

However, f is differentiable on each of $(-\infty, 0]$ and $[0, \infty)$ with

$$f'_{[0,\infty)}(t) = 1 \text{ for all } t \geq 0, \quad \text{and} \quad f'_{(-\infty,0]}(t) = -1 \text{ for all } t \leq 0.$$ ■

The illustrations provided by Examples 1 and 2 are somewhat artificial, because we are mainly concerned with the situation where I is the domain of f. The next example is more indicative of the issues with which we are concerned.

Example 3

Consider the functions graphed in Figures 2abc. In Figure 2a, the function is continuous and differentiable on $[a, b)$. In Figure 2b, the function is continuous on $[a, b]$, but differentiable only on $(a, b]$, since the slope of the graph approaches ∞ as $t \to a^+$. In Figure 2c, the function is continuous and differentiable on $[a, b]$. ■

Figure 2a

Figure 2b

Figure 2c

PROBLEMS

1. Given the "floor" (or greatest integer) function

$$\lfloor t \rfloor = n \text{ if } n \le t < n+1 \text{ and } n \text{ is an integer,}$$

evaluate each of the following.

(a) $\displaystyle\lim_{\substack{t\to 1\\ t\in[0,1]}} \lfloor t \rfloor$ **(b)** $\displaystyle\lim_{\substack{t\to 1\\ t\in[1,2]}} \lfloor t \rfloor$

(c) $\displaystyle\lim_{\substack{t\to -2\\ t\in[-3,-2]}} \lfloor t \rfloor$ **(d)** $\displaystyle\lim_{\substack{t\to -2\\ t\in[-2,-1]}} \lfloor t \rfloor$

2. Let $f(t) = |\sin t|$. Evaluate each of the following.

(a) $f'_{[-\pi,0]}(0)$ **(b)** $f'_{[0,\pi]}(0)$

(c) $f'_{[0,\pi]}(\pi)$ **(d)** $f'_{[\pi,2\pi]}(\pi)$

3. Let $f(t) = |1 - x^2|$. Evaluate each of the following.

(a) $f'_{(-2,-1]}(-1)$ **(b)** $f'_{[-1,1]}(-1)$

(c) $f'_{[0,1]}(1)$ **(d)** $f'_{[1,1.1)}(1)$

4. Let $f(t) = (-1)^{\lfloor t \rfloor}$. (See Problem 1 for the definition of $\lfloor t \rfloor$.) Describe the intervals on which f is differentiable.

5. Let $f(t) = \sqrt{1 - x^2}$.

(a) Show that f is differentiable (and therefore continuous) on $(-1,1)$.

(b) Show that f is continuous but not differentiable on $[-1,1]$.

6. Let f be the function whose graph is shown below. State whether or not f is continuous and whether or not f is differentiable on each of the following intervals.

(a) $(-\infty, a)$ **(b)** $(-\infty, a]$

(c) $(a, b]$ **(d)** $[a, b]$

(e) $[b, c]$ **(f)** $[c, d]$

(g) $[d, \infty)$ **(h)** (d, ∞)

7. Sketch the graph of a function f defined on $(-\infty, \infty)$ that is:

(a) differentiable on $(-\infty, 0]$, but not continuous at 1;

(b) continuous but not differentiable on $(0, 1]$;

(c) differentiable on $[1, 2]$;

(d) continuous on $[1, \infty)$,

(e) differentiable on $[2, 3]$ and $[3, \infty)$, but not differentiable on $[2, \infty)$.

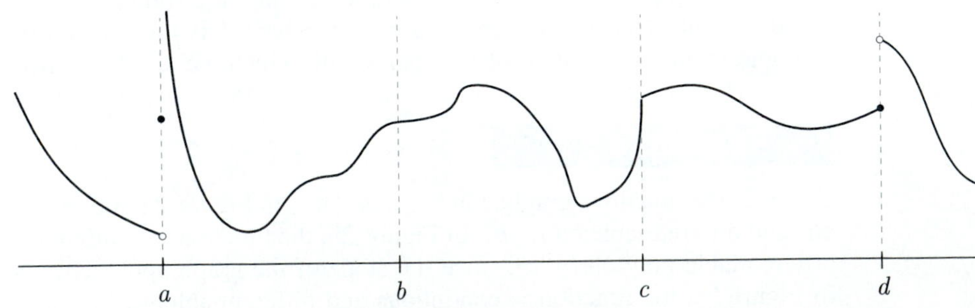

8. Let $f(t) = \begin{cases} e^{-1/t}, & \text{if } t > 0; \\ 0, & \text{if } t = 0. \end{cases}$

(a) Show that f is differentiable on $[0, \infty)$.

(b) Show that f' is continuous on $[0, \infty)$.

9. Let $f(t) = \begin{cases} t \sin \frac{1}{t}, & \text{if } t > 0; \\ 0, & \text{if } t = 0. \end{cases}$

(a) Show that f is differentiable on $[0, \infty)$.

(b) Show that f' is continuous on $(0, \infty)$ but not on $[0, \infty)$.

APPENDIX III

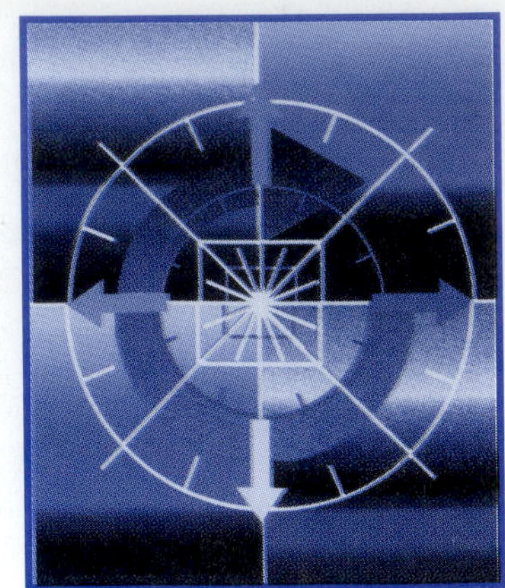

Local Existence and Uniqueness Proofs

This appendix is primarily devoted to the proof of the following local existence and uniqueness theorem. (See Sections 4.1 and 4.2 for definitions of local solution and unique local solution.)

Theorem 1

Suppose that f is continuous on an open region G with the property that, if (t, y_1) and (t, y_2) are in G, then the entire segment joining (t, y_1) and (t, y_2) is contained in G. Suppose further that there are constants K and M such that

i) $|f(t, y)| \le K$ *for all* (t, y) *in* G;
ii) $|f(t, y_1) - f(t, y_2)| \le M|y_1 - y_2|$ *for all* (t, y_1) *and* (t, y_2) *in* G.

Then for any (t_0, y_0) in G the initial-value problem

$$y' = f(t, y), \quad y(t_0) = y_0, \tag{1}$$

has a unique local solution. ▲

Theorem 1 is somewhat more general than the existence-uniqueness theorem stated in Section 4.2, which we restate for convenience here as follows.

Corollary 1 *Suppose that f and $\frac{\partial f}{\partial y}$ are continuous on some closed rectangle*

$$R = [a, b] \times [c, d] = \{(t, y) \mid a \le t \le b, c \le y \le d\}$$

that contains (t_0, y_0) in its interior; that is, $a < t_0 < b$ and $c < y_0 < d$. Then the initial-value problem

$$y' = f(t, y), \quad y(t_0) = y_0,$$

has a unique local solution, which can be uniquely continued for $t < t_0$ and $t > t_0$ at least until $(t, y(t))$ reaches a boundary point of R.

To see how Corollary 1 follows from Theorem 1, let us first suppose that the conditions of Corollary 1 are met. Next we let G be the interior of R. Then, since f and f_y are continuous on R, they are bounded on R and therefore bounded on G. Thus we have constants K and L such that

$$|f(t, y)| \leq K \quad \text{and} \quad |f_y(t, y)| \leq M \text{ for all } (t, y) \text{ in } G.$$

In particular, condition (i) of Theorem 1 is met. Now let (t, y_1) and (t, y_2) be any two points in G. Since the segment joining (t, y_1) and (t, y_2) is in G, the mean value theorem tells us that there is a point (t, y^*) on that segment such that

$$f(t, y_1) - f(t, y_2) = f_y(t, y^*)(y_1 - y_2).$$

Since (t, y^*) is in G, we have

$$|f_y(t, y^*)| \leq M,$$

and so

$$|f(t, y_1) - f(t, y_2)| \leq M|y_1 - y_2|.$$

Thus the conditions of Theorem 1 are met, guaranteeing the existence of a unique local solution. Finally, since the same local existence and uniqueness result is true for any initial point in the interior of R, there can be no point in in the interior of R through which the local solution through (t_0, y_0) cannot be continued.

We remark that condition (ii) of Theorem 1 is often called a *Lipschitz condition* for f on G, and the constant M is called a *Lipschitz constant.*

Before attempting the proof of Theorem 1, we first need to discuss a bit of fundamental background material concerning sequences of continuous functions on a closed interval. This material is usually covered in depth in a course in advanced calculus or real analysis. Here we will only present the definitions and theorems (some without proofs) that are pertinent to—and specialized for—our subsequent proof of Theorem 1. The interested student is encouraged to delve deeper into these and related concepts (and to fill in the gaps in our presentation) with the help of a text such as Goldberg [3] or Ross [4].

Sequences of Continuous Functions and Uniform Convergence Sequences of functions arise naturally in a variety of contexts. For instance, a function defined as power series,

$$\varphi(t) = \sum_{k=0}^{\infty} a_k t^k,$$

may be viewed as the limit of a sequence of polynomials:

$$\varphi_0(t) = a_0, \quad \varphi_1(t) = a_0 + a_1 t, \quad \varphi_2(t) = a_0 + a_1 t + a_2 t^2, \ldots .$$

In our present context of proving Theorem 1, we intend to construct a sequence of functions (by means of the *Picard interation* discussed in Section 4.1) that converges in an appropriate sense to a function that solves (1).

We will consider here only sequences of continuous functions defined on a closed, bounded interval $[a, b]$. Let $\varphi_0, \varphi_1, \varphi_2, \ldots$ be such a sequence. At each number t in $[a, b]$, the sequence of functions gives rise to an ordinary sequence of numbers: $\varphi_0(t), \varphi_1(t), \varphi_2(t), \ldots$. If this numerical sequence converges for each t in $[a, b]$, then the limiting values define a function φ on $[a, b]$, which we call the **pointwise limit** of $\varphi_0, \varphi_1, \varphi_2, \ldots$. This pointwise limit function is defined by the property that

$$\lim_{n \to \infty} \varphi_n(t) = \varphi(t) \text{ for each } t \text{ in } [a, b].$$

A pointwise limit function may fail to share certain crucial properties such as continuity and differentiability with the functions that comprise the terms of the sequence, as illustrated by the following example.

Example 1

Consider the sequence $\varphi_0, \varphi_1, \varphi_2, \ldots$ defined on $[0, 1]$ by

$$\varphi_n(t) = \frac{nt}{1 + nt} \qquad \text{for } n = 0, 1, 2, \ldots .$$

If $0 < t \le 1$, then

$$\lim_{n \to \infty} \frac{nt}{1 + nt} = 1;$$

while at $t = 0$,

$$\lim_{n \to \infty} \frac{nt}{1 + nt} = \lim_{n \to \infty} 0 = 0.$$

Thus, at each t in $[0, 1]$, we have

$$\lim_{n \to \infty} \varphi_n(t) = \varphi(t),$$

where

$$\varphi(t) = \begin{cases} 0, & \text{if } t = 0; \\ 1, & \text{if } 0 < t \le 1. \end{cases}$$

So, even though each φ_n is continuous, the pointwise limit φ is not. ■

Because of the difficulties illustrated by Example 1, we need to define a "stronger" notion of convergence for sequences of continuous functions.

Definition A sequence of continuous functions $\varphi_0, \varphi_1, \varphi_2, \ldots$ on $[a, b]$ is said to **converge uniformly** on $[a, b]$ to φ, if there is a sequence of real numbers $\varepsilon_0, \varepsilon_1, \varepsilon_2, \ldots$ such that

i) for each n, $|\varphi_n(t) - \varphi(t)| \le \varepsilon_n$ for all t in $[a, b]$;

ii) $\lim_{n \to \infty} \varepsilon_n = 0$. ◆

We point out that if $\varphi_0, \varphi_1, \varphi_2, \ldots$ converges uniformly on $[a, b]$ to φ, then φ must also be the pointwise limit of the sequence on $[a, b]$. Thus, the pointwise limit of a sequence (if it exists) is the exclusive candidate to which the sequence may converge uniformly. In particular, if a sequence does not converge pointwise, then it cannot converge uniformly.

Let us now revisit Example 1 and determine whether the sequence there converges uniformly. First, a simple computation reveals that

$$|\varphi_n(t) - \varphi(t)| = \begin{cases} 0, & \text{if } t = 0; \\ \frac{1}{1+nt}, & \text{if } 0 < t \le 1. \end{cases}$$

Next we observe that, for each n, $\lim_{t\to 0^+} |\varphi_n(t) - \varphi(t)| = 1$. Thus, for each n, there is a number t_n in $(0, 1]$ at which $|\varphi_n(t_n) - \varphi(t_n)| > 1/2$ (for instance). Therefore, no sequence of numbers satisfying both (i) and (ii) can exist. Thus $\varphi_0, \varphi_1, \varphi_2, \ldots$ does not converge uniformly.

To cast further light on the definition of uniform convergence, let us consider the situation where $|\varphi_n(t) - \varphi(t)|$ has a maximum value on $[a, b]$ for each n, which is guaranteed when φ is continuous. Then (i) is satisfied if we set

$$\varepsilon_n = \max_{a \le t \le b} |\varphi_n(t) - \varphi(t)|.$$

Thus we have the following theorem.

Theorem 2

Let $\varphi_0, \varphi_1, \varphi_2, \ldots$ be a sequence of continuous functions on a closed, bounded interval $[a, b]$, and suppose that φ is continuous on $[a, b]$. Then the sequence $\varphi_0, \varphi_1, \varphi_2, \ldots$ converges uniformly on $[a, b]$ to φ, if and only if

$$\lim_{n\to\infty} \max_{a \le t \le b} |\varphi_n(t) - \varphi(t)| = 0. \quad \blacktriangle$$

Example 2

Consider the sequence $\varphi_1, \varphi_2, \varphi_3, \ldots$ defined on $[0, 1]$ by

$$\varphi_n(t) = \cos(t/n) \text{ for } n = 1, 2, 3, \ldots .$$

It is easy to show that

$$\lim_{n\to\infty} \cos(t/n) = 1 \quad \text{for each } t \text{ in } [0, 1].$$

Thus, the pointwise limit of the sequence is the continuous function on $[0, 1]$ given by $\varphi(t) = 1$. It is also not difficult to show that, for each n,

$$\max_{0 \le t \le 1} |\cos(t/n) - 1| = |\cos(t/n) - 1|\Big|_{t=1} = 1 - \cos(1/n),$$

and so

$$\lim_{n\to\infty} \max_{0 \le t \le 1} |\cos(t/n) - 1| = \lim_{n\to\infty} (1 - \cos(1/n)) = 0.$$

Therefore, the sequence converges uniformly on $[0, 1]$ to φ. ■

Example 3

Consider the sequence $\varphi_0, \varphi_1, \varphi_2, \ldots$ defined on $[0, \pi]$ by

$$\varphi_n(t) = e^{-nt} \sin nt \quad \text{for } n = 0, 1, 2, \ldots .$$

It is not difficult to show that

$$\lim_{n\to\infty} e^{-nt}\sin nt = 0 \quad \text{for each } t \text{ in } [0,\pi].$$

Thus, the pointwise limit of the sequence is the zero function on $[0, \pi]$, which is continuous. However, a straightforward computation shows that

$$\max_{0\le t\le\pi} |e^{-nt}\sin nt - 0| = \frac{\sqrt{2}}{2}e^{-\pi/4} \quad \text{for all } n.$$

Thus

$$\lim_{n\to\infty}\max_{0\le t\le\pi} |e^{-nt}\sin nt - 0| \neq 0,$$

and so the sequence does not converge uniformly. ■

Example 4

Let $[a, b]$ be any bounded, closed interval containing 0, and define the sequence of polynomials $\varphi_0, \varphi_1, \varphi_2, \ldots$ on $[a, b]$ to be the usual Taylor polynomials for e^t (about $t = 0$):

$$\varphi_n(t) = \sum_{k=0}^{n}\frac{1}{k!}t^k, \quad n = 0, 1, 2, \ldots .$$

The MacLaurin series for e^t,

$$\sum_{k=0}^{\infty}\frac{1}{k!}t^k,$$

is by definition the pointwise limit of $\varphi_0, \varphi_1, \varphi_2, \ldots$, which exists for all t by the ratio test. According to Taylor's theorem, for each t and each $n = 0, 1, 2, \ldots$ there is a number τ between 0 and t such that

$$|\varphi_n(t) - e^t| = \frac{e^\tau}{(n+1)!}t^{n+1}.$$

So we have

$$\max_{a\le t\le b}|\varphi_n(t) - e^t| \le \frac{e^b}{(n+1)!}(b-a)^{n+1}.$$

Since the estimate on the right side approaches 0 as $n \to \infty$, we have

$$\lim_{n\to\infty}\max_{a\le t\le b}|\varphi_n(t) - e^t| = 0.$$

Therefore, the sequence of Taylor polynomials converges uniformly on $[a, b]$ to the function e^t. ■

Two important consequences of uniform convergence are stated in the following theorem, whose proof we omit.

Theorem 3

Let $\varphi_0, \varphi_1, \varphi_2, \ldots$ be a sequence of continuous functions defined on a closed, bounded interval $[a, b]$. If $\varphi_0, \varphi_1, \varphi_2, \ldots$ converges uniformly on $[a, b]$ to φ, then (i) φ is continuous, and

$$(ii) \qquad \lim_{n\to\infty} \int_a^b \varphi_n(t)\,dt = \int_a^b \varphi(t)\,dt. \qquad \blacktriangle$$

It is worth emphasizing that, according to conclusion (i) of Theorem 3, if a sequence converges pointwise to a discontinuous function, then the convergence is not uniform. For instance, the sequence in Example 1 does not converge uniformly, because its pointwise limit is not continuous. Another instance is the sequence t, t^2, t^3,... on [0, 1], which converges pointwise to

$$\varphi(t) = \begin{cases} 0, & \text{if } 0 \le t < 1; \\ 1, & \text{if } t = 1. \end{cases}$$

(Be careful to note that, as Example 3 indicates, a sequence of continuous functions on $[a, b]$ can converge pointwise to a continuous function without converging uniformly.)

Problem 12 at the end of this section provides an example showing that conclusion (ii) of Theorem 3 can fail without uniform convergence; that is, if $\varphi_0, \varphi_1, \varphi_2, \ldots$ converges pointwise on $[a, b]$ but not uniformly, then it is possible that

$$\lim_{n\to\infty} \int_a^b \varphi_n(t)\,dt \neq \int_a^b \lim_{n\to\infty} \varphi_n(t)\,dt,$$

even if the pointwise limit is continuous.

Two More Important Theorems Our next two theorems, as well as Theorem 3, will be crucial in the proof of Theorem 1. The first of these is concerned with sequences of certain compositions of continuous functions, namely, sequences $\psi_0, \psi_1, \psi_2, \ldots$ that arise as

$$\psi_n(t) = f(t, \varphi_n(t)), \quad \text{for } a \le t \le b, \quad n = 0, 1, 2, \ldots,$$

where $\varphi_0, \varphi_1, \varphi_2, \ldots$ is a given sequence of continuous functions on $[a, b]$ and f is a continuous function of two variables. In particular, we need to know what conditions on f will allow us to conclude that $\psi_0, \psi_1, \psi_2, \ldots$ converges uniformly, if $\varphi_0, \varphi_1, \varphi_2, \ldots$ converges uniformly. So suppose that $\varphi_0, \varphi_1, \varphi_2, \ldots$ converges uniformly on $[a, b]$ to φ. The main issue then is how to ensure that

$$\lim_{n\to\infty} \max_{a\le t\le b} |f(t, \varphi_n(t)) - f(t, \varphi(t))| = 0, \quad \text{if } \lim_{n\to\infty} \max_{a\le t\le b} |\varphi_n(t) - \varphi(t)| = 0.$$

One way is to require that there be a constant M such that

$$|f(t, y_1) - f(t, y_2)| \le M|y_1 - y_2|$$

for all t in $[a, b]$ and all numbers y_1 and y_2 within the ranges of $\varphi_0, \varphi_1, \varphi_2, \ldots$. From this we could conclude that

$$\max_{a \le t \le b} |f(t, \varphi_n(t)) - f(t, \varphi(t))| \le M \max_{a \le t \le b} |\varphi_n(t) - \varphi(t)|,$$

and so the left side would converge to 0 as $n \to \infty$, provided that the right side did. Though a more general condition could be stated, we will use this one. All this we state formally as follows.

Theorem 4

Let $\varphi_0, \varphi_1, \varphi_2, \ldots$ be a sequence of continuous functions that converges uniformly on $[a, b]$ to φ, and suppose that c and d are numbers such that

$$c \le \varphi_n(t) \le d \quad \textit{for all } n \textit{ and all } t \textit{ in } [a, b].$$

Suppose further that f is a continuous function of two variables defined on the rectangle

$$\mathcal{R} = \{(t, y) \mid a \le t \le b \textit{ and } c \le y \le d\},$$

and let the sequence $\psi_0, \psi_1, \psi_2, \ldots$ be defined on $[a, b]$ by

$$\psi_n(t) = f(t, \varphi_n(t)), \quad n = 0, 1, 2, \ldots .$$

If there is a constant $M \ge 0$ such that

$$|f(t, y_1) - f(t, y_2)| \le M|y_1 - y_2| \quad \textit{for all } t \in [a, b] \textit{ and } y_1, y_2 \in [c, d],$$

then $\psi_0, \psi_1, \psi_2, \ldots$ converges uniformly on $[a, b]$ to ψ, where

$$\psi(t) = f(t, \varphi(t)) \quad \textit{for } a \le t \le b. \qquad \blacktriangle$$

The next theorem provides a way for us to establish uniform convergence based solely on the difference between consecutive terms of the sequence—without knowing the limit in advance. It is a special case of the *Banach contraction mapping theorem*.

Theorem 5

Suppose that $\varphi_0, \varphi_1, \varphi_2, \ldots$ is a sequence of continuous functions defined on a closed interval $[a, b]$. If there is a number γ in $[0, 1)$ such that

$$\max_{a \le t \le b} |\varphi_{n+1}(t) - \varphi_n(t)| \le \gamma \max_{a \le t \le b} |\varphi_n(t) - \varphi_{n-1}(t)| \quad \textit{for } n = 1, 2, 3, \ldots ,$$

then there is a continuous function φ on $[a, b]$ to which $\varphi_1, \varphi_2, \varphi_3, \ldots$ converges uniformly. Moreover, we have the error estimate

$$\max_{a \le t \le b} |\varphi_n(t) - \varphi(t)| \le \frac{\gamma^n}{1 - \gamma} \max_{a \le t \le b} |\varphi_1(t) - \varphi_0(t)| \quad \textit{for } n = 1, 2, 3, \ldots . \qquad \blacktriangle$$

We omit the complete proof of Theorem 5; however, the subject of Problem 13 is a proof of the error estimate and uniform convergence under the assumption that a pointwise limit exists.

Before we move on to the proof of Theorem 1, let's use a simple specific case to illustrate how we will use Theorem 5 to prove local existence.

Example 5

Let k be any real number, and consider the initial-value problem

$$y' = ky, \quad y(0) = 1. \tag{2}$$

(And pretend that we don't already know the solution!) By integrating each side of the differential equation formally from 0 to t, we arrive at an equivalent integral equation

$$y(t) = 1 + k \int_0^t y(s)\,ds. \tag{3}$$

Our plan is to find a solution of (3) by generating a sequence of functions that converges uniformly to it. So let the sequence $\varphi_0, \varphi_1, \varphi_2, \ldots$ be generated recursively according to

$$\varphi_{n+1}(t) = 1 + k \int_0^t \varphi_n(s)\,ds \qquad \text{for } n = 0, 1, 2, \ldots \tag{4}$$

starting with, say, $\varphi_0(t) = 1$. (In this simple case, it is easy to see that each member of the sequence is actually a polynomial and thus defined for all t.) Our first goal is to use Theorem 5 to show that the sequence converges uniformly on some interval $[-r, r]$, $r > 0$. We begin by letting $n \geq 1$ and noting that

$$\begin{aligned}\varphi_{n+1}(t) - \varphi_n(t) &= \left(1 + k \int_0^t \varphi_n(s)\,ds\right) - \left(1 + k \int_0^t \varphi_{n-1}(s)\,ds\right) \\ &= k \int_0^t (\varphi_n(s) - \varphi_{n-1}(s))\,ds\end{aligned}$$

for any t. Thus, for any $r > 0$,

$$\begin{aligned}\max_{-r\leq t\leq r} |\varphi_{n+1}(t) - \varphi_n(t)| &= \max_{-r\leq t\leq r} \left| k \int_0^t (\varphi_n(s) - \varphi_{n-1}(s))\,ds \right| \\ &\leq |k| \max_{-r\leq t\leq r} \left| \int_0^t |\varphi_n(s) - \varphi_{n-1}(s)|\,ds \right|. \end{aligned} \tag{5}$$

Now, since

$$\left| \int_0^t |\varphi_n(s) - \varphi_{n-1}(s)| ds \right| \leq |t| \max_{s\in J} |\varphi_n(s) - \varphi_{n-1}(s))|,$$

where J is the closed interval with endpoints 0 and t, it follows that, for any t in $[-r, r]$,

$$\left| \int_0^t |\varphi_n(s) - \varphi_{n-1}(s)| ds \right| \leq |t| \max_{-r\leq s\leq r} |\varphi_n(s) - \varphi_{n-1}(s)|.$$

So from (5) we find that, for any $r > 0$,

$$\begin{aligned}\max_{-r\le t\le r} |\phi_{n+1}(t) - \varphi_n(t)| &\le |k| \max_{-r\le t\le r}\left(|t| \max_{-r\le s\le r} |\varphi_n(s) - \varphi_{n-1}(s)|\right)\\ &\le |k| \max_{-r\le t\le r} |t| \max_{-r\le t\le r} |\varphi_n(t) - \varphi_{n-1}(t)|\\ &\le |k|\, r \max_{-r\le t\le r} |\varphi_n(t) - \varphi_{n-1}(t)|.\end{aligned}$$

Now, if we choose r small enough so that $|k|\, r < 1$, then Theorem 4 tells us that $\varphi_0, \varphi_1, \varphi_2, \ldots$ converges uniformly on $[-r, r]$ to some continuous function φ. By Theorem 3 (part (ii)), we conclude that

$$\lim_{n\to\infty} \int_0^t \varphi_n(s)\, ds = \int_0^t \varphi(s)\, ds.$$

Now, by taking the limit as $n \to \infty$ on each side of (4), we conclude that φ satisfies the integral equation (3); that is,

$$\varphi(t) = 1 + k \int_0^t \varphi(s)\, ds \qquad \text{for } t \text{ in } [-r, r].$$

Therefore, by the Fundamental Theorem of Calculus, φ is differentiable with

$$\varphi' = k\varphi \text{ on } [-r, r].$$

Also, since $\varphi_n(0) = 1$ for all n, it follows that $\varphi(0) = 1$. Finally, we conclude that $\varphi_0, \varphi_1, \varphi_2, \ldots$ converges uniformly on $[-r, r]$ to a function that satisfies the initial value problem (2). In particular, (2) *has* a local solution (which of course coincides with $y = e^{kt}$). ■

The Proof of Theorem 1 Suppose that all of the conditions of Theorem 1 are satisfied. Since (1) may be recast as the integral equation

$$y(t) = y_0 + \int_{t_0}^t f(s, y(s))\, ds,$$

we define a sequence of functions $\varphi_0, \varphi_1, \varphi_2, \ldots$ recursively by

$$\varphi_0(t) = y_0, \text{ and } \varphi_{n+1}(t) = y_0 + \int_{t_0}^t f(s, \varphi_n(s))\, ds \qquad \text{for } n = 0, 1, 2, \ldots.$$

(This is the *Picard iteration* discussed in Section 4.1.) Now let $r > 0$ be small enough so that $rM < 1$ and the rectangle

$$\mathcal{R} = \{(t, y) \mid |t - t_0| \le r \quad \text{and } |y - y_0| \le rK\}$$

is contained in G. Then, for $t_0 - r \le t \le t_0 + r$ and all n,

$$|\varphi_{n+1}(t) - y_0| \le \int_{t_0}^t |f(s, \varphi_n(s))|\, ds \le rK$$

by assumption (i) of the theorem. Thus for all n the points $(t, \varphi_n(t))$ remain in the rectangle $\mathcal{R}$ as long as t remains in $[t_0 - r, t_0 + r]$. Now let $n \geq 1$, and note that

$$\begin{aligned}\varphi_{n+1}(t) - \varphi_n(t) &= \left(y_0 + \int_{t_0}^{t} f(s, \varphi_n(s))\, ds\right) - \left(y_0 + \int_{t_0}^{t} f(s, \varphi_{n-1}(s))\, ds\right) \\ &= \int_{t_0}^{t} (f(s, \varphi_n(s)) - f(s, \varphi_{n-1}(s)))\, ds\end{aligned}$$

for all t in $[t_0 - r, t_0 + r]$. Thus we have

$$\begin{aligned}|\varphi_{n+1}(t) - \varphi_n(t)| &\leq \left|\int_{t_0}^{t} |f(s, \varphi_n(s)) - f(s, \varphi_{n-1}(s))|\, ds\right| \\ &\leq M\left|\int_{t_0}^{t} |\varphi_n(s) - \varphi_{n-1}(s)|\, ds\right|\end{aligned}$$

for all t in $[t_0 - r, t_0 + r]$ by assumption (ii) of the theorem. Consequently,

$$\begin{aligned}\max_{|t-t_0|\leq r} |\varphi_{n+1}(t) - \varphi_n(t)| &\leq \max_{|t-t_0|\leq r} M\left|\int_{t_0}^{t} |\varphi_n(s) - \varphi_{n-1}(s)|\, ds\right| \\ &\leq Mr \max_{|t-t_0|\leq r} |\varphi_n(t) - \varphi_{n-1}(t)|.\end{aligned}$$

Since $Mr < 1$, Theorem 5 guarantees that the sequence $\varphi_0, \varphi_1, \varphi_2, \ldots$ converges uniformly on $[t_0 - r, t_0 + r]$ to a continuous function φ. Now we invoke Theorem 4. Because of assumption (ii) of the theorem, the sequence $\psi_0, \psi_1, \psi_2, \ldots$, where

$$\psi_n(t) = f(t, \varphi_n(t)),$$

converges uniformly. So, by Theorem 3,

$$\lim_{n\to\infty} \int_{t_0}^{t} f(s, \varphi_n(s))\, ds = \int_{t_0}^{t} f(s, \varphi(s))\, ds \quad \text{for } t \text{ in } [t_0 - r, t_0 + r].$$

By letting $n \to \infty$ on each side of

$$\varphi_{n+1}(t) = y_0 + \int_{t_0}^{t} f(s, \varphi_n(s))\, ds,$$

we can now conclude that, for t in $[t_0 - r, t_0 + r]$,

$$\varphi(t) = y_0 + \int_{t_0}^{t} f(s, \varphi(s))\, ds.$$

Therefore, by the Fundamental Theorem of Calculus, φ is differentiable with

$$\varphi' = f(t, \varphi) \text{ on } [t_0 - r, t_0 + r].$$

Also, since $\varphi_n(0) = y_0$ for all n, it follows that $\varphi(0) = y_0$. Finally, we conclude that $\varphi_0, \varphi_1, \varphi_2, \ldots$ converges uniformly on $[t_0 - r, t_0 + r]$ to a function that satisfies the initial value problem (1). In particular, (1) *has* a local solution.

In order to prove that this solution is unique, suppose that there were another solution ψ on $[t_0 - r, t_0 + r]$. Then, for t in $[t_0 - r, t_0 + r]$,

$$\varphi(t) = y_0 + \int_{t_0}^{t} f(s, \varphi(s))\, ds \quad \text{and} \quad \psi(t) = y_0 + \int_{t_0}^{t} f(s, \psi(s))\, ds,$$

from which we get

$$\varphi(t) - \psi(t) = \int_{t_0}^{t} (f(s, \varphi(s)) - f(s, \psi(s)))\, ds$$

and then

$$\begin{aligned} |\varphi(t) - \psi(t)| &\le \left| \int_{t_0}^{t} |f(s, \varphi(s)) - f(s, \psi(s))|\, ds \right| \\ &\le M \left| \int_{t_0}^{t} |\varphi(s) - \psi(s)|\, ds \right|. \end{aligned}$$

Consequently,

$$\begin{aligned} \max_{|t-t_0| \le r} |\varphi(t) - \psi(t)| &\le M \max_{|t-t_0| \le r} \left| \int_{t_0}^{t} |\varphi(s) - \psi(s)|\, ds \right| \\ &\le M r \max_{|t-t_0| \le r} |\varphi(t) - \psi(t)|. \end{aligned}$$

Since $Mr < 1$, this implies that

$$\max_{|t-t_0| \le r} |\varphi(t) - \psi(t)| = 0,$$

from which we conclude that $\varphi = \psi$ on $[t_0 - r, t_0 + r]$. Therefore, (1) has a unique solution on $[t_0 - r, t_0 + r]$.

Systems of Equations Theorem 1 has a relatively straightforward extension to systems. We will consider systems of the form

$$\begin{aligned} y_1' &= f_1(t, y_1, y_2, \dots, y_n), \\ y_2' &= f_2(t, y_1, y_2, \dots, y_n), \\ &\vdots \\ y_n' &= f_n(t, y_1, y_2, \dots, y_n), \end{aligned}$$

with specified initial values $y_1(t_0) = y_{0_1}, y_2(t_0) = y_{0_2}, \dots, y_n(t_0) = y_{0_n}$. Each of the functions f_i is a function of $n + 1$ variables and assumed to be continuous on an appropriately chosen region G in $\mathbb{R}^{n+1}$ containing the initial point

$$(t_0, y_{0_1}, y_{0_2}, \dots, y_{0_n}).$$

In order to simplify notation, we define the following symbols:

$$\mathbf{y} = \begin{pmatrix} y_1 \\ y_2 \\ \vdots \\ y_n \end{pmatrix}, \quad (t, \mathbf{y}) = \begin{pmatrix} t \\ y_1 \\ y_2 \\ \vdots \\ y_n \end{pmatrix}, \quad \mathbf{y}_0 = \begin{pmatrix} y_{0_1} \\ y_{0_2} \\ \vdots \\ y_{0_n} \end{pmatrix},$$

$$f_i(t, \mathbf{y}) = f_i(t, y_1, y_2, \ldots, y_n), \quad \text{and} \quad \mathbf{f}(t, \mathbf{y}) = \begin{pmatrix} f_1(t, \mathbf{y}) \\ f_2(t, \mathbf{y}) \\ \vdots \\ f_n(t, \mathbf{y}) \end{pmatrix}.$$

Then the initial-value problem can be expressed compactly as

$$\mathbf{y}' = \mathbf{f}(t, \mathbf{y}), \quad \mathbf{y}(t_0) = \mathbf{y}_0. \tag{6}$$

We assume that each component f_i of $\mathbf{f}$ is continuous on an open set G that contains $(t, \mathbf{y}_0)$ and has the property that, if $(t, \mathbf{y}_1)$ and $(t, \mathbf{y}_2)$ are in G, then the entire segment joining those points,

$$\{(t, \mathbf{y}) \mid \mathbf{y} = \mathbf{y}_1 + \alpha(\mathbf{y}_2 - \mathbf{y}_1), \quad 0 \leq \alpha \leq 1\},$$

is contained in G. One simple type of set that satisfies this condition is an open, $n + 1$-dimensional "box" centered at the initial point defined by

$$G = \left\{ (t, \mathbf{y}) \middle| |t - t_0| + \sum_{i=1}^{n} |y_i - y_{0_i}| < \rho \right\} \text{ for some } \rho > 0.$$

With relatively straightforward technical modifications to the definitions and theorems of this section, the details of which we will not present, the following extension of Theorem 1 can be proven.

Theorem 6

Let G be an open region in $\mathbb{R}^{n+1}$ with the property described previously. Suppose that each f_i is continuous on G and that there are constants K and M such that

i) $\sum_{i=1}^{n} |f_i(t, \mathbf{y})| \leq K$ *for all $(t, \mathbf{y})$ in G;*

ii) $\sum_{i=1}^{n} |f_i(t, \mathbf{u}) - f_i(t, \mathbf{v})| \leq M \sum_{i=1}^{n} |u_i - v_i|$ *for all $(t, \mathbf{u})$ and $(t, \mathbf{v})$ in G.*

Then for any $(t_0, \mathbf{y}_0)$ in G the initial-value problem (6) has a unique local solution. ▲

The following corollary reflects the fact that condition (ii) in Theorem 6 is met, if we require each f_i to have bounded partial derivatives with respect to all variables except t. From this corollary follows local existence and uniqueness for the systems studied in Chapter 9.

COROLLARY 2 *Suppose that each f_i and each partial derivative $\partial f_i/\partial y_j$ is continuous on a closed box*

$$\mathcal{R} = \{(t, \mathbf{y}) \mid a \le t \le b \quad \textit{and} \quad c_i \le y_i \le d_i,\ i = 1, 2, \ldots, n\},$$

where $a < b$ and $c_i < d_i$, $i = 1, 2, \ldots, n$. Then for any initial point $(t_0, \mathbf{y}_0)$ in $\mathcal{R}$ the initial-value problem (6) has a unique local solution, which can be uniquely continued for $t < t_0$ and $t > t_0$ at least until $(t, y(t))$ reaches a boundary point of $\mathcal{R}$.

We now conclude with an application of Corollary 2 to linear systems

$$\mathbf{y}' = A(t)\mathbf{y} + \mathbf{g}(t), \quad \mathbf{y}(t_0) = \mathbf{y}_0, \tag{7}$$

where A is a $n \times n$ matrix-valued function of t with entries a_{ij}, and $\mathbf{g}$ is a vector-valued function with entries g_i. From this corollary follows Theorem 1 in Section 8.1.

COROLLARY 3 *Suppose that each entry of A and each entry of* $\mathbf{g}$ *is continuous on an open interval I. Then for any t_0 in I the initial-value problem (7) has a unique solution on I.*

To see how Corollary 3 follows from Corollary 2, first notice that (7) is a special case of (6) with

$$f_i(t, \mathbf{y}) = \sum_{j=1}^{n} a_{ij}(t)y_i + g_i(t) \quad \text{and} \quad \frac{\partial f_i}{\partial y_j} = a_{ij}(t).$$

Then let $(t_0, \mathbf{y}_0)$ be any initial point with t_0 in I, and let

$$\mathcal{R} = \{(t, \mathbf{y}) \mid a \le t \le b \quad \text{and} \quad c_i \le y_i \le d_i, \quad i = 1, 2, \ldots, n\},$$

where a and b are any numbers in I with $a < t_0 < b$, and where the numbers c_i and d_i satisfy $c_i < y_{0_i} < d_i$. The continuity of each a_{ij} and g_i on I implies that each f_i and each $\frac{\partial f_i}{\partial y_j}$ is continuous on $\mathcal{R}$. So, by Corollary 2, (7) has a solution that can be uniquely continued for $t < t_0$ and $t > t_0$ at least until $(t, y(t))$ reaches a boundary point of $\mathcal{R}$, that is, a point at which either $t = a$, $t = b$, $y_i(t) = c_i$ for some i, or $y_i(t) = d_i$ for some i. Since all of the numbers c_i and d_i may be as large (in absolute value) as we like, we conclude that the solution can be uniquely continued for $t < t_0$ until either $t = a$ or $\lim_{t \to T^+} |y_i(t)| = \infty$ for some i and some T in $[a, t_0)$. Similarly, the solution can be uniquely continued for $t > t_0$ until either $t = b$ or else $\lim_{t \to T^-} |y_i(t)| = \infty$ for some i and some T in $(t_0, b]$. However, it can be shown (we omit the proof) that, each component of the solution of (7) must be bounded on any subinterval of $[a, b]$ on which the solution exists. Therefore, the solution can be continued uniquely throughout the interval $[a, b]$. Finally, since a and b can be any numbers in I with $a < t_0 < b$, we conclude that a unique solution exists on all of I.

PROBLEMS

For each of the functions in Problems 1–3, find constants K and M such that

$$|f(t, y)| \le K \text{ and } |f(t, y_1) - f(t, y_2)| \le M|y_1 - y_2|$$

for all (t, y), (t, y_1), and (t, y_2) in the indicated rectangle.

1. $f(t, y) = \sin ty$, $0 \le t \le 2$, $0 \le y \le \pi$
2. $f(t, y) = ty^3$, $-1 \le t \le 1$, $0 \le y \le 1$
3. $f(t, y) = ye^{-y}$, $-1 \le t \le 1$, $0 \le y \le 3$
4. $f(t, y) = y^2 - ty + t$, $0 \le t \le 4$, $0 \le y \le 2$
5. Suppose that f is continuous on $[a, b]$ and that f is differentiable at all but a finite number of points $c_1, c_2, \ldots, c_\ell$ in $[a, b]$. Suppose further that $|f'(y)| \le M$ for all y in $[a, b]$ where f is differentiable. In this problem you will show that

$$|f(y_1) - f(y_2)| \le M|y_1 - y_2|$$
$$\text{for all } y_1, y_2 \text{ in } [a, b].$$

To begin, let u, v be in $[a, b]$ with $u < v$, and then define

$$g(y) = f(y) - f(u) + M(y - u)$$
$$\text{and } h(y) = f(y) - f(u) - M(y - u)$$

for y in $[u, v]$. (Note that g and h are continuous on $[u, v]$ and differentiable wherever f is.)

(a) Show that $g(u) = 0$ and $g'(y) = f'(y) + M \ge 0$ for all y in $[u, v]$, where f is differentiable. Conclude that g is nondecreasing on $[u, v]$ and therefore $g(v) \ge 0$; that is, $f(v) - f(u) \ge -M(v - u)$. (Why is continuity important here?)

(b) Show that $h(u) = 0$ and $h'(y) = f'(y) - M \le 0$ for all y in $[u, v]$, where f is differentiable. Conclude that h is nonincreasing on $[u, v]$ and therefore $h(v) \le 0$; that is, $f(v) - f(u) \le M(v - u)$.

(c) Put the results of parts (a) and (b) together to obtain

$$-M(v - u) \le f(v) - f(u) \le M(v - u).$$

From this conclude that

$$|f(v) - f(u)| \le M|v - u|.$$

For each of the functions in Problems 6–8, use the result of Problem 5 to find a constant M such that

$$|f(y_1) - f(y_2)| \le M|y_1 - y_2|$$

for all y_1 and y_2 in the indicated interval.

6. $f(y) = 2y + |y|$, $-1 \le y \le 1$
7. $f(y) = e^{|y|}$, $-1 \le y \le 1$
8. $f(y) = |ye^{-y}|$, $-1 \le y \le 1$
9. Let $f(y) = \sqrt[3]{y}$. Show that there is no constant M for which

$$|f(y_1) - f(y_2)| \le M|y_1 - y_2|$$
$$\text{for all } y_1, y_2 \text{ in } [-1, 1].$$

10. Define the sequence $\varphi_0, \varphi_1, \varphi_2, \ldots$ on $[0, 1]$ by

$$\varphi_n(t) = \frac{1}{n+1} t^{n+1}, \quad n = 0, 1, 2 \ldots.$$

(a) Show that $\varphi_0, \varphi_1, \varphi_2, \ldots$ converges uniformly to the zero function on $[0, 1]$.

(b) Show that the sequence of derivatives $\varphi_0', \varphi_1', \varphi_2', \ldots$ converges pointwise to a discontinuous function on $[0, 1]$.

11. Define the sequence $\varphi_0, \varphi_1, \varphi_2, \ldots$ on $[0, 1]$ by

$$\varphi_n(t) = nt^2 e^{-nt}, \ n = 1, 2, 3 \ldots.$$

(a) Show that $\varphi_0, \varphi_1, \varphi_2, \ldots$ converges uniformly to the zero function on $[0, 1]$.

(b) Show that the sequence of derivatives $\varphi_0', \varphi_1', \varphi_2', \ldots$ also converges pointwise to the zero function on $[0, 1]$, but not uniformly.

12. Define the sequence $\varphi_0, \varphi_1, \varphi_2, \ldots$ on $[0, 1]$ by

$$\varphi_n(t) = n^2 t e^{-nt}, \ n = 1, 2, 3 \ldots.$$

(a) Show that $\varphi_0, \varphi_1, \varphi_2, \ldots$ converges pointwise to the zero function on $[0, 1]$, but not uniformly.

(b) Show that

$$\lim_{n\to\infty} \int_0^1 \varphi_n(t)dt \ne \int_0^1 \lim_{n\to\infty} \varphi_n(t)dt.$$

13. Let the sequence $\varphi_0, \varphi_1, \varphi_2, \ldots$ satisfy the hypotheses of Theorem 5, and assume that the sequence converges pointwise to φ on $[a, b]$.

(a) Let $m > n \geq 0$, and let $a \leq t \leq b$. Show that

$$\varphi_m(t) - \varphi_n(t) = \sum_{k=n}^{m-1} (\varphi_{k+1}(t) - \varphi_k(t)),$$

and apply the triangle inequality to obtain the estimate

$$|\varphi_m(t) - \varphi_n(t)| \leq \sum_{k=n}^{m-1} |\varphi_{k+1}(t) - \varphi_k(t)|$$

$$\leq \sum_{k=n}^{m-1} \max_{[a,b]} |\varphi_{k+1} - \varphi_k|.$$

(b) Show that, for $k \geq n$,

$$\max_{[a,b]} |\varphi_{k+1} - \varphi_k| \leq \gamma^{k-n} \max_{[a,b]} |\varphi_{n+1} - \varphi_n|.$$

(c) Show that

$$|\varphi_m(t) - \varphi_n(t)| \leq (1 + \gamma + \gamma^2 + \cdots + \gamma^{m-n-1}) \times \max_{[a,b]} |\varphi_{n+1} - \varphi_n|.$$

(d) Let $n \to \infty$ to show that

$$|\varphi(t) - \varphi_n(t)| \leq \frac{1}{1-\gamma} \max_{[a,b]} |\varphi_{n+1} - \varphi_n|.$$

(e) Show that

$$\max_{[a,b]} |\varphi_{n+1} - \varphi_n| \leq \gamma^n \max_{[a,b]} |\varphi_1 - \varphi_0|.$$

Conclude that

$$|\varphi(t) - \varphi_n(t)| \leq \frac{\gamma^n}{1-\gamma} \max_{[a,b]} |\varphi_1 - \varphi_0|$$

and therefore

$$\max_{[a,b]} |\varphi - \varphi_n| \leq \frac{\gamma^n}{1-\gamma} \max_{[a,b]} |\varphi_1 - \varphi_0|.$$

Finally, observe that

$$\lim_{n\to\infty} \max_{[a,b]} |\varphi - \varphi_n| = 0.$$

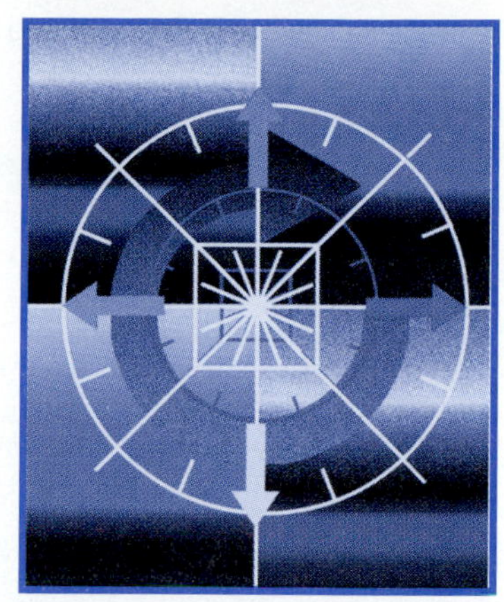

APPENDIX IV

Linear Differential Equations of Arbitrary Order

In this appendix we will describe the basic structure of the solutions of linear differential equations of arbitrary order. The main theme is that everything we have learned about the solutions of second-order linear equations carries over in a fairly straightforward way to higher-order linear equations. We will find out that, in a manner completely analogous to what was seen in Section 5.4 for second-order equations, the general solution of a homogeneous, nth-order differential equation may be constructed in a simple way from n linearly independent particular solutions. For equations with constant coefficients, we will see that the methods of Chapter 6 involving the characteristic polynomial generalize in a rather simple way.

A. Solutions and Linear Independence

Recall from Section 1.3 that a linear nth-order differential equation is one that can be expressed in the form

$$\tilde{p}_n y^{(n)} + \tilde{p}_{n-1} y^{(n-1)} + \cdots + \tilde{p}_1 y' + \tilde{p}_0 y = \tilde{f},$$

where $\tilde{p}_0, \tilde{p}_1, \ldots, \tilde{p}_n$, and $\tilde{f}$ are given functions of the independent variable, which we will assume is t. We will assume also until further notice that $\tilde{p}_0, \tilde{p}_1, \ldots, \tilde{p}_n$ are continuous on some interval I and that $\tilde{p}_n(t) \neq 0$ for all t in I, so that we may divide by $\tilde{p}_n$ to obtain an equation of the form

$$y^{(n)} + p_{n-1} y^{(n-1)} + \cdots + p_1 y' + p_0 y = f, \tag{1}$$

where $p_0, p_1, \ldots, p_{n-1}$ are continuous on I. Note that this equation can be written in operator form as

$$P(\mathcal{D})y = f,$$

where P is defined by

$$P(r) = r^n + p_{n-1}r^{n-1} + \cdots + p_1 r + p_0.$$

This operator $P(\mathcal{D})$ is a linear operator by Theorem 1 in Section 5.3. Thus the principle of superposition gives the following important facts:

i. If $\tilde{y}$ is a particular solution of (1), then $y+\tilde{y}$ is also a solution of (1), where y is any solution of the corresponding homogeneous equation

$$y^{(n)} + p_{n-1}y^{(n-1)} + \cdots + p_0 y = 0. \tag{2}$$

ii. If u and v are any two solutions of (2), then so is $c_1 u + c_2 v$ for any pair of constants c_1, c_2.

We saw in Chapter 5 that, for homogeneous linear second-order equations, every solution could be expressed as a linear combination of any two linearly independent particular solutions. So it is plausible that for homogeneous nth-order linear equations, every solution would be expressible using n particular solutions, say u_1, $u_2, \ldots, u_n$, as $y = c_1u_1 + c_2u_2 + \cdots + c_nu_n$, where $c_1, c_2, \ldots, c_n$ are constants; that is, as a linear combination of $u_1, u_2, \ldots, u_n$. This is indeed true, provided we impose a *linear independence* condition upon $u_1, u_2, \ldots, u_n$.

Definition $u_1, u_2, \ldots, u_n$ are *linearly dependent* on an interval I if there are constants $c_1, c_2, \ldots, c_n$, not all zero, such that $c_1u_1(t)+c_2u_2(t)+\cdots+c_nu_n(t) = 0$ for all t in I. If no such constants exist, then $u_1, u_2, \ldots, u_n$ are *linearly independent* on I. ◆

An alternative way of defining linear independence is to say that $u_1, u_2, \ldots, u_n$ are linearly independent provided that

$$c_1u_1(t) + c_2u_2(t) + \cdots + c_nu_n(t) = 0 \text{ on } I \text{ only if } c_1 = c_2 = \cdots = c_n = 0.$$

In other words, the zero function cannot be obtained as a linear combination of u_1, $u_2, \ldots, u_n$ except in the trivial manner.

The main result in this section is the following theorem, which characterizes the general solutions of (1) and (2).

Theorem 1

Suppose that $u_1, u_2, \ldots, u_n$ are n linearly independent solutions of (2) on the interval I. Then

i) every solution of (2) on I is given by $y = c_1u_1 + c_2u_2 + \cdots + c_nu_n$ for some choice of constants $c_1, c_2, \ldots, c_n$.

If, in addition, $\tilde{y}$ is any particular solution of (1), then

ii) every solution of (1) on I is given by $y = c_1u_1 + c_2u_2 + \cdots + c_nu_n + \tilde{y}$ for some choice of constants $c_1, c_2, \ldots, c_n$. Furthermore, these constants can be chosen so that y satisfies arbitrary initial values $y(t_0) = y_0$, $y'(t_0) = y_1, \ldots, y^{n-1}(t_0) = y_{n-1}$ at any point t_0 in I. ▲

The Wronskian Suppose that we are interested in solving an initial value problem of the form

$$\begin{aligned} &y^{(n)} + p_{n-1}y^{(n-1)} + \cdots + p_1 y' + p_0 y = f \\ &y(t_0) = y_0, \quad y'(t_0) = y_1, \ldots, y^{(n-1)}(t_0) = y_{n-1}. \end{aligned} \tag{3}$$

Further suppose that we have n particular solutions $u_1, u_2, \ldots, u_n$ of (2) and a particular solution $\tilde{y}$ of (1). We would then use superposition to write $y = c_1u_1(t) + c_2u_2(t) + \cdots + c_nu_n(t) + \tilde{y}$ and attempt to solve for the constants $c_1, c_2, \ldots, c_n$. These constants are the solution of the following system of linear algebraic equations:

$$\begin{aligned} c_1u_1(t_0) + c_2u_2(t_0) + \cdots + c_nu_n(t_0) &= y_0 - \tilde{y}(t_0) \\ c_1u_1'(t_0) + c_2u_2'(t_0) + \cdots + c_nu_n'(t_0) &= y_1 - \tilde{y}'(t_0) \\ &\vdots \\ c_1u_1^{(n-1)}(t_0) + c_2u_2^{(n-1)}(t_0) + \cdots + c_nu_n^{(n-1)}(t_0) &= y_{n-1} - \tilde{y}^{(n-1)}(t_0). \end{aligned}$$

This system can be solved for a unique vector of constants $(c_1, c_2, \ldots, c_n)^T$ precisely when the matrix

$$M(t) = \begin{pmatrix} u_1(t) & u_2(t) & \cdots & u_n(t) \\ u_1'(t) & u_2'(t) & \cdots & u_n'(t) \\ & & \vdots & \\ u_1^{(n-1)}(t) & u_2^{(n-1)}(t) & \cdots & u_n^{(n-1)}(t) \end{pmatrix} \tag{4}$$

has a nonzero determinant at $t = t_0$. (See Appendix I.)

The **Wronskian** of n functions $u_1, u_2, \ldots, u_n$ is defined as

$$W[u_1, u_2, \ldots, u_n](t) = \det M(t),$$

where $M(t)$ is defined by (4). Just as in the second-order case, there is a very close connection between the Wronskian and the concept of linear independence. Suppose that $u_1, u_2, \ldots, u_n$ are linearly dependent, differentiable functions on an interval I; that is, $c_1u_1(t) + c_2u_2(t) + \cdots + c_nu_n(t) = 0$ for all t in I, where the coefficients c_i are not all zero. Then, for some k between 1 and n, $c_k \neq 0$ and

$$u_k(t) = \frac{1}{c_k}\left(\sum_{i \neq k} c_iu_i(t)\right) \quad \text{for all } t \in I.$$

Therefore, for each $j = 0, 1, 2, 3, \ldots,$

$$u_k^{(j)}(t) = \frac{1}{c_k}\left(\sum_{i \neq k} c_iu_i^{(j)}(t)\right) \quad \text{for all } t \in I.$$

This implies that the kth column in the matrix $M(t)$ in (4) is a linear combination of the others for all t in I, and therefore $\det M(t) = 0$ on I. That is, if $u_1, u_2, \ldots, u_n$ are $(n-1)$-times differentiable and linearly dependent on I, then $W[u_1, u_2, \ldots, u_n](t)$ is identically zero on I. We have essentially proven half of the following theorem.

Theorem 2

Let $u_1, u_2, \ldots, u_n$ be $(n-1)$-times differentiable on an interval I.

i) If $u_1, u_2, \ldots, u_n$ are linearly dependent on I, then $W[u_1, u_2, \ldots, u_n](t) = 0$ on I.

ii) If $W[u_1, u_2, \ldots, u_n](t) = 0$ on I and one of the u_i is never zero on I, then $u_1, u_2, \ldots, u_n$ are linearly dependent on I. ▲

If $u_1, u_2, \ldots, u_n$ are solutions of a homogeneous, nth-order, linear differential equation, then we can say even more about the Wronskian of $u_1, u_2, \ldots, u_n$. Suppose that $u_1, u_2, \ldots, u_n$ are any n solutions of (2) on an interval I and that there is some point t_0 in I where $W[u_1, u_2, \ldots, u_n](t_0) = 0$. Since $W[u_1, u_2, \ldots, u_n](t_0)$ is the determinant of the matrix $M(t_0)$, where $M(t)$ is as in (4), linear algebra tells us that the matrix equation

$$M(t_0)c = 0,$$

where $c = (c_1, c_2, \ldots, c_n)^T$, has a nontrivial solution. Now, with *this* choice of $c_1, c_2, \ldots, c_n$, the function $y = c_1u_1 + c_2u_2 + \cdots + c_nu_n$ satisfies the initial value problem

$$\begin{aligned} &y^{(n)} + p_{n-1}y^{(n-1)} + \cdots + p_1y' + p_0y = 0 \\ &y(t_0) = y'(t_0) = \ldots y^{(n-1)}(t_0) = 0. \end{aligned} \tag{5}$$

Uniqueness of solutions now tells us that $y(t) = 0$ for all t in I. Now, since $y(t) = c_1u_1(t) + c_2u_2(t) + \cdots + c_nu_n(t) = 0$ for all t in I, we see that $u_1, u_2, \ldots, u_n$ are linearly dependent on I, and consequently $W[u_1, u_2, \ldots, u_n](t) = 0$ for *all* t in I, because of Theorem 2. All of this is summarized in the following theorem.

Theorem 3

If $u_1, u_2, \ldots, u_n$ are any n solutions of (2) on an interval I, and if there is some point t_0 in I where $W[u_1, u_2, \ldots, u_n](t_0) = 0$, then $u_1, u_2, \ldots, u_n$ are linearly dependent on I, and consequently $W[u_1, u_2, \ldots, u_n](t) = 0$ for all t in I. ▲

The following is an easy corollary to Theorem 3.

COROLLARY 1 *If $u_1, u_2, \ldots, u_n$ are any n solutions of (2) on an interval I, then either $W[u_1, u_2, \ldots, u_n](t) = 0$ for all t in I, or else $W[u_1, u_2, \ldots, u_n](t) \neq 0$ for all t in I.*

Finally, we are in a position to prove Theorem 1. So suppose that $u_1, u_2, \ldots, u_n$ are linearly independent solutions of (2) on an interval I, and let $W(t)$ be the Wronskian of $u_1, u_2, \ldots, u_n$. We want to prove that every solution of (2) on I is given by $y = c_1u_1 + c_2u_2 + \cdots + c_nu_n$ for some choice of constants $c_1, c_2, \ldots, c_n$. So suppose that φ is any solution whatever of (2), and pick any point t_0 in I. Because $W(t_0) \neq 0$, we can choose $c = (c_1, c_2, \ldots, c_n)^T$ so that

$$M(t_0)c = \Phi_0,$$

where $\Phi_0 = (\varphi(t_0), \varphi'(t_0), \ldots, \varphi^{(n-1)}(t_0))^T$. With this choice of $c_1, c_2, \ldots, c_n$, we see that

$$\sum_{i=1}^{n} c_i u_i^{(j)}(t_0) = \varphi^{(j)}(t_0) \quad \text{for } j = 0, 1, 2, \ldots, n-1.$$

This means that we have chosen $c_1, c_2, \ldots, c_n$ so that both φ and $c_1u_1 + c_2u_2 + \cdots + c_nu_n$ satisfy the same initial conditions at t_0. Therefore, by uniqueness, we conclude that φ and $c_1u_1 + c_2u_2 + \cdots + c_nu_n$ must be identical on I. This proves that *every* solution of (2) on I can be written as $c_1u_1 + c_2u_2 + \cdots + c_nu_n$.

To prove the second assertion in Theorem 1, suppose again that $u_1, u_2, \ldots, u_n$ are linearly independent solutions of (2) on an interval I and suppose further that $\tilde{y}$ is a particular solution of (1) on I. Let w be *any* solution of (1) on I. Note that $w - \tilde{y}$ satisfies (2). Therefore, the previous argument shows that we can pick $c_1, c_2, \ldots, c_n$ so that $w - \tilde{y} = c_1u_1 + c_2u_2 + \cdots + c_nu_n$ on I; that is, $w = c_1u_1 + c_2u_2 + \cdots + c_nu_n + \tilde{y}$ on I. Thus *every* solution of (1) on I can be written as $c_1u_1 + c_2u_2 + \cdots + c_nu_n + \tilde{y}$. The argument above that motivated the definition of the Wronskian shows that if $W(t) \neq 0$ on I, then $c_1, c_2, \ldots, c_n$ can be chosen so that $y = c_1u_1 + c_2u_2 + \cdots + c_nu_n + \tilde{y}$ satisfies any given initial conditions at any t_0 in I. This completes the proof.

PROBLEMS

In each of Problems 1–3,

(a) use the Wronskian to show that the given functions are linearly independent on any interval I;

(b) show that each of the given functions satisfies the given differential equation;

(c) find the solution of the differential equation that satisfies the specified initial conditions.

1. e^t, te^t, t^2e^t; $(\mathcal{D} - 1)^3 y = 0$; $y(0) = 1$, $y'(0) = y''(0) = 0$

2. $\cos t, \sin t, t\cos t, t\sin t$; $(\mathcal{D}^2 + 1)^2 y = 0$; $y(0) = 1$, $y'(0) = y''(0) = y'''(0) = 0$

3. $e^t, e^{-t}, \cos t, \sin t$; $(\mathcal{D}^4 - 1)y = 0$; $y(0) = 1$, $y'(0) = y''(0) = y'''(0) = 0$

4. Suppose that $u_1, u_2, \ldots, u_n$ satisfy

$$\left.\begin{aligned} &P(\mathcal{D})u_i = 0, && u_i^{(i-1)}(t_0) = 1, \\ &u_i^{(j)}(t_0) = 0, && j \neq i-1,\ j = 1, \ldots, n \end{aligned}\right\} \text{ for } i = 1, 2, \ldots, n,$$

where $P(\mathcal{D})$ is the linear operator arising from (1) and (2). Check that

$$y = y_0u_1 + y_1u_2 + \cdots + y_{n-1}u_n$$

solves the initial value problem (3). We call such functions $u_1, u_2, \ldots, u_n$ the **fundamental solutions** of (2).

5. Suppose that u and v satisfy, respectively,

$$P(\mathcal{D})u = 0, \quad u(t_0) = y_0, \quad u'(t_0) = y_1, \ldots, \quad u^{(n-1)}(t_0) = y_{n-1};$$
$$P(\mathcal{D})v = f, \quad v(t_0) = v'(t_0) = \cdots = v^{(n-1)}(t_0) = 0.$$

where $P(\mathcal{D})$ is the linear operator arising from (1) and (2). Check that $y = u + v$ solves

$$P(\mathcal{D})y = f, \quad y(t_0) = y_0, \quad y'(t_0) = y_1, \dots, \quad y^{(n-1)}(t_0) = y_{n-1}.$$

The function v here, which satisfies zero initial conditions, is called the **rest solution** of $P(\mathcal{D})y = f$ (with respect to the initial point t_0).

The first-order system that arises from a linear nth-order equation can always be expressed in the form

$$\mathbf{y}' = A\mathbf{x} + \mathbf{b},$$

where $\mathbf{y} = (y, y', \dots, y^{(n-1)})^T$, A is an $n \times n$ matrix, and $\mathbf{b}$ is a column vector. If the equation is not autonomous, then some entries in A and/or $\mathbf{b}$ will depend on t. Express each of the equations in 6–10 in this form. Finally, comment on the general form of the matrix A and the vector $\mathbf{b}$ obtained in this way.

6. $y^{(4)} + y'' + y = 0$

7. $y''' + ty' + 2y = t$

8. $y^{(4)} + y = t^3$

9. $y'' + py' + qy = f$

10. $y^{(n)} + p_{n-1}y^{(n-1)} + \dots + p_0 y = f$

B. Homogeneous Equations with Constant Coefficients

Consider a homogeneous nth-order linear equation of the form in the previous section, which we redisplay here for convenience:

$$y^{(n)} + p_{n-1}y^{(n-1)} + \dots + p_1 y' + p_0 y = 0, \tag{1}$$

where $p_{n-1}, p_{n-2}, \dots, p_0$ are constants. We will look for solutions of the form e^{rt} where r is a (possibly complex) constant. Putting $y = e^{rt}$ into the equation yields

$$r^n e^{rt} + p_{n-1}r^{n-1}e^{rt} + \dots + p_1 r e^{rt} + p_0 e^{rt} = 0.$$

Dividing through by e^{rt} produces an nth-degree equation for r:

$$r^n + p_{n-1}r^{n-1} + \dots + p_1 r + p_0 = 0. \tag{2}$$

This equation is called the **characteristic equation** corresponding to the differential equation (1). The point is that if r satisfies the characteristic equation (2), then e^{rt} satisfies the differential equation (1).

Operator Form and the Characteristic Polynomial Notice that we can rewrite equation (1) in operator form as

$$(\mathcal{D}^n + p_{n-1}\mathcal{D}^{n-1} + \dots + p_1\mathcal{D} + p_0\mathcal{I})y = 0$$

or even more succinctly as

$$P(\mathcal{D})y = 0, \text{ where } P(r) = r^n + p_{n-1}r^{n-1} + \dots + p_1 r + p_0.$$

This polynomial P is the **characteristic polynomial** of the operator $P(\mathcal{D})$.

It is easy to see that when P is a polynomial of any degree, these four statements are equivalent:

i. a is a solution of $P(r) = 0$;
ii. $r - a$ is a factor of $P(r)$;
iii. $\mathcal{D} - a\mathcal{I}$ is a factor of the operator $P(\mathcal{D})$;
iv. e^{at} is a solution of $P(\mathcal{D})y = 0$.

A corollary to the *fundamental theorem of algebra* is that every polynomial of degree n with real (or complex) coefficients has precisely n complex (possibly real) roots—if each is counted according to its *multiplicity.* A root r_i of a polynomial P is said to have multiplicity m if m is the largest number such that $(r-a)^m$ is a factor of $P(r)$. Thus, for example, a sixth degree polynomial such as $P(r) = r^2(r-1)^3(r-2)$ has precisely six roots: 0, 0, 1, 1, 1, and 2, when they are listed according to their multiplicities. We are about to see that roots of P with multiplicity m always give rise to m linearly independent solutions of $P(\mathcal{D})y = 0$. Therefore, by knowing all n roots of P, we get n solutions of $P(\mathcal{D})y = 0$, which can be shown to be linearly independent and therefore can be used to assemble the general solution of $P(\mathcal{D})y = 0$.

Real Roots Any real root a of (2) gives rise to a solution e^{at} of (1). Moreover, if $r_1, r_2, \ldots, r_n$ are distinct real roots of (2), then $e^{r_1 t}, e^{r_2 t}, \ldots, e^{r_n t}$ are linearly independent solutions of (1), and so the general solution of (1) is given by

$$y = c_1 e^{r_1 t} + c_2 e^{r_2 t} + \cdots + c_n e^{r_n t}.$$

Complex Roots Just as in the second-order case, a pair of nonreal, complex conjugate roots $\alpha_k \pm \beta_k i$ of (2) gives rise to a pair of linearly independent solutions of (1) of the form

$$e^{\alpha_k t} \cos \beta_k t \text{ and } e^{\alpha_k t} \sin \beta_k t.$$

Repeated Roots If a root a of (2) has multiplicity m (that is, if m is the largest power such that $(r - a)^m$ is a factor of $P(r)$), then

$$e^{at},\ te^{at},\ t^2 e^{at}, \ldots, t^{m-1} e^{at}$$

are m linearly independent solutions of (1). This can be shown very easily by exponential shift. When a pair of nonreal, complex conjugate roots $\alpha \pm \beta i$ have multiplicity m, we get $2m$ linearly independent solutions of the form

$$\begin{array}{llll} e^{\alpha t} \cos \beta t, & te^{\alpha t} \cos \beta t, & t^2 e^{\alpha t} \cos \beta t, \ldots, & t^{m-1} e^{\alpha t} \cos \beta t, \\ e^{\alpha t} \sin \beta t, & te^{\alpha t} \sin \beta t, & t^2 e^{\alpha t} \sin \beta t, \ldots, & t^{m-1} e^{\alpha t} \sin \beta. \end{array}$$

The following examples illustrate how all this is put together to construct general solutions.

Example 1

Consider the third-order equation

$$y''' - 2y' - 4y = 0.$$

The characteristic polynomial is $r^3 - 2r - 4$. Observing that 2 is a root, we are led to the factorization $(r-2)(r^2+2r+2)$. The other roots are then $-1 \pm i$. Therefore, the general solution is

$$y = c_1 e^{2t} + e^{-t}(c_2 \cos t + c_3 \sin t).$$ ■

Example 2

Consider the seventh-order equation

$$(\mathcal{D}^2 + \mathcal{I})^2(\mathcal{D} + 2\mathcal{I})^3 y = 0.$$

The seven roots of the characteristic polynomial are $\pm i$, $\pm i$, -2, -2, -2. Thus the general solution is

$$y = (c_1 + c_2 t)\cos t + (c_3 + c_4 t)\sin t + (c_5 + c_6 t + c_7 t^2)e^{-2t}.$$ ■

PROBLEMS

In Problems 11 and 12, write down the general solution by inspection.

11. $(\mathcal{D} + \mathcal{I})(\mathcal{D} + 2\mathcal{I})^2(\mathcal{D} + 3\mathcal{I})^3 y = 0$

12. $(\mathcal{D} + \mathcal{I})^2(\mathcal{D}^2 + \mathcal{I})^3 y = 0$

In Problems 13–18, find the general solution.

13. $y''' + y'' + 4y' + 4y = 0$

14. $y''' + y'' - y' - y = 0$

15. $y''' + 3y'' + 3y' + y = 0$

16. $y^{(4)} + 2y'' + y = 0$

17. $y^{(4)} - k^4 y = 0$

18. $y^{(4)} + 4y = 0$

19. Assume the fact that $(\mathcal{D} - a\mathcal{I})ue^{kt} = e^{kt}(\mathcal{D} + (k-a)\mathcal{I})u$ is true with a, k any real or complex constants. Show that **exponential shift**

$$P(\mathcal{D})ue^{kt} = e^{kt}P(\mathcal{D} + k\mathcal{I})u$$

is valid for any polynomial P. (*Hint*: Write $P(\mathcal{D})$ as a product of linear factors.)

20. Suppose that r_k is a root of the polynomial P of multiplicity m, so that $P(r) = Q(r)(r - r_k)^m$, where Q is a polynomial of degree $n - m$ with $Q(r_k) \neq 0$.

(a) Use exponential shift to show that $u(t)e^{r_k t}$ is a solution of $P(\mathcal{D})y = 0$ for any function u with $u^{(m)} = 0$.

(b) Conclude that $e^{r_k t}, te^{r_k t}, t^2 e^{r_k t}, \ldots, t^{m-1}e^{r_k t}$ are m solutions of $P(\mathcal{D})y = 0$. Also show that these solutions are linearly independent.

(c) Suppose that $r_k = \alpha_k + \beta_k i$ with $\beta_k \neq 0$. Show that r_k gives rise to the $2m$ linearly independent, real solutions:

$$t^j e^{\alpha_k t}\cos\beta t,\ t^j e^{\alpha_k t}\sin\beta t; \quad j = 0, 1, \ldots, m-1.$$

(Note that $\overline{r}_k = \alpha_k - \beta_k i$ is also a root of $P(r)$ of multiplicity m. However, $\overline{r}_k$ gives rise to precisely the same set of solutions.)

21. This problem is for students who have studied linear algebra and are familiar with the notion of the eigenvalues of a matrix.

(a) Consider the operator $P(\mathcal{D})$, where $P(r) = r^4 + p_3 r^3 + p_2 r^2 + p_1 r + p_0$ with constants $p_3, \ldots, p_0$. Express the equation $P(\mathcal{D})y = 0$ as a first-order system in matrix form

$$\mathbf{y}' = A\mathbf{y},$$

and show that $P(r) = \det(A - r\mathcal{I})$. Conclude that the roots of the polynomial $P(r)$ are precisely the eigenvalues of the matrix A.

(b) Repeat part (a) for the general case $P(r) = r^n + p_{n-1}r^{n-1} + \cdots + p_1 r + p_0$ with constants $p_{n-1}, \ldots, p_0$.

C. Nonhomogeneous Equations

This section is essentially about a connection between the multiplicity of a root a of the polynomial P and the form of a particular solution of the nonhomogeneous equation

$$P(\mathcal{D})y = p(t)e^{at}, \tag{1}$$

where p is a polynomial. To allow simpler statements of the following results, let it be understood that a number a is said to be a root of P of multiplicity $m = 0$ if and only if $P(a) \neq 0$.

First we consider the case where $p(t) = 1$ and state the aforementioned connection as follows.

Proposition 1

Suppose that $p(t) = 1$ and that a is a (not necessarily real) root of the polynomial P with multiplicity $m \geq 0$. Then (1) has the particular solution

$$y = \frac{1}{m!q_0} t^m e^{at},$$

in which q_0 is the constant term in the polynomial $Q(r + a)$, where

$$Q(r)(r - a)^m = P(r). \quad \blacktriangle$$

Proof If a is a root of P with multiplicity m, we may write $P(r) = Q(r)(r - a)^m$, where Q is a polynomial of degree $n - m$ with $Q(r_k) \neq 0$. Thus (1) can be rewritten in the form

$$Q(\mathcal{D})(\mathcal{D} - a\mathcal{I})^m y = e^{at}.$$

We will look for a solution of the form $y = A(t)e^{at}$. Using exponential shift, we find that

$$Q(\mathcal{D} + a\mathcal{I})\mathcal{D}^m A(t) = 1. \tag{2}$$

Let q_j be the coefficient of r^j in $Q(r + a)$, so that $q_j \mathcal{D}^j$ is the jth-order term in $Q(\mathcal{D} + a\mathcal{I})$. Then (2) can be rewritten as

$$\left(q_0\mathcal{I} + q_1\mathcal{D} + q_2\mathcal{D}^2 + \cdots + q_{n-m}\mathcal{D}^{n-m}\right)\mathcal{D}^m A(t) = 1.$$

Since $Q(a) \neq 0$, it follows that $q_0 \neq 0$. (Why?) Thus, in order to find a particular solution, we may simply consider the equation

$$q_0 \mathcal{D}^m A(t) = 1.$$

Repeated antidifferentiation reveals that a particular solution of (2) is $A(t) = t^m/(m!q_0)$. The conclusion of the theorem now follows. ●

Example 1

Consider the equation

$$(\mathcal{D} - 3\mathcal{I})^3(\mathcal{D} - \mathcal{I})^2 y = e^t.$$

Substituting $y = A(t)e^t$ gives

$$(\mathcal{D} - 2\mathcal{I})^3\mathcal{D}^2 A(t) = 1.$$

A particular solution of this is obtained from

$$-8\mathcal{D}^2 A(t) = 1,$$

which yields $A(t) = -t^2/16$. Thus we have $y = -t^2e^t/16$. ■

The following corollary specializes Proposition 1 to the case in which a has nonzero imaginary part.

COROLLARY 2 *Suppose that $\alpha \pm \beta i$ are roots of the polynomial P with multiplicity $m \geq 0$. Then the equations*

$$P(\mathcal{D})x = e^{\alpha t}\cos\beta t, \quad P(\mathcal{D})y = e^{\alpha t}\sin\beta t, \tag{3}$$

have particular solutions of the form

$$x = t^m e^{\alpha t}(a\cos\beta t + b\sin\beta t),$$
$$y = t^m e^{\alpha t}(a\sin\beta t - b\cos\beta t),$$

where a and b are real constants.

Proof We will assemble a solution of (3) in the usual way from the real and imaginary parts of a solution of the complex equation

$$P(\mathcal{D})z = e^{(\alpha+\beta i)t},$$

From Theorem 1, we have a complex solution

$$z = \frac{1}{q_0}t^m e^{(\alpha+\beta i)t} = \frac{1}{q_0}t^m e^{\alpha t}(\cos\beta t + i\sin\beta t)$$

with q_0 the (complex) constant term in $Q(r + \alpha + \beta i)$, where $Q(r)(r - \alpha - \beta i)^m = P(r)$. Thus

$$x = \operatorname{Re} z \quad \text{and} \quad y = \operatorname{Im} z$$

satisfy (3). Computation shows that if $q_0 = \gamma + \sigma i$, then x and y have the forms asserted, where the constants a and b are $\gamma/(\gamma^2 + \sigma^2)$ and $\sigma/(\gamma^2 + \sigma^2)$, respectively. ●

Example 2

Consider the equation

$$(\mathcal{D}^2 + 2\mathcal{D} + 2\mathcal{I})^3 y = e^{-t}(3\cos t - 5\sin t).$$

The roots of P are $-1 \pm i$, each with multiplicity 3. So the complex equation of interest here is

$$(\mathcal{D} + (1+i)\mathcal{I})^3(\mathcal{D} + (1-i)\mathcal{I})^3 z = e^{(-1+i)t}.$$

Substituting $z = A(t)e^{(-1+i)t}$ and using exponential shift gives

$$(\mathcal{D} + 2i\mathcal{I})^3\mathcal{D}^3 A(t) = 1.$$

The zeroth-order term in $(\mathcal{D} + 2i\mathcal{I})^3$ is $-8i$. Therefore, we can obtain a solution from

$$-8i\mathcal{D}^3 A(t) = 1.$$

Antidifferentiation reveals that $A(t) = it^3/48$ is a solution. The real and imaginary parts of $z = it^3 e^{(-1+i)t}/48$ are, respectively,

$$-t^3 e^{-t}\sin t/48 \text{ and } t^3 e^{-t}\cos t/48;$$

so we obtain the real solution

$$y = -t^3 e^{-t}(5\cos t + 3\sin t)/48.$$ ■

We now consider the general case. First we have the following lemma.

Lemma 1

Suppose that $P(0) \neq 0$ and p is a polynomial. Then the equation

$$P(\mathcal{D})y = p(t)$$

has a particular solution $\tilde{p}$, where $\tilde{p}$ is a polynomial of the same degree as p. ▲

Proof First consider the case where $P(\mathcal{D}) = \mathcal{D} - a\mathcal{I}$, $a \neq 0$, and let

$$p(t) = p_n t^n + p_{n-1}t^{n-1} + \cdots + p_1 t + p_0.$$

Substituting

$$y = \tilde{p}(t) = \tilde{p}_n t^n + \tilde{p}_{n-1}t^{n-1} + \cdots + \tilde{p}_1 t + \tilde{p}_0$$

into $(\mathcal{D} - a\mathcal{I})y = p(t)$ and equating coefficients result in

$$-a\tilde{p}_n = p_n, n\tilde{p}_n - a\tilde{p}_{n-1} = p_{n-1}, \ldots, \tilde{p}_1 - a\tilde{p}_0 = p_0.$$

Since $a \neq 0$, this system of equations has a (unique) solution $\tilde{p}_n, \tilde{p}_{n-1}, \ldots, \tilde{p}_0$, from which we get a degree-n polynomial solution $\tilde{p}$ of the differential equation. Thus the result is true when degree $(P) = 1$. We now proceed by induction on the degree of P. Suppose that the result is true if degree $(P) = 1, 2, \dots, k$, and consider the case where degree $(P) = k + 1$. Now we write

$$P(r) = (r - a)Q(r)$$

where Q is of degree k, and we look for a solution of

$$(\mathcal{D} - a\mathcal{I})Q(\mathcal{D})y = p(t).$$

With the substitution $v = Q(\mathcal{D})y$ this becomes

$$(\mathcal{D} - a\mathcal{I})v = p(t),$$

which we have already established has a polynomial solution $v = \tilde{p}$ of the same degree as p. Now, since $v = Q(\mathcal{D})y$, we have the equation

$$Q(\mathcal{D})y = \tilde{p}.$$

But this has a polynomial solution $y = \hat{p}$ of the same degree as $\tilde{p}$ by our induction hypothesis, since degree$(Q) = k$. Therefore, the assertion of the lemma follows by induction. ●

Theorem 1

Suppose that a is a (not necessarily real) root of the polynomial P with multiplicity $m \geq 0$. Then (1) has a particular solution

$$y = t^m \tilde{p}(t)e^{at},$$

where $\tilde{p}$ is a polynomial of the same degree as p. ▲

Proof First, we write $P(r) = Q(r)(r - a)^m$, where degree $(Q) = n - m$ and $Q(r_k) \neq 0$. Thus (1) can be rewritten in the form

$$Q(\mathcal{D})(\mathcal{D} - a\mathcal{I})^m y = p(t)e^{at}.$$

We will look for a solution of the form $y = A(t)e^{at}$. Using exponential shift, the equation becomes

$$Q(\mathcal{D} + a\mathcal{I})\mathcal{D}^m A(t) = p(t),$$

and then after substituting $v = \mathcal{D}^m A(t)$,

$$Q(\mathcal{D} + a\mathcal{I})v = p(t).$$

Lemma 1 tells us that this equation has a polynomial solution $\hat{p}$ of the same degree as p. Now, since $v = \mathcal{D}^m A(t)$, we need a solution to

$$\mathcal{D}^m A(t) = \hat{p}(t).$$

Repeated integration reveals that a particular solution is $A(t) = t^m \tilde{p}(t)$, where $\tilde{p}$ is a polynomial of the same degree as p, thus yielding $y = t^m \tilde{p}(t)e^{-at}$ as a solution of the differential equation (1). ●

We now have the following corollary, whose proof we shall leave to the reader.

COROLLARY 3 *Suppose that $\alpha \pm \beta i$ are roots of the polynomial P with multiplicity $m \geq 0$. Then the equations*

$$P(\mathcal{D})x = p(t)e^{\alpha t}\cos\beta t, \quad P(\mathcal{D})y = p(t)e^{\alpha t}\sin\beta t,$$

where p is a polynomial, have particular solutions

$$\begin{aligned} x &= t^m e^{\alpha t}(Re(\tilde{p}(t))\cos\beta t - Im(\tilde{p}(t))\sin\beta t), \\ y &= t^m e^{\alpha t}(Im(\tilde{p}(t))\cos\beta t + Re(\tilde{p}(t))\sin\beta t), \end{aligned}$$

where $\tilde{p}$ is a (complex) polynomial with the same degree as p.

Example 3

Consider the equation

$$(\mathcal{D} - \mathcal{I})(\mathcal{D} + \mathcal{I})^3 y = 120(t^2 + 1)e^{-t}.$$

According to Theorem 1, there is a solution of the form

$$y = t^3(at^2 + bt + c)e^{-t}.$$

Substituting this into the equation and applying exponential shift result in

$$(\mathcal{D} - 2\mathcal{I})\mathcal{D}^3 t^3(at^2 + bt + c) = 120(t^2 + 1).$$

Computation of the left side gives

$$-120at^2 + (120a - 48b)t + 24b - 12c = 120(t^2 + 1).$$

This produces the coefficients

$$a = -1, \quad b = -5/2, \quad c = -5,$$

which in turn give the solution

$$y = -\tfrac{1}{2}t^3(2t^2 + 5t + 10)e^{-t}.$$ ■

Example 4

Consider the equations

$$(\mathcal{D}^2 + 2\mathcal{D} + 2\mathcal{I})^2 x = 24te^{-t}\cos t \quad \text{and} \quad (\mathcal{D}^2 + 2\mathcal{D} + 2\mathcal{I})^2 y = 24te^{-t}\sin t.$$

The complex equation of interest is

$$(\mathcal{D} + (1+i)\mathcal{I})^2(\mathcal{D} + (1-i)\mathcal{I})^2 z = 24te^{(-1+i)t}.$$

Theorem 1 says that there is a solution of the form

$$z = t^2(at + b)e^{(-1+i)t}.$$

Substituting this into the equation and applying exponential shift, we obtain

$$(\mathcal{D} + 2i\mathcal{I})^2\mathcal{D}^2 t^2(at + b) = 24t.$$

Computation of the left side gives

$$24ai - 8b - 24at = 24t.$$

This produces the coefficients $a = -1$ and $b = -3i$, which in turn give the complex solution

$$z = -t^2(t - 3i)e^{(-1+i)t} = -t^2(t - 3i)e^{-t}(\cos t + i \sin t)$$
$$= -t^2 e^{-t}\,(t \cos t + 3 \sin t + i(-3 \cos t + t \sin t))\,.$$

Thus we have the real solutions

$$x = -t^2 e^{-t}(t \cos t + 3 \sin t) \quad \text{and} \quad y = -t^2 e^{-t}(-3 \cos t + t \sin t),$$

which agrees with the form indicated by Corollary 3. ■

PROBLEMS

For each of the equations in Problems 22–25, give the appropriate form for a trial particular solution.

22. $(\mathcal{D} + \mathcal{I})(\mathcal{D}^2 + 4\mathcal{I})^2 y = te^{-t}$

23. $(\mathcal{D} + \mathcal{I})(\mathcal{D}^2 + 4\mathcal{I})^2 y = (t - 3)e^{-2t}$

24. $(\mathcal{D} - 2\mathcal{I})^3(\mathcal{D} - \mathcal{I})y = t^2 e^{2t}$

25. $(\mathcal{D}^2 + 4\mathcal{D} + 4\mathcal{I})^2 y = (2t^2 - 5)e^{-2t}$

For each of the equations in Problems 26–29, give the corresponding complex equation and an appropriate form for a trial particular solution.

26. $(\mathcal{D} + \mathcal{I})(\mathcal{D}^2 + 4\mathcal{I})^2 y = t \cos 2t$

27. $(\mathcal{D} + \mathcal{I})(\mathcal{D}^2 + 4\mathcal{I})^2 y = (t - 3)e^{-t} \sin 2t$

28. $(\mathcal{D} - 2\mathcal{I})^3(\mathcal{D} - \mathcal{I})y = t^2 \cos t$

29. $(\mathcal{D}^2 + 4\mathcal{D} + 13\mathcal{I})^2 y = (2t^2 - 5)e^{-2t} \sin 3t$

In Problems 30 and 31, find a particular solution of the equation.

30. $(\mathcal{D} + \mathcal{I})^2(\mathcal{D} - \mathcal{I})y = 2(1 - 6t)e^{-t}$

31. $(\mathcal{D} + 2\mathcal{I})^3 y = 6(3 - 4t)e^{-2t}$

In Problems 32 and 33, find a particular solution of each of the pair of equations.

32. $(\mathcal{D}^2 + \mathcal{I})^2 x = 24(t + 1) \cos t,$
$(\mathcal{D}^2 + \mathcal{I})^2 y = 24(t + 1) \sin t$

33. $(\mathcal{D}^2 + 2\mathcal{D} + 2\mathcal{I})^2 x = 24(t + 1)e^{-t} \cos t,$
$(\mathcal{D}^2 + 2\mathcal{D} + 2\mathcal{I})^2 y = 24(t + 1)e^{-t} \sin t$

In Problems 34 and 35, find a particular solution of the equation.

34. $(\mathcal{D}^2 + \mathcal{I})^2 y = 24(t \cos t - \sin t)$

35. $(\mathcal{D}^2 + 2\mathcal{D} + 2\mathcal{I})^2 y = 8e^{-t}(2 \cos t - 3t \sin t)$

36. Show that if 0 is a root of P with multiplicity $m \geq 1$ and p is a polynomial, then the equation $P(\mathcal{D})y = p(t)$ has a particular solution $t^m \tilde{p}(t)$, where $\tilde{p}$ is a polynomial of the same degree as p.

BIBLIOGRAPHY

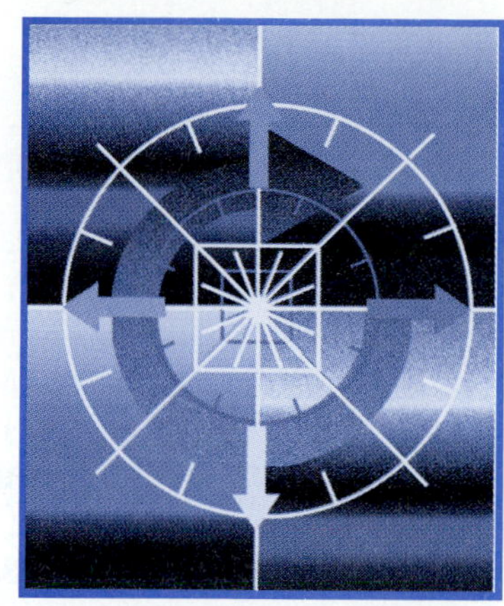

Advanced Calculus and Elementary Analysis

1. Berberian, S. K., *A First Course in Real Analysis*. New York: Springer-Verlag, 1994.
2. Buck, R. C., *Advanced Calculus* (3rd ed.). New York: McGraw-Hill, 1978.
3. Goldberg, R. R., *Methods of Real Analysis* (2nd ed.). New York: John Wiley, 1976.
4. Ross, K., *Elementary Analysis: The Theory of Calculus*. New York: Springer-Verlag, 1980.

Differential Equations

5. Birkhoff, G. and G.-C. Rota, *Ordinary Differential Equations* (2nd ed.). New York: John Wiley, 1969.
6. Boyce, W. E. and R. C. DiPrima, *Elementary Differential Equations and Boundary Value Problems* (7th ed.). New York: John Wiley, 2000.
7. Coddington, E. A., *An Introduction to Ordinary Differential Equations*. Englewood Cliffs, N.J.: Prentice Hall, 1961.
8. Grimshaw, R., *Nonlinear Ordinary Differential Equations*. Oxford: Blackwell Scientific, 1990.
9. Redheffer, R., *Differential Equations: Theory and Applications*. Boston: Jones and Bartlett, 1991.
10. Simmons, G. F., *Differential Equations with Applications and Historical Notes* (2nd ed.). New York: McGraw-Hill, 1991.
11. Walter, W., *Ordinary Differential Equations*. New York: Springer-Verlag, 1998.

Laplace Transforms and Fourier Series

12. Abramowitz, M. and I. A. Stegun, *Handbook of Mathematical Functions*. New York: Dover, 1965.

13. Churchill, R. V., *Operational Mathematics* (3rd ed.). New York: McGraw-Hill, 1972.
14. Churchill, R. V. and J. W. Brown, *Fourier Series and Boundary Value Problems* (6th ed.). New York: McGraw-Hill, 2000.

Linear Algebra

15. Axler, S., *Linear Algebra Done Right* (2nd ed.). New York: Springer-Verlag, 1997.
16. Leon, S. J., *Linear Algebra with Applications* (5th ed.). Upper Saddle River, N.J.: Prentice Hall, 1998.
17. Strang, G., *Linear Algebra and Its Applications* (3rd ed.). New York: Harcourt, 1988.

Mathematical Biology

18. Murray, J., *Mathematical Biology*. New York: Springer-Verlag, 1993.
19. Edelstein-Keschet, L., *Mathematical Models in Biology*. New York: McGraw-Hill, 1988.
20. Perelson, A. S. and P. W. Nelson, Mathematical Analysis of HIV-1 Dynamics in Vivo, *SIAM Review*, vol. 41, no. 1, March 1999.

Numerical Analysis

21. Kincaid, D. R. and E. W. Cheney, *Numerical Analysis: Mathematics of Scientific Computing*. Pacific Grove, Calif.: Brooks/Cole, 1996.
22. Golub, G. H. and J. M. Ortega, *Scientific Computing and Differential Equations: An Introduction to Numerical Methods*. New York: Academic Press, 1991.

Hints and Answers

Chapter 1

Section 1.1, p.1

1. $(5e^{-3t})' + 3(5e^{-3t}) = -15e^{-3t} + 15e^{-3t} = 0$
3. $(t^2 + t^3)''' = (2t + 3t^2)'' = (2 + 6t)' = 6$
4. Suggestion: Show that $x'' + 2x' + x = 0$.
5. $(ce^{-5t} + e^{-2t})' + 5(ce^{-5t} + e^{-2t}) = -5ce^{-5t} - 2e^{-2t} + 5ce^{-5t} + 5e^{-2t} = 3e^{-2t}$
7. $(c_1 \cos 3t + c_2 \sin 3t)'' + 9(c_1 \cos 3t + c_2 \sin 3t) = -9c_1 \cos 3t - 9c_2 \sin 3t + 9c_1 \cos 3t + 9c_2 \sin 3t = 0$
9. If $y = (c_1 + c_2t)e^t$, then $y' = (c_1 + c_2 + c_2t)e^t$ and $y'' = (c_1 + 2c_2 + c_2t)e^t$. Therefore, $y'' - 2y + y = e^t((c_1 + 2c_2 + c_2t) - 2(c_1 + c_2 + c_2t) + (c_1 + c_2t)) = 0$.

Section 1.2, p.4

1. $\ln|y(s)|\Big|_{t_0}^{t} = ks\Big|_{t_0}^{t}$; so $\ln\left|\dfrac{y(t)}{y_0}\right| = k(t - t_0)$; so $y(t) = y_0e^{k(t-t_0)}$. (Absolute values go away because $y(t)$ and y_0 always have the same sign.)
3. $e^{0.03t}$
5. $3e^{-(x-3)/10}$
7. $5e^{-3(z+1)/8}$
9. 99.3%, 93.3%, and 70.7%
11. age $\approx$ 176 years. Answer: No.
12. *Hint*: If $A(t)$ is the number of U^{238} atoms, then the number of Pb atoms is $A_0 - A(t)$.
13. About 49 minutes.
15. About 5.15 hours before.
17. $T_3 = \dfrac{\ln 3}{\ln 2}T_2$
19. Age 10: about 50%. Age 20: about 0.001%.
21. 0.99 m
23. When $k = 1$, the average rate of change over $[a, b]$ is equal to the average value over $[a, b]$.

Section 1.3, p.14

1. **(a)** first order, **(b)** linear, **(c)** nonhomogeneous
2. **(a)** fourth order, **(b)** linear, **(c)** homogeneous
3. **(a)** first order, **(b)** nonlinear, **(c)** nonhomogeneous
5. **(a)** first order, **(b)** nonlinear, **(c)** nonhomogeneous
7. **(a)** third order, **(b)** linear, **(c)** nonhomogeneous
9. **(a)** second order, **(b)** nonlinear, **(c)** homogeneous
15. $y = 1$ and $y = 5$
17. $y = n\pi$, $n = 0, \pm 1, \pm 2, \ldots$
19. With $y = a$, the equation becomes $0 + ta = 0$, which is true on no interval unless $a = 0$.
21. With $y = a$, the equation becomes $0 = a$.
23. With $y = a$, the equation becomes $a = t$, which is true on no interval.
25. e^{2t} and e^{-2t}
27. e^{-2t}
29. t^{-5}
31. t^3
33. $y = \dfrac{k}{2}t^2 + C$, $y = Ce^{kt}$
35. $e^{2t} - 3/2$
37. $(3t + C)e^{2t}$
39. **(a)** $-\dfrac{1}{2}(\cos t + \sin t)$

    (b) $Ce^t - \dfrac{1}{2}(\cos t + \sin t)$
41. $y = 1 - x$
43. $x = \dfrac{1}{t + 1}$
45. *Hint*: By the FTC and the product rule, $y' = -ky_0\, e^{-kt} - ke^{-kt}\displaystyle\int_0^t e^{ks} f(s)\, ds + e^{-kt}e^{kt} f(t)$.
46. $y = e^{2t} + e^{2t}\displaystyle\int_0^t e^{-2s}\cos(s^2)\, ds$. No.
47. In Problem 45, $u = y_0e^{-kt}$ and $v = e^{-kt}\displaystyle\int_0^t e^{ks} f(s)\, ds$.
49. $-3\cos 2t + t + 3$
51. $\dfrac{1}{2}(\sin t - t\cos t)$

53. $1/\sqrt{1-2t}$, $\left(-\infty, \frac{1}{2}\right)$

55. Maximal domain: $(-\pi/2, \pi/2)$

Chapter 2

Section 2.1, p.25

1. $(t+C)e^{-t}$

3. $(\ln|t|+C)t^2$

5. $\frac{2}{3}(t-3)+C(t-3)^{-5}$

7. $\frac{1}{2}\tan t \sin t + C\sec t$

9. $\frac{t+C}{t}\cos t$

11. $\frac{1}{2}\ln t + C/\ln t$

13. $1+C/t$

15. $\frac{t^2+C}{t\ln t}$

17. $e^{-t}\left(y_0+\int_0^t e^s f(s)\,ds\right)$

21. Ce^t+t+1

23. $Ce^{-t}+\cos t+\sin t$

24. To check: $y(1)\approx 0.42884$

27. Not asymptotically stable.

29. Asymptotically stable.

31. The difference between any two solutions satisfies $y'+py=0$ and therefore is $Ce^{-P(t)}$. The assumption guarantees that this approaches 0 as $t\to\infty$.

33. If $y=u^{-1}+\tilde{y}$, then $y'=-u^{-2}u'+\tilde{y}'$. Thus the differential equation is equivalent to $-u^{-2}u'+\tilde{y}'=a(u^{-1}+\tilde{y})^2+b(u^{-1}+\tilde{y})+c$. Simplification (and use of $\tilde{y}'=a\tilde{y}^2+b\tilde{y}+c$) results in $u'=-(2a\tilde{y}+b)u-a$.

35. $u'-3u=2$, $u=Ce^{3t}-2/3$, $y=(Ce^{3t}-2/3)^{-1}+1$.

37. $u'+\left(\frac{1}{2t}-\frac{2}{\sqrt{t}}\right)u=\frac{1}{t}$, $u=\frac{1}{2\sqrt{t}}\left(Ce^{4\sqrt{t}}-1\right)$,

$y=2\sqrt{t}\left(Ce^{4\sqrt{t}}-1\right)^{-1}+\sqrt{t}$.

Section 2.2.1, p.33

1. 10.9 s, −94.7 m/s

3. **(a)** 59.7 m/s, 331 m
(b) 127 m

5. 0.88

7. **(a)** $k_1=4$, $k_2=32$
(b) $mv'=-kv-mg=-32(-40)-160=1120$ lb

8. **(a)** A good choice is $k(t)=\frac{32}{8-7t}$.
(b) $\frac{40}{3}\left(32-28t-35\left(\frac{8-7t}{8}\right)^{32/35}\right)$
(c) 364 lb

10. **(c)** $\frac{\|V_0\|^2}{g}\sin 2\theta$

11. 122 ft/s

13. $x(t)=x_0+\frac{mv_1}{k}(1-e^{-kt/m})$

$y(t)=y_0+\frac{m}{k^2}(mg+kv_2)(1-e^{-kt/m})-\frac{mg}{k}t$

Section 2.2.2, p.37

1. $20k_0e^{-.005t}$, 460.5 minutes

3. $\frac{A}{40}=e^{-.05t}$, 133.7 minutes

5. **(a)** $A'+\frac{2}{30+3t}A=50$, $A(0)=0$

$A=300\left(1+\frac{t}{10}-\left(\frac{10}{10+t}\right)^{\frac{2}{3}}\right)$,

$0\le t\le 10$

(b) $A'+\frac{5}{60}A=50$, $A(0)=150(4-\sqrt[3]{2})$

$A=150(4-\sqrt[3]{2}e^{-t/12})$

7. $A'+\frac{10-t}{100}A=10(10-t)$, $A(0)=0$

$A=1000\left(1-e^{(t-20)t/200}\right)$, $0\le t\le 10$

$\frac{A(10)}{100}=10(1-1/\sqrt{e})\approx 3.935$ grams/gallon

9. $\frac{t}{500}$ grams/ft^3

Section 2.2.3, p.41

1. $Q_0e^{-t/20}$, $20\ln 10\approx 46$ s

3. **(a)** $\frac{1}{10}v_s(1-e^{-100t})$
(b) $\frac{1}{10}v_s$ amps
(c) $(\ln 10)/100\approx 0.023$ s

5. $I_0=-Q_0RC$

7. $RQ' + \frac{1}{C}Q = v_s,\ Q(0) = 0;$
$Q = Cv_s(1 - e^{-t/(RC)})$

9. $\dfrac{\cos(\omega t - \phi)}{\sqrt{R^2 + \omega^2 L^2}} - \dfrac{Re^{-Rt/L}}{R^2 + \omega^2 L^2},\ \phi = \tan^{-1}\dfrac{\omega L}{R}$

11. **(a)** $V_R + V_L + V_C = IR + LI' + Q/C = 0.$ Now differentiate with respect to t.

(b) $R^2 \geq 4L/C$

(c) $R^2 < 4L/C,\quad k = \dfrac{R}{2L},$

$$\omega = \frac{1}{2}\sqrt{\frac{4L - R^2C}{L^2C}}$$

Section 2.3, p.44

1. **(a)**

(b) $y = \begin{cases} -t, & \text{if } t < 0 \\ t & \text{if } t \geq 0 \end{cases}$

(c) *Hint*: Since $p = 0$ and $t_0 = y_0 = 0$, (2) becomes

$$y = \int_0^t f(s)\,ds.$$

3. **(a)**

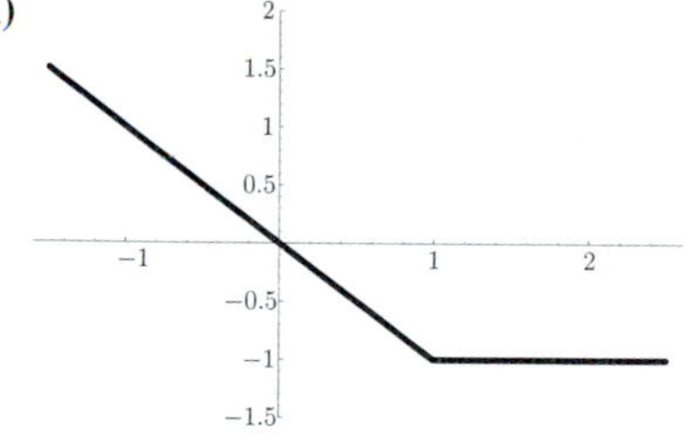

(b) $y = \begin{cases} -t, & \text{if } t < 1 \\ -1 & \text{if } t \geq 1 \end{cases}$

(c) *Hint*: Since $p = 0$ and $t_0 = y_0 = 0$, (2) becomes

$$y = \int_0^t h(s)\,ds.$$

5. $y = \begin{cases} 1 - e^{-(t+1)}, & \text{if } t < -1 \\ 0, & \text{if } -1 \leq t < 1 \\ 1 - e^{-(t-1)}, & \text{if } t \geq 1 \end{cases}$

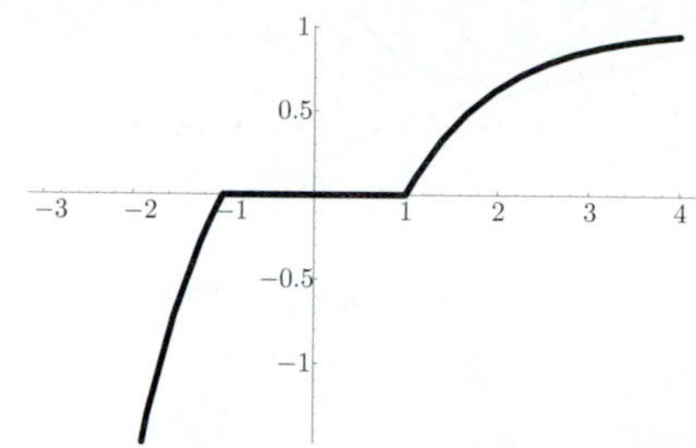

7. $y = \begin{cases} e^t - 1, & \text{if } t < 1 \\ 1 + e(e-2)e^{-t}, & \text{if } t \geq 1 \end{cases}$

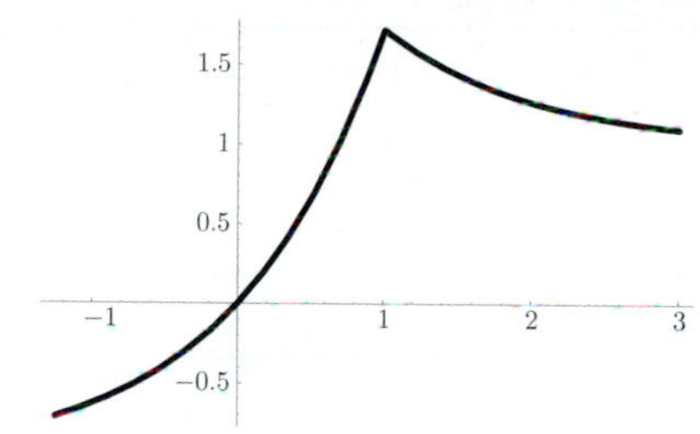

9. $y = \begin{cases} e^t - 1, & \text{if } t < 0 \\ e^{-t} - 1, & \text{if } t \geq 0 \end{cases}$

11. $\dfrac{1}{1000}A = \begin{cases} 1 - e^{-t/5}, & \text{if } 0 \leq t < 5 \\ (e-1)e^{-t/5}, & \text{if } t \geq 5 \end{cases}$

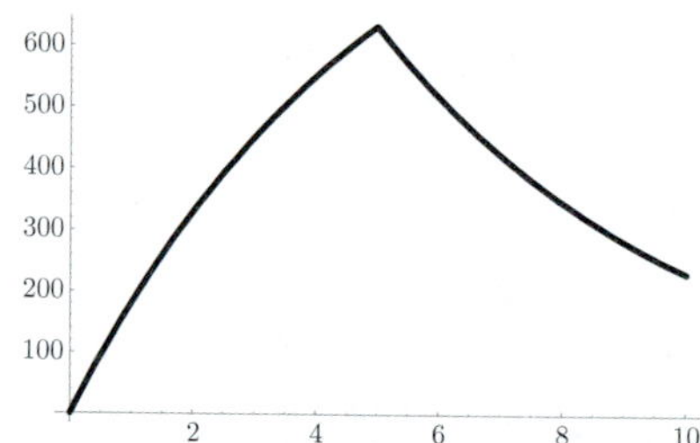

Chapter 3

Section 3.1, p.51

1. **(a)** $Ce^{-t} + t - 1$

(b) $Ce^{-t^2/2}$

3. 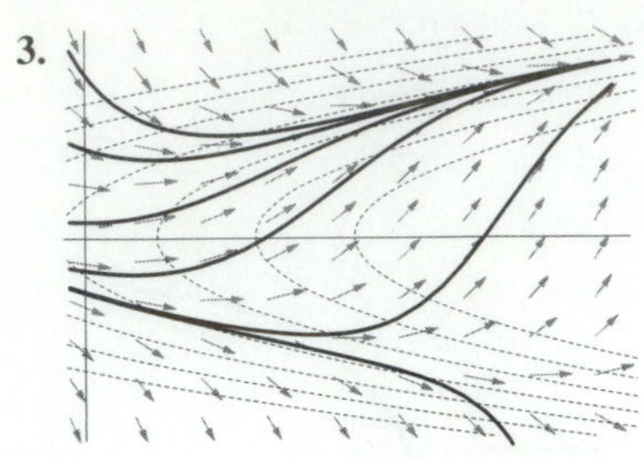

$y' = t - y^2$; isoclines are parabolas.

5. 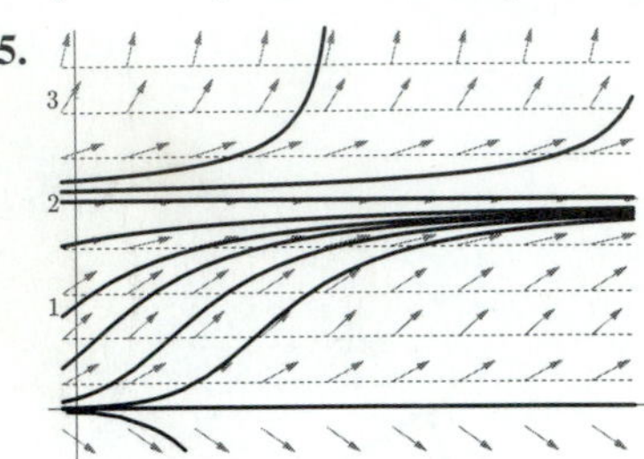

$y' = y(y^2 - 4y + 4)$;
isoclines are horizontal lines.

7. 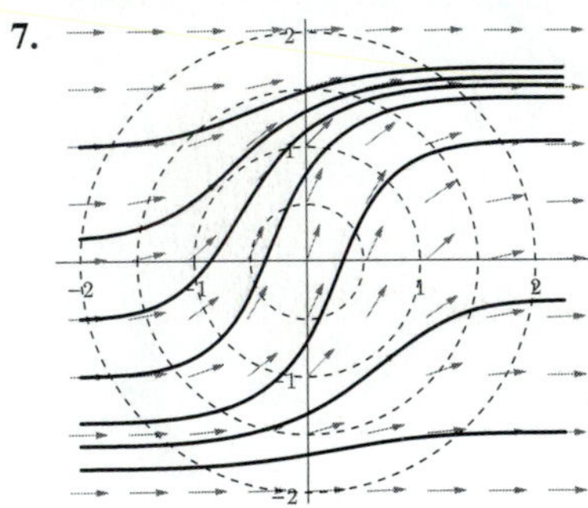

9. 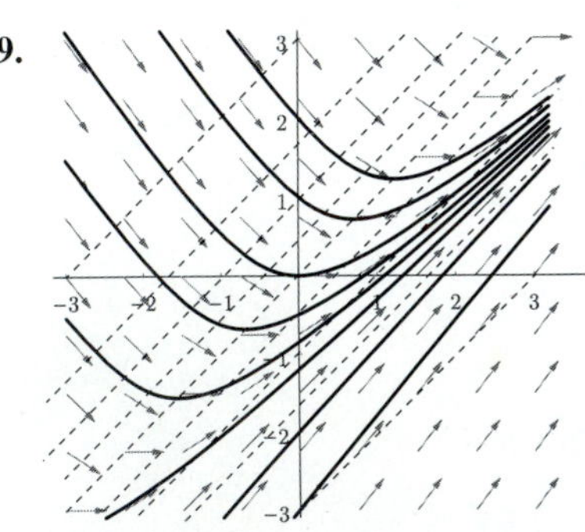

11. With $h = 0.1$: $y(.5) \approx 1.42721$
With $h = 0.05$: $y(.5) \approx 1.42052$
Exact value: $y(.5) = \sqrt{2(.5)+1} = \sqrt{2}$

13. $e^h \approx 1 + h$. Error:
$e^h - (1+h) = \frac{1}{2!}h^2 + \frac{1}{3!}h^3 + \cdots$

15. With $h = 0.1$: $y(.5) \approx 1.41406$
With $h = 0.05$: $y(.5) \approx 1.41418$

17. **(a)** $y(.2) \approx 1.20945$. Error:
$e^{\sin(.2)} - 1.20945 \approx 0.010$

(b) $y(.2) \approx 1.21921$. Error:
$e^{\sin(.2)} - 1.21921 \approx 0.00057$

Section 3.2, p.59

1. $\frac{1}{4-t^2}$, $(-2, 2)$
3. $-\ln(1-t)$, $(-\infty, 1)$
5. e^{e^t}, $(-\infty, \infty)$
7. $\sin y - \sin t = 1/2$
9. **(a)** $y = (1-t/2)^2$ is a solution on $(-\infty, 2]$.

(b) $y = \begin{cases} (1-t/2)^2, & \text{if } t \le 2 \\ 0, & \text{if } t > 2 \end{cases}$

11. $t^2 - 2ty - y^2 = C$
13. $\frac{t}{1-\ln t}$, $(0, \infty)$
15. $t\tan(\pi/4 + \ln t)$, $(0, e^{\pi/4})$
17. $y - \ln|t + y + 1| = C$
19. $\tan^{-1}\left(\frac{y-1}{t-2}\right) - \frac{1}{2}\ln((t-2)^2 + (y-1)^2) = C$
21. $y = \frac{\cos^2 t}{C - \tan t}$

Section 3.3, p.64

1. $m = -1$, $y = \frac{1}{Ce^t + 1}$
3. $m = 1/2$, $y^2 = Ce^{-2t} + 1$
5. $y = \frac{2(t + e^{-t})}{2e^{-t} - t^2 + C}$
7. $y^2 = (2t + C)\ln t$
9. $y = \sqrt[3]{(t^2+1)(C\sqrt{t^2+1} - 6)}$
11. $y^2 = \frac{1}{1 + Ce^{t^2}}$
13. $y^2 = (t^2 + C)e^{-1/t}$
15. $y = \frac{y_0 \sin t}{\sin t_0 - y_0(\cos t - \cos t_0)}$
19. **(b)** $y = e^{-t} + \frac{2}{Ce^{-t} - e^t}$

(c) $y = e^{-t} + \dfrac{2(y_0 - e^{-t_0})}{(1 + y_0 e^{t_0})e^{t_0 - t} + (1 - y_0 e^{t_0})e^{t - t_0}}$

21. (b) $y = e^t + \dfrac{1}{Ce^{-t} - 1}$

(c) $y = e^t + \dfrac{y_0 - e^{t_0}}{e^{t_0} - y_0 - (e^{t_0} - y_0 - 1)e^{t_0 - t}}$

23. (b) $y = \cos t + \dfrac{1 + \sin t}{C - t}$

(c) $y = \cos t - \dfrac{(\cos t_0 - y_0)(1 + \sin t)}{1 + \sin t_0 + (\cos t_0 - y_0)(t - t_0)}$

Section 3.4, p.68

1. $y' = (t^2 + C)^3$

3. $y' = \dfrac{1}{2}e^t + Ce^{-t}$

5. $y' = \sqrt[3]{3\sin t + C}$

7. $y = \tan t$

9. $y = \dfrac{e^{-t} - e^t}{e^{-t} + e^t} = -\tanh t$

11. $y = \dfrac{\sqrt{3}}{2}\tan\left(\dfrac{\sqrt{3}}{2}t + \dfrac{\pi}{6}\right) - \dfrac{1}{2}$

13. $y = \cos^{-1}(e^t/\sqrt{2})$

Section 3.5.1, p.73

1. *Hint*: $\left|\dfrac{v + r}{v - r}\right| = -\dfrac{v + r}{v - r}$, since $-r < v \le 0$.

3. $y_0 - \dfrac{m}{k_q}\ln\left(\cosh(k_q rt/m)\right)$

4. About 17 s and −92 m/s

5. $y_0 +$
$\dfrac{m}{k_q}\ln\left(\sqrt{1 + \dfrac{v_0{}^2}{r^2}}\cos\left(\dfrac{rk_q t}{m} - \tan^{-1}\left(\dfrac{v_0}{r}\right)\right)\right)$

8. (a) 0.019

(b) 0.00051

(c) About 13 s

Section 3.5.2, p.78

1. (a) $a(s) = \pi(s/2)^2$,

$$y = \begin{cases} (6.25(126.5, -0.5602t))^{1/5} & 0 \le t \le 225.8 \\ 0, & t > 225.8 \end{cases}$$

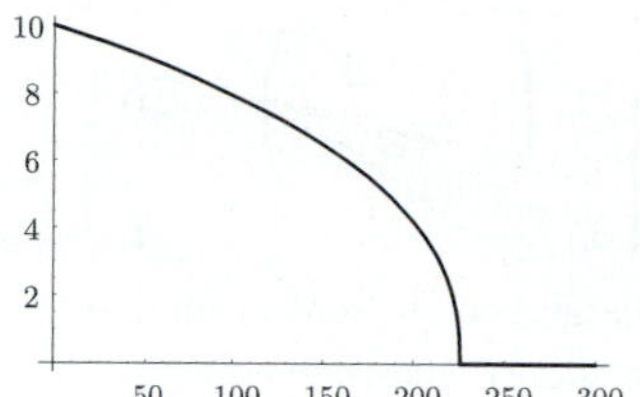

(b) $a(s) = \pi((10 - s)/2)^2$,
$\sqrt{y}\left(1500 - 100y + 3y^2\right) - 2530 = -4.202\,t$
Draining time: $t \approx 602$

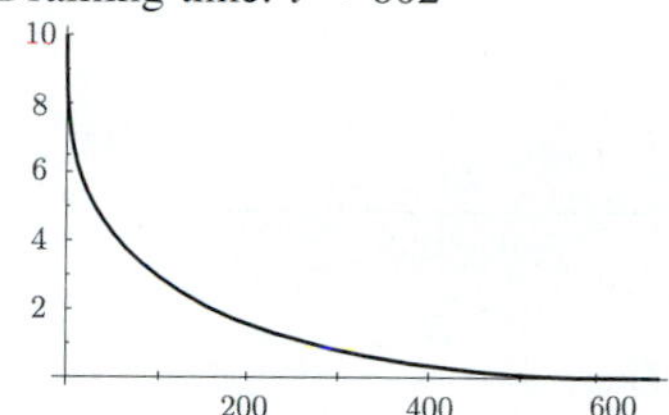

3. (a) $\pi(8^2 - (y - 8)^2)y' = -\rho\sqrt{y}$

$(3y - 80)y^{3/2} + 2028 = \dfrac{15}{2\pi}\rho t$

(b) $T = \dfrac{4096\pi}{15\rho}$

(c) $T \approx 136$ minutes

5. (a) $40\sqrt{3}(\sqrt{8} - 1)/\rho$

(b) 0.3364

9. (a) $\pi R^2 y' = -\rho(\sqrt{y} + \sqrt{y - H/2})$

(b) $\pi R^2 y' = -\rho\sqrt{y}$

11. Any arc of the form $y = kx^4$, $0 \le x \le R$, $k > 0$.

13. (a) $V = aR_{\text{in}}^2/\rho^2$, $A = ak_{\text{in}}R_{\text{in}}^2/\rho^2$

(b) $y = R_{\text{in}}^2/\rho^2$, $A = k_{\text{in}}V(R_{\text{in}}^2/\rho^2)$

15. (a) If $V' = -ka(y)$, then $a(y)y' = -ka(y)$; therefore, $y' = -k$.

(b) $y' = \dfrac{Hq}{Ry} - k$, $y(0) = H/2$

Equilibrium: $y = \dfrac{Hq}{kR}$; meaningful if $q \le kR$.

Section 3.5.3, p.84

1. $Q = Q_0/\sqrt{1 + 2Q_0^2 t}$
The charge decays to zero much more slowly than in the linear case, where the decay is exponential.

3. $Q = \begin{cases} Q_0\left(1 - \dfrac{2t}{3Q_0^{2/3}}\right)^{\frac{3}{2}}, & 0 \le t < \frac{3}{2}Q_0^{2/3} \\ 0, & t \ge \frac{3}{2}Q_0^{2/3} \end{cases}$

The charge goes to zero in finite time, whereas in the linear case, the charge decays exponentially to 0 as $t \to \infty$.

5. $I_\infty = 1 - \sqrt{1 - v_0}$

$$I = I_\infty + \frac{2(v_0 - I_\infty)}{I_\infty + (I_\infty - 2)e^{2(1-I_\infty)t}}$$

7. $I = \dfrac{t}{1+t}$

9. $I = \dfrac{I_0}{\sqrt{1 + (e^{2t} - 1)(1 + I_0^2)}}$

11. $I = \begin{cases} \sin t, & 0 \le t \le \pi \\ \sin t + \frac{1}{2}(t - \pi), & \pi < t \le 5.03709 \\ \sin t + 0.947746, & 5.03709 < t \le 10.6709 \end{cases}$

Chapter 4

Section 4.1, p.87

1. $1 + t + \frac{3}{2}t^2 + \frac{10}{3!}t^3 + \frac{47}{4!}t^4$

3. $e - t + \frac{1}{2e}t^2 + \frac{2}{3!}t^3 - \frac{1 + 2e^2}{4!e^3}t^4$

5. $1,\ 1 + t,\ 1 + t + \frac{1}{2}t^2,\ 1 + t + \frac{1}{2}t^2 + \frac{1}{6}t^3$

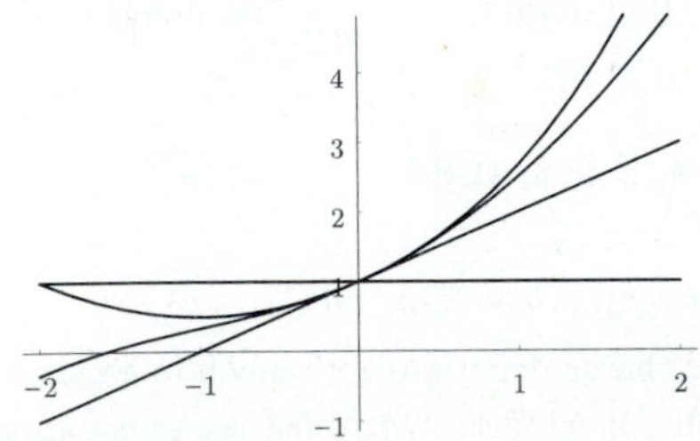

7. $1,\ 2 - \cos t,\ 1 + \frac{1}{2}(3 - 4\cos t + \cos^2 t),$

$1 + \frac{1}{6}(10 - 15\cos t + 6\cos^2 t - \cos^3 t)$

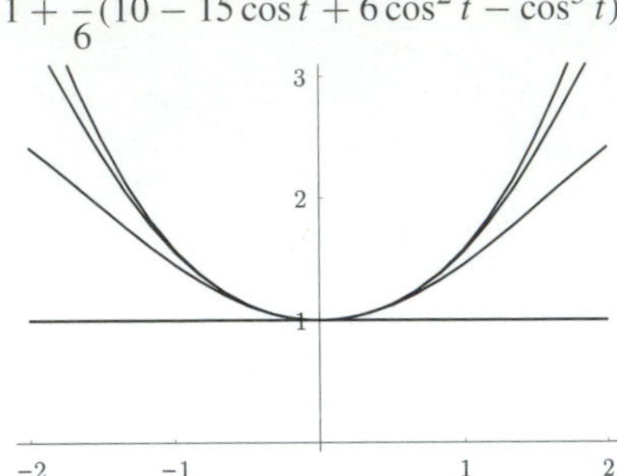

9. One example is $y' = y(\sqrt[3]{t} + \sqrt[3]{y})$, $y(0) = 0$, of which a solution is $y(t) = 0$, $-\infty < t < \infty$.

Section 4.2, p.93

1. $y = Ct$ for any constant C.
$f(t, y) = y/t$ is not continuous at the initial point.

3. **(a)** $y = (2t - 1)^{3/2}$, $a < t < b$, provided that $1/2 < a < 1 < b$.

(b) $y = \begin{cases} 0, & t \le 1/2 \\ (2t - 1)^{3/2}, & t > 1/2 \end{cases}$

Yes, this global solution is unique.

4. **(b)** There are many global solutions. Find two.

5. All (t_0, y_0).

7. All (t_0, y_0) with $y_0 \ne 1$.

9. All (t_0, y_0) with $t_0^2 + y_0^2 \ne \left(n + \frac{1}{2}\right)\pi$, n an integer.

11. (a), (b), and (d)

13. $\left(-\infty, \dfrac{1}{p - 1}\right)$

Section 4.3, p.99

1. Equilibria: $y = 0$ and $y = 3$. $y = 0$ is asymptotically stable; its interval of attraction is $(-\infty, 3)$. $y = 3$ is unstable.

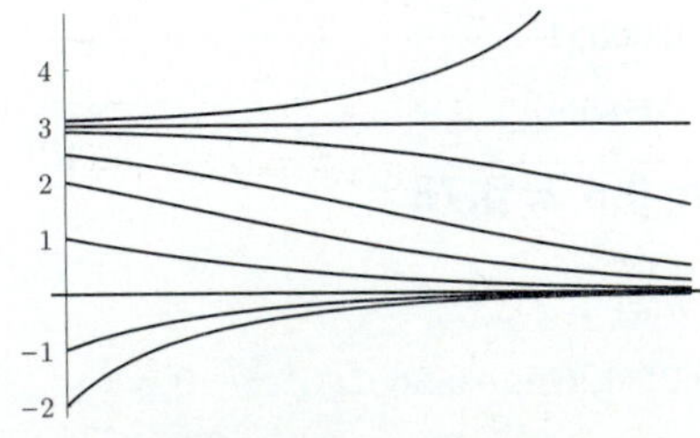

3. Equilibria: $y = 0$ and $y = 3$. $y = 0$ is semistable; its interval of attraction is $(-\infty, 0)$. $y = 3$ is semistable; its interval of attraction is $(0, 3)$.

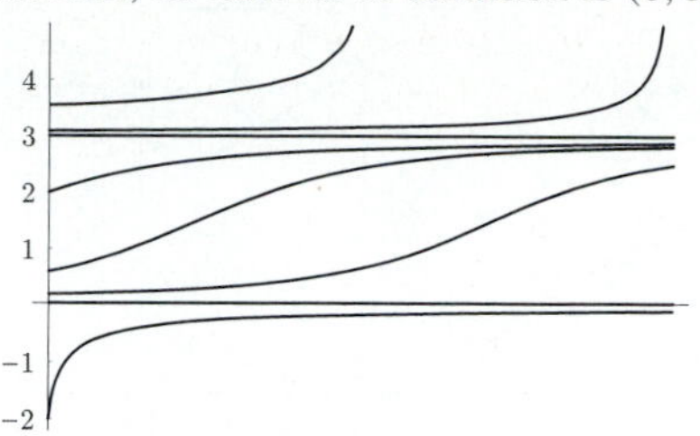

5. Equilibria: $y = 3$ and $y = 5$. $y = 3$ is unstable. $y = 5$ is asymptotically stable; its interval of attraction is $(3, \infty)$.

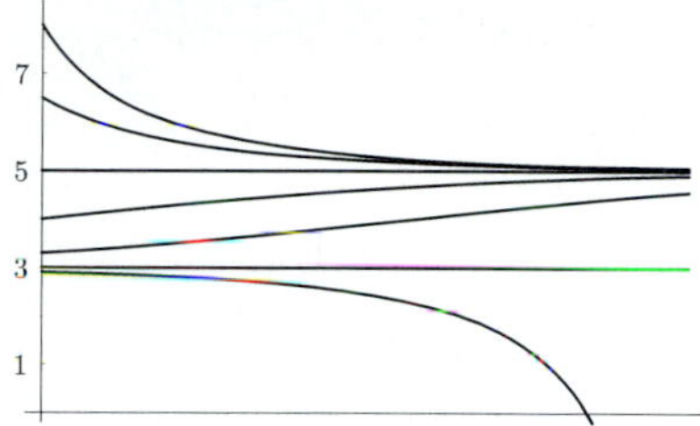

7. Equilibria: $y = n\pi$, $n = 0, \pm 1, \pm 2, \ldots$. For odd n, $y = n\pi$ is asymptotically stable with interval of attraction $((n-1)\pi, (n+1)\pi)$. If n is even, then $y = n\pi$ is unstable.

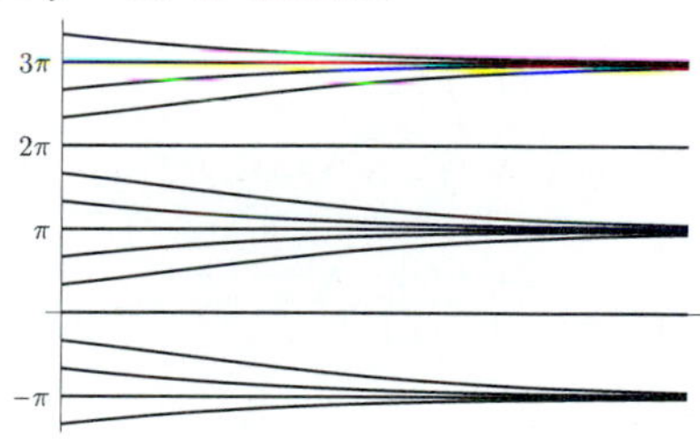

9. Equilibria: $y = \pm 1$. $y = -1$ is asymptotically stable with interval of attraction $(-\infty, 1)$. $y = 1$ is unstable.

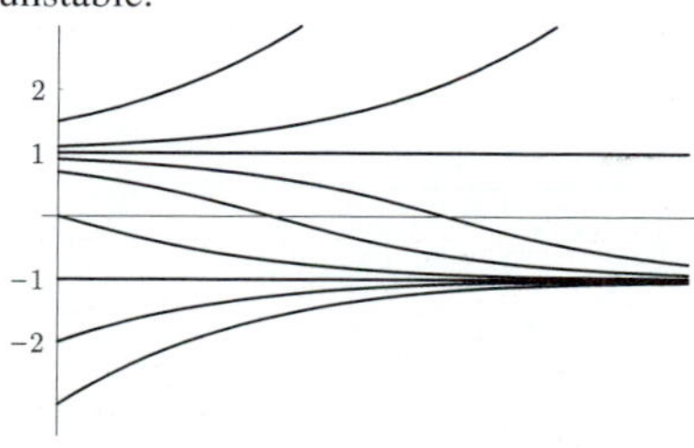

11. **(a)** $k = 1875$

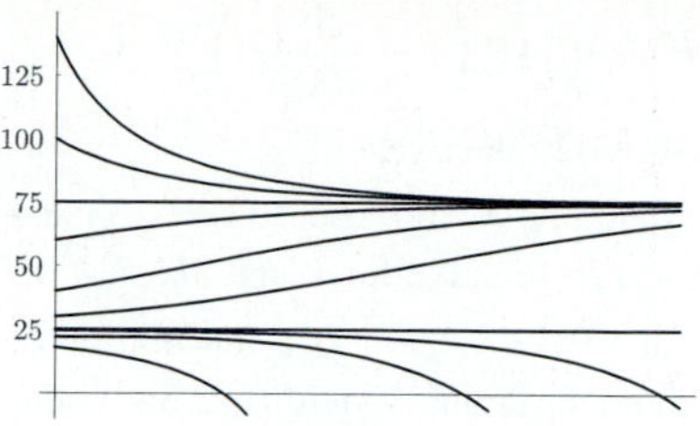

(b) $k < 2500$, $y = b$ for any $b > 50$

13. Since $y' = y(M - y)$, y' has a positive maximum value of $M^2/4$, which occurs when $y = M/2$.

15. **(a)**

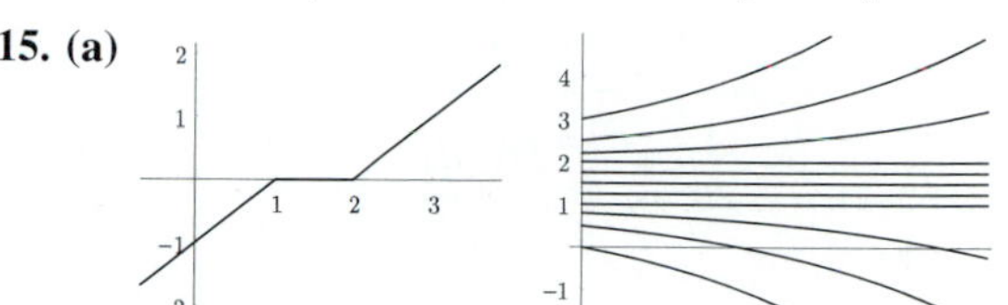

(b) Equilibria: $y = b$ for $1 \le b \le 2$. Unstable if $b = 1, 2$; neutrally stable if $1 < b < 2$.

17. **(a)**

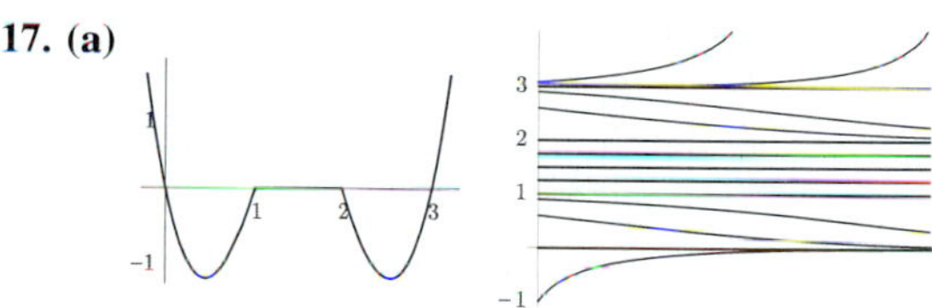

(b) Equilibria: $y = 0$, $y = b$ for $1 \le b \le 2$, and $y = 3$.

$y = 0$ is asymptotically stable;

$y = 1$ and $y = 3$ are unstable;

$y = b$ is neutrally stable if $1 < b < 2$;

$y = 2$ is semistable.

Section 4.4, p.105

2. *Hint*: $M = 1000$

3. **(a)** $P(t) = \dfrac{400(20k - 1)}{20k - e^{(.05-k)t}}$

(b) $k \approx 0.082$

(c) $\lim_{t\to\infty} P(t) \approx 160$

4. *Hint*: With $M = 300$ and $P_0 = 200$, use $P(100) = 280$ to solve for k. Then use $M(1 - \beta/k) = 200$ to find β.

5. **(a)** $P' = 0.14P\left(1 - \dfrac{1}{1000}P\right)$

   (b) About 29 deer/year

6. *Hint*: With $P_0 = 100$, use $P(1) = 133$ and $P(2) = 170$ to solve for k and M.

7. Look at $kP(1 - P/M)$ as a function of P. The maximum sustainable yield (i.e., harvesting rate) is the same as the maximum growth rate of the population. Any greater harvesting rate guarantees that the population will decrease, regardless of its size.

11. **(a)** $y_0 = \dfrac{e^{K(T)} - 1}{\displaystyle\int_0^T e^{K(s)} f(s)\,ds}$

    (b) $\tilde{y}'(t) = y'(t+T)$
    $= k(t+T)y(t+T) - f(t+T)y(t+T)^2$
    $= k(t)\tilde{y}(t) - f(t)\tilde{y}(t)^2$

Section 4.5, p.115

5. **(a)** 0.678213

   (b) 0.666926

   (c) 0.66666661

   (d) $y(1) = 2/3$; so the errors are, respectively (and approximately), 1.2×10^{-2}, 2.6×10^{-4}, and 5×10^{-8}.

7. Euler's method applied to $y' = ky$, $y(0) = y_0$, gives $y(h) \approx y_0(1 + kh)$. The solution of $y' = k_h y$, $y(0) = y_0$, is $y_0 e^{k_h t}$, whose value at $t = h$ is $y_0(1 + kh)$.

   Use l'Hôpital's rule to show $k_h \to k$ as $h \to 0$.

9. *Hint*: $\dfrac{1}{1-h} = 1 + h + h^2 + h^3 + \cdots$

13. $\begin{cases} w_1 + w_2 + w_3 &= 1 \\ w_2\theta_1 + w_3\theta_2 &= 1/2 \\ w_3\theta_1\theta_2 &= 1/6 \end{cases}$

15. **(a)** $w_1 = 2/9$, $w_2 = 1/3$, $w_3 = 4/9$

    (b) $y(1) \approx y_{10} = 0.666673.$

Section 4.6, p.125

3. $(0, 0)$ and $\left(\dfrac{1}{4}, \dfrac{1}{2}\right)$

5. $(2, 3)$ and $(3, 2)$

7. **(ab)**

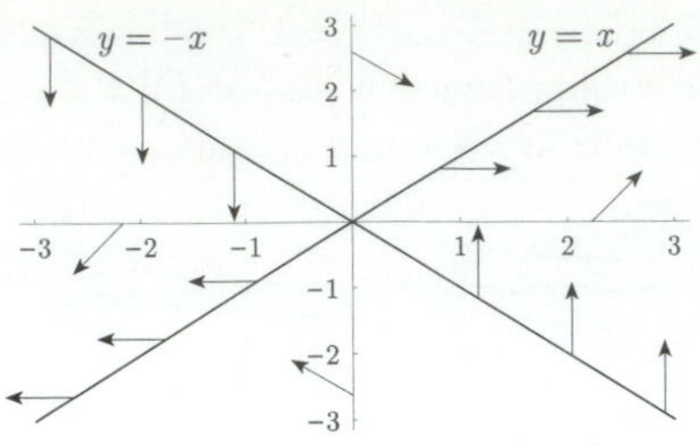

9. **(ab)**

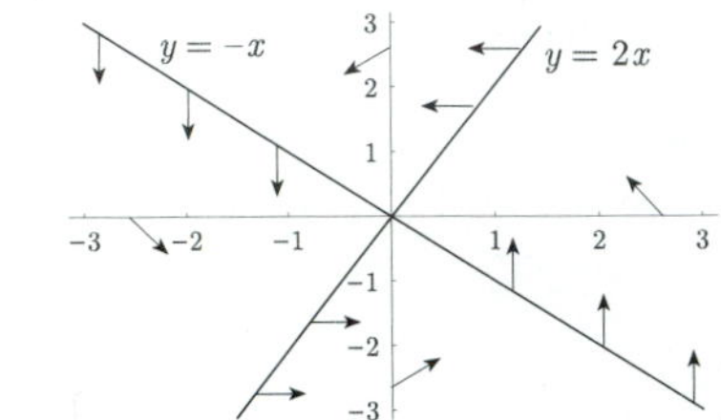

11.

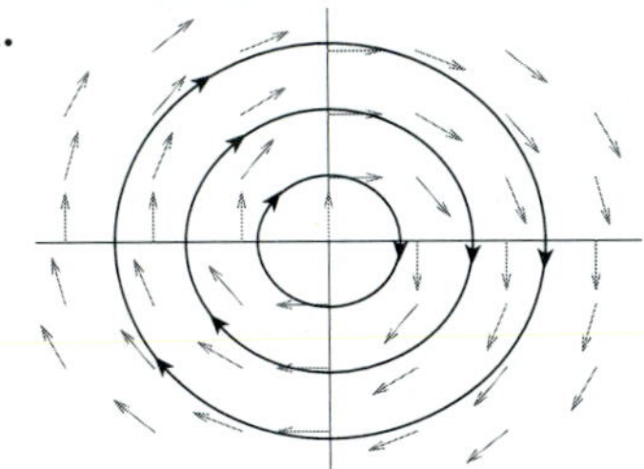

13.

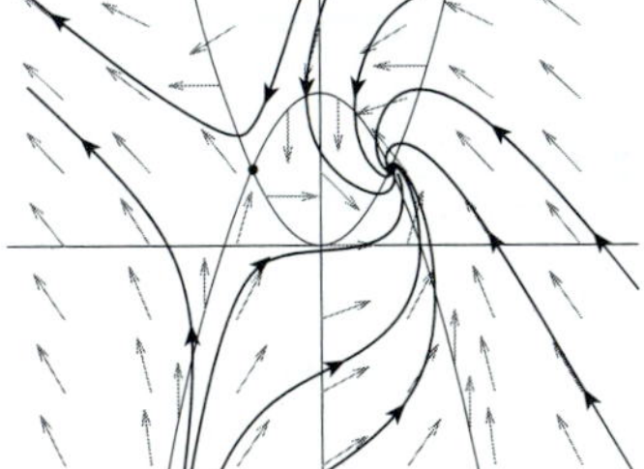

15.

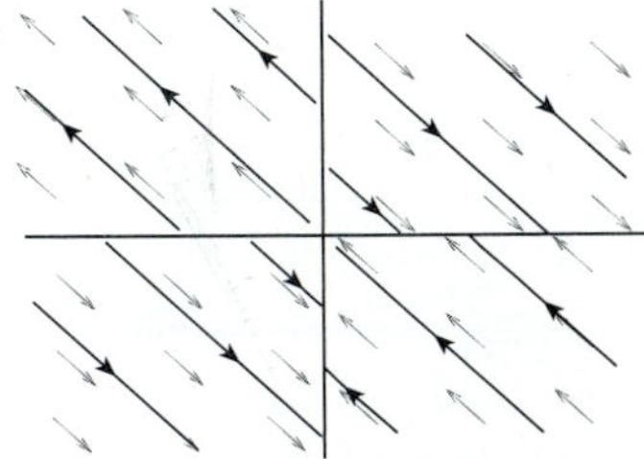

17. $x(.3) \approx 1.0302$, $y(.3) \approx 0.301$

21. $(x^2+y^2)' = 2xx' + 2yy' =$
$2xyp(x, y) - 2xyp(x, y) = 0$
Phase-plane orbits follow circular arcs centered at the origin.

23. The phase-plane orbits are "elliptical spirals" that spiral in toward the origin.

Chapter 5

Section 5.1, p.133

1. **(a)** 0.03115 slugs
(b) 32.1 lb
(c) $128/32.1 \approx 4$ slugs
(d) lb/ft and lb/(ft/s)

3. **(a)** 0.00102 g
(b) 980 dynes
(c) 3 g
(d) dynes/cm and dynes/(cm/s)

4. **(a)** $mg = 10$ and $s = 1/12$; so $k = mg/s = 120$ lb/ft.

5. **(a)** $y'' + 98y = 0$
(b) $y'' + 98y = 0,\ y(0) = .2,\ y'(0) = 0$

7. $2y'' + 4y' + 196y = 4$ or $y'' + 2y' + 98y = 0$

9. The equations turn out to be the same as (2) and (5).

11. $((4 - .01t)y')' + 2y' + y = 0$ or
$(4 - .01t)y'' + 1.99y' + 2y = 0$

13. $((m_0 - \rho t)y')' + ky = T$ or
$(m_0 - \rho t)y'' - \rho y' + ky = T$

14. The answers to (a) and (b) are not the same.

Section 5.2, p.139

3. $y' = v,\ v' = -tv - t^2 y$

5. $y' = v,\ v' = \sin y$

7. $y' = v,\ v' = e^{-y} - v^3$

9. $$\begin{pmatrix} y \\ v \end{pmatrix}' = \begin{pmatrix} 0 & 1 \\ -2 & -t \end{pmatrix}\begin{pmatrix} y \\ v \end{pmatrix} + \begin{pmatrix} 0 \\ t \end{pmatrix}$$

11. **(a)** $$\begin{pmatrix} y \\ v \end{pmatrix}' = \begin{pmatrix} 0 & 1 \\ -q & -p \end{pmatrix}\begin{pmatrix} y \\ v \end{pmatrix} + \begin{pmatrix} 0 \\ f \end{pmatrix}$$

(b) $$\begin{pmatrix} y \\ v \\ w \end{pmatrix}' = \begin{pmatrix} 0 & 1 & 0 \\ 0 & 0 & 1 \\ -r & -q & -p \end{pmatrix}\begin{pmatrix} y \\ v \\ w \end{pmatrix} + \begin{pmatrix} 0 \\ 0 \\ f \end{pmatrix}$$

12. $$\begin{cases} y(.3) \approx 0.99895 \\ y'(.3) \approx -0.032493 \end{cases}$$

13. $$\begin{cases} y(.2) \approx 1.16103 \\ y'(.2) \approx 0.62095 \end{cases}$$

Section 5.3, p.147

1. **(a)** not linear
(b) $\mathcal{T}e^{-t} = 1 - e^{-t}$
(c) $\mathcal{T}(2 - t^2) = 1 - 2t$

3. **(a)** linear
(b) $\mathcal{T}e^{-t} = -e^{-t}(1 + t^2)$
(c) $\mathcal{T}(2 - t^2) = t^4 - 2t^2 - 2t$

5. linear, maps functions to numbers

7. linear, maps functions to numbers

9. $(\mathcal{D}+1)(\mathcal{D}+2)u = (\mathcal{D}+1)(u' + 2u) =$
$u'' + 3u' + 2u = (\mathcal{D}^2 + 3\mathcal{D} + 2)u$

11. **(a)** For any twice-differentiable function u, expansion of each of $(\mathcal{D} + a\mathcal{I})(\mathcal{D} + b\mathcal{I})u$ and $(\mathcal{D} + b\mathcal{I})(\mathcal{D} + a\mathcal{I})u$ gives the same result: $u'' + (a+b)u' + abu$.

(b) A simple example is provided by $a = 0$ and $b = t$. $\mathcal{D}(\mathcal{D} + t\mathcal{I})u = u'' + tu' + u$, while $(\mathcal{D} + t\mathcal{I})\mathcal{D}u = u'' + tu'$.

(c) $(\mathcal{D} + a\mathcal{I})(\mathcal{D} + b\mathcal{I}) =$
$\mathcal{D}^2 + (a+b)\mathcal{D}u + (b' + ab)\mathcal{I}$

12. **(c)** Use the result of part (b) twice:

$$\begin{aligned}(\mathcal{D} + a\mathcal{I})&(\mathcal{D} + b\mathcal{I})(e^{kt}u) \\ &= (\mathcal{D} + a\mathcal{I})(e^{kt}(\mathcal{D} + (k+b)\mathcal{I})u) \\ &= e^{kt}(\mathcal{D} + (k+a)\mathcal{I})(\mathcal{D} + (k+b)\mathcal{I})u\end{aligned}$$

13. **(a)** The statement is true when $n = 1$:

$$\begin{aligned}\mathcal{D}(e^{kt}u) &= e^{kt}u' + ke^{kt}u \\ &= e^{kt}(u' + ku) = e^{kt}(\mathcal{D} + k\mathcal{I})u.\end{aligned}$$

Assuming truth for $n = 1, 2, \ldots, m$, the statement is also true when $n = m + 1$:

$$\begin{aligned}\mathcal{D}^{m+1}(e^{kt}u) &= \mathcal{D}(D^m(e^{kt}u)) \\ &= \mathcal{D}e^{kt}(\mathcal{D} + k\mathcal{I})^m u \\ &= e^{kt}(\mathcal{D} + k\mathcal{I})^{m+1}u.\end{aligned}$$

(b) Let $P(\mathcal{D}) = a_n\mathcal{D}^n + \cdots + a_2\mathcal{D}^2 + a_1\mathcal{D} + a_0\mathcal{I}$. Write out $P(\mathcal{D})(e^{kt}u)$, and apply part (a) to each term.

15. (a)
$$(\mathcal{D}^2 - a^2\mathcal{I})e^{at} = \mathcal{D}^2e^{at} - a^2e^{at}$$
$$(\mathcal{D}^2 - a^2\mathcal{I})e^{at} = \mathcal{D}^2e^{at} - a^2e^{at} = a^2e^{at} - a^2e^{at} = 0$$
$$(\mathcal{D}^2 - a^2\mathcal{I})e^{-at} = \mathcal{D}^2e^{-at} - a^2e^{-at} = a^2e^{-at} - a^2e^{-at} = 0$$

(b) $(\mathcal{D}^2 - a^2\mathcal{I})(c_1e^{at} + c_2e^{-at}) = c_1a^2e^{at} + c_2a^2e^{-at} - a^2c_1e^{at} - a^2c_2e^{-at} = 0$

(c)
$$(\mathcal{D}^2 - a^2\mathcal{I})\left(c_1e^{at} + c_2e^{-at} + \frac{e^{bt}}{b^2-a^2}\right) = 0 + (\mathcal{D}^2 - a^2\mathcal{I})\frac{e^{bt}}{b^2 - a^2} = \frac{1}{b^2 - a^2}(b^2e^{bt} - a^2e^{bt}) = e^{bt}$$

17. If $u' = v'$, then $u - v$ is constant.

19. $(u - v)' - \dfrac{2}{2+t}(u - v) = 0$; so $\left(\dfrac{1}{(2+t)^2}(u - v)\right)' = 0$. Therefore, $u - v = C(2+t)^2$. $C = 1/4$, since $u(0) - v(0) = 1$.

Section 5.4, p.152

1. If $c_1u(x) + c_2v(x) = 0$ for all x in $[0, 1]$, then $c_1u(0) + c_2v(0) = 0 + c_2 = 0$ and $c_1u(1) + c_2v(1) = c_1 + 0 = 0$. Since $c_1 = c_2 = 0$, we conclude that u and v are linearly independent.

3. (a) Eliminating y gives $(ad - bc)x = 0$, and eliminating x gives $(ad - bc)y = 0$. So if $ad - bc \neq 0$, then $x = y = 0$.

(b) If $c_1u + c_2v = 0$ on I, then $c_1u(t_1) + c_2v(t_1) = 0$ and $c_1u(t_2) + c_2v(t_2) = 0$. Since $u(t_1)v(t_2) - u(t_2)\,v(t_1) \neq 0$, $c_1 = c_2 = 0$ by part (a).

5. $W(e^{at}, e^{bt}) = (b - a)e^{(a+b)t}$ is not identically zero, since $a \neq b$.

7. $W(\sin at, \sin bt) = b\sin at\cos bt - a\sin bt\cos at$ is not identically zero unless $a = b$.

9. $W(u, u + c) = -cu'$ is not identically zero, since $c \neq 0$ and u is not constant.

11. $y = c_1\cos t + c_2\sin t + 1$

13. $y = c_1 + c_2t + t^3$

15. $\dfrac{1}{2}(e^t + e^{-t})$ or $\cosh t$

17. $y = 2$

19. $y = t^2$

21. $W(e^t, \cos t) = -e^t(\sin t + \cos t)$ is not constant and has a zero in any interval of length π or greater.

25. $y' = mt^{m-1}u + t^mu'$ and $y'' = m(m-1)t^{m-2}u + 2mt^{m-1}u' + t^mu''$. So
$$t^2y'' - 2mty' + (m(m+1) + \omega^2t^2)y = t^m\left(m(m-1)u + 2mtu' + t^2u'' - 2m^2u - 2mtu' + m(m+1)u + \omega^2t^2u\right) = t^{m+2}(u'' + \omega^2u).$$

29. e^{4t}

33. $4t^2u'' + 5u = 0$

35. $u'' - t^4u = 0$

36. Suggestions: (a) Let $a = \sqrt{-\omega}$. (c) Let $a = \sqrt{\omega}$.

37. (a) $c_1e^{-t} + c_2e^{3t}$

(b) $(c_1 + c_2t)e^{-2t}$

(c) $e^{-2t}(c_1\cos t + c_2\sin t)$

Section 5.5, p.162

1. $\dfrac{1}{2}(\sin t - t\cos t)$

3. $\dfrac{1}{6}(e^t - 3e^{-t} + 2e^{-2t})$

5. (b) $\dfrac{1}{2}e^{3s-t}(e^{-2s} - e^{-2t})$

7. $\dfrac{1}{4}\left(e^{-t}(2t - 1) + e^{-3t}\right)$

9. $\dfrac{1}{8}(t^3 - 2t_0^2t + t_0^4t^{-1})$

11. Verify that $\mathcal{G}(f + g) = \mathcal{G}f + \mathcal{G}g$ and $\mathcal{G}(cf) = c\mathcal{G}f$.

13. Construct G_0 using $\tilde{u} = c_1u + c_2v$ and $\tilde{v} = c_3u + c_4v$; compare with G_0 constructed from u and v.

14. Look for the solution in the form $u(t)e^{-\int p}$.

Section 5.6, p.166

1. (a) $a_0 = 1$, $a_n = \dfrac{1}{n!} + \dfrac{(-1)^{n-1}}{(n-1)!}$ for $n \geq 1$.

(b) $a_n = \dfrac{1}{(n+1)!}$ for $n \geq 0$.

(c) $a_0 = 1$, $a_1 = 0$, $a_n = \dfrac{(n-1)^2}{n!}$ for $n \geq 2$.

3. **(a)** $a_0 = 1$, $a_n = (-1)^n + 2^{n-1}$ for $n \geq 1$.

(b) $a_n = 0$ if $0 \leq n \leq 2$ or n is even;

$a_n = (-1)^{\frac{n-3}{2}} = (-1)^{\frac{n+1}{2}}$ for odd $n \geq 3$.

(c) $a_0 = 1$, $a_n = 2$ for $n \geq 1$.

5. $a_0 = 1$, $a_1 = -1$, and $(n+1)(n+2)a_{n+2} = -2((n+1)\,a_{n+1} + a_n)$ for $n \geq 0$.

$$e^{-t}\cos t = 1 - t + \frac{t^3}{3} - \frac{t^4}{6} + \frac{t^5}{30} - \frac{t^7}{630} + \cdots$$

7. **(a)** $\rho \geq 1$

(b) $(n+1)(n+2)a_{n+2} = n(n+1)a_{n+1} - (n-1)a_n$

(c) $e^t - t$, t, ∞

9.

$$u = 1 - \frac{t^3}{3!\beta} + \frac{(1+3^3)t^6}{6!\beta^2} - \frac{(1+3^3)(1+6^3)t^9}{9!\beta^3} + \frac{(1+3^3)(1+6^3)(1+9^3)t^{12}}{12!\beta^4} - \frac{300979t^{15}}{51321600\beta^5} + \cdots$$

$$v = t - \frac{2t^4}{4!\beta} + \frac{2(1+4^3)t^7}{7!\beta^2} - \frac{2(1+4^3)(1+7^3)t^{10}}{10!\beta^3} + \frac{2(1+4^3)(1+7^3)(1+10^3)t^{13}}{13!\beta^4} - \frac{87763t^{16}}{18662400\beta^5} + \cdots$$

11. **(a)** $a_{n+2} = \dfrac{1}{n+2}a_n$

(b) $u = 1 + \dfrac{t^2}{2} + \dfrac{t^4}{2\cdot 4} + \dfrac{t^6}{2\cdot 4\cdot 6} + \dfrac{t^8}{8!!} + \cdots$

$v = t + \dfrac{t^3}{3} + \dfrac{t^5}{3\cdot 5} + \dfrac{t^7}{3\cdot 5\cdot 7} + \dfrac{t^9}{9!!} + \cdots$

(c) ∞

13. **(a)** $a_{n+2} = \dfrac{n^2 - n + 1}{4(n+1)(n+2)}a_n$

(b) $u = 1 + \dfrac{t^2}{8} + \dfrac{t^4}{128} + \dfrac{13t^6}{15360} + \dfrac{403t^8}{3440640} + \cdots$

$v = t + \dfrac{t^3}{24} + \dfrac{7t^5}{1920} + \dfrac{7t^7}{15360} + \dfrac{301t^9}{4423680} + \cdots$

(c) $\rho = 2$

15. **(a)** $a_{n+2} = -\dfrac{n^2+1}{(n+1)(n+2)}a_n$

(b)

$$u = 1 - \frac{t^2}{2} + \frac{(2^2+1)t^4}{4!} - \frac{(2^2+1)(4^2+1)t^6}{6!} + \frac{(2^2+1)(4^2+1)(6^2+1)t^8}{8!} - \cdots$$

$$v = t - \frac{2t^3}{3!} + \frac{2(3^2+1)t^5}{5!} - \frac{2(3^2+1)(5^2+1)t^7}{7!} + \cdots$$

(c) $\rho = 1$

17. **(a)** $a_{n+2} = \dfrac{(n+3)(n-2)}{(n+1)(n+2)}a_n$

(b) $u = 1 - 3t^2$,

$v = t - \dfrac{2t^3}{3} - \dfrac{t^5}{5} - \dfrac{4t^7}{35} - \dfrac{5t^9}{63} - \dfrac{2t^{11}}{33} - \cdots$

(c) $\rho = 1$ for nonterminating power-series solutions.

19. $\dfrac{t^3}{3\cdot 2} - \dfrac{t^6}{6\cdot 5\cdot 3\cdot 2} + \dfrac{t^9}{9\cdot 8\cdot 6\cdot 5\cdot 3\cdot 2} - \dfrac{t^{12}}{12\cdot 11\cdot 9\cdot 8\cdot 6\cdot 5\cdot 3\cdot 2} + \cdots$

21. $\dfrac{t^2}{2} + \dfrac{t^4}{4!} + \dfrac{3t^6}{6!} + \dfrac{5\cdot 3t^8}{8!} + \dfrac{7\cdot 5\cdot 3t^{10}}{10!} + \dfrac{9!!t^{12}}{12!} + \cdots$

23. $\dfrac{t^4}{4\cdot 3} + \dfrac{t^8}{8\cdot 7\cdot 4\cdot 3} + \dfrac{t^{12}}{12\cdot 11\cdot 8\cdot 7\cdot 4\cdot 3} + \cdots$

25. $y' - 2ty = t^3 e^{t^2}$, $y(0) = 1$; $y = \dfrac{1}{4}(t^4 + 4)e^{t^2}$

27. **(a)** Every solution produced by the recurrence formula is a multiple of the solution satisfying $y(0) = 1$ and $y'(0) = -1/2$.

(d) $v = 0$ produces an ordinary power-series solution:

$$1 - \frac{t}{2} + \frac{t^2}{2^2\cdot 3\cdot 2} - \frac{t^3}{2^3\cdot 5\cdot 3\cdot 3!} + \frac{t^4}{2^4\cdot 7\cdot 5\cdot 3\cdot 4!} - \cdots;$$

$\nu = 1/2$ produces

$$\sqrt{t}\left(1 - \frac{t}{3\cdot 2} + \frac{t^2}{2^2\cdot 5\cdot 3\cdot 2} - \frac{t^3}{2^3\cdot 7\cdot 5\cdot 3\cdot 3!} + \cdots\right)$$

28. $1 + 2t + \frac{2t^2}{3} + \frac{4t^3}{45} + \frac{2t^4}{315} + \frac{4t^5}{14175}\cdots,$
$\sqrt{t}\left(1 + \frac{2t}{3} + \frac{2t^2}{15} + \frac{4t^3}{315} + \frac{2t^4}{2835} + \frac{4t^5}{155925}\cdots\right)$

29. $t^{1/3}\left(1 - \frac{3t^2}{16} + \frac{9t^4}{896} - \frac{9t^6}{35840} + \cdots\right),$

$t^{-1/3}\left(1 - \frac{3t^2}{8} + \frac{9t^4}{320} - \frac{9t^6}{10240} + \cdots\right)$

31. $t^{-1/3}, t^{2/3}\left(1 - \frac{t^2}{6} + \frac{t^4}{40} - \frac{t^6}{336} + \cdots\right)$

Section 5.7, p.179

1. **(a)** $(n+1)(n+2)a_{n+2} = -(m-n)(m+n+1)a_n$

(b)
$$1 - \frac{m(m+1)t^2}{2} + \frac{(m-2)m(m+1)(m+3)t^4}{4!} - \frac{(m-4)(m-2)m(m+1)(m+3)(m+5)t^6}{6!} + \cdots,$$
$$t - \frac{(m-1)(m+2)t^3}{3!} + \frac{(m-3)(m-1)(m+2)(m+4)t^5}{5!} - \cdots$$

(c) $1, t, \frac{1}{2}(3t^2-1), \frac{1}{2}t(5t^2-3), \frac{1}{8}(35t^4 - 30t^2 + 3)$

(d) $\rho = 1$

3. **(a)** $(n+1)(n+2)a_{n+2} = -2(m-n)a_n$

(b) $1 - mt^2 + \frac{2^2 m\,(m-2)\,t^4}{4!} - \frac{2^3 m\,(m-2)\,(m-4)\,t^6}{6!} + \cdots,$

$$t - \frac{2\,(m-1)\,t^3}{3!} + \frac{2^2\,(m-1)\,(m-3)\,t^5}{5!} - \frac{2^3\,(m-1)\,(m-3)\,(m-5)\,t^7}{7!} + \cdots$$

(c) $1, 2t, 4t^2 - 2, 8t^3 - 12t, 16t^4 - 48t^2 + 12$

(d) $\rho = \infty$

7. **(b)** $W = -m/\sqrt{1-t^2}$

9.
$$\begin{aligned} T_{m+1}(t) &= \cos((m+1)\cos^{-1}t) \\ &= \cos\left(\cos^{-1}t + m\cos^{-1}t\right) \\ &= t\cos(m\cos^{-1}t) - \sin(\cos^{-1}t)\sin\left(m\cos^{-1}t\right) \\ &= tT_m(t) - \sin(\cos^{-1}t)\sin\left(m\cos^{-1}t\right) \end{aligned}$$
$$\begin{aligned} T_{m-1}(t) &= \cos((m-1)\cos^{-1}t) \\ &= \cos\left(m\cos^{-1}t - \cos^{-1}t\right) \\ &= \cos(m\cos^{-1}t)t + \sin\left(m\cos^{-1}t\right)\sin(\cos^{-1}t) \\ &= tT_m(t) + \sin\left(m\cos^{-1}t\right)\sin(\cos^{-1}t) \end{aligned}$$
Therefore, $T_{m+1}(t) + T_{m-1}(t) = 2tT_m(t)$.

Chapter 6

Section 6.1, p.185

1. $r^2 + 5r + 4 = 0$; $r = -4, -1$; $c_1e^{-4t} + c_2e^{-t}$

3. $r^2 - 2r - 8 = 0$; $r = -2, 4$; $c_1e^{-2t} + c_2e^{4t}$

5. $r^2 + 5r + 6 = 0$; $r = -3, -2$; $c_1e^{-3t} + c_2e^{-2t}$

7. $r^2 + 2r + 1 = 0$; $r = -1, -1$; e^{-t}

9. $r^2 - 4r + 4 = 0$; $r = 2, 2$; e^{2t}

11. $\mathcal{D}^2 + 3\mathcal{D} + 2\mathcal{I} = (\mathcal{D} + \mathcal{I})(\mathcal{D} + 2\mathcal{I})$, $r = -1, -2$

13. $\mathcal{D}^2 + 2\mathcal{D} + \mathcal{I} = (\mathcal{D} + \mathcal{I})^2$, $r = -1, -1$

15. $\mathcal{D}^2 + 2\mathcal{D} + 2\mathcal{I}$, $r = -1 \pm i$

17. $c_1e^t + c_2$

19. **(a)** $r + p = 0$

(b) $(\mathcal{D} + p\mathcal{I})y = 0$

(c) $r = -p$, $y = ce^{-pt}$

20. *Hints*: **(a)** $r^3 + r^2 - 4r - 4 = (r+1)(r^2-4)$

(b) $r^4 - 16 = (r^2+4)(r^2-4)$

22. **(a)** The choice of $c_1 = 0$ produces $y' = 3y$; the choice of $c_2 = 0$ produces $y' = 2y$. For any nontrivial orbit,

$$\begin{aligned} \lim_{t\to\infty}\frac{y'(t)}{y(t)} &= \lim_{t\to\infty}\frac{2c_1e^{2t} + 3c_2e^{3t}}{c_1e^{2t} + c_2e^{3t}} \\ &= \lim_{t\to\infty}\frac{2c_1e^{-t} + 3c_2}{c_1e^{-t} + c_2} = 3, \\ \lim_{t\to-\infty}\frac{y'(t)}{y(t)} &= \lim_{t\to-\infty}\frac{2c_1e^{2t} + 3c_2e^{3t}}{c_1e^{2t} + c_2e^{3t}} \\ &= \lim_{t\to-\infty}\frac{2c_1 + 3c_2e^t}{c_1 + c_2e^t} = 2. \end{aligned}$$

So orbits become parallel to the line $y = 3x$ as $t \to \infty$ and tangent to the line $y = 2x$ as $t \to -\infty$.

Section 6.2, p.191

1. $(c_1 + c_2 t)e^{3t}$
3. $(c_1 + c_2 t)e^{-t}$
5. $(t^2 + 4t + 6)e^{-t}$
7. $-\frac{1}{8}(2t^2 + 2t + 1)e^{-t}$
9. $\frac{1}{36}(6t - 5)e^{t}$
11. $\left(c_1 + c_2 t + \frac{1}{6}t^3\right) e^{-2t}$
13. **(b)** $\lim_{t\to\infty} \frac{y'(t)}{y(t)} = \lim_{t\to\infty} \frac{(c_2 - 2c_1 - 2c_2 t)e^{-2t}}{(c_1 + c_2 t)e^{-2t}}$ $= -2$. The limit as $t \to -\infty$ is the same.

Section 6.3, p.196

1. $c_1 \cos 3t + c_2 \sin 3t$
3. $e^{-3t}(c_1 \cos 3t + c_2 \sin 3t)$
5. $e^{-t}(c_1 \cos 2t + c_2 \sin 2t + t)$
7. $e^{-t} - \cos t + \sin t$
9. $e^{-t}(1 - \cos t)$
15. **(a)** $(e^{bit})' = (\cos bt + i \sin bt)' = -b \sin bt + bi \cos bt = bi(\cos bt + i \sin bt) = bie^{bit}$
19. **(a)** $\frac{1}{2}(1+i)$ **(b)** $\frac{1}{5}(1+3i)$
    (c) $-i$
21. **(a)** $\sqrt{2}e^{i\pi/4}$ **(b)** $2e^{i\pi/2}$
    (c) $3e^{i\pi}$ **(d)** $e^{-i\pi/2}$
22. **(a)** $\frac{1}{2}(1 - i\sqrt{3})$
23. First expand $r_1(\cos\theta_1 + i \sin\theta_1)r_2(\cos\theta_2 + i \sin\theta_2)$.
25. $c_1 t + c_2 t^3$
27. $c_1 \cos(\ln t) + c_2 \sin(\ln t)$
29. $\frac{1}{t}(c_1 \cos(\ln t) + c_2 \sin(\ln t))$

Section 6.4, p.202

1. $z'' + z = e^{it}$, $z = -\frac{i}{2}te^{it}$,
   $x = \frac{1}{2}t \sin t$, $y = -\frac{1}{2}t \cos t$
3. $z'' + z' + 3z = 5e^{it}$, $z = e^{it}(2 - i)$,
   $x = 2\cos t + \sin t$, $y = 2 \sin t - \cos t$
5. $z'' - z = e^{-1+2it}$, $z = \frac{1}{8}(-1+i)e^{-1+2it}$,
   $x = -\frac{1}{8}e^{-t}(\cos 2t + \sin 2t)$,
   $y = \frac{1}{8}e^{-t}(\cos 2t - \sin 2t)$
7. **(a)** $\frac{1}{4}((1-2t)\cos t + (1+2t)\sin t)$
   (b) $\frac{1}{4}((3-2t)\sin t - (1+6t)\cos t)$
9. **(a)** $\frac{1}{5}(19\cos t + 12 \sin t)$
   (b) $5\sin t - 5\cos t$
11. **(a)** $e^{-t}(\cos 2t - 3\sin 2t)$
    (b) $e^{-t}(5\cos 2t - \sin 2t)$
13. $e^{-t}(c_1 \cos t + c_2 \sin t) + \cos t + 2\sin t$
15. $c_1 \cos t + c_2 \sin t - \frac{1}{2}t \cos t$
17. $\cos 2t + 4\sin 2t - \frac{1}{2}e^{-t}(2\cos 2t + 9 \sin 2t)$
21. $A - 3B = 0$, $3A + B = 10$; $A = 3$, $B = 1$

Section 6.5, p.206

1. **(a)** $\omega = 2\ \text{s}^{-1}$, $\frac{\omega}{2\pi} = \frac{1}{\pi}$ cps
   (b) $0.75\cos 2t$, $A = 0.75$
   (c) $\sin 2t$, $A = 1$
   (d) $0.75\cos 2t + \sin 2t$, $A = 1.25$
3. **(a)** $r^2 > 4km$ **(b)** $r^2 = 4km$
   (c) $r^2 < 4km$
5. **(a)** $CR^2 > 4L$ **(b)** $CR^2 = 4L$
   (c) $CR^2 < 4L$
7. $y = e^{-pt/2}\left(y_0 \cos\omega t + \frac{2v_0 - py_0}{2\omega}\sin\omega t\right)$
   $= \frac{1}{2\omega}\sqrt{4\omega^2 y_0^2 + (2v_0 + py_0)^2}\cos(\omega t - \phi)$,
   where $\omega = \sqrt{q - p^2/4}$ and $\tan\phi = \frac{2v_0 + py_0}{2\omega y_0}$.
9. **(a)** $y'' + py' + qy = 0$, where $p = \frac{2}{3}\ln 2 \approx 0.46$ and $q = \frac{1}{9}(324 + (\ln 2)^2) \approx 36$.

(b) $m = k/q = \dfrac{9}{324 + (\ln 2)^2} \approx 0.028,$

$r = mp = \dfrac{6 \ln 2}{324 + (\ln 2)^2} \approx 0.013$

11. (a) $5(3e^{-2t/3} - e^{-2t})$
(b) $10e^{-t}(1 + t)$
(c) $10e^{-4t/5}\left(\cos\left(\frac{2}{5}t\right) + 2\sin\left(\frac{2}{5}t\right)\right)$

13. (a) $\frac{1}{3}\sin 3t$ (b) $\frac{1}{2}e^{-\sqrt{5}t}\sin 2t$
(c) te^{-3t} (d) $\frac{1}{8}(e^{-t} - e^{-9t})$

Section 6.6, p.211

1. $4(\cos 3t + \sin t)$
3. $\frac{9}{16}(2\cos t + \cos 5t)$
5. (a) $-3t\cos 2t$ (b) $-\frac{3}{2}\cos 2t$
7. $\cos(3t - \phi)$, $\phi = \tan^{-1}\left(\frac{12}{5}\right)$, $G = \frac{14}{13}$
9. $\cos(2t - \phi)$, $\phi = \tan^{-1}\left(\frac{4}{-3}\right) + \pi$, $G = \frac{1}{5}$
11. (a) $x = 2(t\cos t + t^2\sin t)$,
$y = 2(-t^2\cos t + t\sin t)$
(b) $(c_1 + 2t(1 - t))\cos t + (c_2 + 2t(1 + t))\sin t$
13. (b) When $p, q > 0$, the roots of the characteristic polynomial will have negative real parts, and so u will have an exponential factor that causes $u(t)$ to approach 0 as $t \to \infty$. Since $y = u + w$, it follows that $y(t) \approx w(t)$ for large t.
15. $k/\sqrt{(k - m\omega_0^2)^2 + r^2\omega_0^2}$
17. $p = 20$, $q = 300$
21. $q \approx 48$, $p \approx 5.8$, $\dfrac{\omega_0}{2\pi} \approx 0.89$

Chapter 7

Section 7.1, p.220

1. $\dfrac{-3 - 2s}{(s - 1)s}$
3. $\dfrac{s}{s^2 - b^2}$
5. $\dfrac{1}{s^2 + 4s + 3}$
7. $\dfrac{1}{s^2 + s + 1}$
9. $\dfrac{s^2 + 2s + 2}{(s + 1)(s^2 - 4)}$
11. $\dfrac{F(s)}{s^2 + s + 1}$
13. $\dfrac{F(s)}{s^3 + 1}$
15. $\dfrac{1}{(s - a)^2}$
17. $\dfrac{2\omega s}{(s^2 + \omega^2)^2}$
19. $\dfrac{A_1}{s} + \dfrac{A_2}{s^2} + \dfrac{Bs + C}{s^2 + 2s + 2}$
21. $\dfrac{A_1}{s - 1} + \dfrac{A_2}{s + 1} + \dfrac{B_1 s + C_1}{s^2 + 1}$
23. $\dfrac{2}{s - 2} - \dfrac{1}{s - 1}$
25. $\dfrac{1}{s - 1} + \dfrac{1}{s + 1}$
27. $\dfrac{1}{(s - 1)^2} + \dfrac{1}{(s - 1)^3}$
29. $\dfrac{1}{(s - 1)^2} + \dfrac{2}{(s - 1)^3} + \dfrac{1}{(s - 1)^4}$
31. $\dfrac{1}{s - 1} + \dfrac{1 - s}{s^2 + 1}$
33. $\dfrac{1}{s - 1} - \dfrac{2}{s^2 + 2s + 2}$
35. $Y(s) = \dfrac{1/2}{s - 1} - \dfrac{1/2}{s + 1}$,
$y(t) = \frac{1}{2}(e^t - e^{-t}) = \sinh t$
37. $Y(s) = \dfrac{1}{s - 2} + \dfrac{1}{s + 2} - \dfrac{2}{s}$,
$y(t) = e^{2t} + e^{-2t} - 2 = 2\cosh 2t - 2$
39. $Y(s) = \dfrac{2}{s - 1} + \dfrac{9}{3s + 2} - \dfrac{5}{s}$,
$y(t) = 2e^t + 3e^{-2t/3} - 5$
41. (a) The integrand is $e^{-st}e^{t^2} = e^{t^2 - st}$, which (for fixed s) does not approach zero as $t \to \infty$.
(b) The integrand is $e^{-st}t^t = e^{t\ln t - st}$, which (for fixed s) does not approach zero as $t \to \infty$.

43. **(a)** $\int_0^\infty e^{-st} f(ct)\, dt =$

$\int_0^\infty e^{-(s/c)\tau} f(\tau)\frac{1}{c} d\tau = \frac{1}{c} F\left(\frac{s}{c}\right)$

(b) Replace c with $1/c$ in part (a).

Section 7.2, p.227

1. $\frac{1}{s} - \frac{1}{s+2} = \mathcal{L}[1 - e^{-2t}]$,

$\frac{2}{(s+1)^2 - 1} = \mathcal{L}[e^{-t}(2\sinh t)]$

3. $\frac{5}{s-5} - \frac{1}{s-1} = \mathcal{L}[5e^{5t} - e^t]$,

$\frac{4s}{(s-3)^2 - 4} = \frac{4(s-3)}{(s-3)^2 - 4} + \frac{12}{(s-3)^2 - 4}$

$= \mathcal{L}[e^{3t}(4\cosh 2t) + e^{3t}(6\sinh 2t)]$

5. $e^{-3t}(\cos 2t - \sin 2t)$

7. $e^{-t}(2\cos 2t - \sin 2t)$

9. $(t^2 - 4t + 2)e^{-t}$

11. $t(t+1)e^t$

13. $\sin 2t - 2t\cos 2t$

15. $\cos 2t - t\sin 2t$

17. $e^{-t}(2t\cos 2t + (4t-1)\sin 2t)$

19. $\frac{1}{2}\left(3e^{-3t} - e^{-t}\right)$

21. te^{-2t}

23. $\frac{1}{338}\left(26t - 12 + e^{-3t}(12\cos 2t + 5\sin 2t)\right)$

25. $2\sin t - \sin 2t$

27. $\frac{1}{2} t \sin t$

29. Let $y = \int_0^t f(\tau)\, d\tau$. Then $y' = f$ and $y(0) = 0$. Therefore, $Y(s) = F(s)/s$.

31. $\ln\left(\frac{s-a}{s}\right)$

33. $\tan^{-1}(\beta/s)$ **35.** $\sqrt{\frac{\pi}{s}}$

37. $\lim_{s\to\infty} \mathcal{L}[f'(t)] = \lim_{s\to\infty}(sF(s) - f(0)) = 0$; therefore, $\lim_{s\to\infty} sF(s) = f(0)$.

39. $1 - \cos t$

40. $e^{-t} + 2e^{t/2}\cos\left(\frac{\sqrt{3}}{2}t\right)$

Section 7.3, p.234

1. $\begin{cases} 0, & t < 0; \\ 1, & 0 \le t < 1; \\ 2, & 1 \le t < 2; \\ 3, & 2 \le t < 3; \\ 2, & 3 \le t < 4; \\ 1, & 4 \le t < \infty. \end{cases}$

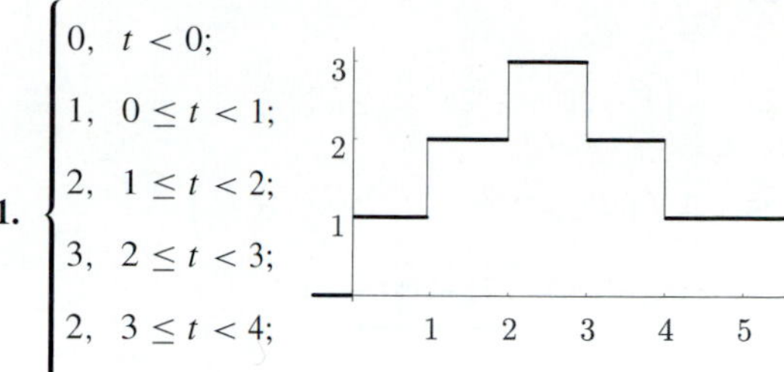

3. $\begin{cases} 0, & t < 0; \\ t, & 0 \le t < 1; \\ 1, & 1 \le t < 2; \\ 0, & 2 \le t < \infty. \end{cases}$

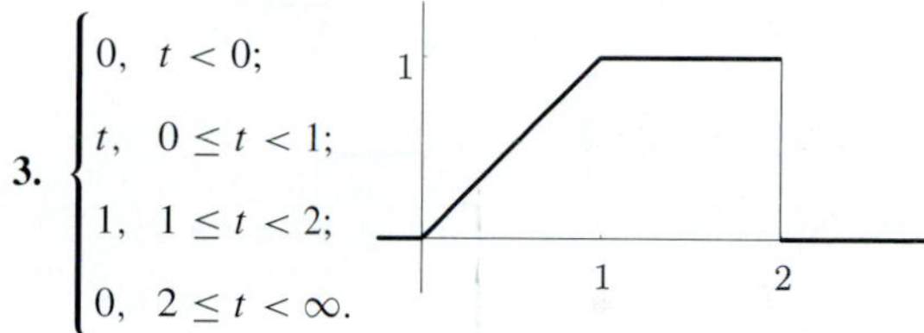

5.

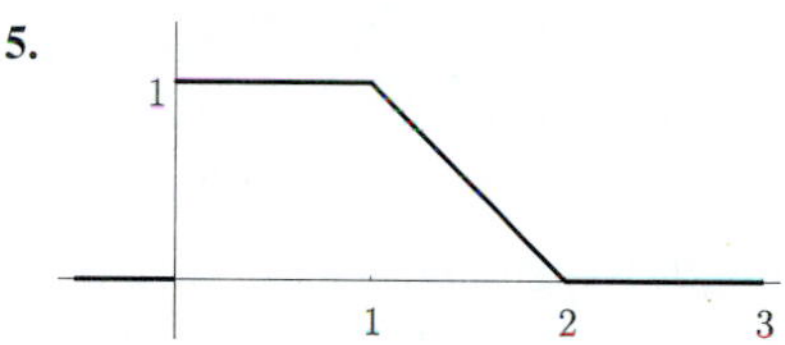

7. $th(t) + (1-2t)h(t-1) - (1-t)h(t-2)$

9. $2h(t) - h(t-1) + h(t-2) - 2h(t-3)$

11. $th(t) - h(t-1) - h(t-2) - (t-2)h(t-3)$

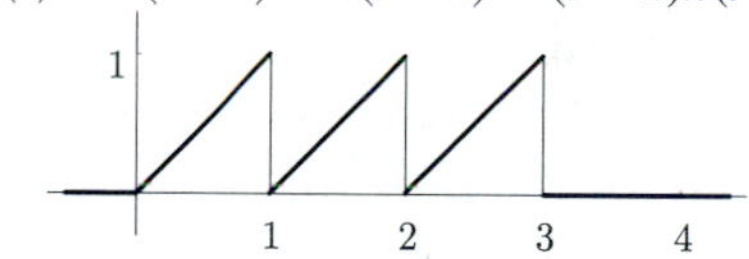

13. $3h(t) - h(t-1) - h(t-2) - h(t-3)$

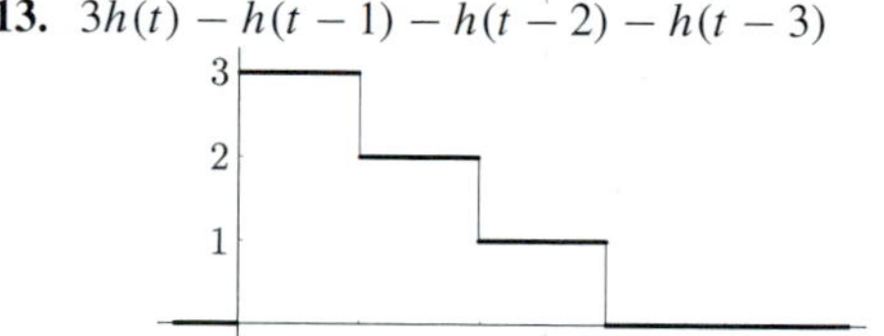

15. $4\mathcal{L}\left[(t - t^2)h(t) + ((t-1) + (t-1)^2)h(t-1)\right]$

$= 4\left(\frac{1}{s^2} - \frac{2}{s^3} + \left(\frac{1}{s^2} + \frac{2}{s^3}\right)e^{-s}\right)$

17. $\mathcal{L}[3h(t) - h(t-1) - h(t-2) - h(t-3)]$

$= \frac{1}{s}\left(3 - e^{-s} - e^{-2s} - e^{-3s}\right)$

19. $\frac{1}{s}\left(1 - 2e^{-s} + e^{-2s} + e^{-3s} - 2e^{-4s} + e^{-5s}\right)$

21. $\frac{1}{s^2}\left(se^{-s} - e^{-2s} + e^{-3s}\right)$

23. $\frac{1}{s}\left(1 - 2e^{-s} + 2e^{-2s} - e^{-3s}\right)$

25. $\frac{1}{s^2}\left(1 - se^{-s/2} - se^{-3s/2} - e^{-2s}\right)$

27. $h(t) - 2h(t-2) + h(t-3)$

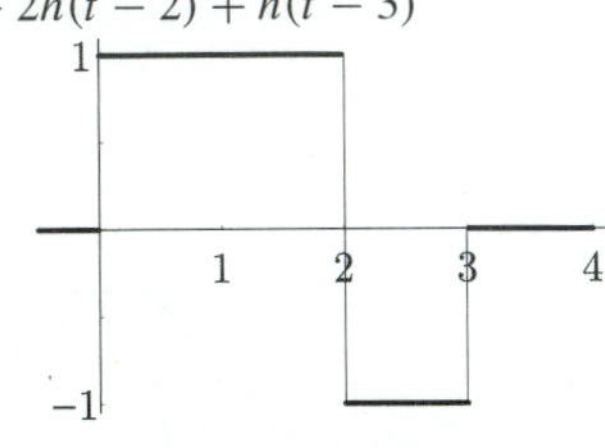

29. $h(t) - 2(t-3)h(t-3) + (t-2)h(t-2)$

31. $2\sin^2 t\,(1 - 2h(t-\pi) + h(t-2\pi))$

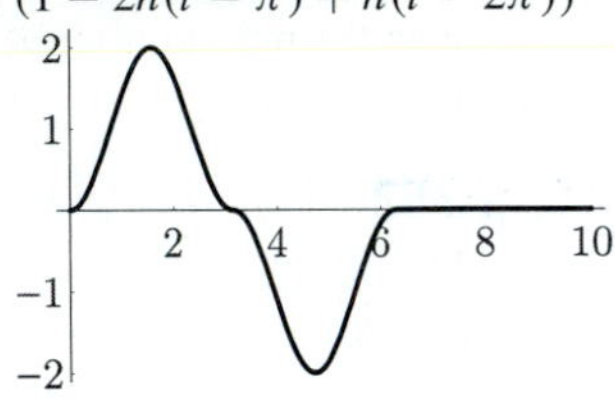

33. $e^t + (e - e^t)h(t-1)$

35. $\cos \pi t + (1 + \cos \pi t)h(t-1) + (-1 + \cos \pi t)h(t-2)$

37. $\int_0^{\infty} e^{-st} f(t)\,dt = \int_1^2 e^{-st}\,dt$

39. $\int_0^{\infty} e^{-st} f(t)\,dt = \int_0^1 e^{(1-s)t}\,dt$

Section 7.4, p.241

1. $\dfrac{1 - (s+2)e^{-s} + (s+1)e^{-2s}}{s^2(1 - e^{-4s})}$

3. $\dfrac{2 - e^{-s} + e^{-2s} - 2e^{-3s}}{s(1 - e^{-4s})}$

5. $f_0(t) = t, 0 \le t < 1, p = 1$

$F_0(s) = \dfrac{1}{s^2}$

$F(s) = \dfrac{1}{s^2(1 - e^{-s})}$

7. $\cos t \sum_{n=0}^{\infty} (-1)^n h(t - n\pi)$

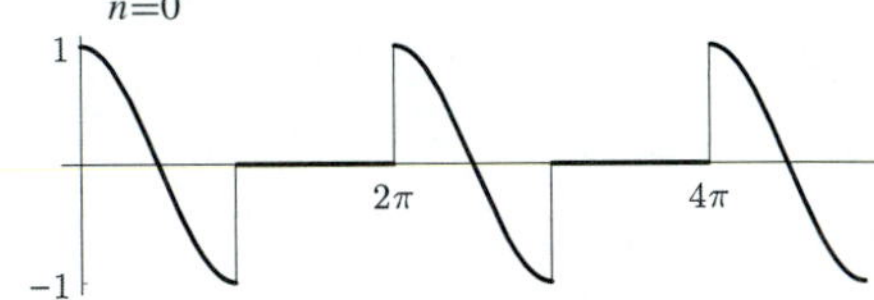

9. $\sum_{n=0}^{\infty} (-1)^n (t-n)h(t-n)$

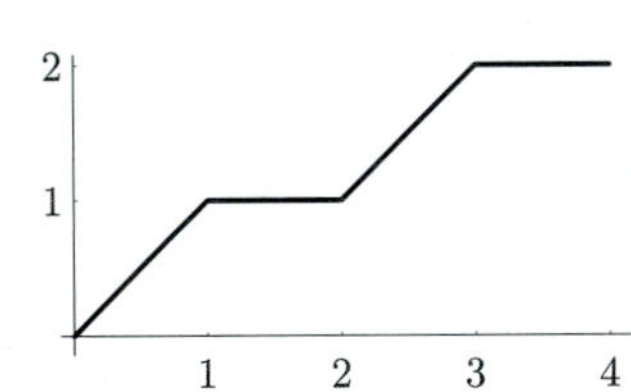

11. $\sum_{n=0}^{\infty} (2 - (t-n))\,h(t-n)$

$+\left(\frac{1}{2\pi}\sin(2\pi t) - 2\cos(2\pi t)\right)\sum_{n=0}^{\infty} h(t-n)$

13. $(1 - \cos(2\pi t))(h(t) - 2h(t-1) + h(t-2)$
$+h(t-4) - 2h(t-5) + h(t-6) + \cdots)$

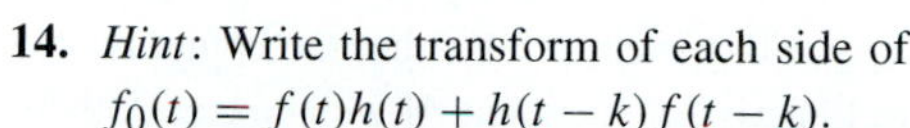

14. *Hint*: Write the transform of each side of $f_0(t) = f(t)h(t) + h(t-k)f(t-k)$.

16. (b) *Hints*: g is periodic with period $2k$, and $1 - e^{-2ks} = (1 - e^{-ks})(1 + e^{-ks})$.

19. *Hint*: $|f(t)| = h(t)$.

21. $\dfrac{1}{(1 - e^{-\pi(s+1)})((s+1)^2 + 1)}$

Section 7.5, p.247

1. 7

3. -1

5. 1

7. $th(t) - (t-1)h(t-1)$

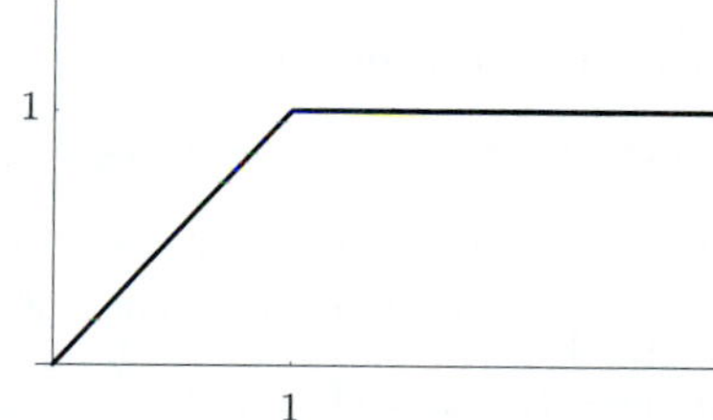

9. $\sin(t-1)h(t-1)$

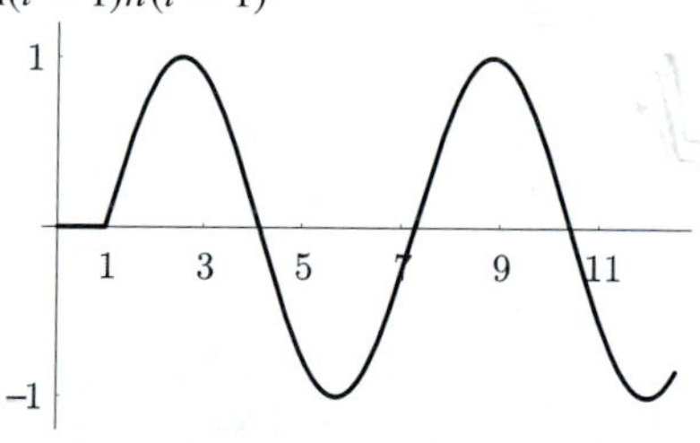

11. $(e^{-t} - e^{-2t})h(t) + 3(3e^{-2t} - e^{-t})h(t - \ln 3)$

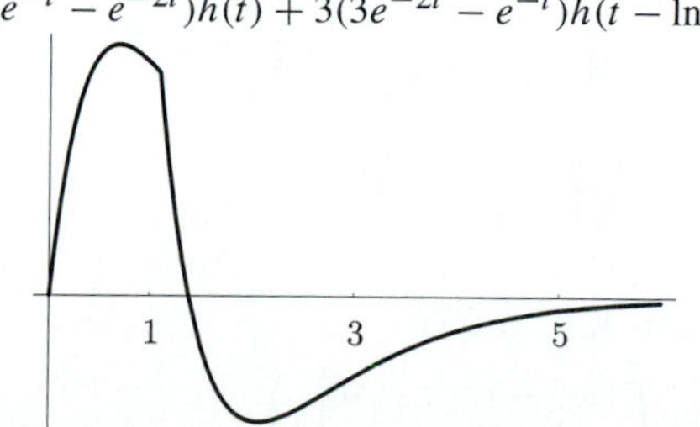

13. $1 + h(t-1)$

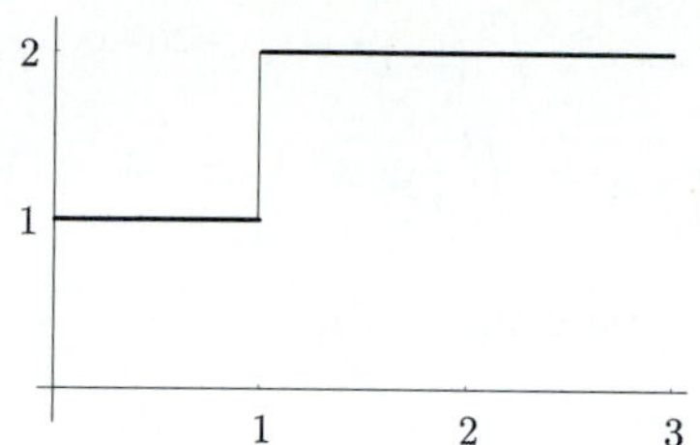

15. $e^{-t}(1 - 2h(t - \ln 2))$

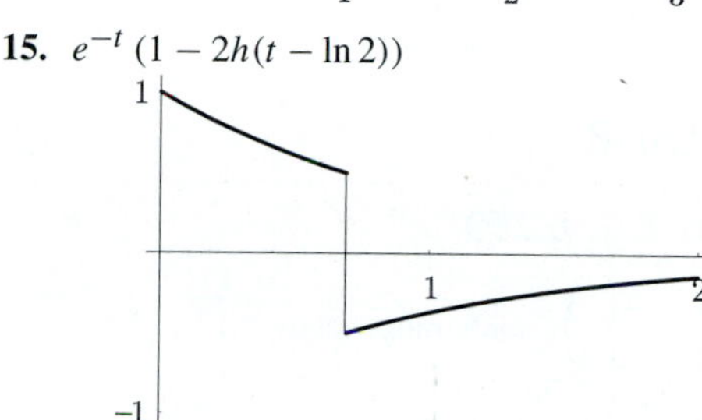

19. (a) $\dfrac{1}{a}((1 - \cos t)h(t) - (1 - \cos(t-a))h(t-a))$

(c) We may as well take the limit as $a \to 0^+$ of the result of (b), because $0 < a < t$ if a is sufficiently small.

(d) The limit as $a \to 0^+$ of the solution from (a) is the same as the rest solution of $y'' + y = \delta(t)$ for all t.

Section 7.6, p.252

1. $\dfrac{4}{15}t^{5/2}$

3. $\sin t$

5. $\dfrac{2}{3}(\cos t - 1)\sin t$

7. $\dfrac{1}{2}(e^t - \cos t + \sin t)$

9. $\dfrac{1}{2}(1 + e^{-2t})$

11. $\sin t$

13. $h(t)\cosh t - h(t-1)\cosh(t-1)$

15. $\left(2 - (t+2)e^{-t}\right)h(t)$

17. $\left(t + \dfrac{1}{6}t^3\right)h(t)$

19. $h(t) + h(t-\pi) + h(t-2\pi) + \cdots$

23. $\dfrac{1}{a^2}\left(e^{at} - at - 1\right)$

25. (a) Begin with $u * f = u * y'' + pu * y' + qu * y$.

(b) $y + (1 - e^{-t}) * y = \cosh t - 1$

27. $\sinh t$, $\displaystyle\int_0^t \sinh(t-x)f(x)\,dx$

29. $e^{-t} - e^{-2t}$, $\int_0^t \left(e^{-(t-x)} - e^{-2(t-x)}\right) f(x)\,dx$

31. $2e^{-t} - e^{-2t}$, $\int_0^t \left(2e^{-(t-x)} - e^{-2(t-x)}\right) f(x)\,dx$

33. $\frac{1}{4}(1+3\cos 2t)$,
$\frac{1}{4}\int_0^t (1+3\cos(2(t-x)))\, f(x)\,dx$

Chapter 8

Section 8.1, p.259

1. $\begin{pmatrix} 1 & -1 \\ t & 1 \end{pmatrix}$, nonhomogeneous

3. $\begin{pmatrix} 1 & -1 & 1 \\ 1 & -1 & -1 \\ 1 & 1 & -1 \end{pmatrix}$, homogeneous

5. $y_1' = ty_1 - y_2$, $y_2' = y_1 - t^2 y_2$

7. $x'' - 3x' + x = 0$

13. $\begin{cases} a_1' = r_0c_0 - \dfrac{r_0+r_1}{V}a_1 + \dfrac{r_1}{V}a_3 \\ a_2' = \dfrac{r_0+r_1}{V}(a_1 - a_2) \\ a_3' = \dfrac{r_0+r_1}{V}a_2 - \dfrac{r_1}{V}a_3 \end{cases}$

15. $\begin{cases} i_1' = \dfrac{1}{L_2}v \\ v' = -\dfrac{1}{C}i_1 - \dfrac{R^2}{L_1}i_2 - \dfrac{R}{L_2}v \\ i_2' = -\dfrac{R}{L_1}i_2 - \dfrac{1}{L_2}v \end{cases}$

Section 8.2, p.265

1. $a_1 = 2(6 - e^{-3t/2} - 5e^{-3t/10})$,
$a_2 = 3(4 + e^{-3t/2} - 5e^{-3t/10})$

5. $\begin{pmatrix} e^{-t} \\ e^{-t} \end{pmatrix}$

7. $\begin{pmatrix} e^{-t}\sin t \\ e^{-t}\cos t \end{pmatrix}$

9. $\begin{pmatrix} (3t-1)e^{-t} + e^{-4t} \\ (6t+1)e^{-t} - e^{-4t} \\ 4(3t+2)e^{-t} + e^{-4t} \end{pmatrix}$

11. $\begin{pmatrix} \sinh t \\ e^t \\ -\sinh t \end{pmatrix}$

Section 8.3, p.269

1. Suppose that $c_1\begin{pmatrix} p \\ k \end{pmatrix} + c_2\begin{pmatrix} h \\ q \end{pmatrix} = 0$. This is equivalent to the pair of equations $pc_1 + hc_2 = 0$ and $kc_1 + qc_2 = 0$. Eliminating c_2 gives $(pq - hk)\ c_1 = 0$, and eliminating c_1 gives $(pq - hk)\ c_2 = 0$. So if $pq - hk \neq 0$, then $c_1 = c_2 = 0$, which implies that the columns of X are linearly independent. If $pq - hk = 0$, then a nontrivial solution c_1, c_2 can always be found (why?), which implies that the columns of X are linearly dependent.

3. The determinant is 0; so the columns are linearly dependent.

5. $3\begin{pmatrix} 1 \\ -1 \end{pmatrix} + 2\begin{pmatrix} 2 \\ 3 \end{pmatrix} = \begin{pmatrix} 7 \\ 3 \end{pmatrix}$

7. $0\begin{pmatrix} 5 \\ 2 \\ 3 \end{pmatrix} + 1\begin{pmatrix} 1 \\ -3 \\ -1 \end{pmatrix} + 0\begin{pmatrix} 3 \\ 1 \\ 2 \end{pmatrix} = \begin{pmatrix} 1 \\ -3 \\ -1 \end{pmatrix}$

9. Let $U = (\mathbf{u}\,\mathbf{v}\,\mathbf{w})$. Then $|U(t)| = t$, which is nonzero somewhere in every interval.

11. **(b)** The entries of A are continuous on $\left(-\frac{\pi}{4}, \frac{\pi}{4}\right)$ but not on any interval containing $\pm\frac{\pi}{4}$.

(c) $|Y(t)| = -\cos 2t$, which is never zero in $\left(-\frac{\pi}{4}, \frac{\pi}{4}\right)$. Therefore, the columns of $Y(t)$ are linearly independent on $\left(-\frac{\pi}{4}, \frac{\pi}{4}\right)$.

(d) $\begin{pmatrix} c_1\sin t + c_2\cos t \\ c_1\cos t + c_2\sin t \end{pmatrix}$

13. **(b)** The entries of A are continuous on all of $\mathbb{R}$.

(c) $|Y(t)| = -5e^{-3t}$, which is never zero. Therefore, the columns of $Y(t)$ are linearly independent on $\mathbb{R}$.

(d) $\begin{pmatrix} c_1e^{-t} + 3c_2e^{-2t} \\ 2c_1e^{-t} + c_2e^{-2t} \end{pmatrix}$

15. $Y(t) = \begin{pmatrix} -e^{-4t} & e^{-t} \\ e^{-4t} & 2e^{-t} \end{pmatrix}$, $|Y(t)| = -3e^{-5t}$

17. $A = \frac{1}{10}\begin{pmatrix} -7 & 6 \\ 6 & 2 \end{pmatrix}$

Section 8.4, p.276

1. $\frac{1}{2t^2}\begin{pmatrix} 8t^3 - t & t - 4t^3 \\ 8t^3 - 1 & 1 - 4t^3 \end{pmatrix}$

3. $\tan t \begin{pmatrix} \sqrt{2}\sec t - 1 & \sqrt{2} - \sec t \\ \sqrt{2} - \sec t & \sqrt{2}\sec t - 1 \end{pmatrix}$

5. $\begin{pmatrix} 4t - 1/t \\ 4 - 1/t^2 \end{pmatrix}$

7. $\tan t \begin{pmatrix} \sqrt{2}\sec t - 1 & \sqrt{2} - \sec t \\ \sqrt{2} - \sec t & \sqrt{2}\sec t - 1 \end{pmatrix} \times \begin{pmatrix} \sqrt{2} - 4\cos^3 t \\ 3\cos^4 t - 3/4 \end{pmatrix}$

9. $\begin{pmatrix} 6e^t - e^{-2t} - 5 \\ 6e^t + e^{2t} - 7 \end{pmatrix}$

11. $(1+t)^{3/2}\ln(1+t)\begin{pmatrix} 1 \\ -1 \end{pmatrix}$

13. Since the entries of A are constants, it follows that $\Psi' = A + tA^2 + \frac{t^2}{2!}A^3 + \cdots$, which is the same as $A\Psi$.

Section 8.5, p.283

1. i) $e^0 = 1$, ii) $e^{a(t+s)} = e^{at}e^{as}$, iii) $\frac{1}{e^{at}} = e^{-at}$, iv) $(e^{at})' = ae^{at}$, v) $e^{at} = 1 + at + \frac{1}{2!}a^2t^2 + \frac{1}{3!}a^3t^3 + \cdots$

3. *Hint*: To verify property (iii), compute $e^{At}e^{A(-t)}$.

5. (a) $\begin{pmatrix} 1+t^2 & t & \frac{1}{2}t^2 \\ 2t & 1 & t \\ -2t^2 & -2t & 1-t^2 \end{pmatrix}$ **(b)** $\begin{pmatrix} \frac{1}{2}t^2 \\ t \\ -t^2 \end{pmatrix}$

7. $\begin{pmatrix} \sin t & -\cos t \\ \cos t & \sin t \end{pmatrix}$, addition formulas for sine and cosine.

9. (a) ± 2 **(b)** $\begin{pmatrix} -3e^{-2t} & e^{2t} \\ e^{-2t} & e^{2t} \end{pmatrix}$

(c) $\frac{1}{4}\begin{pmatrix} 3e^{-2t} + e^{2t} & -3e^{-2t} + 3e^{2t} \\ -e^{-2t} + e^{2t} & e^{-2t} + 3e^{2t} \end{pmatrix}$

11. (a) $-2, -1$ **(b)** $\begin{pmatrix} -e^{-2t} & -3e^{-t} \\ e^{-2t} & 2e^{-t} \end{pmatrix}$

(c) $\begin{pmatrix} -2e^{-2t} + 3e^{-t} & -3e^{-2t} + 3e^{-t} \\ 2e^{-2t} - 2e^{-t} & 3e^{-2t} - 2e^{-t} \end{pmatrix}$

13. (a) $-4, -4, 0$ **(b)** $\begin{pmatrix} e^{-4t} & e^{-4t} & -1 \\ 0 & e^{-4t} & 2 \\ e^{-4t} & 0 & 1 \end{pmatrix}$

(c) $\frac{1}{4}\begin{pmatrix} 1 + 3e^{-4t} & -1 + e^{-4t} & -1 + e^{-4t} \\ -2 + 2e^{-4t} & 2 + 2e^{-4t} & 2 - 2e^{-4t} \\ -1 + e^{-4t} & 1 - e^{-4t} & 1 + 3e^{-4t} \end{pmatrix}$

15. (a) $-1, -1 \pm i$

(b) $e^{-t}\begin{pmatrix} 0 & \cos t + \sin t & \sin t - \cos t \\ -1 & \cos t + \sin t & \sin t - \cos t \\ 1 & \cos t & \sin t \end{pmatrix}$

(c) $e^{-t} \cdot \begin{pmatrix} 0 & 0 & 0 \\ -1 & 1 & 0 \\ 1 & -1 & 0 \end{pmatrix} + e^{-t}\cos t \begin{pmatrix} 1 & 0 & 0 \\ 1 & 0 & 0 \\ -1 & 1 & 1 \end{pmatrix} + e^{-t}\sin t \begin{pmatrix} -3 & 2 & 2 \\ -3 & 2 & 2 \\ -2 & 1 & 1 \end{pmatrix}$

17. $\frac{1}{5}\begin{pmatrix} 4 + e^{-5t} & 2 - 2e^{-5t} \\ 2 - 2e^{-5t} & 1 + 4e^{-5t} \end{pmatrix}$

19. $e^{-4t}\begin{pmatrix} \cos 6t & -\frac{3}{2}\sin 6t \\ \frac{2}{3}\sin 6t & \cos 6t \end{pmatrix}$

21. $\frac{1}{8}\begin{pmatrix} 11e^{-7t} - 3e^t & 33(e^t - e^{-7t}) \\ e^{-7t} - e^t & 11e^t - 3e^{-7t} \end{pmatrix}$

23. $\begin{pmatrix} e^t & e^t - e^{-t} & e^{-t} - e^t \\ e^{-t} - e^t & 2e^{-t} - e^t & e^t - e^{-t} \\ e^{-t} - e^t & e^{-t} - e^t & e^t \end{pmatrix}$

25. $e^{-4t}\begin{pmatrix} 1 + 12t & -9t \\ 16t & 1 - 12t \end{pmatrix}$

27. $e^{-t}\begin{pmatrix} 1 - t & -t + 2t^2 & t - t^2 \\ -t & 1 + 3t + 2t^2 & -t - t^2 \\ -2t & 6t + 4t^2 & 1 - 2t - 2t^2 \end{pmatrix}$

29. $\begin{pmatrix} 1 & 2 & -2 & 1 \\ 1 & 2 & -2 & 1 \\ 1 & 2 & -2 & 1 \\ 0 & 0 & 0 & 0 \end{pmatrix} - e^t \begin{pmatrix} 0 & 0 & 0 & 0 \\ 1 & 0 & -1 & 0 \\ 1 & 0 & -1 & 0 \\ 0 & 0 & 0 & 0 \end{pmatrix}$

$- e^{-t} \begin{pmatrix} 0 & 2+t & 2+t & 1+t \\ 0 & 1 & -1 & 1 \\ 0 & 2+t & -2-t & 1+t \\ 0 & t & -t & t-1 \end{pmatrix}$

31. (a) $\begin{cases} (A-\lambda\mathcal{I})^j \mathbf{v}_j & = (A-\lambda\mathcal{I})^{j-1}\mathbf{v}_{j-1} \\ & = \cdots = (A-\lambda I)\mathbf{v}_1 \\ & = p \neq 0 \\ (A-\lambda\mathcal{I})^{j+1}\mathbf{v}_j & = (A-\lambda\mathcal{I})^j \mathbf{v}_{j-1} \\ & = \cdots = (A-\lambda\mathcal{I})p \\ & = 0. \end{cases}$

(b) $(A-\lambda\mathcal{I})\mathbf{v}_j = \mathbf{v}_{j-1}$,
$(A-\lambda\mathcal{I})^2\mathbf{v}_j = \mathbf{v}_{j-2}, \ldots,$
$(A-\lambda\mathcal{I})^j\mathbf{v}_j = p.$

33. $p = \begin{pmatrix} 1 \\ 0 \\ 1 \end{pmatrix}, \mathbf{v}_1 = \frac{1}{2}\begin{pmatrix} 0 \\ 1 \\ 0 \end{pmatrix}, \mathbf{v}_2 = \frac{1}{8}\begin{pmatrix} 2 \\ 1 \\ 0 \end{pmatrix},$

$Y(t) = \frac{1}{8}e^{-t}\begin{pmatrix} 8 & 8t & 2+4t^2 \\ 0 & 4 & 1+4t \\ 8 & 8t & 4t^2 \end{pmatrix}$

35. $e^t \begin{pmatrix} 0 & 1 & 1+t & 2+t+t^2/2 \\ 1 & t & t^2/2 & t^3/6 \\ 0 & 1 & 2+t & 6+2t+t^2/2 \\ 0 & 1 & 2+t & 5+2t+t^2/2 \end{pmatrix}$

37. $\begin{pmatrix} 0 & -e^{-t} & 0 & -1 \\ e^{-t} & te^{-t} & -1 & -t \\ e^{-t} & te^{-t} & 0 & 1 \\ 0 & 0 & 1 & t \end{pmatrix}$

Section 8.6, p.293

1. (a) $\ell_1(x) = \frac{x-\lambda_2}{\lambda_1-\lambda_2}$ and $\ell_2(x) = \frac{x-\lambda_1}{\lambda_2-\lambda_2}$;

so $\gamma(x) = e^{\lambda_1 t}\frac{x-\lambda_2}{\lambda_1-\lambda_2} + e^{\lambda_2 t}\frac{x-\lambda_1}{\lambda_2-\lambda_2}$.

(b) $\frac{e^{\alpha t}}{2\beta i}\big((\cos\beta t + i\sin\beta t)(A-(\alpha-\beta i)\mathcal{I})$
$-(\cos\beta t - i\sin\beta t)(A-(\alpha+\beta i)\mathcal{I})\big) = \cdots$

(c) $\gamma(x) = te^{\lambda_1 t}(x-\lambda_1) + e^{\lambda_1 t} =$
$e^{\lambda_1 t}(tx+1-\lambda_1 t)$

3. $\frac{1}{2}\begin{pmatrix} e^{4t}+e^{-2t} & e^{4t}-e^{-2t} \\ e^{4t}-e^{-2t} & e^{4t}+e^{-2t} \end{pmatrix}$

5. $\frac{1}{2}\begin{pmatrix} e^{-3t}+e^{-5t} & e^{-3t}-e^{-5t} \\ e^{-3t}-e^{-5t} & e^{-3t}+e^{-5t} \end{pmatrix}$

7. $e^{-t}\begin{pmatrix} \cos 2t + t\sin 2t \\ t\cos 2t - \sin 2t \end{pmatrix}$

9. $\frac{1}{2}(e^{-t}(5\mathcal{I}+4A+A^2) - e^{-2t}\ (\cos t(3\mathcal{I}+4A+$
$A^2) + \sin t(\mathcal{I}+A)^2))$

11. $e^{-2t}\left(\mathcal{I}+t(A+2\mathcal{I})+\frac{1}{2}t^2(A+2\mathcal{I})^2\right)$

13. $\mathcal{I}+(e^t-1)A$

15. $\cos t\,\mathcal{I}+\sin t\,A$

Section 8.7, p.301

1. (a) $\frac{t^2+2t+C}{t+1}$

(b) Asymp. stable;

solutions $\approx \frac{t^2+2t}{t+1}$ as $t\to\infty$.

3. (a) $2\sqrt{t+1}-1+Ce^{-2\sqrt{t+1}}$

(b) Asymp. stable;

solutions $\approx 2\sqrt{t+1}-1$ as $t\to\infty$.

5. (a) $\frac{2t+8\sin t+\sin 2t+C}{4(2+\cos t)}$

(b) not asymp. stable.

7. $\gamma < 1/4$

9. $|\gamma| > 1$

11. (a) Eigenvalues: -1, -3

(b) $\begin{pmatrix} 4\sin t - 3\cos t \\ \sin t - 2\cos t \end{pmatrix}$

13. *Hint*:
$\int te^t\cos t\,dt = \frac{1}{2}e^t(t\cos t+(t-1)\sin t)+C.$

The general solution is $y = \frac{Ce^{-t}-\sin t}{2t} + w(t)$,

where $w(t) = \frac{1}{2}(\cos t+\sin t)$.

16. Recall from Section 4.4 that the solution with $y(0)=y_0$ is $y = \frac{My_0}{y_0+(M-y_0)e^{-kt}}$. Also, negative solutions exist only for $t < \frac{1}{k}\ln\left(\frac{y(0)-M}{y(0)}\right)$.

17. Solutions with $0 < y(0) < 2$ are asymptotically stable.

19. $\frac{4+2\sin t}{4-\cos t+\sin t}$

Chapter 9

Section 9.1, p.309

1.

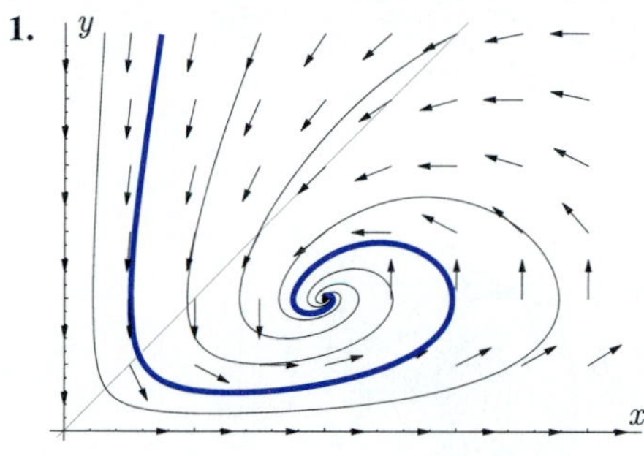

3.

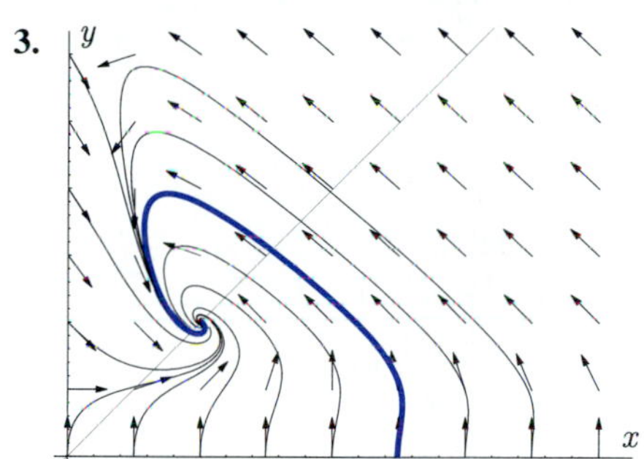

5.

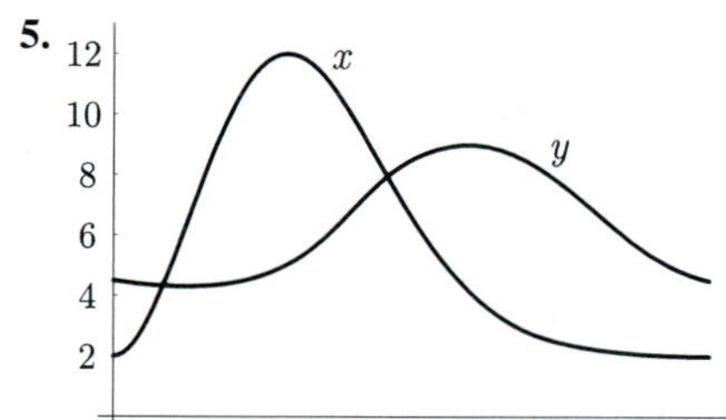

7.

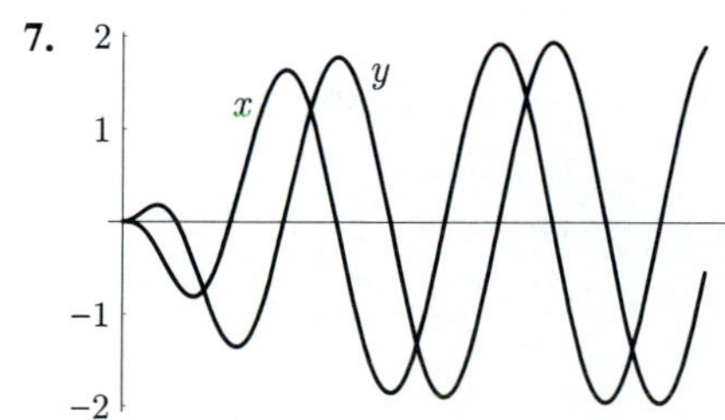

9.

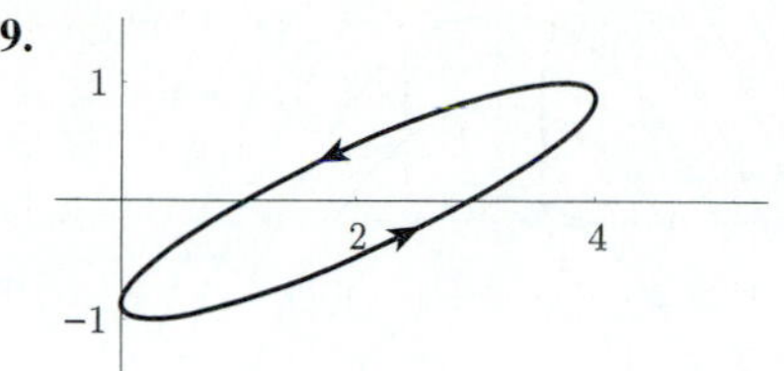

11.

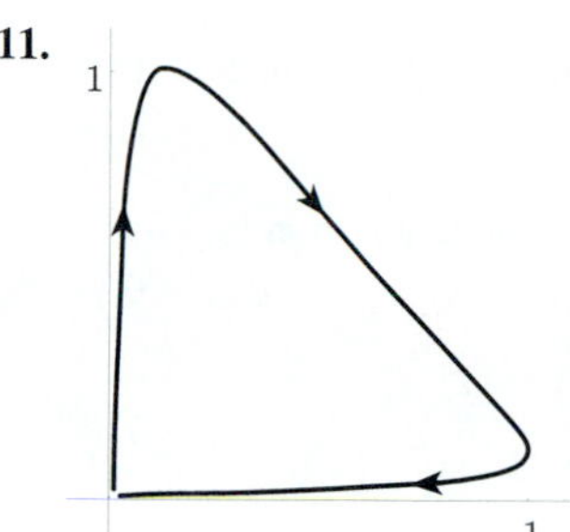

13.

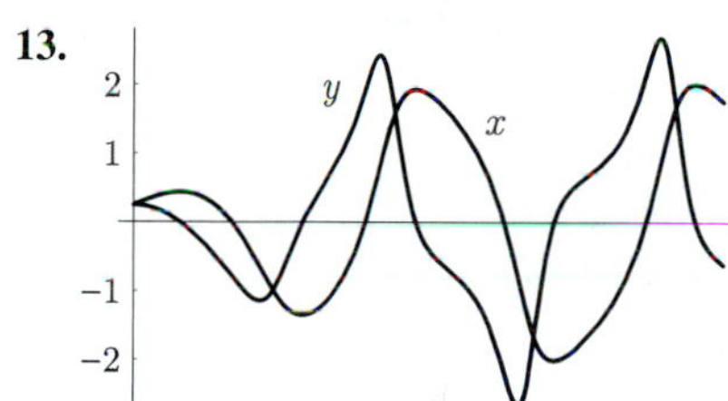

15. (0, 0), (1, 0), and (3, 0) are unstable; (2, 0) is neutrally stable.

17. In brief: The first two plots in Figure 11 correspond to the two inner, periodic orbits in Figure 10, respectively. The third plot corresponds to the clockwise orbit comprising the upper portion of outer circle, and the fourth corresponds to the counterclockwise orbit comprising the lower portion of outer circle.

19. $y^2 - x^2 = C$

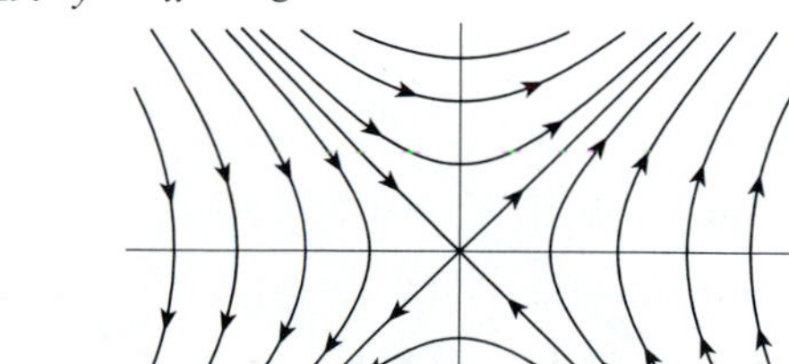

21. $2y + x^2 = C$

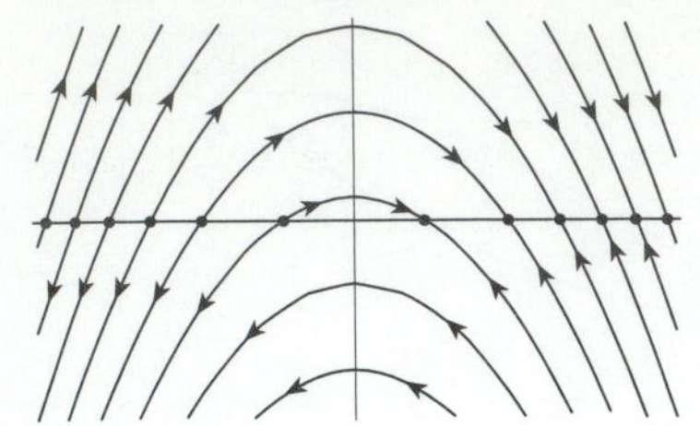

23. $y^2 = 2\cos x + C$

24. Show that the slope equation is satisfied when $x^2 + y^2 = 1$ and $\dfrac{dy}{dx} = -x/y$.

25. *Hint*: Rearrange the slope equation into $(2x + x^2)\,dx + 2y\,dy - (x\,dy + y\,dx) = 0$.

Section 9.2, p.320

1. **(a)** unstable saddle point;
   (b) real, opposite sign
2. **(a)** stable spiral point;
   (b) nonreal, negative real part
3. **(a)** stable biaxial node;
   (b) real, negative
4. **(a)** stable coaxial node;
   (b) real, repeated, negative
5. **(a)** unstable biaxial node;
   (b) real, positive
6. **(a)** unstable spiral point;
   (b) nonreal, real part > 0

7.

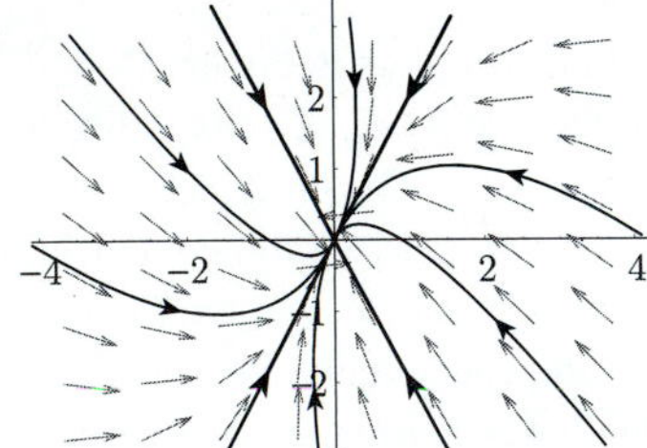

9.

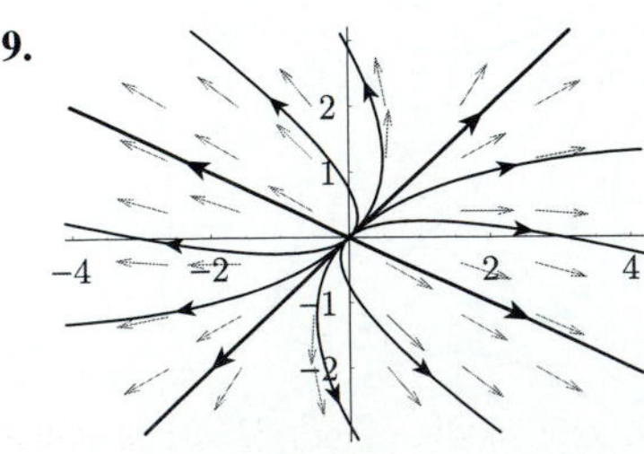

11.

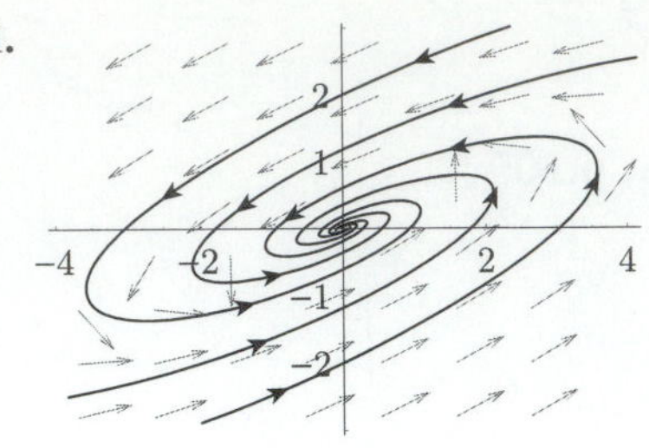

13.

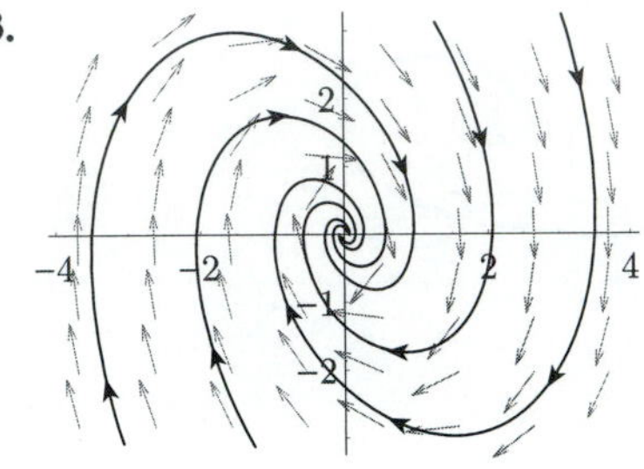

15. (a)

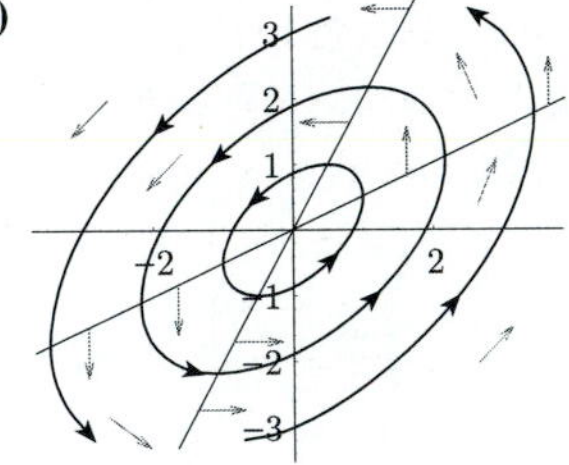

(b)

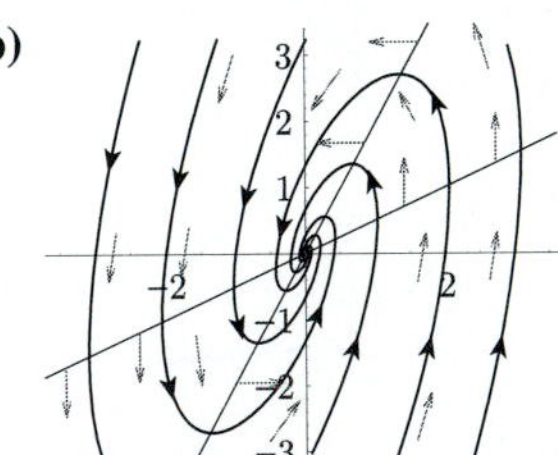

(c)

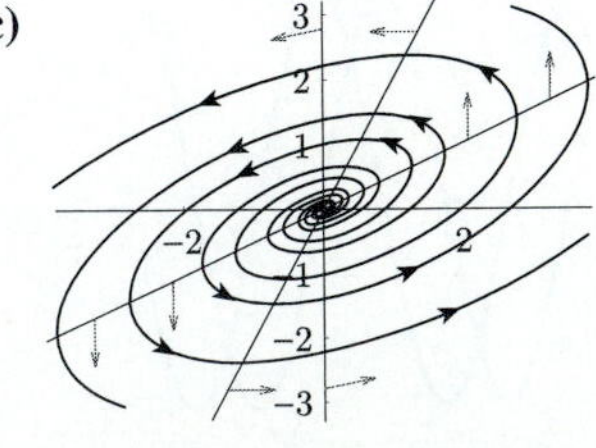

(d)

17. (a)

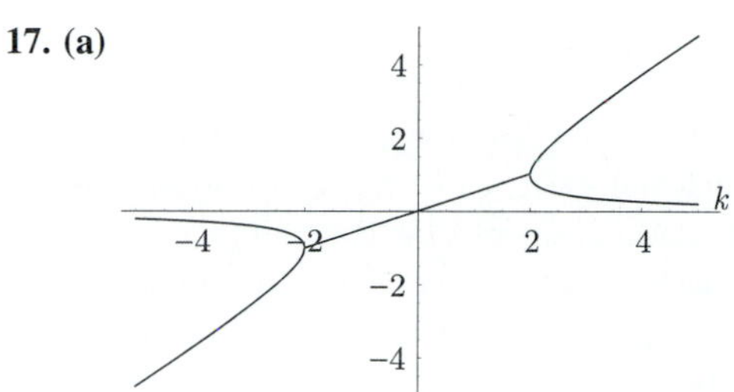

(b) Real and distinct when $|k| > 2$; real and repeated when $k = \pm 2$; imaginary when $k = 0$; nonreal complex when $|k| < 2$.

(c) $k < -2$: stable biaxial node. $k = -2$: stable coaxial node. $-2 < k < 0$: stable spiral point. $k = 0$: center. $0 < k < 2$: unstable spiral point. $k = 2$: unstable coaxial node. $k > 2$: unstable biaxial node.

21. Eigenvalues: -2, -1.
Straight-line orbits: $y = -x$, $y = -2x$.

23. Eigenvalues: $-1 \pm 2i$.
Straight-line orbits: none.

25. Eigenvalues: 0, 2. Nullclines coincide to form a line ($y = x$) of equilibrium points. All nontrivial orbits are straight lines: $x + y = C$.

Section 9.3, p.334

1. (a) (0, 0), (−1, 1), (2, 4)

(b) $\mathcal{J}(0,0) = \begin{pmatrix} -2 & 0 \\ 0 & -1 \end{pmatrix}$,
eigenvalues: $-2, -1$

$\mathcal{J}(-1,1) = \begin{pmatrix} 1 & -1 \\ -2 & -1 \end{pmatrix}$,
eigenvalues: $\pm\sqrt{3}$

$\mathcal{J}(2,4) = \begin{pmatrix} -2 & 2 \\ 4 & -1 \end{pmatrix}$,
eigenvalues: $\frac{1}{2}(-3 \pm \sqrt{33})$

(c) (0,0) is a stable node; (−1, 1) and (2,4) are saddle points.

(e)

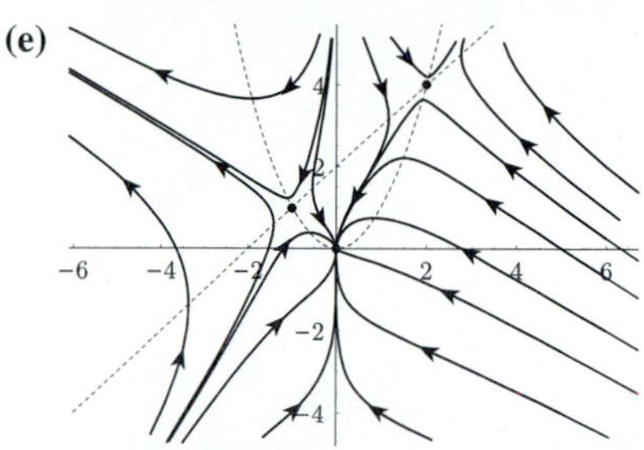

3. (a) (0, 0), (1, 1)

(b) $\mathcal{J}(0,0) = \begin{pmatrix} 1 & 0 \\ 1 & -1 \end{pmatrix}$, eigenvalues: -1, 1

$\mathcal{J}(1,1) = \begin{pmatrix} 0 & -1 \\ 1 & -1 \end{pmatrix}$,
eigenvalues: $\frac{1}{2}(-1 \pm i\sqrt{3})$

(c) (0, 0) is a saddle point; (1, 1) is a stable spiral point.

(e)

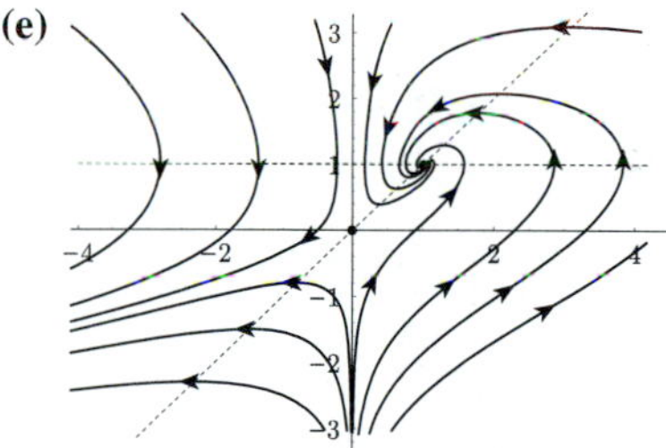

5. (a) (−1, −1), (1, 1)

(b) $\mathcal{J}(-1,-1) = \begin{pmatrix} 2 & 2 \\ 1 & -1 \end{pmatrix}$,
eigenvalues: $\frac{1}{2}(-1 \pm \sqrt{17})$

$\mathcal{J}(1,1) = \begin{pmatrix} -2 & -2 \\ 1 & -1 \end{pmatrix}$,
eigenvalues: $\frac{1}{2}(-3 \pm i\sqrt{7})$

(c) (−1, −1) is a saddle point; (1, 1) is a stable spiral point.

(e)

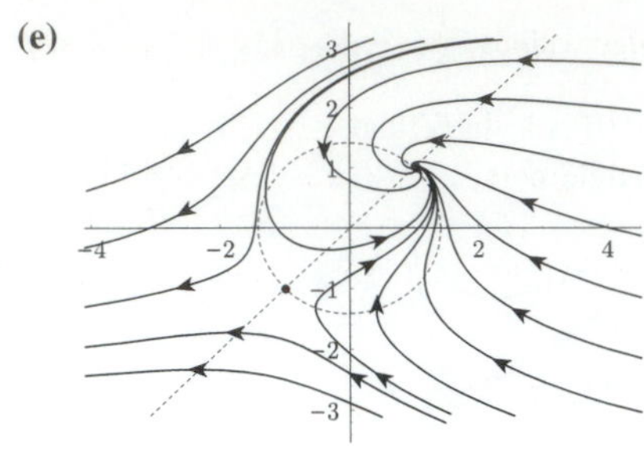

7. (a) $(-1,-1)$, $(-1,1)$, $(1,-1)$, $(1,1)$

(b) $\mathcal{J}(-1,-1)=\begin{pmatrix} -2 & -2 \\ -2 & 2 \end{pmatrix}$,
eigenvalues: $\pm\sqrt{8}$

$\mathcal{J}(-1,1)=\begin{pmatrix} -2 & 2 \\ -2 & -2 \end{pmatrix}$,
eigenvalues: $-2\pm 2i$

$\mathcal{J}(1,-1)=\begin{pmatrix} 2 & -2 \\ 2 & 2 \end{pmatrix}$,
eigenvalues: $2\pm 2i$

$\mathcal{J}(1,1)=\begin{pmatrix} 2 & 2 \\ 2 & -2 \end{pmatrix}$, eigenvalues: $\pm\sqrt{8}$

(c) $(-1,-1)$ and $(1,1)$ are saddle points; $(-1,1)$ is a stable spiral point; $(1,-1)$ is an unstable spiral point.

(e)

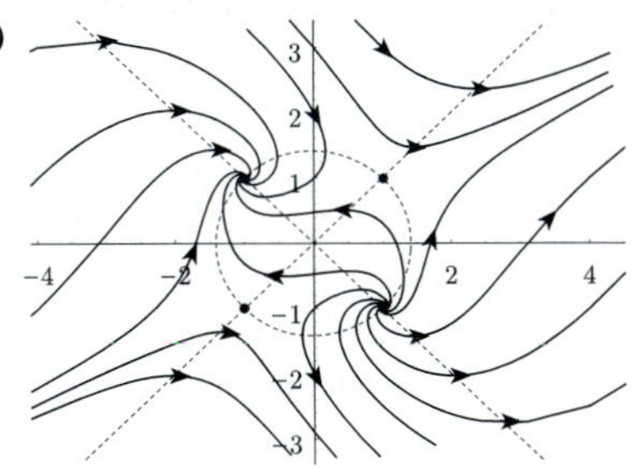

9. (a) $\left(1,\frac{1}{2}\right)$

(b) $\mathcal{J}\left(1,\frac{1}{2}\right)=\begin{pmatrix} 1 & 2 \\ 0 & -2 \end{pmatrix}$, eigenvalues: -2,1

(c) $\left(1,\frac{1}{2}\right)$ is a saddle point.

(e)

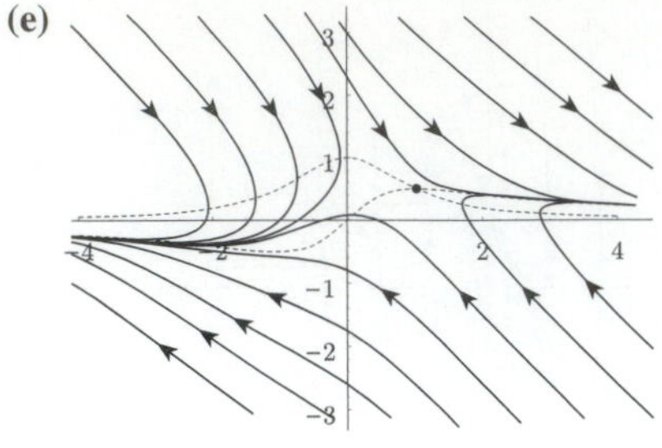

11. As in the undamped case, each odd n corresponds to the unstable position at the top of the pendulum's arc, and each even n corresponds to the stable position at the bottom of the pendulum's arc. When $\rho'(0)$ is small relative to k, damped oscillations occur, corresponding to a stable spiral point. For larger values of $\rho'(0)$, damping prevents oscillation and causes the equilibrium point to be a stable node.

13. Equilibrium point: $\left(k^2,\frac{1}{k}\right)$.

Stable node if $0<k^3<3-2\sqrt{2}$.

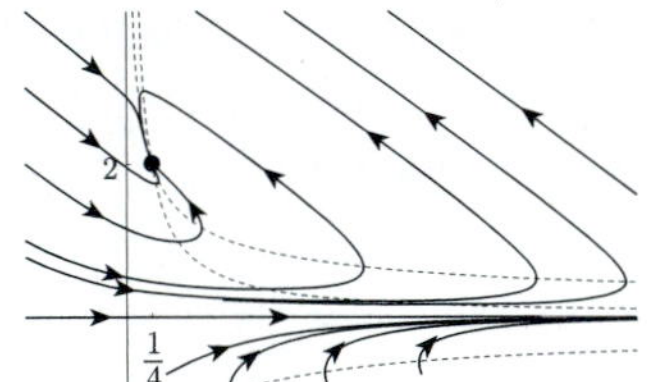

Stable spiral point if $3-2\sqrt{2}<k^3<1$.

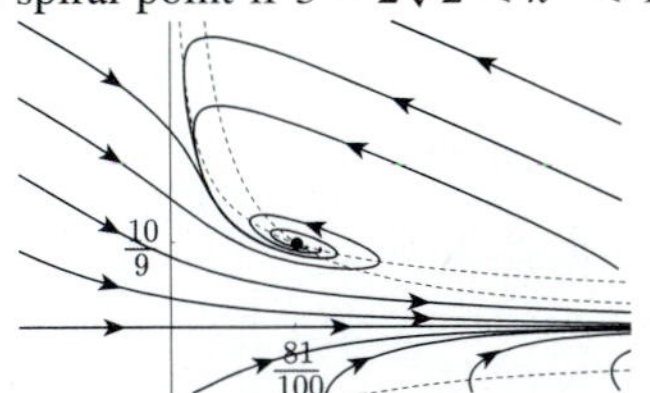

Possible center if $k=1$.

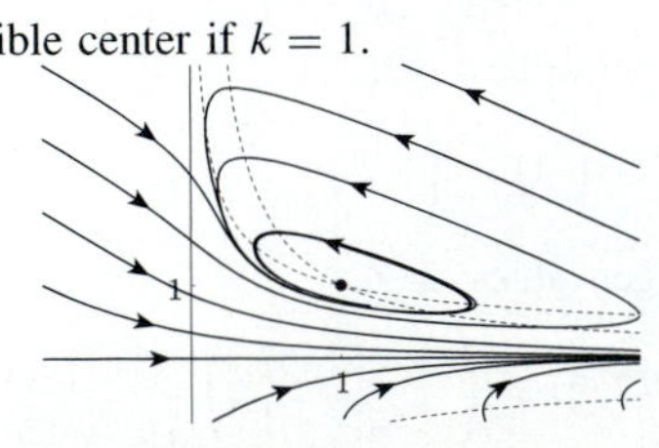

Unstable spiral point if $1 < k^3 < 3 + 2\sqrt{2}$.

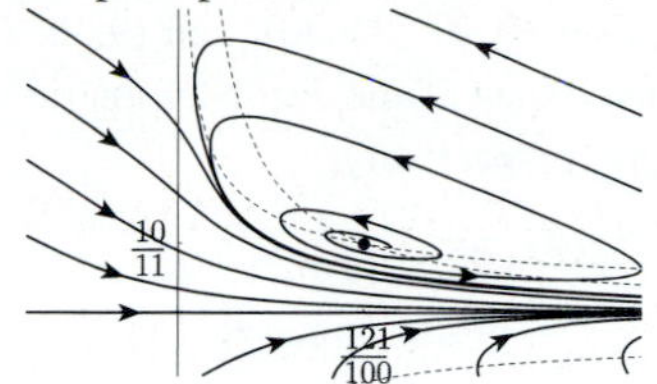

Unstable node if $k^3 > 3 + 2\sqrt{2}$.

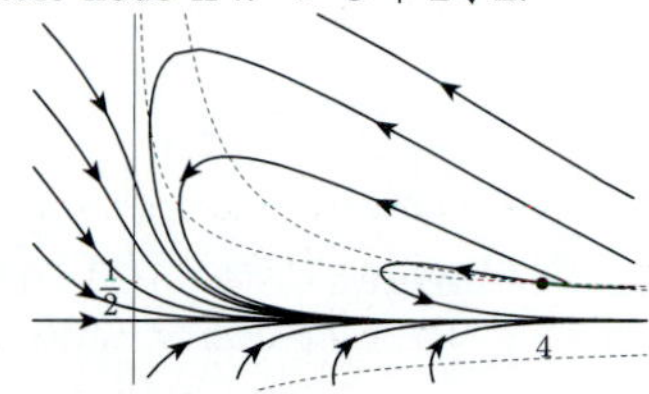

15. (0, 0): unstable node
(1, 1) and (0, 2): saddle points
(0.2956, 1.839): stable node

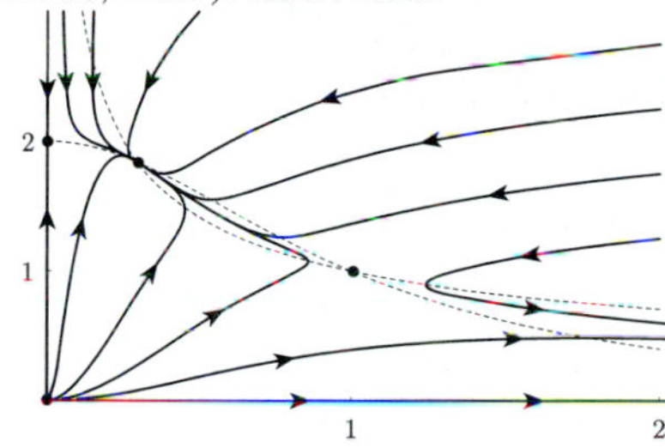

17. (0, 0) and (3, 0) are stable spiral points, and (2, 0) is a saddle point. (1, 0) has characteristics of a stable node and a saddle point, consistent with the fact that the eigenvalues of $\mathcal{J}(1, 0)$ are -1 and 0.

19. (0, 0): stable spiral point; $\left(\dfrac{5}{2}, 0\right)$: saddle point.

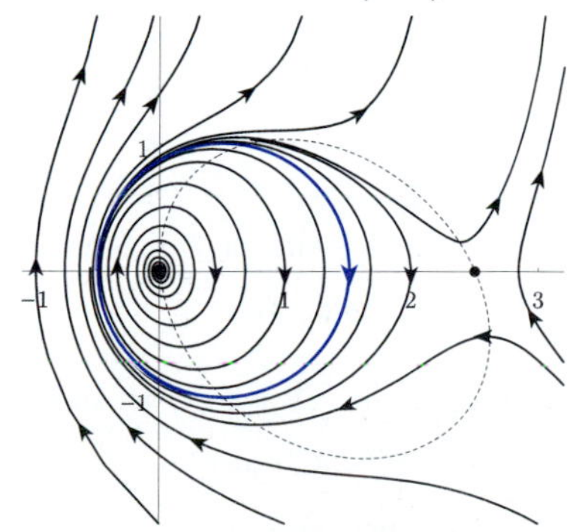

21. **(a)** Orbits follow the contours (level curves) of the surface.

(b) Nearby orbits trace closed curves around (x_0, y_0). Therefore, (x_0, y_0) is a center.

23. **(a)** $\varphi(x, y) = C$, where $\varphi(x, y) = \dfrac{1}{2}y^2 - \cos x$.

(b) If n is even, then
$(\varphi_{xx}\varphi_{yy} - \varphi_{xy}^2)|_{(\pm n\pi, 0)} = \cos(\pm n\pi) = 1 > 0$ and $\varphi_{xx}|_{(\pm n\pi, 0)} = \cos(\pm n\pi) = 1 > 0$. Nearby orbits are periodic; each $(\pm n\pi, 0)$ is a center.

Section 9.4, p.344

1. **(b)** If $r^2 \le p(0) \le 1$, then $p(t)$ is nondecreasing for all $t \ge 0$, and $p(t) \to 1$ as $t \to \infty$. If $1 \le p(0) \le R^2$, then $p(t)$ is nonincreasing for all $t \ge 0$, and $p(t) \to 1$ as $t \to \infty$.

(c) If $0 < p(0) < 1$, then $p(t)$ is increasing for all $t \ge 0$, and $p(t) \to 1$ as $t \to \infty$. Thus all orbits near (0, 0) move strictly away from (0, 0).

(d) The annulus $\mathcal{A}_{r,R}$ contains a periodic orbit. (Letting $r \to 1^-$ and $R \to 1^+$ shows that a periodic orbit traces out the unit circle, as is easily verified.)

3. **(b)** If $R^2 \le p(t) \le r^2$, then $p'(t) < 0$.

(c) $p'(t) > 0$ when $(x(t), y(t))$ is near the origin.

(d) A periodic orbit exists inside the disk $x^2 + y^2 \le R^2$.

5. **(a)** If $x(t) \ge 0$ and $y(t) = 0$, then $y'(t) = bx(t) + \beta \ge 0$. If $x(t) = 0$ and $y(t) \ge 0$, then $x'(t) = py(t) + \alpha \ge 0$. Therefore, no orbit can leave the first quadrant.

(b) *Hint*: Show that the first-quadrant portion of the line $x + y = k$ lies above the line $(a + b)x + (p + q)y + \alpha + \beta = 0$.

(c) The triangle $\{(x, y)|x \ge 0, y \ge 0, x + y \le k\}$ is forward-invariant for any $k \ge 4/3$. Also, the only equilibrium point in the first quadrant is (1/2, 2/3), which is a stable spiral point. This suggests, and a phase portrait confirms, that *every* first quadrant orbit approaches (1/2, 2/3) as $t \to \infty$.

7. The only equilibrium point is (1, 3), which is an unstable spiral point. A forward-invariant region containing (1, 3) is

$$\left\{(x, y)|x \ge 0, y \ge 0, y \le \frac{9}{4}(x + 5), x + y \le \frac{29}{2}\right\}.$$

Hints: $(x + y)' \le 0$ if $x \ge 1$;

$$y' \le \frac{9}{4y} \text{ if } 0 \le x \le 1 \text{ and } y \ge 3;$$

$$x' \ge 1 - \frac{4}{y} \text{ if } 0 \le x \le 1 \text{ and } y > 4;$$

$$\frac{dy}{dx} \le \frac{9}{4} \text{ if } 0 \le x \le 1 \text{ and } y \ge 5.$$

11. Every periodic orbit must enclose the origin. Therefore, since the first quadrant is a forward-invariant region (by Theorem 2), no periodic orbit can exist.

13. Every periodic orbit must enclose a point on the line $y = x$, which consists entirely of equilibrium points. This implies that a periodic orbit must pass through an equilibrium point, which is impossible. Therefore, no periodic orbit can exist.

17.

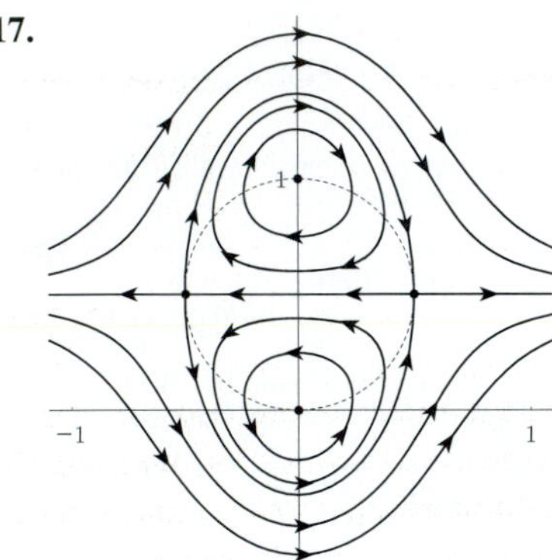

19. $\varphi_x f + \varphi_y g = 2xy - 2yx = 0$. Therefore, disks centered at $(0, 0)$ are forward-invariant. (In fact, orbits are circular.)

21. $\varphi_x f + \varphi_y g = 2x(y^2 - x^3) - 2xy^2 = -2x^4 \le 0$. Therefore, disks centered at $(0, 0)$ are forward-invariant. Also, $(0, 0)$ is asymptotically stable.

Section 9.5, p.355

1. $(0, 0, 0)$ is the only equilibrium point. The eigenvalues of the Jacobian matrix there are $-c$ and $\frac{1}{2}\left(-(a+1) \pm \sqrt{(a-1)^2 + 4ab}\right)$. Since $a > 0$ and $0 < b \le 1$, it follows that

$$\frac{1}{2}\left(-(a+1) + \sqrt{(a-1)^2 + 4ab}\right) < 0.$$

Therefore, all three eigenvalues are real and negative.

3. The equilibrium points are (approximately) $(0,0,0)$, $(-9.30, 27, -9.30)$, and $(9.30, 27, 9.30)$. The eigenvalues of the Jacobian matrix at these points are, respectively,

$$-22.8, 11.8, -3.2;$$
$$-14.2, 0.00200 \pm 11.0i;$$
$$-14.2, 0.00200 \pm 11.0i.$$

5. $\frac{d}{dt}\ell(x, y, z) = -2\left(y^2z^2 + x^2(1 + z^2)\right)$.

(a) no. **(b)** yes.

7. $\frac{d}{dt}\ell(x, y, z) = 2x(x - y)(1 + z)$.

(a) no. **(b)** no.

Chapter 10

Section 10.1, p.366

1. (a) k_1, k_2: time^{-1}; c_1, c_2: "individuals;" α_1, α_2: individuals$^{-1}\cdot$ time^{-1}

3. $x' = 5x(1 - x - 2y)$, $x(0) = 1$

$y' = 20y(1 - \frac{5}{2}x - y)$, $y(0) = 1$

The phase portrait is similar to Figure 2c with $\left(\frac{1}{4}, \frac{3}{8}\right)$ an unstable node. Also, the phase portrait shows that $(x(t), y(t)) \to (1, 0)$ as $t \to \infty$. Thus $(p(t), q(t)) \to (500, 0)$ as $t \to \infty$.

5. $x' = 10x(1 - x - \frac{1}{2}y)$, $x(0) = \frac{4}{5}$

$y' = 20y(1 - \frac{1}{2}x - y)$, $y(0) = \frac{1}{2}$

The phase portrait is similar to Figure 2d with $\left(\frac{2}{3}, \frac{2}{3}\right)$ a stable node. Thus $(x(t), y(t)) \to \left(\frac{2}{3}, \frac{2}{3}\right)$ as $t \to \infty$, and so $(p(t), q(t)) \to (667, 67)$ as $t \to \infty$.

7. *Hint*: Use the slope equation

$$\frac{dy}{dx} = \frac{k_2 y(1 - y - bx)}{k_1 x(1 - x - ay)}.$$

Also, when $k_1 = k_2$, the straight-line orbit has slope $m = (b - 1)/(a - 1)$.

9.

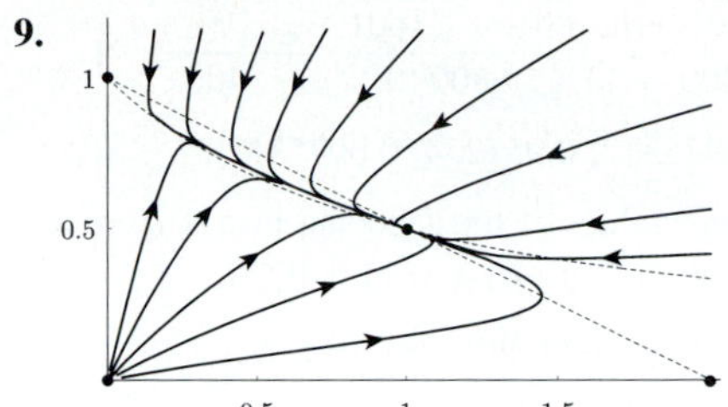

11. $x' = x(1 - \frac{2}{5}x - y)$, $x(0) = \frac{5}{2}$

$y' = -20y(1 - x)$, $y(0) = 10$

The phase portrait is similar to Figure 4c with $\left(1, \frac{3}{5}\right)$ a stable spiral point. Thus $(x(t), y(t)) \to \left(1, \frac{3}{5}\right)$ as $t \to \infty$, and so $(p(t), q(t)) \to (200, 60)$ as $t \to \infty$.

13. $x' = 50x(1 - \frac{1}{2}x - y)$, $x(0) = \frac{8}{5}$

$y' = -5y(1 - x)$, $y(0) = \frac{1}{20}$

The phase portrait is similar to Figure 4b with $\left(1, \frac{1}{2}\right)$ a stable node. Thus $(x(t), y(t)) \to \left(1, \frac{1}{2}\right)$ as $t \to \infty$, and so $(p(t), q(t)) \to (500, 500)$ as $t \to \infty$.

15.

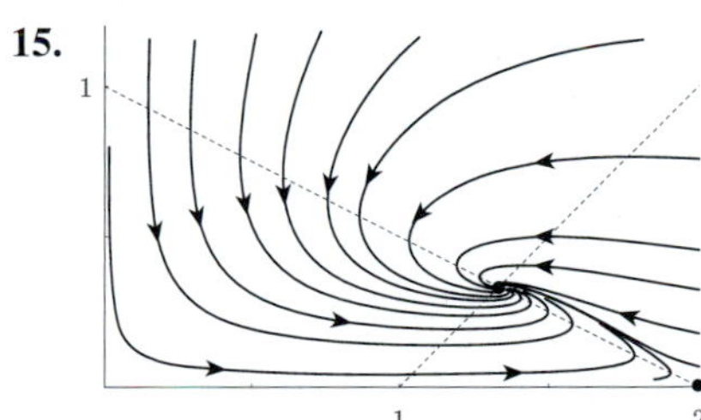

17.

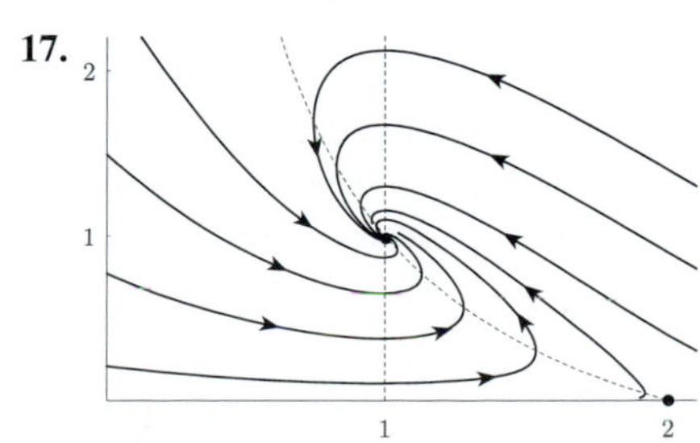

19.

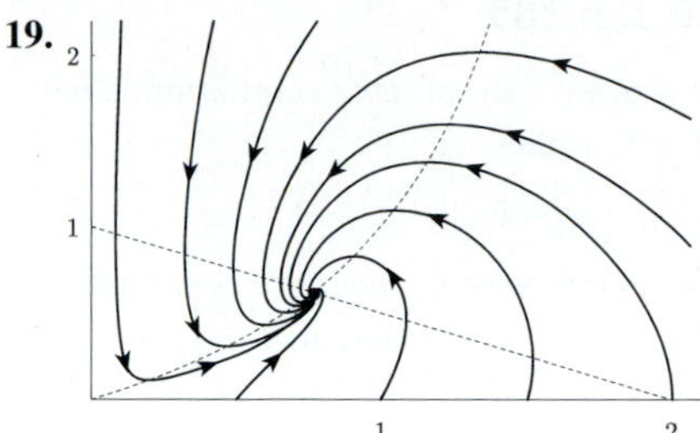

21. Equilibria: $(0, 0)$, $(\varepsilon, 0)$, $(0, \beta)$, and $\left(\frac{\varepsilon(1+\beta)}{1-\beta\varepsilon}, \frac{\beta(1+\varepsilon)}{1-\beta\varepsilon}\right)$.

The first three are always unstable. The fourth is of interest only if $\beta\varepsilon < 1$, and in that case it is stable. If $\beta\varepsilon \geq 1$, then $x(t) \to \infty$ and $y(t) \to \infty$ as $t \to \infty$ for any positive initial values.

23. **(a)** y preys on x, x preys on z, and z preys on y.

(b) $(0, 0, 0)$ is stable. $(5, 8, \frac{9}{2})$ is unstable.

Section 10.2, p.380

1. **(a)** The total population obeys a logistic equation in the absence of the disease (i.e., if $\delta = 0$).

(b) The term $-bI(S + I)$ reflects the facts that (1) the birth rate of infectives is zero, and (2) the per capita death rate due to natural causes is the same as in the general population (i.e., it's proportional to $S + I$). The term $(a - bS)(S + I)$, or $a(S + I) - bS(S + I)$, indicates that the birth rate of susceptibles is proportional to the total population and that the per capita death rate due to natural causes is is the same as in the general population.

(c) $\left(\frac{5}{2}, \frac{5}{2}\right)$ is a stable spiral point.

5. **(a)** $I = -S + \rho \ln S + C$, where $\rho = (\gamma + \delta)/\beta$.

(b) $I(0) + S(0) - \rho\,(1 - \ln \rho + \ln S(0))$

9. **(b)** The equilibrium point is $\left(\frac{q}{a}, 0, 0\right)$.

(c) $0 < b < 1$ and $(1 - b)q > abc$

(d) With $b = \frac{1}{2}$, $a = c = 1$, and $q = 2$ (for example), the equilibrium point of interest is $(1, 2, 1)$, and it is asymptotically stable. The infection is not eliminated, but it is kept under control by the treatment.

Section 10.3, p.385

3. $(0, 0, q)$ is always an unstable equilibrium point. If $\beta \geq 1 + 1/q$, then $\left(\frac{\beta}{\beta - 1}(q\beta - q - 1), 0, \frac{1}{\beta - 1}\right)$ is an equilibrium point with nonnegative coordinates, which is asymptotically stable if $\beta > \gamma$ and unstable if $\beta < \gamma$. If $\gamma \geq 1 + 1/q$, then $\left(0, \frac{\gamma}{\gamma - 1}(q\gamma - q - 1), \frac{1}{\gamma - 1}\right)$ is an equilibrium point with nonnegative coordinates, which is asymptotically stable if $\gamma < \beta$ and unstable if $\gamma > \beta$.

5. There is a single equilibrium point (x^*, y^*) at the intersection of the nullclines. The eigenvalues of $\mathcal{J}(x^*, y^*)$ are $-1 \pm i\frac{\sqrt{2ay^*}}{b + (y^*)^2}$. Thus (x^*, y^*) is an asymptotically stable spiral point. A phase-portrait sketch makes it clear that all nonnegative solutions approach (x^*, y^*) as $t \to \infty$.

7. *Hints*: **(a)** v^* is a root of $f(v) = bv^3 + 3(1 - b)v - 3a$. Look at $f'(v)$.
(b) You want $f(1) > 0$. (Why?)
(c) The eigenvalues of $\mathcal{J}(v^*, w^*)$ are $\frac{1}{2k}\left(\alpha \pm \sqrt{\alpha^2 - 4k^2(1 - b + bv^{*2})}\right)$ where $\alpha = k^2(1 - v^{*2}) - b$.

Section 10.4, p.392

1. $a' = -k_1ab^2 + k_2c$, $b' = -2k_1ab^2 + 2k_2c$, $c' = k_1ab^2 - k_2c$

3. $a' = -k_1ab + k_2c - k_3ac$, $b' = -k_1ab + k_2c$, $c' = k_1ab - k_2c - k_3ac$, $p' = k_3ac$

5. $a' = -k_1abc + k_2p$, $b' = -k_1abc + k_2p$, $c' = -k_1abc + k_2p$, $p' = k_1abc - k_2p$

7. Asymptotically stable if $\beta < 1 + \alpha^2$, unstable if $\beta > 1 + \alpha^2$.

9. **(a)** $(\gamma, 1/\gamma)$ is unstable if $0 < \gamma < 1$ and asymptotically stable if $\gamma > 1$.

(b) *Hint*: Look at the direction field between the x-nullclines $x = 0$ and $xy = 1$.

(c) $\left(\gamma + \rho, \gamma/(\gamma + \rho)^2\right)$ is unstable if $(\gamma + \rho)^3 < \gamma - \rho$ and asymptotically stable if $(\gamma + \rho)^3 > \gamma - \rho$.

11. Equilibrium point:

$$\left(\frac{\sqrt{1601} - 1}{400}, \frac{801 - \sqrt{1601}}{800}, \frac{\sqrt{1601} - 1}{400}\right),$$

or (0.0975312, 0.951234, 0.0975312).

The eigenvalues of the Jacobian matrix there are (approximately) -16.4, $0.22 \pm 0.55i$.

13. You need to show the following:

If $x = q$, $\frac{q}{q+1} \leq y \leq \frac{1}{2q}$, and $q \leq z \leq 1$, then $x' \geq 0$.

If $x = 1$, $\frac{q}{q+1} \leq y \leq \frac{1}{2q}$, and $q \leq z \leq 1$, then $x' \leq 0$.

If $q \leq x \leq 1$, $y = \frac{q}{q+1}$, and $q \leq z \leq 1$, then $y' \geq 0$.

If $q \leq x \leq 1$, $y = \frac{1}{2q}$, and $q \leq z \leq 1$, then $y' \leq 0$.

If $q \leq x \leq 1$, $\frac{q}{q+1} \leq y \leq \frac{1}{2q}$, and $z = q$, then $z' \geq 0$.

If $q \leq x \leq 1$, $\frac{q}{q+1} \leq y \leq \frac{1}{2q}$, and $z = 1$, then $z' \leq 0$.

Section 10.5, p.399

1. $x_1' = v_1$, $v_1' = \frac{k}{m}(-2x_1 + x_2 + x_3)$

$x_2' = v_2$, $v_2' = \frac{k}{m}(x_1 - 2x_2 + x_3)$

$x_3' = v_3$, $v_3' = \frac{k}{m}(x_1 + x_2 - 2x_3)$

Eigenvalues: 0, $\pm i\sqrt{3k/m}$

2. *Hint*: $KE = \frac{1}{2}(m_1 + m_2)\|P'(t)\|^2$, where $P(t)$ is the center of mass of the system. Express $P(t)$ in rectangular coordinates. Also, the potential energy of the pivot is constant and so may be ignored.

3. Let ℓ be the equilibrium length of the spring, and let $\ell + z(t)$ be its length at time t. Let θ be the usual angle. Then the Lagrangian is

$$L = \frac{1}{2}m\left(\ell^2\theta'^2 + z'^2 - 2g\left(\ell - (\ell + z)\cos\theta\right) - \frac{k}{m}z^2\right),$$

and the equations of motion are

$$\ell^2\theta'' = -g(\ell + z)\sin\theta, mz'' = mg\cos\theta - kz.$$

5. Let the position of the first mass be given by the usual polar-coordinate angle θ, and let x be the position of the second mass. Then the Lagrangian is

$$L = \frac{1}{2}\left(m_1\theta'^2 + m_2x'^2\right) + \frac{m_1m_2G}{\sqrt{1 - 2x\cos\theta + x^2}},$$

and the equations of motion are

$$m_1\theta'' = \frac{m_1m_2Gx\sin\theta}{(1 - 2x\cos\theta + x^2)^{3/2}},$$
$$m_2x'' = \frac{m_1m_2G(x - \cos\theta)}{(1 - 2x\cos\theta + x^2)^{3/2}}.$$

7. Partial answer:

$$L = \tfrac{1}{2}m\ell^2(\theta'^2 + \phi'^2) - mg\ell(2 - \cos\theta - \cos\phi)$$
$$-k\ell\left(\sqrt{\begin{array}{l}(3 + \sin\phi - \sin\theta)^2\\ +(\cos\phi - \cos\theta)^2\end{array}} - 2\right)^2$$

Chapter 11

Section 11.1, p.408

13. $13 + 12x - x^3$

15. There is no solution.

Section 11.2, p.417

1. $e^{-9\pi^2kt}\sin(3\pi x)$

3. $2e^{-\pi^2kt}\sin(\pi x) + e^{-4\pi^2kt}\sin(2\pi x)$

5. $3e^{-kt}\sin x + e^{-4kt}\sin(2x) - \frac{1}{2}e^{-9kt}\sin(3x)$

7. **(b)** $2 + 5e^{-\pi^2kt}\cos(\pi x) - e^{-9\pi^2kt}\cos(3\pi x)$

(c) $1 + e^{-4kt}\cos(2x) + e^{-9kt}\cos(3x)$

(d) $\sum_{n=0}^{\infty} a_ne^{-\sigma_n^2kt}\cos(\sigma_nx)$

9. **(b)**

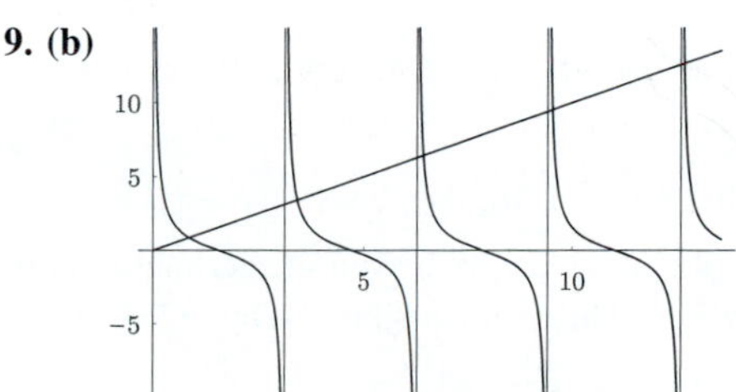

(c) $\sum_{n=1}^{\infty} a_ne^{-\sigma_n^2kt}\cos(\sigma_nx)$

11. **(a)** $1 - x^3$

(b) $w_t - \frac{1}{2}w_{xx} = 0,\ 0 < x < 1,\ t > 0,$

$w(0, x) = x^3 + x - 1,\ 0 \le x \le 1,$

$w(t, 0) = w(t, 1) = 0,\ t > 0;$

12. **(b)** Partial answer: $w = -e^{-t}\sin x + e^{-4t}\sin 2x$

15. **(a)** Integrate each side of $-kz_{xx} = f$ over $[0, \ell]$.

(b) Partial answer: One solution is

$$z = -\frac{1}{k}\int_0^x\int_0^\xi f(\eta)\,d\eta\,d\xi.$$

17. *Hint*: $|\sin\lambda_nx| \le 1$ and $|\cos\lambda_nx| \le 1$.

18. *Suggestion*: Consider, in turn, the nth term, the Nth partial sum, and the limit of the partial sums.

Section 11.3, p.427

1. **(a)** Coefficients: $a_0 = \frac{1}{2}$; $a_n = -\frac{1}{n\pi}\sin(\frac{n\pi}{2})$,

$$b_n = \frac{1}{n\pi}\left(\cos(\frac{n\pi}{2}) - (-1)^n\right),\ n \ge 1$$

(b)

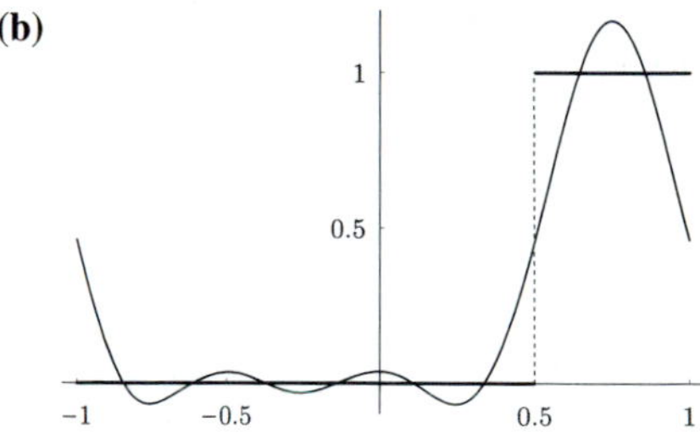

(c)

3. **(a)** $\frac{1}{2} + \sum_{n=1}^{\infty}\frac{2}{n\pi}\sin\left(\frac{n\pi}{2}\right)\cos(n\pi x)$

(b)

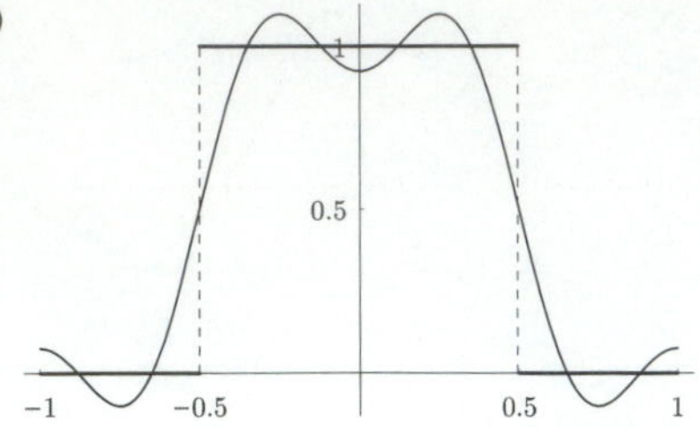

(c)

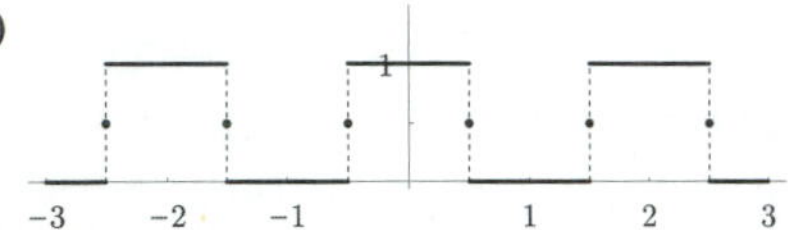

5. **(a)** $\sum_{n=1}^{\infty} \frac{2(-1)^{n+1}}{n\pi} \sin(n\pi x)$

(b)

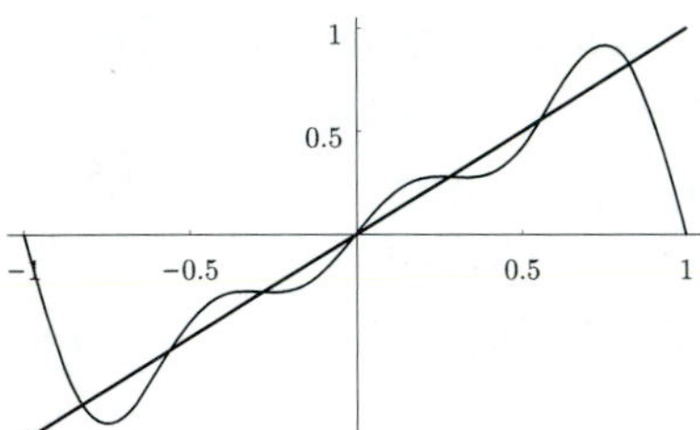

(c)

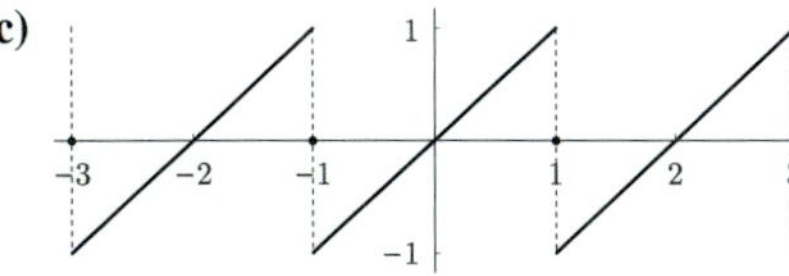

7. **(a)** $2\sum_{n=1}^{\infty} \frac{1-(-1)^{n}}{n\pi} \sin(n\pi x)$

(b)

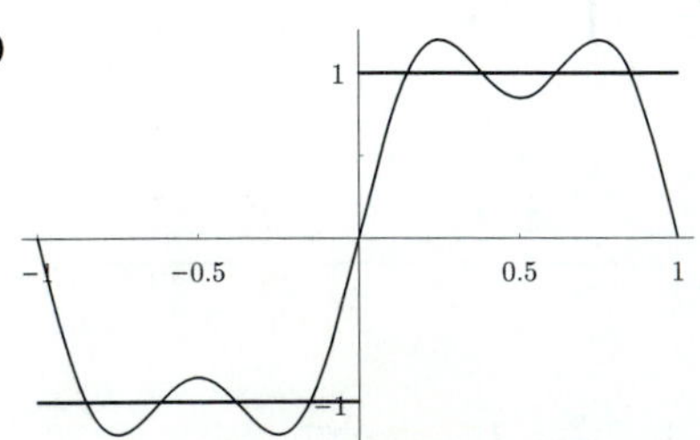

(c)

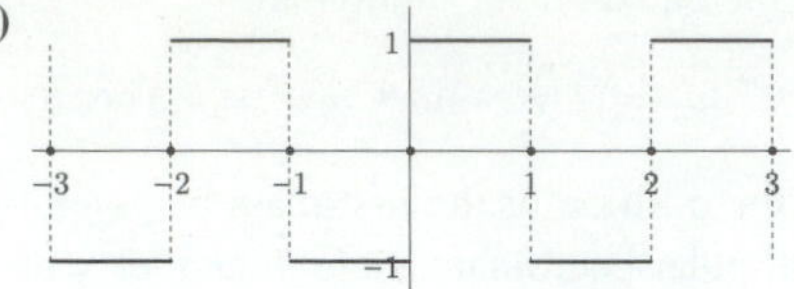

9.

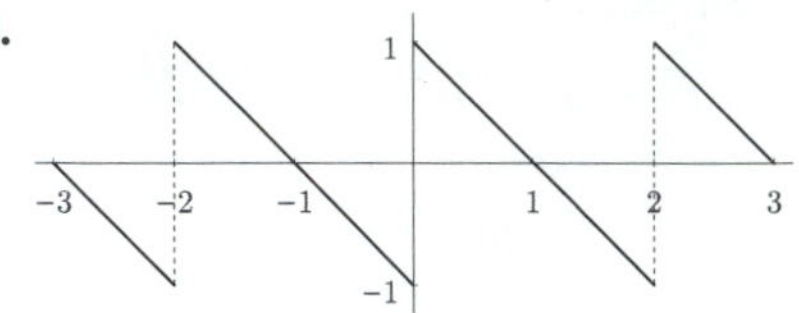

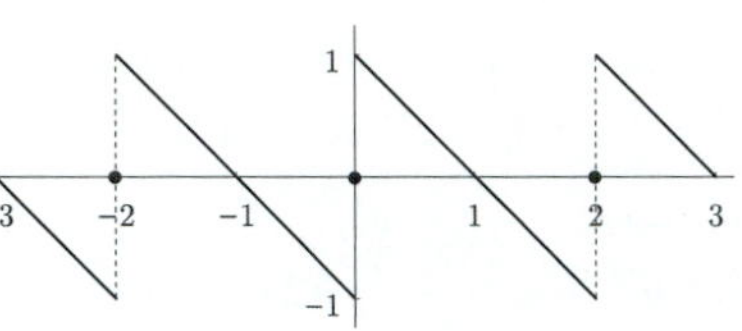

11.

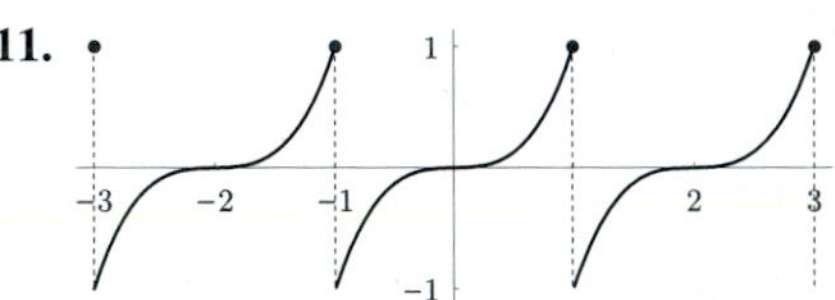

13. *Hint*: Use the "product to sum" trigonometric identities.

15. **(a)** *Hint*:

$$\frac{f(x+p+h)-f(x+p)}{h} = \frac{f(x+h)-f(x)}{h}$$

(c) *Hint*: If $mp = nq$, then

$$f(x+mp)+g(x+mp) = f(x+mp)+g(x+nq).$$

(d) 22π

(e) If $f(p) = f(0)$, then $\cos p + \cos \pi p = 2$.

Thus p and πp are each even-integer multiples of π. (Why?) This is impossible. (Why?)

17. $|x| = \frac{1}{2} + \frac{2}{\pi^2}\sum_{n=1}^{\infty} \frac{(-1)^n - 1}{n^2} \cos(n\pi x)$

$$= \frac{1}{2} - \frac{4}{\pi^2} \sum_{k=1}^{\infty} \frac{1}{(2k-1)^2} \cos((2k-1)\pi x)$$

With $x = 1$, this is $1 = \frac{1}{2} + \frac{4}{\pi^2} \sum_{k=1}^{\infty} \frac{1}{(2k-1)^2}$, which gives the desired result.

Section 11.4, p.437

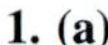

1. (a)

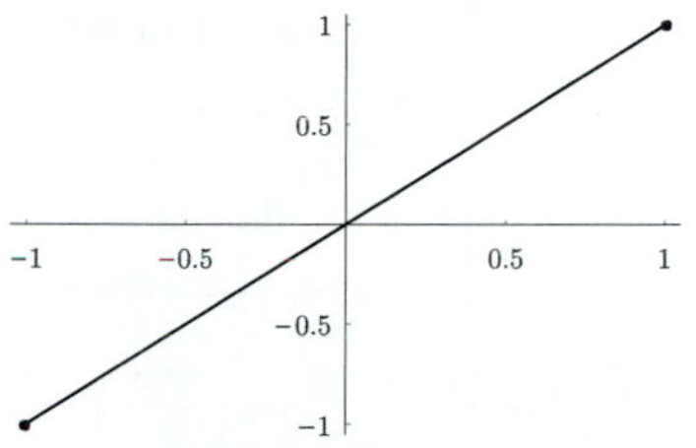

(b) $-\frac{2}{\pi} \sum_{n=1}^{\infty} \frac{(-1)^n}{n} \sin(n\pi x)$

(c)

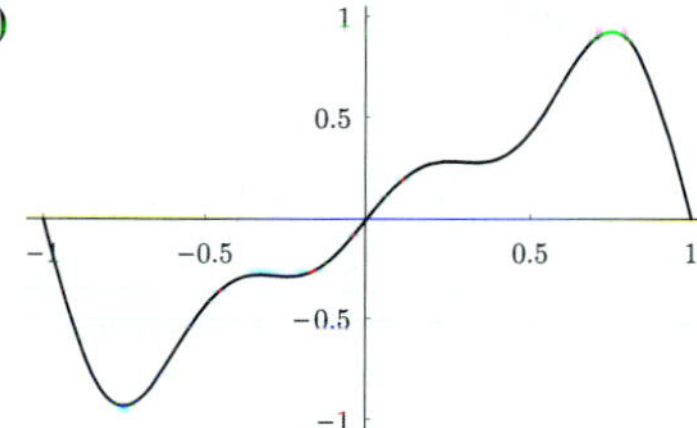

(d)

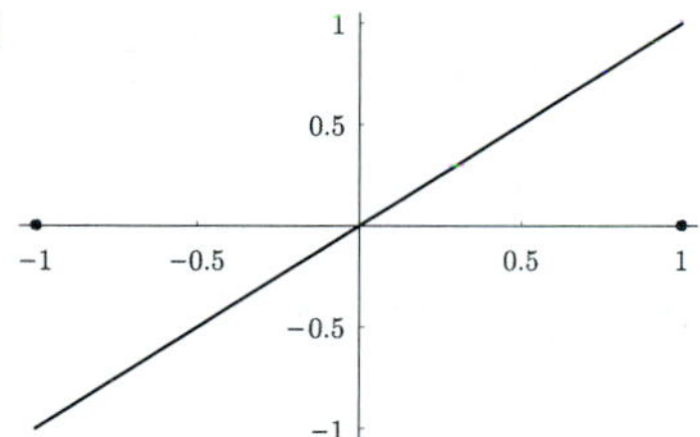

3. (a)

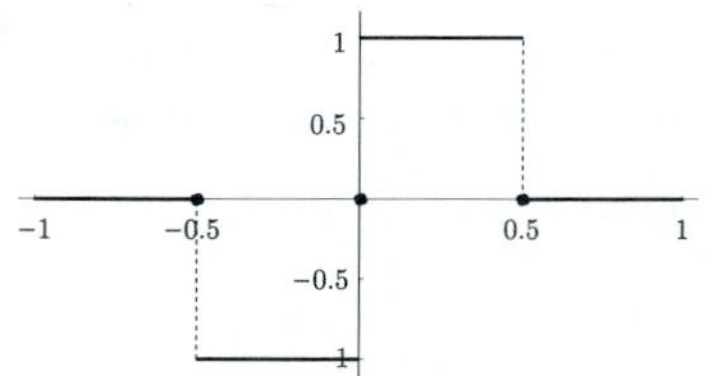

(b) $\frac{4}{\pi} \sum_{n=1}^{\infty} \frac{1}{n} \sin^2\left(\frac{n\pi}{4}\right) \sin(n\pi x)$

(c)

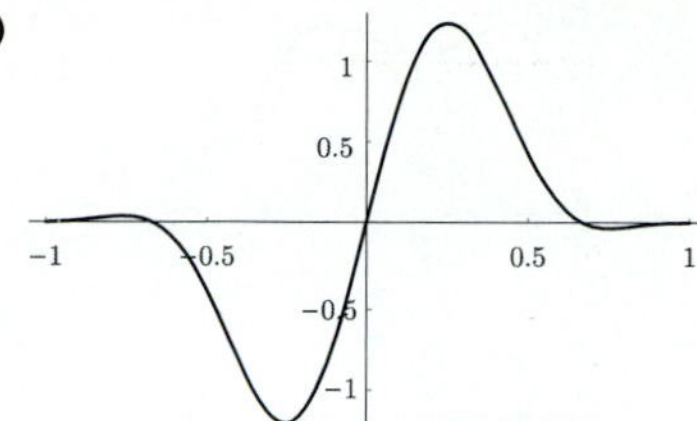

(d)

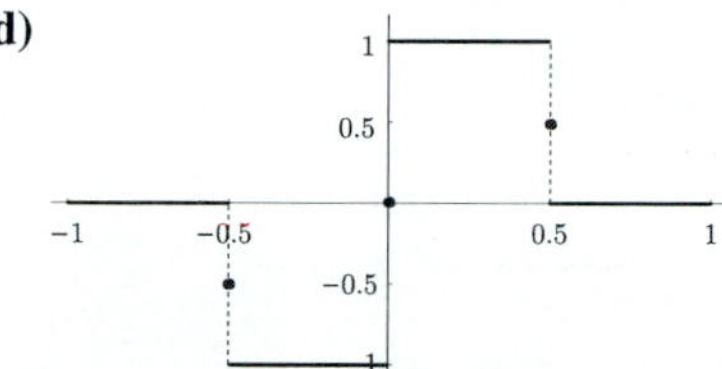

5. (a)

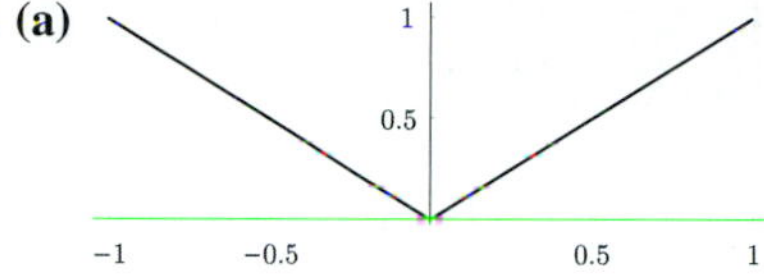

(b) $\frac{1}{2} + \frac{2}{\pi^2} \sum_{n=1}^{\infty} \frac{(-1)^n - 1}{n^2} \cos(n\pi x)$

(c)

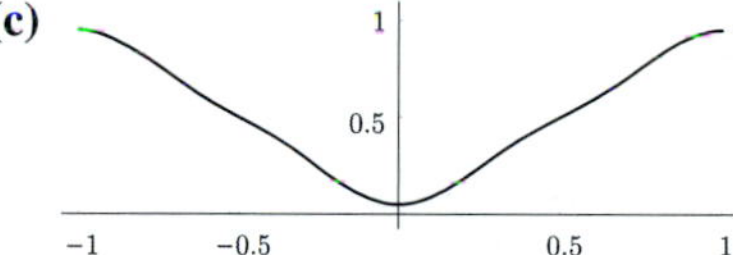

(d) Same as (a).

7. (a)

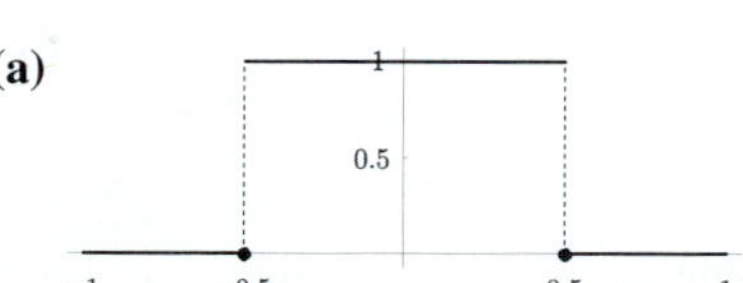

(b) $\frac{1}{2} + \frac{2}{\pi} \sum_{n=1}^{\infty} \frac{1}{n} \sin\left(\frac{n\pi}{2}\right) \cos(n\pi x)$

(c)

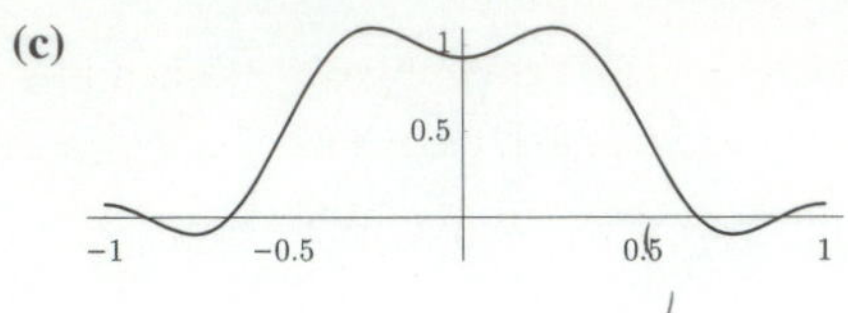

(d)

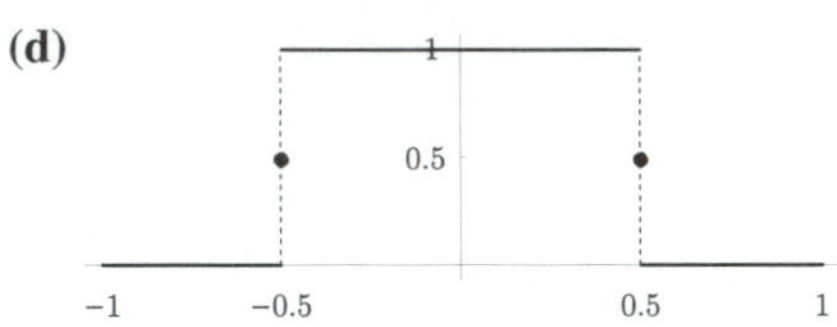

9. $\dfrac{2}{\pi}\sum_{n=1}^{\infty}\dfrac{(1+(-1)^n)n}{n^2-1}\sin nx$

11. $\sin x$

13. **(a)** $w(t,x) =$
$$\frac{2}{\pi}\sum_{n=1}^{\infty}\frac{1-(-1)^n}{n}e^{-n^2\pi^2 t}\sin(n\pi x)$$

(b) $w(0.01, x) \approx$
$$1.1536\sin(\pi x) + 0.1746\sin(3\pi x) + 0.0216\sin(5\pi x)$$

$w(0.1, x) \approx 0.4745\sin(\pi x)$

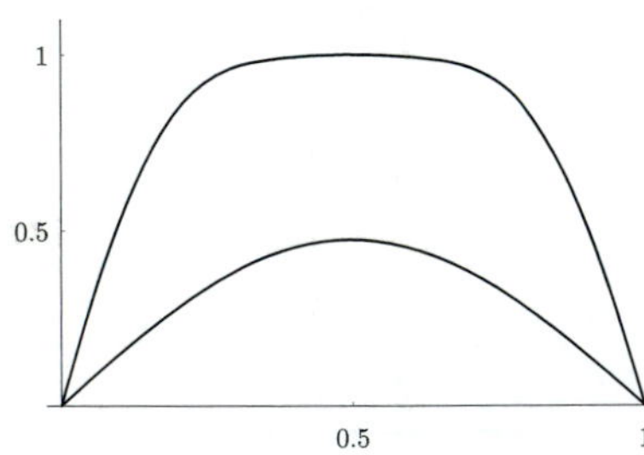

15. The substitution $x = -y$ is used in (i) and (ii):

i) $$\int_{-\ell}^{\ell} f(x)\,dx = \int_{-\ell}^{0} f(x)\,dx + \int_{0}^{\ell} f(x)\,dx = \int_{\ell}^{0} f(y)\,dy + \int_{0}^{\ell} f(x)\,dx = 0$$

ii) $$\int_{-\ell}^{\ell} f(x)\,dx = \int_{-\ell}^{0} f(x)\,dx + \int_{0}^{\ell} f(x)\,dx = -\int_{\ell}^{0} f(y)\,dy + \int_{0}^{\ell} f(x)\,dx = 2\int_{0}^{\ell} f(x)\,dx$$

iii) $$f(-x)g(-x) = \begin{cases} f(x)g(x), & \text{if } f, g \text{ are either both odd or both even,} \\ -f(x)g(x), & \text{if } f \text{ is odd and } g \text{ is even (or vice versa).} \end{cases}$$

17. $$w(t,x) = \sum_{n=1}^{\infty} b_n e^{-n^2\pi^2 t/\ell^2}\sin\left(\frac{n\pi x}{\ell}\right) = e^{-\pi^2 t/\ell^2}\sum_{n=1}^{\infty} b_n e^{-(n^2-1)\pi^2 t/\ell^2} \times \sin\left(\frac{n\pi x}{\ell}\right)$$

Therefore,
$$|w(t,x)| \le e^{-\pi^2 t/\ell^2}\sum_{n=1}^{\infty}\left|b_n\sin\left(\frac{n\pi x}{\ell}\right)\right|,$$
and so $|w(t,x)| \to 0$ as $t \to \infty$ (for each x in $[0,\ell]$).

19. *Hint*: Compute $\varphi_{\cos}\left(\frac{1}{2}\right)$

Section 11.5, p.445

1. $\mathcal{S}\psi = e^{-t}\left((e^t\psi')'\right)$

3. $\lambda_0 = 0$, $\lambda_n = -\dfrac{1}{4} - n^2\pi^2$, $n \ge 1$

$\psi_0 = 1$,
$$\psi_n = e^{-x/2}\left(\cos(n\pi x) + \frac{1}{2n\pi}\sin(n\pi x)\right)$$

5. **(a)** $\lambda_n = -k\omega_n^2$, $n = 1, 2, 3, \ldots$, where $\omega_1, \omega_2, \ldots$ are the positive solutions of
$$\frac{\omega^2-1}{\omega} = 2\cot(\omega\pi).$$
$$\psi_n = \cos(\omega_n x) + \frac{1}{\omega_n}\sin(\omega_n x)$$

(b) $\psi_n =$
$$\sqrt{\frac{2}{2+(1+\omega_n^2)\pi}}\left(\omega_n\cos(\omega_n x) + \sin(\omega_n x)\right)$$

7. **(a)** $\mathcal{S}\psi = (\psi')' + \gamma\psi$

(b) $\lambda_n = -n^2\pi^2 + \gamma$, $n = 1, 2, 3, \ldots$
$\psi_n = \sin(n\pi x)$

9. $(x^2+1)^{-1}\left((x^2+1)^2\psi'\right)'$

13. **(c)** $\lambda_n = -\dfrac{1}{4} - n^2\pi^2$, $n = 1, 2, 3, \ldots$

$$\psi_n = \frac{1}{\sqrt{1+x}} \sin\left(\sqrt{-\frac{1}{4} - \lambda_n} \ln(1+x)\right)$$

15. **(a)** $\mathcal{S}\psi = e^{-x^2/2}\left(e^{x^2/2}\psi'\right)'$

(b) $u_\lambda = 1 + \dfrac{\lambda}{2}x^2 + \dfrac{\lambda(\lambda-2)}{4!}x^4 + \dfrac{\lambda(\lambda-2)(\lambda-4)}{6!}x^6 + \cdots,$

$$v_\lambda = x + \frac{\lambda-1}{3!}x^2 + \frac{(\lambda-1)(\lambda-3)}{5!}x^5 + \cdots$$

(c) $\psi_n = v_{\lambda_n}$, where $v_{\lambda_n}(1) = 0$; that is,

$$1 + \frac{\lambda-1}{3!} + \frac{(\lambda-1)(\lambda-3)}{5!} + \frac{(\lambda-1)(\lambda-3)(\lambda-5)}{7!} + \cdots = 0$$

(d) Truncating after five terms gives $\lambda_0 \approx -11$ and

$$\psi_0(x) \approx x - 2x^3 + \frac{7}{5}x^5 - \frac{8}{15}x^7 + \frac{2}{15}x^9.$$

Truncating after six terms gives $\lambda_0 \approx -10.4$ and

$$\psi_0(x) \approx x - 1.9x^3 + 1.27x^5 - 0.467x^7 + 0.113x^9 - 0.0199x^{11}.$$

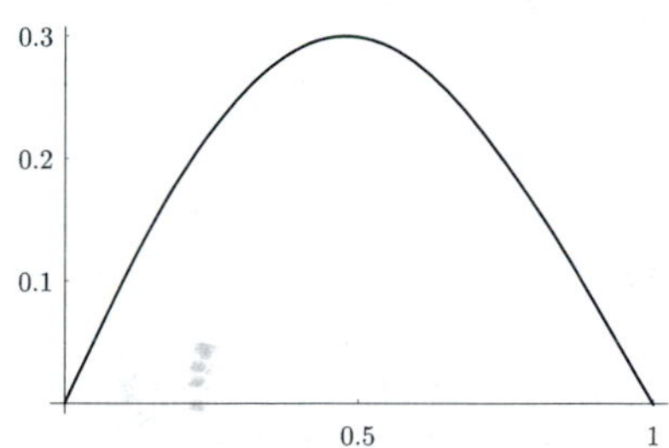

17. If $\mathcal{S}\psi_n = \lambda_n\psi_n$, then $\mathcal{S}_c\psi_n = (\lambda_n + c)\psi_n$.

19. **(a)** $\lambda_n = -4n^2$, $n = 0, 1, 2, \ldots$

(b) $\cos(2nx)$ and $\sin(2nx)$ are each eigenfunctions corresponding to λ_n.

Section 11.6, p.458

1. $\lambda_n = -(2n-1)^2$, $\psi_n = \cos\left((2n-1)\cos^{-1}x\right)$

3. **(a)** $\mathcal{S}\psi = ((1-x^2)\psi')'$

(b) Use the result of Problem 1, Section 5.7.

(c) Use the fact that the radius of converges is 1.

5. **(b)** $\lambda > 0$: $\psi(x) = \sinh\left(\sqrt{\lambda}\left(\dfrac{1}{x} - \pi\right)\right)$

$\lambda = 0$: $\psi(x) = \dfrac{1}{x} - \pi$

$\lambda < 0$: $\psi(x) = \sin\left(\sqrt{-\lambda}\left(\dfrac{1}{x} - \pi\right)\right)$

Section 11.7, p.462

1. **(a)** $\lambda_n = -n^2$, $\psi_n = \sqrt{\dfrac{2}{\pi}}e^x \sin nx$, $n = 1, 2, 3, \ldots$

(b) $\sqrt{\dfrac{2}{\pi}} \displaystyle\sum_{n=1}^{\infty} \frac{1-(-1)^n}{n}\psi_n(x)$

3. **(a)** $\lambda_0 = \omega_0^2$, where ω_0 is the unique positive solution of $\omega = \tanh \pi\omega$, and $\lambda_n = -\omega_n^2$, $n \geq 1$, where $\omega_1, \omega_2, \ldots$ are the positive solutions of $\omega = \tan \pi\omega$.

$$\psi_0 = \sqrt{\frac{2}{\pi\omega_0^2 - \pi + 1}} \times e^x(\omega_0 \cosh \omega_0 x - \sinh \omega_0 x),$$ and for $n \geq 1$,

$$\psi_n = \sqrt{\frac{2}{\pi\omega_n^2 + \pi - 1}} \times e^x(\omega_n \cos \omega_n x - \sin \omega_n x).$$

(b) $\displaystyle\sum_{n=0}^{\infty} a_n\psi_n(x)$, where

$$a_0 = \sqrt{\frac{2}{\pi\omega_0^2 - \pi + 1}} \times \left(1 + (\omega_0^2 - 1)\cosh \pi\omega_0\right),$$

$$a_n = \sqrt{\frac{2}{\pi\omega_n^2 + \pi - 1}} \times \left(-1 + (\omega_n^2 + 1)\cos \pi\omega_n\right), \ n \geq 1.$$

5. **(a)** $\lambda_0 = 0$, $\psi_0 = \dfrac{2}{\sqrt{e^{4\pi} - 1}}$;

$n \geq 1$: $\lambda_n = -(n^2 + 1)$,

$$\psi_n = \sqrt{\frac{2}{(4+n^2)\pi}}\, e^{-2x}(n\cos nx + 2\sin nx)$$

(b) $\displaystyle\sum_{n=0}^{\infty} a_n\psi_n(x)$, where $a_0 = \dfrac{e^{2\pi} - 1}{\sqrt{e^{4\pi} - 1}}$ and

$$a_n = \frac{2(1-(-1)^n)}{n}\sqrt{\frac{2}{(4+n^2)\pi}}, n \geq 1.$$

7. (a) $\lambda_n = -(n^2\pi^2+1)$, $\psi_n = \sqrt{2}e^{-x}\sin(n\pi x)$, $n \geq 1$

(b) $e^{-x}(1.73081\sin\pi x - 0.323145\sin 2\pi x + 0.0985087\sin 3\pi x - 0.0429401\sin 4\pi x)$

9. $T_0 - \frac{1}{4}T_1 + \frac{1}{4}T_3$

11. (a) $\lambda_0, \ldots, \lambda_4$ are
$-0.620, -2.79, -6.84, -12.9, -20.9.$
Eigenfunctions:
$\psi_0 = 0.729\sin(0.788x)$,
$\psi_1 = 0.766\sin(1.67x)$,
$\psi_2 = 0.782\sin(2.62x)$,
$\psi_3 = 0.789\sin(3.59x)$,
$\psi_4 = 0.792\sin(4.57x)$.

(b) $w \approx 1.65e^{\lambda_0 t}\psi_0 + 0.223e^{\lambda_1 t}\psi_1 + 0.406e^{\lambda_2 t}\psi_2 + 0.161e^{\lambda_3 t}\psi_3 + 0.210e^{\lambda_4 t}\psi_4$

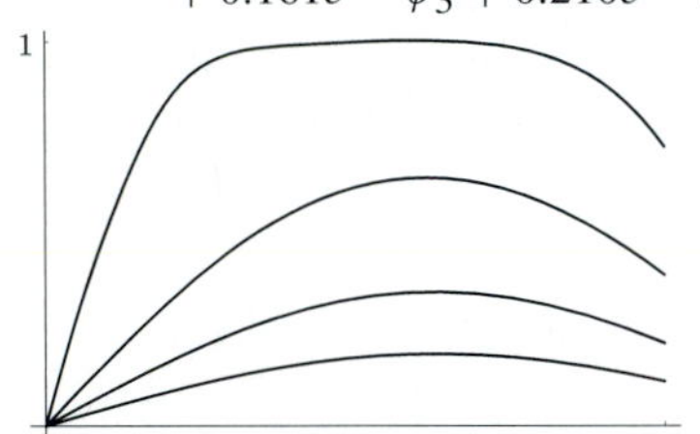

13. (a) $\lambda_0, \ldots, \lambda_4$ are
$-1.62, -3.79, -7.84, -13.9, -21.9.$
Eigenfunctions:
$\psi_0 = 0.729e^x\sin(0.788x)$,
$\psi_1 = 0.766e^x\sin(1.67x)$,
$\psi_2 = 0.782e^x\sin(2.62x)$,
$\psi_3 = 0.789e^x\sin(3.59x)$,
$\psi_4 = 0.792e^x\sin(4.57x)$.

(b) $w \approx 0.355e^{\lambda_0 t}\psi_0 + 0.338e^{\lambda_1 t}\psi_1 + 0.261e^{\lambda_2 t}\psi_2 + 0.204e^{\lambda_3 t}\psi_3 + 0.165e^{\lambda_4 t}\psi_4$

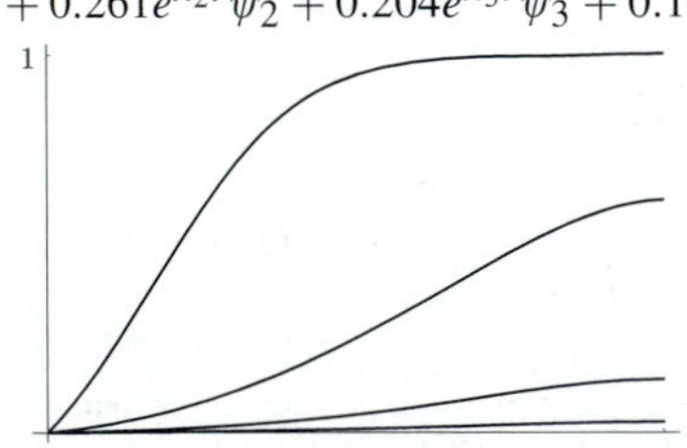

15. (a) $\lambda_n = 1-(n+1)^2$,
$\psi_n = \frac{2}{\sqrt{\pi}}\sin((n+1)\pi x)$

(b) $w = \sum_{n=0}^{\infty} a_n e^{\lambda_n t}\psi_n$, where $a_n = \frac{(-1)^n\sqrt{2\pi}}{n+1}$

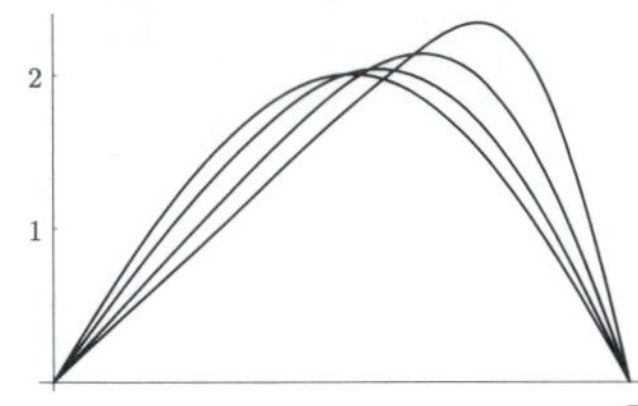

$$\lim_{t\to\infty} w(t,x) = 2\sin x.$$

17. (a) $\frac{1}{5}P_0 + \frac{2}{5}P_1 + \frac{4}{7}P_2 - \frac{2}{5}P_3 + \frac{8}{35}P_4$

(b) $\frac{1}{5} + \frac{2}{5}e^{-2t}P_1 + \frac{4}{7}e^{-6t}P_2 - \frac{2}{5}e^{-12t}P_3 + \frac{8}{35}e^{-20t}P_4$

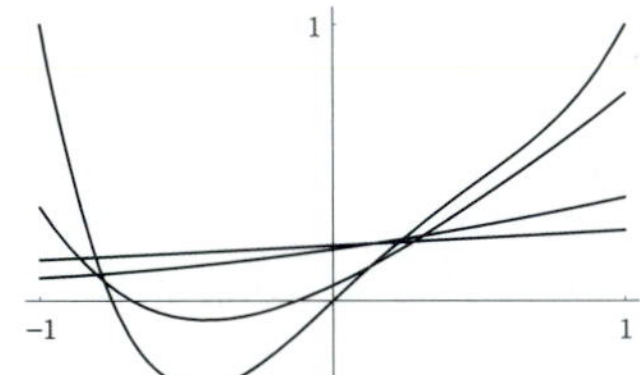

19. (b) $\lambda_n = -\frac{(n+1)^2\pi^2+1}{4}$,
$\psi_n = \frac{1}{\sqrt{1+x}}\sin\left(\frac{(n+1)\pi}{2}\ln(1+x)\right)$

(c) $e^{-t/4}\left(e^{-\pi^2 t/4}\psi_0 - \frac{1}{3}e^{-9\pi^2 t/4}\psi_2 + \frac{1}{2}e^{-25\pi^2 t/4}\psi_4\right)$

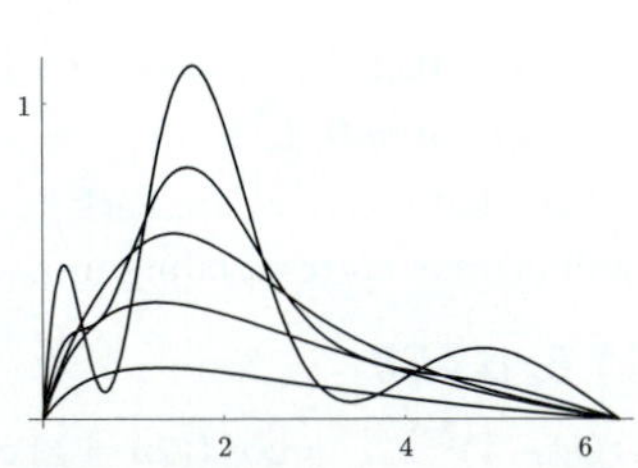

Chapter 12

Section 12.1, p.472

1.

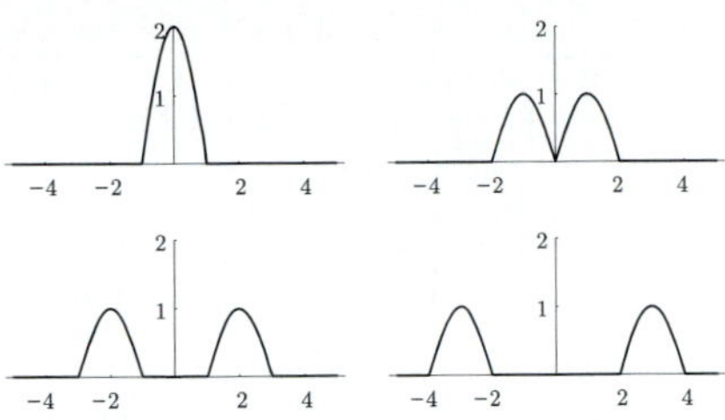

3.

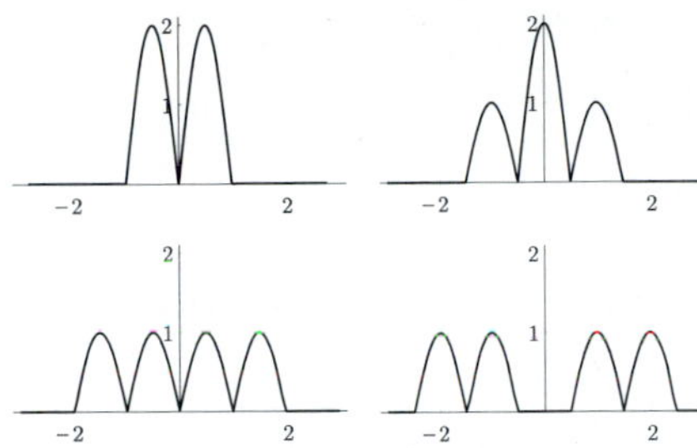

5.

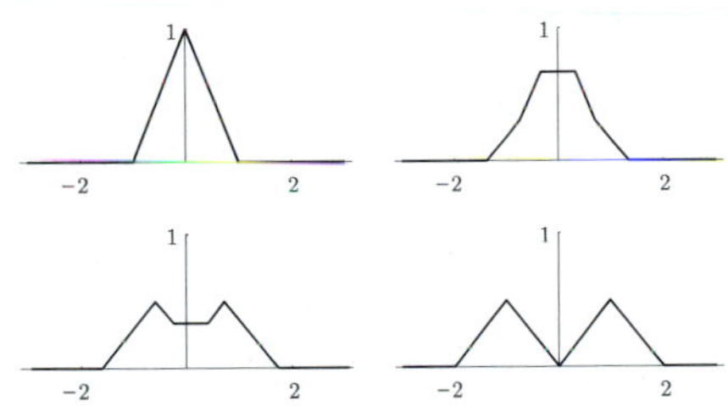

7. $\displaystyle\sum_{n=1}^{\infty} \frac{2^n}{n!} \cos(2nt)\sin(nx)$

9. $\displaystyle\sum_{n=1}^{\infty} b_n \cos(nt)\sin(\frac{nx}{3})$, where

$$b_n = \begin{cases} \frac{6}{(9-n^2)\pi}\left(\sin(\frac{n\pi}{3}) + \sin(\frac{2n\pi}{3})\right), & n \neq 3, \\ -\frac{1}{3}, & n = 3. \end{cases}$$

Note that $\sin\left(\frac{n\pi}{3}\right) + \sin\left(\frac{2n\pi}{3}\right)$ cycles as follows: $\sqrt{3}$, 0, 0, 0, $-\sqrt{3}$, 0, $\sqrt{3}$, 0, 0, 0, $-\sqrt{3}$, 0, ...

11. $\displaystyle\sum_{n=1}^{\infty} b_n \cos\left(\frac{nt}{3}\right)\sin\left(\frac{nx}{3}\right)$, where

$$b_n = \frac{2}{n\pi}\left(\cos\left(\frac{n\pi}{3}\right) - \cos\left(\frac{2n\pi}{3}\right)\right).$$

Note that $\cos\left(\frac{n\pi}{3}\right) - \cos\left(\frac{2n\pi}{3}\right)$ cycles as follows: 1, 0, -2, 0, 1, 0, 1, 0, -2, 0, 1, 0, ...

13. $\displaystyle\sum_{n=1}^{\infty} b_n \sin\left(\frac{n\pi ct}{\ell}\right)\sin\left(\frac{n\pi x}{\ell}\right)$, where

$$b_n = \frac{2}{n\pi c}\int_0^{\ell} v(x)\sin\left(\frac{n\pi x}{\ell}\right).$$

15. $\sin(x + ct)$

21. (a)

$$\sum_{n=1}^{\infty} \frac{b_n}{2\omega_n} e^{-\frac{\beta t}{2}} (2\omega_n \cos\omega_n t + \beta\sin\omega_n t) \times \sin\left(\frac{n\pi x}{\ell}\right),$$

where $\omega_n = \frac{1}{2\ell}\sqrt{4n^2\pi^2c^2 - \beta^2\ell^2}$.

(b)

$$\frac{1}{2}b_1\left(1 + \frac{\beta\ell}{\sqrt{\beta^2 - 4\pi^2c^2}}\right)\left(e^{r_1 t} - \frac{r_1}{r_2}e^{r_2 t}\right) \times \sin\left(\frac{\pi x}{\ell}\right) + \sum_{n=2}^{\infty} \frac{b_n}{2\omega_n} e^{-\beta t/2} (2\omega_n \cos\omega_n t + \beta\sin\omega_n t) \sin\left(\frac{n\pi x}{\ell}\right),$$ where

$$\begin{aligned} r_1 &= -\beta + \frac{1}{\ell}\sqrt{\beta^2\ell^2 - 4\pi^2c^2}, \\ r_2 &= -\beta - \frac{1}{\ell}\sqrt{\beta^2\ell^2 - 4\pi^2c^2}, \\ \omega_n &= \frac{1}{2\ell}\sqrt{4n^2\pi^2c^2 - \beta^2\ell^2}. \end{aligned}$$

Section 12.2, p.484

1. $\sin(2\pi y)\operatorname{csch}(2\pi)\left(\sinh(2\pi(1-x)) - \sinh(2\pi x)\right)$

3. $\displaystyle\frac{2}{\pi}\sum_{n=1}^{\infty} \frac{1-(-1)^n}{n} \sin(n\pi x)\operatorname{csch}(n\pi)\sinh(n\pi y) - \frac{2}{\pi}\sum_{n=1}^{\infty} \frac{(-1)^n}{n} \sin(n\pi y)\operatorname{csch}(n\pi)\sinh(n\pi x)$

5. (b) $a_{mn} = -\dfrac{\ell^2 h^2}{(\ell^2 n^2 + h^2 m^2)\pi^2} c_{mn}$

7. $1 + r\sin\theta$

9. $\frac{1}{2} + \frac{1}{\pi} \sum_{n=1}^{\infty} \frac{1-(-1)^n}{n} r^n \sin n\theta$

13. $\sum_{n=0}^{\infty} a_n \cos(\omega_n x) \operatorname{sech}(\omega_n h) \cosh(\omega_n (h-y))$,

where $\omega_n = \frac{(2n+1)\pi}{2\ell}$ and

$a_n = 2\int_0^{\ell} g_1(x) \cos(\omega_n x)\, dx.$

15. $\frac{a_0}{2} + \sum_{n=1}^{\infty} a_n \cos(\omega_n x) \operatorname{sech}(\omega_n h) \times \cosh(\omega_n (h-y))$,

where $\omega_n = \frac{n\pi}{\ell}$ and

$a_n = 2\int_0^{\ell} g_1(x) \cos(\omega_n x)\, dx.$

17. $a_{00} = \frac{1}{\alpha} c_{00}$;

$a_{mn} = \frac{h^2\ell^2}{h^2\ell^2\alpha - m^2\pi^2 h^2 - n^2\pi^2\ell^2} c_{mn}$

if $n \geq 1$ or $m \geq 1$.

Section 12.3, p.497

1. $5e^{-2kt} \sin x \sin y - e^{-13kt} \sin 2x \sin 3y$

3. $\sum_{n=1}^{\infty} \sum_{m=1}^{\infty} \frac{16c_{mn}}{mn\pi^2} e^{-(m^2+n^2)kt} \sin mx \sin ny$,

where $c_{mn} = \sin^2\left(\frac{m\pi}{4}\right) \sin^2\left(\frac{n\pi}{4}\right)$.

5. $\sum_{m=1}^{\infty} b_{mn} \sin\left(\frac{m\pi x}{\ell}\right) =$

$\frac{2}{h}\int_0^h \varphi(x, y) \sin\left(\frac{m\pi y}{h}\right) dy$; therefore,

$$b_{mn} = \frac{2}{\ell}\int_0^{\ell} \left(\frac{2}{h}\int_0^h \varphi(x, y) \sin\left(\frac{n\pi y}{h}\right) dy\right) \times \sin\left(\frac{m\pi x}{\ell}\right) dx$$

$$= \frac{4}{\ell h}\int_0^{\ell}\int_0^h \varphi(x, y) \sin\left(\frac{n\pi y}{h}\right) \times \sin\left(\frac{m\pi x}{\ell}\right) dy\, dx.$$

9. $\sum_{n=1}^{\infty} \sum_{m=0}^{\infty} a_{mn} e^{-\left(\left(m+\frac{1}{2}\right)^2+n^2\right)kt} \cos \left(\left(m + \frac{1}{2}\right) x\right) x \sin ny$, where

$$a_{mn} = \frac{4}{\pi^2}\int_0^{\pi}\int_0^{\pi} \varphi(x, y) \cos\left(\left(m + \frac{1}{2}\right) x\right) \times \sin(ny)\, dx\, dy.$$

13. $\alpha'_{mn} + (m^2+n^2)k\alpha_{mn} = 1$, $\alpha_{mn}(0) = 0$; therefore,

$\alpha_{mn} = \frac{1}{(m^2+n^2)k}\left(1 - e^{-(m^2+n^2)kt}\right)$,

and so $u = \frac{1}{(m^2+n^2)k} \times \left(1 - e^{-(m^2+n^2)kt}\right) \sin mx \sin ny.$

21. $\sum_{n=1}^{\infty} a_n \cos(x_n ct) J_0(x_n r)$, where

$x_1, x_2, x_3, \ldots$ are the positive zeros of J_0

and $a_n = \frac{\int_0^1 \gamma(r) J_0(x_n r) r\, dr}{\int_0^1 (J_0(x_n r))^2 r\, dr}$.

Appendix I

Appendix I.A, p.512

1. $x_1 = 2$, $x_2 = 2$, $x_3 = 1$

3. $x_1 = 2$, $x_2 = 1$, $x_3 = 2$

5. $x_1 = -1$, $x_2 = 2$, $x_3 = -2$

7. $x_1 = 7$, $x_2 = -5$, $x_3 = 1$

9. $x_1 = 1$, $x_2 = 2$, $x_3 = 0$

11. $x_1 = 1$, $x_2 = 2$, $x_3 = -1$

13. $x_1 = 2 - 2x_2$, $x_2 =$ any real number, $x_3 = 3$

15. No solution

17. $x_1 = -2x_3$, $x_2 = 1 - x_4$, $x_3 =$ any real number, $x_4 =$ any real number

19. Infinitely many solutions

21. No solution

23. Unique solution

25. $x_1 = 0$, $x_2 = 0$, $x_3 = 0$

27. $x_1 = -\frac{5}{11}x_3$, $x_2 = -\frac{7}{11}x_3$, $x_3 =$ any real number

29. $x_1 = 8$, $x_2 = -1$, $x_3 = 3$

31. $x_1 = -2$, $x_2 = 0$, $x_3 = 5$, $x_4 = 2$, $x_5 = -3$

Appendix I.B, p.523

1. $\begin{pmatrix} 1 \\ 12 \end{pmatrix}$

3. $\begin{pmatrix} 11 \\ 4 \\ 11 \\ 6 \end{pmatrix}$

5. $\begin{pmatrix} 7 & -1 & 13 \\ 2 & 0 & 3 \\ 4 & 2 & 9 \end{pmatrix}$

7. $\begin{pmatrix} 13 & 2 \\ 14 & 4 \end{pmatrix}$

9. $\begin{pmatrix} -12 & -2 \\ -5 & 3 \end{pmatrix}$

11. $\begin{pmatrix} 2 & -8 \\ 8 & -2 \end{pmatrix}$

17. $\begin{pmatrix} 1 & 0 & -1 \\ 0 & 0 & 1 \\ -1 & 1 & 1 \end{pmatrix}$

19. $\frac{1}{4}\begin{pmatrix} 3 & 2 & 1 \\ 2 & 4 & 2 \\ 1 & 2 & 3 \end{pmatrix}$

21. The inverse doesn't exist.

23. $\begin{pmatrix} 1 & -3 & 9 \\ 0 & 1 & -3 \\ 0 & 0 & 1 \end{pmatrix}$

25. **(b)** $\begin{vmatrix} 1 & 3 & 1 \\ 2 & 3 & 4 \\ 1 & 0 & 3 \end{vmatrix} = 0$; the vectors are linearly dependent.

27. **(b)** $\begin{vmatrix} -2 & 4 \\ -3 & 6 \end{vmatrix} = 0$; the vectors are linearly dependent.

29. **(b)** $\begin{vmatrix} 2 & -1 & 5 \\ -1 & 2 & 3 \\ 3 & 4 & -2 \end{vmatrix} = -89$; the vectors are linearly independent.

31. -9

33. 4

35. $0, \frac{1}{2}$

37. $\pm 1, \pm 4$

39. **(a)** *Hint*: Let A be any invertible matrix and $B = -A$.

(b) One simple example is $A = B = \mathcal{I}$.

Appendix I.C, p.543

3. $\left(-1, \begin{pmatrix} -1 \\ 1 \end{pmatrix}\right), \left(3, \begin{pmatrix} 1 \\ 1 \end{pmatrix}\right)$

5. $\left(-3, \begin{pmatrix} 2 \\ 3 \end{pmatrix}\right), \left(-2, \begin{pmatrix} 1 \\ 2 \end{pmatrix}\right)$

7. $\left(5 - 3i, \begin{pmatrix} 1 - 3i \\ 2 \end{pmatrix}\right), \left(5 + 3i, \begin{pmatrix} 1 + 3i \\ 2 \end{pmatrix}\right)$

9. $\left(-2, \begin{pmatrix} -1 \\ 1 \\ 0 \end{pmatrix}\right), \left(1, \begin{pmatrix} 2 \\ 0 \\ 1 \end{pmatrix}\right), \left(3, \begin{pmatrix} 1 \\ 2 \\ 1 \end{pmatrix}\right)$

11. $\left(-i, \begin{pmatrix} 1 \\ -1 \\ i \end{pmatrix}\right), \left(i, \begin{pmatrix} 1 \\ -1 \\ -i \end{pmatrix}\right), \left(1, \begin{pmatrix} 0 \\ 1 \\ 1 \end{pmatrix}\right)$

13. $\left(0, \begin{pmatrix} -2 \\ -1 \\ 2 \end{pmatrix}\right), \left(2, \begin{pmatrix} 0 \\ 1 \\ 0 \end{pmatrix}\right)$

15. $-4, 3$

17. $-3 + 2i, -2$

19. Similar

21. Not similar

23. Not similar

25. $\begin{pmatrix} 3 & 1 & -2 \\ -6 & -4 & 6 \\ -2 & -2 & 3 \end{pmatrix}$

26. *Hint*: $a^2 - 2ac + c^2 = (a - c)^2$

27. *Hint*: Try an upper-triangular matrix.

29. *Hint*: $A\mathbf{x} = \lambda\mathbf{x}$ is equivalent to $\mathbf{x} = \lambda A^{-1}\mathbf{x}$.

31. One 2×2 example is $\frac{1}{5}\begin{pmatrix} 3 & 4 \\ -4 & 3 \end{pmatrix}$.

One 3×3 example is $\frac{1}{5}\begin{pmatrix} 0 & -3 & 4 \\ 5 & 0 & 0 \\ 0 & 4 & 3 \end{pmatrix}$.

$$\mathbf{a}_{:i}^T \mathbf{a}_{:j} = \begin{cases} 0, & \text{if } i \neq j \\ 1, & \text{if } i = j \end{cases}$$

Appendix II

1. **(a)** 0 **(b)** 1 **(c)** -3 **(d)** -2

3. **(a)** -2 **(b)** 2 **(c)** -2 **(d)** 2

5. **(a)** $f'(x) = \dfrac{-x}{\sqrt{1 - x^2}}$ for $-1 < x < 1$

(b) $\lim_{x \to -1^+} f(x) = f(-1)$ and $\lim_{x \to 1^-} f(x) = f(1)$;

neither $f'_{[-1,1]}(-1)$ nor $f'_{[-1,1]}(1)$ exists.

6. (a) Continuous and differentiable

(c) Continuous and differentiable

(e) Continuous and not differentiable

(g) Not continuous and not differentiable

7. One of many posibilities:

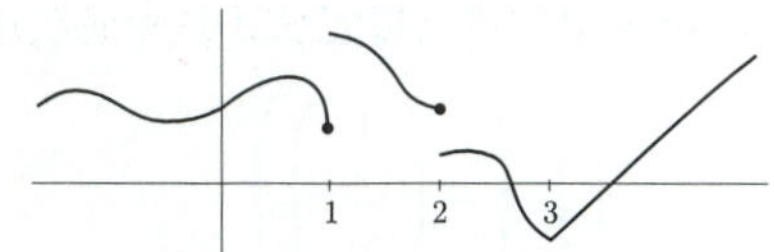

9. *Hint*: $f'_{[0,\infty)}(0) = \lim_{h \to 0^+} \dfrac{h^2 \sin \frac{1}{h} - 0}{h}$

Appendix III

1. $K = 1$, $M = 2$

3. $K = e^{-1}$, $M = 1$

7. $M = e$

9. *Hint*: Let $y_2 = 0$ and look at $\lim_{y_1 \to 0^+} \dfrac{f(y_1) - f(0)}{y_1 - 0}$.

Appendix IV

1. (a) $W(t) = 2e^{3t} \neq 0$ for all t

(c) $e^t \left(1 - t + \frac{1}{2}t^2\right)$

3. (a) $W(t) = 2(-3 + \cos 2t) \neq 0$ for all t

(c) $\frac{1}{4}\left(e^t + e^{-t} + 2\cos t\right)$

7. $$\frac{d}{dt}\begin{pmatrix} y \\ y' \\ y'' \end{pmatrix} = \begin{pmatrix} 0 & 1 & 0 \\ 0 & 0 & 1 \\ -2 & -t & 0 \end{pmatrix}\begin{pmatrix} y \\ y' \\ y'' \end{pmatrix} + \begin{pmatrix} 0 \\ 0 \\ t \end{pmatrix}$$

9. $$\frac{d}{dt}\begin{pmatrix} y \\ y' \end{pmatrix} = \begin{pmatrix} 0 & 1 \\ -q & -p \end{pmatrix}\begin{pmatrix} y \\ y' \end{pmatrix} + \begin{pmatrix} 0 \\ f \end{pmatrix}$$

11. $c_1 e^{-t} + (c_2 + c_3 t)e^{-2t} + (c_4 + c_5 t + c_6 t^2)e^{-3t}$

13. $c_1 e^{-t} + c_2 \cos 2t + c_3 \sin 2t$

15. $(c_1 + c_2 t + c_3 t^2)e^{-t}$

17. $c_1 e^{kt} + c_2 e^{-kt} + c_3 \cos kt + c_4 \sin kt$

21. (a) $$\frac{d}{dt}\begin{pmatrix} y \\ y' \\ y'' \\ y''' \end{pmatrix} = \begin{pmatrix} 0 & 1 & 0 & 0 \\ 0 & 0 & 1 & 0 \\ 0 & 0 & 0 & 1 \\ -p_0 & -p_1 & -p_2 & -p_3 \end{pmatrix}\begin{pmatrix} y \\ y' \\ y'' \\ y''' \end{pmatrix},$$

$$\det\begin{pmatrix} r & -1 & 0 & 0 \\ 0 & r & -1 & 0 \\ 0 & 0 & r & -1 \\ p_0 & p_1 & p_2 & r + p_3 \end{pmatrix} = p_0 + p_1 r + p_2 r^2 + (r + p_3)r^3$$

23. $(at + b)e^{-2t}$

25. $t^4(at^2 + bt + c)e^{-2t}$

27. $(\mathcal{D} + \mathcal{I})(\mathcal{D}^2 + 4\mathcal{I})^2 z = (t - 3)e^{(-1+2i)t}$, $(at + b)e^{(-1+2i)t}$

29. $(\mathcal{D}^2 + 4\mathcal{D} + 13\mathcal{I})^2 z = (2t^2 - 5)e^{(-2+3i)t}$, $t^2(at^2 + bt + c)e^{(-2+3i)t}$

31. $t^3(3 - t)e^{-2t}$

33. $x = -t^2 e^{-t}\,((t + 3)\cos t - 3\sin t)$, $y = -t^2 e^{-t}\,(3\cos t + (t + 3)\sin t)$

35. $t^2 e^{-t}(\cos t + t \sin t)$

INDEX

I

J

K

L

M

N

O

P

Q

R

S

T

U

V

W

Y

Z